SWINE SCIENCE
(Animal Agriculture Series)

Cover pictures.
Orange ginger chef's prime™ pork loin and Symbol II
were provided by the National Pork Producers Council,
Des Moines, Iowa

ABOUT THE AUTHORS

Marion Eugene Ensminger completed B.S. and M.S. degrees at the University of Missouri, and the Ph.D. at the University of Minnesota. Dr. Ensminger served, in order, as Manager of the Dixon Springs Agricultural Center (University of Illinois), Simpson, Illinois; and on the staffs of the University of Massachusetts, the University of Minnesota, and Washington State University. Dr. Ensminger also served as Consultant, General Electric Company, Nucleonics Department, and as the first President of the American Society of Agricultural Consultants. Since 1964, Dr. Ensminger has served as President of Agriservices Foundation, Clovis, California, a nonprofit foundation serving world agriculture in the area of World Food, Hunger, and Malnutrition.

Among Dr. Ensminger's many honors and awards are: Distinguished Teacher Award, American Society of Animal Science; the "Ensminger Beef Cattle Research Center" at Washington State University, Pullman, named after him in recognition of his contributions to the University; Faculty-Alumni Award of the University of Missouri; Outstanding Achievement Award of the University of Minnesota; Distinguished Service Award of the American Medical Association (with Mrs. Ensminger); Honorary Professor, Huazhong Agricultural College, Wuhan, China; Doctor of Laws (LL.D.) conferred by the National Agrarian University of Ukraine; and an oil portrait of him was placed in the 300-year-old gallery of the famed Saddle and Sirloin Club, Lexington, Kentucky.

DR. M. E. ENSMINGER

In 1995, Cuba honored Dr. Ensminger by making him an Honorary Member of the Cuban Association of Animal Production; presenting him the 30th anniversary Gold Medal of the Institute of Animal Science, at Havana; making him an Honorary Guest Professor of the Agricultural University (ISCAH), at Havana; and making him an Honorary Guest Professor of the University of Camaguey, Camaguey, Cuba.

In 1995, Dr. Ensminger received the Distinguished Teacher Award, the highest honor of the National Association of Colleges and Teachers of Agriculture (NACTA).

In 1996, Iowa State University awarded Dr. Ensminger the honorary degree, Doctor of Humane Letters for "extraordinary achievements in animal science, education, and international agriculture."

In 1996, Dr. Ensminger was the recipient of the International Animal Agriculture Bouffault Award, Paris, France, and the American Society of Animal Science.

Dr. Ensminger founded the International Ag-Tech Schools, which he directed for more than 50 years. He has directed schools, lectured, and/or conducted seminars in 70 countries. Dr. Ensminger is the author of more than 500 scientific articles, bulletins, and feature articles; and he is the author or co-author of 22 books, which are in several languages and used all over the world. He waives all royalties on the foreign editions of his books in order to help the people. The whole world is Dr. Ensminger's classroom.

Dr. Richard O. Parker grew up on a general livestock farm in Idaho; graduated from Minidoka County High School, Rupert, Idaho; obtained the B.S. degree from Brigham Young University; completed the Ph.D. degree at Iowa State University; and pursued postdoctorate training at the University of Alberta, Edmonton, and at the University of Wyoming, Laramie.

From 1980 to 1984, Dr. Parker served as author's assistant to Dr. M. E. Ensminger, Clovis, California, with writing assignments on *Foods & Nutrition Encyclopedia*, and co-authorship of *Swine Science* and *Sheep & Goat Science*. From 1981 to 1984, Dr. Parker served as an Adjunct Professor in Animal Science at California State University, Fresno.

From January 1984 to July 1985, Dr. Parker was self-employed at Rupert, Idaho, in farming, consulting, computer instruction, and equipment sales.

From February to May, 1985, Dr. Parker served as part-time instructor in the agricultural department, College of Southern Idaho, Twin Falls. Since 1985, he has been a full-time staff member of College of Southern Idaho; serving as coordinator and instructor in the agricultural department and part-time computer trainer from 1985 to December 1986; Division Director for agriculture and part-time computer trainer since 1987; and Affiliate Faculty since July 1988.

DR. R. O. PARKER

SWINE SCIENCE
(Animal Agriculture Series)

by

M. E. Ensminger, B.S., M.A., Ph.D.   **R. O. Parker**, B.S., Ph.D.

SIXTH EDITION

INTERSTATE PUBLISHERS, INC.
Danville, Illinois

Editions:

First 1952
Second 1957
Third 1961
Fourth 1970
Fifth 1984
Sixth 1997

Library of Congress Catalog Card No. 96-79328

ISBN 0-8134-3108-5

1 2 3
4 5 6
7 8 9

Other books by M. E. Ensminger
available from Interstate Publishers:

Animal Science
Animal Science Digest
Beef Cattle Science (with R. Perry)
Dairy Cattle Science
Feeds & Nutrition (with J. Oldfield & W. Heinemann)
Feeds & Nutrition Digest (with J. Oldfield & W. Heinemann)
Horses and Horsemanship
Poultry Science
Sheep and Goat Science
Stockman's Handbook, The
Stockman's Handbook Digest

Animal Science presents a perspective or panorama of the far-flung livestock industry; whereas each of the other books presents specialized material pertaining to the specific class of farm animals indicated by its title.

Feeds & Nutrition and *Feeds & Nutrition Digest* bring together both the art and the science of livestock feeding, narrow the gap between nutrition research and application, and assure more and better animals in the current era of biotechnology.

The Stockman's Handbook presents the "why" as well as the "how." It contains, under one cover, the pertinent things that a stockman needs to know in the daily operation of a farm or ranch. It covers the broad field of animal agriculture, concisely and completely, and wherever possible in tabular and outline form.

Dedicated

to

the incredible pig who, from the remote day of its domestication forward, has artfully mirrored the world around it. Do you know—

■ **Where "Wall Street" got its name?**

The residents of colonial New York protected their precious fields from free-roaming swine by erecting a long, permanent wall on the northern edge of what is now Lower Manhattan. The street that came to border this wall was aptly named, *Wall Street*.

■ **Where the saying "living high on the hog" came from?**

It originated from poor people eating low-cost pigs feet, while those who could afford to ate more costly hams, pork roasts, and pork chops from high up on the carcass.

■ **The origin of the saying "a pig in a poke"?**

It has reference to a common trick in 17th century England to palm off on some unsuspecting greenhorn a cat in a poke (bag) instead of a suckling pig. Subsequently, when the poke was opened the trick was revealed.

■ **What President Harry Truman had to say about pigs?**

No man should be allowed to be a U.S. President who does not understand pigs.

■ **What living legacy the humble pig left to a hungry world?**

Root hog or die.

PREFACE TO THE SIXTH EDITION

Since 1970, changes in U.S. pork production, processing, and marketing have paced the whole field of agriculture. The changes came as a result of a shift from a labor-based economy to a capital-based economy; as a result of a shift from predominantly individual ownership, outdoor production, and the sweat of the brows of many caretakers, to corporate pigs, much confinement and environmentally controlled production, and maximum automation. It is expected that this trend will be slowed and resisted in the decades to come, but that science and technology will continue to be the great multipliers in the pork industry.

Present Status

The present status of the U.S. pork industry is indicated by the following facts:

1. **Fewer but bigger hog farms.** In 1994, there were only 208,780 hog farms in the U.S., down from 2.39 million hog farms in 1954; and the number of hogs per farm increased tenfold in the same 40-year period.

2. **Pork production is big business.** Pork producers are no longer their father's pig slopper. Today, pork production is big business.

3. **Per capita pork production has increased.** Pork consumption has moved to a higher per capita plateau.

4. **Big pork producers are industrialized.** Modern mega-pork producers are industrialized—hogs are produced in specialized factory-like facilities staffed with specialized labor. The driving forces back of it have been new technologies, new management techniques, lower cost of production, and increased quality control.

5. **Records are on computer.** Big and complex swine operations have outgrown hand record keeping. It is too time consuming. Today, there are many types of software (programs) available for use in swine production.

6. **Systems have evolved.** Pork production systems have evolved. Basically, there are three systems: (a) farrow-to-finish, (b) feeder pig production, and (c) grower-finisher.

What's Ahead for Pork

The U.S. pork team—producers, marketers, processors, and retailers—needs to project what's ahead. Here is what the senior author's crystal ball shows:

1. **Fewer and bigger hog farms.** Although this trend will continue, it will be impacted by (a) the ability and willingness of mega-operations to handle wastes in an environmentally responsible manner; (b) the concern given to the natural habitat/behavior of swine, and to breeding swine for adaptation; (c) the adverse consequences of displaced hog producers, displaced agribusinesses, and deteriorating rural communities; (d) the support accorded by big hog operators to colleges, hospitals, churches, and other charities in the states in which they operate, and (e) the increased business acumen required by big hog farms.

2. **Small producers will band together.** For survival, small producers will band together in co-ops and networks.

3. **Fewer, but bigger, integrators.** Big vertical integrators will increase their control of pork production and distribution through (a) joint complete ownership, or (b) contracts.

4. **Fat will continue to be an "ugly word."** Markets will pay a premium of $1.00 to $1.50 per cwt for each $1/10$ in. less backfat over the last rib.

5. **Outdoor breeding, gestation, and finishing will remain competitive.** Outdoor swine production will be reinvented as "outdoor intensive swine production," in which some of the efficiencies of complete confinement are incorporated.

6. **Multi-site production will become mainstream.** Having breeding, gestating, and farrowing at one site, nursery pigs at a second site, and growing-finishing pigs at a third site, with a mile or more between the three units, will increase for pig health reasons, accompanied by greater rate and efficiency of gains.

7. **Segregated early weaning (SEW) and all-in/all-out (AIAO) will increase.** For the control of infectious diseases, segregated early weaning—weaning at 10 to 17 days of age; and all-in/all-out management—in which pigs are moved in groups through (a) farrowing, (b) nursery, and (c) growing-finishing, with the barn vacated, washed and disinfected between each group—will increase.

8. **Big breeding companies will dominate the sales of swine breeding stock.** Each mega-hog producer will continue to hook up with a breeding company which can supply at one time a large number of healthy, high-quality breeding animals, free of the "stress" gene.

9. **Slaughter weights will increase.** Leaner genetics and growth enhancers will make for heavier slaughter weights. Symbol II (see cover of this book and Fig. 19-8 for pictures of Symbol II, the U.S. pork industry's graphic image of the ideal market hog, prepared by the National Pork Producers Council) boasts (a) a 260-lb live weight, with a 195 lb carcass; (b) a gilt that is marketed at 164 days, with a fat-free lean index of 52.2%; (c) a barrow that is marketed at 156 days, with a fat-free lean index of 49.8%; (d) produced free of the stress gene; and (e) the result of a terminal crossbreeding program.

10. **Fewer, but larger, feed companies will sell complete diets direct to producers.** With this trend, feed companies will bypass dealers and become more involved in joint financing hog farms.

11. **Pork may become number one in per capita meat consumption.** Pork, the other white meat, may overtake beef and poultry, in U.S. per capita meat consumption.

12. **The exciting age of biotechnology will arrive.** Biotechnology will reshape every facet of pork from breeding to meat on the table, including genetic makeup, physiology, stress tolerance, disease resistance, feed efficiency, the quality and quantity of pork produced, and food safety.

13. **The family farm will make a comeback.** Modern environmentally controlled swine buildings use a large variety of high-energy items—for heating, cooling, ventilation, etc. During the first half of the 21st century, fossil fuels will become scarcer and more costly, favoring sustainable agriculture, bio-mass products, labor intensive operations, soil conservation, and the comeback of the family farm.

National Live Stock & Meat Board, We Shall Miss Thee

Since its formation in 1922—75 years ago—the National Live Stock and Meat Board has imparted class, style, and elegance to the entire livestock industry. It has made a great difference. Its works will live on. Hopefully, the glory days of the National Live Stock and Meat Board will return soon.

I am Grateful to Many People

I am grateful to all those who contributed so richly to this major revision of *Swine Science*. Audrey H. Ensminger (Mrs. E) provided invaluable professional help, encouragement, book design, and layout. The following swine specialists provided a critique of the current edition of *Swine Science*, and made invaluable suggestions for this new sixth edition: Lee J. Johnston, Ph.D., West Central Experiment Station, University of Minnesota, Morris, Minnesota; William G. Luce, Extension Swine Specialist, Oklahoma State University, Stillwater, Oklahoma; and Wayne L. Singleton, Ph.D., Purdue University, West Lafayette, Indiana. Lawrence A. Duewer, Ph.D., Agricultural Economist, USDA/ERS provided many of the statistics. Wilton W. Heinemann, Ph.D., Animal Scientist and Nutrition Consultant, Yakima, Washington, made a final cover-to-cover reading of the camera-ready copy of the book and applied our standard checklist. Randall and Susan Rapp were par excellence in typesetting and proofreading. Additionally, a host of individuals, associations, and companies provided pictures and made other notable contributions, which are gratefully acknowledged throughout the book.

M. E. Ensminger
Clovis, California
1997

CONTENTS

Pigs were first brought to America by Hernando de Soto. The energetic Spanish explorer arrived at Tampa Bay (now Florida) in 1539. On his several vessels, he had 13 head of hogs. (Courtesy, The Bettmann Archive, Inc., New York, NY)

1

HISTORY AND DEVELOPMENT OF THE SWINE INDUSTRY

Nomadic peoples could not move swine about with them as easily as they could cattle, sheep, or horses. Moreover, close confinement was invariably accompanied by the foul odors of the pig sty. For this reason, the early keepers of swine were often regarded with contempt. This may have been the origin of the Hebrew and Moslem dislike of swine, later fortified by religious precept. As swine do not migrate great distances under natural conditions and the early nomadic peoples could not move them about easily, there developed in these animals, more than in most stock, a differentiation into local races that varied from place to place. It also appears that swine were domesticated in several different regions and that each region or country developed a characteristic type of hog.

ORIGIN AND DOMESTICATION OF SWINE

Wild pigs of the present day represent as many as 6 genera and 31 species. But the ancestors of the domestic pig are traceable to the genus and specie, *Sus scrofa*, which is recognized under three common names: pigs, hogs, and swine—terms which are used interchangeably.

Sus scrofa contributed to American breeds of swine, primarily through the following subspecies, or races: (1) the East Indian pig (largely, *Sus scrofa vittatus* and *Sus scrofa christatus*), and (2) the European wild boar (*Sus scrofa Eurasia*).

Archeological evidence indicates that swine were first domesticated in the East Indies and southeastern Asia, in the Neolithic Period, or New Stone Age, (1) beginning about 9000 B.C., in the eastern part of New Guinea (now known as Papua New Guinea), an island in the Pacific Ocean just north of Australia; and (2) about 7000 B.C., in Jericho, which lies in Jordan, north of the Dead Sea.

The domestication of the European wild boar came independently and later than the East Indian pig.

The East Indian pig was taken to China about 5000 B.C. Thence, Chinese pigs were taken to Europe in the last century, where they were crossed on the descendants of the European wild boar, thereby fusing the European and Asiatic strains of *Sus scrofa* and forming the foundation of present-day breeds.

In their wild state, pigs were gregarious animals, often forming large herds. Their feed consisted mostly of roots, mast (especially acorns and beechnuts), and such forage as they could glean from the fields and forests. While their diet was primarily vegetarian, they ate carrion (dead animals), wounded small animals, young birds, eggs, lizards, snakes, frogs, and fish. Because of their roving nature, diseases and parasites were almost unknown. Swine seem to have been especially variable under domestication and especially amenable to human selection. But, when given the opportunity, pigs promptly revert within only a few generations to a wild or feral state in which they acquire the body form and characteristics of their wild progenitors many generations removed. The self-sustaining razorback of the United States is an example of this reversion.

EURASIAN WILD BOAR (Sus scrofa)

The Eurasian[1] wild boar *(Sus scrofa)* is distributed in Europe, North Africa, and Asia. Although much reduced in numbers in the last few hundred years, it appears unlikely that the famous wild boar will become extinct like the Aurochs (the chief progenitor of domestic cattle). In comparison with the domestic pig, this race of hogs is characterized by its coarser hair (with an almost manelike crest along the back), larger and longer head, larger feet, longer and stronger tusks, narrower body, and greater ability to run and fight. The color of mature animals is nearly black, with a mixture of gray and rusty brown on the body. Very young pigs are striped. The ears are short and erect.

These sturdy ancestors of domestic swine are extremely courageous, are stubborn fighters, and are able to drive off most of their enemies, except humans. If attacked, they will use their tusks with deadly effect,

---

[1]The term *Eurasian* refers to Europe and Asia as a whole.

Fig. 1-1. Eurasian wild boar *(Sus scrofa)*, progenitor of the domestic breeds. Note the coarse hair, long head and snout, large feet, and long tusks. (Courtesy, New York Zoological Society, Bronx, NY)

although normally they are as shy as most animals and prefer to avoid people. The wild boar hunt has been regarded as a noble sport throughout history. Custom decrees that the hunt shall be on horseback and with dogs and that the quarry shall be killed with a spear.

The Eurasian wild boar will cross freely with domestic swine, and the offspring are fertile.

SOUTHEAST ASIAN PIG

At one time, this pig was considered to be a separate species which included a number of wild stocks of swine native to the East Indies and southeastern Asia. In appearance, these pigs are smaller

Fig. 1-2. The southeast Asian pig, *Sus scrofa vittatus*, was long considered to be a separate species. It is now classified as an eastern subspecies of the wild boar. (Courtesy, Field Museum of Natural History, Chicago, IL)

and more refined than the Eurasian wild boar. Also, they have a white streak along the sides of the face and the crest of hair on the back is absent. Today, these pigs are classified primarily as *Sus scrofa vittatus* and *Sus scrofa christatus*, eastern subspecies of the wild boar, since all intermediates between western and eastern specimens have been determined.

POSITION OF THE HOG IN THE ZOOLOGICAL SCHEME

The following outline shows the basic position of the domesticated hog in the zoological scheme:

Kingdom *Animalia*: Animals collectively; the animal kingdom.

Phylum *Chordata*: One of approximately 21 phyla of the animal kingdom, in which there is either a backbone (in the vertebrates) or the rudiment of a backbone (in the cephalochordates).

Class *Mammalia*: Mammals or warm-blooded, hairy animals that produce their young alive and suckle them for a variable period on a secretion from the mammary glands.

Order *Artiodacttyla*: Even-toed, hoofed mammals.

Family *Suidae*: The family of nonruminant, artiodactyl ungulates, consisting of wild and domestic swine but, in modern classifications, excluding the peccaries, which belong to the family *Tayassuidae*.

Genus *Sus*: The typical genus of swine, formerly comprehensive but now restricted to the European wild boar and its allies, with the domestic breeds derived from them.

Species *Sus scrofa*: *Sus scrofa* is the wild boar (hog) of Europe and Asia from which most domestic swine have been derived. Subspecies, therefore, include (1) the Central European wild boar, *Sus scrofa scrofa*, (2) the Japanese wild boar, *Sus scrofa leucomystax*, (3) the Southeast Asian pig, *Sus scrofa vittatus* and *Sus scrofa christatus*, and (4) the domestic pig, *Sus scrofa domestica*.

Other major species in the *Sus* genus include the pygmy hog, *Sus salvanis*; the bearded pig, *Sus barbatus*; and the Javan pig, *Sus verrucosus*.

Within the pig family, there are other genera which include the warthog, *Phacochoerus aethiopicus*; the babirusa, *Babyrousa babyrussa*; the giant forest hog, *Hylochoerus meinertzhageni*; and the bush pigs, *Potamochoerus porcus*.

HOW BACON GOT ITS NAME

The word *bacon* is said by one authority to have been derived from the old German word *baec*, which means *back*. However, others express the opinion that the word may have been derived from the noted Englishman, Lord Bacon, for the reason that the crest of Lord Bacon depicts a pig. In support of the latter story, it can be said that bacon has for many years been a favorite item in the English menu. It has also been suggested that the expression, "bringing home the bacon" is of English origin, having grown out of an ancient English ceremony, which annually took place

Fig. 1-3. The crest of Lord Bacon (1561–1626)—noted English Viscount, lawyer, statesman, politician. (Courtesy. Picture Post Library, London, England)

in a village about 40 miles from London. In this ceremony, it was the custom to award a "flitch of bacon" to each couple who, after a year of married life, could swear that they had been happy and had not wished themselves unwed. This 700-year-old tradition was revived in 1949 (see Fig. 1-4).

Fig. 1-4. Trial of the Dunmow flitch (or side) of bacon. Custom decreed that the married couple kneel on pointed stones while swearing that they had told the truth about their happy marital life. The traditional English ceremony of Dunmow, Essex, England, is—so the story goes— responsible for the origin of the word *bacon*. It was the custom to award a flitch of bacon to each couple who, after a year of marital life, could prove to the satisfaction of a judge and jury (composed of spinsters and bachelors) that they had been happy and had not wished themselves unwed. This homey 700-year-old tradition was started in the 12th century by a young lord who married a commoner, and, having proved after a year that he was happy with her, was given the flitch of bacon by the Prior of Dunmow. (Courtesy, Picture Post Library, London, England)

INTRODUCTION OF SWINE TO AMERICA

Although many wild animals were widely distributed over the North American continent prior to the coming of Europeans, the wild boar was unknown to the native American Indian.

Columbus first brought hogs to the West Indies on his second voyage, in 1493. According to historians, only eight head were landed as foundation stock. However, these hardy animals must have multiplied at a prodigious rate, for, 13 years later, the settlers of this same territory found it necessary to hunt the ferocious wild swine with dogs; they had grown so numerous that they were killing cattle.

Although swine were taken to other Spanish settlements following the early explorations of Columbus, pigs first saw America when touring the continent with Hernando De Soto. The energetic Spanish explorer arrived in Tampa Bay (now Florida), in 1539. Upon his several vessels (between 7 and 10), he had 600 or more soldiers, some 200 or 300 horses, and 13 head of hogs.

This hardy herd of squealing, scampering pigs traveled with the army of the brave Spanish explorer. The hazardous journey stretched from the Everglades of Florida to the Ozarks of Missouri. In spite of battles with hostile Indians, difficult travel, and other hardships, the pigs thrived so well that at the time of De Soto's death on the upper Mississippi, 3 years after the landing at Tampa, the hog herd had grown to 700.

De Soto's successor, Moscoso, then ordered that the swine be auctioned off among the men.

It is reasonable to assume, therefore, that the cross-country tour of De Soto's herd of pigs was the first swine enterprise in America. No doubt some of De Soto's herd escaped to the forest, and perhaps still others were traded to the Indians. At any rate, this sturdy stock served as foundation blood for some of the early American razorbacks.

Fig. 1-5. A typical Arkansas razorback. This two-year-old sow weighed 180 lb. (Courtesy, United Duroc Swine Registry, Peoria, IL)

COLONIAL SWINE PRODUCTION IN THE UNITED STATES

Sir Walter Raleigh brought sows to the Jamestown Colony in 1607. Their semi-wild progeny made such rampages in New York colonists' grain fields that every owned hog more than 14 in. high had to have a ring in its nose. On Manhattan Island, a long solid wall was constructed on the northern edge of the colony to control the roaming herds of hogs. This area is now known as *Wall Street*. One of the historical documents of 1633 refers to their innumerable swine. It is also noted that John Pynchon, founded the first meat packing plant in Springfield, Massachusetts, in 1641. He packed salted pork in barrels for shipment to the West Indies, most of which was exchanged for sugar and rum. Pynchon's records reveal something about the hogs in the early New England Colonies. They were described as black or sandy in color, with razorback build; and they were said to be speedy runners (as might be inferred from the fact that they ranged the woods in a half-wild state). With their huge tusks, the boars were believed to be quite capable of taking care of any wolves that might attack them. Pynchon's book records the weight of one lot of 162 hogs as 27,409 lb, an average of only 170 lb per animal. Sixteen of these weighed less than 120 lb. Twenty-five weighed more than 200 lb; and the 2 heaviest tipped the scales at 270 and 282 lb, respectively. Others followed Pynchon's lead in packing pork, and, in the year of 1790, it was reported that 6 million lb of pork and lard were exported from the United States.

Unlike the cattle, sheep, and horses—which were largely confined to the town commons for pasturage—the hogs of early New England roamed the surrounding countryside. Many of them were caught by hound dogs at marketing time. Usually a dog held on to each ear of the hog until the animal could be tied and pitched into a properly enclosed wagon. In order to prevent too much damage from rooting, some of the New England towns designated a "hog ringer," whose duty it was to ring all swine above a certain height (Hadley, Massachusetts, drew the line at 14 in.). Few hogs were marked, for these animals were so numerous that nobody minded the theft of a pig. The only care ever given the hogs, and then infrequently, was at farrowing time, at which time the sow might be given the privilege of using a shelter or be allowed to crawl under the barn or house.

Fig 1-6. Old-time farm slaughter scene. When nearly all the people lived on the land—prior to the growth of cities and the rise of the town butcher—each family did its own slaughtering and consumed the fresh meats or dried, smoked, or salted them for later use. (Courtesy, Swift and Company)

THE CREATION OF AMERICAN BREEDS OF SWINE

The most thoroughly American domestic animal is the hog. In no other class of animals have so many truly American breeds been created. These facts probably result from (1) the suitability of native maize or Indian corn as a swine feed, (2) the ease with which pork could be cured and stored prior to the days of refrigeration, and (3) the need for fats and high-energy foods for laborers engaged in the heavy development work of a frontier country.

Unlike the beef and dairy producers, who sent their native cattle to slaughter and imported whole herds of blooded cattle from England, the American hog raiser was content to use the mongrel sow descended from colonial ancestry as a base, upon which were crossed imported Chinese, Neapolitan, Berkshire, Tamworth, Russian, Suffolk Black, Byfield, and Irish Grazier boars. These importations began as early as the second quarter of the 19th century. Out of the various crosses, which varied from area to area, were created the several genuinely American breeds of swine.

Structurally, the creation of the modern hog has been that of developing an animal that would put flesh on the sides and quarters, instead of running to bone and a big head. Physiologically, breed improvement has resulted in an elongation of the intestine of the hog, thus enabling it to consume more feed for conversion into meat. According to naturalists, the average length of the intestine of a wild boar compared with his body is in the proportion of 9 to 1; whereas, in the improved American breeds, it is in the proportion of 13.5 to 1.

THE RISE OF CINCINNATI AS A PORK-PACKING CENTER

With the expansion of farming west of the Allegheny Mountains, corn was marketed chiefly through hogs and cattle. At the time of the first United States census, taken in 1840, the important hog-production centers were in what were then the corn growing areas of Tennessee, Kentucky, and Ohio.

Cincinnati became the earliest and foremost pork-packing center in the United States. By 1850, it was known throughout the length and breadth of the land as "Porkopolis." Cincinnati originated and perfected the system that packed 15 bushels of corn into a pig, packed that pig into a barrel, and sent it over the mountains and over the ocean to feed humanity.

Some idea relative to the rapid rise in pork slaughtering in Cincinnati and the price fluctuations of the period may be gained from Table 1-1.

Cincinnati was favored as an early-day, pork-packing center because (1) it was then the center of the finest hog-raising region in the world, and (2) it was strategically located from the standpoint of shipping, large quantities of cured pork being shipped to

TABLE 1-1
NUMBER OF HOGS SLAUGHTERED AND PRICES ON THE
CINCINNATI MARKET, 1833–67

Year	Number of Hogs Slaughtered	Year	Price/Cwt
1833	85,000	1855	$ 5.75
1838	182,000	1860	6.21
1843	250,000	1862	3.28
1853	360,000	1865	14.62
1863	606,457	1866	11.97
		1867	6.95

southern points in flat boats via the Ohio and Mississippi Rivers. Both pork and lard were exported to the West Indies, to England, and to France.

The inflated price of 1865 and the slump of two years later were the result of the demands caused by the Civil War. History repeated itself during World Wars I and II.

CHICAGO AS THE PORK-PACKING CENTER

By 1860, the center of pork production and packing had again shifted westward, and Chicago was the foremost packing center. Many railroads converged in Chicago, bringing hogs from the productive regions of the East and South. Initially, each of the five major railroads of the time built yards as an enticement of business. Then in 1865, the Illinois legislature incorporated the Union Stockyards and Transit Company—a single facility to accommodate all rail lines. Between 1865 and 1907, the Union Stockyards received about 241,000,000 hogs, or an average of about 6 million pigs per year.[2] For many years, Chicago held the position of the major market center. In 1914, the poet and native son of Illinois, Carl Sandburg, penned these words to describe Chicago:

"Hog Butcher for the World,
Tool Maker, Stacker of Wheat,
Player with Railroads, and the Nation's
  Freight Handler;
Stormy, husky, brawling,
City of the Big Shoulders...."

Until after World War I, centralized terminal marketing remained the dominant method of marketing. Gradually, however, marketing decentralized. Packing plants were built nearer the areas of production, and

---

[2]Coburn, F. D., *Swine in America*, Orange Judd company, New York, NY, 1910, p. 5.

producers began to market hogs direct. After World War II, the number of hogs received at Chicago declined from about 3.5 million in 1945 to about 1.4 million in 1965, and to only about 300,000 the first part of 1970. In May, 1970, the "hog butcher for the world," stopped accepting hogs at the Union Stockyards thus ending another era. Other public stockyards—namely, Omaha, National Stockyards (East St. Louis, IL), South St. Joseph (MO), South St. Paul, and Sioux City—continue receiving hogs.

GROWTH OF THE UNITED STATES SWINE INDUSTRY

The growth of hog production has paralleled very closely the production of corn in the north central or Corn Belt states, and these states produce nearly three-fourths of the corn grown in the country. Thus, when corn yields are down, the price of feed is up, and swine production decreases. The opposite occurs when corn yields are high.

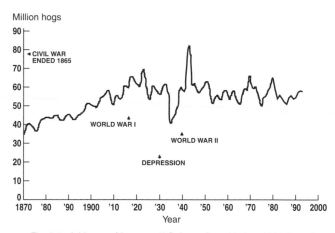

Fig. 1-7. A history of hogs on U.S. farms from 1870 to 1994, based on USDA data.

As shown in Fig. 1-7, hog numbers change sharply from year to year. A sharp increase occurred during the war years, with an all-time peak of 83,741,000 head on January 1, 1944. In recent years, however, hog numbers have oscillated below 60 million. Changes have probably been the result, in part at least, from increased per capita consumption of poultry. It is noteworthy, however, that hog numbers increased in the early 1990s.

QUESTIONS FOR STUDY AND DISCUSSION

1. Why didn't nomadic peoples move swine with them to the same extent that they moved cattle, sheep, and horses?

2. Trace the origin, domestication, and fusion of the East Indian pig and the European wild boar.

3. What prompted such early explorers as Columbus and De Soto to take swine with them?

4. More truly American breeds were created in swine than was the case with cattle, sheep, and horses. How do you explain this situation?

5. Discuss the reasons back of the rise and fall of Cincinnati as an early day pork-producing center.

6. Why did Chicago become "the hog butcher for the world," and then gradually lose this distinction?

7. Fig. 1-7 shows that there has been a tendency for hog numbers to level off since 1982, but to increase in the early 1990s. Why has this been so?

SELECTED REFERENCES

Title of Publication	Author(s)	Publisher
Ancestor for the Pigs: Taxonomy and Phylogeny of the Genus "Sus"	C. Groves	Australian National University Press, Canberra, Australia, 1982
Domesticated Animals from Early Times	J. Clutton-Brock	British Museum, London, and University of Texas Press, Austin, TX, 1981
Encyclopaedia Britannica		Encyclopaedia Britannica, Chicago, IL
Grzimek's Animal Life Encyclopedia, Vol. 13, *Mammals IV*	Ed. by B. Grzimek	Van Nostrand Reinhold Company, New York, NY, 1972
History of Livestock Raising in the United States 1607–1860	J. W. Thompson	Agri. History Series No. 5, U.S. Department of Agriculture, Washington, DC, November 1942
Natural History of the Pig, The	I. M. Mellen	Exposition Press, New York, NY, 1952
Our Friendly Animals and Whence They Came	K. P. Schmidt	M. A. Donohue & Co., Chicago, IL, 1938
Pigs from Cave to Corn Belt	C. W. Towne E. N. Wentworth	University of Oklahoma Press, Norman, OK, 1950
Pork Facts 1995–1996	Staff	National Pork Producers Council, Des Moines, IA, 1995–1996
Swine Nutrition	E. R. Miller D. E. Ullrey A. J. Lewis	Butterworth-Heinemann, Stoneham, MA, 1991

From whence they came. The European Wild Boar, after a painting by Ernest Griset. (Courtesy, Smithsonian Institution)

CHINA	🐖	🐖	🐖	🐖	🐖	🐖	393,865,000
U.S.A.	🐖	🐖	🐖	57,904,000			
RUSSIAN FED.	🐖	🐖	31,520,000				
BRAZIL	🐖	🐖	31,050,000				
GERMANY	🐖	26,466,000					
POLAND	🐖	18,860,000					
SPAIN	🐖	18,000,000					
MEXICO	🐖	16,832,000					
UKRAINE	🐖	16,175,000					
VIETNAM	🐖	14,861,000					

TEN LEADING SWINE PRODUCING COUNTRIES OF THE WORLD

2

WORLD AND U.S. SWINE AND PORK— PAST, PRESENT, AND FUTURE[1]

[1]The authors wish to express their appreciation to Mr. Kenneth Johnson and Dr. Terence R. Dockerty, National Live Stock and Meat Board, Chicago, Illinois, for their authoritative review of this chapter.

Swine are produced most numerously in the temperate zones and in those areas where the population is relatively dense. There is reason to believe that these conditions will continue to prevail. But the future can often be more clearly determined by studying some historical trends.

WORLD SWINE DISTRIBUTION AND PRODUCTION

Fig. 2-1 shows how the world swine population is distributed over the globe.

Most of the pigs in Asia are in China, which has long had the largest hog population of any nation. But,

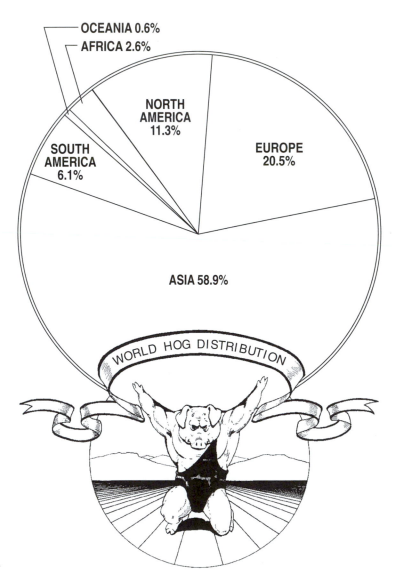

Fig. 2-1. World distribution of swine, by major areas. (Based on estimates from the *FAO Production Yearbook*, FAO/UN, Rome, Italy, 1994, pp. 192–194, Table 90)

because of the large human population, production in that country is largely on a domestic basis, with very negligible quantities of pork entering into world trade. Also, it must be remembered that in China pigs are primarily scavengers, and that the value of the manure produced is one of the main incentives for keeping them.

In South America, hog numbers have advanced at a rapid pace since the late 1950s. Brazil accounts for most of the production in that area.

In general, in the European countries, hog numbers are closely related to the development of the dairy industry and the production of barley and potatoes—in much the same manner as the distribution of swine in the United States is closely related to the acreage of corn.

Corn is raised extensively in the La Plata region of South America and in the Danube Basin of southern Europe. In these corn-growing areas, hog production is a dominant type of farming.

Dairy byproducts—skim milk, buttermilk, and whey—have long been important swine supplements in Denmark, the Netherlands, Ireland, and Sweden. In Germany and Poland, potatoes have always been extensively used in swine feeding.

Most North American swine are produced in the United States, where the hog and corn combination go together.

Table 2-1 provides additional details relative to the numbers of swine in specific countries, along with their relationship to the human population and country size.

From 1923 to the 1950s, except for the increases occurring during World War II, there was a downward trend in the exports of pork and lard from the United States. This was due to a marked increase in production in Canada and in the European countries, particularly in Denmark, Germany, and Ireland, and to various trade restrictions imposed by the importing countries. In no sense was this decrease due to any lack of capacity to produce on the part of the American farmer. Moreover, the general trend in exports of pork has been upward since the early 1960s, but lard production and exportation has declined due to changes in consumer preference.

Hog numbers fluctuate rather sharply on the basis of available feed supplies. Also, the annual per capita consumption of pork in different countries of the world varies directly with production and availability, cost, the taste preference of the people, and in some cases with the religious beliefs that bar the use of pork as a food.

TABLE 2-1
SIZE AND DENSITY OF HOG POPULATION OF 10 LEADING HOG PRODUCING COUNTRIES OF THE WORLD, BY RANK

Country	Population		Size of Country		Hogs per Capita	Hogs per Unit of Area	
	Hogs[1]	Humans[2]					
			(sq mi)	(sq km)		(sq mi)	(sq km)
China	393,965,000[3]	1,169,619,000	3,696,100	9,572,899	0.34	106.6	41.2
United States	57,904,000	248,709,873	3,618,770	9,372,614	0.23	16.0	6.2
Russian Fed.	31,520,000[3]	149,527,000	8,649,538	17,075,352	0.21	4.8	1.8
Brazil	31,050,000	158,000,000	3,286,470	8,511,957	0.20	9.4	3.6
Germany	26,466,000	80,387,000	137,838	357,000	0.33	192.0	74.1
Poland	18,860,000	38,385,000	120,727	312,683	0.49	156.0	60.3
Spain	18,000,000	39,118,000	194,896	504,781	0.46	92.4	35.7
Mexico	16,832,000	92,380,000	761,604	1,972,554	0.18	22.1	8.5
Ukraine	16,175,000	51,994,000	233,100	602,729	0.31	69.4	26.8
Vietnam	14,861,000	68,964,000	127,330	329,785	0.22	116.7	45.1

[1] *FAO Production Yearbook*, FAO/UN, Rome, Italy, 1993, Vol. 47, p. 192, Table 90.

[2] *The World Almanac* 1994.

[3] *Pork Facts 1995/1996*, National Pork Producers Council, Des Moines, Iowa.

U.S. PRODUCTION AND DISTRIBUTION

The contribution of the humble pig to American agriculture is expressed by its undisputed title as the "mortgage lifter." No other animal has been of such importance to the farmer, but the number of hogs and their value fluctuate.

The point and purpose of swine in the United States is the production of meat. Table 2-2 shows how pork production compares with the production of beef/veal and lamb/mutton. The most recent data available show that pork production accounts for about 42% of all red meat produced in the United States.

The sections that follow show where hogs are distributed in the United States, and give pertinent information.

AREAS OF SWINE PRODUCTION

The geographical distribution of swine in the United States coincides closely with the acreage of corn, the principal swine feed. Normally, one half of the corn crop is fed to hogs. It is not surprising, therefore, to find that about 60% of the hog production is centered in the seven Corn Belt states: Iowa, Illinois, Indiana, Ohio, Missouri, Nebraska, and Kansas.2 From this it should not be concluded that sections other than

TABLE 2-2
U.S. PRODUCTION OF RED MEAT[1]

Meat	Year	Production		Production as Percentage of All Red Meats
		- - - - - - (million) - - - - - -		
		(lb)	(kg)	(%)
Beef & Veal	1975	24,847	11,268	68
	1980	22,044	9,997	57
	1985	24,242	11,006	62
	1990	23,070	10,474	59
	1993	23,335	10,594	57
Pork (excluding lard)	1975	11,503	5,218	31
	1980	16,615	7,535	43
	1985	14,805	6,721	38
	1990	15,353	6,970	40
	1993	17,087	7,757	42
Lamb & Mutton	1975	410	186	1.1
	1980	318	145	0.8
	1985	357	162	0.9
	1990	362	164	0.9
	1993	337	153	0.8

[2] *Agricultural Statistics, 1994*, p. 239, Table 401.

[1] *Agricultural Statistics*, USDA, 1981, p. 342, Table 504; 1994, p. 262, Table 436.

the Corn Belt are not well adapted to pork production. As a matter of fact, any area that produces small grains is admirably adapted to the production of pork of the highest quality. Fig. 2-2 shows swine numbers and distribution by geographical areas in the United States.

Since 1920, there has been a significant increase in pork production in the northwestern Corn Belt and in the northern and western Great Plains area. This has been attributed to the increased corn and barley production in these areas. But the most phenomenal recent increase in swine production has occurred in North Carolina, which now ranks second only to Iowa in hog numbers (see Fig. 2-3).

The eastern, New England, and western states of the United States are pork deficit areas. Despite the greatly expanded human population of the Pacific Coast, it is noteworthy that hog numbers in this area have remained about the same since 1900.

In the past, many of the live hogs slaughtered in the West Coast plants were shipped distances of 1,500 to 2,000 miles. They were being transported greater distances as live animals than was necessary a century ago when the eastern packers were prompted to move their slaughtering plants from the East Coast to Chicago. Today, packing plants are located near the areas of production and hogs are trucked shorter distances; and many of them sold direct from producer to packer, with the price determined by carcass merit.

CORN-HOG RELATIONSHIP

With the opening of the central Mississippi Valley region, it soon became evident that here was one of the greatest corn countries in the world. The fertile soil, relatively long growing season, ample moisture, and warm nights were ideal for corn production. Also, it was soon realized that corn was unsurpassed as a hog feed. Here appeared to be an invincible combination, but, unfortunately, the early-day hog proved quite incompetent in the efficient conversion of corn into meat and lard. Undaunted, the swine producers of America promptly set about improving the existing swine, eventually developing several new and distinctly American breeds. Thus, "King Corn" and the American hog have played no small part in the development of American agriculture and in the prosperity of our farmers. The hog created a channel of disposal, or market,

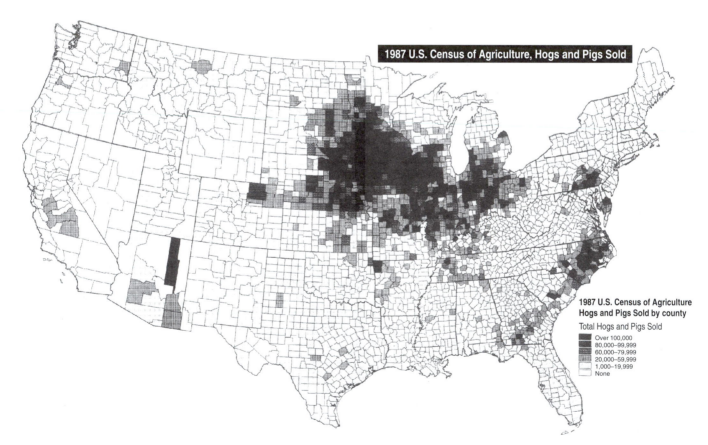

1987 U.S. Census of Agriculture, Hogs and Pigs Sold

1987 U.S. Census of Agriculture
Hogs and Pigs Sold by county
Total Hogs and Pigs Sold
Over 100,000
80,000–99,999
60,000–79,999
20,000–59,999
1,000–19,999
None

Fig. 2-2. The distribution of hogs in the United States. (Courtesy, *Pork Facts 1995/1996*, National Pork Producers Council, Des Moines, IA. Source: *National Hog Farmer*)

for corn and supplied the people of this and other countries with highly palatable and nutritious meat at a moderate price.

HOG AND BEEF COMBINATION

Farmers in the central states long ago recognized the advantages of combining beef cattle and hogs. Regardless of the system of beef production—cow and calf production, the growing of stockers and feeders, finishing steers, dual-purpose production, or a combination of two or more of these systems—the beef cattle and swine enterprises complement each other in balanced farming.

The cattle are able to utilize effectively great quantities of roughages, both dry forages (hay, fodder, etc.) and pastures; whereas the pig is fed primarily on concentrates. In brief, the beef cattle-hog combination makes it possible to market efficiently all the forages and grains through livestock. Also, such a combination makes for excellent distribution of labor. The largest labor requirements for both beef cattle and hogs come in the winter and early spring. During the growing and harvesting seasons, therefore, most of the labor is released for attention to the crops.

SWINE PRODUCTION IN THE SOUTH

In recent years, there has been a concerted effort to diversify the agriculture of the South away from the straight cotton, peanut, and tobacco farming traditional to the area. In part, this movement has been motivated by the recognized need for greater attention to improved soil conservation practices and the unprofitable prices sometimes encountered from the sale of cotton, peanuts, and tobacco. Also, vertical integration has speeded the movement through providing necessary capital and know-how.

The southern states are by no means uniform in the crops that they grow. A great variety of suitable swine feeds is produced from area to area. The common hog feeds of the South include corn, peanuts, soybeans, velvet beans, sweet potatoes, molasses, cottonseed meal, some small grains, and numerous grazing crops. From the standpoint of pastures, the South has the unique advantage of possible year-round grazing, particularly when permanent pastures are properly supplemented with adapted temporary grazing crops.

SWINE PRODUCTION IN THE WEST

As has already been pointed out, the West is a swine deficit area. This is largely due to the fact that other enterprises have been more remunerative; and, over a long period of time, farmers and ranchers do those things which are most profitable to them. Moreover, in the West, wheat, the leading grain crop of the area, is frequently too high in price to use profitably as a swine feed. This is due to the fact that wheat has always been considered primarily a human cereal, and that it frequently has been federally subsidized.

Expansion of swine production in the West in the 1980s and 1990s was driven primarily by large industrial swine operations seeking wide open spaces (1) to escape the increasing legislation and litigation against odor and contamination nuisances in the Corn Belt, and (2) to isolate their big operations from other hogs and swine diseases.

LEADING STATES IN SWINE PRODUCTION

All 50 states have some hogs, but the state of Iowa has held undisputed lead in hog numbers since 1880. The rank of the other states has shifted considerably. A ranking of the 10 leading states, based on recent data, is shown in Fig. 2-3.

Growing corn and producing pork have contributed largely toward making the farmers of the upper Mississippi Valley the wealthiest agricultural people on the globe.

TEN LEADING PORK PRODUCING STATES

IOWA	14.2
NORTH CAROLINA	7.0
ILLINOIS	5.35
MINNESOTA	4.85
INDIANA	4.5
NEBRASKA	4.35
MISSOURI	3.45
OHIO	1.8
SOUTH DAKOTA	1.74
KANSAS	1.31

MILLIONS OF HOGS ON FARMS ON A SPECIFIC DATE

Fig. 2-3. Ten leading states in hog numbers, by rank. (Based on USDA data)

DECLINE IN HOG FARMS; INCREASE IN SIZE

The number of U.S. farms reporting hogs has declined sharply in recent years. Table 2-3 shows that during the 26-year period 1968 to 1994, the number of hog farms dropped from 967,580 to 208,780—a whopping 78%. It is expected that this trend will continue, but with less momentum.

Table 2-4 presents data for (1) five different size hog operations, ranging from 1–99 head to 2,000+ head; (2) 10 states; (3) 16 states; (4) the other 34 states; and (5) the entire U.S. Clearly this shows that many small hog farms are going out of business, and that the big farms are getting bigger. It is expected that this trend will continue, but at a slower pace and with a tendency to level off.

TABLE 2-3
NUMBER OF U.S. HOG OPERATIONS[1]

Year	No.	Year	No.
1968	967,580	1982	482,190
1969	873,840	1983	462,110
1970	871,200	1984	429,580
1971	869,000	1985	391,000
1972	778,200	1986	348,000
1973	735,700	1987	331,620
1974	733,100	1988	326,600
1975	661,700	1989	306,210
1976	658,300	1990	275,440
1977	647,000	1991	253,890
1978	635,300	1992	248,700
1979	653,600	1993	225,210
1980	670,350	1994	208,780
1981	580,060		

[1]From: *Pork Facts 1995/96*, National Pork Producers Council, Des Moines, IA. Source: USDA *Hogs and Pigs*, December, 1994.

TABLE 2-4
SIZE OF U.S. SWINE OPERATIONS[1]

	Operations Having									
	1–99 Head		100–499 Head		500–999 Head		1,000–1,999 Head		2,000+ Head	
	1993	1994	1993	1994	1993	1994	1993	1994	1993	1994
Georgia	4,500	3,900	1,100	930	220	190	100	100	80	80
Illinois	4,000	3,800	4,400	4,000	1,800	1,900	840	820	460	480
Indiana	6,000	5,500	3,800	3,500	1,400	1,200	600	600	400	400
Iowa	8,000	7,000	15,000	12,300	6,500	6,000	2,600	2,700	900	1,000
Kansas	2,900	2,300	1,800	1,600	390	360	120	140	90	100
Minnesota	5,900	6,000	5,200	5,200	1,800	1,700	760	700	340	400
Missouri	5,500	5,300	3,800	3,500	1,100	1,000	450	500	150	200
Nebraska	4,000	3,650	5,200	5,150	1,550	1,450	500	500	250	250
North Carolina	5,700	5,000	580	550	280	280	320	350	620	820
Ohio	10,200	9,200	2,400	3,100	760	730	300	320	40	50
10 States	56,700	51,650	43,280	39,830	15,800	14,810	6,590	6,730	3,330	3,780
Kentucky	4,500	3,200	930	900	200	230	110	110	60	60
Mississippi	3,400	3,400	1,100	1,100	220	230	160	140	120	130
Pennsylvania	4,500	4,300	930	920	280	280	180	190	110	110
South Dakota	2,700	2,300	3,300	3,100	750	760	250	230	100	110
Tennessee	6,000	4,900	600	600	200	200	60	70	40	30
Wisconsin	5,500	5,000	2,300	2,000	430	420	120	130	50	50
16 States	83,300	74,750	52,440	48,540	17,880	16,930	7,470	7,600	3,810	4,270
Other 34 States	54,200	50,500	4,500	4,600	750	800	510	520	35	360
United States	137,500	125,250	56,940	53,050	18,630	17,730	7,980	8,120	4,160	4,630

[1]From: *Pork Facts 1995/96*, National Pork Producers Council, Des Moines, IA.

CASH RECEIPTS

Table 2-5 shows the annual cash receipts from U.S. animal commodities and total farm income from 1973 to 1994. The column on the far right includes the total farm receipts from both animals and crops.

The pie diagram that follows (Fig. 2-4), shows the proportion of animals and their products derived from each class of livestock in 1994.

It is noteworthy that swine accounted for 12.4% of the cash receipts from animals and animal products in 1994, and that income from swine was exceeded by beef, dairy, and poultry.

As would be expected, the proportions are somewhat changeable from year to year, depending upon the relative value of the various farm products and the amount produced.

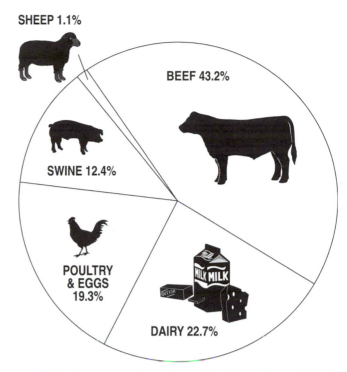

SHEEP 1.1%
BEEF 43.2%
SWINE 12.4%
POULTRY & EGGS 19.3%
DAIRY 22.7%

Fig. 2-4. The proportion of cash income derived from each class of animals and their products, in 1994.

TABLE 2-5
CASH RECEIPTS FROM ANIMAL COMMODITIES AND TOTAL FARM INCOME[1]

Year	Poultry	Beef	Dairy	Swine	Sheep	Total Income From Animals and Products	Total Income From Animals and Crops
					(mil. $)		
1973	6.9	22.3	8.3	7.5	0.52	45.9	87.1
1974	6.4	17.8	9.7	6.9	0.45	41.5	92.4
1975	6.8	17.5	10.2	7.9	0.45	43.3	88.9
1976	7.2	19.3	11.8	7.5	0.49	46.4	95.4
1977	7.2	20.2	11.8	7.3	0.48	47.5	96.2
1978	8.0	28.2	12.7	8.8	0.56	59.0	112.9
1979	8.7	34.4	14.7	9.0	0.61	68.1	133.8
1980	8.9	31.5	16.6	8.9	0.59	67.8	142.0
1981	9.9	29.6	18.1	9.8	0.55	63.3	144.1
1982	9.5	29.9	18.2	10.6	0.54	70.3	147.1
1983	10.0	28.7	18.8	9.8	0.52	69.4	141.1
1984	12.2	30.6	17.9	9.7	0.47	72.9	150.7
1985	11.2	29.0	18.1	9.0	0.50	69.8	151.9
1986	12.7	28.9	17.8	9.7	0.50	71.0	147.0
1987	11.5	33.6	17.7	10.3	0.60	76.0	141.8
1988	12.9	36.8	17.6	9.2	0.50	79.4	161.1
1989	15.4	36.9	19.4	9.5	0.50	84.1	160.9
1990	15.2	39.9	20.2	11.6	0.40	89.9	169.9
1991	15.1	39.6	18.0	11.1	0.40	86.7	168.7
1992	15.5	37.9	19.8	10.1	0.50	86.3	171.2
1993	17.2	40.0	19.3	10.9	0.50	90.6	175.1
1994	17.0	38.0	20.0	10.9	1.00	88.0	178.0

[1]From: *Pork Facts 1995/96*, National Pork Producers Council, Des Moines, IA. Source: USDA.

WORLD PORK CONSUMPTION

In general, pork consumption (and production) is highest in the temperate zones of the world, and in those areas where the population is relatively dense. In many countries, such as China, pigs are primarily scavengers; in others, hog numbers are closely related to corn, barley, potato, or dairy production. As would be expected, the per capita consumption of pork in different countries of the world varies directly with its production and availability (see Table 2-6). Food habits and religious restrictions also affect the amount of pork consumed. For example, Islam and Judaism both prohibit the consumption of pork.

In 1993, the United States produced 18.5 million metric tons of red meat, or about 15.8% of the total world production. China is now the largest meat producer with 32.3 million metric tons, or 27.6% of the total red meat production.

Table 2-6 gives a summary of per capita meat consumption in the leading meat-eating countries of the world and shows the position of pork. It is noteworthy that pork is in an especially favored position in Spain, Germany, France, and Italy. However, the greatest per capita consumers of pork are not necessarily big eaters of all meats.

Fig. 2-5 shows the leading countries in per capita consumption of pork.

TABLE 2-6
MEAT PRODUCTION AND PER CAPITA CONSUMPTION IN 10 LEADING COUNTRIES OF THE WORLD

Country[1] (leading countries, by rank, of all meats)	Production			Per Capita Consumption[2]			
	Total Red Meat	Pork	Percent Pork of Red Meat	Total Red Meat		Pork	
	- - - - - (1,000 metric tons) - - - - -		(%)	(lb)	(kg)	(lb)	(kg)
China	32,255	28,544	88	62	28	53	24
United States	18,488	7,751	42	167	76	67	30
Russian Fed.	6,273	2,551	41	90	41	37	17
Brazil	5,864	1,250	21	73	33	18	8
Germany	4,817	3,095	64	150	68	108	49
France	4,018	2,151	54	158	72	84	38
Spain	2,817	2,088	74	161	73	119	54
Australia	2,810	328	12	166	76	40	18
Mexico	2,718	870	32	71	32	22	10
Italy	2,642	1,371	52	141	64	77	35

[1]*Agricultural Statistics 1994*, USDA, p. 263, Table 437 (1993 data).

[2]*Livestock and Poultry World Markets and Trade*, USDA FAS, October 1994, pp. 53-55.

PORK CONSUMPTION

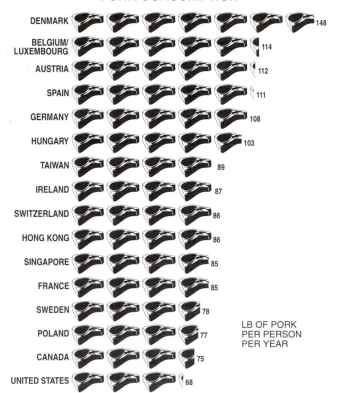

DENMARK	148
BELGIUM/ LUXEMBOURG	114
AUSTRIA	112
SPAIN	111
GERMANY	108
HUNGARY	103
TAIWAN	89
IRELAND	87
SWITZERLAND	86
HONG KONG	86
SINGAPORE	85
FRANCE	85
SWEDEN	78
POLAND	77
CANADA	75
UNITED STATES	68

LB OF PORK PER PERSON PER YEAR

Fig. 2-5. Leading countries in per capita pork consumption, carcass-weight basis, 1994. (Based on data from *Pork Facts 1995/1996*, National Pork Producers Council, p. 18.)

TABLE 2-7
TOP 10 COUNTRIES IN TOTAL PORK CONSUMPTION[1]

Country	Pork Consumed
	(1,000 metric tons)
China	30,000
United States	7,929
Germany	2,985
Russia	2,300
France	2,240
Spain	2,088
Netherlands	1,730
Denmark	1,557
Japan	1,410
Italy	1,340

[1]*Pork Facts, 1995/1996*, National Pork Producers Council, Des Moines, IA, p. 21.

It is noteworthy that the countries eating the most pork per individual (Fig. 2-5) are not the countries consuming the most total pork (Table 2-7), because the latter figure is determined by per capita consumption and population.

U.S. PORK AND OTHER ANIMAL PRODUCT CONSUMPTION

From Tables 2-8 and 2-9, and Fig. 2-6, the following conclusions can be drawn:

1. Pork ranks second as the preferred red meat in the United States, beef having replaced pork in the first position in the early 1950s.

2. Since 1960, pork consumption has fluctuated between 60 and 80 lb per capita. In the mid-1990s it showed signs of moving to a slightly higher plateau.

TABLE 2-8
U.S. PER CAPITA RED MEAT AND PORK CONSUMPTION
(carcass weight basis)[1]

Year	All Red Meats		Pork		% Pork of All Red Meats
	(lb)	(kg)	(lb)	(kg)	(%)
1984	176.0	80.0	66.1	30.0	38
1985	177.4	80.6	66.5	30.2	37
1986	174.7	79.4	62.9	28.6	36
1987	170.6	77.5	63.2	28.7	37
1988	173.9	79.0	67.5	30.7	39
1989	168.4	76.5	67.0	30.4	40
1990	163.2	74.2	64.1	29.1	39
1991	163.1	74.1	64.9	29.5	40
1992	166.1	75.5	68.4	31.1	41
1993	163.0	74.1	67.5	30.7	41
10-year avg.	169.6	77.1	65.8	29.9	39

[1]*Agricultural Statistics 1994*, p. 267, Table 443.

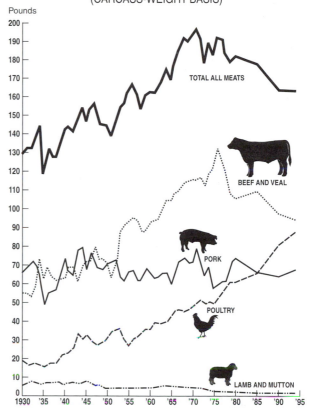

MEAT CONSUMPTION PER PERSON
(CARCASS-WEIGHT BASIS)

Fig. 2-6. A history of the per capita meat and poultry consumption (carcass-weight basis) in the United States. Poultry consumption has tripled since 1960, while pork consumption has hovered between 60 and 80 lb per capita. Beef consumption has gone down and down since 1976. (Based on data from various years of *Agricultural Statistics*, USDA)

TABLE 2-9
U.S. PER CAPITA CONSUMPTION OF MEAT, EGGS, AND MILK[1]

Animal Products	1960	1970	1980	1990	1993
Meats:					
Beef (carcass wt. equivalent) . . . . . (lb)	85.2	113.5	103.4	96.1	93.0
Pork (carcass wt. equivalent) . . . . . (lb)	65.2	72.6	73.5	64.1	67.5
Chicken (ready to cook) . . . . (lb)	27.8	40.5	49.7	71.9	70.1
Turkey (ready to cook) . . . . (lb)	6.2	8.0	10.5	18.1	17.8
Fish . . . . (lb)	10.3	11.8	12.8	15.0	14.9
Eggs . . . . (no.)	334	311	272	234	234
All milk equivalent . . . . (lb)	653	561	543	570	565

[1]USDA sources.

3. Since 1960, chicken consumption has shown a 2.5-fold increase, turkey consumption has shown a 2.9-fold increase, and fish consumption has shown a 1.4-fold increase.

4. Since 1960, the consumption of eggs and the consumption of dairy products on a milk equivalent basis have decreased.

5. Although not shown in Tables 2-8, 2-9, or Fig. 2-6, the per capita consumption of fruits, vegetables, and cereal products has increased in recent years.

U.S. PORK IMPORTS AND EXPORTS

Hog producers are prone to ask why the United States, with a successful productive swine industry, buys pork abroad. Conversely, consumers sometimes wonder why we export pork. Occasionally, there is justification for such fears, on a temporary basis and in certain areas, but as shown in Table 2-10, this nation neither imports nor exports large quantities of pork.

Table 2-10 also reveals that the United States imports slightly more pork than it exports, but total pork imports actually constitute a very small percentage of the available U.S. pork.

The amount of pork imported depends to a substantial degree on (1) the level of U.S. meat production, (2) consumer buying power, (3) hog prices, and (4) tariffs. The current rate of duty for pork ranges from 1 to 3¢/lb for the most-favored nations to 3.25¢/lb for those nations not accorded most-favored status.[3] But because tariffs have always been in politics and are subject to change, pork producers need to increase efficiency of production as a means of meeting foreign competition.

---

[3] Tariff Schedules of the United States Annotated (1987), U.S. Govt. Printing Office, Washington, DC.

The amount of pork exported from this country is dependent upon (1) the volume of meat produced in the United States, (2) the volume of meat produced abroad, and (3) the relative vigor of international trade, especially as affected by buying power and trade restrictions.

Our pork imports come principally from Brazil, Canada, Denmark, Hungary, the Netherlands, Poland, and Sweden. Neither swine nor the fresh, chilled, or frozen meat from hogs can be imported from any country in which it has been determined that rinderpest or foot-and-mouth disease exists.

The United States exports pork to Brazil, Canada, Colombia, EU-12, Hong Kong, Japan, Korea, Mexico, the Netherlands-Antilles, and Russia.

We export more lard than any other nation. Our leading countries of lard destination are: Asia, Canada, Mexico, the Netherlands, and Spain. Mexico buys more than half of our lard.

Some types of pork products and the amounts the United States imports and exports are listed in Table 2-11.

FUNCTIONS OF SWINE

The average person is aware, at least in part, of the basic utility function of swine in contributing food. Few recognize, however, that—because of their added functions—swine are an integral part of a sound, mature, and permanent agriculture.

The primary functions of swine are to—

1. Contribute food.
2. Provide profitable returns from available labor.
3. Convert inedible feeds into valuable products.
4. Aid in maintaining soil fertility.
5. Serve as an important companion of grain production.

TABLE 2-10
U.S. IMPORTS, EXPORTS, AND NET IMPORTS OF PORK (carcass weight basis)[1]

Year	Production		Imports		Exports		Net Imports		Imports	Exports	Net Imports
	- - - - - - - - - - - - - - (million lb [kg] carcass equivalent) - - - - - - - - - - - - - - -								- - - - - - - (percent of U.S. production) - - - - - - -		
	(lb)	(kg)	(lb)	(kg)	(lb)	(kg)	(lb)	(kg)			
1989	15,759	7,163	895	407	262	119	633	288	5.7	1.7	4.0
1990	15,300	6,954	898	408	239	109	659	299	5.9	1.6	4.3
1991	15,948	7,249	775	352	283	129	492	224	4.9	1.8	3.1
1992	17,184	7,811	646	294	407	185	239	109	3.8	2.4	1.4
1993	17,030	7,741	740	336	435	198	305	139	4.4	2.6	1.8
5 yr. avg.	16,244	7,384	791	359	325	148	466	212	4.9	2.0	2.9

[1] From: Meat & Poultry Facts, 1994, p. 48, American Meat Institute, Washington, DC.

TABLE 2-11
QUANTITIES OF U.S. PORK IMPORTS AND EXPORTS, BY TYPE OF PRODUCT[1]

Product	Year									
	1984	1985	1986	1987	1988	1989	1990	1991	1992	1993
Import	- (metric tons) -									
Fresh and frozen	207,703	254,538	263,488	302,392	282,728	225,304	232,253	215,933	185,672	207,652
Canned[2]	142,423	160,802	151,730	145,464	139,847	107,267	98,479	72,666	54,114	70,577
Other prepared or preserved[3] . . .	3,372	6,312	8,221	9,387	10,214	8,651	10,055	11,760	13,104	14,295
Sausage, all types[4]	2,243	2,192	2,640	2,688	2,906	2,656	3,421	2,144	2,453	2,695
Export										
Fresh and frozen	46,098	34,394	20,969	29,145	54,598	79,318	66,756	76,193	116,496	129,240
Hams and shoulders, cured or cooked	1,474	1,175	650	1,227	2,138	6,101	5,567	4,702	8,181	5,208
Bacon	621	450	474	617	1,045	3,788	4,518	5,443	7,396	7,092
Other pork, pickled, salted or otherwise cured	3,837	4,066	4,796	3,597	4,924	2,204	4,310	6,133	5,812	4,579
Other pork, canned	513	638	349	376	268	1,395	1,036	1,278	2,352	2,349

[1]Agricultural Statistics, USDA, 1994, pp. 264-265.

[2]Includes canned hams, shoulders, and bacon.

[3]Includes pickled and cured.

[4]Includes fresh and cured sausage.

6. Supplement other enterprises.
7. Provide other functions.

Each of these functions will be detailed in a separate section that follows.

CONTRIBUTE FOOD

Table 2-12 shows the United States per capita consumption of selected foods and products. It is noteworthy that the U.S. per capita consumption of pork ranks second only to beef in terms of meat consumption and seventh in overall consumption of foods and products.

Furthermore, the food supplied by pork is of the highest quality. The protein of pork provides all of the essential amino acids, including lysine and methionine, along with certain minerals and vitamins.

Additionally, pork is highly digestible and very palatable. These factors are important to peoples of the world, for how they live and how long they live are determined in large part by the diet. In the United States, pork makes a significant contribution toward meeting the nutritional needs of people, as shown in Table 2-13. The pork available in the American diet

TABLE 2-12
ANNUAL PER CAPITA CONSUMPTION OF SELECTED FOODS AND PRODUCTS[1]

Product	Per Capita per Year	
	(lb)	(kg)
Dairy products (all milk equiv.)	572	260
Vegetables (fresh, canned, and frozen) . . .	244	111
Grains (excluding corn syrup and sugar) . .	156	71
Potatoes and sweet potatoes	122	55
Fruits (fresh and processed)	110	50
Beef (carcass wt. equiv.)	93	42
Pork (carcass wt. equiv.)	67	30
Fats and oils	64	29
Sugar (refined)	63	29
Poultry (ready-to-cook)	61	28
Fish (edible weight)	15	7
Lamb and mutton	1	0.5
Veal	1	0.5
All red meat	162	74

[1]Data for 1993. Agricultural Statistics 1994, USDA, p. 267, Table 443; and p. 431, Table 653.

supplies about 21% of the protein, 16% of the phosphorus, 16% of the iron, 20% of the thiamin, and 13% of the niacin that is required for good health. Also, pork supplies 100% of the requirement for vitamin B-12—a vitamin which occurs only in animal food sources and fermentation products. It is noteworthy, too, that the availability of iron in meat is about twice as high as in plants. (For further details regarding the nutritive qualities of pork, refer to Chapter 18.)

TABLE 2-13
PERCENTAGE OF RECOMMENDED DAILY DIETARY ALLOWANCES
(RDA) SUPPLIED BY PORK IN THE UNITED STATES

Nutrient	RDA[1]	Amount Supplied Daily by Pork[2]	Percent of RDA
			(%)
Food energy	2,900 kcal	208 kcal	7.2
Protein	63 g	13 g	20.6
Calcium	800 mg	6 mg	0.8
Phosphorus	800 mg	130 mg	16.3
Iron	10 mg	1.6 mg	16.0
Magnesium	350 mg	14 mg	4.0
Vitamin A	1,000 mcg RE	0	0
Thiamin	1.5 mg	0.3 mg	20.0
Riboflavin	1.7 mg	0.12 mg	7.1
Niacin	19 mg	2.5 mg	13.2
Vitamin B-6	2.0 mg	0.2 mg	10.0
Vitamin B-12	2 mcg	2 mcg	100.0
Vitamin C	60 mg	0 mg	0

[1]Recommended Dietary Allowances, National Research Council, 1989, for a 25- to 50-year-old male.

[2]Computed using data from Agricultural Statistics, USDA, 1994, and assuming that the pork available for consumption represents about 26% of that supplied by meat, poultry, and fish (2.1 oz pork eaten/day).

PROVIDE PROFITABLE RETURNS FROM AVAILABLE LABOR

Consciously or unconsciously, most farmers keep hogs simply because they find them remunerative. Primary factors contributing to a profitable swine enterprise are: (1) a relatively high labor return in comparison with other types of livestock production, and (2) a fairly uniform labor requirement throughout the year.

CONVERT INEDIBLE FEEDS INTO VALUABLE PRODUCTS

Swine are better adapted than any other class of livestock to utilizing many wastes and byproducts that are not suited for human consumption.

Among what would otherwise be waste feeds that are fed to swine are garbage, bakery wastes, garden waste, cull or damaged grain, and root crops and fruit.

Pigs also convert to edible foods the numerous byproducts of the meat packing, fishery, grain milling, and vegetable oil processing industries. Some of these residues (or wastes) have been used for animal feeds for so long, and so extensively, that they are commonly classed as feed ingredients, along with such things as the cereal grains, without reference to their byproduct origin. Most of these processing residues have little or no value as a source of nutrients for human consumption.

Fig. 2-7. Chinese hogs on Ping Chou People's Commune, Kwangtung Province, in China. Their ration consisted of two byproducts—rice millfeed and bagasse (the pith of sugarcane), along with water hyacinth—all of which the pigs ate with relish. In China, swine utilize millions of tons of otherwise wasted crop residues and byproducts. (Photo by Audrey H. Ensminger)

AID IN MAINTAINING SOIL FERTILITY

Cash crops, whether they be grains or forages, result in the marketing of soil fertility. Although it is possible to use fertilizers and green-manure crops to maintain soil fertility, usually it is more practical to attain part of this end through feeding the grains and forages to animals. On the average general farm, with various classes and ages of animals, probably 80% of the fertilizing value of the feed is excreted in the feces and urine. Historical support of this situation is found in the fact that, despite the predominantly cereal grain diet of the people, China has the largest swine population of any country in the world. Every Chinese peasant recites the following teaching: "The more pigs, the more manure; and the more manure, the more grain." Indeed, animal manure is very precious in China; it is carefully conserved and added to the land. Manure is used as a way in which to increase yields of farmland already under cultivation. With proper conservation, therefore, this fertility value may be returned to the soil (see Chapter 14 for further details).

SERVE AS AN IMPORTANT COMPANION OF GRAIN PRODUCTION

Swine provide a large and flexible outlet for the year-to-year changes in grain supplies. When there is a large production of grain, (1) more sows can be bred to farrow, and (2) market hogs can be carried to heavier weights. On the other hand, when grain prices are high, (1) pregnant sows can be marketed without too great a sacrifice in price, (2) market hogs can be slaughtered at lighter weights, and (3) the breeding herd can be maintained by reducing the grain that is fed and increasing the pasture or ground alfalfa. Thus, swine give elasticity and stability to grain production.

The hog-corn ratio is an example of how grain production determines pork production. The hog-corn ratio is calculated simply by dividing the price received for live hogs per hundredweight by the current price for corn per bushel at that time.

Corn is the primary ingredient of the pig's diet in the majority of feeding areas in the United States, and feed constitutes a major portion of the total costs of hog production.

Therefore, when corn is cheap, relative to hog prices, many farmers opt to add value to it by feeding it to hogs rather than selling it outright. When the corn price is high, relative to hog prices, farmers may choose to sell the corn. Therefore, above a certain ratio, farmers will more often expand production of hogs, whereas below that level, they will usually cut back production.

In this way, the hog-corn ratio serves as a general indicator of future trends.

Additionally, the following points should be recognized, when considering grain production companion to pork (swine) production:

1. Such well-known grains as corn and barley would have only limited value if restricted solely to direct human consumption, but, because eventually they can ride to market as animal products, their value is immensely greater. A distinction needs to be made, therefore, between food grains for people and feed grains for livestock.

2. Finally, there is a saving in market transportation costs, because the 12 to 15 bushels of grain consumed by the pig in growing to market weight require only about one-fourth the space on the four legs of the pig as would be needed in marketing the grains. Even with animal transportation rates about double that of grain per hundredweight, the cost of marketing the grain through animals is reduced by one-half.

SUPPLEMENT CROP PRODUCTION

Swine supplement crop production through hogging-down certain crops. In addition to doing their own harvesting, the maximum fertility value of the manure is conserved. This contribution of pigs is valuable especially where crops have been damaged or lodged, where harvesting labor is not available, or where crop prices are disastrous.

Fig. 2-8. Hogging down small grains. Sometimes small grains that have been badly lodged or otherwise damaged are harvested by hogs (Courtesy, Miss. Agr. Exp. Sta.)

PROVIDE OTHER FUNCTIONS

Swine have other values. Every day millions of people use swine products for their health, enjoyment, amusement, beautification, and general happiness. Among these products are:

1. Replacement heart valves for humans.

2. Pig intestines for sutures.

3. Paint brushes made with bristles, the short stiff hair of the hog.

4. Adrenal glands to provide the extract which is used to treat Addison's disease, and to provide the drug epinephrine (adrenaline) which is used to treat bronchial asthma and whooping cough.

5. Pigskin which is used for gloves, shoes, hats, billfolds, coats, pants, suits, vests, and topcoats, and as an aid in the treatment of severe burns.

6. Hog pancreases which provide insulin for the treatment of diabetes.

7. Gelatin, for use in ice cream mixes and other products, which is derived from pigskin.

8. The enzyme pepsin, used in chewing gums, is derived from pig's stomachs.

Indeed, the list is long, but it could be made longer by just citing the hog's medical usefulness, which would include its usefulness as an experimental animal to study conditions affecting humans, since physiologically hogs and humans are quite similar.

These innumerable byproducts, many of which would not be available without a swine industry, contribute to the quality of human life.

FACTORS FAVORABLE TO SWINE PRODUCTION

The important position that the hog occupies in American agriculture is due to certain factors and economic conditions favorable to swine production. These are enumerated as follows:

1. Hogs are well adapted to the practice of self-feeding, thereby minimizing labor.

2. Compared with many other agricultural enterprises, the initial investment for a beginning farmer to get into the business is small, and the returns come quickly. A gilt may be bred at 6 to 8 months of age and the pigs marketed 5 to 6 months after farrowing.

3. Swine excel in dressing percentage, yielding 65 to 80% of their liveweight when dressed packer style—with head, leaf fat, kidneys, and ham facings removed. On the other hand, cattle dress only 50 to 60%, and sheep and lambs 45 to 55%. Moreover, because of the small proportion of bone, the percentage of edible meat in the carcass of hogs is greater.

4. The spread in price in market hogs is relatively small—much smaller, for example, than the spread which usually exists between the price of slaughter lambs and ewes. Hogs may be sold at weights ranging from 150 to 250 lb without any great penalty in price. Also, old sows that have outlived their usefulness in the breeding herd may be disposed of without difficulty.

5. Swine are prolific, commonly farrowing from 8 to 14 pigs, and producing two litters per year.

6. The pig is adapted to both diversified and intensified agriculture.

7. Hogs are efficient converters of wastes and byproducts into pork.

8. Swine excel other red-meat–producing animals (beef cattle and sheep) in converting feed to food (see Table 2-14), though they are not as efficient as dairy cattle, fish, or poultry.

FACTORS UNFAVORABLE TO SWINE PRODUCTION

It is not recommended that hogs be raised under any and all conditions. There are certain limitations that should receive consideration if the venture is to be successful. Some of these reservations follow:

1. Because of the nature of the digestive tract, the growing-finishing pig must be fed a maximum of concentrates and a minimum of roughages. Where or when grains are scarce and high in price, this may result in high production costs.

2. Because of the nature of their diet and their rapid growth rate, hogs are extremely sensitive to unfavorable rations and to careless management.

3. Swine are very susceptible to numerous diseases and parasites.

4. Fences of a more expensive kind are necessary in hog raising.

5. Sows should have skilled attention at farrowing time.

6. Because of their rooting and close-grazing habits, hogs are hard on pasture.

7. Hogs are not adapted to a frontier type of agriculture where grazing areas are extensive and vegetation is sparse. Neither are they best suited to the utilization of permanent-pasture areas.

8. Increasing energy costs are unfavorable to modern confinement swine production.

THE FUTURE OF THE AMERICAN SWINE INDUSTRY

Some of the factors that will affect the future of the American swine industry are:

1. Competition for grain.
2. Efficiency of feed conversion.
3. Foreign competition.
4. The lard situation.
5. Increased pork consumption.
6. Competition between geographic areas.
7. High energy cost.
8. Pork's image.
9. Animal welfare and animal rights.
10. Industrialization of U.S. hog production.
11. Handling wastes in an environmentally-responsible manner.

Each of these factors will be discussed in a separate section which follows.

COMPETITION FOR GRAIN

Many countries of the world will experience grain shortages and require imported grain to feed their growing populations. By the year 2020, it is estimated that the United States and Canada, and possibly Latin America and Southeast Asia, will be the only areas producing substantially more grain than they consume. Hogs will be in more direct competition with humans for food.

■ **Grain shortages favor ruminants**—On the average, 95.7% of the feed of swine is derived from concentrates and 4.3% from roughages, whereas beef cattle are almost the opposite—84.5% of their feed is derived from roughages and 15.5% from concentrates. Thus, grain shortages place hogs in a less favorable position than ruminants.

■ **Grain shortages favor cereal grain diet for humans and lessening grain for animals**—Cereal grain is the most important single component of the world's food supply, accounting for between 30 and 70% of the food produced in all world regions. It is the major, and sometimes almost exclusive, source of food for many of the world's poorest people, supplying 60 to 75% of the total calories many of them consume. With sporadic food shortages and famine in different parts of the world, the competition for grain will increase and people will ask more frequently: Who should eat grain— people or animals? Shall there be feed or food?

Forgetting for a moment the high nutritive value of meats, milk, and eggs, there can be no question that more hunger can be alleviated with a given quantity of grain by completely eliminating animals. About 2,000 lb of concentrates (mostly grain) must be supplied to livestock in order to produce enough meat and other livestock products to support a human for a year, whereas 400 lb of grain (corn, wheat, rice, soybeans, etc.) eaten directly will support a human for the same period of time. Thus, a given quantity of grain eaten directly will feed five times as many people as it will if it is first fed to livestock and then is eaten indirectly by humans in the form of livestock products. This inefficiency is the result of unavoidable nutrient losses in all animal feeding and the fact that no return is received from that portion of the animal's feed which goes for maintenance (which amounts to approximately one-half). This is precisely the reason why the people of the Orient have become vegetarians.

EFFICIENCY OF FEED CONVERSION

On a feed, caloric, or protein conversion basis, it's not efficient to feed grain to animals and then consume the livestock products. This fact is pointed up in Table 2-14. Moreover, as Table 2-14 indicates, other food animals are more efficient converters of feed to food than swine; namely, dairy cows, broilers, layers, rabbits, and fish.

Thus, in the developing countries, where the population is growing rapidly, virtually all grain is eaten directly by people. Precious little of it is converted to animal products. Then, as a nation becomes more affluent, more grain is used, but most of it is converted into animal products.

FOREIGN COMPETITION

Several European and Scandinavian countries are pork-exporting nations. Some of the South American countries are potential pork-producing and pork-exporting nations, an encouraging market being the only needed incentive. Canada is making great progress in swine production, in both quality and quantity. Only tariffs, quotas, and embargoes enacted by our federal government can prevent future and serious competition from foreign imports. However, with our huge corn production and improved swine-production methods, it is not anticipated that pork will ever have the potential foreign competition that exists with beef.

THE LARD SITUATION

Lard was a "drug on the market" prior to 1941 and soon after World War II this status returned. Satisfactory vegetable oils can now be produced at lower cost.

TABLE
FEED TO FOOD EFFICIENCY RATING BY SPECIES OF ANIMALS,
(Based on Energy as TDN or DE and Crude Protein in Feed Eaten by Various Kinds of

Species	Unit of Production (on foot)	Feed Required to Produce One Production Unit				Dressing Yield	
		Pounds	TDN[1]	DE[2]	Protein	Percent	Net Left
		(lb)	(lb)	(kcal)	(lb)	(%)	(lb)
Broiler	1 lb chicken	2.1[7]	1.7[8]	3,400	0.21[8]	72[13]	0.72
Fish	1 lb fish	1.6[9]	0.98	1,960	0.57	65[10]	0.65
Dairy cow	1 lb milk	1.11[7]	0.9[8]	1,800	0.1[8]	100	1.0
Turkey	1 lb turkey	5.2[7]	4.21[8]	8,420	0.46[8]	79.7[13]	0.797
Layer	1 lb eggs (8 eggs)	4.6[7]	3.73[8]	7,460	0.41[8]	100	1.0
Hog (birth to market weight)	1 lb pork	4.0[18]	3.2	6,400	0.36	70[16]	0.7
Rabbit	1 lb fryer	3.0[19]	2.2	4,400	0.48	55[19]	0.55
Beef steer (yearling finishing period in feedlot)	1 lb beef	9.0[18]	5.85	11,700	0.90	58[16]	0.58
Lamb (finishing period in feedlot)	1 lb lamb	8.0[18]	4.96	9,920	0.86	47[16]	0.47

[1]TDN pounds computed by multiplying pounds feed (column to left) times percent TDN in normal rations. Normal ration percent TDN taken from M. E. Ensminger's books and rations, except for the following: dairy cow, layer, broiler, and turkey from *Agricultural Statistics 1974*, p. 358, Table 518. Fish based on averages recommended by Michigan and Minnesota Stations and U.S. Fish and Wildlife Service.

[2]Digestible Energy (DE) in this column given in kcal, which is 1 Calorie (written with a capital C), or 1,000 calories (written with a small c). Kilocalories computed from TDN values in column to immediate left as follows: 1 lb TDN = 2,000 kcal.

[3]From *Lessons on Meat*, National Live Stock and Meat Board, 1965.

[4]Feed efficiency as used herein is based on pounds of feed required to produce 1 lb of product. Given in both percent and ratio.

[5]Kilocalories in ready-to-eat food = kilocalories in feed consumed, converted to percentage. Loss = kcal in feed ÷ kcal in product.

[6]Protein in ready-to-eat food = protein in feed consumed, converted to percentage. Loss = pounds protein in feed ÷ pounds protein in product.

[7]*Agricultural Statistics 1974*, p. 358, Table 518. Pounds feed per unit of production is expressed in equivalent feeding value of corn.

[8]Pounds feed (column No. 2) per unit of production (column No. 1) is expressed in equivalent feeding value of corn. Therefore, the values for corn were used in arriving at these computations. No. 2 corn values are TDN, 81%; protein, 8.9%. Hence, for the dairy cow 81% × 1.11 = 0.9 lb TDN; and 8.9% × 1.11 = 0.1 lb protein.

[9]Data from report by Dr. Philip J. Schaible, Michigan State University, *Feedstuffs*, April 15, 1967.

[10]*Industrial Fishery Technology*, edited by Maurice E. Stansby, Reinhold Pub. Corp., 1963, Ch. 26, Table 26-1.

2-14
RANKED BY PROTEIN CONVERSION EFFICIENCY
Animals Converted into Calories and Protein Content of Ready-to-Eat Human Food)

Ready-to-Eat; Yield of Edible Product (meat & fish deboned & after cooking)				Feed Efficiency[4]		Efficiency Rating			
As % of Raw Product (carcass)	Amount Remaining from One Unit of Production	Calorie[3]	Protein[3]	(lb feed to produce one lb product)		Calorie Efficiency[5]		Protein Efficiency[6]	
(%)	(lb)	(kcal)	(lb)	(%)	(ratio)	(%)	(ratio)	(%)	(ratio)
54[14]	0.39	274	0.11	47.6	2.1:1	8.1	12.4:1	52.4	1.9:1
57[11]	0.37	285	0.27	62.5	1.6:1	14.5	6.9:1	47.6	2.1:1
100	1.0	309	0.037	90.0	1.11:1	17.2	5.8:1	37.0	2.7:1
57[15]	0.45	446	0.146	19.2	5.2:1	5.3	18.9:1	31.7	3.2:1
100[12]	1.0[12]	616	0.106	21.8	4.6:1	8.3	12.1:1	25.9	3.9:1
44[17]	0.31	341	0.088	0.25	4.0:1	5.3	18.8:1	24.4	4.1:1
79[19]	0.43	301	0.08	35.7	2.8:1	6.8	14.6:1	16.7	6.0:1
49[17]	0.28	342	0.085	11.1	9.0:1	2.9	34.2:1	9.4	10.6:1
40[17]	0.19	225	0.052	12.5	8.0:1	2.3	44.1:1	6.0	16.5:1

[11]*Ibid.* Reports that "Dressed fish averages about 73% flesh, 21% bone, and 6% skin." In limited experiments conducted by A. Ensminger, it was found that there was a 22% cooking loss on filet of sole. Hence, these values—73% flesh from dressed fish, minus 22% cooking losses—give 57% yield of edible fish after cooking, as a percent of the raw, dressed product.

[12]Calories and protein computed basis per egg; hence, the values herein are 100% and 1.0 lb, respectively.

[13]*Marketing Poultry Products*, 5th Ed., by E. W. Benjamin, *et al.*, John Wiley & Sons, 1960, p. 147.

[14]*Factors Affecting Poultry Meat Yields*, University of Minnesota Sta. Bull. 476, 1964, p. 29, Table 11 (fricassee).

[15]*Ibid.* Page 28, Table 10.

[16]Ensminger, M. E., *The Stockman's Handbook*, 6th Ed., Sec. XII.

[17]Allowance made for both cutting and cooking losses following dressing. Thus, values are on a cooked, ready-to-eat basis of lean and marbled meat, exclusive of bone, gristle, and fat. Values provided by National Live Stock and Meat Board (personal communication of June 5, 1967, from Dr. Wm. C. Sherman, Director, Nutrition Research, to the author), and based on the data from *The Nutritive Value of Cooked Meat*, by Ruth M. Leverton and George V. Odell, Misc. Pub. MP-49, Appendix C, March 1958.

[18]Estimates by the senior author.

[19]Based on information in *Commercial Rabbit Raising*, Ag. Hdbk. No. 309, USDA, 1966, and *A Handbook on Rabbit Raising*, by H. M. Butterfield, Washington State University Ext. Bull. No. 411.

When processed lard sells for less than the price of hogs on foot, it should be perfectly evident that the product is lacking in demand. In order to alleviate the surplus lard situation, the soundest approach consists of: (1) breeding a type of hog that is less lardy in conformation, (2) feeding so as to produce less excess fat, (3) marketing at lighter weights, and (4) purchasing hogs on a quality basis (preferably rail-graded).

U.S. LARD PRODUCTION AND EXPORTS

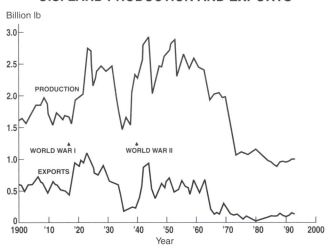

Fig. 2-9. Production and exports of lard from the United States, 1900 to 1993. Note that lard exports increased sharply during both World War I and World War II, but they were very small between 1935 and 1940. Most authorities agree that future lard exports will be negligible. (Based on USDA data)

INCREASED PORK CONSUMPTION

Increased U.S. consumption of pork has accrued from two sources: (1) increased population, and (2) increased per capita consumption of pork brought about through the production of a higher quality, more versatile product, receiving more exposure. One such promising avenue is the introduction of pork products to the fast-food industry.

COMPETITION BETWEEN GEOGRAPHIC AREAS

The Corn Belt will remain the major hog producing area, though there will be certain shifts in production. Thus, it is noteworthy that three states—North Carolina, Minnesota, and South Dakota—not in the Corn Belt, were among the top 10 pork-producing states of the U.S. in 1994 (see Fig. 2-3).

Some expansion of swine production in the West occurred in the 1980s and 1990s, driven primarily by

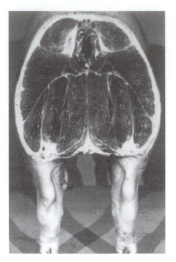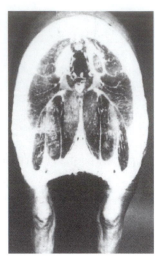

Fig. 2-10. Breeding made the difference! The hogs received the same ration and were slaughtered at the same weight. Note the difference in the amount of lean meat, with the hog on the left being superior.

mega-operations seeking wide open spaces (1) to escape the increasing legislation and litigation against odor and contamination nuisances in the Corn Belt, and (2) to isolate their operations from other hogs and swine diseases.

HIGH ENERGY COST

The future of the American swine industry will be affected by energy costs. This is so because a modern swine farm has a large variety of energy-consuming items—heating, cooling, ventilation, etc. It follows that more efficient energy consumption will be important to the success or failure of a swine operation. The energy crisis of the early 1970s caused producers and the world in general to seek alternatives and more efficient use of energy. Some values for energy consumption are given in Table 2-15. These values will vary depending upon location, season, and type of operation. To produce feeder pigs, energy costs represent 8 to 10% of the total cost, while energy costs represent about 2 to 3% of the total costs of farrow-to-finish operations, but only about 1% in feeder pig finishing operations. Thus, continually increasing energy costs (see Fig. 2-11) are unfavorable to modern swine production, particularly high investment confinement systems.

Looking at energy costs from a different perspective, in the U.S. it takes an average input of 13.1 calories to produce, process, market, and cook every calorie of food consumed.

Swine producers will likely focus on other forms of energy such as solar, methane, geothermal, as well as construction and management techniques which will conserve energy.

TABLE 2-15
ENERGY CONSUMPTION[1]

Animal	Farrowing		Nursery	
	Electricity	LP Gas	Electricity	LP Gas
	(kwh)	(gal)	(kwh)	(gal)
Per sow maintained	234.6	18.1	260.0	39.6
Per litter farrowed	118.2	9.2	130.6	20.1
Per pig farrowed, live	11.6	0.9	12.8	2.0
Per pig weaned	13.5	1.0	14.9	2.3
Per pig marketed	14.6	1.1	16.2	2.2

[1]North Carolina Swine Development Center, 1976–79, as reported in *Yearbook of Agriculture*, USDA, 1980, p. 65.

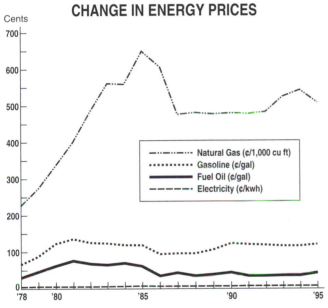

Fig. 2-11. Since 1978 the costs for all forms of energy have spiraled, causing many individuals and businesses to conserve and to seek alternatives. (Data from *Monthly Energy Review*, September 1995, U.S. Dept. of Energy)

A COMEBACK OF THE FAMILY FARM

In the 1980s and 1990s, there was a growing dominance of giant hog corporations linking up tightly with their suppliers. Many of these mega-operations integrated backward for control of their inputs. Corporate hog farms will likely dominate swine production and marketing until about 2015 to 2020. During this period, more and more family farms will be squeezed out.

From about 2015 to 2020 to the end of the 21st century, fossil fuels will become scarcer and more costly, favoring organic farming, agroforestry, integrated pest management, and sustainable agriculture. As a result of high energy prices, from about 2015 to 2020, agriculture will be characterized by bio-mass products, labor intensive operations, soil conservation, and the comeback of the family farm.

PORK'S IMAGE

Despite the excellent nutritive qualities of pork, it has been portrayed as being an unhealthy food by some of the media and by some self-proclaimed nutritionists. Producers will need to continue to battle this false image through education of consumers. But one fact is certain: Consumers do not want a fat product.

ANIMAL WELFARE AND ANIMAL RIGHTS

To all animal caretakers, the principles and application of animal behavior and environment depend on understanding; and on recognizing that they should provide as comfortable an environment as feasible for their animals, for both animal welfare and economic reasons. This requires that attention be paid to environmental factors that influence the behavioral welfare of their animals as well as their physical comfort, with emphasis on the two most important influences of all in animal behavior and environment—feed and confinement.

INDUSTRIALIZATION OF U.S. HOG PRODUCTION

Industrialization of hog production is the production of hogs in specialized factory-like facilities staffed with specialized labor. Formerly, hog production per unit varied with the acres of associated corn land. Today, the number of hogs per unit varies with the size of facilities.

The trend to industrialized hog production started in about 1970 with a rapid movement of hog production into partial or total confinement. The driving forces back of it have been new technologies, new managerial techniques, lower cost of production, and increased quality control. Adverse consequences are being felt by displaced hog producers, displaced agribusinesses, and deteriorating rural communities. Also, big industrialized hog operations are not noted for their liberal support of colleges, hospitals, churches, and other charities in the states in which they operate.

Fig. 2-12. A 1,200-head sow farm and lagoon on Murphy Family Farms, Rose Hill, North Carolina. Murphy Farms, the nation's largest pork producer, produces 4 million pigs a year. (Courtesy, Dr. Garth W. Boyd and Ms. Rhonda Campbell, Murphy Farms, Rose Hill, NC)

HANDLING WASTES IN AN ENVIRONMENTALLY-RESPONSIBLE MANNER

The future of the U.S. swine industry will be impacted by the ability and willingness of producers to handle wastes in an environmentally-responsible manner.

If not handled properly, swine may produce the following pollutants in troublesome quantities: manure, gases/odors, dust, flies and other insects. Also, they may pollute water supplies.

In 1995, the animal industry of North Carolina received three rude wake-up calls. On June 21, 25 million gallons of manure from a 10,000 hog operation broke out of a lagoon, pouring into nearby gullies and streams feeding the New River. That same day, a lagoon on a smaller hog farm ruptured in North Carolina. Then, on July 3, a four-acre poultry lagoon broke

As hog production is being concentrated in fewer and bigger production units, communities and states belatedly debate whether they shall welcome or try to bar such industrialization. Although trying to save family farms by anti-incorporation legislation at the state level does not work, a new force for their impediment is the growing trend of neighbors and communities legislating and litigating against odor and contamination nuisances. NIMBY (not in my back yard) opposition has become very real in the Corn Belt. This is driving some hog operations to the wide open spaces of the West, which may accelerate a more rapid concentration of ownership and vertical integration.

The 1995 level of vertical integration by four animal species is shown in Fig. 2-13. Over the next decade, the level of vertical integration of hog production may reach 50%, but the farmer-hog producer will always be an integral part of U.S. pork production.

Fig. 2-14. In ground concrete-sided manure storage structure provides safe waste storage. (Courtesy, Iowa State University, Ames, IA)

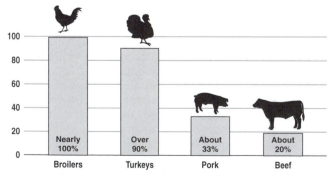

LEVEL OF INTEGRATION BY SPECIES

100			
80			
60			
40			
20	Nearly 100%	Over 90%	About 33%	About 20%
0	Broilers	Turkeys	Pork	Beef

Fig. 2-13. Level of integration of four species in 1995.

Fig. 2-15. Injecting liquid manure into the ground disposes of waste in an environmentally-responsible manner. (Courtesy, Iowa State University, Ames, IA)

in North Carolina, spewing 8.6 million gallons of waste into the tributaries of the Northeast Cape Fear River.

In the past, many environmentalists have spent much time trying to save the world from plastic plates while ignoring the meat that was served on them. Now, family farmers and environmentalists are uniting to halt the proliferation of factory hog farms. United, they will have an enormous impact on mega hog farms in the future.

QUESTIONS FOR STUDY AND DISCUSSION

1. Why are there so many hogs in Asia (see Fig. 2-1)?

2. Discuss the factors which account for each of the five leading swine producing countries (see Table 2-1) holding their respective ranks.

3. How do you account for the fact that about 60% of the hog production is centered in the seven Corn Belt states?

4. Select a certain farm or ranch (your home, farm, or ranch, or one with which you are familiar). Then discuss the relation of swine production on this farm or ranch to the type of agriculture.

5. Assuming that a young person had no "roots" in a particular location, in what area of the United States would you recommend that he/she establish a swine enterprise? Justify your answer.

6. Iowa has a commanding lead as a swine producing state (see Fig. 2-3). Why is this so?

7. North Carolina ranks a solid second as a U.S. swine producing state (see Fig. 2-3). How do you account for this state which is not in the Corn Belt having so many hogs?

8. Why has there been such a big decline in number of U.S. hog farms, accompanied by an increase in size (see Table 2-3 and 2-4)? Is this good or bad?

9. The pie diagram (Fig. 2-4) shows that swine rank fourth in U.S. cash receipts from animal commodities. Do you foresee the ranking of swine moving higher than beef, dairy, or poultry? Justify your answer.

10. As a food, what position does pork hold around the world (see Fig. 2-5)? What position does it hold in the United States (see Table 2-8, Table 2-9, and Fig. 2-6)? What factors and changes have contributed to this position?

11. What is the importance of pork as an export and as an import, and what determines the amount of pork that the United States exports and imports (see Table 2-10)?

12. Compare the type and amounts of pork products that are imported and exported by the United States (see Table 2-11).

13. List and discuss the functions of swine. What do you consider to be the most important function of swine?

14. Looking into the future, do you feel that the factors favorable to swine production outweigh the unfavorable factors? Justify your answer.

15. How would you answer the question, "Who should eat grain—people or animals?"

16. Compared to other food animals, how efficient is the hog in converting feed to food (calories and protein) (see Table 2-14)?

17. When lard sells for less than the price of live hogs, it should be perfectly evident that the product is lacking in demand. How may producers alleviate the surplus lard situation?

18. How may the cost of energy affect swine production (see Table 2-15 and Fig. 2-11)?

19. Discuss the animal welfare and animal rights movement.

20. Discuss the industrialization of U.S. hog production. Has this been good or bad?

21. How will handling wastes in an environmentally-responsible manner impact the future of the U.S. swine industry?

22. On the whole, do you feel that the future of U.S. swine production warrants optimism or pessimism? Justify your answer.

SELECTED REFERENCES

Title of Publication	Author(s)	Publisher
Feeds & Nutrition, second edition	M. E. Ensminger J. E. Oldfield W. W. Heinemann	The Ensminger Publishing Company, Clovis, CA, 1990
Feeds & Nutrition Digest	M. E. Ensminger J. E. Oldfield W. W. Heinemann	The Ensminger Publishing Company, Clovis, CA, 1990
Food and Fiber for a Changing World	G. W. Thomas S. E. Curl W. F. Bennett, Sr.	The Interstate Printers & Publishers, Inc., Danville, IL, 1982
Global 2000 Report to the President, The		U.S. Government Printing Office, Washington, DC, 1980
Pork Facts 1995–1996	Staff	National Pork Producers Council, Des Moines, IA, 1995–1996
Swine Production	C. E. Bundy R. V. Diggins V. W. Christensen	Prentice-Hall, Inc., Englewood Cliffs, NJ, 1976
Swine Production	J. L. Krider J. H. Conrad W. E. Carroll	McGraw-Hill Book Co., New York, NY, 1982
Swine Production in Temperate and Tropical Environments	W. G. Pond J. H. Maner	W. H. Freeman and Co., San Francisco, CA, 1974
World Food and Nutrition Study	National Research Council	National Academy of Sciences, Washington, DC, 1977
World Food Production, Demand, and Trade	L. L. Blakeslee E. O. Heady C. F. Framingham	Iowa State University Press, Ames, IA, 1973

Many large commercial hog producers now buy breeding stock from *Specialized Breeding Stock Suppliers*, who usually breed and market one or more of the traditional purebreds, along with their synthetic strains. (Courtesy, Lone Willow Genetics, Roanoke, IL)

3

BREEDS OF SWINE[1]

In the hands of skilled caretakers, swine are the most plastic of any species of farm animals. This is due to their early maturity, multiple rate of reproduction, and the short time between generations. A farmer who produces a total of 100 spring-farrowed pigs yearly needs only 14 gilts to raise a crop of the same size the following spring. There will be approximately 50 gilts from which he may select the 14 brood sows needed. He has a wide choice of keeping meaty-type gilts or of picking others that are shorter and thicker. Continued selection each year with emphasis

---

[1]Sometimes people construe the write-up of a breed of livestock in a book or in a U.S. Department of Agriculture bulletin as an official recognition of the breed. Nothing could be further from the truth, for no person or office has authority to approve a breed. The only legal basis for recognizing a breed is contained in the Tariff Act of 1930, which provides for the duty-free admission of purebred breeding stock provided they are registered in the country of origin. But the latter stipulation applies to imported animals only.

In this book, *no official* recognition of any breed is intended or implied. Rather, the authors have tried earnestly, and without favoritism, to present the factual story of the breeds in narrative and picture.

upon the same characteristics and the purchase of boars of the same type can completely alter the conformation of the hogs in a herd in the short period of 4 to 5 years. Despite this fact, progress in producing meat-type hogs was often slow and painful throughout the 1930s and 1940s. As a result, (1) pork gradually lost its place as the preferred meat, with beef taking the lead in the early 1950s, and (2) a number of new American breeds of swine evolved, most of them carrying some Landrace breeding.

TYPES OF HOGS

Swine types are the result of three contributing factors: (1) the demands of the consumer, (2) the character of the available feeds, and (3) the breeding and pursuit of type fads by breeders.

Historically, three distinct types of hogs have been recognized; namely (1) lard type, (2) bacon type, and (3) meat type. At the present time, however, the goal for all U.S. swine breeds is for a meat-type hog, although it is obvious that some breeds have more nearly achieved this than others.

LARD TYPE

Originally, breeders of hogs stressed immense size and scale and great finishing ability. This general type persisted until the latter part of the 19th century. Beginning about 1890, breeders turned their attention to the development of early maturity, great refinement, and a very thick finish. In order to obtain these desired qualities, animals were developed that were smaller in size, thick, compactly built, and very short of leg. In the Poland China breed, this fashionable fad was carried to the extreme. It finally culminated in the development of the "hot bloods." Hogs of this chuffy

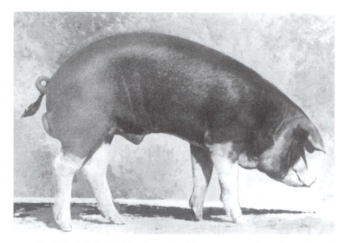

Fig. 3-2. A Poland China boar pig of the rangy type. Long legged, weak loined, "cat hammed" animals of this type dominated the American show-ring from 1915 to 1925.

type were notoriously lacking in prolificacy. They often farrowed twins and triplets; and, when they were carried to weights in excess of 200 lb, their gains were very expensive. Small, refined animals of this type dominated the show-ring from about 1890 to 1910.

In order to secure increased utility qualities, breeders finally, about 1915, began the shift to the big-type strains. Before long, the craze swept across the nation, and again the pendulum swung too far. Breeders demanded great size, growthiness, length of body, and plenty of bone. The big-type animal was rangy in conformation and slow in maturity. Many champions of the show-ring included as their attributes long legs, weak loins, and "cat hams." One popular champion of the day was advertised as being "so tall that it makes him dizzy to look down." Inasmuch as this type failed most miserably in meeting the requirements of either the packer or the producer— being too slow to reach maturity and requiring a heavy weight in order to reach market finish—another shift in ideals became necessary.

BACON TYPE

Bacon-type hogs are more common in those areas where the available feeds consist of dairy byproducts, peas, barley, wheat, oats, rye, and root crops. As compared with corn, such feeds are not so fattening. Thus, instead of producing a great amount of lard, they build sufficient muscle for desirable bacon. The countries of Denmark, Canada, and Ireland have long been noted for the production of high-quality bacon. In the past, the surplus pork produced in these countries has found a ready market in England, largely selling as Wiltshire sides.

In emphasizing the importance of character of

Fig. 3-1. A Poland China gilt of the chuffy type. Small, refined animals of this type dominated the American show-ring from 1890 to 1910. (Courtesy, University of Illinois)

feeds as a factor influencing the production of bacon-type hogs, it is not to be inferred that there is no hereditary difference. That is to say, when bacon-type hogs are taken into the Corn Belt and fed largely on corn, they never entirely lose their bacon qualities.

MEAT TYPE

Since about 1925, American swine breeders have been striving to produce meat-type hogs—animals that are intermediate between the lard and bacon types. The best specimens of the meat type combine muscling, length of body, balance, and the ability to reach market weight and finish without excess fat. In achieving the meat type, the selection and breeding programs of producers have been stoutly augmented by meat certification programs, livestock shows, and swine type conferences.

BREEDS OF HOGS IN THE UNITED STATES

Worldwide, there are over 400 breeds of swine.[2] But, presently, there are active registries for only the following nine breeds in the United States: Berkshire, Chester White, Duroc, Hampshire, Hereford, Landrace, Poland China, Spotted, and Yorkshire. In the past, several other breeds, such as the Tamworth, Ohio Improved Chester (OIC), Conner Prairie, English Large Black, Kentucky Red Berkshire, and Wessex Saddleback, maintained registries, but these do not appear to be active at the present time.

With the exception of the Berkshire, Landrace, and Yorkshire, the breeds of swine common to the United States are strictly American creations. This is interesting in view of the fact that only one of our breeds of draft horses and few of our better known breeds of sheep and beef cattle were American creations. With the exception of the Hereford breed and some of the newer hybrid breeds of swine, the American breeds came into being in the period from 1800 to 1880—an era which was characterized by the production of an abundance of corn for utilization by hogs and by consumer demand for fat, heavy cuts of pork. The European breeds did not seem to meet these requirements.

It must be remembered, however, that the American breeds of swine were not developed without recourse to foreign stock. Prior to De Soto's importation, no hogs were found on the continent. The offspring of De Soto's sturdy razorbacks, together with subsequent importations of European and Oriental hogs, served as the foundation stock for the American breeds which followed. Out of these early-day, multiple-colored and conglomerate types of swine, the swine producers of different areas of the United States, through the tools of selection and controlled matings, gradually molded uniform animals, later to be known as breeds. It is to be noted, however, that these foundation animals carried a variable genetic composition. This made them flexible in the hands of breeders and accounted for the radical subsequent shifts in swine types that have been observed within the pure breeds.

RELATIVE POPULARITY OF BREEDS OF SWINE

Table 3-1 shows the 1994 registrations of the common breeds of hogs. Although trends are not shown in this table, the 1994 registrations are indicative of the current popularity of each of the breeds.

TABLE 3-1
ANNUAL REGISTRATIONS OF SWINE BY
UNITED STATES BREED ASSOCIATIONS

Breed	Annual Registrations
Berkshire (litters)	1,792
Chester White (individuals)	27,342
Duroc	130,625
Hampshire (individuals)	122,002
Hereford (individuals)	518
Landrace (litters)	4,714
Poland China (individuals)	14,872
Spotted (litters)	26,930
Yorkshire (individuals)	208,662

While there are many breed differences and most breed associations are constantly extolling the virtues of their respective breeds, it is perhaps fair to say that there is more difference within than between breeds from the standpoint of efficiency of production and carcass quality. Without doubt, the future and enduring popularity of each breed will depend upon how well it fulfills these two primary requisites.

BERKSHIRE

The Berkshire is one of the oldest of the improved breeds of swine. The striking style and carriage of the Berkshire has made it known as the aristocrat among the breeds of swine.

_____
[2]*Van Nostrand's Scientific Encyclopedia*, 5th ed., Van Nostrand Reinhold Company, New York, 1976, p. 2116.

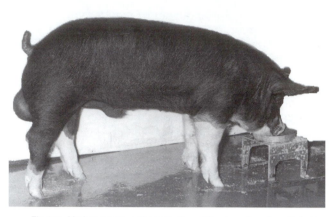

Fig. 3-3. Modern meat type Berkshire boar. (Courtesy, American Berkshire Assn., West Lafayette, IN)

ORIGIN AND NATIVE HOME

The native home of the Berkshire is in south central England, principally in the counties of Berkshire and Wiltshire. The old English hog, a descendant of the wild boar, served as foundation stock; and these early animals were improved by introducing Chinese, Siamese, and Neapolitan blood. In 1789, the Berkshire was described as follows: reddish brown in color, with black spots; large drooping ears; short legs; fine bone; and the disposition to fatten at an early age.

EARLY AMERICAN IMPORTATIONS

The earliest importation of Berkshire hogs into the United States, of which there is authentic record, was made by John Brentnall of New Jersey, in 1823. This importation was followed by those of Bagg and Wait, of Orange County, New York, in 1839, and A. B. Allen of Buffalo, New York, in 1841.

Although there have been many constructive Berkshire breeders in this country, certainly the name of N. H. Gentry, of Sedalia, Missouri, is among the immortals. Few breeders either here or abroad achieved the success which was his. He was truly a master breeder. The majority of the best Berkshires of today trace to Gentry breeding, particularly to the great and prepotent boar, Longfellow.

BERKSHIRE CHARACTERISTICS

The distinct peculiarity of the Berkshire breed is the short, and sometimes upturned, nose. The face is somewhat dished, and the ears are erect but inclined slightly forward. The color is black with six white points—four white feet, some white in the face, and a white switch on the tail. However, splashes of white may be located on any part of the body.

The conformation of the Berkshire may be described as excellent meat type. Fortunately, the big-type craze did not gain great momentum among Berkshire breeders. Thus, there seems to be pronounced uniformity within the breed at the present time.

The typical Berkshire is long-bodied, with a long, deep side; moderately wide across the back; smooth throughout; well balanced; and medium in length of leg. The meat is exceptionally fine in quality, well streaked with lean, and has no heavy covering of fat. The breed has established an enviable record in the barrow and carcass contests of the country.

There was a time when Berkshire sows were considered lacking in prolificacy. Judicious breeding and selection, however, have largely eliminated this criticism.

The Berkshire is not as large as most of the other breeds. However, because of its great length, depth,

Fig. 3-4. Berkshire gilt exhibited in the 1994 Natioinal Barrow Show by Phenotypic Acres, Ames, Iowa. (Courtesy, American Berkshire Assn., West Lafayette, IN)

and balance, it is likely to be underestimated in weight.

Animals possessing (1) a swirl on the upper half of the body, (2) any color other than black or white, (3) a hernia, (4) undescended testes (cryptorchid), (5) absence of anal opening, (6) a rectal or uteral prolapse, or (7) fewer than 12 teats are disqualified from registry.

Overall, Berkshires are hardy, rugged hogs and able to withstand cold weather. Berkshires are often used in commercial herds because they are not as closely related to most other breeds.

CHESTER WHITE

The Chester White breed is very popular on the farms of the northern part of the United States, and

Fig. 3-5. Chester White boar. (Courtesy, Chester White Swine Record, Peoria, IL)

barrows of this breed have an enviable reputation in the barrow classes of livestock shows.

ORIGIN AND NATIVE HOME

The Chester White had its origin in the fertile agricultural section of southeastern Pennsylvania, principally in Chester and Delaware counties. Both of these counties border on Lancaster County, one of the most noted agricultural and livestock counties in the United States. The breed seems to have originated early in the 19th century from the amalgamation of several breeds, most of which were white in color. The foundation stock included imported pigs of English Yorkshire, Lincolnshire, and Cheshire breeding. In 1818, Captain James Jeffries of Chester County imported a pair of white pigs from Bedfordshire, England. The boar in this importation was destined to exert a marked refining influence on the foundation stock of the Chester White breed. By 1848, the breed had reached such a degree of uniformity and purity that it was named Chester County White. The word *county* was soon dropped, and the present name became established.

Fig. 3-6. Chester White gilt. (Courtesy, Chester White Swine Record, Peoria, IL)

CHESTER WHITE CHARACTERISTICS

As the name indicates, the breed is white in color. Although small bluish spots, called freckles, are sometimes found on the skin, such spots are to be discriminated against.

In general, type changes within this breed have followed those of the Poland China and the Duroc, although they have been less radical in nature. No doubt, some Yorkshire blood was infused into certain herds during the big-type craze. If so, perhaps it can be added that this blood brought real improvement and benefit to the breed.

Chester White sows are very prolific and are exceptional mothers. The pigs adapt well to a variety of conditions; they mature early; and the finished barrows are very popular on the market.

Any of the following are disqualifications: Not ⅔ big enough for age; upright ears; off-colored hair; spots on skin larger than a silver dollar; cryptorchidism in males; hernia in males or females; or swirls on body above the flanks.

DUROC

Fig. 3-7. A high performing Duroc herd sire. (Courtesy, *Seedstock*, West Lafayette, IN)

Currently, the Duroc ranks second in annual registrations, being exceeded only by Yorkshires.

ORIGIN AND NATIVE HOME

The Duroc breed of swine originated in the northeastern section of the United States. Although several different elements composed the foundation stock, it is reasonably certain that the Durocs, of New York, and the Jersey Reds, of New Jersey, contributed most to the ancestry.

The Jersey Reds were large, coarse, prolific red hogs that were bred in New Jersey early in the 19th century. The Durocs were smaller in size, more compact, and possessed great refinement. The red hogs of uncertain origin, which were found in New York, were named after the famous stallion, Duroc, a noted horse of that day. Beginning about 1860, these two strains of red hogs were systematically blended together, and thus there was formed the breed that is known at the present time as the Duroc.

DUROC CHARACTERISTICS

Fig. 3-8. A long, meaty, productive Duroc female. (Courtesy, *Seedstock*, West Lafayette, IN)

The Duroc is red in color, with the shades varying from light to dark. Although a medium cherry red is preferred by the majority of breeders, there is no particular discrimination against lighter or darker shades so long as they are not too extreme. Duroc ears are medium sized and tipped forward.

During the big-type era, the Duroc also had its change in type. The show-ring was dominated by tall, rangy individuals—animals possessing narrow loins, light hams, shallow bodies, and slow maturity. Although some lack of uniformity still exists within the breed as a result of this radical shift, today the best representatives are of the most approved meat type. The popularity of the breed may be attributed to the valuable combination of size, feeding capacity, prolificacy, and hardiness.

Any of the following are disqualifications for registry: white feet or white spots on any part of the body; any white on the end of the nose; black spots larger than 2 in. diameter on the body; swirls on upper half of the body or neck; or ridgeling (one testicle) boars or fewer than six udder sections on either side.

HAMPSHIRE

The Hampshire is one of the youngest breeds of swine, but its rise in popularity has been rapid. It is

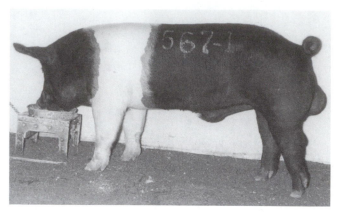

Fig. 3-9. A high cutability Hampshire herd sire prospect. (Courtesy, *Seedstock*, West Lafayette, IN)

widely distributed throughout the Corn Belt and the South.

ORIGIN AND NATIVE HOME

The Hampshire breed of swine originated in Boone County, Kentucky, just across the Ohio River from Cincinnati. The foundation stock consisted of 15 head of belted hogs, generally known as Thin Rinds and Ring Middles. This original herd was purchased in Pennsylvania in 1835 by Major Joel Garnett, who had the animals driven to Pittsburgh and then sent by boat to Kentucky. Years later, in 1893, six Boone County (Kentucky) farmers organized the Thin Rind Association, which, in 1904, became the Hampshire Swine Association.

The origin of the foundation herd purchased by Major Garnett is clouded in obscurity. However, belted hogs were reported in Massachusetts and New York between 1820 and 1830. Moreover, the Essex and Wessex Saddleback breeds of England possess the same color pattern as the present-day Hampshire.

Fig. 3-10. A modern Hampshire female. (Courtesy, *Seedstock*, West Lafayette, IN)

HAMPSHIRE CHARACTERISTICS

The most striking characteristic of the Hampshire is the white belt around the shoulders and body, including the front legs. The black color with the white belt constitutes a distinctive trademark; breed enthusiasts refer to it as the million dollar trademark.

Hampshire breeders have always stressed great quality and smoothness. The jowl is trim and light, the head refined, the ears erect, the shoulders smooth and well set, and the back well arched. An effort is now being made to secure bigger-framed, leaner, deeper-bodied, and sounder boars and sows to carry to heavier weights, yet remaining lean and efficient. In general, the breed has not been subjected to radical type fads. The trimness and freedom from excess lardiness give promise of a bright future for the Hampshire. Animals of this breed are active, and the sows have a reputation of raising a high percentage of the pigs farrowed. Also, they are of medium size and adapt well to confinement or outdoor environments.

The Hampshire Swine Registry lists the following disqualifications: white belt that does not entirely encircle the body, white on the head and snout, white on the hind legs that extends above the knob of hock, any red on any part of the animal, swirl on the upper half of the body or neck, or evidence of an extra dewclaw.

HEREFORD

The Hereford is one of the newer and less widely distributed breeds of swine. Among many swine growers, it has enjoyed a special attraction because the color markings emulate those of Hereford cattle

Fig. 3-11. Hereford boar. Grand Champion at the 1994 Hereford Hog Show and Sale, Reynolds, Indiana. (Courtesy, National Hereford Hog Record Assn., Flandreau, SD)

ORIGIN AND NATIVE HOME

The Hereford breed of hogs was founded by R. U. Webber, of LaPlata, Missouri. He conceived the idea of producing a white-faced hog with a cherry-red body color that would resemble the markings of Hereford cattle. The foundation stock included Chester Whites, OICs, Durocs, and pigs of unknown origin. By inbreeding and selection, the breed was further developed over a period of 20 years. In 1934, the National Hereford Hog Association was organized, under the sponsorship of the Polled Hereford Cattle Registry Association.

HEREFORD CHARACTERISTICS

Fig. 3-12. Hereford gilt. Grand Champion at the 1994 Hereford Hog Show and Sale, Reynolds, Indiana. (Courtesy, National Hereford Hog Record Assn., Flandreau, SD)

The most distinctive characteristic of the Hereford breed of hogs is their color marking, which is similar to that of Hereford cattle. In order to be eligible for registry, animals must be at least 2/3 red in color, either light or dark, and must have faces which are 4/5 white; the ears can be red or white, or red and white; and white must appear on at least three feet and extend at least an inch above the hoof, all the way around the leg. The ideal colored Hereford has a white head and ears, four white feet, white switch, and white markings on the underline.

In size, the Hereford is smaller than the other breeds of swine. In the past, many specimens of the breed were considered too lardy, heavy shouldered, and rough; but great improvement has been made through selection.

Any of the following disqualify an animal from registry or exhibiting: absence of some white in the face and not having a minimum of 2/3 red; a white belt extending over the shoulders, back, or rump; less than two white feet; a swirl; no marks of identification (ear notches or tattoo required); boar with one testicle; permanent deformities of any kind; underlines of fewer than six teats on each side; or any animal whose dam was under 10 months of age at time of farrowing.

LANDRACE

Fig. 3-13. Landrace boar. (Courtesy, Cedar Ridge Farms, Inc., Red Bud, IL)

As was originally true of Spain with its Merino sheep, Denmark long held a monopoly on the Landrace breed of swine. In 1934, Landrace hogs were shipped to the United States and Canada, but, by government agreement, for several years thereafter they could not be released as purebreds; their use being restricted to crossbreeding. Subsequently, an agreement was reached with Denmark whereby surplus purebred Landrace swine could be released. Thereupon (in 1950), the American Landrace Association, Inc., was organized, and the breed became known as the Landrace in this country

ORIGIN AND NATIVE HOME

The Landrace breed of hogs is native to Denmark, where it has been bred and fed to produce the highest quality bacon in the world. With the aid of government testing stations, the Landrace breed has long been selected for improved carcass quality and efficiency of pork production. In addition to improved breeding, the feeds common to Denmark—small grains and dairy byproducts—are also conducive to the production of high-quality bacon. For many years, the chief outlet for the surplus pork of Denmark has been the London market, where it is sold primarily in the form of Wiltshire sides.

EARLY AMERICAN IMPORTATIONS

The United States Department of Agriculture made

an importation of Danish Landrace swine from Denmark in 1934. Under the terms of the original agreement, purebred Landrace stock could not be raised by private individuals; only by certain experiment stations. Subsequently, Denmark approved of the release of surplus breeding stock that was already in the United States. Then, in 1954, 38 head of boars and gilts were imported from Norway by four individual breeders. The latter animals carried Norwegian, Danish, and Swedish Landrace blood. Other importations followed.

LANDRACE CHARACTERISTICS

The Landrace breed is white in color, although black skin spots or freckles are acceptable. The breed is characterized by its very long side, level top, well defined underline, deep flanks, trim jowl, straight snout, and medium lop ears. It is noted for prolificacy and for efficiency of feed utilization.

The breed registry association lists the following disqualifications: black in the hair coat; fewer than six teats on either side; erect ears, with no forward break; and swirls on topline (head to tail).

Fig. 3-14. Landrace sow. (Courtesy, Cedar Ridge Farms, Inc., Red Bud, IL)

POLAND CHINA

No other breed of swine has been subjected to such radical shifts in type as the Poland China. Likewise, no other breed has swung from such heights of popularity or fallen so low in disrepute.

ORIGIN AND NATIVE HOME

The Poland China breed of swine originated in southwestern Ohio in the fertile area known as the Miami Valley, particularly in Warren and Butler counties. In the early part of the 19th century, the Miami Valley was the richest corn-producing section of the

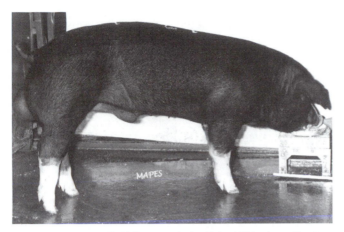

Fig. 3-15. 1994 World Expo top placing Poland China boar. (Courtesy, Poland China Record, Peoria, IL)

United States. Moreover, prior to the Civil War, Cincinnati was the pork-packing metropolis of the country ("Porkopolis"). Thus, the conditions were ideal for the development of a new breed of hogs.

The common stock kept by the early settlers of the Miami Valley were described as being of mixed color, breeding, and type. The foundation animals were crossed with the Russian and Byfield hogs that were introduced into the valley early in the 19th century. In 1816, the Shaker Society, a religious sect, introduced the Big China breed. This breeding and improvement gave rise to the so-called Warren County hog, which gained considerable prominence as a result of its huge size and great fattening ability. Later, hogs of Berkshire and Irish Grazer breeding were introduced into the area and were crossed with the Warren County hogs. This improved the quality and refinement of the stock and resulted in earlier maturity. It is generally agreed that no outside blood was brought into the Miami Valley after about 1845.

The name, *Poland China*, was established in 1872 by the National Swine Breeders in convention at Indianapolis, Indiana. In view of the fact that it seems to be definitely established that no breed known as Poland hogs was ever used as foundation stock, it is of interest to know how the name *Poland China* was selected. A Mr. Asher, a prominent Polish farmer in the Valley, was supposedly responsible for the word *Poland*. As China hogs had been introduced by the Shaker Society, the name of *Poland China* was adopted.

POLAND CHINA CHARACTERISTICS

Modern Poland Chinas are black in color with six white points—the feet, nose, and tip of tail—but prior to 1872, they were generally mixed black and white and spotted. Absence of one or two of the six white points is of small concern, and a small white spot on the body is not seriously criticized.

Until the latter part of the 19th century, Poland China hogs were noted for their immense size and scale. Beginning about 1890, breeders turned their attention to the development of greater refinement, earlier maturity, and smaller size. This fashionable fad finally culminated in the development of the "hot bloods," an extremely small, compact, short-legged type which dominated the show-ring from 1900 to 1910.

Finally, in order to secure increased utility qualities, breeders began the shift to the big-type strains. The craze swept across the nation, and again the pendulum swung too far. However, since 1925, the efforts of breeders have been toward development of the more conservative meat type. Poland Chinas now yield a high quality carcass with a high percentage of lean.

Many producers pick Polands as the sire breed in a terminal cross because their dark color is dominated by the color of the second breed.

The following disqualify an animal from registry: fewer than six teats on a side, a swirl on the upper half of the body, hernia, or cryptorchidism.

Fig. 3-16. Poland China gilt exhibited at the 1995 National Barrow Show by Roger Waller, Riveria Polands, Bennett, Iowa. (Courtesy, *Poland China Record*, Peoria, IL)

SPOTTED

The popularity of the "Spots" is chiefly attributed to the success of breeders in preserving the utility value of the old Spotted Polands while making certain improvements in the breed.

ORIGIN AND NATIVE HOME

The Spotted was developed in the north central part of the United States, principally in the state of Indiana. As has been indicated, the foundation stock

Fig. 3-17. Spotted boar. (Courtesy, National Spotted Swine Record, Peoria, IL)

of the Miami Valley Poland Chinas was frequently of a black and white spotted color. Moreover, these animals had a reputation for size, ruggedness, bone, and prolificacy. During the era of the "hot bloods," many breeders forsook the Poland China breed and attempted to revive the utility qualities of the original spotted hogs of the Miami Valley. Thus, the Spotted was established. In addition to utilizing selected strains of the Poland China, breeders introduced, in 1914, the blood of the Gloucester Spotted hog of England. Although the Spotted breed had its official beginning with the organization of record associations in 1914, liberal infusion of big-type Poland China blood was continued until 10 years later.

SPOTTED CHARACTERISTICS

Fig. 3-18. Spotted sow. (Courtesy, National Spotted Swine Record, Peoria, IL)

At the present time, there is no great difference between the most approved type of Poland China and the Spotted. The former has a little more size, but in general the conformation is much alike.

The most desired color in the Spotted is spotted black and white—about 50% of each. Females must have at least six prominent teats on each side to be eligible for show or sale.

The following constitute disqualifications and bar animals from registry: brown or sandy spots; less than 20% or more than 80% white on body; boar with a swirl; small upright ears; not over half normal size; cramped or deformed feet; seriously diseased, barren, or blind; or if scoring less than 60 points.

TAMWORTH

Fig. 3-18a. Tamworth boar; Junior Champion at the Ohio State Fair. He was tested at the Ohio Boar Station. (Courtesy, Tamworth Swine Assn., Winchester, OH)

The Tamworth is one of the oldest and probably one of the purest of all breeds of hogs. It is also recognized as the most extreme bacon type of any breed.

ORIGIN AND NATIVE HOME

The Tamworth originated in Ireland where they were known as *Irish Grazers*, since they were such good foragers. About 1812, Sir Robert Peel imported some of them to his estate at Tamworth, England, from whence they derived their name.

There is record of pure breeding and careful selection dating back more than 100 years. Tamworths are thought to be one of the purest breeds in Britain.

EARLY AMERICAN IMPORTATIONS

The earliest importation of Tamworths into the United States, of which there is authentic record, was made by Thomas Bennett, of Rossville, Illinois, in 1882.

TAMWORTH CHARACTERISTICS

The color of the breed is red, varying from light to dark. The conformation may be described as that of extreme bacon type.

Swirls, fewer than 12 teats, and more than 5% black disqualify pigs from registration.

YORKSHIRE

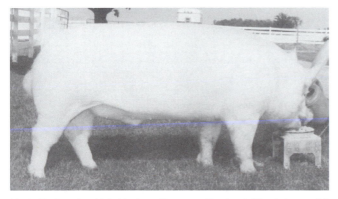

Fig. 3-19. A modern Yorkshire boar. (Courtesy, *Seedstock*, West Lafayette, IN)

In its native home, England, the Yorkshire breed is known as the Large White.

ORIGIN AND NATIVE HOME

The Yorkshire is a popular English bacon breed which had its origin nearly a century ago in Yorkshire and neighboring counties in northern England.

EARLY AMERICAN IMPORTATIONS

Although Yorkshires were probably first brought to the United States early in the 19th century, representatives of the modern type were first introduced by Wilcox and Liggett, of Minnesota, in 1893. A great many importations were made into Canada at an early date, and the breed has always enjoyed a position of prominence in that country.

Fig. 3-20. A performance proven Yorkshire brood sow. (Courtesy, *Seedstock*, West Lafayette, IN)

YORKSHIRE CHARACTERISTICS

Yorkshires should be entirely white in color. Although black pigment spots, called "freckles," do not constitute a defect, they are frowned upon by breeders. The face is slightly dished, and the ears are erect.

Yorkshire sows are noted as good mothers. They not only farrow and raise large litters, but they are great milkers. The pigs are excellent foragers and compare favorably with those of any other breed in economy of gains.

The following disqualify animals from registry: the presence of swirls on the upper third of the body; hernia; hair color other than white; cryptorchidism; hermaphrodite; blind or inverted teats; total blindness; fewer than six teats on each side; permanent lameness; excessive amounts of black and/or dark pigment of the skin; or extra dew claws.

INBRED BREEDS OF SWINE

Beginning in the 1930s and continuing through the 1950s, there was interest in developing new breeds of swine that would be prolific, gain rapidly and efficiently, and produce a high quality carcass. New breeds were developed at the USDA Agricultural Research Center at Beltsville, Maryland, at several of the state agricultural experiment stations, by private breeders, and in Canada. As the numbers of these inbred (new breeds) swine increased, a demand for registration arose. So, in 1946, the Inbred Livestock Registry was organized.

While most of these newer breeds of swine did not survive, it is to their credit that they shook their older counterparts out of their lethargy.

SPECIALIZED BREEDING STOCK SUPPLIERS

Until about 1950, typical U.S. commercial swine producers of market hogs selected most replacement female breeding stock from their own herds, while almost all boars were purchased from purebred breeders. But the entry of *specialized breeding stock suppliers* has changed many breeding programs. A recent survey showed that specialized breeding stock suppliers were the source of 14.1% of the nation's gilts, and of 27.7% of the nation's boars.[3] Also, the survey revealed a trend of large commercial hog producers to buy breeding stock from these new specialized breeding stock suppliers.

---

[3]Survey conducted by Michigan State University, financed by *National Hog Farmer*, reported by *National Hog Farmer*, September 15, 1989, pp. 52–58.

SPECIALIZED BREEDING STOCK

Fig. 3-21. Hampshire boar. *Note:* Most of the Specialized Breeding Stock suppliers breed and market one or more of the traditional purebred breeds. (Courtesy, *Seedstock Edge*, West Lafayette, IN)

Fig. 3-24. #14 gilt. Note the muscling and the good feet and legs. (Courtesy, American Diamond Swine Genetics, Prairie City, IA)

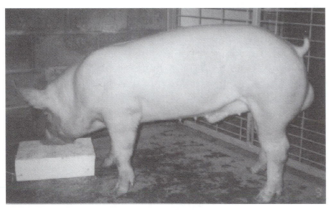

Fig. 3-22. Hybrid boar. This line is being used to develop new White Diamond Blue females. (Courtesy, American Diamond Swine Genetics, Prairie City, IA)

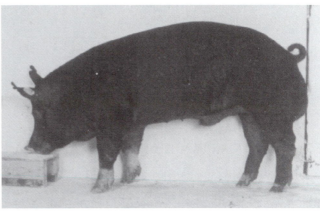

Fig. 3-25. FHC#1 Best Black, F-5 boar. (Courtesy, Farmers Hybrid Co., Inc., West Des Moines, IA)

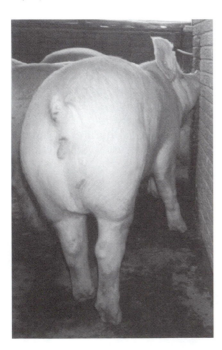

Fig. 3-23. Large White gilt. (Courtesy, American Diamond Swine Genetics, Prairie City, IA)

Fig. 3-26. FHC#1 Best E_2 boar. (Courtesy, Farmers Hybrid Co., Inc., West Des Moines, IA)

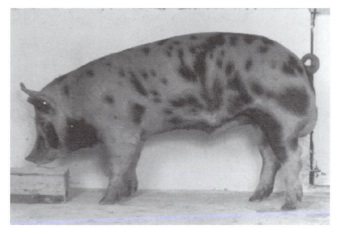

Fig. 3-27. FHC#1 Best Red & Black Elite Boar. (Courtesy, Farmers Hybrid Co., Inc., West Des Moines, IA)

Fig. 3-30. YXL Line. 50% inbred Yorkshire and 50% inbred Lacombe. (Courtesy, McLean County Hog Service, Inc., LeRoy, IL)

Fig. 3-28. A Babcock Grandparent gilt, performance tested and indexed for growth rate and leanness. (Courtesy, Babcock Swine, Inc., Plainview, MN)

Fig. 3-31. MCHS sow. (Courtesy, McLean County Hog Service, Inc., LeRoy, IL)

Fig. 3-29. 50% inbred Hampshire X 50% Large White boar. (Courtesy, McLean County Hog Service, Inc., LeRoy, IL)

Fig. 3-32. Duroc boar, class winner at the Nebraska State Fair, bred and exhibited by Waldo Farms. (Courtesy, Waldo Farms, DeWitt, NE)

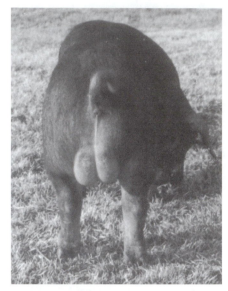

Fig. 3-33. Modern muscular Duroc. (Courtesy, Lone Willow Genetics, Roanoke, IL)

Fig. 3-35. Modern high lean Hampshire boar. (Courtesy, Lone Willow Genetics, Roanoke, IL)

Fig. 3-34. Modern muscular Yorkshire. (Courtesy, Lone Willow Genetics, Roanoke, IL)

Fig. 3-36. Spot X 1 Boar. 50% Minnesota No. 1, and 50% Spotted. (Courtesy, McLean County Hog Service, Inc., LeRoy, IL)

Most of the specialized breeding stock suppliers breed and market one or more of the traditional pure-bred breeds. In addition, they usually produce and market, often under their own genetic names, one or more hybrids, crossbreds, inbreds, maternal female lines, and/or terminal sire lines. They have well-developed sales and testing programs, which emphasize performance and carcass quality. Some of them also provide financial assistance to their customers.

Among the specialized breeding stock suppliers are the following:[4]

American Diamond Swine Genetics
Prairie City, Iowa

Babcock Swine, Inc.
Plainview, Minnesota

Compart's Boar Store
Nicollet, Minnesota

DeKalb Swine Breeders, Inc.
DeKalb, Illinois

Farmers Hybrid
West Des Moines, Iowa

Genetipork
Morris, Minnesota

Lone Willow Genetics
Roanoke, Illinois

McLean County Hog Service
LeRoy, Illinois

Newsham Hybrids
Colorado Springs, Colorado

NPD Carroll Foods
North Carolina

Pig Improvement Company (PIC)
Franklin, Kentucky

Segher's Hybrids
Muscoda, Wisconsin

Waldo Farms, Inc.
DeWitt, Nebraska

---

[4]No claim is made that this list is complete. Neither should a listing be construed as an endorsement by the authors.

MINIATURE SWINE

During the 1950s, the need developed for a smaller breed of swine for use in biomedical studies. For example, an awareness of the need arose from some of the work of the U.S. Atomic Energy Commission. Swine were selected as the animals of choice for certain studies. But the large size and the amount of waste (feces) produced by standard swine created a disposal problem and discouraged their use for long-term studies. In 1949, the Hormel Institute of the University of Minnesota initiated the development of genetically small pigs. These were used in some of the first Atomic Energy Commission studies. Later, other small strains were developed such as the Pitman-Moore, the Hanford Miniature Swine, and the Gottingen Miniature.[5] These miniatures are much smaller than standard swine. They are good experimental animals for biomedical studies, offering economic and convenience advantages over standard swine; they require less housing space, eat less, are easier to handle, and produce less waste (feces).

■ **Vietnamese pot-bellied pigs**—These native Asian pigs, which are normally less than one-fifth the size of traditional U.S. hogs, were first brought into the United States in 1988 for zoos. But they very quickly became novelty pets, selling for as much as $5,000 or more. Normally, they are 18 to 24 in. tall; and they have a pot-belly, a swayed back, and a straight tail which they wag when they are happy.

By 1994, it was estimated that the nation's pot-bellied pig population numbered 200,000, and prices plummeted to as low as $100.

Although the novelty has worn off, many people need pets; so, pot-bellied pigs will be around for a very long time. Thus, it is important that the basics of owning and caring for these pets, along with their behavior, be understood. To this end, the following brief is presented.

1. **Local ordinances restricting the keeping of livestock within the city or corporate limits.** Pot-bellied pigs belong to the same species as traditional U.S. hogs. So, local governing bodies will have to decide if citizens can or cannot keep them within corporate limits.

2. **Pot-bellied pigs are subject to the same state veterinary regulations as U.S. commercial swine.** For example, owners need to be aware that pot-bellied pigs are subject to brucellosis and tuberculosis, which can be transmitted from pigs to people. So, keepers of pot-bellied pigs should contact their veterinarian for assistance in the testing and health care of these pets.

3. **Feed.** Pot-bellied pigs should be fed similar feeds to those fed to their bigger counterparts, but in much smaller quantities. Thus, the starter diet should contain 18 to 20% crude protein, the grower diet 14 to 16% crude protein, and the maturity diet 14% crude protein. Feed ½ to ¾ lb of feed per 50 lb of body weight, depending on body condition.

4. **Size.** The best way to avoid ending up with a big hog is to see both of the parents before buying.

5. **Spoiled and destructive.** Their herd instinct tends to make pot-bellied pigs aggressive, which may cause them to harm children. Also, their rooting and foraging instinct may cause them to lift up the carpet and to eat everything chewable. And they must be housebroken!

---

[5]Dr. M. E. Ensminger, the co-author of this book, served as Consultant in the development and use of the Hanford Miniature at the Hanford Laboratories operated by the General Electric Company, Nucleonics Department (Atomic Energy Commission).

QUESTIONS FOR STUDY AND DISCUSSION

1. Must a new breed of swine be approved by someone, or can anyone start a new breed? Justify your answer.

2. Trace shifting of swine types throughout the years, including the factors that prompted such shifts.

3. Why have U.S. swine types shifted more rapidly and more radically than cattle and sheep types?

4. Most American-created breeds of swine evolved during two periods; namely, (a) 1800 to 1880, and (b) since 1940. What is the explanation for this?

5. Table 3-1 gives the annual registrations of breeds of swine. These figures show that, currently, three breeds—Yorkshire, Duroc, and Hampshire—dominate the swine breeds. Why is this?

6. List the place of origin, and the distinguishing characteristics for the nine major U.S. breeds of swine. Discuss the importance of each.

7. Obtain breed registry association literature and a sample copy of a magazine of your favorite breed of swine. Evaluate the soundness and value of

the material that you receive. (See the Appendix for addresses.)

8. Justify any preference that you may have for one particular breed of swine.

9. During the 1930s and 1940s, several new American inbred breeds of swine evolved, most of them carrying some Landrace breeding. These new breeds did not survive. What was the reasoning behind their development, why did they fail to survive? Was their development and cost justified?

10. How do the specialized breeds of swine differ from the inbred breeds and from the traditional breeds?

11. Why are miniature swine more suitable for biomedical studies than the standard breeds of swine?

12. Would you recommend Vietnamese pot-bellied pigs as a family pet?

13. How important are breed characteristics? Can breed differences in hams and pork chops be detected?

SELECTED REFERENCES

Title of Publication	Author(s)	Publisher
Breeds of Livestock, The	C. W. Gay	The Macmillan Company, New York, NY, 1918
Breeds of Livestock in America	H. W. Vaughan	R. G. Adams and Company, Columbus, OH, 1937
Breeds of Swine		Farmers' Bul. 1263, U.S. Department of Agriculture, Washington, DC, 1968
History of Yorkshires and American Yorkshire Club	W. L. Plager	American Yorkshire Club, Inc., West Lafayette, IN, 1975
Modern Breeds of Livestock	H. M. Briggs D. M. Briggs	The Macmillan Company, New York, NY, 1980
Pigs from Cave to Corn Belt	C. W. Towne E. N. Wentworth	University of Oklahoma Press, Norman, OK, 1950
Stockman's Handbook, The, Seventh Edition	M. E. Ensminger	Interstate Publishers, Inc., Danville, IL, 1992
Story of Durocs, The	B. R. Evans G. G. Evans	United Duroc Record Association, Peoria, IL, 1946
Study of Breeds in America, The	T. Shaw	Orange Judd Company, New York, NY, 1912
Swine Production	J. L. Krider J. H. Conrad W. L. Carroll	McGraw-Hill Book Company, New York, NY, 1982
Types and Breeds of Farm Animals	C. S. Plumb	Ginn and Company, Boston, MA, 1920
World Dictionary of Breeds Types and Varieties of Livestock, The	I. L. Mason	CAB International, Wallingford, Oxon, 0X10 8DE UK, 1988

Also, breed literature pertaining to each breed may be secured by writing to the respective breed registry associations (see Appendix for the name and address of each association).

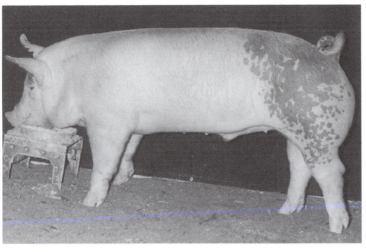

A modern lean, meat-type barrow, a Hampshire X Yorkshire X Duroc cross. (Courtesy, *Seedstock*, West Lafayette, IN)

4

ESTABLISHING THE HERD; SELECTING AND JUDGING SWINE

The problems encountered and the principles employed in establishing the swine herd and in selecting and judging hogs are very similar to those for beef cattle and sheep. In general, however, one can establish a swine herd at a lower cost and more quickly than is possible with other classes of farm animals. Until recently, hog markets were less discriminating than cattle or sheep markets, with the result that the average commercial producer gave far less attention to becoming proficient in selecting and judging swine. But, today, it is recognized that selecting and judging swine by the "eyeballing" technique is only part of the process.

FACTORS TO CONSIDER IN ESTABLISHING THE HERD

At the outset, it should be recognized that the vast majority of the swine producers of this nation keep hogs simply because they expect them to be profitable. That hogs have usually lived up to this expectation is attested by their undisputed claim to the title of the "mortgage lifter." For maximum profit and satisfaction in establishing the herd, the individual swine producer must give consideration to the type, breeding, breed, size of herd, uniformity, health, age, price, and suitability of the farm

TYPE: LEAN AND MEATY

Historically, there were two types of hogs: the lard type and the bacon type. The lard type was a thick-bodied hog carrying a large amount of fat, while the bacon type was developed to meet the demand for

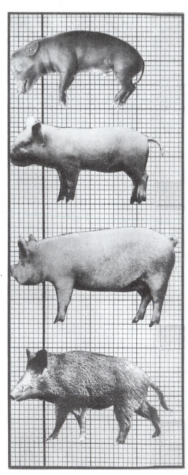

**DOMESTIC PIG FETUS
9 WEEKS — ½ LB**

**DOMESTIC PIG
3 WEEKS — 15 LB**

**DOMESTIC PIG
15 WEEKS — 100 LB**

**WILD BOAR
ADULT — 300 LB**

ALL SCALED TO THE SAME HEAD SIZE

Fig. 4-1. This figure illustrates how years of selection have changed the conformation of the pig to meet market requirements. Each animal is reduced to the same head size. As shown, when an improved breed matures, the proportion of loin to head and neck increases greatly; but an unimproved type such as the wild boar matures without much change in body proportions. (From: *Farm Animals*, by John Hammond; published by Edward Arnold & Co., London; courtesy, Sir John Hammond)

high quality lean bacon in the British Isles and Scandinavia.

With lard becoming an unwanted product, often selling for less per pound than the price of slaughter hogs on foot, breeders of meat-type breeds stressed leanness, a minimum amount of fat, and the maximum cut-out value of primal cuts.

Today, the vast majority of market hogs are lean and meaty. However, through the years, most purebred breeds of swine have run the gauntlet in types, producing animals of the chuffy, rangy, and medium types. It is evident even today that these breeds possess the necessary store of genes through which such shift in types may be made by breeding and selection. Even so, most pork producers prefer the medium or intermediate type to either the chuffy or rangy type. Chuffy-type animals lack in prolificacy and rapidity of gains, whereas the rangy hog must be carried to too heavy a weight in order to reach market finish. The packers and consumers object to the chuffy animals because of their excess lardiness and to the rangy ones because of their large cuts. It may be concluded, therefore, that most successful swine producers of the present day favor hogs that are lean and meaty.

PUREBREDS, SEEDSTOCK, OR CROSSBREDS

Generally speaking, only the experienced breeder should undertake the production of purebreds with the intention of eventually furnishing foundation or replacement stock to other purebred breeders, seedstock producers, or commercial producers.

Seedstock producers are discussed in Chapter 3, under the heading "Specialized Breeding Stock Suppliers"; hence, the reader is referred thereto.

Crossbreeding of swine is fully discussed in Chapter 6. At this point, it is sufficient to say that this type of breeding program is more widely used in the production of hogs than with any other class of livestock.

SELECTION OF THE BREED OR STRAIN

No one breed or strain of hogs can be said to excel all others in all points of swine production and for all conditions. It is true, however, that particular breed or strain characteristics may result in a certain breed or strain being better adapted to given conditions; for example, hogs of light color are subject to sunburn if raised outdoors in hot, sunny areas. Usually, however, there is a greater difference among individuals within the same breed or strain than between the different breeds or strains; this applies both to type and efficiency of production.

SIZE OF HERD

Hogs multiply more rapidly than any other class of farm animals. They also breed at an early age, produce twice each year, and bear litters. It does not take long, therefore, to get into the hog business.

The eventual size of the herd is best determined by the following factors: (1) proximity to the neighbors, (2) suitable manure handling facilities, (3) adequate land for manure application, (4) confinement facilities, (5) kind and amount of labor, (6) the disease and parasite situation, including isolation, (7) the market(s), and (8) comparative profits from hogs and other types of enterprises.

UNIFORMITY

Uniformity of type and ancestry gives assurance of the production of high-quality pigs that are alike and true to type. This applies both to the purebred and the commercial herd. Uniform offspring sell at a premium at any age, whether they are sold as purebreds for foundation stock, as feeder pigs, or as slaughter hogs. With a uniform group of sows, it is also possible to make a more intelligent selection of herd boars.

HEALTH

Producers should match the health status of incoming stock to that of their existing herd. Incoming breeding animals should be in thrifty, vigorous condition, and should have been raised by suppliers who have exercised great care in the control of diseases and parasites. All purchases should be made subject to the animals being free from contagious diseases.

AGE

In establishing the herd, the beginner may well purchase a few bred gilts that are well grown, uniform in type, and of good ancestry and that have been mated to a proven sire. Although less risk is involved in the purchase of tried sows, the cost is likely to be greater in relation to the ultimate value of the sows on the market.

Then, too, with limited capital, it may be necessary to consider the purchase of a younger boar. Usually a wider selection is afforded with this procedure, and,

in addition, the younger animal has a longer life of usefulness ahead.

PRICE

The beginner should always start in a conservative way. However, this should never be cause for the purchase of poor individuals—animals that are high at any price.

SUITABILITY OF THE FARM

A swine producer may have one neighbor who is a cattle feeder, a second who operates a dairy, a third who keeps a sizeable farm flock of sheep, a fourth who produces light horses for recreation and sport, and a fifth whose chief source of income is from poultry. All may be successful and satisfied with their respective livestock enterprises. This indicates that several types of livestock farming may be equally well adapted to an area or region. Therefore, the selection of the dominant type of livestock enterprise should be analyzed from the standpoint of the individual farm.

Usually a combination of several factors suggests the livestock enterprise or enterprises best adapted to a particular farm and farmer. Some of the things that characterize successful major swine enterprises are:

1. Adequate distance from the neighbors.
2. Suitable manure handling facilities.
3. Swine knowledge, interest, and skill of the operator.
4. A plentiful supply of grains or other high energy feeds in the immediate area.
5. Available labor skilled in caring for swine, especially at farrowing time.
6. A satisfactory market outlet.
7. Adequate and convenient, but not elaborate, buildings and equipment.

SELECTION BASES

Generally, the selection of foundation hogs, boars, and replacement gilts, is made on the basis of one or more of the following considerations: (1) type or individuality, (2) mechanical measures, (3) pedigree, (4) show-ring winnings, (5) production testing, (6) STAGES, EBV, EPD, and/or BLUP, or (7) genetically-free of Porcine Stress Syndrome.

SELECTION BASED ON TYPE OR INDIVIDUALITY

Selection based on type or individuality implies the selection of those animals that approach the ideal or standard of perfection most closely and the culling out of those that fall short of these standards.

The ideal lean/meaty hog is properly balanced with a blending of all parts. The head and neck are clean-cut. The back is moderately and evenly arched, and of adequate width. The loins are wide and strong, and the hams are deep, thick, slightly bulging and meated well down to the hocks. The shoulders are well-laid in and smooth, while the sides are long, deep, and smooth. The legs are straight and squarely on the corners of the body, and the pasterns are strong. The walk is free and easy, not stiff. Additionally, the ideal meat-type hog possesses adequate size for age.

SELECTION BASED ON MECHANICAL MEASURES

Methods of measuring backfat and loin muscle area in swine are very important due to value-based marketing and demand for lean breeding stock.

The backfat probe, by which backfat is measured with a stainless steel backfat ruler, has been used for years; and it is still the most economical, objective means of measuring backfat. Lean meters are used by some meat packers to determine carcass leanness.

Ultrasound, to determine the body composition of people, has been used by the medical profession for more than 30 years. Due to the improvements made in scanners and transducers utilized in medicine, the animal/meat industry has been able to use ultrasound technology for the past 10 years.

Today, the following methods are available and are used in the swine industry to measure body composition: (1) backfat probe, (2) lean meters, and (3) ultrasonics.

■ **Backfat probe**—Fig. 4-2 shows the probing sites on the live hog. These three locations correspond to the three locations where backfat determinations are made on the carcass.

The only equipment needed for probing is a snare to restrain the hog, a sharp knife or scalpel blade, and a narrow 6-in. metal ruler with $\frac{1}{10}$-in. graduations. The steps and technique in probing are:

1. Wrap the knife or scalpel with several layers of tape about $\frac{3}{8}$ in. from the tip to prevent the blade from going too deep.
2. Weigh the hog.
3. Restrain the hog with a nose snare.
4. Jab the knife through the skin at a right angle

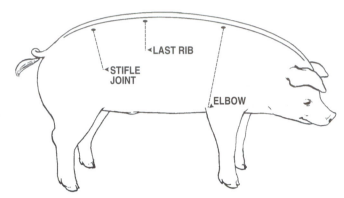

Fig. 4-2. Probe each hog at three locations: (1) midpoint of shoulder above elbow, (2) middle of back where last rib joins the vertebrae, and (3) rump, straight above the stifle joint.

to the hog's body at 1.5 to 2 in. to one side of the midline.

5. Insert the probe in the cut and slant it so that it points toward the center of the hog's body.
6. Force the probe through the fat down to the loin muscle. When the probe reaches the loin muscle, a firm resistance will be noted.
7. Push the clip on the probe snugly down to the skin line. Remove the ruler and read the measurement.

In order to make valid comparisons and selections, backfat thickness should be adjusted to a common liveweight basis, commonly 230 lb.

Where pinpoint accuracy is not considered essential, measuring backfat at a single probe site is suggested. The recommended site where one probe only is taken is the seventh rib, located at a distance approximately four fingers wide behind the shoulder-probing location and 1 in. off the midline of the back.

■ **Lean meters**—Some pork packers are now using lean meters to evaluate pork carcasses. One such system (and there are others), known as TOBEC (total body electrical conductivity), is targeted for use in packing plants where both speed and accuracy are important. Basically, it consists of a huge, precisely wound coil of wire through which the carcass passes. TOBEC measures lean rather than fat, hence, it rewards heavily muscled carcasses, rather than carcasses that are just lean.

■ **Ultrasonics**—Electronic equipment employs the "pulse echo" technique. Basically, this is the generation of very short bursts of high-frequency (nondestructive, inaudible) sound into the animal, detecting the reflection of the pulses, and measuring the elapsed time between introduction of the sound pulse into the animal and the return of the reflected pulse. When the machine is properly calibrated, the fat depth can be read directly from the scale.

Basically, two types of ultrasound equipment are

in use in the swine industry—the A-mode machine (see Fig. 4-3), and the B-mode (real time) machine; and there are two or more manufacturers of each.

Both types of machines use reflected sound waves to measure depth of backfat and muscle. But A-mode machines use a measurement of loin-eye depth to estimate loin-eye area, whereas B-mode (real time) machines actually measure loin-eye area.

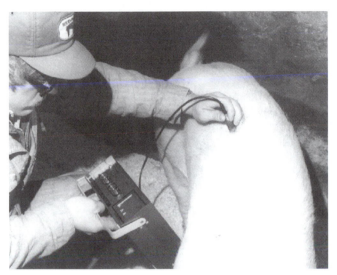

Fig. 4-3. Ultrasonic measurement of backfat. (Courtesy, International Livestock Improvement Services Corp., Ames, IA)

Of course, the level of sophistication adds greatly to the cost of the machine and its attachments. The A-mode scanners sell for between $495 and $1,400, whereas the B-mode scanners sell for $10,000 to $25,000.

One of the greatest variables in ultrasound accuracy is the technician effect. Technicians must be consistent in finding the 10th or last rib locations and interpreting the information.

■ **Current recommendations relative to ultrasonics**—Since there is considerable differential between A-mode and B-mode equipment, producers must determine the level of accuracy that is desired in measuring fat thickness and loin muscle area. A-mode equipment can be effectively used (1) to sort hogs for backfat when marketing on a grade and yield basis, and (2) to select replacement gilts for fat thickness. But the use of A-mode equipment to predict loin muscle area is not as reliable.

Specialized seedstock producers should consider B-mode equipment, since it makes for greater accuracy in both fat thickness and loin muscle area.

When data is used across herds, reliability is important; hence, B-mode equipment by trained technicians is recommended. Also, B-mode equipment is recommended for scan data collected at central test stations, and for on-farm data that is used by national breed registries.

SUMMARY: SWINE BREEDERS MUST MEASURE

Mechanical methods provide necessary adjuncts to visual appraisal and scales in the selection of meaty-type breeding animals and the production of higher quality pork carcasses. Increasingly, they will be used in the selection programs of purebred, seedstock, and commercial producers, and in value-based marketing.

■ **Comparing animals on the basis of backfat and loin area measurements**—In order to make valid backfat and loin area comparisons of animals, all measurements should be adjusted to a 230 lb weight basis. Participants in purebred swine association STAGES and/or EPD programs can secure adjusted measures from their respective breed registries. However, swine producers can easily adjust their measurements. The following formulas can be used to adjust backfat:

$$\text{Adjusted backfat} = \text{avg. of 3 backfat probes}$$
$$\text{(see Fig. 4 - 3)} + \left[(230 - \text{weight}) \times \frac{(\text{Avg. Backfat})}{\text{weight} - 25} \right]$$

Example: A pig had an average backfat depth of 1.0 in. when probed at 220 lb.

$$1 + (10 \times 0.005)$$

$$1 + \left[230\,\text{lb} - 220 \times \frac{1}{220 - 25} \right]$$

$$1 + 0.05 = 1.05$$

In addition to including backfat thickness, and possibly loin eye area, a sound selection program should include other important economic traits such as sow productivity, post weaning growth (days to 230 lb), and structural soundness.

(See subsequent section in this chapter headed "Selection Based on STAGES, EBV, EPD, and BLUP.")

SELECTION BASED ON PEDIGREE

In the selection of breeding animals, the pedigree is a record of the individual's heredity or inheritance. If the ancestry is good, it lends confidence in projecting how well young animals may breed. It is to be emphasized, however, that mere names and registration numbers are meaningless. A pedigree may be considered as desirable only when the ancestors close up in the lineage—the parents and grandparents—were superior individuals and outstanding producers. Too often, purebred hog breeders are prone to play up one or two outstanding animals back in the third or fourth generations. If pedigree selection is to be of any help, one must be familiar with the individual animals listed therein.

The boar should be of known ancestry. This alone is not enough, for he should also be a good representative of the breed or strain selected. Likewise, it is important that the sows be of good ancestry, regardless of whether they are purebreds or crossbreds. Such ancestry and breeding give more assurance of the production of high quality pigs that are uniform and true to type.

Also, a registration certificate will be greatly enhanced by being a performance pedigree, presenting the following economically important traits: backfat thickness, days to 230 lb, number born alive, and 21-day litter weight.

SELECTION BASED ON SHOW-RING WINNINGS

Until about 1970, most swine producers looked favorably upon using show-ring winnings as a basis of selection. Purebred breeders were quick to recognize this appeal and to extol their champions through advertising. In most instances, the selection of foundation or replacement hogs on the basis of show-ring winnings and standards was for the good. On some occasions, however, purebred and commercial breeders alike came to regret selections based on show-ring winnings. This was especially true during the eras when the chuffy or rangy type was sweeping shows from one end of the country to the other. This would indicate that some scrutiny should be given to the type of animals winning in the show, especially to ascertain whether such animals are of a type that is efficient from the standpoint of the producer, and whether, over a period of years, they will command a premium on a discriminating market.

Perhaps the principal value of selections based on show-ring winnings lies in the fact that shows direct the attention of the amateur to those types and strains of hogs that at the moment are meeting with the approval of the better breeders and judges.

Swine shows have lost favor with many producers. But properly conducted junior livestock shows will continue to have values for our best products, boys and girls—values which cannot be measured in dollars and cents. When FFA, 4-H Club, and other junior exhibitors are showering their love and affection on their animals, they are not being subjected to the temptations of today's society. Moreover, competing in junior shows is great preparation for a lifetime of competition. The senior author of this book will ever be grateful for his experiences in fitting and showing animals, including swine in the nation's FFA, 4-H Club, and other junior livestock shows.

SELECTION BASED ON PRODUCTION TESTING

No criterion that can be used in selecting an animal is so accurate or important as past performance.

Most progressive purebred and seedstock producers now have production records. Without doubt, swine producers of the future will make increasing use of such records as a basis for selection.

Production testing systems and the relative merits of each are presented in Chapter 6. Also, suggested record forms are reproduced in the same chapter.

SELECTION BASED ON STAGES, EBV, EPD, AND BLUP

■ **STAGES,** which stands for **Swine Testing and Genetic Evaluation System**, is a system of the breed registries recording performance data provided by swine producers. In 1994, the combined swine breeds had 260,000 sow productivity records, and 120,000 growth records.

■ **EBV,** which stands for **Estimated Breeding Value**, is the entire worth of the parent as a source of genetic material. Stated differently, the EBVs represent the value of an individual as a source of genetic material for the herd.

■ **EPD**, which stands for **Estimated Progeny Difference**, is half the genetic worth of each parent and is an estimate of how much of their performance will be passed on to their offspring. EPD is a prediction of the progeny performance of an animal compared to the progeny of an average animal in the population, based on all information currently available.

Accuracy (ACC) of the EPD can range from 0 to 1.0; it is an expression of the reliability of the EPD. The values for accuracy are more reliable if they exceed 0.5. The decision as to which boar to use should be based on the EPD of each individual boar. The accuracy value should be used to determine how extensively the animal should be used. Boars with favorable EPDs and high accuracies can be used with confidence; they will contribute to the genetic improvement of the herd. Accuracy is not constant because considerable new information is added for each subsequent analysis.

Note: Positive EPDs are more desirable for number born alive and 21-day weight. Negative EPDs are more desirable for days/230 and backfat.

EPDs are available for the following:

EPD Days/230
EPD Backfat
EPD Terminal Sire Index (TSI)
EPD Number Born Alive (NBA)
EPD Litter Weight (LW)
EPD Sow Productivity Index (SPI)
EPD Maternal Sire Index (MSI)

■ **BLUP,** which stands for **Best Linear Unbiased Prediction,** is a statistical procedure that can be used to analyze swine performance data. Some of the special features of this procedure are:

1. BLUP uses information from all known relatives, past and present, in assessing a pig's genetic merit.

2. BLUP uses information from other associated traits through genetic correlations in determining the genetic merit of a pig for a particular trait.

3. BLUP removes most of the biases from the genetic analysis introduced by systematic environmental influences such as—

    a. Age differences of animals.

    b. Management group differences.

    c. Litter effects.

    d. Parity effects.

    e. Preferential or "corrective" mating.

Thus, genetic trends free of these environmental effects can be estimated.

BLUP makes it possible for breeders (1) to make a much more accurate assessment of the animals that they are considering for selection and culling; and (2) to evaluate the genetic progress that they are making in their breeding programs, which is particularly important when market conditions are changing.

BLUP is the most accurate technology available for the estimation of genetic merit. This procedure develops an Expected Progeny Deviation for each individual based on its own record, record on ancestors in the individual's pedigree, and collateral relatives with performance records (littermates, half-sibs, etc.), as well as any progeny of the individual that has a performance record. Using all of these sources of information (through a genetic relationship matrix for the population) accounts for genetic trends and non-random matings, thus resulting in maximum prediction accuracy.

SELECTION BASED ON GENETICALLY-FREE PORCINE STRESS SYNDROME (PSS)

Prevent PSS by selecting replacement gilts and boars that are genetically-free from PSS, which may (1) cause sudden death when animals are stressed, and (2) produce pale, soft, exudative pork, by requiring that replacement gilts and boars pass a Halothane or DNA test.

CHOOSING THE SEEDSTOCK SUPPLIER

The role of the seedstock industry is to provide the commercial swine industry with healthy, genetically superior breeding animals. If the commercial industry is to make genetic progress, it must come through the seedstock producers. Thus, seedstock producers must

Fig. 4-4. Nursery owned by Keith Rhoda, Prairie City, Iowa. (Courtesy, American Diamond Swine Genetics, Prairie City, IA)

Fig. 4-5. Nursery facilities of Farmers Hybrid Co., a seedstock producer. (Courtesy, Farmers Hybrid Co., Inc., Des Moines, IA)

concentrate their selection efforts on economically important traits, base their selection on measured performance, and maintain the use of visual appraisal where it is appropriate and effective. Therefore, a planned, effective breeding program is essential to genetic progress. Furthermore, seedstock producers must maintain sufficient production volume to meet the needs of their customers as well as provide the genetic diversity between lines that will allow the commercial producer to maximize heterosis and utilize the superior characteristics of each breed or strain through systematic use of a crossbreeding program.

From the above, it may be concluded that when choosing the seedstock supplier the commercial producer should evaluate both the hogs and the supplier.

SELECTING REPLACEMENT GILTS

The productivity of the sow herd is the foundation of commercial pork production. Also, the sow herd contributes half of the genetic makeup of growing-finishing hogs. Together, these factors indicate the importance of careful selection of replacement gilts.

When selecting replacement gilts, consideration should be given to their EPDs and the following traits:

1. **Performance.** In general, the fastest growing gilts which are from large litters should be saved for replacement gilts. This requires identification at birth and a good set of records. Gilts should be from litters of 10 to 12 pigs which demonstrate uniform pig weights. Standardization of weight for age may be accomplished by assuming a daily gain of 2 lb at the time of evaluation. For example, the data could be adjusted to a 200-lb standard by adding one-half day to the gilt's age for each pound below 200 lb or deducting one-half day for each pound over 200 lb.

Fig. 4-6. M₃ gilts. (Courtesy, Farmers Hybrid Co., Inc., Des Moines, IA)

Fig. 4-7. Bred gilts. (Courtesy, American Diamond Swine Genetics, Prairie City, IA)

2. **Backfat.** Replacement gilts should be lean, having 1.2 in. or less of backfat, adjusted to a 230 lb basis.

3. **Feed efficiency.** Feed efficiency is favored indirectly by selecting fast-growing, low backfat gilts.

4. **Well-developed underline.** Replacement gilts should possess a sufficient number of functional teats to nurse a large litter. Six or more functional and uniformly spaced teats on each side is a good standard. Gilts with inverted or scarred nipples should not be saved.

5. **Reproductive soundness.** Most anatomical defects of the reproductive system are internal and hence not visible. However, gilts with small vulvas are likely to possess infantile reproductive tracts and should not be kept.

6. Feet and legs. A gilt should have legs that are set wide out on the corners of the body and the legs should be heavy boned with a slight angle to the pasterns.

SELECTING BOARS

Greater selection reach is possible with boars than with gilts since in most herds one boar is selected for every 15 to 20 gilts.

When selecting boars, some traits are apparent

Fig. 4-8. Hampshire boar. (Courtesy, Lone Willow Farm, Roanoke, IL)

from the records of relatives, while other traits are apparent, and may be selected for, from the boar's own record. In selecting boars, the following traits or standards should be examined:

1. **Behavior.** Behavioral traits are those characteristics that express themselves as docileness, temperament, sex characteristics, maturity, and aggressiveness. These are associated with reproductive potential.

2. **Dam productivity.** Dam productivity traits include such things as reproductive ability, litter size, milking ability, and mothering ability. The number of pigs farrowed and weaned and the average pig birth weight in litter are the most common measures. Litter weight at 21 days is probably the best single measure of dam productivity. Boars should be selected only from those litters of 10 or more pigs farrowed and 8 or more pigs weaned.

Dam behavioral and productivity traits are very important in the financial returns of the swine enterprise. Therefore, when selecting a boar for these traits, use records of the sire and dam, litter records, records of other relatives, and any records available on the boar being selected. A cross-breeding program will maximize improvement of these traits.

3. **Performance.** Performance traits include (a) growth rate measured as gain per day from weaning to 230 lb, and (b) feed conversion. These traits are above average in economic value. As guidelines, boars should (a) reach 230 lb at 155 days, or less, of age; (b) consume about 275 lb of feed, or less, per 100 lb of weight gain, between the weights of 60 and 230 lb; and (c) gain 2 lb, or more, per day during this same time. When selecting for these items, one should place more emphasis on the boar's own record and less emphasis on records of relatives.

4. **Backfat.** Carcass merit is best evaluated by taking measurements of backfat thickness, loin eye area, or the muscle in the animal. These traits have

very high heritability. As a guide, the backfat of a boar adjusted to 230 lb should be 1.0 in. or less.

5. **Reproductive soundness.** Characteristics associated with soundness include: the spacing, number, and presentation of the teats; genetic abnormalities such as hernia and cryptorchidism; and mating ability. Boars should possess 12 or more well spaced teats. Genetic abnormalities and mating ability traits have a very high economic importance. For these traits, insist that relatives of these selected boars be free of these defects and rely on the breeder's integrity. Physical soundness of the feet and legs, and bone size and strength, are also important. Feet and legs should demonstrate medium to large bone; wide stance both front and rear; free in movement; good cushion to both front and rear feet; and equal size toes.

6. **Conformation.** This includes body length, depth, height, and skeletal size; muscle size and shape; boar masculinity characteristics and testicular development. Conformation traits such as length and height have high heritability values. It is important to select boars on the basis of their own records for these characteristics.

Fig. 4-9. Montana X Duroc boar. Fifty percent inbred Montana No. 1 and 50% inbred Duroc. (Courtesy, McLean County Hog Service, Inc., LeRoy, IL)

Boars should be selected and purchased at 6 to 7 months of age for use beginning at a minimum of 8 months of age. It is recommended that all replacement boars be purchased at least 60 days before the breeding season. This allows them to be isolated and checked for health, conditioned, and test mated or evaluated for reproductive performance.

The primary consideration of producers is to select only boars that will increase the present production level of the herd and at the same time lessen weaknesses in the herd.

■ **What's a good boar worth?**—Saving $100 or more when buying a boar may be costly in the long run. Here's why: Assume a boar breeds four sows a week. With 90% conception and weaning eight pigs per litter, that boar will sire 187 litters, or 1,496 pigs per year.

Now, let's assume that the producer used a superior boar, and the offspring are marketed at 240 lb:

A 5% improvement in feed efficiency will save 37.5 lb of feed per pig, or over 28 tons of feed per year. At $125 a ton, that's a savings of $3,500.

A 5% improvement in daily gain will cut nine days off the time it takes pigs to reach market weight. At 25¢ per pig per day, that's a savings of $2.25 per pig, or $3,366 a year.

A $1.50/cwt premium for improved carcass quality is worth $3.60 a pig, or $5,385 a year.

An extra 0.1 pig sold per litter accounts for an extra 150 pigs per year. After variable costs are figured, that still leaves an extra $3,750.

Most people would agree that these are fairly attainable goals. When these four improvements in the boar's progeny are added together, a superior boar can increase profitability by $16,000 during one year! Also, this same boar may be used for 2 or 3 years.

This doesn't mean that a producer should buy the most expensive boar available. Rather, performance and carcass data should justify the higher price.

JUDGING SWINE

As previously indicated, until recently the small price spread in market classes and grades of swine offered little incentive to commercial swine producers to become proficient in judging. It is to the everlasting credit of purebred swine breeders, however, that they have been very progressive in this respect. The swine-type conferences sponsored by the various breed associations have made a unique contribution. Through bolstering live-animal work with a liberal amount of carcass data, these contests have set fashions for both the producer and the packer.

The discussion that follows represents a further elucidation of the first point discussed under selection—individuality. In addition to individual merit, the word judging implies the comparative appraisal or placing of several animals.

Judging swine, like all livestock judging, is an art, the rudiments of which must be obtained through patient study and long practice. The master breeders throughout the years have been competent livestock judges. Shrewd traders have also been masters of the art, even to the point of deception.

The essential qualifications that a good judge of swine must possess, and the recommended procedure to follow in the judging assignment are as follows:

1. **Knowledge of the parts of an animal.** This consists of mastering the language that describes and locates the different parts of an animal (see Fig. 4-10). In addition, it is necessary to know which of these parts are of major importance; that is, what comparative evaluation to give the different parts.

2. **A clearly defined ideal or standard of perfection.** The successful swine judge must know what

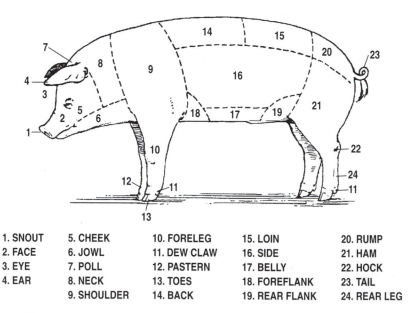

1. SNOUT	5. CHEEK	10. FORELEG	15. LOIN	20. RUMP
2. FACE	6. JOWL	11. DEW CLAW	16. SIDE	21. HAM
3. EYE	7. POLL	12. PASTERN	17. BELLY	22. HOCK
4. EAR	8. NECK	13. TOES	18. FOREFLANK	23. TAIL
	9. SHOULDER	14. BACK	19. REAR FLANK	24. REAR LEG

Fig. 4-10. Parts of a hog. The first step in preparation for judging hogs consists of mastering the language that describes and locates the different parts of the animal.

to look for; that is, the judge must have in mind an ideal or standard of perfection.

3. **Keen observation and sound judgment.** The good judge possesses the ability to observe both good conformation and defects, and to weigh and evaluate the relative importance of the various good and bad features.

4. **Honesty and courage.** The good judge of any class of livestock must possess honesty and courage, whether it be in making a show-ring placing or conducting a breeding and marketing program. For example, it often requires considerable courage to place a class of animals without regard to: (a) placings in previous shows, (b) ownership, and (c) public applause. It may take even greater courage and honesty to discard from the herd a costly animal whose progeny has failed to measure up.

5. **Logical procedure in examining.** There is always great danger of beginners making too close an inspection; oftentimes getting "so close to the trees that they fail to see the forest." Good judging procedure consists of the following three separate steps: (a) observing at a distance and securing a panoramic view where several animals are involved, (b) using close inspection, and (c) moving the animal in order to observe action.

Since a pig will neither stand still nor remain in the same vicinity for long, it is not possible to arrive at a set procedure for examining swine. In this respect, the judging of hogs is made more difficult than the judging of other classes of livestock. Where feasible, however, the steps for examining as illustrated in Fig. 4-11 are very satisfactory, and perhaps as good as any.

6. **Tact.** In discussing either (a) a show-ring class, or (b) animals on a producer's farm or ranch, it is important that the judge be tactful. The owners are likely to resent any remarks that imply that their animals are inferior.

Having acquired this knowledge, the judge must spend long hours in patient study and practice in comparing animals. Even this will not make expert and proficient judges in all instances, for there may be a grain of truth in the statement that "the best judges are born and not made." Nevertheless, training in judging and selecting animals is effective when it is directed by a competent instructor or an experienced producer.

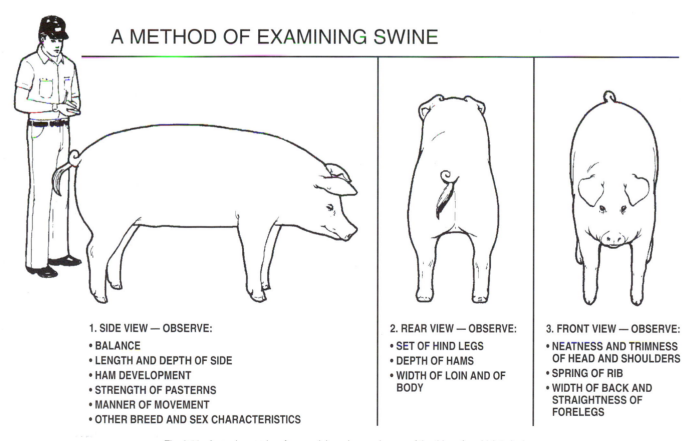

A METHOD OF EXAMINING SWINE

1. SIDE VIEW — OBSERVE:
- BALANCE
- LENGTH AND DEPTH OF SIDE
- HAM DEVELOPMENT
- STRENGTH OF PASTERNS
- MANNER OF MOVEMENT
- OTHER BREED AND SEX CHARACTERISTICS

2. REAR VIEW — OBSERVE:
- SET OF HIND LEGS
- DEPTH OF HAMS
- WIDTH OF LOIN AND OF BODY

3. FRONT VIEW — OBSERVE:
- NEATNESS AND TRIMNESS OF HEAD AND SHOULDERS
- SPRING OF RIB
- WIDTH OF BACK AND STRAIGHTNESS OF FORELEGS

Fig. 4-11. A good procedure for examining a hog and some of the things for which to look.

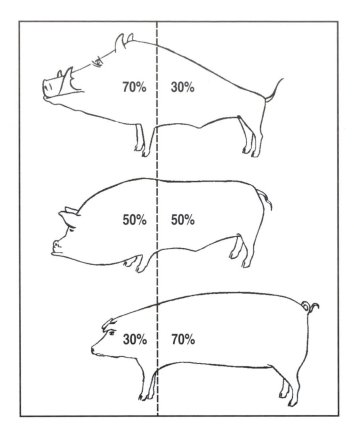

Fig. 4-12. Accompanying this line drawing was this caption by the late Sir John Hammond of Cambridge, England: "The object of breeders should be to strive for a light fore-end." According to Sir John, the European wild boar has a very heavy fore-end and light hams; so, unless constant selection is carried out for these characters there tends to be reversion to the primitive type. (From: "The Growth of the Pig," printed in *Pig Progress*, July, 1957)

IDEAL TYPE AND CONFORMATION

A major requisite in judging or selection is to have clearly in mind a standard or ideal. Presumably, this ideal should be based on a combination of (1) the efficient performance of the animal from the standpoint of the producer, and (2) the desirable carcass characteristics of the market animals as determined by the consumer.

The most approved meat-type breeding animals combine size, smoothness, and quality, and the offspring possess the ability to finish during the growing period without producing an excessive amount of lard. The head and neck should be trim and neat; the back well arched and of ample width; the sides long, deep, and smooth; and the hams well developed and deep. The legs should be of medium length, straight, true, and squarely set; the pasterns should be short and strong; and the bone should be ample and show plenty of quality. With this splendid meat type, there should be style, balance, and symmetry and an abundance of quality and smoothness.

The most approved bacon-type breeding hogs differ from meat-type animals chiefly in that greater emphasis is placed on length of side and the maximum development of the primal cuts with the minimum of lard. Also, bacon-type hogs generally have less width over the back and have a squarer type of ham, and show more trimness throughout.

With both meat- and bacon-type animals, the brood sows should show great femininity and breediness; and the udder should be well developed, carrying from 10 to 12 teats. The herd boar should show great masculinity as indicated by strength and character in the head, a somewhat crested neck, well-developed but smooth shoulders, a general ruggedness throughout, and an energetic disposition. The reproductive organs of the boar should be clearly visible and well developed. A boar with one testicle should never be used.

Fig. 4-13 shows the ideal meat-type hog versus some of the common faults. Since no animal is perfect, the proficient swine judge must be able to recognize, weigh, and evaluate both the good points and the common faults. In addition, the judge must be able to arrive at a decision as to the degree to which the given points are good or bad.

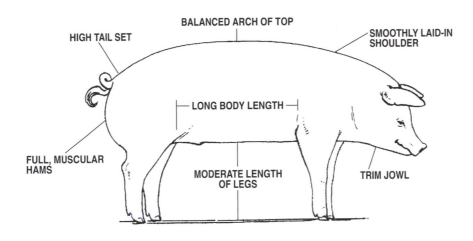

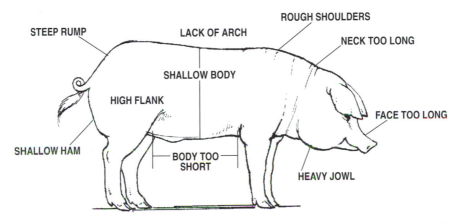

Fig. 4-13. Ideal meat type (above) vs common faults (below). The successful hog judge must know what to look for and be able to recognize and appraise both the good points and the common faults.

QUESTIONS FOR STUDY AND DISCUSSION

1. In establishing the herd and in selecting and judging, what primary differences exist in swine as compared to beef cattle and sheep?

2. Select a certain farm or ranch (your home, farm, or ranch, or one with which you are familiar). Assume that there are no hogs on this establishment at the present time. Then outline, step by step, (a) how you would go about establishing a herd, and (b) the factors that you would consider. Justify your decisions.

3. In establishing a herd, why are the following factors important: type, breeding, breed or strain, size of herd, uniformity, health, age, price, and suitability of the farm.

4. Fig. 4-12 shows that selection has changed the conformation of improved swine so that they have greater proportion of loin to head than the European wild boar. Why have producers made such a change?

5. Discuss each of the seven bases of selection of swine. Why must the swine industry measure?

6. Why are backfat and muscling such important considerations when selecting swine?

7. Cite examples of how purebred and commercial breeders alike have come to regret selections based on show-ring winnings.

8. What do STAGES, EBV, EPD, and BLUP stand for? Detail how you should use EPDs in selection.

9. What factors should a commercial swine producer consider when choosing a seedstock supplier?

10. What important traits or standards would you look for in (a) replacement gilts, and (b) replacement boars?

11. Why is it important to know the parts of a hog?

12. Why is it difficult to arrive at a set procedure for examining a pig?

SELECTED REFERENCES

Title of Publication	Author(s)	Publisher
Breeding & Improvement of Farm Animals	E. J. Warwick J. E. Legates	McGraw-Hill Book Co., New York, NY, 1979
Genetic Evaluation	Bob Uphoff, Chairman of the NPPC Genetic Program Committee	National Pork Producers Council, 1995
Genetics of Livestock Improvement	J. F. Lasley	Prentice-Hall, Inc., Englewood Cliffs, NJ, 1978
Livestock Judging, Selection and Evaluation, Third Edition	R. E. Hunsley W. M. Beeson	Interstate Publishers, Inc., Danville, IL, 1988
Pork Industry Handbook		Cooperative Extension Service, Purdue University, West Lafayette, IN
Selecting, Fitting and Showing Swine	J. E. Nordby H. E. Lattig	The Interstate Printers & Publishers Inc., Danville, IL, l961 (out of print)
Stockman's Handbook, The, Seventh Edition	M. E. Ensminger	Interstate Publishers, Inc., Danville, IL, 1992

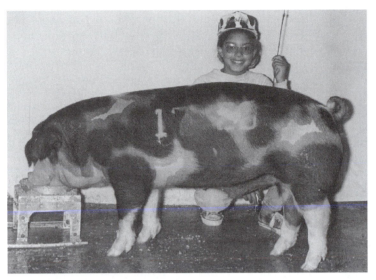

Champion Market Hog in the Junior Barrow Show at National Spotted Type Conference, 1994, exhibited by Dawn Symonds, Dahinda, Illinois. (Courtesy, National Spotted Swine Record, Peoria, IL)

5

SELECTING, FITTING, AND SHOWING SWINE

Through the years show-ring fashions in swine have fluctuated more radically than those in any other class of livestock. This story indicates better than voluminous words that, over a period of years, show-ring standards survive only when based on such utilitarian considerations as efficiency of production and selling price on a discriminating market. Breed fancy points and decisions made by judges are but passing fads when they conflict with the primary objective of producing swine, which is pork over the counter.

Note well: Comply with the rules of the show. Today, some major shows do not permit the use of oil, paint, powder, or other dressing on market swine. Also, the use of unapproved drugs (drugs not approved by the FDA and/or the USDA) is forbidden. Exhibitors violating these rules will be barred from showing.

ADVANTAGES OF SHOWING

Though not all exhibitors share equally in the advantages which may accrue from showing hogs, in general the following reasons may be advanced for exhibiting swine:

1. It serves as the best available medium for molding breed type.

2. It gives the breeders an opportunity to observe the impartial appraisal by a competent judge of their entries in comparison with others.

3. It offers an opportunity to study the progress being made within other breeds and classes of livestock.

4. It provides an excellent medium of advertising.

5. It gives breeders an opportunity to exchange ideas, thus serving as an educational event.

6. It offers an opportunity to sell a limited number of breeding animals.

7. It sets sale values for the animals back home, such values being based on the sale of show animals.

SELECTING SHOW ANIMALS

Fig. 5-1. Selection of the prospective show animal is the first and most important assignment. Since this requires a projection into the future, no judging assignment is quite so difficult as that of selecting an animal for further development.

The first and most important assignment in preparation for the show is the selection of the prospective show animals. Unless the exhibitor has had considerable experience and is a good judge of hogs, it is well to secure the assistance and advice of a competent judge when selecting the animals that are to be fitted.

Selections should be made as far in advance of the show as is possible. In fact, the show-ring objective is usually kept in mind at the time matings are made and farrowing dates arranged. Selection is really a year-round job for the person who desires to exhibit a full herd. In general, all breeding animals intended for

show, except those in the junior pig classes, should be selected at least 4 to 6 months in advance, thus allowing ample time for fitting and training. There is an unavoidable delay in selecting junior pigs and barrows.

As some animals may not develop properly, it is advisable to begin the fitting work with larger numbers in each class than it is intended to show, especially in the younger groups. In this manner, those animals which fail to respond may be culled out from time to time and a stronger show herd assembled.

Exhibitors may show (1) breeding animals or (2) barrows. When possible and when the classifications are available, it is desirable to show in both groups. In recent years, great emphasis has been placed on barrow shows, and winning in a strong barrow show is looked upon as a fine accomplishment for the herd or breed/strain which produced the champion. As the type of winning barrows generally reflects consumer demands, this is a good thing.

Exhibitors can enhance their chances of winning and of securing sufficient premium money to cover expenses by filling as many of the individual classes and groups as possible, and this applies to both breeding animals and barrows. Provided that the animals are good, it costs little more and is usually good economy to have a sizable and well balanced show herd.

TYPE

The most approved breeding animals in modern shows possess adequate size for age and are of lean meat type. They should be clean-cut about the head and neck; the back should be moderately and evenly arched, and of adequate width; the loin should be wide and strong; the hams should be deep, thick, slightly bulging and meated well down to the hocks; the shoulders should be well laid in and smooth; the sides should be long, deep, and smooth; the legs should be set well apart, straight, and squarely on the corners; and the pasterns should be strong. With this special meat type, there should be a proper balance and blending of all parts, and the animals should be stylish and showy. An alert, active walk is a decided asset. Sex character, breediness, and adherence to distinctive breed characteristics are important in boars and sows.

In show barrows, special emphasis should be placed on trimness of middle and quality throughout as well as on the other characteristics mentioned. The ideal barrow when ready for the show must possess superior conformation, finish, and quality. There should be the maximum development in the high-priced cuts and a minimum of lardiness.

Group classes in both breeding and barrow classes should be as uniform as possible, for group placings are determined by the merits and uniformity of the entire group rather than on the basis of 1 or 2 outstanding individuals therein.

Further information relative to breed characteristics and other factors of importance in making selections can be obtained from Chapters 3 and 4.

SHOW CLASSIFICATIONS

It is also desirable to select hogs as old as possible within the respective age or weight classifications in order that they may show to the best possible advantage. Classifications for breeding animals are based on age. Barrow classifications are usually based on weight rather than age.

The swine classifications of four major livestock shows are presented in Table 5-1. As noted, shows vary greatly in classifications.

Many shows are giving less emphasis to the on-foot classes and are adding classes for barrows that are judged on-foot and in carcass. Placings are based on such measures as backfat thickness, size of loin area, and primal cuts.

The weight divisions of different barrow shows vary considerably, thus making it imperative that the exhibitor study these carefully. In fact, the breeding program should be planned with this information in mind, because animals farrowed at the proper time may be better fitted and may reach the proper bloom for different weight divisions, if the breeding and feeding programs are properly synchronized.

Besides the breeding swine and barrow (market hog) classifications, a few fairs may have classes for groups such as (1) produce of one sow, (2) get of sire, and (3) premier exhibitor.

TABLE 5-1
SWINE SHOW CLASSIFICATIONS OF FOUR MAJOR LIVESTOCK SHOWS

Show	Breeds	Open Class Breeding Swine	Junior Market Barrows	Junior Barrow Carcass Contest
Houston Livestock Show and Rodeo, Houston, TX, 1995	**Open Class Breeds:** Berkshire, Chester White, Duroc, Hampshire, Poland China[1], Spotted[1], Yorkshire. **Junior Market Barrow Breeds:** Same as in Open Class, with Landrace and Crossbreds added.	Four breeding animals either sex produce of one sow farrowed July 1–Oct. 15, 1993, bred and owned by exhibitor. Four breeding animals either sex get of one boar farrowed July 1–Oct. 15, 1993. **Age Classes, Sows:** Sept. 1–Oct. 15, 1994; Aug. 1–Aug. 31, 1994; July 1–July 31, 1994. **Age Classes, Boars:** Sept. 1–Oct. 31, 1994; Aug. 1–Aug. 31, 1994; July 1–July 31, 1994.	Only 660 barrows accepted and allowed to compete in the Houston Show. Barrows must be within wt. range of 220–260 lb. **Awards:** Grand Champion; Res. Grand Champion; Breed Champion; Res. Breed Champion; Multiple placings: first place, second place, third place, etc., to 12th place, ending with "All Remaining Placing Barrows."	The first and second place barrows of each breed in the Junior Market Barrow Show will be slaughtered, and the following carcass measurements will be made: Adjusted 170 lb hot carcass wt. (approx. 230 lb live wt.) Min. carcass length 29.5 in. Max. backfat thickness at last rib 1.1 in. Min. loin area at 10th rib 4.5 sq. in.
Iowa State Fair, Des Moines, IA, 1995	**Open Class Breeds:** Berkshire, Chester White, Duroc, Hampshire, Landrace, Poland China, Spotted, Yorkshire	**Classification for Each Breed:** Boars farrowed Dec. 1, 1994–March 31, 1995; Gilts farrowed Dec. 1, 1994–March 31, 1995. **Awards:** Grand Champion Boar; Grand Champion Female; Premiere Sire; Premier Exhibitor. **Truckload:** Barrow Show, 6 head, farrowed on or after Feb. 1, 1995; Purebreds 220–270 lb; Crossbreds 220–270 lb. **Awards:** Grand Champ. Truckload; Res. Champ. Truckload; Grand Champ. Barrow; Res. Grand Champ. Barrow; Breed Champ. Barrow; Res. Breed Champ. Barrow		

(Continued)

TABLE 5-1 (Continued)

Show	Breeds	Open Class Breeding Swine	Junior Market Barrows	Junior Barrow Carcass Contest
National Western Stock Show & Rodeo, Denver, CO, 1995	**Classifications for each:** Berkshire Chester White Duroc Hampshire Poland China Spotted Yorkshire Other Purebreds Crossbreds		Min. wt. 210 lb Max. wt. 260 lb **Awards (for each breed through 10th place):** Grand Champ. Barrow Res. Grand Champ. Barrow	
North American International, Louisville, KY, 1994	All breeds will show together, divided by wts.		Wt. 210–260 lb Classes will be divided into 3 divisions—light, medium, and heavy, with a Division Champ. and Reserve Champ. in each. **Awards:** In addition to division awards, there will be a Champ. and Res. Champ. of Show Market Carcass Contest Loin eye and fat cover at 10th rib will be measured. **Awards:** Novice, Champ. & Reserve Junior, Champ. & Reserve Senior, Champ. & Reserve	

[1]At the Houston Livestock Show, Poland China and Spotted are shown together in the Open Show, but separately in the Junior Show.

BREEDING

Animals selected for the show should always be of good ancestry, this being added assurance of satisfactory future development. Breeding animals should show the distinctive breed characteristics, but certain fancy points may be overlooked in selecting barrows. In fact, there are usually provisions for showing crossbred barrows. Consideration is also given to breeding from the standpoint of filling the get-of-sire and produce-of-dam classes.

FEEDING AND HANDLING FOR THE SHOW

All animals intended for show purposes must be placed in proper condition, a process requiring great attention to details.

KEEP THE ANIMALS CONTENTED AND HEALTHY AND PROVIDE EXERCISE

Uncomfortable quarters, filthy wallows, annoyance by parasites (internal or external), improper handling, and unnecessary noise are the most common causes of discontentment. Prospective show animals should have the best quarters on the farm, and, above all, they should be kept cool, clean, and free from parasites. A certain amount of exercise is necessary in order to promote good circulation and to increase the thrift and vigor of the animal. Exercise tends to stimulate the appetite, making for greater feed consumption; keeps the animals sound on their feet and legs; and promotes firmness of fleshing and trimness of middle. Mature boars, sows, and barrows may need to be walked, thus forcing exercise. During warm weather, it is best to exercise show animals in the early morning or late evening, avoiding unnecessary handling in the heat of the day.

SEGREGATE THE BOARS AND SOWS

Usually more individual attention can be given, especially in the matter of feeding, if show hogs are handled in small groups. With junior pigs, the boars should be separated from the gilts when the pigs are 4 months old. At the time young show boars begin to rant, usually when 4 to 7 months of age, it is desirable that they be placed in isolated lots, preferably where they cannot see or hear other hogs. For the most part, show boars of all ages must be kept separate, but several females of about the same age can be run together.

SOME SUGGESTED RATIONS

Any of the rations listed in Chapter 9 for the respective classes and ages of swine are suitable for use in fitting show animals of similar classification. In general, however, instead of self-feeding most experienced caretakers feel that they can get superior bloom and condition by either (1) hand-feeding or (2) using a combination of hand-feeding and self-feeding (hand-feeding twice daily and allowing free access to a self-feeder). When hand-feeding, they also prefer mixing the ration with skimmed milk, buttermilk, or condensed buttermilk and feeding the entire ration in the form of a slop.

Adding milk to a ration that is already properly balanced does make for a higher protein content than necessary. On the other hand, most experienced caretakers prefer using rations of higher protein content for fitting purposes. They feel they get more bloom that way. In general, however, when skimmed milk or buttermilk is used in slop-feeding, the protein feeds of the ration may be reduced by one-half without harm to the animal.

In fitting show barrows, it may be necessary to decrease or discontinue slop-feeding 2 to 4 weeks before the show to avoid paunchiness and lowering the dressing percentage.

When oatmeal (hulled oats) is not too high priced, many successful hog exhibitors, replace up to 50% of the grain (corn, wheat, barley, oats, and/or sorghum) in the ration with oatmeal. They do this especially when fitting hogs—both breeding animals and barrows—in the younger age groups. Oatmeal is highly palatable, lighter, and less fattening than corn.

RULES OF FEEDING

The general principles and practices of swine feeding are fully covered in Chapters 8 and 9. The feeding of show hogs differs from the feeding of the rest of the herd primarily in that the former are fed for maximum development and bloom, with less attention being paid to economy from the standpoint of both feed and labor. It is to be noted, however, that it is a disadvantage to have the hogs excessively fat for modern shows, thus rendering possible harm to breeding animals and making barrows too lardy. Rather, a firm, smooth finish is desired.

The most successful producers have worked out systems of their own as a result of years of practical experience and close observation. The beginner may well emulate their methods. Some rules of feeding show hogs as practiced by experienced producers are:

1. **Practice economy, but avoid false economy.** Although the ration should be as economical as possible, it must be remembered that proper condition is the primary objective, even at somewhat additional expense. Perhaps the most common mistake made in the fitting ration, especially of breeding animals, is the heavy feeding of corn or other grains.

2. **Hand-feeding versus self-feeding.** Barrows are more often self-fed than breeding animals. Most experienced producers feel, however, that they can get superior bloom and condition by hand-feeding show animals. The majority prefer mixing the grain ration with skimmed milk, buttermilk, or condensed buttermilk, and feeding the entire ration in the form of a slop.

3. **Provide a variety of feeds.** A good variety of feeds increases the palatability of the ration, thus increasing feed consumption. A small amount of molasses adds to the palatability of a ration.

4. **Feed a balanced ration.** A balanced ration will be more economical and will result in better growth and finish. Experienced producers usually prefer a ration that is on the narrow side (high in proteins). For assistance in selecting a ration, the reader is referred to Chapter 9.

5. **The ration must not be too bulky.** Hogs cannot handle a great amount of bulk. Consumption of too much bulk will cause the animal to become paunchy and will lower the dressing percentage, a condition severely criticized in a show animal, especially a barrow.

6. **Feed regularly.** Animals intended for show purposes should be fed with exacting regularity. In the early part of the fitting period, two feedings per day may be adequate. Later, the animals should be fed 3 to 5 times a day, especially when fitting rapidly growing pigs.

7. **Avoid sudden changes.** Sudden changes in either the kind or amount of feed are apt to cause digestive disturbances. Any necessary changes should be gradual.

8. **Provide minerals.** When proteins in the ration are of animal origin (tankage, fish meal, milk, etc.), there may be no deficiency of minerals other than salt. However, access to a mineral mixture can be arranged at little cost and may be good protection with animals that are being crowded, especially junior pigs being fitted for the show. The chapter on feeding swine lists satisfactory mineral supplements.

EQUIPMENT FOR FITTING AND SHOWING HOGS

In the fitting, training, and grooming operations, the essential equipment consists of brushes, clipper, rasp, sharp knife, soap, canes or whips, and hurdles. In loading for the fair, all of these items should be included in the show box. In addition, most exhibitors take water buckets, light troughs, a sprinkling can, a fork and a broom, oil (and powder when white hogs or hogs with white spots are being shown), a saw, hammer, hatchet, a few nails, some rope, flashlight, blankets or a sleeping bag, and a limited and permissible supply of feed and bedding.

The show box, in which all smaller equipment is stored, is usually of durable wood construction, freshly painted, with a neat sign on the top or front giving the name and address of the exhibitor. The feed taken to the show should be identical to the mix which was used at home.

TRAINING AND GROOMING THE ANIMAL

Most show-rings are cursed with the presence of too many squealing, scampering, unmanageable, and poorly groomed pigs. Such animals make an adverse impression on ringside spectators, annoy other exhibitors, and fail to catch the eye of the judge. When competition is keen, there may be several well-bred and beautifully fitted individuals of the right type in each class, with the result that the winner will be selected by a very narrow margin. Under such circumstances, proper training, grooming, and showing is often a deciding factor.

Fig. 5-2. A junior swine show. (Courtesy, William G. Luce, Oklahoma State University, Stillwater, OK)

TRAINING THE PIG

Proper training of the pig requires time, patience, and persistence. Such schooling makes it possible for the judge to see the animal to best advantage. Some exhibitors prefer to use the whip, whereas others use the conventional stockman's cane. The pig should be trained to respond to either one or the other (cane or whip) but not to both. With mature boars, it is good protection to use a hand hurdle. However, the use of a hurdle in showing young animals is usually an admission of either a mean disposition or a lack of training.

Long before the show, the exhibitor should study the individual animal, arriving at a decision as to the most advantageous pose. Then the animal should be trained to perfection so that it will execute the proper pose at the desired time. It is important that the pig be trained to stop when necessary, perhaps when the whip or cane is placed gently in front of his face. The direction of walking should be easily guided by merely placing the cane or whip alongside the animal's head.

The exhibitor should avoid making a pet of the pig. Such an animal may display a nasty disposition when placed in the show-ring, or it may slouch down when anyone comes near, including the judge. Also, animals should be trained to stand squarely on their feet and to keep their backs up and heads down.

TRIMMING THE FEET

In order that the animal may stand squarely and walk properly and that the pasterns appear straight and strong, the toes and dew claws should be trimmed regularly. Moreover, long toes and dew claws are unsightly in appearance. Trimming can best be done with the animal lying on its side. Usually this position can best be acquired by merely stroking the animal's belly. With this procedure, tying and possible injury therefrom is unnecessary.

The practice of standing the animal on a hard surface and cutting off the ends of the toes with a hammer and chisel gives only temporary relief and is not recommended. The bottoms of the toes should be trimmed. Proper trimming can best be accomplished by using a small rasp and a knife.

The outside toes of the rear feet grow faster than the inside toes. In extreme cases this condition may result in crooked hind legs. This condition may be corrected by trimming the outside toes more frequently.

The toes should be trimmed regularly, with some animals as often as every six weeks. Too much trimming at any one time, however, may result in lameness. For this reason, it is not advisable to work on the feet

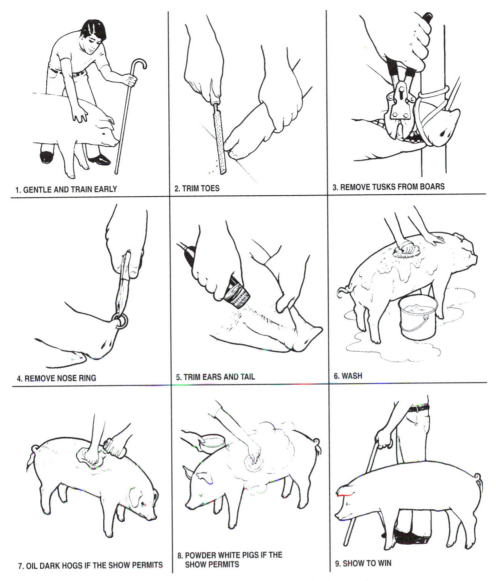

1. GENTLE AND TRAIN EARLY

2. TRIM TOES

3. REMOVE TUSKS FROM BOARS

4. REMOVE NOSE RING

5. TRIM EARS AND TAIL

6. WASH

7. OIL DARK HOGS IF THE SHOW PERMITS

8. POWDER WHITE PIGS IF THE SHOW PERMITS

9. SHOW TO WIN

Fig. 5-3. Some essentials in training and grooming hogs for show.

within two weeks before the show. In no case should trimming be so severe as to draw blood.

The dew claws should be cut back and dressed down neatly, making the pasterns appear short and straight.

REMOVING TUSKS AND RINGS

Boars over one year of age usually have tusks of considerable size. These should be removed a month or two before the show. To remove the tusks, tie the animal to a post with a strong rope, draw up the upper jaw, and use a hoof parer.

Rings should also be removed well in advance of fair time, allowing all soreness to disappear. Rings may be removed by securely tying the animal, as for the removal of tusks, and by cutting the rings with wire pliers or nippers.

CLIPPING

Clipping is usually done with hand clippers. It should be done a few days before the show. Many successful exhibitors prefer to use the clippers twice, about two weeks before the show and a second time the day before entering the ring. The usual practice is to clip the ears, both inside and outside, and to clip the tail. Clipping of the tail should begin above the switch and extend up to the tail head. The exact amount of switch to leave will vary somewhat with each individual, requiring judgment on the part of the exhibitor.

It may also help to remove long hairs from about the head and jowl. The udder of a gilt will often show to better advantage if some of the hairs on the belly are carefully removed. In all trimming, it is important that the hair be gradually tapered off so that the clipped and unclipped areas blend nicely.

WASHING

An occasional washing and brushing keeps the animal clean and makes the skin smooth and mellow. It also assists materially in shedding the coat of older animals. Lukewarm water, plenty of tar soap, and a fairly stiff brush will do the job. After the skin and hair have been dampened thoroughly, rub the hair with tar soap until a suds is formed and then work this lather into the hide with the hands and brush. Clean all parts of the body thoroughly, using care not to get any of the water into the ears. A small amount of bluing in the water is helpful in bleaching out stained spots on white hogs. Following washing, the animal should be rinsed off in order to remove all traces of soap from the hair and skin. The animal should then be placed in a clean pen to dry.

OILING

Oiling softens the skin and hair and gives the necessary bloom to the coat. It is most important that the oil be used sparingly and that it be evenly distributed over the body. For best results a light application of oil should be given the night before the show, using an oiled cloth or brush. Then give the animal a thorough grooming with a brush and a woolen rag just before entering the ring. This gives the necessary bloom. When the pig enters the ring, surplus oil should not be in evidence. Any clear, light vegetable oil is satisfactory for oiling purposes. Paraffin oil is very satisfactory when mixed with small proportions of rubbing alcohol.

Oiling mature boars and sows that are in high condition may cause them to overheat more easily during warm weather. Under such conditions, some experienced exhibitors use no oil but sprinkle water over the animals before entering the ring, and then follow with the brush.

POWDERING

White hogs or white spots on dark hogs are usually powdered with talcum powder or corn starch. The animal should first be thoroughly washed and allowed to dry. The powder should be dusted on before entering the ring. In order to be most effective, it should be evenly distributed. Some major shows have, however, eliminated powdering, and one should check carefully before using powder.

MAKING FAIR ENTRIES

Well in advance of the show, the exhibitor should request that the fair manager or secretary forward a premium list and the necessary entry blanks. Usually, entries close from 2 to 4 weeks prior to the opening date of the show. All rules and regulations of the show should be studied carefully and followed to the letter—including requirements relative to entrance, registration certificates, vaccination, health certificates, pen fees, exhibition and helper's tickets, and other matters pertaining to the show. Most entry blanks for purebred hogs call for the following information: breed, sex, name and registration number of the entry, date farrowed, name and registration number of the sire and dam, a description of markings (such as ear notches), and the class in which the animal is to be shown. Entries should be made in all individual and group classes but not in breed specials or championship classes. It is not necessary to specify the identity of individuals constituting herds or groups, because, when the exhibitor has a choice, the winnings in the individual class will largely determine this.

PROVIDING HEALTH CERTIFICATES

Most fairs require that swine must be accompanied by a health certificate when they enter the exposition grounds. Although the stipulations vary, perhaps the following provisions of the Iowa State Fair are rather typical:

■ **Health certificate**—All swine must be individually identified on a Certificate of Veterinary Inspection and originate from **herds not under quarantine**.

■ **Brucellosis**—

1. **Native Iowa swine.** No brucellosis test required for exhibition purposes.

2. **Swine from out of state.** All breeding swine six months of age or older must either:

　　a. Originate from a Brucellosis Class "free" state; or

　　b. Originate from a brucellosis validated herd

with herd certification number and date of last test listed on the certificate; or

c. Have a negative brucellosis test conducted within 60 days prior to the show and confirmed by a state-federal laboratory.

■ **Aujeszky's disease (pseudorabies)—**

1. Exhibitor must present *a test record and Certificate of Veterinary Inspection* that indicates that all swine have had a negative test for pseudorabies within 30 days prior to the show, regardless of the status of the herd. Also, animals must have a test tag in their ear along with the official test results chart.

2. Swine returning from an exhibition to its home herd or moved to a purchaser's herd, following an exhibition or consignment sale, must be isolated and retested negative for pseudorabies not less than 30 and not more than 60 days after reaching the swine's destination. (Code of Iowa 166 D.13(2).)

SHIPPING TO THE FAIR

Show hogs should be shipped so that they arrive within the limitations imposed by the fair and a minimum of two days in advance of showing. Because of the greater convenience and speed, most hauls are made by truck. Public conveyances should always be thoroughly disinfected before loading hogs. Show hogs should not be crowded, and those of different age groups and sexes should be separated by suitable partitions. During warm weather, properly wetted sand makes the best bedding. In hot weather, sand should be wetted down before loading and while en route. Also, hogs may be drenched when necessary, but water should never be applied to the backs of hot hogs. Regardless of the method of transportation, animals should be fed lightly (about a half ration) just prior to and during shipping. A heavy fill is likely to result in digestive disturbances and overheating in warm weather.

PEN SPACE, FEEDING, AND MANAGEMENT AT THE FAIR

Most swine pens at fairs or exhibitions range from 6 to 8 feet square. When it is not too hot and the hogs are used to each other, about the following numbers of the same sex can be accommodated in one pen: 3 to 5 junior pigs or barrows, 2 seniors, and 2 each of the older age groups—except that boars older than junior pigs had best be kept in separate pens. Sufficient pens should be obtained in order to avoid overcrowding, especially during warm weather. It is easier to keep the animals clean when there is ample space.

The advertising value of the exhibit will be enhanced through displaying a neat and attractive sign over the pens, giving the name and address of the breeder, farm, or ranch. It is also important that the pens and alleys be kept neat and attractive at all times, thus impressing spectators and showing management your desire to cooperate to put on a good show.

Following a day of rest and light feeding after unloading at the show, normal feeding may be resumed provided that it is accompanied by exercise. Usually, the animals are fed twice daily, and, if possible, the exhibitors should feed outside the pens, preferably behind the exhibition building. Most exhibitors prefer to clean the pens thoroughly in the early morning, while the animals are confined within hurdles for feeding or are being taken for exercise. It is important that all animals receive sufficient and regular exercise—at least a ½-hour walk daily—while they are on the fairgrounds.

The final washing may be given a day or two after arrival, but the coat should be brushed daily and oiled or powdered as the breed may require.

SHOWING THE PIG

Expert showmanship cannot be achieved through reading any set of instructions. Each show and each ring will be found to present unusual circumstances. However, there are certain guiding principles that are always adhered to by the most successful exhibitors including the following:

1. Train the animal well, long before you enter the ring.
2. Have the animal carefully groomed and ready for the parade before the judge.
3. Dress neatly for the occasion.
4. Enter the ring promptly when the class is called.
5. Do not crowd the judge.
6. Be courteous and respect the rights of other exhibitors .
7. Work in close partnership with the animal.
8. Keep your animal in view at all times.
9. Do not allow your hog to fight or bite other animals .
10. Always be showing. Keep one eye on the judge and the other on the pig. The animal may be under observation of the judge when you least suspect it.
11. Keep calm, confident, and collected. Remember that the nervous exhibitor creates an unfavorable impression.
12. Be a good sport. Win without bragging and lose without squealing.

AFTER THE FAIR IS OVER

Before an exhibitor can leave the show grounds, it is customary to require a signed release from the superintendent of the show. Immediately prior to that time, all of the equipment should be loaded, followed by loading the hogs. The same care and precautions that applied in travel to the show should prevail in the return trip.

Because of the possible disease and parasite hazard resulting from contact with other herds and through transportation facilities, it is good protection to quarantine the show herd for a period of three weeks following return from the fair. There is also the problem of letting them down in condition or reducing from the usual heavily fitted condition to breeding condition. To do the latter operation properly, especially with mature animals, requires great skill; for the manner in which animals are reduced in flesh often determines their fertility and future usefulness in the breeding herd as well as their possibility of "coming back" for future shows. Those who are most successful in this phase of management usually employ three methods—namely, (1) plenty of exercise, (2) added bulk in the ration, and (3) a gradual reduction of the concentrate allowance. In addition, the concentrate ration should be made more bulky by adding such feeds as oats, wheat bran, and ground alfalfa, and the feed allowance should be gradually reduced. With active, growing junior pigs, a slight reduction in the grain allowance will usually suffice in placing them in breeding condition.

QUESTIONS FOR STUDY AND DISCUSSION

1. Do you approve or disapprove of the rules of some shows not permitting the use of oil, paint, powder, or other dressing on market swine; and of forbidding the use of unapproved drugs (drugs not approved by the FDA and/or the USDA)? Defend your position.

2. List each (a) the similarities, and (b) the differences in the swine show classifications of the four major livestock shows presented in Table 5-1. Discuss your listings.

3. Under what circumstances would you recommend that each, a purebred swine producer, a seedstock producer, a commercial swine producer, and an FFA or 4-H swine club member, (a) should show, and (b) should not show hogs?

4. Defend either the affirmative or the negative position of each of the following statements:

a. Fitting and showing does not harm animals.

b. Livestock shows have been a powerful force in swine improvement.

c. Too much money is spent on livestock shows.

d. Unless all animals are fitted, groomed, and shown to the same degree of perfection, show-ring winnings are not indicative of the comparative quality of animals.

5. How may livestock shows be changed so that they (a) more nearly reflect consumer preference, and (b) make for greater swine improvement?

6. It is generally agreed that livestock shows abetted the radical shifts in swine types of the past—shifts which later proved to be detrimental. How could this have been averted?

7. How does feeding for show differ from commercial feeding for market?

8. Why do the health certificates of most fairs detail the requirements relative to brucellosis and pseudorabies?

9. What advice would you give someone showing swine for the first time?

SELECTED REFERENCES

Title of Publication	Author(s)	Publisher
Selecting, Fitting and Showing Swine	J. E. Nordby H. E. Lattig	The Interstate Printers & Publishers, Inc., Danville, IL, 1961 (out of print)
Stockman's Handbook, The, Seventh Edition	M. E. Ensminger	Interstate Publishers, Inc., Danville, IL, 1992

Today's genetic improvements determine tomorrow's profits. This is a 50% inbred Montana and 50% inbred Duroc. (Courtesy, McLean County Hog Service, LeRoy, IL)

PRINCIPLES OF SWINE GENETICS

The objective of swine breeding is to mate individuals whose offspring will possess the necessary heritability to (1) produce the maximum amount of pork, (2) develop the desired body type, and (3) perform at the desired level. These animals should then be fed and managed so that their maximum genetic potential will be expressed, since swine are the products of heredity and environment, as are all other animals.

The economic justification for improved breeding is that good pigs make more money. Through the years, the swine industry has been responsive to this motivating force. For example, economics caused breeding programs to shift from lard-type to meat-type hogs. In addition to changes in body type and cutout value, it is important that producers initiate breeding and selection programs designed to improve the performance traits of economic importance, such as litter size, rapid growth, and feed efficiency. The bottom line to all of it: With each improvement wrought through breeding programs, the cost of producing pork declines.

Today, purebred hogs are generally raised to be sold to commercial producers as seed stock for crossbreeding programs. These crossbreeding programs strive to combine the best traits of several breeds to produce superior offspring with the sought after characteristics of leanness, meatiness, good feed efficiency, fast growth, and durability.

Each individual swine producer is interested in herd improvement. Any herd improvement means more profit. Also, any permanent herd improvements made by an individual swine producer inevitably contribute to permanent breed improvement.

Swine breeders of the future should have clear goals in mind and a definite breeding program, for, despite the remarkable progress of the past, much remains to be done. A casual glance at the daily receipts of any hog market is convincing evidence of the task ahead. The challenge is primarily that of improving the great masses of animals in order that more of them may approach the too few nearly perfect specimens. Also, in this computer age, there must be greater efficiency of production; and this means more rapid growth, less feed to produce 100 lb of meat, and lifting of the percentage pig crop well above the present United States average. With the experience of the pioneers to guide us and with our present knowledge of genetics and physiology of reproduction, our progress should now be much more certain and rapid. In the past, animal breeding was an art. In the future, it should be both an art and a science.

EARLY ANIMAL BREEDERS

Until very recent times, the general principle that like begets like was the only recognized law of heredity.

That the application of this principle over a long period of time has been effective in modifying animal types in the direction of selection is evident from a comparison of present-day types and breeds of swine.

There can be little doubt that men like Bakewell, the English patriarch, and other 18th century breeders had made a tremendous contribution in pointing the way toward livestock improvement before Mendel's laws became known to the world in the early part of the 20th century. Robert Bakewell's use of progeny testing through his ram letting was truly epoch making, and his improvement of Shire horses and Longhorn cattle was equally outstanding. He and other pioneers had certain ideals in mind, and, according to their standards, they were able to develop some nearly perfect specimens. These men were intensely practical, never overlooking the utility value or the market requirements. No animal met with their favor unless such favor was earned by meat on the block, milk in the pail, weight and quality of wool, pounds gained for pounds of feed consumed, draft ability, or some other performance of practical value. Their ultimate goal was that of furnishing better animals for the market and/or lowering the cost of production. It must be just so with the master breeders of the present and future.

Others took up the challenge of animal improvement where Bakewell and his contemporaries left off, slowly but surely molding animal types. Armed with a better understanding of genetics, during the past 100 years remarkable progress has been made in breeding better meat animals—animals that are more efficient and, at the same time, that produce cuts of meat more nearly meeting the exacting requirements of the consuming public. The wild boar and the Arkansas razorback have been replaced by modern meat-type swine.

The laws of heredity apply to swine breeding exactly as they do to all classes of farm animals. But the breeding of swine is more flexible because: (1) hogs normally breed at an earlier age, thus making for a shorter interval between generations, and (2) they are litter-bearing animals. Because of these factors, together with the available feeds and the type of pork products demanded by the consumer, the American swine producer has created more new breeds and made more rapid shifts in hog types than in any other class of farm animals.

MENDEL'S CONTRIBUTION TO GENETICS

Modern genetics was really founded by Gregor Johann Mendel, a cigar-smoking Austrian monk, who conducted breeding experiments with garden peas from 1857 to 1865, during the time of the Civil War in the United States. In his monastery at Brunn (now

Fig. 6-1. Gregor Johann Mendel (1822–1884), a cigar-smoking Austrian monk, who founded modern genetics through breeding experiments with garden peas. (Courtesy, The Bettmann Archive, Inc., New York, NY)

and animals follows the biological laws discovered by Mendel.

SOME FUNDAMENTALS OF HEREDITY IN SWINE

In the sections that follow, no attempt will be made to cover all of the diverse field of genetics. Rather, the authors present a condensation of the pertinent facts in regard to the field and briefly summarize their application to swine.

GENES AND CHROMOSOMES

Chromosomes carry all the hereditary characteristics of animals, from the body type to the color of the hair. Hundreds and even thousands of genes are carried on a single chromosome. Genes are the functional unit of inheritance.

The bodies of all animals are made up of millions or even billions of tiny cells, microscopic in size. Each cell contains a nucleus in which there are a number of pairs of bundles called chromosomes. The central

Brno, in Czechoslovakia), Mendel applied a powerful curiosity and a clear mind to reveal some of the basic principles of hereditary transmission. In 1866, he published in the proceedings of a local scientific society a report covering 8 years of his studies, but for 34 years his findings went unheralded and ignored. Finally, in 1900, 16 years after Mendel's death, three European biologists independently duplicated his findings, and this led to the dusting off of the original paper published by the monk 34 years earlier.

The essence of Mendelism is that inheritance is by particles or units (called genes), that these genes are present in pairs—one member of each pair having come from each parent—and that each gene maintains its identity generation after generation. Thus, Mendel's work with peas laid the basis for two of the general laws of inheritance: (1) the law of segregation, and (2) the independent assortment of genes. Later, genetic principles were added; yet all the phenomena of inheritance, based upon the reactions of genes, are generally known under the collective term, *Mendelism*.

Thus, modern genetics is really unique in that it was founded by an amateur who was not trained as a geneticist and who did his work merely as a hobby. During the years since the rediscovery of Mendel's principles (in 1900), many additional genetic principles have been added, but the fundamentals as set forth by Mendel have been proved correct in every detail. It can be said, therefore, that inheritance in both plants

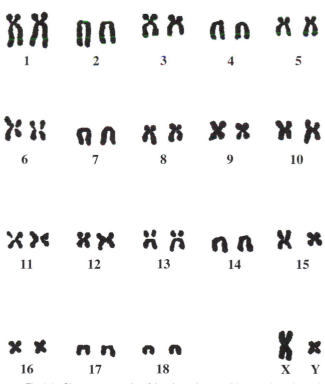

Fig. 6-2. Chromosome pairs of the pig as they would appear in a photomicrograph made from chromosomes of the metaphase of cell division when they can be stained and observed. There are 18 pairs of autosomes and one pair of sex chromosomes (XY)—all the genetic information necessary to make a male pig.

inner portion of each chromosome contains a long double helical molecule called deoxyribonucleic acid, or DNA for short. The DNA molecule is the genetic material—the genetic code. Genes form a portion of each DNA molecule, each in a fixed or special position, called a locus, of each chromosome. Since chromosomes occur in pairs, the genes are also in pairs. The nucleus of each body cell of swine contains 19 pairs of chromosomes,[1] or a total of 38, whereas there are perhaps thousands of pairs of genes. These genes determine all the hereditary characteristics of living animals via the coded information in their DNA. Thus, inheritance is transmitted by units (chromosomes) rather than by the blending of two fluids, as our grandfathers thought.

■ **DNA (deoxyribonucleic acid)**—In recent years, the molecular basis of heredity has become much better understood. The most important genetic material in the nucleus of the cell is deoxyribonucleic acid (DNA), which serves as the genetic information source. It is composed of nucleotides containing adenine, guanine, cytosine, and thymine. The sequence of these four bases in DNA acts as a code in which messages can be transferred from one cell to another during the process of cell division. The code can be translated by cells (1) to make proteins and enzymes of specific structure that determine the basic morphology and functioning of a cell; (2) to control differentiation, which is the process by which a group of cells become an organ; or (3) to control whether an embryo will become a pig, a cow, or a human.

But DNA is far more than a genetic information center, or master molecule. The recent development of the recombinant DNA technique ushered in a new era of genetic engineering—with all its promise and possible hazard.

Recombinant DNA techniques are of enormous help to scientists in mapping the positions of genes

---

[1]Cattle have 60 chromosomes; horses have 64 chromosomes, and sheep have 54 chromosomes.

and learning their fundamental nature. It may lead to new scientific horizons—of allowing introduction of new genetic material directly into the cells of an individual to repair specific genetic defects or to transfer genes from one species to another. On the other hand, the opponents of tinkering with DNA raise the specter (1) of reengineered creatures proving dangerous and ravaging the earth, and (2) of moral responsibility in removing nature's "evolutionary barrier" between species that do not mate. Nevertheless, molecular biologists are working ceaselessly away in recombinant DNA studies; hence, the swine geneticist should keep abreast of new developments in this exciting field.

The modern breeder knows that the job of transmitting qualities from one generation to the next is performed by the germ cells—a sperm from the male and an ovum or egg from the female. All animals, therefore, are the result of the union of two such tiny cells, one from each of its parents. These two germ cells contain all the anatomical, physiological, and psychological characters that the offspring will inherit.

In the body cells of an animal, the chromosomes occur in pairs (diploid numbers), whereas in the formation of the sex cells, the egg and the sperm, a reduction division occurs and only one chromosome and one gene of each pair goes into a sex cell (haploid number). This means that only half the number of chromosomes and genes present in the body cells of the animal go into each egg and sperm, but each sperm or egg cell has genes for every characteristic of its species. As will be explained later, the particular half that any one germ cell gets is determined by chance. When mating and fertilization occur, the single chromosomes from the germ cell of each parent unite to form new pairs, and the genes are again present in duplicate in the body cells of the embryo.

With all possible combinations in 19 pairs of chromosomes (the species number in swine) and the genes that they bear, any boar or sow can transmit over one billion different samples of its own inheritance; and the combination from both parents makes possible one billion times one billion genetically different offspring.

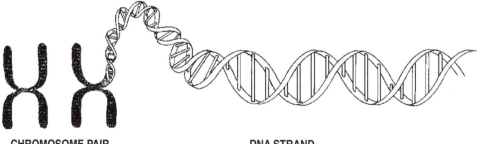

CHROMOSOME PAIR   **DNA STRAND**

Fig. 6-3. Chromosomes consist of deoxyribonucleic acid (DNA) and a type of protein coiled into a tightly packaged structure. The DNA molecule (uncoiling left to right) has a double helical structure. *Note:* The concept is greatly exaggerated in this diagram.

It is not strange, therefore, that no two animals within a given breed are exactly alike (except identical twins from a single egg which split after fertilization). Rather, we can marvel that the members of a given breed bear as much resemblance to each other as they do.

Even between such closely related individuals as full sisters, it is possible that there will be quite wide differences in size, growth rate, temperament, conformation, and in almost every conceivable character. Admitting that many of these differences may be due to undetected differences in environment, it is still true that in such animals much of the variation is due to hereditary differences. A boar, for example, will sometimes transmit to one offspring much better inheritance than he does to most of his get, simply as the result of chance differences in the genes that go to different sperm at the time of the reduction division. Such differences in inheritance in offspring have been called both the hope and the despair of the livestock breeder.

Generally there are two genes for each trait. For each locus in one of the members of a chromosome pair there is a corresponding locus in the other member of that chromosome pair. These genes located at corresponding loci in chromosome pairs may be identical in the way they affect a trait, or they may contrast in the way they affect a trait. If the genes are identical in the way they affect a trait, then the individual is said to be homozygous at that locus; if they contrast in the way they affect a trait, the individual is said to be heterozygous at that locus. It follows that homozygous individuals are genetically pure, since the genes for a trait passed on in the egg or sperm will always be the same. But heterozygous individuals produce sperm or eggs bearing one of two types of genes. Homozygous and heterozygous may also refer to a series of hereditary factors. However, few if any animals are entirely homozygous—producing totally uniform offspring as a result of one sperm or egg being just like any other. Rather, animals are heterozygous, and there is often wide variation within the offspring of any given sire or dam. The wise and progressive breeder recognizes this fact, and he insists on the production records of all offspring rather than that of a few meritorious individuals.

Variation between the offspring of animals that are not pure or homozygous, to use the technical term, is not to be marveled at, but is rather to be expected. No one would expect to draw exactly 20 sound apples and 10 rotten ones every time a random sample of 30 is taken from a barrel containing 40 sound ones and 20 rotten ones, although on the average—if enough samples were drawn—about that proportion of each may be expected. Individual drawings would of course vary rather widely. Exactly the same situation applies to the relative number of "good" and "bad" genes that may be present in different germ cells from the same animal. Because of this situation, the mating of a sow with a fine production record to a boar that on the average transmits relatively good offspring will not always produce pigs of merit equal to that of their parents. The pigs could be markedly poorer than the parents or, happily, they could in some cases be better than either parent.

Selection and closebreeding are the tools through which the swine producer can obtain boars and sows whose chromosomes and genes contain similar hereditary determiners—animals that are genetically more homozygous.

MUTATIONS

Gene changes are technically known as mutations. *A mutation may be defined as a sudden variation that results from changes in a gene or genes which is later passed on through inheritance.* Mutations are not only rare, but they are prevailingly harmful. For all practical purposes, therefore, the genes can be thought of as unchanged from one generation to the next. The observed differences between animals are usually due to different combinations of genes being present rather than to mutations. Each gene probably changes only about once in each 100,000 to 1,000,000 animals produced.

Once in a great while a mutation occurs in a farm animal, and it produces a visible effect in the animal carrying it. These animals are sometimes called *sports*. Mutations are occasionally of practical value. The occurrence of the polled characteristic within the horned Hereford and Shorthorn breeds of cattle is an example of a mutation of economic importance.[2] Out of this has arisen polled Hereford and polled Shorthorn cattle.

Gene changes can be accelerated by exposure to ionizing radiation (x-rays, radium), a wide variety of chemicals (mustard gas, LSD), and ultraviolet light rays.

Geneticists have always dreamed of producing specific or directed mutations thereby creating new varieties. Whether this will be achieved in farm animals is uncertain, but it is an enticing possibility, and recombinant DNA techniques suggest its eventuality.

SIMPLE GENE INHERITANCE
(Qualitative Traits)

In the simplest type of inheritance, only one pair of genes is involved. This type of inheritance can be used to demonstrate how genes segregate in the sperm and egg at random. Therefore, the possible

---

[2]The horned gene mutates to the polled gene at a fairly high frequency—about 1 in 20,000.

gene combinations are governed by the laws of chance (probability) operating in much the same manner as the results obtained from flipping coins. Relatively large numbers are required for certain proportions to be evident. For example, if a penny is flipped often enough, the number of heads and tails will come out about even. However, with the laws of chance in operation, it is possible that out of any four tosses one might get all heads, all tails, or even three to one.

In swine, a pair of genes is responsible for erect and lop ears, hair color, eye color, type of blood, and lethals.

DOMINANT AND RECESSIVE FACTORS

When genes at corresponding loci on chromo-some pairs are unlike, one of the genes often over-powers the expression of the other. This gene is referred to as the dominant gene and the gene whose expression is prevented is called the recessive gene.

The pair of genes responsible for erect and lop ears in swine may be used as an example of dominance and recessiveness, and as an example to illustrate how the law of probability operates in inheritance. This situation is illustrated in Fig. 6-4. In this example, the gene for lop ears has its full effect regardless of whether it is present with another just like itself or is paired with a recessive gene. An erect eared boar possesses only recessive genes for lop ears and can only form sperm bearing recessive genes. A lop ear sow bearing only the dominant genes for lop ears can only produce eggs bearing the dominant. All the first cross progeny from these animals are hybrid lop eared; that is, they appear lop eared but possess a dominant gene for lop ears and a recessive gene for erect ears.

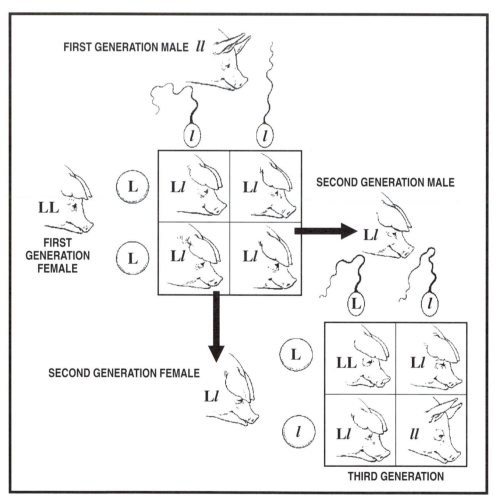

Fig. 6-4. Initially, a boar pure (homozygous) for erect ears (ll) is bred to a sow pure (homozygous) for lop ears (LL). Since lop ears are dominant, all of their offspring will be lop eared (Ll). When these lop-eared offspring are mated to each other, they produce germ cells (spermatozoa and eggs) for lop ears and erect ears in equal proportion. Their offspring will, on the average, consist of three lop ears to one erect ear.

In scientific terms, their phenotype (how they appear) is lop ear while their genotype is lop-erect. If mated, these offspring produce sperm and eggs bearing either a dominant or a recessive gene for lop ear. When large numbers are involved, the results of this mating would be that phenotypically, ¾ of the offspring would have lop ears and ¼ would have erect ears. Genotypically, ¼ would be lop-lop (homozygous dominant); ½ would be lop-erect (heterozygous); and ¼ would be erect-erect (homozygous recessive).

The mating of two erect-eared pigs will always produce erect-eared offspring, but the mating of the lop-eared pigs may produce lop or erect ears depending on the genetic makeup—whether the animals are homozygous or heterozygous.

It is clear, therefore, that a dominant character will cover up a recessive. Hence a hog's breeding performance cannot be recognized by its phenotype (how it looks), a fact which is of great significance in practical breeding.

As can be readily understood, dominance often makes the task of identifying and discarding all animals carrying an undesirable recessive factor a difficult one. Recessive genes can be passed on from generation to generation, appearing only when two animals both of which carry the recessive factor happen to mate. Even then, only one out of four offspring produced will, on the average, be homozygous for the recessive factor and show it.

COLOR INHERITANCE

In swine, white is dominant to colored hair, with the result that when white hogs are crossed on colored hogs, the first crosses are white, though the colored breeds sometimes transmit some of their skin pigmentation.

Some examples of color inheritance in swine follow:

1. **Black breed (Poland China) X white breed (Chester White, Yorkshire, or Landrace).** The offspring are usually white with small black spots, although there may be roans in some cases.
2. **Black breed (Poland China) X red breed (Duroc).** The offspring will be black-and-red spotted.
3. **Red breed X white breed.** The offspring are usually white, although there may be roans in some cases.
4. **Belted breed X red or black breed.** The offspring are generally colored with white belts.
5. **Belted (Hampshire) X white.** The offspring are usually white with some black spots, or there may be some degree of roan. Ghost patterns often show.

RECESSIVE DEFECTS

Examples of undesirable recessives in animals are: scrotal hernia and inverted nipples (blind teats) in pigs. When these conditions appear, one can be very certain that both the sire and dam contributed equally to the condition and that each of them carries the recessive gene therefor. This fact should be given consideration in the culling program.

Assuming that a hereditary defect or abnormality has occurred in a herd and that it is recessive in nature, the breeding program to be followed to prevent or minimize the possibility of its future occurrence will depend somewhat on the type of herd involved—especially on whether it is a commercial or purebred herd. In an ordinary commercial herd, the breeder can usually guard against further reappearance of the undesirable recessive simply by using an outcross (unrelated) sire within the same breed or by crossbreeding with a sire from another breed. With this system, the breeder is fully aware of the recessive being present, but has taken action to keep it from showing up.

On the other hand, if such an undesirable recessive appears in purebred or seedstock herds, the action should be more drastic. A reputable breeder has an obligation not only to himself/herself but to customers among both the purebred and commercial herds. Purebred and seedstock animals must be purged of undesirable genes and lethals. This can be done by:

1. Eliminating those sires and dams that are known to have transmitted the undesirable recessive character.
2. Eliminating both the abnormal and normal offspring produced by these sires and dams, since approximately half of the normal animals will carry the undesirable character in the recessive condition.
3. In some instances, breeding a prospective herd sire to a number of females known to carry the factor for the undesirable recessive, thus making sure that the new sire is free from the recessive.

Such action in a purebred or seedstock herd is expensive, and it calls for considerable courage. Yet it is the only way in which the purebred swine of the country can be freed from such undesirable genes.

MULTIPLE GENE INHERITANCE (Quantitative Traits)

Relatively few characters of economic importance in farm animals are inherited in as simple a manner as the ears described. Important characters—such as meat production, milk and butterfat production, egg production, and wool production—are due to many genes; thus, they are called multiple-factor characters or multiple-gene characters. Because such characters show all manner of gradation—from high to low performance, for example—they are sometimes referred to as quantitative traits. Still the mechanism of inheritance is the same whether a few or multiple genes are involved.

In quantitative inheritance, the extremes (either good or bad) tend to swing back to the average. Thus, the offspring of a high producing boar and a high producing sow are not apt to be as good as either parent. Likewise, and happily so, the progeny of two very mediocre parents will likely be superior to either parent.

Estimates of the number of pairs of genes affecting each economically important characteristic vary greatly, but the majority of geneticists agree that for most such characters ten or more pairs of genes are involved. Growth rate in swine, for example, is affected by: (1) the animal's appetite or feed consumption; (2) the efficiency of assimilation—that is, the proportion of the feed eaten that is absorbed into the blood stream; and (3) the use to which the nutrients are put after assimilation—growth or finishing. This example should indicate clearly enough that such a characteristic as growth rate is controlled by many genes and that it is difficult to determine the mode of gene action of such characters.

HEREDITY AND ENVIRONMENT

A sleek hog, with an ideal body conformation, is undeniably the result of two forces—heredity and environment. If turned to the forest, a littermate to the sleek pig would present an entirely different appearance. By the same token, optimum environment could never make an attractive animal out of a pig with scrub ancestry.

These are extreme examples, and they may be applied to any class of farm animals; but they do emphasize the fact that any particular animal is the product of heredity and environment. Stated differently, heredity may be thought of as the foundation, and environment as the structure. Heredity has already made its contribution at the time of fertilization, but environment works ceaselessly away until death.

Admittedly, after looking over an animal, a breeder cannot with certainty know whether it is genetically a high or a low producer; and there can be no denying the fact that environment—including feeding, management, and disease—plays a tremendous part in determining the extent to which hereditary differences that are present will be expressed in animals.

Experimental work has long shown conclusively enough that the vigor and size of animals at birth are dependent upon the environment of the embryo from the minute the ovum or egg is fertilized by the sperm, and some evidence indicates that newborn animals are affected by the environment of the egg and sperm long before fertilization has been accomplished. In other words, perhaps due to the storage of substances, the kind and quality of the ration fed to young females may later affect the quality of their progeny. Generally speaking, then, environment may inhibit the full expression of potentialities from a time preceding fertilization until physiological maturity has been attained.

It is generally agreed, therefore, that maximum development of characters of economic importance—growth, body form, milk production, etc.—cannot be achieved unless there are optimum conditions of nutrition and management. However, the next question is whether a breeding program can make maximum progress under conditions of suboptimal nutrition (such as is often found under some farm conditions). One school of thought is that selection for such factors as body form and growth rate in animals can be most effective only under nutritive conditions promoting the near maximum development of those characters of which the animal is capable. The other school of thought is that genetic differences affecting usefulness under suboptimal conditions will be expressed under such suboptimal conditions, and that differences observed under forced conditions may not be correlated with real utility under less favorable conditions. Those favoring the latter thinking argue, therefore, that the production and selection of breeding animals for suboptimal nutritive conditions should be under less favorable conditions and that the animals should not be highly fitted.

 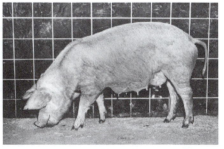

Fig. 6-5. Feed made the difference! The two sows were of the same age and breeding, but the sow shown in the picture at left received all she could eat from birth, whereas the gaunt sow shown in the picture at right was limited to 70% of the ration consumed by the better fed animal. This 10-year experiment, conducted at Washington State University, was designed to study the effect of plane of nutrition on meat animal improvement. (Courtesy, Washington State University)

In general, the results of a long-time experiment conducted at Washington State University[3] support the contention that selection of breeding animals should be carried on under the same environmental conditions as those under which commercial animals are produced.

Within the pure breeds of swine—managed under average or better than average conditions—it has been found that, in general, only 15 to 30% of the observed variation in a characteristic is actually brought about by hereditary variations. To be sure, if we contrast animals that differ very greatly in heredity—for example, a top producing hog and a scrub—90% or more of the apparent differences in type may be due to heredity. The point is, however, that extreme cases such as the one just mentioned are not involved in the advancement within improved breeds of livestock. Here the comparisons are between animals of average or better than average quality, and the observed differences are often very minor.

The problem of progressive breeders is that of selecting the very best animals available genetically for parents of the next generation of offspring in their herds. Since only 15 to 30% of the observed variation is due to differences in inheritance, and since environmental differences can produce misleading variations, mistakes in the selection of breeding animals are inevitable. However, if the purebred or seedstock breeder has clearly in mind a well-defined ideal and adheres rigidly to it in selecting breeding stock, very definite progress can be made.

SEX DETERMINATION

On the average, and when considering a large population, approximately equal numbers of males and females are born in all common species of animals. To be sure, many notable exceptions can be found in individual herds or flocks.

Sex is determined by the chromosomal makeup of the individual. One particular pair of the chromosomes is called the sex chromosomes. In farm animals, the female has a pair of similar sex chromosomes, called X chromosomes; the male has a pair of unlike sex chromosomes, called X and Y chromosomes. In the bird, this condition is reversed, the female having the unlike pair and the male having the like pair.

The pairs of sex chromosomes separate when the germ cells are formed. Thus, each of the ova or eggs produced by the sow contains the X chromosomes; whereas the sperm of the boar are of two types, one-half containing the X chromosome and the other half the Y chromosome. Since, on the average, the

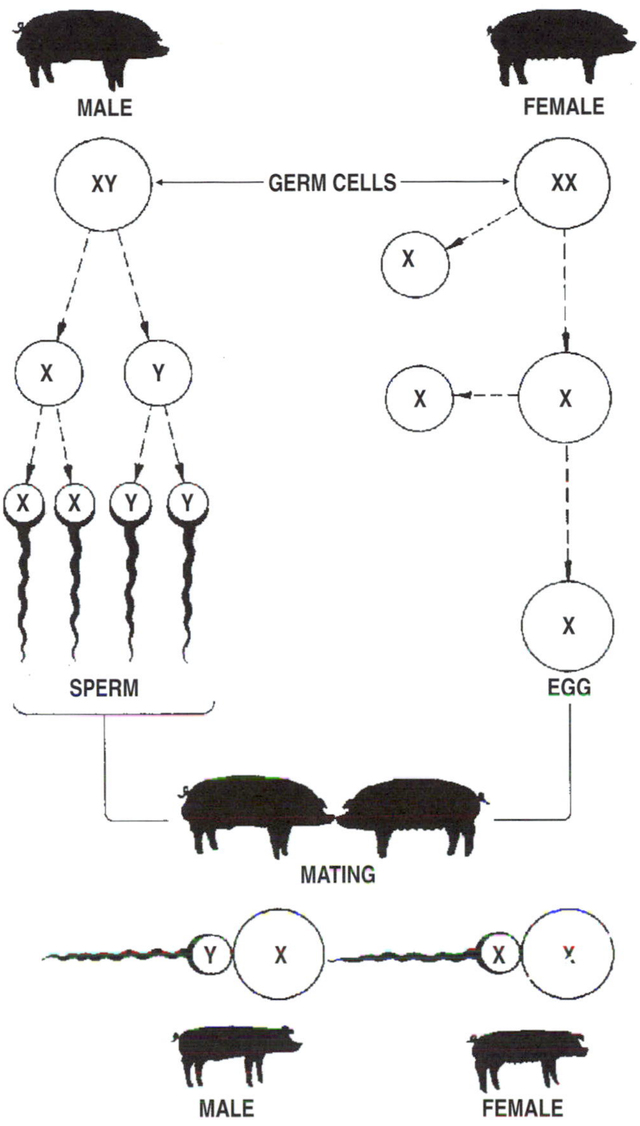

Fig. 6-6. Diagrammatic illustration of the mechanism of sex determination in hogs, showing how sex is determined by the chromosomal make-up of the individual. The sow has a pair of like sex chromosomes, called X chromosomes; the boar has a pair of unlike sex chromosomes, called X and Y chromosomes. Thus, if an egg and a sperm of like sex chromosomal make-up unite, the offspring will be a female, and if an egg and a sperm of unlike sex chromosomal make-up unite, the offspring will be a male.

eggs and sperm unite at random, it can be understood that half of the progeny will contain the chromosomal make up XX (females) and the other half XY (males).[4]

---

[3]Wash. Ag. Exp. Sta. Bul. 34, January, 1961.

[4]The scientists' symbols for the male and female, respectively, are: ♂ (the sacred shield and spear of Mars, the Roman god of war), and ♀ (the looking glass of Venus, the Roman goddess of love and beauty).

SEX RATIO CONTROL

Through the ages humans have desired to select the sex of their offspring. Rulers were always anxious to have sons, while some rulers desired that their servants have fewer sons. For the producers, controlling the sex of the offspring of farm animals is of great interest because of the economic advantage of being able to produce all males or females depending on the need. The normal sex ratio (males to females) is about 50:50, since whether an X- or Y-bearing sperm will fertilize an egg is decided by chance just as flipping a coin for heads or tails.

Many approaches have been tried to alter the sex ratio. Some of these approaches are sophisticated and scientific while others are myths that develop from casual observations. Most scientific efforts have focused upon eliminating or altering the X or Y sperm in semen before artificial insemination. The three main approaches are (1) sedimentation, in which it is hoped that the heavy X sperm will settle to the bottom and the Y sperm float to the top; (2) electrical charge, in which it is assumed that X and Y sperm have a different net charge and chemical or electrical methods could bind or attract either negatively or positively charged sperm; and (3) treatment of sperm with a substance deadly to X or Y sperm. Potential methods of sex ratio control include (1) immunological in which antibodies would prevent Y-bearing sperm from fertilizing an egg;

(2) embryo transfer where only embryos of the desired sex were transferred; and (3) cloning, where the whole genetic makeup including sex could be preselected.

It appears that nature is about to yield to sex control. Predetermination of the sex of 6- to 12-day-old embryos is a reality, and progress is being made in the separation of sperm cells containing X chromosomes from those containing Y chromosomes.

Research and theories will continue because the stakes are high if any workable method can be found.

LETHALS AND OTHER HEREDITARY DEFECTS IN SWINE

The term "lethal" refers to a genetic factor that causes death of the young, either during prenatal life or at birth. Other defects occur which are not sufficiently severe to cause death but which do impair the usefulness of the affected animals. Some of the lethals and other abnormalities that have been reported in swine are summarized in Table 6-1.

Many such abnormal animals are born on the nation's farms and ranches each year. Unfortunately, the purebred or seedstock breeders, whose chief business is that of selling breeding stock, are likely to "keep mum" about the appearance of any defective

TABLE 6-1
SOME HEREDITARY LETHALS AND OTHER HEREDITARY ABNORMALITIES
WHICH HAVE BEEN REPORTED IN SWINE

Type of Abnormality	Description of Abnormality	Probable Mode of Inheritance
A. Lethals		
Atresia ani	No anal opening. Pigs born alive.	Undetermined
Bent legs	Legs bent at right angle and stiff.	Recessive
Brain hernia	Pigs usually born alive but skull openings present inovlving frontal and parietal bones.	Probably recessive
Catlin mark	Incomplete development of the skull.	Recessive
Cleft palate	Pigs born alive but unable to nurse.	Recessive
Excessive fatness	Pigs become excessively fat at 70 to 150 lb *(32 to 68 kg)* and die.	Undetermined
Fetal mortality	Born dead or are reabsorbed.	Recessive
Hydrocephalus	Fluid on the brain, head enlarged; often accompanied by short tail.	Recessive
Legless	Pigs born alive but without legs.	Recessive
Muscle contracture	Usually only forelegs affected, but sometimes hindlegs are involved. Forelegs rigid. Animals usually stillborn or live only a short time.	Recessive
Paralysis	Complete paralysis of hindlegs. Born alive but starve unless given special care.	Recessive
Split ears	Ears split usually associated with cleft palate and deformed hindlegs.	Probably recessive
Thickened forelimbs	Thickening of forelegs caused by infiltration of connective tissues which replace the muscle fibers. Pigs usually born alive.	Recessive

(Continued)

TABLE 6-1 (Continued)

Type of Abnormality	Description of Abnormality	Probable Mode of Inheritance
B. Non Lethals		
Cryptorchidism	One or both testicles retained in body cavity.	Recessive
Hair whorls (swirls)	Hair at a given point in back region flanges out in all directions.	Mode of inheritance not certain but probably due to complementary action of two dominant factors
Hairlessness	Animals born with little or no hair (not to be confused with hairlessness caused by an iodine deficiency).	Recessive
Hemophilia	Blood fails to clot promptly when wounds are inflicted.	Recessive
Hermaphrodites	Animals that possess characteristics of both sexes.	Sex-limited recessive
Inverted nipples (blind teats)	Teats inverted and nonfunctional.	Undetermined
Kinky tail	Rigid angles in the tail at birth.	Recessive
Polydactyl	Extra toes on forefeet.	Undetermined
Porcine Stress Syndrome (PSS)	Sudden death of heavily muscled pigs and/or production of pale, soft, exudative musculature of carcass.	Autosomal recessive
Red eyes	Observed in Hampshires. Affected animals also have light brown hair coat.	Probably recessive
Scrotal hernia	Ruptured; intestines extending into scrotum.	Result of two pairs of recessive factors
Syndactyl (mule foot)	Only one toe instead of two.	Dominant
Umbilical hernia	Weakness at umbilicus; intestines protrude.	Dominant
Wattles	Skinlike flaps hanging from throat near lower jaw.	Dominant
Wooly	Kinky hair.	Dominant

animals in their herds because of the justifiable fear that it may hurt their sales. With the commercial producers, however, the appearance of such lethals is simply so much economic loss, with the result that they generally, openly and without embarrassment, admit the presence of the abnormality and seek correction.

The embryological development—the development of the young from the time that the egg and the sperm unite until the animal is born—is very complicated. Thus, the oddity probably is that so many of the offspring develop normally rather than that a few develop abnormally.

Many such abnormalities (commonly known as monstrosities or freaks) are hereditary, being caused by certain "bad" genes. Moreover, the bulk of such lethals are recessive and may, therefore, remain hidden for many generations. The prevention of such genetic abnormalities requires that the germ plasm be purged of the "bad" genes. This means that, where recessive lethals are involved, the producer must be aware of the fact that both parents carry the gene. For the total removal of the lethals, test matings and rigid selection must be practiced. The best test mating to use for a given sire consists of mating him to some of

his own daughters. Thus, where there is suspicion of scrotal hernia, it is recommended that a boar be bred back to some of his daughters.

In addition to hereditary abnormalities, there are certain abnormalities that may be due to nutritional deficiencies, or to accidents of development—the latter

Fig. 6-7. Pig with "thick forelegs," a lethal condition caused by a simple recessive genetic factor. (Courtesy, Department of Veterinary Pathology and Hygiene, College of Veterinary Medicine, University of Illinois)

including those which appear to occur sporadically and for which there is no well-defined reason. When only a few defective individuals occur within a particular herd, it is often impossible to determine whether their occurrence is due to: (1) defective heredity, (2) defective nutrition, (3) viral infection of the dam, or (4) accidents of development. If the same abnormality occurs in any appreciable number of animals, however, it is probably either hereditary or nutritional. In any event, the diagnosis of the condition is not always a simple matter.

The following conditions would tend to indicate a hereditary defect:

1. If the defect had previously been reported as hereditary in the same breed of livestock.

2. If it occurred more frequently within certain families or when there had been inbreeding.

3. If it occurred in more than one season and when different rations had been fed.

The following conditions might be accepted as indications that the abnormality was due to a nutritional deficiency:

1. If previously it had been reliably reported to be due to a nutritional deficiency.

2. If it appeared to be restricted to a certain area.

3. If it occurred when the ration of the mother was known to be deficient.

4. If it disappeared when an improved ration was fed.

If there is suspicion that the ration is defective, it should be improved, not only from the standpoint of preventing such deformities, but from the standpoint of good and efficient management.

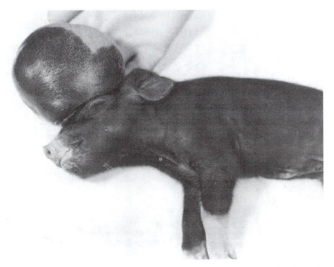

Fig. 6-8. Hydrocephalic (literally meaning "water in the head"). Pigs affected with this condition die soon after birth. It is inherited as a simple recessive. (Courtesy, Purdue University)

Fig. 6-9. A six-legged Duroc gilt. Perhaps this condition can best be described as an "accident of development."

If there is good and sufficient evidence that the abnormal condition is hereditary, the steps to be followed in purging the herd of the undesirable gene are identical to those already outlined for ridding the herd of an undesirable recessive factor. An inbreeding program, of course, is the most effective way in which to expose hereditary lethals in order that purging may follow.

RELATIVE IMPORTANCE OF THE BOAR AND THE SOW

As a boar can have so many more offspring during a given season or a lifetime than a sow, he is from a hereditary standpoint a more important individual than any one sow so far as the whole herd is concerned, although both the boar and the sow are of equal importance so far as concerns any one offspring. Because of their wider use, therefore, boars are usually culled more rigidly than sows, and the breeder can well afford to pay more for an outstanding boar than for an equally outstanding sow.

Experienced swine producers have long felt that boars often resemble their daughters more closely than their sons, whereas sows resemble their sons. Some boars and sows, therefore, enjoy a reputation based almost exclusively on the merit of their sons, whereas others owe their prestige to their daughters. Although this situation is likely to be exaggerated, any such phenomenon that may exist is due to sex-linked inheritance which may be explained as follows: The genes that determine sex are carried on one of the chromosomes. The other genes that are located on the same chromosome will be linked or associated with sex and will be transmitted to the next generation in combination with sex. Thus, because of sex linkage, there are more color-blind men than color-blind

Fig. 6-10. Boars at Murphy Farms, Inc., Rose Hill, North Carolina. (Courtesy, Dr. Garth W. Boyd, and Ms. Rhonda Campbell)

women. In poultry breeding, the sex-linked factor is used in a practical way for the purpose of distinguishing the pullets from the cockerels early in life, through the process known as *sexing* the chicks. Thus, when a black cock is crossed with barred hens, all the cocks come barred and all the hens come black. It should be emphasized, however, that under most conditions it appears that the influence of the sire and dam on any one offspring is about equal. Most breeders, therefore, will do well to seek excellence in both sexes of breeding animals.

PREPOTENCY

Prepotency refers to the ability of the animal, either male or female, to stamp its own characteristics on its offspring. The offspring of a prepotent boar, for example, resemble both their sire and each other more closely than usual. The only conclusive and final test of prepotency consists of the inspection of the get.

From a genetic standpoint, there are two requisites that an animal must possess in order to be prepotent: (1) dominance, and (2) homozygosity. Every offspring that receives a dominant gene or genes will show the effect of that gene or genes in the particular character or characters which result therefrom. Moreover, a perfectly homozygous animal would transmit the same kind of genes to all of its offspring. Although entirely

homozygous animals probably never exist, it is realized that a system of inbreeding is the only way to produce animals that are as nearly homozygous as possible.

Popular beliefs to the contrary, there is no evidence that prepotency can be predicted by the appearance of an animal. To be more specific, there is no reason why a vigorous, masculine-appearing boar will be any more prepotent than one less desirable in these respects.

It should also be emphasized that it is impossible to determine just how important prepotency may be in animal breeding, although many sires of the past have enjoyed a reputation for being extremely prepotent. Perhaps these animals were prepotent, but there is also the possibility that their reputation for producing outstanding animals may have rested upon the fact that they were mated to some of the best females of the breed.

In summary, it may be said that if a given boar or sow possesses a great number of genes that are completely dominant for desirable type and performance and if the animal is relatively homozygous, the offspring will closely resemble the parent and resemble each other, or be uniform. Fortunate, indeed, is the breeder who possesses such an animal.

NICKING

If the offspring of certain matings are especially outstanding and in general better than their parents, breeders are prone to say that the animals nicked well. For example, a sow may produce outstanding pigs to the service of a certain boar, but when mated to another boar of apparent equal merit as a sire, the offspring may be disappointing. Or sometimes the mating of a rather average boar to an equally average sow will result in the production of a most outstanding individual both from the standpoint of type and performance.

So-called successful nicking is due, genetically speaking, to the fact that the right combination of genes for good characters are contributed by each parent, although each of the parents within itself may be lacking in certain genes necessary for excellence. In other words, the animals "nicked" well because their respective combinations of good genes were such as to complement each other.

The history of animal breeding includes records of several supposedly favorable nicks. Because of the very nature of successful nicks, however, outstanding animals arising therefrom must be carefully scrutinized from a breeding standpoint; because, with their heterozygous origin, it is quite unlikely that they will breed true.

FAMILY NAMES

In animals, depending upon the breed, family names are traced through either the males or females. Unfortunately, the value of family names is generally grossly exaggerated. Obviously, if the foundation boar or sow, as the case may be, is very many generations removed, the genetic superiority of this head of a family is halved so many times by subsequent matings that there is little reason to think that one family is superior to another. The situation is often further distorted by breeders placing a premium on family names of which there are few members, little realizing that, in at least some cases, there may be unfortunate reasons for the scarcity in numbers.

Such family names have about as much significance as human family names. Who would be so foolish as to think that the Joneses as a group are alike and superior to the Smiths? Perhaps, if the truth were known, there have been many individuals with each of these family names who have been of no particular credit to the clan, and the same applies to all other family names.

Family names lend themselves readily to speculation. Because of this, the history of livestock breeding has often been blighted by instances of unwise pedigree selection on the basis of not too meaningful family names. Fortunately, for swine producers, there has been less worshipping of family names in hogs than in certain other classes of livestock.

Of course, certain linebred families—linebred to a foundation sire or dam so that the family is kept highly related to it—do have genetic significance. Moreover, if the programs involved have been accompanied by rigid culling, many good individuals may have evolved, and the family name may be in good repute.

SYSTEMS OF BREEDING

The many diverse types and breeds among each class of farm animals in existence today originated from only a few wild types within each species. These early domesticated animals possessed the pool of genes, which, through controlled matings and selection, proved flexible in the hands of breeders.

Perhaps at the outset it should be stated that there is no one best system of breeding or secret of success for any and all conditions. Each breeding program is an individual case, requiring careful study. The choice of the system of breeding should be determined primarily by the size and quality of the herd, by the finances and skill of the operator, and by the ultimate goal ahead.

PUREBREEDING

A purebred animal may be defined as a member of a breed, the animals of which possess a common ancestry and distinctive characteristics; and it is either registered or eligible for registry in that breed. The breed association consists of a group of breeders banded together for the purposes of: (1) recording the lineage of their animals, (2) protecting the purity of the breed, and (3) promoting the interest of the breed.

The term *purebred* refers to animals whose entire lineage, regardless of the number of generations removed, traces back to the foundation animals accepted by the breed or to animals which have been subsequently approved for infusion.

The terms pure breeding and homozygosity may bear very different connotations. Yet there is some interrelationship between purebreds and homozygosity. Because most breeds had a relatively small number of foundation animals, the unavoidable inbreeding and linebreeding during the formative state resulted in a certain amount of homozygosity. Moreover, through the normal sequence of events, it is estimated that purebreds become more homozygous by from 0.25 to 0.5% per animal generation.

It should be emphasized that being a purebred animal does not necessarily guarantee superior type or high productivity. That is to say, the word purebred is not, within itself, magic, nor is it sacred. Many a person has found, to much sorrow, that there are such things as purebred scrubs. Yet, on the average, purebred animals are superior to non-purebreds.

For the producer with experience and adequate capital, the breeding of purebreds may offer unlimited opportunities. It has been well said that honor, fame, and fortune are all within the realm of possible realization of the purebred breeder; but it should also be added that only a few achieve this high calling.

Purebred breeding is a highly specialized type of production. Generally speaking, only the experienced breeder should undertake the production of purebreds with the intention of furnishing foundation or replacement stock to other purebred breeders, or purebred boars for crossbreeding programs. Although we have had many constructive swine breeders and great progress has been made, it must be remembered that only a few achieve sufficient success to classify as master breeders.

INBREEDING

Inbreeding is rarely practiced among present-day swine producers, though it was common in the foundation animals of most of the breeds.

Inbreeding is the mating of animals more closely

related than the average of the population from which they came.

Most scientists divide inbreeding into various categories, according to the closeness of the relationship of the animals mated and the purpose of the matings. There is considerable disagreement, however, as to both the terms used and the meanings that it is intended they should convey. For purposes of this book and the discussion which follows, the following definitions will be used:

Closebreeding is the mating of closely related animals: sire to daughter, son to dam, and brother to sister.

Linebreeding is the mating of animals more distantly related than in closebreeding and in which the matings are usually directed toward keeping the offspring closely related to some highly admired ancestor; such as half-brother to half-sister, female to grandsire and cousins.

CLOSEBREEDING

In closebreeding there are a minimum number of different ancestors. In the repeated mating of a brother with his full sister, for example, there are only 2 grandparents instead of 4, only 2 great-grandparents instead of 8, and only 2 different ancestors in each generation farther back—instead of the theoretically possible 16, 32, 64, 128, etc. The most intensive form of inbreeding is self-fertilization. It occurs in some plants, such as wheat and garden peas, and in some lower animals; but domestic animals are not self-fertilized. Closebreeding is rarely practiced by present-day producers, though it was common in the foundation animals of most of the breeds.

The reasons for practicing closebreeding are:

1. It increases the degree of homozygosity within animals, making the resulting offspring pure or homozygous in a larger proportion of their gene pairs than in the case of linebred or outcross animals. In so doing, the less desirable recessive genes are brought to light so that they can be more readily culled. Thus, closebreeding, together with rigid culling, affords the surest and quickest method of fixing and perpetuating a desirable character or group of characters.

2. If carried on for a period of time, it tends to create lines or strains of animals that are uniform in type and in other characteristics.

3. It keeps the relationship to a desirable ancestor highest.

4. Because of the greater homozygosity, it makes for greater prepotency. That is, selected inbred animals are more homozygous for desirable genes (genes which are often dominant), and they, therefore, transmit these genes with greater uniformity.

5. Through the production of inbred lines or families by closebreeding and the subsequent crossing of certain of these lines, it affords a modern approach to livestock improvement. Moreover, the best of the inbred animals are likely to give superior results in outcrosses.

6. Where a breeder is in the unique position of having a herd so far advanced that to go on the outside for seed stock would merely be a step backward, it offers the only sound alternative for maintaining existing quality or making further improvement.

The precautions in closebreeding may be summarized as follows:

1. As closebreeding greatly enhances the chances that recessives will appear during the early generations in obtaining homozygosity, it is almost certain to increase the proportion of worthless breeding stock produced. This may include such so-called degenerate characteristics as reduction in size, fertility, and general vigor. Lethals and other genetic abnormalities often appear with increased frequency in closebred animals.

2. Because of the rigid culling necessary in order to avoid the "fixing" of undesirable characters, especially in the first generations of a closebreeding program, it is almost imperative that this system of breeding be confined to a relatively large herd and to instances when the owner has sufficient finances to stand the rigid culling that must accompany such a program.

3. It requires skill in making planned matings and rigid selection, thus being most successful when applied by "master breeders."

4. It is not adapted for use by the breeder with average or below average stock because the very fact that the animals are average means that a goodly share of undesirable genes are present. Closebreeding would merely make the animals more homozygous for undesirable genes and, therefore, worse.

Judging from outward manifestations alone, it might appear that closebreeding is predominantly harmful in its effects—often leading to the production of defective animals lacking in the vitality necessary for successful and profitable production. But this is by no means the whole story. Although closebreeding often leads to the production of animals of low value, the resulting superior animals can confidently be expected to be homozygous for a greater than average number of good genes and thus more valuable for breeding purposes. Figuratively speaking, therefore, closebreeding may be referred to as "trial by fire," and the breeder who practices it can expect to obtain many animals that fail to measure up and that have to be culled. On the other hand, if closebreeding is handled properly, the breeder can also expect to secure animals of exceptional value.

Although closebreeding has been practiced less during the past century than in the formative period of the different pure breeds of livestock, it has real merit when its principles and limitations are fully understood. Perhaps closebreeding had best be confined to use by the skilled master breeder who is in a sufficiently sound financial position to endure rigid and intelligent culling and delayed returns and whose herd is both large and above average in quality.

LINEBREEDING

From a biological standpoint, closebreeding and linebreeding are the same thing, differing merely in intensity. In general, closebreeding has been frowned upon by swine producers, but linebreeding (the less intensive form) has been looked upon with favor in some quarters.

In a linebreeding program, the degree of relationship is not closer than half-brother and half-sister or matings more distantly related: cousin matings, grandparent to grand offspring, etc.

Linebreeding may be practiced in order to conserve and perpetuate the good traits of a certain outstanding boar or sow. Because such descendants are of similar lineage, they have the same general type of germ plasm and therefore exhibit a high degree of uniformity in type and performance.

In a more limited way, a linebreeding program has the same advantages and disadvantages of a closebreeding program. Stated differently, linebreeding offers fewer possibilities both for good and harm than closebreeding. It is a more conservative and safer type of program, offering less probability to either "hit the jackpot" or "sink the ship." It is a middle-of-the road program that the vast majority of average and small breeders can follow safely to their advantage. Through it, reasonable progress can be made without taking any great risk. A greater degree of homozygosity of certain desirable genes can be secured without running too great a risk of intensifying undesirable ones.

Usually a linebreeding program is best accomplished through breeding to an outstanding sire rather than to an outstanding dam because of the greater number of offspring of the former. If a swine breeder is in possession of a great boar—proved great by the production records of a large number of his get—a linebreeding program might be initiated in the following way: Select two of the best sons of the noted boar and mate them to their half-sisters, balancing all possible defects in the subsequent matings. The next generation matings might well consist of breeding the daughters of one of the boars to the son of the other, etc. If, in such a program, it seems wise to secure some outside blood (genes) to correct a common defect or defects in the herd, this may be done through

selecting a few outstanding proved sows from the outside—animals whose get are strong where the herd may be deficient—and then mating these sows to one of the linebred boars with the hope of producing a son that may be used in the herd.

The small operator—the owner of a few sows—can often follow a linebreeding program by breeding sows to a boar purchased from a large breeder who follows such a program—thus in effect following the linebreeding program of the larger breeder.

Naturally, a linebreeding program may be achieved in other ways. Regardless of the actual matings used, the main objective in such a system of breeding is that of rendering the animals homozygous—in desired type and performance—to some great and highly regarded ancestor, while at the same time weeding out homozygous undesirable characteristics. The success of the program, therefore, is dependent upon having desirable genes with which to start and an intelligent intensification of these good genes."

Uncle" Nick Gentry of Sedalia, Missouri— whose name is among the immortals as a Berkshire swine breeder—linebred and closebred with great skill and success to his outstanding boar, Longfellow. That Mr. Gentry was very cautious, however, is attested by the fact that he often spent an hour walking back and forth between his herd boars before deciding which one to mate to a particular sow.

It should be emphasized that there are some types of herds that should almost never closebreed or linebreed. These include herds of only average quality.

The owners of grade or commercial herds run the risk of undesirable results, and, even if successful as commercial breeders, they cannot sell their stock at increased prices for breeding purposes.

With purebred herds of only average quality, more rapid progress can usually be made by introducing superior outcross sires. Moreover, if the animals are of only average quality they must have a preponderance of "bad" genes that would only be intensified through a closebreeding or linebreeding program.

OUTCROSSING

Outcrossing is the mating of animals that are members of the same breed, but which show no relationship close up in the pedigree (for at least the first 4 to 6 generations).

Most of our purebred animals of all classes of livestock are the result of outcrossing. It is a relatively safe system of breeding, for it is unlikely that two such unrelated animals will carry the same "undesirable" genes and pass them on to their offspring.

Perhaps it might well be added that the majority of purebred breeders with average or below average herds had best follow an outcrossing program, because, in such herds, the problem is that of retaining a heterozygous type of germ plasm with the hope that genes for undesirable characters will be counteracted by genes for desirable characters. With such average or below average herds, a closebreeding program would merely make the animals homozygous for the less desirable characters, the presence of which already makes for their mediocrity. In general, continued outcrossing offers neither the hope for improvement nor the hazard of retrogression in linebreeding or closebreeding programs.

Judicious and occasional outcrossing may well be an integral part of linebreeding or closebreeding programs. As closely inbred animals become increasingly homozygous with germ plasm for good characters, they may likewise become homozygous for certain undesirable characters even though their general overall type and performance remains well above the breed average. Such defects may best be remedied by introducing an outcross through an animal or animals known to be especially strong in the character or characters needing strengthening. This having been accomplished, the wise breeder will return to the original closebreeding or linebreeding program, realizing full well the limitations of an outcrossing program.

GRADING UP

Grading up is that system of breeding in which a purebred or seedstock sire is mated to a native or grade female. Its purpose is to impart quality and to increase performance in the offspring.

Naturally, the greatest single step toward improved quality and performance occurs in the first cross. The first generation from such a mating results in offspring carrying 50% of the hereditary material of the purebred or seedstock parent (or 50% of the "blood" of the purebred or seedstock parent, as many producers speak of it). The next generation gives offspring carrying 75% of the "blood" of the purebred or seedstock parent, and in subsequent generations the proportion of inheritance remaining from the original scrub parent is halved with each cross. Later crosses usually increase quality and performance still more, though in less marked degree. After the third or fourth cross, the offspring compare very favorably with purebred or seedstock in conformation, and only exceptionally good sires can bring about further improvement. This is especially so if the boars used in grading up successive generations are derived from the same strain.

CROSSBREEDING

Crossbreeding is the mating of different breeds.

In a broad sense, crossbreeding also includes the mating of purebred sires of one breed with high grade females of another breed.

Today, there is renewed interest in crossbreeding swine, and increased research is underway on the subject. Crossbreeding is being used by swine producers to (1) increase productivity over straightbreds, because of the resulting hybrid vigor or heterosis; (2) produce commercial hogs with a desired combination of traits not available in any one breed; and (3) produce foundation stock for developing new breeds.

The motivating forces back of increased crossbreeding in farm animals are (1) more artificial insemination, thereby simplifying the rotation of sires of different breeds; and (2) the necessity for swine producers to become more efficient in order to meet their competition, both from within their industry and from without.

Crossbreeding will play an increasing role in the production of market animals in the future, because it offers the several advantages discussed in the sections which follow.

HYBRID VIGOR OR HETEROSIS

Heterosis, or hybrid vigor, is the name given to the biological phenomenon which causes crossbreds to outproduce the average of their parents. For numerous traits, the performance of the cross is superior to the average of the parental breeds. This phenomenon has been well known for years and has been used in many breeding programs. The production of hybrid seed corn by developing inbred lines and then crossing them is probably the most important attempt to take advantage of hybrid vigor. Today, heterosis is also being used extensively in commercial swine, sheep, layer, and broiler production. An estimated 80% of market lambs are crossbred; 95% of broilers are crosses; and about 90% of the hogs raised for slaughter are crossbred.

The genetic explanation for the hybrid's extra vigor is basically the same, whether it be cattle, hogs, sheep, broilers, hybrid corn, hybrid sorghum, or whatnot. Heterosis is produced by the fact that the dominant gene of a parent is usually more favorable than its recessive partner. When the genetic groups differ in the frequency of genes they have and dominance exits, then heterosis will be produced.

Heterosis is measured by the amount the crossbred offspring exceeds the average of the two parent breeds or inbred lines for a particular trait, using the following formula for any one trait:

$$\frac{\text{Crossbred average} - \text{Purebred average}}{\text{Purebred average}} \times 100 = \text{Percent hybrid vigor}$$

Thus, if the average of the two parent populations for litter weaning weight is 284 lb and the average of their crossbred offspring is 336 lb, application of the above formula shows that the amount of heterosis is 52 lb or 18%.

Traits high in heritability—like carcass length, backfat thickness, and loin eye area—respond consistently to selection but show little response to hybrid vigor. Traits low in heritability—like litter size, litter weaning weight, and survival rate—usually demonstrate good response to hybrid vigor.

COMPLEMENTARY

Complementary refers to the advantage of a cross over another cross or over a purebred, resulting from the manner in which two or more characters combine or complement each other. It is a matching of breeds so that they compensate each other, the objective being to get the desirable traits of each. Thus, in a crossbreeding program, breeds that complement each

Fig. 6-11. Hampshire boar. On the paternal side, Hampshire and Duroc boars excel in crossbreeding. (Courtesy, Lone Willow Farm, Roanoke, IL)

Fig. 6-12. Yorkshire sow and litter. On the maternal side, Yorkshire and Large White sows excel in crossbreeding. (Courtesy, Land O Lakes, Ft. Dodge, IA)

other should be selected, thereby maximizing the desirable traits and minimizing the undesirable traits. Since breeds which are selected because they tend to express a maximum of some trait will have some undesirable traits, different breeds must be selected for different purposes.

No one breed or strain has a monopoly on all the desired characteristics. Therefore, producers must study their operation and the merits of different breeds or seedstock before choosing a breed.

INTRODUCE NEW GENES QUICKLY

Crossbreeding provides a way in which to introduce new and desired genes quickly—at a faster rate than can be achieved by selection within a breed. As undesirable qualities are often recessive, crossbreeding offers the best way in which to improve certain characteristics merely by hiding them with dominants.

GET HYBRID VIGOR EXPRESSED IN THE FEMALE

Except for a two-breed cross, crossbreeding offers an opportunity to have hybrid vigor expressed in breeding females. This is most important in the swine herd where it results in increased fertility, survivability of piglets, litter weaning size, and pig growth rate—all factors that mean more profit for the producer.

FACTORS AFFECTING MAGNITUDE OF ADVANTAGES FROM CROSSBREEDING

Many other examples of each of the advantages of crossbreeding could be cited. It should be noted, however, that the total magnitude of the advantage of these factors—achieving the 15 to 25% potential immediate increase in yield per female unit through continuous crossbreeding compared to continuous straight breeding—depends upon the following:

1. Making wide crosses. The wider the cross, the greater the heterosis.

2. Selecting breeds that are complementary. A crossbreeding program should involve breeds that possess the favorable expression of traits desired in the crossbred offspring that will be produced.

3. Using high-performing stock. Once a crossbreeding program is initiated, further genetic improvement is primarily dependent upon the use of superior production tested boars.

4. Following a sound crossbreeding system. For a continuous high expression of heterosis and maximum output per female, a sound system of crossbreeding must be followed. This should include the use of crossbred females, for research clearly indicates that

over one-half the higher profits from a crossbreeding program results therefrom.

5. Tapping purebreds constantly. Purebreds must be constantly tapped to renew the vigor of crossbreds; otherwise, the vigor is dissipated.

DISADVANTAGES OF CROSSBREEDING

Crossbreeding does, however, possess certain disadvantages which should be understood:

1. Generally speaking, a crossbred hog lacks the uniformity in color and general attractiveness of purebreds.

2. Desirable boars of the two or three breeds or strains must be located and purchased.

3. It should not be assumed that the virtues of crossbreeding are sufficiently powerful to alleviate the necessity of selecting outstanding boars.

4. Crossbreeding should not be looked upon as a panacea for neglect of sound practices of breeding, feeding, management, and sanitation.

SWINE CROSSBREEDING SYSTEMS[5]

Crossbreeding can be a very useful means for the pork producer to increase the efficiency and profit of

Fig. 6-13. A crossbred litter of pigs. (Courtesy, Dr. Garth Boyd and Ms. Rhonda Campbell, Murphy Farms, Inc., Rose Hill, NC)

[5]The section on Swine Crossbreeding Systems, including the Figs., was adapted by the authors from *Oklahoma Extension Facts, No. 3603*, by David S. Buchanan, William G. Luce, and Archie C. Cutter.

an operation. Full benefits from crossbreeding can be gained only by careful combination of available breeds and selection of outstanding breeding animal replacements from within those breeds. Crossbreeding programs must be systematic and well planned to take full advantage of heterosis and breed differences.

Crossbreeding enables the producer to take advantage of heterosis and to combine desirable characteristics of different breeds. Desirable characteristics of different breeds can be utilized if some breeds can be identified as good maternal breeds and others as good paternal breeds. A system where males from paternal breeds (superior growth and carcass) are mated to females from maternal breeds (superior reproductive and mothering ability) can take advantage of the strengths of both breeds while minimizing some of the weaknesses.

Carefully designed crossbreeding programs can be used to enhance improvement for selection in the purebreeds. Crossbreeding does not change the genes that are present in a population, but it arranges them in more favorable combinations. It follows that the initial boost from crossbreeding can be maintained by continued crossing. Permanent improvement can result only through selection.

Two basic systems of crossbreeding exist; namely, the rotational cross system and the terminal cross system. Also, the two systems can be used in combination. Rotational cross systems combine two or more breeds, where the breed of boar used is different from the previous generation and replacement crossbred females are retained from each cross. In the terminal cross, female replacements are usually purchased or produced by maintaining purebred herds that emphasize reproductive performance.

ROTATION CROSS SYSTEMS

The two-breed rotational cross uses boars of two different breeds in alternate generations, and retains crossbred females for maternal stock (Fig. 6-14). This system is fairly simple to follow once the producer chooses two breeds. Breeds used in rotation should be productive since, over time, each breed contributes equally to both the production traits of the market offspring and the reproductive traits of replacement females. Therefore, reproductively sound breeds with adequate growth and carcass characteristics should be used.

In this system purebred boars are mated to sows with a certain percentage of the same breed as the boars. Therefore, the maximum response from heterosis will not be realized (see Table 6-2). In fact, the actual heterosis retained changes a little each generation until the sixth generation, after which the two-

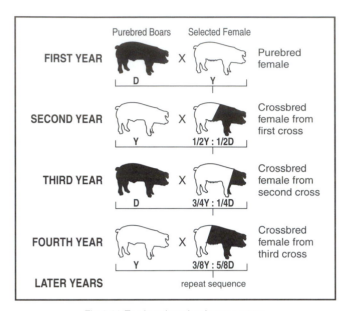

Fig. 6-14. Two-breed rotational cross system.

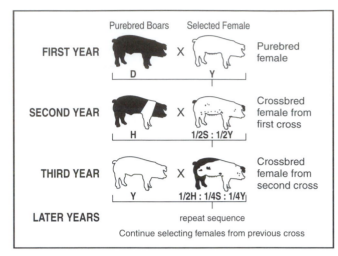

Fig. 6-15. Three-breed rotational cross system.

breed rotation realizes about two-thirds of the total advantage obtained from crossbreeding.

More heterosis can be realized with the addition of a third breed to the rotation (Table 6-2 and Fig. 6-15). A three-breed rotation realizes about 86% of the advantage obtained from crossbreeding. Again, each breed contributes as both a sire and a dam so reproductively sound breeds with adequate growth and carcass characteristics should be used. Three-breed rotations are recommended over two-breed rotations, since a higher percentage of the total advantage obtained from crossbreeding is realized.

A fourth breed could be added to the rotation, in

which case about 92% of the total advantage from crossbreeding would be realized. Even though a higher percentage of the total heterosis is realized with a four-breed rotation, it is generally not recommended over a three-breed rotation because of the difficulty in finding a fourth breed with a higher average level of productivity. The number of breeding groups that need to be maintained also increases with the number of breeds included in the rotation since females of different parities will be at different stages of the rotation and will require different breeds of boars or mates.

The rotation cross system is more popular and thought by many to be a more practical program than the terminal cross system. This is because the only outside breeding stock that needs to be purchased once the program is established are boars. Thus, a

TABLE 6-2
BREED COMPOSITION AND PERCENT OF MAXIMUM HETEROSIS EXPECTED FROM TWO-BREED AND THREE-BREED ROTATIONAL CROSSING PROGRAMS

Generation Number	Two-Breed Crosses				Three-Breed Crosses				
	% Blood		Expected Heterosis		% Blood			Expected Heterosis	
	A	B	Offspring	Dam	A	B	C	Offspring	Dam
1	50	50	100	0	50	50[1]	0	100	0
2	75[1]	25	50	100	25	25	50[1]	100	100
3	38	62[1]	75	50	63[1]	12	25	75	100
4	69[1]	31	62	75	31	56[1]	12	88	75
5	34	66[1]	69	62	16	28	56[1]	88	88
6	67[1]	33	66	69	58[1]	14	28	84	88
7	33	67[1]	67	66	29	57[1]	14	86	84
8	67[1]	33	67	67	14	29	57[1]	86	86

[1]Breed of sire used to produce offspring.

producer does not have the difficulty of obtaining replacement females at a reasonable cost. Also, there is less risk of introducing disease to the herd since only boars are purchased.

Combinations of desirable traits of breeds cannot be fully utilized in a rotational system. Ideally, market hogs should be out of sows with good mothering ability and sired by boars that excel in growth and carcass characteristics. This is difficult to achieve in the framework of a rotational crossing system.

TERMINAL CROSS SYSTEMS

Terminal cross systems may involve two, three, or four breeds (Table 6-3). A two-breed cross will utilize purebred boars on one breed mated to purebred sows of another breed. For example, Hampshire boars and Yorkshire females could be used to produce every pig crop. This system allows the producer to combine a dam breed superior for reproductive performance with a sire breed superior for growth and carcass characteristics. All pigs produced are crossbred, therefore this system reaps all the advantages obtained by having crossbred pigs. The sows, however, are purebred and one of the superiority obtained from crossbred sows will be realized.

A three-breed terminal cross will utilize purebred boars of one breed mated to crossbred sows of two other breeds. For example, Duroc boars and Landrace X Yorkshire cross females could be used to produce each pig crop. In this system, sows that are a cross between two breeds superior for maternal characteristics are mated to a third breed of sire that is superior for growth and carcass characteristics. All pigs produced are crossbreds. Crossbred sows are utilized; so, this system maximizes heterosis and is one of the most productive systems.

A four-breed terminal cross has all of the same advantages on the maternal side as the three-breed cross. In addition, there are some advantages in conception rate associated with use of crossbred boars. A disadvantage is that crossbred boars may be more difficult to obtain than purebred boars. However, since some seedstock producers maintain purebred herds of two or more breeds, they could easily produce some crossbred boars if there should be a demand for them.

In general, terminal crosses capitalize on the strengths of each breed and realize maximum gains from heterosis. Example terminal crossbreeding systems are shown in Table 6-3. A big disadvantage of terminal cross can be the difficulty of obtaining replacement females. If they are purchased, there is a health risk associated with the introduction of outside breeding stock. If they are raised within the herd, the commercial producer will need at least one purebred herd to produce replacement stock.

A COMBINATION SYSTEM

The advantages of both systems can be utilized if all replacement females are produced in a rotational cross of prolific, highly productive breeds and the market hogs are then sired by a boar of another breed (Fig. 6-16). The producer would maintain a small portion of the herd (10–15%) in the two-breed rotational cross. The best females would be kept in the rotation while most of the remaining rotation females would be mated to the terminal cross sire. The rotation breeds should be ones with good maternal abilities and the terminal sire should be from a breed with good growth and carcass characteristics and rank highly for those traits as an individual. All pigs are from crossbred dams so much of the maternal heterosis can be used and 100% of the individual heterosis utilized in the terminal cross pigs. Also, only boars need to be brought in from the outside, so disease problems are minimized.

TABLE 6-3
PERCENTAGE HETEROSIS MAINTAINED BY TERMINAL CROSSES
AND ITS EFFECT ON ONE TRAIT

Sire Breed	Dam Breed	% Heterosis			Litter Weight (lb) at 21 Days per Female Exposed
		Pig	Dam	Sire	
Purebred					
A	A	0	0	0	72.0
Two-breed cross					
A	B	100	0	0	80.1
Three-breed cross					
C	AB	100	100	0	93.8
Four-breed cross					
CD	AB	100	100	100	97.0

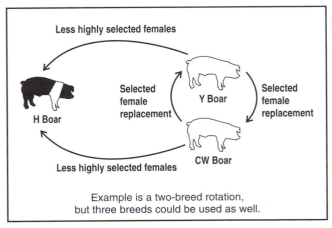

Fig. 6-16. Terminal sire on rotation female system.

This system has the disadvantages of being rather complicated and requires large numbers to allow it to operate efficiently. Unless a producer farrows at least 200 litters per year the terminal sire on rotation female system will be hard to maintain and will make inefficient use of the boars, particularly those in the rotation breeds.

When considering a crossbreeding system there are many choices one can make. Each has several advantages and disadvantages. Some programs are simple but do not make maximum use of heterosis. Others are more complex, require more time to manage, but should have higher average levels of performance because they take more advantage of the strengths of breeds and retain a higher percentage of heterosis. No one system will be best for every producer.

Larger producers who spend considerable time managing their swine operations can utilize more complicated systems that have higher expected levels of performance. In addition, facilities, source of breeding stock, disease control programs, and perhaps other factors will be important when deciding which system is best adapted to a particular production unit.

PRODUCTION TESTING SWINE

Fig. 6-17. A recent addition to the New Ulm Swine Evaluation Station. This facility was added to evaluate the National Barrow Show pigs. The capacity of the New Ulm station makes the Minnesota Swine Evaluation Station a national center for swine testing. (Courtesy, University of Minnesota)

As a basis for selection, increasing emphasis is being placed on production testing which embraces both (1) performance testing (sometimes called individual merit testing), and (2) progeny testing. The distinction between and the relationship of these terms are set forth in the following definitions:

1. Performance testing is the practice of evaluating and selecting animals on the basis of their individual merit or performance.
2. Progeny testing is the practice of selecting animals on the basis of the merit of their progeny.

3. Production testing is a more inclusive term, including performance testing and/or progeny testing.

Production testing involves the taking of accurate records—from birth on—rather than casual observation. Also, in order to be most effective, the accompanying selection must be based on characteristics of economic importance and high heritability (see Table 6-4), and an objective measure or "yardstick" (such as pounds, inches, etc.) should be placed upon each of the traits to be measured. Finally, those breeding animals that fail to meet the high standards set forth must be removed from the herd promptly and unflinchingly. Breeding animals can only transmit desirable qualities unfailingly to all their offspring when they themselves have been rendered relatively homozygous or pure for the necessary genes. This process can be made more rigid and certain through securing and intelligently using production records, but a knowledge of what records to keep and how to use them is necessary.

Fig. 6-18. Production testing necessitates weighing, preferably litter weight at 21 days of age, and individual off test weight at 230 lb. (Courtesy, Farmland Industries, Inc., Kansas City, MO)

CENTRAL TESTING STATIONS

Boar test stations, first established in Denmark in 1907, have been in operation in the United States since 1954. They have provided uniform environments

Fig. 6-19. Iowa swine Testing station showing regular test boar unit. Progeny tests consist of three boars and one barrow. Pigs remain in these units from the time of entry at about 8 to 10 weeks of age until they are auctioned at 6 to 7 months of age. (Courtesy, Iowa Swine Testing station, Ames, IA)

for evaluating the genetic merit of swine for three of the most important economic traits: growth rate, feed efficiency, and backfat thickness. The number of central testing stations has declined since 1980. Some test boars only. Others test boars, gilts, and market hogs. Most central test stations follow the Guidelines for Uniform Swine Improvement Programs, of the National Swine Improvement Federation. These follow:[6]

1. Entry requirements.

Individual pigs must weigh between 40 and 65 lb and have a weight per day of age between 0.6 and 1.2 lb.

All pigs entered must have an interstate health paper stating that pigs are free from designated communicable diseases outlined by the participating test station. Some stations may require a negative PRV test before or shortly after pigs arrive at the station.

If the pigs are purebreds, they must be eligible for registry, and the purebred nomination papers must be submitted to the test station manager before the time of the 35-day report.

2. Testing procedures.

All pigs should be grown under uniform management.

It is recommended that a high energy, corn-soy diet with a minimum of 18% protein be fed for the first 35 days after test initiation, and that a 16% protein be fed for the remainder of the test period.

The number of animals per pen should be determined by the size of the pen (3 boars per 5' × 15' pen). All animals should be the progeny of one sire.

---

[6]*Guidelines for Uniform Swine Improvement Programs,* USDA, 1987, pp. 8-1 and 8-2.

Test animals should be grouped by sex. Boars, gilts, or barrows are to be tested in separate groups or pens.

There will be a 5- to 7-day adjustment period from entry into the station to on-test weight.

The average minimum on-test pen weight will be 70 lb and the maximum weight 80 lb.

Pigs will be taken off test when the pen averages between 230 and 240 lb. On-test average daily gain will be used to adjust days to 230 lb. Test average daily gain should be adjusted to a standard on-test weight, using an adjustment of 0.004 lb per pound for each pound over or under the standard on-test weight.

Feed efficiency should be computed at the off-test average weight of 230 to 240 lb. Report whether a meal or pelleted ration is fed.

Backfat is to be measured between 220 and 240 lb. The current method of measurement is an average of three locations: above the point of the elbow, the last rib, and the stifle joint. All measurements are taken 1½ to 2 in. off the midline. If reported, 10th rib backfat should be taken 3 in. off the midline. All measurements should be adjusted and reported at a 230-lb basis.

3. Adjustment factors and selection indexes.

For details relative to computing these, see Guidelines for Uniform Swine Improvement Programs, National Swine Improvement Federation, USDA, 1987.

Central testing stations are used primarily to (1) acquaint and educate producers with performance records, and (2) compare individual pigs' performance for rate of gain, feed conversion, and backfat. The most a central testing station can offer is reliable records for comparisons among individuals within test groups. Ideally, central testing stations and on-farm testing programs should be linked and complementary.

Central testing stations provide an excellent source of replacement boars with the ultimate in test records—compiled under a common environment.

ON-FARM TESTING

Performance tests can be conducted on the farm—regardless of the facilities. The prime requirement is that all animals be handled similarly. Giving a small group of pigs preferential treatment leads to false conclusions and stymies, if not regresses, genetic improvement. A good on-farm testing program promotes rapid genetic improvement, since it permits testing of a larger sample than is possible in central testing stations. The on-farm testing program is designed to assist breeders in evaluating their herds in a systematic manner. The program will (1) identify superior individuals, strains, lines, or breeds; (2) assist breeders in the selection of boars and gilts; (3) provide a means of following up on the breeding value of boars

Fig. 6-20. Production testing facilities for litters on a purebred establishment. Each litter is on test from weaning to 230 lb. The owner obtains (1) for each pig, a daily grain record and a probed backfat reading; and (2) for each litter, the feed consumed. (Courtesy, Hampshire Swine Registry)

purchased from central test stations or seedstock producers; and (4) allow breeders to use common terminology and guidelines in selection of breeding stock.

ON-FARM TESTING ESSENTIALS

The objectives of the on-farm testing program can be met completely only if the whole herd is tested. Testing a selected sample of the herd yields limited and biased information. Comparisons are most meaningful if based on all the pigs produced so trait ratios and indexes are not distorted.

Accuracy is an extremely important part of any testing program. Most producers have the ability to conduct performance tests adequately. However, professional assistance from breed associations, testing organizations, Cooperative State Extension Services, and private commercial concerns is available to aid producers in the mechanics, record processing, and reporting of data connected with a testing program.

Pigs should be evaluated within test groups and divided by farrowing group, month, or season. These test groups should be managed and fed uniformly. All pigs in the test group should be given an equal opportunity.

Each record should be expressed as a ratio of the test group or herd average. The individual's index should be used when ranking possible herd replacements.

ON-FARM TESTING PROCEDURES

The National Swine Improvement Federation recommends the following on-farm testing procedures:[7]

---

[7]*Guidelines for Uniform Swine Improvement Programs*, National Swine Improvement Federation, USDA, 1987, p. 6-1.

1. Identification of all pigs in herd. It is recommended that the ear notching system identify the litter in the right ear and the individual pig in the left ear.

2. Birth record. Within three days of birth, all pigs should be individually ear notched, sex noted, and the birth date and parents recorded in an appropriate record book or file kept by the breeder.

3. Sow productivity. The number of pigs farrowed alive and dead should be recorded. Litters should be standardized to between 8 and 10 pigs per litter within 24, but not later than 48 hours after birth. After standardization, record the number of pigs in the litter.

At 14 to 28 days of age, adjusted to a 21-day basis, record litter weight and the number of pigs alive—all pigs raised by the sow, including foster pigs. An individual breeder may wean at any time, but litter weight should be collected before weaning and as near 21 days as possible.

4. Growth. Pigs should be placed on test at a weight of approximately 70 lb, and weighed off test near 230 lb, adjusted to 230 lb. These weights may be used to estimate days to 230 lb.

5. Backfat and loin eye area. All pigs will be measured for backfat thickness. The current method of measurement is an average of three locations: (1) above the point of the elbow, (2) the last rib, and (3) the stifle joint. The data should be averaged and adjusted to 230 lb or a constant weight.

Loin muscle area should be measured on pigs within a range of ±30 lb of the desired constant weight. Loin muscle area is not included in any recommended selection index.

6. Feed efficiency. If possible, feed efficiency should be measured on an individual basis. If group fed, pigs should be tested by progeny groups. With group feeding, the number of pigs per pen, sex, and relationship among pigs in pen should be noted.

Fig. 6-21. Ultrasonic scanning at the termination of test conducted at the Minnesota Swine Evaluation Station. (Courtesy, University of Minnesota)

ON-FARM TESTING INDEXES

Recommended indexes for on-the-farm testing are given in the Guidelines for Uniform Swine Improvement Programs, National Swine Improvement Federation, USDA, 1987.

ON-FARM CARCASS EVALUATION

Swine producers can obtain carcass evaluation data on their hogs through sources such as meat processing plants, locker plants, fairs, shows, and home slaughter. An impartial and experienced individual should collect or advise in the collection of these data.

The procedures to evaluate market hogs include: (1) identification by tattoo, (2) inspection, (3) hot carcass weight, and (4) ribbing the carcass.

■ Quantitative characteristics—To determine the proportionate amount of lean or muscle, one of the following two methods is recommended, with the specific methods often depending on circumstances in the cooperating slaughtering facility.

1. Method one includes hog carcass weight, fat depth, and loin eye area at the tenth rib.
2. Method two combines hot carcass weight, last rib fat, and carcass muscling score. It should be used when carcasses cannot be ribbed.

■ Qualitative characteristics—A carcass evaluation also includes the estimation of (1) muscle color, (2) muscle firmness, and (3) loin muscle marbling. Loin muscle color, firmness, and marbling are given scores of 1 to 3.

ON-FARM TESTING REPORTS

All reports should include ratios (see earlier section on Ratios) for the traits measured. Actual measurements are optional. Reports should include:

1. Number born alive ratio.
2. Adjusted 21-day litter weight ratio.
3. Adjusted days to 230 lb ratio.
4. Adjusted backfat at 230 lb ratio.
5. Index.

Standardized, uniform on-farm test records may be computerized.

PRODUCTION TESTING BY SWINE RECORD ASSOCIATIONS

Today, STAGES, EBV, EPD, and BLUP overshadow the once thriving production testing by breed registries.

Nevertheless, some breed registries still conduct Production Testing Programs for their members. It is suggested that purebred breeders who are interested in production testing by their swine registry, contact the executive secretary or a field representative of their breed registry.

(Also see Chapter 4)

THE DANISH SWINE PRODUCTION TESTING SYSTEM

Fig. 6-22. Interior view of the world's first swine production testing station established in 1907 at Elsesminde, Denmark. In Denmark four pigs of each litter are tested under standard conditions, with evaluation based on (1) rate and economy of gain, and (2) carcass quality at 200 lb. The most outstanding example in the world of production testing meat animals is, without doubt, the swine breeding work of Denmark. (Courtesy, Danish Embassy)

Without doubt, the outstanding example of production testing work with meat animals is the swine breeding work of Denmark. This work was started in 1907, and since then, it has operated continuously, except for three years during World War I when shortage of feed forced the suspension of all testing. In the Danish system, the registration of swine is supervised by a national committee in charge of swine breeding. Only animals bred at organized swine-breeding centers are eligible for registration, because these are the farms where the breeders have complied with certain regulations, including sending each year to the testing stations half as many litters as they have sows in their herd. At the testing stations, these test litters of four pigs each are fed under standard conditions, and the rates and economies of gain are recorded. When each pig reaches a weight of 200 lb, it is slaughtered at a nearby bacon factory, and its dressing percentage and the type, conformation, and quality of its carcass are measured and scored. Twice annually, each breeding center is inspected by a committee that scores it for: (1) management and general appearance of the farm, (2) conformation of breeding animals, (3) fertility of the breeding animals, (4) efficiency in the use of feed by the test pigs from this center, and (5) slaughter quality of the test pigs. The advance made in the carcass qualities (body length, belly and backfat thickness), plus efficiency of feed utilization, have been phenomenal. Swine selection in Denmark is strictly based on utility considerations.

Fig. 6-23. Landrace boar. (Courtesy, Cedar Ridge Farms, Red Bud, IL)

USING HERD RECORDS IN SELECTION

It is important that all animals be evaluated for their performance in terms of economically important traits. This requires a good record keeping system in which an animal's performance becomes part of a permanent record. Consistently good production is to be desired. It is all too easy for a breeder to remember the good individuals produced by a given sow or boar and to forget those which are mediocre or culls.

A prerequisite for any production data is that each animal be positively identified—by means of ear notches. For purebred breeders, who must use a system of animal identification anyway, this does not constitute an additional detail. But the taking of weights, grades, and notes does require additional time and labor—an expenditure which is highly worthwhile, however.

Information on the productivity of close relatives (the sire and the dam and the brothers and sisters) can supplement that on the animal itself and thus be a distinct aid in selection. The production records of more distant relatives are of little significance, because individually, due to the sampling nature of inheritance, they contribute only a few genes to an animal many generations removed.

Finally, it should be recognized that swine are raised primarily for profit, and profit is dependent upon efficiency of production and market price. Fortunately, the factors making for efficiency of production—including litter size and survival, growth rate, and feed efficiency—do not change with type fads. For this reason, emphasis should be placed on proper balance of the production factors. It might be added that type changes, quite likely, would not be so radical as in the past if they were guided by market demands based on carcass values.

In order not to be burdensome, the record forms should be relatively simple. Fig. 6-24 is an individual sow record designed for use in recording the lifetime production record of one sow; whereas Fig. 6-25 is a litter record form for use in recording detailed information on one litter.

A good plan for progeny testing boars consists of retaining and mating one or more boar pigs—the numbers depending upon the size of the herd—to a limited number of females during their first season of breeding. The progeny are then tested and evaluated, and only those boars that prove to be best on the basis of their progeny are retained for further breeding purposes. If boar pigs are each mated to 6 or 8 sows, pigs should be born 114 days later, and the progeny can be tested. Thus, with good fortune, it is possible to have progeny data on a boar when he is approximately 12 months of age.

Animals too young to progeny test may be evaluated by performance testing.

Herd records are, however, of little value unless they are intelligently used in culling operations and in deciding upon replacements. Also, most swine producers can and should use production records for purposes of estimating the rate of progress and for determining the relative emphasis to place on each character.

PORK QUALITY

In addition to measuring efficiency in terms of producing large, healthy litters that gain rapidly on minimum of feed, producers also should be concerned about how much lean, edible pork is produced and how desirable it is for consumption.

Desirable lean quality in pork is reddish-pink in color, slightly firm in texture and practically free of surface exudation (watery). Quality variations from this ideal result in less desirable conditions of pale, soft, and exudative (PSE) lean, and dark, firm, and dry (DFD) lean. It is not unusual to find a wide range of lean quality in pork cuts displayed in a retail meat case, including varying degrees of the PSE and DFD condition.

SWINE
Individual Sow Record

Breed _____ Name and registration no._____

Date farrowed _____ Identification _____
(ear notch, tattoo)

Bred by _____
(Name and address)

Sow's pedigree _____ { _____
(Sire)   _____

_____ { _____
(Dam)   _____

Record of litter of which the sow was a member

No. in litter _____ No. of pigs weaned _____

Weaning wt. at _____ days of age
(fill in)

Her own wt. _____ Avg. wt. of litter _____

Litter mate carcass record, if any:

No. carcasses _____ ; avg. backfat _____ ; loin eye _____ ; length _____
(in.)   (sq in.)   (in.)

Number of teats _____

Production Record of Sow

	1	2	3	4	5	6	7	8
Litter no.								
Sire								
No. services								
Farrowing data								
Date								
Temperament of sow (gentle, nervous, cross)								
No. pigs born: Alive								
Dead								
Mummies								
Total								
Avg. birth wt.								
No. functiong teats								
Weaning data: Age								
No. weaned								
Avg. weaning wt.								
Offspring saved for breeding: No. gilts								
No. boars								

Disposal of Sow

Date _____ Reasons _____

Sold to _____
(Name and address)

Price $ _____

Fig. 6-24. Individual sow record form.

SWINE
Litter Record

Breed _____ Litter No. _____
                                                                                   (notch, tattoo)

Data on Dam:

    Pedigree _____ { _____
            (name, reg. no., and ear notch)   (Sire)

    Birth date _____ _____
                    (date and year)   (Dam)

    Litter mate carcass data, if any:

        No. carcasses _____ ; avg. backfat _____ ; loin eye _____ ; length _____
                                (in.)   (sq in.)   (in.)

    Sow's _____ litter
           (1st, 2nd, etc.)

Data on Sire:

    Pedigree _____ { _____
            (name, reg. no., and ear notch)   (Sire)

    Birth date _____ _____
                    (date and year)   (Dam)

    Litter mate carcass data, if any:

        No. carcasses _____ ; avg. backfat _____ ; loin eye _____ ; length _____
                                (in.)   (sq in.)   (in.)

Date of Birth _____   Health Services:

No. Pigs Born: Date cholera vaccinated _____

    Alive _____   Date erysipelas vaccinated _____

    Dead _____   Date wormed _____

    Mummies _____   Other, including iron pills or shots (list) _____

    Total _____   _____

No. Pigs Weaned _____   _____

Individual Pig Record

Pig's No.	Sex	No. Teats	Birth Wt.	Off-Color Markings	Defects & Abnormalities	Weaning Wt. _____ days (fill in)	Date Castrated	Date & Cause of Death	Disposal Date & To Whom	Remarks

Fig. 6-25. Litter record form.

LEAN PORK GAIN[8]

For the most complete and comprehensive evaluation of market hogs, both live-hog production and carcass merit should be included. Also, wherever possible producers should determine the pounds of quality lean pork gain per day on test.

Procedure to determine pounds of quality lean pork gain per day on test follows—

1. Production Test in a Central Testing Station.

The rather standard procedure followed in most test stations is detailed in an earlier section headed "Central Testing Stations"; hence, the reader is referred thereto.

The specific steps to determine pounds of quality lean pork gain per day on test follow:

2. Tattoo hog or use ear tag to identify prior to slaughter.

3. Visually evaluate warm carcass and eliminate from consideration if it is condemned or excessively trimmed (greater than 5%).

4. Obtain carcass weight (pounds), adjust for minor trim losses (if necessary) and express on a warm, skin-on basis. Eliminate from consideration if weight is less than 150 lb.

5. Measure carcass length (inches) and eliminate from consideration if less than 29.5 in.

6. Cut carcass between the 10th and 11th ribs and measure fat depth (inches) and eliminate from consideration if greater than 1.3 in.

7. Determine loin muscle area (square inches) and eliminate from consideration if less than 4.5 sq in.

8. Assess color score of loin muscle and eliminate from consideration if score is either (a) pale pinkish gray or (b) dark purplish red.

9. Assess firmness/wetness score of loin muscle and eliminate from consideration if score is either (a) very soft and very watery or (b) soft and watery.

10. Assess marbling score of loin muscle and eliminate from consideration if score is (a) devoid to practically devoid or (b) moderately abundant or greater.

11. Assess quality of fat throughout the carcass and eliminate from consideration if it is soft and oily.

12. For a carcass meeting minimum qualifications established above, and for purposes of ranking, determine pounds of quality lean pork gain per day on test. Use the formula which follows.

---

[8]This section was adapted by the authors from: *Procedures to Evaluate Market Hogs*, Third Edition, 1991, pp. 4-5, published by the National Pork Producers Council.

POUNDS OF ACCEPTABLE QUALITY LEAN PORK (containing 5% fat) GAIN PER DAY ON TEST

$$= \frac{(\text{lb of lean in carcass}) - (\text{lb of lean in feeder pig})}{(\text{days on test})}$$

$$= \frac{\left[\begin{array}{l} 7.231 + 0.437 \times \text{adj. warm carc. wt., lb} \\ - 18.746 \times \text{10th rib fat depth, in.} \\ + 3.877 \times \text{10th rib LMA, sq in.} \end{array} \right] - \Big[(0.418 \times \text{live wt., lb}) - 3.650 \Big]}{(\text{days on test})^3}$$

Pounds of lean in carcass[1]   *Pounds of lean in feeder pig*[2]

[1]This is the preferred equation. Substitute others if working with live ultrasonics, carcass electronic probe or unribbed observations.

[2]Equation is from Brannaman, *et al.*

[3]Days include both first and last days of test.

ECONOMICALLY IMPORTANT TRAITS AND THEIR HERITABILITY

That swine show variation in economically important traits is generally recognized. The problem is to measure these differences from the standpoint of discovering the most desirable genes and then increasing their concentration and, at the same time, to purge the herd of the less desirable traits.

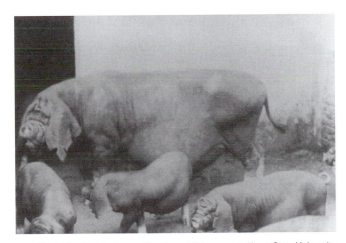

Fig. 6-26. Litter size is a profit indicator! This prompted Iowa State University to import this very prolific Chinese breed of swine. (Courtesy, Iowa State University, Ames, IA)

Table 6-4 gives the economically important traits in swine and their estimated heritability. It should be understood that a heritability estimate indicates the percentage of a trait due to heredity while the remaining portion of the variance is due to the environment. For example, a heritability estimate of 15% for number of pigs born alive (Table 6-4) indicates that about 15% of larger than average litters selected is due to genetic influences, but it also says that 85% is due to environmental influences—whatever they may be.

TABLE 6-4
ECONOMICALLY IMPORTANT TRAITS IN SWINE AND THEIR HERITABILITY[1]

Economically Important Characters	Approximate Heritability of Characters[2]	Comments
	(%)	
1. No. of pigs born alive	10	On the average, a sow will have consumed a total of ¾ to 1 ton of feed during the period between breeding and the date her litter is weaned. Thus, if this quantity of feed must be charged against a litter of 4 or 5 pigs, the chance of eventual profit is small.
2. Birth weight of pigs	5	Very light pigs usually lack vigor.
3. 21-day litter weight	15	21-day litter weight is important, for it has been shown that pigs that are heaviest at 21 days reach market weight more quickly. The low heritability of this factor indicates that it is largely a function of the nursing ability of the sow rather than genetic.
4. Litter size at weaning	12	Although greatly influenced by herdsmanship, litter survival to weaning is a measure of the mothering ability of the sow.
5. Age at puberty	35	Early puberty makes it possible to get animals in production at younger ages, thereby lowering production costs.
6. Days to 230 lb	30	Days to 230 lb is important because (1) it is highly correlated with efficiency of gain, and (2) it makes for a shorter time in reaching market weight and condition, thus effecting a saving in labor, making for less exposure to risk and disease, and allowing for a more rapid turnover in capital.
7. Average daily gain	30	High daily gains make for early market weight and greater feed efficiency.
8. Lb feed/lb gain	30	The most profitable animals generally require less feed to make 100 lb *(45 kg)* of gain.
9. Nipple number	18	There should be a minimum of 12 functional nipples.
10. Confirmation score	29	This heritability figure is likely to be considered higher in a herd of low quality.
11. Carcass characteristics:		
a. Length	60	Carcass length is perhaps the most highly hereditary trait in hogs. This accounts for the rapid shifts that frequently have been observed; for example, in changing from chuffy to rangy hogs.
b. Backfat thickness	40	The probe, lean meter, or ultrasonic equipment can be used to measure backfat thickness on prospective breeding animals.
c. Loin lean area	50	Loin area is an indication of muscling or red meat.
d. Percent ham, based on carcass weight	50	Ham is a high-priced cut; hence, the aim is to get as large a ham as possible.
e. Percent lean cuts, based on carcass weight	50	A high yield of lean cuts means trimmable fat and more edible meat.
12. Eating quality of pork[3]	20	Eating quality of pork is indicated by tenderness, reddish-pink color, marbling, slightly firm texture, and palatability.

[1]These heritability estimates apply to within herd and within breed variations. Variations between breeds are much higher in heritability than the variations within breeds.

[2]The rest is due to environment. The heritability figures given herein are averages based on large numbers; thus, some variation from these may be expected in individual herds.

[3]From: *Pigs-Misset*, Aug. 1994, pp. 14–15, article entitled "Pork Quality: Only 20% Genetic Influence," by deVries, A. G., P. G. van der Wal, G. E. Eikelenboom, J. W. M. Merks, DLO Research Inst. for An. Prod., Zeist, Netherlands.

APPRAISING CHANGES IN PRODUCTION DUE TO HEREDITY AND ENVIRONMENT

Swine producers are well aware that there are differences in litter size, in weaning weight, in body type, etc. If those animals which excel in the desired traits would, in turn, transmit without loss these same improved qualities to their offspring, progress would be simple and rapid. Unfortunately, this is not the case. Such economically important characters are greatly

affected by environment (by feeding, care, management, etc.). Thus, only part of the apparent improvement in certain animals is hereditary, and can be transmitted on to the next generation.

As would be expected, improvements due to environment are not inherited. This means that if most of the improvement in an economically important character is due to an improved environment, the heritability of that character will be low and little progress can be made through selection. On the other hand, if the character is highly heritable, marked progress can be made through selection. Thus, color of hair in swine is a highly heritable character, for environment appears to have little or no part in determining it. On the other hand, such a character as weight per pig at weaning is of low heritability because, for the most part, it is affected by environment (by the nursing ability of the sow).

There is need, therefore, to know the approximate amount of percentage of change in each economically important character which is due to heredity and the amount which is due to environment. Table 6-4 gives this information for swine in terms of the approximate percentage heritability of each of the economically important characters. The heritability figures given therein are averages based on large numbers; thus some variations from these may be expected in individual herds. Even though the heritability of many of the economically important characters listed in Table 6-4 is disappointingly small, it is gratifying to know that much of it is cumulative and permanent.

RATIOS

To compare animals from different environments, and to determine if an animal is better because of genetics or environment, ratios are often used. Ratios simply involve the expression of an animal's performance relative to the herd or test group average. They are calculated as follows:

$$\frac{\text{Animal's performance} \times 100}{\text{Average performance of all animals in the group}}$$

For example, a boar gaining 2.20 lb per day from a group averaging 2.00 lb per day has a daily gain ratio of $(2.20 \times 100) \div 2.00 = 110$. A ratio of 110 implies that boar is 10% above the test group average for that trait. Similarly, a ratio of 90 would indicate the boar performed 10% below the average of contemporary pigs tested. A ratio of 100 indicated the animal is average.

Ratios remove differences in average performance levels among groups (and generally among traits). Thus, they should allow for a more unbiased comparison among individuals that were tested in different groups. This is only true to the extent that average differences among groups are not genetic. Most differences among groups are due to feeding, weather, housing, management, etc. Ratios allow individuals to be compared relative to contemporary groups. They are a useful method for comparing individuals that are not contemporaries, such as those in different tests, herds, or for comparing different traits.

The National Swine Improvement Federation (NSIF) recommends that in production testing programs, the level of performance for various traits should be expressed as a ratio.

ESTIMATING RATE OF PROGRESS

For purposes of illustrating the way in which the heritability figures in Table 6-4 may be used in practical breeding operations, the following example is given:

In a certain herd of swine, the litters in a given year average 7 pigs each, with a range of 4 to 15 pigs. There are available sufficient of the larger litters (averaging 12 pigs) from which to select replacement breeding stock. What amount of this larger litter size (5 pigs above the average) is likely to be transmitted to the offspring of these pigs?

Step by step, the answer to this question is secured as follows:

1. $12 - 7 = 5$ pigs, the number by which the selected litter size exceeds the average from which they arose.

2. By referring to Table 6-4, it is found that the number of pigs born alive is 10% heritable. This means that 10% of the 5 pigs can be expected due to the superior heredity of the stock saved as breeders, and that the other 90% is due to environment (feed, care, management, etc.).

3. $5 \times 10\% = 0.5$ pig; which means that for litter size the stock saved for the breeding herd is 0.5 pig per litter superior, genetically, to the stock from which it was selected.

4. $7 + 0.5 = 7.5$ pigs per litter; which is the expected performance of the next generation.

It is to be emphasized that the 7.5 pigs per litter is merely the expected performance. The actual outcome may be altered by environment (feed, care, management, etc.) and by chance. Also, it should be recognized that where the heritability of a character is lower, less progress can be made. The latter point explains why the degree to which a character is heritable has a very definite influence on the effectiveness of mass selection.

Using the heritability figures given in Table 6-4, and assuming certain herd records, the progress to be expected from one generation of selection in a given herd of swine might appear somewhat as summarized in Table 6-5. Naturally, the same procedure can be applied to each of the traits listed in Table 6-4.

TABLE 6-5
ESTIMATING RATE OF PROGRESS IN SWINE

Economically Important Characters	Average of Herd	Selected Individuals for Replacements	Average Selection Advantage	Heritability Percent	Expected Performance Next Generation
1. No. pigs born alive	7	12	5	10	7.5
2. No. pigs weaned	6	10	4	12	6.48
3. 21-day litter weight	180	400	220	15	213
4. Average daily gain	1.2	1.6	0.4	30	1.32
5. Lb feed/lb gain	450	375	75	30	427.5
6. Conformation score[1] . . .	3	7	4	29	4.16

[1]The type grades used herein are as follows: Excellent = 9, Good = 7, Medium = 5, Fair = 3, and Inferior = 0. Naturally, the principle herewith illustrated may be applied to any measurable system of grading.

FACTORS INFLUENCING RATE OF PROGRESS

Swine producers need to be informed relative to the factors which influence the rate of progress that can be made through selection. They are:

1. The heritability of the character. When heritability is high, much of that which is selected for will appear in the next generation, and marked improvement will be evident.

2. The number of characters selected for at the same time. The greater the number of characters selected for at the same time, the slower the progress in each. In other words, greater progress can be attained in one character if it alone is selected for. For example, if selection of equal intensity is practiced for four independent traits, the progress in any one will be only one-half of that which would occur if only one trait were considered; whereas selection for nine traits will reduce the progress in any one to one-third. This emphasizes the importance of limiting the traits in selection to those which have greatest importance as determined by economic value and heritability. At the same time, it is recognized that it is rarely possible to select for one trait only, and that income is usually dependent upon several traits.

3. The genotypic and phenotypic correlation between traits. The effectiveness of selection is lessened by (a) negative correlation between two desirable traits, or (b) positive correlation of desirable with undesirable traits.

4. The amount of heritable variation measured in such specific units as pounds, inches, numbers, etc. If the amount of heritable variation—measured in such specific units as pounds, inches, or numbers—is small, the animals selected cannot vary much above the average of the entire herd, and progress will be slow. For example, there is much less spread, in pounds, in the birth weights of pigs than in the 154-day weights (usually there is less than 2 lb spread in weights at birth, whereas a spread of 30 to 40 lb is common at 154 days of age). Therefore, more marked progress in selection can be made in the older weights than in birth weights of pigs, when measurements at each stage are in pounds.

5. The accuracy of records and adherence to an ideal. It is a well established fact that a breeder who maintains accurate records and selects consistently toward a certain ideal or goal can make more rapid progress than one whose records are inaccurate and whose ideals change with fads and fancies.

6. The number of available animals. The greater the number of animals available from which to select, the greater the progress that can be made. In other words, for maximum progress, enough animals must be born and raised to permit rigid culling. For this reason, more rapid progress can be made with swine than with animals that have only one offspring, and more rapid progress can be made when a herd is either being maintained at the same numbers or reduced than when it is being increased in size.

7. The age at which selection is made. Progress is more rapid if selection is practiced at an early age. This is so because more of the productive life is ahead of the animal, and the opportunity for gain is then greatest.

8. The length of generation. Generation interval refers to the period of time required for parents to be succeeded by their offspring, from the standpoint of reproduction. The minimum generation interval of farm animals is about as follows: horses, 4 years; cattle, 3 years; sheep, 2 years; and swine, 1 year. By way of comparison, the average length of a human generation is 33 years.

Shorter generation lengths will result in greater progress per year, provided the same proportion of animals is retained after selection.

Usually it is possible to reduce the length of the

generation of sires, but it is not considered practical to reduce materially the length of the generation of females. Thus, if progress is being made, the best young males should be superior to their sires. Then the advantage of this superiority can be gained by changing to new generations as quickly as possible. To this end, it is recommended that the breeder change to younger sires whenever their records equal or excel those of the older sires. In considering this procedure, it should be recognized, however, that it is very difficult to compare records made in different years or at different ages.

9. The caliber of the sires. Since a much smaller proportion of males than of females is normally saved for replacements, it follows that selection among the males can be more rigorous and that most of the genetic progress in a herd will be made from selection of males. Thus, if 2% of the males and 50% of the females in a given herd become parents, then about 75% of the hereditary gain from selection will result from the selection of males and 25% from the selection of females, provided their generation lengths are equal. If the generation lengths of males are shorter than the generation lengths of females, the proportion of hereditary gain due to the selection of males will be even greater.

DETERMINING RELATIVE EMPHASIS TO PLACE ON EACH TRAIT

A replacement animal seldom excels in all of the economically important characters. The producer must decide, therefore, how much importance shall be given to each factor. Thus, the swine producer will have to decide how much emphasis shall be placed on litter size, litter survival, rate of gain, efficiency of feed utilization, and carcass characteristics.

Perhaps the relative emphasis to place on each character should vary according to the circumstances. Under certain conditions, some characters may even be ignored. Among the factors which determine the emphasis to be placed on each character are the following:

1. The economic importance of the character to the producer. Table 6-4 lists the economically important characters in swine, and summarizes (see comments column) their importance to the producer.

By economic importance is meant their dollars and cents value. Thus, those characters which have the greatest effect on profits should receive the most attention.

2. The heritability of the character. It stands to reason that the more highly heritable characters should receive higher priority than those which are less heritable, for more progress can be made thereby.

3. The amount of variation in each character.

Obviously, if all animals were exactly alike in a given character, there could be no selection for that character. Likewise, if the amount of variation in a given character is small, the selected animals cannot be very much above the average of the entire herd, and progress will be slow.

4. The level of performance already attained. If a herd has reached a satisfactory level of performance for a certain character, there is not much need for further selection for that character.

5. The genetic correlation between traits. One trait may be so strongly correlated with another that selection for one automatically selects for the other. For example, rate of gain and economy of gain are correlated to the extent that selection for rate of gain tends to select for the most economical gains as well. Conversely, one trait may be negatively correlated with another so that selection for one automatically selects against the other.

SYSTEMS OF SELECTION

Hand in hand with the breeding system and production testing, the swine producer needs to follow a system of selection which will result in maximum total progress over a period of several years or animal generations.

Among the several selection systems used by swine producers are the following (each of which will be detailed): Selection Based on Tandem, Minimum Standards, and Indexes; Selection Based on STAGES, EBV, EPD, and BLUP; and/or Selection Based on Genetically-free Porcine Stress Syndrome (PSS).

SELECTION BASED ON TANDEM, MINIMUM STANDARDS, AND INDEXES

1. Tandem selection. This refers to that system in which there is selection for only one trait at a time until the desired improvement in that particular trait is reached, following which selection is made for another trait, etc. This system makes it possible to make rapid improvement in the trait for which selection is being practiced, but it has two major disadvantages: (a) usually it is not possible to select for one trait only, and (b) generally income is dependent on several traits.

Tandem selection is recommended only in those rare herds where one character only is primarily in need of improvement; for example, where a certain herd of swine needs improving primarily in litter size.

2. Establishing minimum standards for each character, and selecting simultaneously but independently for each character. This system, in which several of the most important characters are selected for simultaneously, is without doubt the most common system

of selection. It involves establishing minimum standards for each character and culling animals which fall below these standards. For example, it might be decided to cull all pigs in litters of fewer than seven pigs, or weighing less than 40 lb at weaning, or gaining less than 2 lb per day from weaning to 230 lb. Of course, the minimum standards may have to vary from year to year if environmental factors change markedly (for example, if pigs average light at weaning time due to a disease).

The chief weakness of this system is that an individual may be culled because of being faulty in one character only, even though he is well nigh ideal otherwise.

3. Selection index. Selection indexes combine all important traits into one overall value or index. Theoretically, a selection index provides a more desirable way in which to select for several traits than either (a) the tandem method, or (b) the method of establishing minimum standards for each character and selecting simultaneously but independently for each character.

Selection indexes are designed to accomplish the following:

a. To give emphasis to the different traits in keeping with their relative importance.

b. To balance the strong points against the weak points of each animal.

c. To obtain an over-all total score for each animal, following which all animals can be ranked from best to poorest.

d. To assure a constant and objective degree of emphasis on each trait being considered, without any shifting of ideals from year to year.

e. To provide a convenient way in which to correct for environmental effects, such as feeding differences, etc.

Selection indexes have been devised by the National Swine Improvement Federation and are given in the Guidelines for Uniform Swine Improvement Programs, published by the U.S. Department of Agriculture.

SELECTION BASED ON STAGES, EBV, EPD, and BLUP

STAGES stands for Swine Testing and Genetic Evaluation System. EBV stands for Estimated Breeding Value. EPD stands for Expected Progeny Difference. BLUP stands for Best Linear Unbiased Prediction. These systems of selection are covered in Chapter 4 under the heading "Selection Based on STAGES, EBV, EPD, and BLUP"; hence, the reader is referred thereto.

SELECTION BASED ON GENETICALLY-FREE PORCINE STRESS SYNDROME (PSS)[9]

The porcine stress syndrome (PSS) is characterized by the sudden death of heavily muscled pigs when stressed and/or the production of pale, soft, exudative (PSE) musculature of their carcasses. PSS is inherited as an autosomal recessive. Birth of a PSS pig incriminates both parents as carriers of the recessive gene (heterozygous), and possibly having two copies of the recessive gene (homozygous).

PSS animals may appear shorter-bodied and smaller than their normal herd mates. Also, they may be more muscular in appearance than normal animals, with groove-shaped loins, indentations in their rumps, and circular shape to the hams. A separation between the major muscles of the hams is often evident. Dilation of the pupils and tremor of the tail are often observed following exposure to physical stress. Caution: Not all heavily-muscled pigs are affected by PSS.

The homozygous mutant form (NN) of the stress gene can be devastating. If the physical stress doesn't cause death, it will likely cause pale, soft, and exudative (PSE) pork under the best of handling and processing conditions. Two questions remain: (1) Will the presence of the normal gene prevent PSE meat? (2) Is the PSS carrier or heterozygous form (Nn) intermediate to the two homozygotes (nn and NN) for lean composition?

Several reports from Iowa State University and one from Denmark indicate that carriers are 1.5 to 4% higher than normal pigs in lean percentage, and that carriers are equal, if not superior, to normal pigs in growth rate. All of these studies, however, indicate that the muscle quality of carrier animals is generally undesirable and that the frequency of PSE is in the 30–50% range.

The National Genetic Evaluation Program (NGEP) initiated in 1990, and completed in 1995, was the largest and most comprehensive swine industry program ever undertaken. Over 500 people participated in the program.

Pigs began the test at 65 lb and finished it at 250 lb. They were evaluated for feed intake, growth, and leg straightness.

Carcass backfat measurements were made. Loin muscle samples were evaluated for the meat quality traits of color, marbling, firmness, pH, water holding capacity, drip loss, and the eating quality traits of tenderness, cooking loss, juiciness, chewiness, and cooked moisture content.

All test pigs were classified for halothane (stress) genotype by the DNA probe test. About 12% of the

---

[9]Adapted by the authors from the *Special Report, National Genetic Evaluation Program*, by *National Hog Farmer*, June 1, 1995.

test pigs were carriers (heterozygotes, Nn) of the halothane gene. The Genetics Program Committee estimated that about 10% of the dams of all test pigs were carriers of the halothane gene.

■ Impact of the Stress Gene—The National Genetic Evaluation Program reported the following impact of the stress gene:

1. In the homozygous mutant form (nn), the stress gene can have some devastating effects. If physical stress doesn't cause death, it is almost sure to cause pale, soft, and exudative (PSE) pork.

2. The presence of the normal allele (N) will usually prevent stress death and the rigidity response to the anesthetic, halothane.

3. All barrows and gilts were classified by their stress genotype, using the Halothane DNA probe test. Of the 3,261 pigs finishing the test, 7 were mutant pigs (nn), 391 were mutant-carrier pigs (Nn), and 2,863 were normal (NN).

A comparison of the 2,863 normal (NN) pigs and the 391 carriers (Nn) revealed the following:

1. Growth rate, leg soundness, and backfat thickness were about the same.

2. The 391 stress gene carrier group had a 0.29 sq in. advantage in loin muscle area, a 0.4% higher dressing percentage, and slightly more lean gain. But these pigs produced paler and less tender meat which possessed less intramuscular fat and greater drip loss. Note: A 5% shrink per carcass, a conservative estimate for a PSS carcass, results in approximately $10.00/pig loss. Based on this study, it is clearly evident that the gene should be eliminated from the U.S. pig population.

3. Even without the PSS stress gene, 22.5% of the 2,863 loins were disqualified because of lack of quality. So the PSS gene doesn't account for all of the industry's muscle quality problems. Therefore, a program to monitor pork quality, and to select for the improvement of quality traits is needed.

Prior to the relatively new DNA probe test,[10] known as the HAL-1843 DNA test, used in the National Genetic Evaluation Program, now being licensed, the following two tests were used to evaluate pigs objectively for PSS, and they are still used:

1. The creatine phosphokinase (CPK) test—This involves catching a small drop of blood from the pig's ear on a special card and sending the card to a chemical laboratory where it is analyzed for the activity of creatine phosphokinase, a serum enzyme that is abnormally high in PSS swine.

2. The Halothane anesthesia test—In this test, the pig is put to sleep using the anesthetic Halothane. PSS animals respond to the Halothane anesthesia by showing signs of extreme muscle rigidity within five minutes from the start of the treatment. This test provides immediate results, but the equipment is expensive and must be used under the direction of a trained technician.

---

[10]Peter O'Brien, University of Guelph, and David MacLennan, University of Toronto, developed the DNA probe, which is 100% accurate, barring human error.

QUESTIONS FOR STUDY AND DISCUSSION

1. Why is swine breeding more flexible in the hands of humans than cattle and sheep breeding?

2. What unique circumstances surrounded the founding of genetics by Mendel?

3. Explain the relationships between chromosomes, genes, and DNA.

4. Under what conditions might a theoretically completely homozygous state in hogs be undesirable and unfortunate?

5. Give two examples each of (a) dominance, and (b) recessive phenomena in swine.

6. In order to make intelligent selections and breed progress, how important is the environment?

7. What did the Washington State University longtime swine experiment show relative to selection and environment?

8. Explain how sex is determined. Is it possible to alter the sex ratio?

9. When abnormal pigs are born, what conditions tend to indicate each: (a) a hereditary defect, or (b) a nutritional deficiency?

10. The "sire is half the herd"! Is this an understatement or an overstatement?

11. Define (a) prepotency, (b) nicking, (c) outcrossing, and (d) grading up.

12. What system of breeding do you consider to be best adapted to your herd, or to a herd with which you are familiar? Justify your choice.

13. Define and explain heterosis or hybrid vigor.

14. Discuss the advantages and the disadvantages of crossbreeding swine.

15. Which of the following crossbreeding systems

would you recommend: Two-breed rotational cross, three-breed rotational cross, or terminal cross? Justify your choice.

16. Wherein are the following production tests (a) alike, and (b) different: (1) central testing stations, (2) on-farm testing, and (3) swine registry testing?

17. Why is pork quality important? Describe high quality pork.

18. What is lean pork gain? How is it determined?

19. Why is it important that a modern swine producer keep good records?

20. Based on (a) heritability, and (b) dollars and cents value, what characteristics should receive greatest emphasis in a swine production testing program?

21. What is the ratio of a boar that has an average daily gain of 2.3 lb per day where the test group average is 2.1 lb per day?

22. List and discuss some factors influencing the rate of progress that can be made through selection.

23. What system of selection—(a) tandem, minimum standards, and indexes; (b) STAGES, EBV, EPD, and BLUP; and/or (c) genetically-free Porcine Stress Syndrome (PSS) would you recommend and why?

24. Is the Porcine Stress Syndrome (PSS) good or bad? Justify your answer.

SELECTED REFERENCES

Title of Publication	Author(s)	Publisher
Animal Breeding	L. M. Winters	John Wiley & Sons, Inc., New York, NY, 1954
Animal Breeding Plans	J. L. Lush	Collegiate Press, Inc., Ames, IA, 1963
Animal Science	M. E. Ensminger	Interstate Publishers Inc., Danville, IL, 1991
Breeding and Improvement of Farm Animals	E. M. Warwick J. E. Legates	McGraw-Hill Book Co., New York NY, 1979
Developmental Anatomy	L. B. Arey	W. B. Saunders Co., Philadelphia, PA, 1940
Embryology of the Pig, The	B. M. Patten	P. Blakiston's Son & Co., Inc., Philadelphia, PA, 1931
Fifty Years of Progress in Swine Breeding	W. A. Craft	Journal of Animal Science, Vol. 17 No. 4, November, 1958
Genetic Basis of Selection, The	I. M. Lerner	John Wiley & Sons, Inc., New York, NY, 1958
Genetic Evaluation	Bob Uphoff, Chairman of NPPC Genetic Program Committee	National Pork Producers Council, 1995
Genetics of Livestock Improvement	J. F. Lasley	Prentice-Hall, Inc., Englewood Cliffs, NJ, 1978
Hammond's Farm Animals	J. Hammond, Jr. I. L. Mason T. J. Robinson	Edward Arnold & Co., London, England, 1971
Improvement of Livestock	R. Bogart	The Macmillan Company, New York, NY, 1959
Livestock Improvement	J. E. Nichols	Oliver and Boyd, Tweeddale Court, Edinburgh, 1957
Modern Developments in Animal Breeding	I. M. Lerner H. P. Donald	Academic Press, London and New York, 1966
Pork Industry Handbook		Cooperative Extension Service, Purdue University, West Lafayette, IN
Tables for Coefficients of Inbreeding in Animals	C. D. Mueller	Agri. Exp. Sta. Tech. Bul. 80, Kansas State University, Manhattan, KS

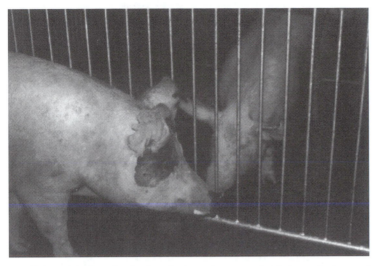

Estrous detection, showing important nose-to-nose contact between the boar and the sow. (Courtesy, Dr. D. G. Levis, University of Nebraska, Lincoln, NE)

7

REPRODUCTION IN SWINE

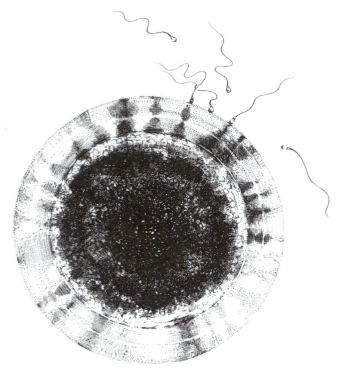

Fig. 7-1. The ultimate objective of swine breeding is the union of an egg or ovum from the female and a sperm from the male, which transmit to the offspring all the inheritance from each parent. Each egg contains 19 half pairs (haploid numbers) of chromosomes as does each sperm; hence, the offspring will have 38 pairs (diploid number) of chromosomes. The sperm and egg are drawn to scale.

Swine producers have many reproductive problems, a reduction of which calls for a full understanding of reproductive physiology and the application of scientific practices therein. In fact, it may be said that reproduction is the first and most important requisite of swine breeding, for if animals fail to reproduce the breeder is soon out of business.

Many outstanding individuals are disappointments because they are either sterile or reproduce poorly. From 5 to 30% of the fertilized eggs do not develop normally, resulting in embryonic mortality or death; and from 10 to 20% of the live pigs farrowed die within the first 7 to 10 days of life. The subject of physiology of reproduction is, therefore, of great importance.

REPRODUCTIVE ORGANS OF THE BOAR

The boar's functions in reproduction are: (1) to produce the male reproductive cells the *sperm* or *spermatozoa*, and (2) to introduce sperm into the female reproductive tract at the proper time. In order that these functions may be fulfilled, swine breeders should have a clear understanding of the anatomy of the reproductive system of the boar and of the functions of each of its parts. Fig. 7-2 is a schematic drawing of the reproductive organs of the boar. A description of each part follows:

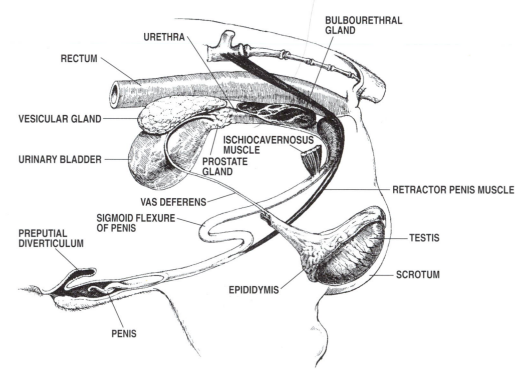

Fig. 7-2. Diagram of the reproductive organs of the boar, showing their location in the body. The testis, its ducts, the vesicular gland, and the bulbourethral gland are paired but, for simplicity, only those on the left side have been reproduced.

1. **Testes (testicles).** The primary function of the testes (singular testis) is to produce sperm. They are enclosed in the scrotum, a diverticulum of the abdomen.

The chief function of the scrotum is thermoregulatory; to maintain the testes at temperatures several degrees lower than that of the body proper.

Cryptorchids are males one or both of whose testes have not descended to the scrotum. The undescended testicle(s) is usually sterile because of the high temperature in the abdomen.

The testes communicate through the inguinal canal with the pelvic cavity, where accessory organs and glands are located. A weakness of the inguinal canal, which is heritable in swine, sometimes allows part of the viscera to pass out into the scrotum—a condition called *scrotal hernia.*

The following structures within the testes (see Fig. 7-3) contribute to their function:

a. **Seminiferous tubule.** This is the germinal portion of the testis, in which are situated the spermatogonia (sperm-producing cells). If laid end to end, it has been estimated that the seminiferous tubules of one testicle of the boar would be nearly 2 miles long. Each gram of testis is capable of producing about 27 million sperm each day.

Around and between the seminiferous tubules are the *interstitial (Leydig) cells,* which produce the male sex hormone, *testosterone.* Testosterone is necessary for the mature development and continued function of the reproductive organs, for the development of the male secondary sexual characteristics, and for sexual drive.

b. **Rete testis.** The rete testis is formed from the union of several seminiferous tubules.

c. **Vasa efferentia (efferent ducts).** The efferent ducts carry the sperm cells from the rete testis to the head of the epididymis. Also, it is thought that their secretions are vital to the nutrition and maturing of the sperm cells.

2. **Epididymis.** The efferent ducts of each testis unite into one duct, thus forming the epididymis. This greatly coiled tube consists of three parts.

a. **The head.** Which includes several tubules that are grouped into lobules.

b. **The body.** The part of the epididymis which passes down along the sides of the testis.

c. **The tail.** The part located at the bottom of the testis.

The epididymis has four functions; namely, (a) as a passageway for sperm from the seminiferous tubules, (b) the storage of sperm, (c) the maturation of the sperm, and (d) concentration of sperm.

3. **Vas deferens (Ductus deferens).** This slender tube, which is lined with ciliated cells, leads from the tail of the epididymis to the pelvic part of the urethra. Its primary function is to move sperm into the urethra at the time of ejaculation.

The vas deferens—together with the longitudinal strands of smooth muscle, blood vessels, and nerves; all encased in a fibrous sheath—make up the spermatic cord (two of them) which pass up through an opening in the abdominal wall, the inguinal canal, into the pelvic cavity.

The cutting or closing off of the ductus deferens, known as *vasectomy,* is the most usual operation performed to produce sterility, where sterility without castration is desired.

4. **Vesicular glands.** This pair of glands flanks the vas deferens near its point of termination. They are the largest of the accessory glands of reproduction in the male, and are located in the pelvic cavity.

The vesicular glands secrete a fluid which provides a medium of transport, energy substrates, and buffers for the spermatozoa. They are not a store house for sperm.

5. **Prostate gland.** This is located at the neck of the bladder, surrounding or nearly surrounding the urethra and ventral to the rectum. The prostate gland contributes fluid and salts (inorganic ions) to semen. It provides bulk and a suitable medium for the transport of sperm.

6. **Bulbourethral gland (Cowper's gland).** These two glands, which often reach a diameter of 1.5 in. in the boar, are located on either side of the urethra in the pelvic region. They communicate with the urethra by means of a small duct.

These glands produce the prominent gel-like component of boar semen, which forms a plug in the vagina of the female.

7. **Urethra.** This is a long tube which extends from the bladder to the end of the penis. The vas deferens

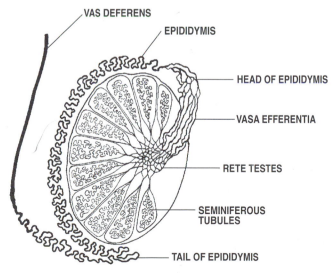

VAS DEFERENS

EPIDIDYMIS

HEAD OF EPIDIDYMIS

VASA EFFERENTIA

RETE TESTES

SEMINIFEROUS TUBULES

TAIL OF EPIDIDYMIS

Fig. 7-3. A schematic drawing of the structures within the testis (sagittal section).

and vesicular glands open to the urethra close to its point of origin.

The urethra serves for the passage of both urine and semen.

8. **Penis.** This is the boar's organ of copulation. It is composed essentially of fibrous tissue. At the time of erection, cavernous spaces in the penis become engorged with blood.

In total, the reproductive organs of the boar are designed to produce semen and to convey it to the female at the time of mating. The semen consists of two parts; namely (1) the sperm which are produced by the testes, and (2) the liquid portion, or seminal plasma, which is secreted by the seminiferous tubules, the epididymis, the vas deferens, the vesicular glands, the prostate, and the bulbourethral glands. Actually, the sperm make up only a small portion of the ejaculate. On the average, at the time of each service, a boar ejaculates 150 to 250 ml, with a range from less than 50 to well over 500 ml.

REPRODUCTIVE ORGANS OF THE SOW

The sow's functions in reproduction are: (1) to produce the female reproductive cells, the *eggs* or *ova*, (2) to develop the new individual, the *embryo*, in the uterus, (3) to expel the fully developed young at the time of *birth* or *parturition*, and (4) to produce milk for the nourishment of the young. Actually, the part played by the sow in the generative process is much more complicated than that of the boar. It is imperative, therefore, that modern swine producers have a full understanding of the anatomy of the reproductive organs of the sow and the functions of each part. Figs. 7-4 and 7-5 show the reproductive organs of the sow: and a description of each part follows:

1. **Ovaries.** The two irregular-shaped ovaries of the sow are suspended in the abdominal cavity near the backbone and just in front of the pelvis.

The ovaries have three functions: (a) to produce the female reproductive cells, the *eggs* or *ova*, (b) to secrete the female sex hormone, *estrogen*, and (c) to form the *corpora lutea*, which secrete the hormone progesterone. The ovaries may alternate somewhat irregularly in the performance of these functions.

The ovaries differ from the testes in that eggs are produced in very limited numbers and at intervals, during or shortly after heat. Each miniature egg is contained in a *follicle*, which surrounds the egg with numerous small cells. Large numbers of follicles are scattered throughout the ovary. Generally, the follicles

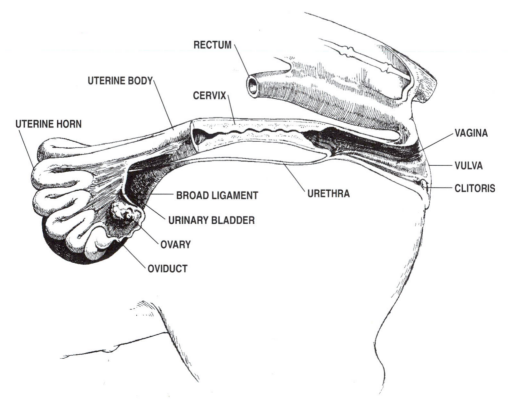

Fig. 7-4. The reproductive organs of the sow, showing their location in the body.

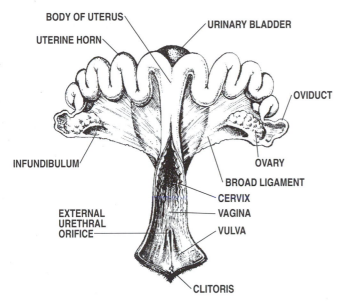

Fig. 7-5. The reproductive organs of the sow, as viewed from above. The vagina and cervix are slit open.

remain in an unchanged state until the advent of puberty, at which time some of them begin to enlarge through an increase in the follicular liquid within. Toward the end of heat (estrus), the follicles (which at maturity measure about 0.33 in. in diameter) rupture and discharge the egg. This process is known as *ovulation*. As soon as the eggs are released the corpora lutea form at the site from which the eggs were released. The corpora lutea secrete a hormone called *progesterone*, which (a) acts on the uterus so that it implants and nourishes the embryo, (b) prevents other eggs from maturing and keeps the animal from coming in heat during pregnancy, (c) maintains the animal in a pregnant condition, and (d) assists estrogen and other hormones in the development of the mammary glands. If the eggs are not fertilized, however, the corpora lutea atrophy and allow new follicles to ripen and a new heat period to begin, Occasionally, cystic ovaries develop, thus inducing temporary sterility. Actually, there are three different types of disturbed conditions which are commonly called cystic ovaries; namely (a) cystic follicles, (b) cystic corpus luteum, and (c) persistent corpus luteum. All three conditions prevent normal ovulation and constitute major causes of sterility in sows.

The egg-containing follicles also secrete into the blood the female sex hormone, *estrogen*. Estrogen is necessary for the development of the female reproductive system, for the mating behavior or *heat* of the female, for the development of the mammary glands, and for the development of the secondary sex characteristics, or femininity, in the sow.

From the standpoint of the practical hog breeder,

the ripening of the first Graafian follicle in a gilt generally coincides with puberty, and this marks the beginning of reproduction.

2. **Oviducts (fallopian tubes).** These small, cilia-lined tubes or ducts lead from the ovaries to the horns of the uterus. They are about 10 in. long in the sow and the end of each tube nearest the ovary, called *infundibulum*, flares out like a funnel. They are not attached to the ovaries but lie so close to them that they seldom fail to catch the released eggs.

At ovulation, the eggs pass into the infundibulum where, within a few minutes, the ciliary movement within the tube, assisted by the muscular movements of the tube itself, carries them down into the oviduct. If mating has taken place, the union of the sperm and eggs usually takes place in the upper third of the oviduct. Thence, the fertilized eggs move into the uterine horn. All this movement from the ovary to the uterine horn takes place in 3 to 4 days.

3. **Uterus.** The uterus is the muscular sac, connecting the fallopian tubes and the vagina, in which the fertilized eggs attach themselves and develop until expelled from the body of the sow at the time of parturition. The uterus consists of the two horns, the body, and the neck or cervix. In the sow, the horns are about 4 to 5 ft long, the body about 2 in. long, and the cervix about 6 in. long.

In swine, the fetal membranes that surround the developing embryo are in contact with the entire lining of the uterus; and there are no *buttons* or *cotyledons* as in the cow and ewe.

4. **Cervix.** Although it is technically part of the uterus, the cervix is often discussed as a distinct organ. It is a thick-walled, inelastic structure about 6 in. long. The canal of the cervix communicating between the body of the uterus and the vagina is funnel-shaped, with ridges in the canal having a corkscrew configuration which conforms to that of the end of the boar's penis. The main function of the cervix is to prevent microbial contamination of the uterus. In swine, semen is deposited directly into the cervix during natural mating. When birth occurs, the cervix must dilate to allow passage of the piglets.

5. **Vagina.** The vagina is the canal which admits the penis of the boar at the time of service. At the time of birth, it expands and serves as the final passageway for the fetus.

6. **Vulva (or urogenital sinus).** The vulva is the external opening of both the urinary and genital tracts. It is about 3 in. in length. During sexual excitement, the vulva of the sow becomes swollen, thus producing one of the signs of heat.

MATING

Mating is a prolonged process in swine, varying

from 3 to 20 minutes in which waves of high and low sperm concentration exist in the flow of the ejaculate. For this reason, it is important that copulation take place without disturbance.

An ejaculation consists of the following three phases:

1. The first, or pre-sperm phase, which lasts 1 to 5 minutes, consists of a watery fluid in which there are tapioca-like pellets but no sperm, and comprises 5 to 20% of the ejaculate.

2. The second, or sperm-containing phase, which lasts 2 to 5 minutes, consists of a whitish, uniform fluid which contains the sperm, and comprises 30 to 50% of the ejaculate.

3. The last phase, which lasts 3 to 8 minutes, contains very few sperm, helps form a gelatinous plug in the female tract, and comprises 40 to 60% of the total volume.

Fig. 7-6. The white breeds—Yorkshire, Large White, Landrace, and Chester White—excel in maternal characteristics. (Courtesy, Land O Lakes, Ft. Dodge, IA)

BREEDING METHODS

Three breeding methods are used in swine: (1) hand-mating, (2) pen-mating, and (3) artificial insemination. Regardless of the method of breeding, the all-important factor in achieving a high conception rate and good litter size is to get sperm into the female's reproductive tract at the right time when pregnancy rate and litter size will be maximized.

Hand-mating is practiced more in swine than with either cattle or sheep. It lends itself to better record keeping, since (1) exact breeding dates are known, (2) there is assurance that each female is bred, and (3) more is known about the breeding performance of the boar.

Hand-mating is usually practiced by purebred swine breeders who must be able to identify the sire of each litter.

Pen-mating is usually practiced by smaller commercial swine producers. They may (1) split the sow herd so as to have 1 young boar per group of 10 to 12 females, or (2) alternate boars in the sow herd—that is, use one boar or set of boars one day and another boar or set of boars the next day. Pen-mating requires less labor and lower cost facilities than hand-mating, but the resulting farrowing rate may be lower than for sows that are individually mated.

Hand-mating or AI are commonly used by industrialized swine operations, with AI increasing.

NORMAL REPRODUCTIVE TRAITS OF FEMALE SWINE

The pig lends itself very well to experimental study in confined conditions. It is reasonable to expect,

therefore, that we should have a considerable store of knowledge relative to the normal breeding habits of swine, perhaps more than we have of any other class of farm animals. Fig. 7-7 shows the average reproductive traits of female swine.

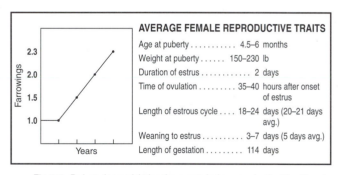

AVERAGE FEMALE REPRODUCTIVE TRAITS

Age at puberty	4.5–6 months
Weight at puberty	150–230 lb
Duration of estrus	2 days
Time of ovulation	35–40 hours after onset of estrus
Length of estrous cycle	18–24 days (20–21 days avg.)
Weaning to estrus	3–7 days (5 days avg.)
Length of gestation	114 days

Fig. 7-7. By knowing and timing the events in the reproductive life of female swine, producers increase the number of farrowings per sow per year.

AGE OF PUBERTY

The age of puberty in swine varies from 4.5 to 6 months. This rather wide range is due to difference in breeds and strains, sex, and environment—especially nutrition. In general, boars do not reach puberty quite so early as gilts.

AGE TO BREED GILTS

Reasonably early breeding has the advantages of establishing regular and reliable breeding habits and reducing the cost of the pigs at birth. Although gilts may come into heat at 4 or 5 months of age, the general recommendation is to breed at the third heat, primarily to take advantage of any increase in ovulation rate. Therefore, gilts are usually bred at 7 to 8 months of age and farrow at 11 to 12 months of age.

Mixing pens of confinement-reared gilts and then regrouping them in direct boar contact is likely to start heat periods earlier. Also this may help synchronize the first heat and the second heat.

HEAT PERIOD (Estrus)

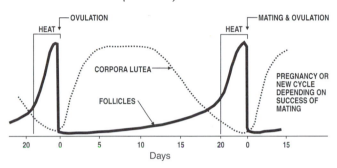

Fig. 7-8. The estrous cycle of the sow—follicles form, ovulate, corpora lutea form, regress, and more follicles form. Unless mating and conception occur, the cycle is normally repeated about every 20–21 days. Progesterone is secreted by the corpora lutea while the follicles secrete estrogen. In the event of pregnancy, the corpora lutea do not regress but remain functional throughout the period of gestation. Ultimately the control of the estrous cycle resides with the hypothalamus and the pituitary in the brain.

The heat period or estrus is the time during which the sow will accept the boar. It lasts from 1 to 5 days, with an average of 2 days. Older sows generally remain in heat longer than gilts.

Ovulation occurs 38 to 42 hours after the start of estrus. Live sperm must be in the female reproductive tract a few hours before ovulation occurs; otherwise, litter size will be reduced. Optimal breeding is based on the number of times per day that a producer checks the females for standing estrus. With once-a-day detection, females should be bred each day they will accept the boar. With twice-a-day detection, females should be bred at 12 and 24 hours after they are first detected in heat. Gilts will sometimes have heat periods less than 2 days long and may have to be bred as soon as they are detected in heat and then each succeeding 12 hours they will stand for the boar. Heat detection should always be accomplished in the presence of a boar, since his presence maximizes the chances of detecting all possible females in heat.

When the heat period lasts longer than 3 days, continued breeding is likely a waste of boar power since conception is doubtful. If not bred, the heat period normally recurs at intervals of 16 to 25 days, with an average of 20–21 days.

The external signs of heat in the sow are restless activity, swelling and pink-red coloring of the vulva, frequent mounting of other sows, frequent urination, and occasional loud grunting. All signs are not always present.

OVULATION RATE

The ovulation rate—the number of eggs (ova) released—is associated with genetic background, age at breeding, weight at breeding, and nutrition. Cross-breeding increases ovulation rate. In gilts, the ovulation rate may increase 1 or 2 eggs with each successive heat period. Following the first litter, the ovulation rate increases until about the fifth or sixth litter when the rate plateaus. Increasing the energy level of the diet of gilts increases the ovulation rate.

FERTILIZATION

It is believed that the eggs are usually liberated about 40 hours after the onset of estrus (standing heat). During standing heat, the boar mates the female. The sperm (or male germ cells) are deposited in the cervix and uterus at the time of service and from there ascend the female reproductive tract. Under favorable conditions, they meet the eggs, and one of them fertilizes each egg in the upper part of the oviduct near the ovary.

A series of delicate time relationships must be met, however, or the eggs will not be fertilized. The sperm cells live only 24 to 48 hours in the reproductive tract of the female. Moreover, the eggs are viable for an even shorter period of time, between 8 and 10 hours after ovulation. For conception, therefore, breeding must take place at the right time. Furthermore, when aged sperm or eggs are involved in fertilization, developmental abnormalities may result; hence. timing is critical.

GESTATION PERIOD

The average gestation period of sows is 114 days, though extremes of 98 to 124 days have been reported. An easy way to remember the gestation period is that it is 3 months, 3 weeks, and 3 days. Table 7-1 provides approximate due dates for pigs when the breeding date is known.

TABLE 7-1
GESTATION TABLE BASED ON 114 DAYS

Date Bred	Date Due	Date Bred	Date Due
Jan. 1	April 25	July 5	Oct. 27
Jan. 6	April 30	July 10	Nov. 1
Jan. 11	May 5	July 15	Nov. 6
Jan. 16	May 10	July 20	Nov. 11
Jan. 21	May 15	July 25	Nov. 16
Jan. 26	May 20	July 30	Nov. 21
Jan. 31	May 25	Aug. 4	Nov. 26
Feb. 5	May 30	Aug. 9	Nov. 31
Feb. 10	June 4	Aug. 14	Dec. 6
Feb. 15	June 9	Aug. 19	Dec. 11
Feb. 20	June 14	Aug. 24	Dec. 16
Feb. 25	June 19	Aug. 29	Dec. 21
Mar. 2	June 24	Sept. 3	Dec. 26
Mar. 7	June 29	Sept. 8	Dec. 31
Mar. 12	July 4	Sept. 13	Jan. 5
Mar. 17	July 9	Sept. 18	Jan. 10
Mar. 22	July 14	Sept. 23	Jan. 15
Mar. 27	July 19	Sept. 28	Jan. 20
April 1	July 24	Oct. 3	Jan. 25
April 6	July 29	Oct. 8	Jan. 30
April 11	Aug. 3	Oct. 13	Feb. 4
April 16	Aug. 8	Oct. 18	Feb. 9
April 21	Aug. 13	Oct. 23	Feb. 14
April 26	Aug. 18	Oct. 28	Feb. 19
May 1	Aug. 23	Nov. 2	Feb. 24
May 6	Aug. 28	Nov. 7	Mar. 1
May 11	Sept. 2	Nov. 12	Mar. 6
May 16	Sept. 7	Nov. 17	Mar. 11
May 21	Sept. 12	Nov. 22	Mar. 16
May 26	Sept. 17	Nov. 27	Mar. 21
May 31	Sept. 22	Dec. 2	Mar. 26
June 5	Sept. 27	Dec. 7	Mar. 31
June 10	Oct. 2	Dec. 12	April 5
June 15	Oct. 7	Dec. 17	April 10
June 20	Oct. 12	Dec. 22	April 15
June 25	Oct. 17	Dec. 27	April 20
June 30	Oct. 22		

BREEDING AFTER FARROWING

Sows will often come in heat during the first few days after farrowing, but they rarely conceive if bred at this time—for the reason that they fail to ovulate.

The trend of large commercial operators is to wean at 3 weeks of age, with weaning to estrus of the sow following in 3 to 7 days, with an average of 5 days. So, if pigs are weaned at 21 days of age, sows may be bred 26 days after farrowing.

FERTILITY AND PROLIFICACY IN SWINE

Under domestication and conditions of good care, a high degree of fertility is desired. The cost of carrying a litter of 10 or 12 pigs to weaning time is little greater than the cost of producing a litter of only 5 or 6. In

Fig. 7-9. Prolificacy in swine is much desired, for the cost of keeping a pregnant sow and of carrying a litter of 10 or 12 pigs to weaning time is little greater than the cost of producing a litter of only 5 or 6. (Courtesy, Murphy Farms, Rose Hill, NC; photo by Dr. Garth Boyd and Rhonda Campbell)

other words, the maintenance costs on both the sow and the boar remain fairly constant. It must be remembered, however, that in the wild state high fertility may not have been characteristic of swine. Survival and natural selection were probably in the direction of smaller litters, but nature's plan has been reversed through planned matings and selection.

Low fertility in swine is most commonly attributed to hereditary and environmental factors. Maximum prolificacy depends upon having a large number of eggs shed at the time of estrus, upon adequate viable sperm present for fertilization at the proper time, and upon a minimum of embryonic and fetal mortality.

It is a well-known fact that some breeds and strains of swine are much more prolific than others. Litters of 12 or more are considered normal rather than exceptional among Chinese swine. Also, through selection, more prolific strains of swine can be developed. As shown in Table 6-4, however, the heritability of litter size at birth and at weaning is only 10 and 12%, respectively. Yet, because such gains are cumulative, they are highly worthwhile.

The number of pigs produced increases with the age of the sow. All in all, it appears that more can be accomplished through proper management and environment to increase litter size than can be done through selection.

Practical swine producers generally associate type with prolificacy. To substantiate their theory, they point out that the fat, chuffy hogs in vogue during the early part of the twentieth century were not prolific, but were prone to farrow twins and triplets. Experimental work substantiates this opinion.

Even though many eggs may be shed and fertilized, the size of the litter may be affected materially by embryonic and fetal mortality, which ranges from 5 to 30%. Exact causes of this are uncertain, but embryonic and fetal mortality have been attributed to (1) hereditary factors, perhaps recessive lethals; (2) overcrowding resulting from a large number of pigs and a consequent limited uterine surface area available for the nourishment of the individual embryos; (3) nutrition of the dam prior to and during gestation; (4) aged sperm or eggs at fertilization; (5) diseases or parasites; (6) accidents or injuries; or (7) hormone imbalances, Since some sows carry all embryos to term without loss, additional studies need to be made relative to the cause and prevention of embryonic and fetal mortality.

Certainly, the boar cannot affect the number of eggs shed, and under usual circumstances fertilization is very much an all or none phenomenon. Still, some evidence suggests that the boar can have a marked effect on litter size, primarily because embryos and fetuses sired by certain boars are less likely to survive to term. This further supports the contention that embryonic and fetal mortality are due to genetic and fertilization errors (aged sperm or eggs). In this regard, conception rate and litter size can be increased by using more than one boar on each female. Again, additional research is needed to determine the extent of the influence of the boar.

It is recommended that, in advance of the regular use, new boars should be test mated to a few gilts. During these test matings, the producer should observe the boar for aggressiveness and desire to mate, and if necessary, give the boar assistance the first service or two. Also, the producer should check the boar's ability to enter the gilt, which may be hindered by a limp, infantile, or tied penis. Most importantly, serviced gilts should become pregnant. A semen evaluation conducted by a veterinarian or qualified technician will complement the test matings. While there is no absolute test for fertility, test matings and semen evaluation can often detect a sterile boar or one of questionable fertility.

As Table 7-2 indicates, an English study revealed that infertility—failure to breed—is the chief reason given by producers for the disposal of sows; and for good reason, since with each 21-day delay the sow must produce about two extra pigs just to pay for the time and feed she consumed.

FLUSHING SOWS AND GILTS

The practice of feeding sows and gilts more liberally so that they gain in weight from 1 to 1.5 lb daily from 1 to 2 weeks before the opening of the breeding season until they are safely in pig is known as *flushing*. This is usually accomplished by increasing the concentrate ration by about 2 lb per animal per day. Formerly, the following beneficial effects were attributed to this practice: (1) more eggs are shed, and this

TABLE 7-2
ANALYSIS OF REASONS FOR SLAUGHTERING SOWS IN ENGLAND[1]

Reasons for Slaughter	Percent of Total
	(%)
Failure to breed	21.4
Piglet mortality	17.8
Old age	15.0
Low fertility	10.1
Disease	7.5
Milk failure and udder troubles	6.1
Uneven or unthrifty litters	4.6
Foot-and-mouth disease restrictions	3.2
Injury	2.6
Giving up pig breeding	2.6
Too fat or too big	2.1
Labor difficulties	2.0
Miscellaneous	5.0
Total	100.0

[1]Pomeroy, R. W., "Infertility and Neonatal Mortality in the Sow," *Journal of Agricultural Science*, Vol. 54, No. 1, 1960.

results in larger litters; (2) the sows come in heat more promptly; and (3) conception is more certain.

In modern swine production, experiments and experiences do not show any benefits from flushing second and third litter sows, probably because of the short period between weaning and mating. However, flushing does seem to be effective for gilts, especially limited-fed gilts, perhaps due to the higher energy needs of these younger animals.[1]

PREGNANCY DIAGNOSIS

Open females are expensive to maintain!

With ultrasonic detectors, pregnancy diagnosis is a reality. Producers can determine with a 90 to 95% accuracy the number of females that have settled. These detectors are most accurate when they are used between 30 and 45 days after mating. Their accuracy drops off rapidly after 45 days.

The principle of ultrasonic pregnancy detectors is an ultrasonic echo from fluid in the uterus. Uterine fluid increases rapidly following conception and reaches detectable levels 25 to 30 days after breeding. It remains detectable for 80 to 90 days after breeding,

[1]Tribble, L. F., and D. E. Orr, Jr., "Effect of Feeding Level After Weaning on Reproduction in Sows," *Journal of Animal Science*, 1982, Vol. 55, No. 3, p. 608.

following which the mass of pigs in the uterus exceeds the fluid content.

Among the several advantages of early pregnancy detection in sows and gilts are (1) it makes it possible to cull or rebreed nonpregnant, feed-wasting females; (2) it allows closer grouping of a number of sows for a farrowing period; (3) it gives early warning of breeding troubles, such as infertile boars and cystic ovaries of sows; (4) it enables the producer to make more effective use of breeding facilities and to plan more adequately for farrowing, nursing, and finishing; and (5) it makes it possible to guarantee pregnancy on females that are for sale.

MEASURING SOW PRODUCTIVITY

Because of the prolificacy and fecundity of swine, the output of sows is one of the most important economic traits used to evaluate efficiency of production.

Sow productivity is usually measured by the following economically important traits:

1. Number of pigs weaned per sow per year.
2. Number of litters per year.
3. Number of pigs born alive.
4. Litter size at weaning.
5. Nonproductive sow days.

Since these traits have low heritability, emphasis on good management and nutrition is needed in order to minimize the effect of environmental differences on sow productivity. If environmental factors are standardized and optimized, the measured differences between litters can be regarded as genetic and can be more effectively used in selection programs.

The most important single measure of sow productivity is the number of pigs weaned per sow per year. The two main components affecting it are (1) the number of litters per year, and (2) the litter size at weaning. Basically, the number of litters per sow per year depends on the interval between farrowings. Since the gestation period is a constant component of farrowing interval, the lactation length (or weaning age) and the weaning-to-remating interval become the main factors that can influence the farrowing interval.

Litter size at birth depends on ovulation rate, fertilization rate, embryo losses, and fetal losses.

Litter size at weaning is affected by (1) the number of pigs born alive, and (2) the preweaning mortality. Preweaning losses can range from 10 to 20%, caused primarily by overlaying and starvation.

To optimize sow productivity, producers need to take good care of the components affecting it.

To ensure a high number of pigs at weaning, the producer should do the following:

1. Select sows with proven genetic ability to be prolific.
2. Use boars with maximum proven fertility.
3. Provide optimum housing and management throughout the reproductive cycle.
4. Provide a proper feeding program for sows consisting of (a) low plane feeding throughout the gestation period, and (b) rather liberal feeding during lactation.
5. Optimize management at farrowing and during early lactation.
6. Keep records of each sow and litter, and use this data in making selections for high weaning performance.
7. Maintain a sound herd health program.

Reducing lactation length is the most effective way in which to increase sow productivity. With an adequate management program, weaning pigs at 3 to 4 weeks of age should be the regular practice. The empty days herd average should be 10 to 15 days or fewer. To achieve this, the following practices are important:

1. Follow the progress of individual sows from weaning onward.
2. Evaluate the performance of each sow, and determine the best time for culling.
3. Use boars of high fertility and service each sow twice during the heat period.
4. Use proper procedure to check heat, and to check returns to service.
5. Use an accurate pregnancy diagnostic technique.
6. Check breeding sows and boars regularly for reproductive diseases.

CARE AND MANAGEMENT OF BOARS

Proper care and management of boars makes for long and productive service. This generally involves the proper consideration of the following: (1) the purchase and induction of boars, (2) test mating and semen evaluation, (3) housing and feeding, (4) ranting problem, (5) age and service of the boar, (6) maximizing fertility, and (7) keeping and using adequate records.

PURCHASE AND INDUCTION OF BOARS

Boars should be purchased at least 45 to 60 days before they are needed for breeding. If any purchase adjustments are necessary, the "Code of Fair Practices" adopted by the National Association of Swine

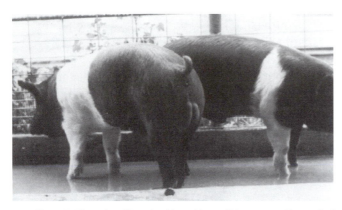

Fig. 7-10. The colored breeds—Hampshire and Duroc—excel in terminal characteristics. (Courtesy, National Pork Producers Council, Des Moines, IA)

Breeders contains good guidelines to follow. Also, boar buyers should purchase boars from sellers who can provide good health records; and all new boars should be isolated for at least 30 days prior to use.

TEST-MATING AND SEMEN EVALUATION

Records indicate that about 1 in 10 young untried boars has a fertility problem which renders them either sterile or subfertile. The simple practice of test-mating to identify a problem boar before the breeding starts can avert lost time and interrupted pig flow. Boars should be test-mated following their isolation period, and at 7 months of age or older.

HOUSING AND FEEDING BOARS

Fig. 7-11. Boars in breeding crates at Murphy Farms, Rose Hill, North Carolina. (Courtesy, Dr. Garth W. Boyd and Ms. Rhonda Campbell)

When using the hand-mating system (individual mating system), boars should be penned separately, commonly in crates which are approximately 28 in. wide by 7 ft long, or in pens about 6 ft × 6 ft. Individual housing of boars eliminates fighting, riding, and competition for feed. Groups of boars used in a pen-mating system should be penned together when they are removed from a sow group.

During the initial isolation period, boars should be fed a ration similar to that fed by the seller. This reduces stress associated with relocation. Gradually, the diet should be changed to match the diet used by the buyer. Young boars that are still growing should be fed at a level which allows for moderate weight gain. Depending upon the diet, boar age, boar condition, and housing/climate conditions, the boar should be fed 5 to 6.5 lb of a balanced 14% crude protein diet per day—5 to 5.5 lb for younger boars, 5 to 6 lb for mature boars. Boars with exceptionally high sex drive, or that are penned in outside facilities during cold weather, have a higher than normal feed requirement; hence, their feed level should be adjusted accordingly. Overfeeding boars can lead to reproductive problems and decrease the length of service in the herd.

When limit-feeding any diet to reduce energy intake, be certain that adequate levels of protein (amino acids), vitamins, and minerals are present to meet the boar's requirements for these nutrients.

Formulation of special boar rations may be justified where several boars are to be maintained. However, in most herds, a well balanced gestation record is satisfactory.

RANTING PROBLEM

Some boars chop their jaws and slobber. Such action is called ranting. Young boars that take to excessive ranting may go off feed, become "shieldy" (hard, leather shoulders), and fail to develop properly. Although this condition will not affect their breeding ability, it is undesirable from the standpoint of appearance. Isolation from other boars or from the sow herd is usually an effective means of quieting such boars. Should the boar remain off feed, placing a barrow or a bred sow in the pen with him will help to get him back on feed.

AGE AND SERVICE OF THE BOAR

The number of services allowed will vary with the age, development, temperament, health, breeding condition, distribution of services, system of mating (hand-breeding, pen-breeding, or use of artificial insemination), and farrowing schedule. No standard number of services can be recommended for any and all conditions. Yet the practices followed by good swine producers are not far different. Such practices are summarized in Table 7-3.

TABLE 7-3
BOAR USE GUIDE

| Age | Services | | Females per Boar | Comments |
	Daily	Weekly		
Young (8 to 12 months)	1	5	8 to 10	Turning a young untried boar into a group of just weaned sows coming into heat may be disastrous.
Mature (over 12 months)	2	7	10 to 12	If the estrous cycle of a group of sows tend to be synchronized, more boar power is needed.

For best results, the boar should be at least eight months old and well grown before being put into service. Even then, he should be limited to one service per day and a maximum of five per week.

The number of boars necessary should be thought of in terms of the services required rather than sows per boar since sows or gilts will breed more than once during a heat period, though hand breeding controls the number of services by a boar. In pen breeding systems, more boar power will be needed if sows are weaned and bred back by groups.

When fed and cared for by an experienced producer, a strong, vigorous boar from 1 to 4 years of age (the period of most active service) may serve two sows per day during the breeding season provided a system of hand coupling is practiced. Excessive service will result in the release of a decreased concentration of sperm as well as immature sperm. With pen mating, fewer sows can be bred.

A boar should remain a vigorous and reliable breeder up to 6 or 8 years of age or older, provided he has been managed properly throughout his lifetime.

MAXIMIZING FERTILITY

Producers should maximize boar fertility by providing adequate boar power, rotating boars or individual-mating, providing an adequate breeding area, keeping them cool during the summer months, and using other sound management practices.

KEEPING AND USING ADEQUATE RECORDS

In order to utilize boar power more efficiently and identify breeding problems early, records are necessary. Producers should keep records of the frequency of boar services, and if artificial insemination is used, the date and volume of each ejaculate should be recorded. Computer programs are available for use in keeping, organizing, and analyzing records.

CARE AND MANAGEMENT OF PREGNANT SOWS

Without attempting to duplicate the discussion on feeding the gestating sow as found in Chapter 9, it may be well to emphasize that there are two cardinal principles which the producer should keep in mind when feeding sows during the pregnancy period. These are: (1) to provide a ration which will insure the complete nourishment of the sow and her developing fetal litter, and (2) to choose the feeds and adopt a method of feeding which will prove economical and adaptable to local conditions. Contrary to a common practice, a laxative ration is necessary at farrowing time only if constipation is a problem; and constipation is seldom a problem.

In the past, the shelter for bred sows was not elaborate or expensive. The chief requirements for sow shelter were that it be tight overhead, that it provide protection from inclement weather, that it be well drained and dry; and that it be of sufficient size to allow the animals to move about and lie down in comfort. Moreover, except during the most inclement weather, sows were encouraged to run outdoors for exercise, fresh air, and sunshine. But times have changed! An increasing number of hog producers are confining bred sows throughout gestation.

High land values, environmental problems, and efficiency in handling and managing large numbers of sows have caused producers to turn to a confinement system of housing. The type of system is, however, dependent upon the manager's skill, ability, and finances. Chapter 14 discusses production systems, including confinement.

Some **advantages** of confinement sow housing include: (1) better control of mud, dust, and manure; (2) reduced labor for feeding, breeding, and moving to farrowing house; (3) improved control of internal and external parasites; (4) smaller land requirements; (5) better supervision of herd at breeding time; (6) use of existing buildings; (7) improved operator comfort and convenience; and (8) opportunities for better all-around management.

Some **disadvantages** of confinement sow housing

include: (1) higher initial investment; (2) possible delayed sexual maturity and breeding age, lower conception rates in gilts, and lower rebreeding efficiency in sows; (3) requirement of better management and daily attention to details; and (4) increase in feet and leg problems.

CARE OF SOWS AT FARROWING TIME

The careful and observant producer realizes the importance of having everything in readiness for farrowing time. If pregnant sows have been so fed and managed as to give birth to a crop of strong, vigorous pigs, the next problem is that of saving the pigs at farrowing time.

It is estimated that from 10–20% of the pigs farrowed never reach weaning age, and an additional loss of 5–10% occurs after weaning.

SIGNS OF APPROACHING PARTURITION

The immediate indications that the sow is about to farrow are extreme nervousness and uneasiness, an enlarged vulva, a possible mucous discharge, and presence of milk in the teats.

PREPARATION FOR FARROWING

About two weeks prior to farrowing, the sow should be dewormed. She should also be treated for external

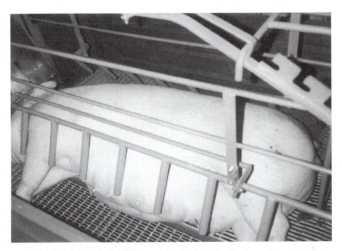

Fig. 7-12. Sow in last stages of gestation, in the farrowing room. Note underline and placement of teats. (Courtesy, American Diamond Swine Genetics, Prairie City, IA)

parasites at least twice within a few days of moving to the farrowing facility.

When farrowing in crates or pens, sows should be moved to these no later than the 110th day of gestation.

If constipation is a problem, substitute 20% wheat bran or 10% dehydrated alfalfa meal or beet pulp for grain in the diet upon moving the sow into the farrowing unit. Some producers avoid the constipation problem by adding 20 lb of magnesium sulfate (Epsom salts) or 15 lb of potassium chloride per ton of farrowing-lactation ration.

SANITARY MEASURES

Performance of pigs can be improved by developing an "all-in, all-out" management system. This system entails scheduling the breeding of the sows so that a clean farrowing facility can be filled with sows within a short time. This allows a cleanup period before another group of sows is brought into the same facility.

Before being moved into the farrowing quarters, sows should be washed with warm water and soap and rinsed with a mild disinfectant. This is necessary to minimize contamination of the clean farrowing quarters.

Cleaning the farrowing facility can be accomplished by scraping, high pressure cleaners, steam cleaners, and/or a stiff scrub brush. A complete job is necessary; otherwise, the use of a disinfectant is useless. Many good commercial disinfectants are available, including the quaternary ammonium compounds, iodoform compounds, and lye. Chapter 15 provides information on disinfectants.

THE QUARTERS

Hogs are sensitive to extremes of heat and cold and require more protection than any other class of farm animals. This is especially true at the time of parturition. It is recommended that the temperature in the sow area be in the range of 55° to 75°F, and the temperature in the baby pig area have supplemental heat so as to maintain the temperature at 90° to 95°F for the first few days. After this time, the temperature for piglets can be decreased to 70° to 80°F until weaning. Along with this temperature, there should be adequate ventilation at all times.

FARROWING CRATES OR PENS

Many producers use farrowing crates because they reduce the number of piglets crushed by the sow. An additional advantage is that the operator is protected. Most farrowing crates are 5 ft wide by 7 ft long. The width includes an 18-in. piglet area on both sides of the 24-in. sow stall. Commercial crates adjust to accommodate very large or very small females. Crates may have slotted or solid floors.

When open pen farrowing is practiced, a guard rail around the farrowing pen is an effective means of preventing sows from crushing their pigs. The importance of this simple protective measure may be emphasized best by pointing out that approximately one-half of the young pig losses are accounted for by those pigs that are over-laid by their mothers. The rail should be raised 8 to 10 in. from the floor and should be 8 to 12 in. from the walls. It may be constructed of two-by-fours, two-by-sixes, or strong poles or steel pipe.

BEDDING

In open pen farrowing, quarters should be lightly bedded with clean, fresh material. Any good absorbent that is not too long and coarse is satisfactory. Wheat, barley, rye, or oat straw; short or chopped hay; ground corncobs; peanut hulls; cottonseed hulls; shredded corn fodder; or shavings are most commonly used. Additional information regarding bedding is given in Chapter 14, Swine Management.

ATTENDANT

Attending sows at farrowing decreases the incidence of stillborn pigs that die during the birth process, and the incidence of pigs dying within the first few hours after birth. Moreover, care given during this time improves survival the first few days after farrowing. The caretaker, therefore, should be on the job.

On the average, the interval between the birth of pigs is approximately 15–20 minutes unless a problem develops. Normal presentation is either head first or tail first. An attendant can (1) free piglets from the membranes, (2) help piglets reach a teat, (3) possibly revive some piglets which are not breathing, and (4) treat the navel cord with tincture of iodine. If farrowing is proceeding normally but slow, oxytocin may be used to speed the rate of delivery. Oxytocin should not be used if there is any suggestion that a pig is blocking the birth canal.

Continued strong labor for an extended period without the birth of piglets indicates the need for manual assistance by the attendant. A well lubricated gloved hand and arm should be inserted in the vulva and up the vagina as far as is needed to find the piglet blocking the birth canal. Then the piglet should be grasped and gently but firmly pulled. Since manual assistance increases the chances of complications, it is advisable to use an antibacterial solution as a lubricant and as an infusion following farrowing.

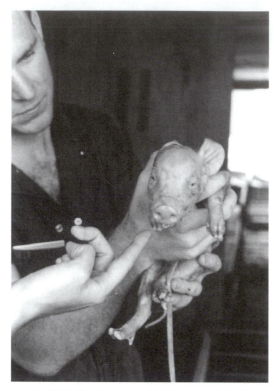

Fig. 7-13. Attending sows at birth decreases the incidence of (1) pigs that die during the birth process, and (2) pigs that die within the first few hours after birth. (Courtesy, Murphy Farms, Rose Hill, North Carolina. Photo by Dr. Garth Boyd and Ms. Rhonda Campbell)

Some producers choose to induce farrowing in sows in order to control when parturition will occur. Induced farrowing avoids parturition at inconvenient times such as late hours or over the weekend. It also (1) results in more efficient use of farrowing crates and barns, (2) facilitates cross-fostering of pigs from very large litters to sows that delivered smaller litters, (3) makes it possible to wean pigs together that are more uniform in age and size, and (4) results in sows coming in heat (estrus) about the same time after weaning.

Several different hormonal injections will induce farrowing. Lutalyse (Upjohn Co.) is the most commonly used product for this purpose. All label instructions should be followed when using Lutalyse. Sows should not be treated earlier than 110 days of gestation, usually animals will farrow within 48 hours after treatment. Another product that will induce farrowing, and which some producers favor, is prostaglandin F22.

As soon as the afterbirth is expelled, it should be removed from the pen and burned or buried in lime. This prevents the sow from eating the afterbirth and prevents the development of bacteria and foul odors. Many good swine producers are convinced that eating the afterbirth encourages the development of the pig-eating vice. Dead pigs should be removed for the same reason.

If bedding is used, it is also well to work over the bedding; remove wet, stained, or soiled bedding and provide clean, fresh material.

CHILLED AND WEAK PIGS

Pigs arriving in a cold environment are easily chilled. If they are born in an unheated building during cold weather, it may be advisable to take the pigs from the mother as they are born and to place them in a half-barrel or basket lined with straw or rags. A few hot bricks or a jug of warm water (properly wrapped to prevent burns) may be placed in the barrel or basket; or the pigs may be taken to a warm room until they are dry and active.

One of the most effective methods of reviving a chilled pig is to immerse the body, except the head, in water as warm as one's elbow can bear. The pig should be kept in this for a few minutes, then removed and rubbed vigorously with cloths.

Of course, modern pig nurseries avert chilled pigs.

ORPHAN PIGS

Pigs may be orphaned either through sickness or death of their mother. In either event, the most satisfactory arrangement for the orphans is to provide a foster mother. When it is impossible to transfer the pigs to another sow, they may be raised on cow's milk or milk replacer. The problem will be simplified if the pigs have received a small amount of colostrum (the first milk) from their mother. Colostrum can be hand milked if necessary.

If cow's milk is used, do not add cream or sugar; however, skim milk powder, at the rate of a tablespoonful to a pint of fluid milk may be added, if available. Sow milk replacer should be mixed according to the directions found on the container. The first 2 or 3 days the orphans should be fed regularly every 2 hours, and the milk should be at 100°F. Thereafter, the intervals may be spaced farther apart. All utensils (pan feeding or a bottle and nipple may be used) should be clean and sterilized.

Orphan pigs should be started on a prestarter or starter ration when they are one week old. Also, a source of iron should be provided (in keeping with instructions given in Chapter 9).

RUNTS

Small pigs or "runts" present a problem for pro-

ducers. Larger litters have more runts, and these runts are often some of the last pigs born; thus, forcing them to compete for a teat with the larger earlier born pigs. Since runts often perish, in the past many producers have sacrificed them rather than try to save them. With today's economy, the effort to save these pigs may be worthwhile. Some research indicates that supplemental feeding of underweight (under 2 lb) newborn pigs (runts) can reduce their mortality. Supplemental feedings may consist of a commercial milk replacer or a mixture of 1 qt milk, ½ pt "half and half," and 1 raw egg, which is administered once or twice daily in 15 to 20 ml portions with a soft plastic tube attached to a syringe.[2]

ARTIFICIAL HEAT

Raising the air temperature of the farrowing unit is important to prevent the chilling of newborns.

Artificial heat usually must be provided, especially for pigs farrowed in the northern United States. Most nursery houses are equipped with a heating unit for use in winter farrowing, designed to maintain the temperature at 70° to 75°F.

Furthermore, providing an area of supplemental heat is necessary to maintain baby pigs in their thermoneutral zone. This heat zone should be 90° to 95°F. A major factor adversely affecting piglet survival is its difficulty in maintaining body temperature due to the high ratio between body surface area and size, sparse hair covering, and limited body fat. Even at thermoneutral temperatures, substantial amounts of metabolic energy are required to maintain body functions. A few days after birth, age-related changes occur which markedly improve the thermostability of the piglet.

SOW AND LITTER

The care and management given the sow and litter should be such as to get the pigs off to a good start. As is true of other young livestock, young pigs make more rapid and efficient gains than older hogs. Strict sanitation and intelligent feeding are especially important for the well being of the young pig. This and other management practices pertinent to the well being of young pigs—including adjusting litter size, removal of the needle teeth, ear notching, castrating, etc.—are fully covered in Chapter 14.

■ **Feeding the nursing sow**—High-producing sows

---

[2]Moody, N. W., *et al.*, "Effects of Supplemental Milk on Growth and Survival of Suckling Pigs," *Journal of Animal Science*, Vol. 25, 1966, p. 1250.

Fig. 7-14. Sow and litter. (Courtesy, Babcock Swine, Inc., Fairview, MN)

weaning 9–11 pigs per litter and averaging 2.2 litters per year, or more, have a much higher nutritional requirement than the average sow weaning 7–8 pigs per litter and averaging 2 or fewer litters per year.

High-producing sows should be on full feed by day four after lactation.

Throughout the lactation period, sows should be fed liberally with feeds that will stimulate milk production. The most essential ingredients fed in the brood sow's ration during this period are an ample amount of protein (amino acids), vitamins, and minerals (see Chapter 9).

NORMAL BREEDING SEASON AND TIME OF FARROWING

Sows will breed any time of the year, but as in other farm animals the conception rate is much higher during those seasons when the temperature is moderate and the nutritive conditions are good. For the country as a whole, spring pigs are preferred, as is shown by the size of the spring pig crop in comparison with the fall pig crop. However, big operations and environmentally controlled buildings have made for more year-round production.

ARTIFICIAL INSEMINATION OF SWINE[3]

Artificial insemination (AI) is the deposition of

---

[3]In the preparation of the section on Artificial Insemination of Swine, the authors adapted much of the material from the following two state-of-the-art sources: *1995 Proceedings Nebraska Whole Hog Days*, article on "Artificial Insemination," by D. G. Levis, pp. 155–175; and *The Swine AI Book*, by Billy Flowers, *et al.*, North Carolina State University, published by Southern Cross, Raleigh, NC, 1994.

spermatozoa in the female genitalia by artificial rather than by natural means.

Swine AI has long been widely used in the former Soviet Union, China, Japan, the European and Scandinavian countries, and in other areas having large swine populations. But the U.S. swine industry has been slower to adopt AI than the dairy and turkey industries. However, increasing numbers of the large U.S. hog producers are using AI. By the mid-1990s, it was estimated that 10 to 15% of the bred sows and gilts in the U.S. were artificially inseminated.

The **advantages** of AI are:

1. It makes it possible to use genetically superior boars more widely than would be possible with natural service.

2. It provides a way to bring in new genetic material with a minimum of disease risk.

3. Fewer boars are needed. For example, a mature boar should not breed more than two females per day. However, in a large operation where sows are weaned in groups, AI makes it possible to breed 10 or more sows in one day from one ejaculation.

4. There is less risk of injury to the sows, the boar, and the people handling them when AI is used.

5. The number of genetically superior boars and sows for sale is increased.

The two most common **misconceptions** relative to AI in swine are:

1. That the farrowing rate and litter size will be lower with AI than with natural service.

Fact: If AI is performed properly, reproductive performance will be as good as the level achieved with natural service.[4]

2. That AI matings require more labor than natural matings.[5]

Fact: A study in North Carolina showed that when more than three females were bred in a day, the labor was significantly less with AI than with natural service.[4]

Several AI businesses operating worldwide have added top quality boars. Also, there is an increasing number of strictly swine service AI organizations in the United States. Generally, swine service businesses sell mostly fresh semen (but some frozen semen), provide custom processing and on-site AI training, sell AI supplies, and sell semen throughout the U.S. and abroad.

In Canada, many years ago the Ontario Swine AI Assn., Woodstock, Ontario, a farmer cooperative, assembled a battery of outstanding boars, representing five breeds, to provide semen throughout Canada. It

---

[4]Levis, D. G., *1995 Proceedings Nebraska Whole Hog Days*, p. 163, Table 2.

[5]*Ibid.*, p. 166, Fig. 1.

has improved swine for lean meat, both domestically and internationally.

SELECTING AI BOARS

When choosing AI boars, producers may select from within their herds, or buy new animals from outside their herds. AI gives them a third choice: They may purchase semen from outside the herd, rather than purchase boars. Purchased boars or semen from outside sources may be obtained from either independent purebred breeders, or from seedstock producers. EPDs can be used when selecting boars or semen from outside herds. (See Chapter 4, section headed "Selection Based on STAGES, EBV, EPD, and/or BLUP.")

In addition to genetic value, when selecting boars they should be evaluated on those traits that allow them to produce sperm and mate (see Chapter 4, section on "Selecting Boars").

AI LABORATORY

Where AI is planned, a modest, but adequate, AI laboratory should be arranged. Preferably, it should be adjacent to the semen collection pen, with hand-delivery access between them by a sliding window.

The laboratory should be equipped for (1) preparing semen collection equipment, (2) examining semen, (3) preparing semen dilutents, (4) diluting semen, (5) storing semen, and (6) cleaning and storing equipment.

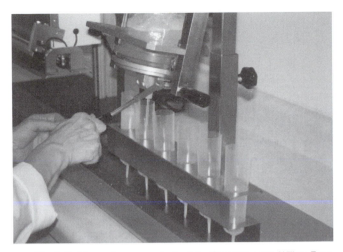

Fig. 7-16. Dispensing semen in the lab. (Courtesy, Lone Willow Farm, Roanoke, IL)

Fig. 7-17. Preparing semen for shipping. (Courtesy, Lone Willow Farm, Roanoke, IL)

SEMEN COLLECTION AREA

The collection room should be well lighted and weather-comfortable. The collection pen should be approximately 10 ft × 10 ft; and it should have safety escape corners, or be surrounded by escape posts, for the protection of the collectors if a boar should become hostile. Also, keep a hurdle (stockboard) nearby when handling boars.

The collection pen should be equipped with a dummy sow securely fastened to the floor or a pen partition. Swine producers can construct their own dummy (Fig. 7-18) or purchase a dummy from a company that specializes in AI equipment (Fig. 7-19).

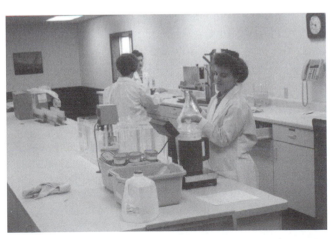

Fig. 7-15. Modern AI laboratory. (Courtesy, Lone Willow Farm, Roanoke, IL)

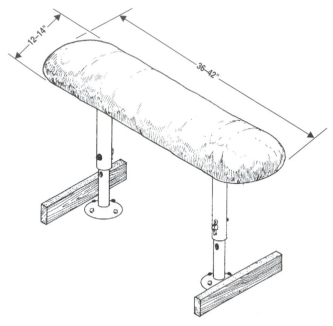

Fig. 7-18. A simple dummy for semen collection from boars.

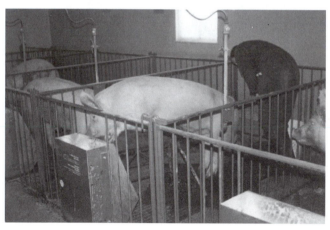

Fig. 7-19. Mounting dummy. *Note:* This dummy looks like a Yorkshire or Large White. (Courtesy, Lone Willow Farm, Roanoke, IL)

TRAINING BOARS

Training boars for collection of semen requires great patience.

As soon as the young boar starts to rant, or at approximately 7–8 months of age, his training to mount the dummy sow should start. The following techniques have been used to train a boar to mount a dummy sow: (1) introduce the trainee to the dummy sow immediately after another boar has worked on it; (2) pour urine, semen, or preputial fluid from a mature boar on the rear of the dummy, (3) sexually stimulate the boar by exposing him to a strange boar or gilt adjacent to the collection area, then have the strange boar or gilt exit in a manner which leads the trainee-boar to the dummy; (4) allow the trainee to mount, but not breed, an estrous female near the dummy; then, exit the female to lead the trainee to the dummy; (5) collect from the trainee when he is mounted on an estrous female standing near the dummy; or (6) allow the trainee to mount an in-heat female standing adjacent to the dummy, ejaculate for about one minute, then lift him off the female and place him on the dummy.

COLLECTING SEMEN

Before collecting semen, trim the hair surrounding the preputial orifice.

Experienced technicians collect boar semen, step by step, as follows:

1. Wash and dry hands.
2. Introduce the boar to the dummy; squeeze the preputial area to empty the prepuce of any urine, and clean the surrounding area with a paper towel; and put on disposable, nonspermicidal vinyl gloves.
3. After boar mounts the dummy, his penis will emerge. Grasp the penis with the gloved hand. Remember that the semen collector's hand must simulate the sow's cervix, and that the penis, which forms a rigid spiral, locks into it. Apply firm pressure on the first and second ridges of the spiral portion of the penis; and do not loosen grip until ejaculation is finished. Once the lock is formed, the boar will be begin to ejaculate.
4. Three different fractions of semen will be ejaculated in the following order:
     a. The gel fraction, which should not be collected as it contains few sperm.
     b. The sperm-rich fraction, which is opaque and milky.
     c. The final gel fraction, which should not be collected.
Caution: Avoid contaminating the ejaculate with preputial fluid, which will kill sperm.
5. Collection will take 5 to 20 minutes. Do not let hold of the boar's penis until the boar has finished ejaculating.

For adequate semen volume, sperm concentration, and doses per ejaculate, a collection frequency of 48 to 72 hours is recommended.

During collection, and in handling immediately following collection, the following precautions should be observed in order to maximize the fertility of sperm cells:

1. Protect sperm cells from extreme heat or cold.
2. Prevent contact with water or harmful chemicals.

Fig. 7-20. Over-working a boar can cause him to produce an ejaculate without sperm cells. Beaker on the right contains an ejaculate without sperm cells from a boar that was over-worked. (Courtesy, D. G. Levis, University of Nebraska, Lincoln)

3. Avoid exposure of semen to direct sunlight; and minimize exposure to air.

4. Gently stir or swirl semen, but do not shake.

ASSESSING SEMEN QUALITY

Semen should be evaluated promptly after collection. Routinely, the first few ejaculates of a new boar should be examined, followed by an examination monthly thereafter and additional examinations if there are fertility problems. Semen should be evaluated on the following bases:

1. **Volume.** Volume of semen can be measured by pouring it into a measuring beaker. Normally, a boar will ejaculate 150–250 ml, with a range of 50–500 ml. Alternatively, the semen can be weighed and the volume computed: 1 g = 1 ml.

2. **Motility.** Good motility indicates good viability of semen. Motility should be assessed on the proportion of sperm moving forward in a fairly straight line (progressive motility).

3. **Morphology (abnormal cells).** This is a measure of abnormal sperm, including: coiled tails; crooked or bent tails; tails without heads; heads without tails; giant heads; and many others.

4. **Sperm density (sperm concentration or sperm count).** Sperm density may be determined with a microscope, a hemacytometer, or a photometer. Unless sperm density is known, doses should be limited to 6 to 10 per collection.

5. **Color.** Chalky, milky appearance indicates high density; opalescent, watery appearance indicates lack of density; pink indicates that the semen is contaminated with blood and should be discarded; yellow indicates that the semen is contaminated with preputial secretions, urine, and/or pus and should be discarded.

6. **Odor.** A clean ejaculate has little odor, whereas ejaculate contaminated with preputial fluid has a distinctive odor.

EXTENDING (Diluting) SEMEN

A semen extender is a liquid preparation that is used to increase volume, protect against cold shock, serve as a buffer against high pH, provide nutrients for sperm, and lengthen the viability of sperm cells.

The following three-day extenders are being used with or without antibiotics (antibiotics are not needed if semen is used immediately after dilution): *Androhep, Beltsville Thawing Solution (BTS), Guelph, Kiev, Merck III,* and *Modena.*

Note: Extended semen should be used as soon as possible. Pregnancy rate starts to decline when the three-day extenders are stored longer than 48 hours.

More complex dilutents, such as *Reading* and *Zorpva,* can extend semen life to five days.

INSEMINATING SOWS AND GILTS

The actual process of AI is relatively easy since the anatomy of the female's vagina and cervix guides the inseminating catheter to the cervix without the aid of sight or some manipulation to ensure the passage into the cervix.

The major factor in achieving maximum farrowing rate and litter size is to inseminate females at the right time. While the technique is simple, being certain it is carried out correctly is difficult. When more frequent heat detections are conducted, it is more likely that insemination will occur at the proper time. The best heat check is not to allow boar-to-sow contact for one hour before actual time of heat checking. This separation causes the females quickly to exhibit a strong immobilization response when encountering a boar. Where possible, the latter is best accomplished by driving a boar through the passageway in front of the sows, thereby providing nose-to-nose contact.

It is generally agreed that ovulation in swine begins 36–44 hours after the onset of estrus and lasts for 1 to 3 hours. However, the onset and duration of ovulation are very variable.

Confining the in-heat female to a small pen with a boar nearby, preferably with nose-to-nose contact, and applying hand pressure to the back of the female to bring about an immobile stance, or gently rubbing a nervous sow's flank, is all that is necessary for females to stand during insemination.

Both the rubber catheter, shaped like the boar's penis, and the plastic catheter will work for AI. The major advantage to the plastic catheter is that it is disposable.

Artificial insemination using either the rubber catheter or the plastic catheter follows these general steps:

1. Wipe vulva with paper towel.

2. Insert the tip of the catheter into the vulva with the tip pointing upward to prevent entrance into the ureteral orifice.

3. Slide the catheter along the top of the vagina until firm resistance of the cervix is felt, usually about 8 to 10 in.

4. Rotate catheter counterclockwise to lock into the cervix.

5. When the catheter is in position, the semen container is attached and semen is squeezed slowly into the catheter. It normally takes 5 to 10 minutes to complete the insemination.

6. After semen is deposited, rotate the catheter clockwise and slowly remove.

7. Reinseminate the female in 18–24 hours.

CLEANING AND STORING EQUIPMENT

In order to minimize breakdown in sanitation, disposable equipment is recommended, including plastic catheters, plastic insemination bottles, and plastic bags in collection vessels.

If rubber catheters are used, they should be scrubbed with hot water to remove debris, rinsed thoroughly with cold water, and boiled in distilled water for 10 minutes. When removed from the sterilizer, shake them to remove excess water and place them spiral-end up in a drying cabinet. After they are dry, store them in a sealed plastic storage bag.

FROZEN SEMEN

Frozen semen is available in both pellets and macro tubes. But boar semen does not freeze well. Frozen semen can be stored longer than fresh (liquid) semen. However, in comparison with fresh semen, frozen semen (1) produces fewer doses, (2) lowers farrowing rates an average of 30 to 40%, and (3) reduces litter size by as much as one pig.

Before use, frozen semen must be thawed, which should be according to supplier's directions; and it should be used immediately after thawing.

REGISTRATION OF PIGS PRODUCED BY AI

Table 7-4 summarizes the rules and attitudes of registries toward AI.

SUMMARY

AI requires a greater managerial input, a minimum amount of specialized equipment, and specialized training. However, by following a few precautions, litter size and conception rates will be equal to natural service, and the genetic level of the entire herd will be improved.

TABLE 7-4
AI RULES OF SWINE REGISTRY ASSOCIATIONS

Breed	Pertinent Rules or Attitude of Each Registry Association Relative to Artificial Insemination (AI)
Berkshire	Accepted provided AI certificate is signed and dated by both the inseminator and the owner of the boar.
Chester White	Accepted. Seller must purchase $15 AI certificate from Chester White Record Assn., and furnish it to buyer of semen, who submits it with registration application.
Duroc	AI-sired pigs accepted, but information must be furnished by inseminator; and the Registry charges a fee of $25 per AI litter.
Hampshire	Accepted. But boar used in AI must be tested for "red gene factor" by mating to two Duroc sows, with all pigs from these matings belted or black. If any pigs are red or red belted, the boar's pedigree is pulled, and no pigs sired by this boar can be registered.
Hereford	Accepted, but must be approved by the Board of Directors.

(Continued)

TABLE 7-4 (Continued)

Breed	Pertinent Rules or Attitude of Each Registry Association Relative to Artificial Insemination (AI)
Landrace	Accepted. But Landrace boars used in AI must be certified for white color purity; and there is an AI certificate fee of $15 per litter.
Poland China	Accepted. Applicant must send breeding certificate, date of insemination, identity of dam, and name of inseminator.
Spotted	Accepted. Semen seller must purchase $15 AI certificate from Spotted Assn., and furnish it to buyer of semen, who submits it with registration application.
Yorkshire	Accepted. Breeding certificate signed by owner of the sire, if different than the owner of the litter.

QUESTIONS FOR STUDY AND DISCUSSION

1. Diagram and label the reproductive organs of the boar, and briefly describe the function of each organ.

2. Diagram and label the reproductive organs of the sow, and briefly describe the function of each organ.

3. Differentiate between hand-mating, pen-mating, and artificial insemination. List some advantages and disadvantages of each.

4. Describe the estrous cycle of the sow.

5. In order to synchronize ovulation and insemination, when should sows be bred with relation to the heat period?

6. What are the signs of heat?

7. Sows will often come in heat during the first few days after farrowing. If bred at this time, why do they usually fail to conceive?

8. When should sows be bred after farrowing?

9. From 5 to 30% of the fertilized eggs do not develop normally, resulting in embryonic mortality or death; and from 10 to 20% of the live pigs farrowed die within the first 7 to 10 days of life. Discuss the possible causes and economics of this situation.

10. In your opinion, is pregnancy diagnosis a valuable asset to a swine operation? Defend your answer.

11. List the five economically important traits which are usually used in measuring sow productivity. Discuss the importance of each of them.

12. Discuss each of the following aspects of the care and management of boars: (1) purchase and induction of boars; (2) test mating of boars; (3) housing and feeding; (4) ranting problem; (5) age and service of the boar; (6) maximizing fertility; and (7) keeping and using adequate records.

13. How often should a mature boar be used when hand-mating is practiced? What guidelines would you follow if pen-mating is practiced?

14. Outline some practices that you feel would ensure the survival of a maximum number of pigs at farrowing and the first few days afterwards.

15. Describe a sow that shows signs of farrowing and describe the actual birth process.

16. Why must an area of supplemental heat be provided for piglets?

17. What advantages could accrue from the practical and extensive use of artificial insemination in swine?

18. Will AI result in the farrowing rate and litter size being lower than natural service? Will AI require more labor than natural service?

19. How would you go about purchasing a boar for AI in your herd?

20. Discuss training boars for AI.

21. List and discuss six bases for evaluating boar semen.

22. List the steps that should be followed in using either the rubber catheter or plastic catheter in inseminating sows.

23. Compare fresh (liquid) boar semen and frozen semen.

24. Discuss the rules of the swine registries relative to registering pigs produced by AI.

SELECTED REFERENCES

Title of Publication	Author(s)	Publisher
Animal Reproduction—Principles and Practices	A. M. Sorensen, Jr.	McGraw-Hill Book Co., New York, NY, 1979
Applied Animal Reproduction	H. J. Bearden J. W. Fuquay	Reston Publishing Co., Inc., Reston, VA, 1996
Biology of the Pig, The	W. G. Pond K. A. Houpt	Cornell University Press, Ithaca, NY, 1978
Embryology of the Pig, The	B. M. Patten	Maple Press Co., York, PA, 1931
Managing Swine Reproduction	L. H. Thompson	University of Illinois, Urbana-Champaign, IL, 1981
Pork Industry Handbook		Cooperative Extension Service, Purdue University, West Lafayette, IN
Reproduction in Domestic Animals	Ed. by P. T. Cupps	Academic Press, Inc., New York, NY, 1991
Reproduction in Farm Animals	Ed. by E. S. E. Hafez	Lea & Febiger, Philadelphia, PA, 1993
Stockman's Handbook, The	M. E. Ensminger	Interstate Publishers, Inc., Danville, IL, 1992
Swine AI Book, The	Billy Flowers, et al.	Southern Cross, Raleigh, NC, 1994

As evidenced by the whiskers of this baby pig, milk is good, and good for baby pigs. (Courtesy, Murphy Farms, Rose Hill, NC; Dr. Garth W. Boyd and Ms. Rhonda Campbell)

8

FUNDAMENTALS OF SWINE NUTRITION

Efficient and profitable swine production depends upon an understanding and application of the fundamentals of swine nutrition for two primary reasons:

1. Feed represents 60 to 75% of the total cost of production. For this reason, swine producers should endeavor to provide diets that are both satisfactory and as low cost as possible, and that make for the maximum production of quality pork per unit of feed consumed.

2. Today, many of the hogs produced in the United States are raised in confinement. As a result, they have less choice in their selection of feed than any other class of four-footed animals. For the most part, they are able to consume only what the caretaker provides. This consists largely of concentrated feeds with only a small proportion of forage. These conditions are made more critical because hogs grow much faster in proportion to their body weight than the larger farm animals, and they produce young at an earlier age—factors which have been accentuated under modern forced production.

Thus, a knowledge of the nutritional needs of swine is especially important.

PERSPECTIVE OF NUTRITION

Nutrition is more than just feeding. Correctly speaking, *nutrition is the science of the interaction of a nutrient with some part of a living organism*. It begins with a knowledge of the fertility of the soil and the composition of plants; and it includes the ingestion of feed, the liberation of energy, the elimination of wastes, and all the syntheses essential for maintenance, growth, reproduction, lactation, and finishing.

Several factors affect the nutritive requirements of swine; among them, the following:

1. Environment (temperature, weather, housing, and competition for feed).
2. Breed or strain, sex, and genetics of pigs.
3. Health status of the herd.
4. Presence of molds, toxins, or inhibitors in the diet.
5. Availability and absorption of dietary nutrients.
6. Variability of nutrient content, and availability of the feed.
7. Level of feed additives or growth promotants.
8. Energy concentration of the diet.
9. Level of feeding, such as limit feeding versus *ad libitum*.

BODY COMPOSITION OF SWINE

Nutrition encompasses the various chemical and

Fig. 8-1. Environmentally controlled nursery units. (Courtesy, Farmers Hybrid Co., Inc., Des Moines, IA)

physiological reactions which change feed elements into body elements. It follows that body composition is useful in understanding swine response to nutrition.

In 1843, Lawes and Gilbert, famed English scientists, initiated at the Rothamsted Station the pioneering and laborious task of analyzing the entire bodies of farm animals—studies which extended for over a half century. Other similar studies involving pigs, cattle, and sheep followed throughout the world.

These studies were summarized and published by Reed, J. T., *et al.*, under the auspices of the National Academy of Sciences. The data for 714 pigs are presented in Table 8-1.

TABLE 8-1
RANGE IN CHEMICAL COMPOSITION OF BODIES OF PIGS
FROM BIRTH TO MATURITY[1]

Species	Number in Group	Range in Body Composition of Ingesta-Free (Empty) Body			
		Water	Fat	Protein	Ash
		(%)	(%)	(%)	(%)
Pigs	714	30.7–80.8	1.1–61.5	8.3–19.6	1.3–6.6

[1]Reid, J. T., *et al.*, *Body Composition of Animals and Man*, National Academy of Sciences, 1968, p. 20, Table 1.

The data in Table 8-1 represented 8 distinct breeds of swine and 1 crossbred group, ranging in age from 1 to 923 days (2.5 years), and included both barrows and females. It showed the following:

1. **Water.** On a percentage basis, swine showed a marked decrease in water content, ranging from 80.8% in newborn pigs to 30.7% in mature and very fat animals.

2. **Fat.** The percentage of fat increased with growth and fatness, ranging from 1.1% in the newborn pig to 61.5% in mature and very fat animals.

3. **Fat and water.** As the percentage of fat increased, the percentage of water decreased.

4. **Protein.** The percentage of protein increased with growth, ranging from 8.3% in newborn pigs to 19.6% during growth. The percentage of protein remained rather constant during growth, but decreased as pigs fattened. On the average, there are 3 to 4 lb of water per 1 lb of protein in the pig's body.

5. **Ash.** The percentage of ash showed the least change. However, it decreased as pigs fattened because fat tissue contains less mineral than lean tissue.

6. **Composition of gain.** The gain in weight of pigs tells nothing about the composition of gain. The composition of gain is important from a nutrition standpoint because efficiency of feed utilization (pounds feed per pound body gain) is greatly influenced by the proportion of fat and lean produced.

Body composition as a parameter—Although the general relationships presented in Table 8-1 remain, genetic selection has shifted the tissue growth of modern swine, with less fat and more protein being produced. In addition to genetic background, the following factors influence the body composition of today's pigs: weaning age, breed, sex, additives, temperature, diet, and compensatory growth. Cost has been the primary factor limiting the continued and widespread use of body composition as a parameter.

Also, the chemical composition of the body varies widely between organs and tissues and is more or less localized according to function. Thus, water is an essential of every part of the body, but the percentage composition varies greatly in different body parts; blood plasma contains 90 to 92% water, muscle 72 to 78%, bone 45%, and the enamel of the teeth only 5%. Proteins are the principal constituents, other than water, of muscles, tendons, and connective tissues. Most of the fat is localized under the skin, near the kidneys, and around the intestines. But it is also present in the muscles (known as marbling in a carcass), bones, and elsewhere.

Table 8-1 does not reveal the very small amount of carbohydrates (mostly glucose and glycogen) present in the bodies of animals and found principally in the liver, muscles, and blood. Although these carbohydrates are very important in animal nutrition, they account for less than 1% of the body composition. It is noteworthy, too, that the carbohydrate content is one of the fundamental differences between the composition of plants and animals. In animals, the walls of the body cells are made chiefly of protein, whereas in plants they are composed of cellulose and other carbohydrates. Also, in plants most of the reserve food is stored as starch, another carbohydrate, whereas in animals nearly all the reserve is stored in the form of fat.

DIGESTIVE SYSTEM OF THE PIG

To grow rapidly and efficiently, swine must receive a high-energy, concentrated grain diet, low in fiber. Cattle and sheep, on the other hand, can digest large quantities of fibrous feeds such as hay and pasture. This is largely due to differences in their digestive tracts.

Meat animals can be divided into two broad classifications—ruminants and nonruminants. Pigs are nonruminants. They have a single stomach, in contrast to ruminants which have a stomach divided into four compartments. Cattle and sheep are ruminants.

The digestive tract (or gastrointestinal tract) can be considered a continuous hollow tube—open at both ends—with the body built around it. It is a factory assembly line in reverse; instead of building some-

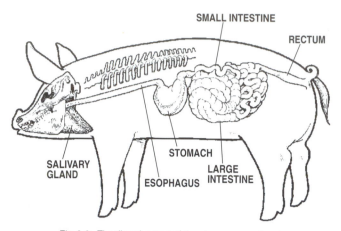

Fig. 8-2. The digestive tract of the pig—a nonruminant.

thing, it takes things apart. The digestive tract of the pig includes five main parts: the mouth, esophagus, stomach, small intestine, and large intestine.

Although ruminants and nonruminants differ in their physical makeup, the job of the digestive tract is the same in all animals. It breaks down feedstuffs into simple chemical components so that the animal can absorb and rearrange them into its own characteristic body composition.

CLASSIFICATION OF NUTRIENTS

Pigs do not utilize feeds as such. Rather, they use those portions of feeds called *nutrients* that are released by digestion, then absorbed into the body fluids and tissues.

Nutrients are those substances, usually obtained from feeds, which can be used by the animal when made available in a suitable form to its cells, organs, and tissues. They include carbohydrates, fats, proteins, minerals, vitamins, and water. (More correctly speaking, the term *nutrients* refers to the more than 40 nutrient chemicals, including amino acids, minerals, and vitamins.) Energy is frequently listed with nutrients, since it results from the metabolism of carbohydrates, proteins, and fats in the body.

Knowledge of the basic functions of nutrients in the animal body, and of the interrelationships between various nutrients and other metabolites within the cells of the animal, is necessary before one can make practical scientific use of the principles of nutrition.

FUNCTIONS OF NUTRIENTS

Of the feed consumed, a portion is digested and absorbed for use by the pig. The remaining undigested portion is excreted and constitutes the major portion of the feces. Nutrients from the digested feed are used for a number of different body processes, the exact usage varying with the class, age, and productivity of the pig. All pigs use a portion of their nutrients to carry on essential functions, such as body metabolism and maintaining body temperature and the replacement and repair of body cells and tissues. These uses of nutrients are referred to as *maintenance*. That portion of digested feed used for growth, finishing, or the production of milk is known as *production requirements*. Another portion of the nutrients is used for the development of the fetus and is referred to as *reproduction requirements*.

Based on the quantity of nutrients needed daily for different purposes, nutrient demands may be classed as high, low, variable, or intermediate. Requirements for milk are considered *high-demand uses*, whereas hair growth is a *low-demand use*. The last stage of pregnancy has *variable requirements*. Growth and finishing may be classed as intermediate in nutrient demands. Each of these needs will be discussed in more detail.

MAINTENANCE

Pigs, unlike machines, are never idle. They use nutrients to keep their bodies functioning every hour of every day, even when they are not being used for production.

Maintenance requirements may be defined as the combination of nutrients which are needed by the animal to keep its body functioning without any gain or loss in body weight or any productive activity. Although these requirements are relatively simple, they are essential for life itself. A mature pig must have (1) heat to maintain body temperature, (2) sufficient energy to keep vital body processes functional, (3) energy for minimal movement, and (4) the necessary nutrients to repair damaged cells and tissues and to replace those which have become nonfunctional. Thus, energy is the primary nutritive need for maintenance. Even though the quantity of other nutrients required for maintenance is relatively small, it is necessary to have a balance of the essential amino acids, minerals, and vitamins.

No matter how quietly a pig may be lying in a pen, it requires a certain amount of fuel and other nutrients. The least amount on which it can exist is called its *basal maintenance requirement*. With the exception of horses, most animals require about 9% more fuel (calories) when standing than when lying, and even more is needed when they walk or run.

There are only a few times in the normal life of a pig when only the maintenance requirement needs to be met. Such a status is closely approached by mature males not in service and by mature, dry, nonpregnant females. Nevertheless, maintenance is the standard bench mark or reference point for evaluating nutritional needs.

Even though maintenance requirements might be considered an expression of the nonproduction needs of a pig, there are many factors which affect the amount of nutrients necessary for this vital function; among them, (1) exercise, (2) weather, (3) stress, (4) health, (5) body size, (6) temperament, (7) individual variation, (8) level of production, and (9) lactation. The first four are *external factors*—they are subject to control to some degree through management and facilities. The others are *internal factors*—they are part of the animal itself. Both external and internal factors influence requirements according to their intensity. For example, the colder or hotter it gets from the most comfortable (optimum) temperature, the greater will be the maintenance requirements.

GROWTH

Growth may be defined as the increase in size of bones, muscles, internal organs, and other parts of the body. It is the normal process before birth, and after birth until the pig reaches its full, mature size. Growth is influenced primarily by nutrient intake. The nutritive requirements become increasingly acute when young animals are under forced production, such as when pigs are fed to reach market weight by 160 to 190 days of age. The growth of pigs in relation to age, daily feed intake, daily gain, and feed efficiency is graphed in Fig. 8-3.

Growth is the very foundation of swine production. Young swine will not make the most economical finishing gains unless they have been raised to be thrifty and vigorous. Likewise, breeding females may have their reproductive ability seriously impaired if they have been improperly grown.

Generally speaking, organs vital for the maintenance of life—*e.g.*, the brain, which coordinates body activities, and the gut, upon which the rest of the postnatal growth depends—are early developing; and the commercially more valuable parts such as muscle develop later.

Knowledge of normal growth and development is useful for a variety of purposes. From a nutritional standpoint, growth curves are used primarily as standards against which to gauge the adequacy of nutrient allowances. In fact, such curves are often the entire basis for the allowances set down in dietary and feeding standards. Also, they provide a basis for comparisons of breeding groups and serve as a reference point from which to establish breeding and management objectives. Economically, growth is important, for young gains are cheap gains—less feed for gain (see Table 8-2). This is generally so because, in comparison

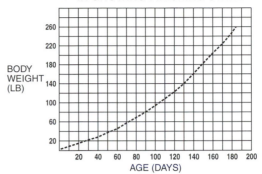

RELATIONSHIP OF AGE TO BODY WEIGHT IN GROWING-FINISHING SWINE

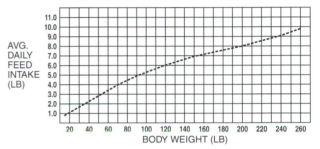

RELATIONSHIP OF FEED INTAKE TO BODY WEIGHT IN GROWING-FINISHING SWINE

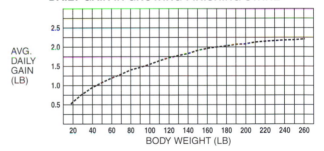

RELATIONSHIP OF BODY WEIGHT AND AVERAGE DAILY GAIN IN GROWING-FINISHING SWINE

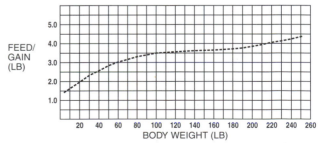

RELATIONSHIP OF FEED REQUIRED PER POUND OF GAIN AND BODY WEIGHT IN GROWING-FINISHING SWINE

Fig. 8-3. The growth of pigs—body weight changes—in relation to age, daily feed intake, daily gain, and feed efficiency (feed/gain).

TABLE 8-2
TARGET PERFORMANCE FOR IMPROVED PIGS

Stage	Age/Weight	Feed Intake Per Day		Daily Gains		Feed Conversion
		(lb)	*(kg)*	*(lb)*	*(kg)*	
Pre-nursery	3–6 weeks	0.5–0.7	1.1–1.5	0.5–0.6	1.1–1.3	1.0–1.25
Nursery	6–9 weeks	2.3–2.5	5.1–5.5	1.2–1.4	2.6–3.1	1.6–1.7
Grower	50–120 lb	3.6–4.2	7.9–9.2	1.6–1.8	3.5–4.0	2.3–2.6
Finisher	120–240 lb	6.5–7.5	14.3–16.5	1.9–2.2	4.2–4.8	3.4–3.8

with older animals, young animals (1) consume more feed according to size, (2) use a smaller proportion of their feed for maintenance, and (3) form relatively more muscle tissue which has a lower caloric value than fat. Still, the nutritive needs for growth vary with breed, sex, rate of growth, and health.

REPRODUCTION

Being born and born alive are the first and most important requisites of pork production, for if pigs fail to reproduce, the breeder is soon out of business. A "mating of the gods," involving the greatest genes in the world, is of no value unless these genes result in (1) the successful joining of the sperm and egg, and (2) the birth of live offspring. Still, research shows that embryonic mortality claims 5 to 30% of the swine embryos; of the pigs born alive 10 to 20% die before weaning, and that 15% of all sows bred fail to produce litters. Certainly, there are many causes of reproductive failure, but scientists agree that nutritional inadequacies play a major role.

Since swine producers largely determine their own destiny when it comes to feeding, it is important that they know the causes of reproductive failure and how to rectify them.

A review of the literature clearly points to three reproductive difficulties: (1) females failing to show signs of heat, (2) the low conception rate at first service, and (3) the excessive losses at birth or within the first two weeks of age.

Research gives ample evidence that the real cause of most reproductive failure is a deficiency of one or more nutrients just before or immediately following parturition—nutritive deficiencies during the critical period when life begins—a deficiency of energy, protein, minerals, and/or vitamins.

With all mammalian species, most of the growth of the fetus occurs during the last third of pregnancy. Additionally, the female must store body reserves during pregnancy, for the demands for milk production are generally greater than can be supplied by the diet fed during early lactation. Hence, the nutrient requirements are very critical during this period, especially for young pregnant females.

It is also known that the ration exerts a powerful effect on sperm production and semen quality. Too fat a condition can lead to temporary or permanent sterility. Moreover, there is abundant evidence that greater fertility of herd sires exists under conditions where a well-balanced diet is provided.

LACTATION

The lactation requirements of females of all mammalian species for moderate to heavy milk production are much more rigorous than the maintenance or pregnancy requirements. Fortunately, females can store up body reserves during pregnancy and then draw upon them during lactation. But, if there has not been proper body storage, something must "give"—and that something will be the mother, for nature ordained that growth of the fetus, and the lactation that follows, shall take priority over the maternal requirements. Hence, when there is a nutrient deficiency, the female's body will be deprived, or even stunted if she is young, before the developing fetus or milk production will be materially affected.

FINISHING

Finishing refers to the phase in the life cycle of market hogs from approximately 120 lb to market weight of 230 to 250 lb.

Finishing is usually achieved through the use of high-energy feeds, primarily carbohydrates, along with adequate and balanced amino acids, minerals, and vitamins.

The objective of swine producers is to finish animals to carcass weight desired by consumers, with a maximum of lean meat and a minimum of fat, at a maximum of profit for their efforts.

■ **Fitting**—*Fitting is the conditioning of animals, usually for show or sale, through careful feeding and grooming to enhance their bloom and attractiveness.*

Fitting animals for show or sale involves the application of similar principles and practices to those followed in finishing hogs for market. Animals intended for show or sale should be fed so as to achieve a certain amount of finish or bloom, but they should not be fat. In general, most fitting rations are similar to the rations used in commercial finishing operations for animals of comparable ages, except that they are usually higher in protein content; experienced herdsmen feel that they get more bloom by use of high-protein diets. Also, it is common practice to feed a palatable milk replacer to young animals that are being fitted for show or sale.

NUTRIENTS

Nutrients are utilized in one of two metabolic processes: (1) for anabolism, or (2) for catabolism. *Anabolism is the process by which nutrient molecules are used as building blocks for the synthesis of complex molecules. Anabolic reactions are endergonic—that is, they require the input of energy into the system. Catabolism is the oxidation of nutrients, liberating energy (exergonic reaction) which is used to fulfill the body's immediate demands.*

ENERGY

Energy is required for practically all life processes—for the action of the heart, maintenance of blood pressure and muscle tone, transmission of nerve impulses, ion transport across membranes, reabsorption in the kidneys, protein and fat synthesis, and the production of milk.

A deficiency of energy is manifested by slow or stunted growth, body tissue losses, and/or lowered production of meat rather than by specific signs, such as those which characterize many mineral and vitamin deficiencies. For this reason, energy deficiencies often go undetected and unrectified for extended periods of time.

It is common knowledge that a diet must contain carbohydrates, fats, and proteins. Although each of these has specific functions in maintaining a normal body, all of them can be used to provide energy for maintenance, for growth, and for reproduction. From the standpoint of supplying the normal energy needs of pigs, however, the carbohydrates are by far the most important, more of them being consumed than any other compound, whereas the fats are next in importance for energy purposes. Carbohydrates are usually more abundant and cheaper, and most of them are very easily digested, absorbed, and transformed into body fat. Also, carbohydrate feeds may be more easily stored than fats in warm weather and for longer periods of time.

CARBOHYDRATES

Carbohydrates are organic compounds composed of carbon, hydrogen, and oxygen—formed in plants by the process of photosynthesis. They constitute about 75% of the dry weight of plants and grain and make up a large part of the swine ration. They serve as a source of heat and energy in the pig's body, and a surplus of them is transformed into fat and stored.

No appreciable amount of carbohydrate is found in the animal body at any one time, the blood supply being held rather constant at about 0.05 to 0.1% for most animals. However, this small quantity of glucose in the blood, which is constantly replenished by changing the glycogen of the liver back to glucose, serves as the chief source of fuel with which to maintain the body temperature and furnish the energy needed for all body processes. The storage of glycogen (so-called animal starch) in the liver amounts to 3 to 7% of the weight of that organ.

FATS

Lipids (fat and fatlike substances), like carbohydrates, contain the three elements—carbon, hydrogen, and oxygen.

As feeds, fats function much like carbohydrates in that they serve as a source of heat and energy and for the formation of fat. Because of the larger proportion of carbon and hydrogen, however, fats liberate more heat than carbohydrates when digested, furnishing approximately 2.25 times as much heat or energy per pound on oxidation as do the carbohydrates. A smaller quantity of fat is required, therefore, to serve the same function.

Research indicates that the addition of 3 to 5% fat to growing-finishing swine diets will improve feed conversion, and often improve average daily gain. However, recent data indicates that adding fat to *ad libitum* fed diets tends to increase backfat thickness.

Research with sows indicates that the addition of 5% fat to the diet about 10 days before farrowing may improve pig survivability, perhaps due to increased milk yield and milk fat content.

■ **Feed fats affect body fats**—Thus, swine consuming soft fat such as vegetable oils may produce soft pork.

■ **Fatty acids**—These are the key components of fats (lipids). Their length and degree of saturation (amount of hydrogen) determine many of the physical aspects—melting point and stability—of fats (lipids).

There is evidence that linoleic, linolenic, and arachidonic acids are dietary essentials. Deficiency symptoms of these fatty acids have been observed in swine, mice, poultry, dogs, guinea pigs, and human

Fig. 8-4. Fat deficiency. Littermate pigs. *Top:* Pig received fat in the diet (5.0% ether extract). *Bottom:* Pig received little fat in the diet (0.06% ether extract). Note loss of hair and scaly dandrufflike dermatitis—especially on the feet and tail. (Courtesy, Dr. W. M. Beeson, Purdue University)

infants. Depending on the type of animal, numerous manifestations of these deficiencies are seen; among them, dermatitis, reduced growth, increased water consumption and retention, impaired reproduction, and increased metabolic rate. Two functions of the fatty acids have been postulated: (1) precursors of prostaglandins, and (2) structural components of cells.

The evidence of the first theory is conclusive. The second theory has substantial support inasmuch as the essential fatty acids are in highest concentrations in phospholipids, a type of lipid that plays an important role in the structural integrity of the cell.

MEASURING AND EXPRESSING ENERGY VALUE OF FEEDSTUFFS

One nutrient cannot be considered as more important than another, because all nutrients must be present in adequate amounts if efficient production is to be maintained. Yet, historically, feedstuffs have been compared or evaluated primarily on their ability to supply energy to animals. This is understandable because (1) energy is required in larger amounts than any other nutrient, and (2) energy is the major cost associated with feeding swine.

Our understanding of energy metabolism has increased through the years. With this added knowledge, changes have come in both the methods and terms used to express the energy value of feeds.

Two methods of measuring energy are employed in the United States—the total digestible nutrient (TDN) system, and the calorie system.

ENERGY DEFINITIONS AND CONVERSIONS

Some pertinent definitions and conversions of energy terms follow:

■ **Calorie (cal)**—The amount of energy as heat required to raise the temperature of 1 g of water 1°C (precisely from 14.5° to 15.5°C). It is equivalent to 4.184 joules. Although *not preferred*, it is also called a *small calorie* and so designated by being spelled with a lower case "c." *Note well:* In popular writings, especially those concerned with human caloric requirements, the term *calorie* is frequently used erroneously for the kilocalorie.

■ **Kilocalorie (kcal)**—The amount of energy as heat required to raise the temperature of 1 kg of water 1°C (from 14.5° to 15.5°C). Equivalent to 1,000 calories. In human nutrition, it is referred to as a kilogram calorie or as a "large Calorie" and is so designated by being spelled with a capital "C" to distinguish it from the "small calorie."

■ **Megacalorie (Mcal)**—Equivalent to 1,000 kcal or 1,000,000 calories. Also, referred to as a *therm*, but the term *megacalorie* is preferred.

■ **British Thermal Unit (Btu)**—The amount of energy as heat required to raise 1 lb of water 1°F; equivalent to 252 calories. This term is seldom used in animal nutrition.

■ **Joule**—A proposed international unit (4.184J = 1 calorie) for expressing mechanical, chemical, or electrical energy, as well as the concept of heat. In the future, energy requirements and feed values will likely be expressed by this unit.

■ **Converting TDN to Mcal**—One lb of TDN = 2.0 Mcal or 2,000 kcal. It is recognized, however, that the roughage component in a ration affects its energy value.

CALORIE SYSTEM OF ENERGY EVALUATION

To measure calories, an instrument known as the

bomb calorimeter is used, in which the feed (or other substance) tested is placed and burned in the presence of oxygen.

Through various digestive and metabolic processes, numerous losses of the energy in feed occur as it passes through the pig's digestive system. These losses are illustrated in Fig. 8-5.

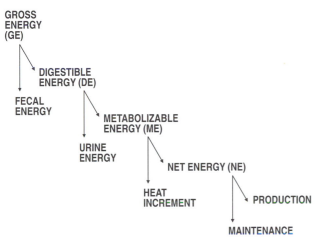

Fig. 8-5. Utilization of energy. Digestible energy is roughly comparable to total digestible nutrients (TDN).

As shown in Fig. 8-5, energy losses occur in the digestion and metabolism of feed. Measures that are used to express energy requirements and the energy content of feeds differ primarily in the digestive and metabolic losses that are included in their determination. Thus, the following terms are used to express the energy value of feeds:

■ **Gross energy (GE)**—Gross energy represents the total combustible energy in a feedstuff. It does not differ greatly between feeds, except for those high in fat.

■ **Digestible energy (DE)**—Digestible energy is that portion of the GE in a feed that is not excreted in the feces.

■ **Metabolizable energy (ME)**—Metabolizable energy represents that portion of the GE that is not lost in the feces, urine, and gas. The losses of energy as gas produced in the digestive tract of swine are small; therefore, the ME values are not corrected for this energy loss. Although ME more accurately describes the useful energy in the feed than does GE or DE, it does not take into account the energy lost as heat. ME values are calculated from the formula—

$$ME = DE \frac{(96 - 0.2 \times \% \text{ crude protein})}{100}$$

■ **Net energy (NE)**—Net energy represents the energy fraction in a feed that is left after the fecal, urinary, gas, and heat losses are deducted from the GE, or in other words, the ME minus the heat losses. The net energy, as a fraction of the ME, varies from 27 to 69% for common swine feeds. While the NE may be the best measure of the energy available for maintenance and production, it is difficult to measure. At present, NE requirements for maintenance and production are not available. Therefore, the energy requirements of swine and the energy values of the feedstuffs for swine are presented in DE and ME values.

TOTAL DIGESTIBLE NUTRIENTS (TDN)

Total digestible nutrients (TDN) is the sum of the digestible protein, fiber, nitrogen-free extract, and fat × 2.25. It has been the most extensively used measure for energy in the United States.

Back of TDN values are the following steps:

1. **Digestibility.** The digestibility of a particular feed for a specific species is determined by a digestion trial.

2. **Computation of digestible nutrients.** Digestible nutrients are computed by multiplying the percentage of each nutrient in the feed (protein, fiber, nitrogen-free extract [NFE], and fat) by its digestion coefficient. The result is expressed as digestible protein, digestible fiber, digestible NFE, and digestible fat. For example, if No. 2 corn contains 8.9% protein of which 77% is digestible, the percent of digestible protein is 6.9.

3. **Computation of total digestible nutrients (TDN).** The TDN is computed by use of the following formula:

$$\% \text{ TDN} = \frac{\text{DCP} + \text{DCF} + \text{DNFE} + (\text{DEE} \times 2.25)}{\text{feed consumed}} \times 100$$

where DCP = digestible crude protein; DCF = digestible crude fiber; DNFE = digestible nitrogen-free extract; and DEE = digestible ether extract.

TDN is ordinarily expressed as a percent of the ration or in units of weight (lb or kg), not as a caloric figure.

The main **advantage** of the TDN system is that it has been used for a very long time and many people are acquainted with it.

The main **disadvantages** of the TDN system are:

1. It is really a misnomer, because TDN is not an actual total of the digestible nutrients in a feed. It does not include the digestible mineral matter (such as salt, limestone, and defluorinated phosphate—all of which are digestible); and the digestible fat is multiplied by the factor 2.25 before being included in the TDN figure,

because its energy value is higher than carbohydrates and protein. As a result of multiplying fat by the factor 2.25, feeds high in fat will sometimes exceed 100 in percentage TDN (a pure fat with a coefficient of digestibility of 100% would have a theoretical TDN value of 225% − 100% × 2.25).

2. It is an empirical formula based upon chemical determinations that are not related to actual metabolism of the animal.

3. It is expressed as a percent or in weight (lb or kg), whereas energy is expressed in calories.

4. It takes into consideration only digestive losses; it does not take into account other important losses, such as losses in the urine, gases, and increased heat production (heat increment).

5. It overevaluates roughages in relation to concentrates when fed for high rates of production, due to the higher heat loss per pound of TDN in high-fiber feeds.

Because of these several limitations, in the United States the TDN system is gradually being replaced by other energy evaluation systems. However, many of the available data, both for the energy requirements of animals and for the energy composition of feeds, are reported as TDN. Hence, TDN will be used for a long time to come. In this book, the energy requirements of swine are given in terms of digestible energy (DE) and metabolizable energy (ME). Nevertheless, the TDN composition of feeds is included along with DE and ME energy values.

PROTEINS

Proteins are complex organic compounds made up chiefly of amino acids, which are present in characteristic proportions for each specific protein. This nutrient always contains carbon, hydrogen, oxygen, and nitrogen; and in addition, it usually contains sulfur and frequently phosphorus. Proteins are essential in all plant and animal life as components of the active protoplasm of each living cell.

Crude protein refers to all the nitrogenous compounds in a feed. It is determined by finding the nitrogen content and multiplying the result by 6.25. The nitrogen content of protein averages about 16% (100 ÷ 16 = 6.25).

In plants, the protein is largely concentrated in the actively growing portions, especially the leaves and seeds. Plants also have the ability to synthesize their own proteins from such relatively simple soil and air compounds as carbon dioxide, water, nitrates, and sulfates, using energy from the sun. Thus, plants, together with some bacteria which are able to synthesize these products, are the original sources of all proteins.

Proteins are much more widely distributed in animals than in plants. Thus, the proteins of the animal body are primary constituents of many structural and protective tissues—such as bones, ligaments, feathers, skin, and the soft tissues which include the organs and muscles.

Pigs of all ages require adequate amounts of protein of suitable quality for maintenance, growth, and reproduction. Of course, the protein requirements for growth are the greatest and most critical.

From a nutritional standpoint, the requirement for protein is not for protein *per se* but rather for certain of the amino acids which cannot be synthesized by the pig but are necessary for normal growth and development. These amino acids which must be supplied by the diet are referred to as essential or indispensable amino acids. An amino acid is nonessential (dispensable) if it can be synthesized in the body. This requires nitrogen which is a function of protein in the diet.

In the body, the amino acids function as building blocks for new protein, as well as functioning in some other specific metabolic roles such as the formation of neurotransmitters, hormones, purines, and urea. The essential and nonessential amino acids are identified as follows:

Essential (Indispensable)	Nonessential (dispensable)
Arginine	Alanine
Histidine	Asparagine
Isoleucine	Aspartic acid
Leucine	Cysteine
Lysine	Cystine
Methionine	Glutamic acid
Phenylalanine	Glutamine
Threonine	Glycine
Tryptophan	Hydroxyproline
Valine	Proline
	Serine
	Tyrosine

It is becoming increasingly important to specify lysine levels when formulating and evaluating swine diets.

If a diet is inadequate in any essential amino acid, protein synthesis cannot proceed beyond the rate at which that amino acid is available. This is called the limiting amino acid.

The lysine and other essential amino acid requirements for swine differ as follows:

1. They are higher for young, rapidly growing pigs.
2. They are higher for boars and barrows than for gilts.
3. They are higher for high lean growth pigs.
4. They are higher in the hot summer months,

when the appetite and feed consumption of pigs decreases.

When selecting feeds, producers should be aware that all proteins are not created equal. Plant proteins often contain insufficient quantities of lysine, methionine and cystine, tryptophan, and/or threonine. Hence, the term *quality of protein* is often used to describe the amino acid balance of a protein. A protein is said to be of good quality when it contains all the essential amino acids in proper proportions and amounts, and to be of poor quality when it is deficient in either content or balance of essential amino acids. From this it is evident that the usefulness of a protein source depends upon its amino acid composition, because the real need of the pig is for amino acids and not for protein as such.

Although it is common practice to refer to "percent protein" in a ration, this term has little significance in swine nutrition unless there is information about the amino acids present. For swine, quality is just as important as quantity. It is possible for pigs to perform better on a 12% protein diet, well balanced for amino acids, than on a 16% diet having a poor amino acid balance.

From a practical standpoint, the problem of building a balanced diet for swine is centered around correcting the deficiencies of the cereal grains. Although corn, wheat, and barley may contain from 8 to 12% protein, their protein is seriously deficient in the essential amino acid, lysine. Corn is also deficient in tryptophan, as is meat and bone meal. Since protein supplements are more expensive than grain, the tendency is to feed too little of them.

Previously, it was stated that when an excess of energy is consumed by the pig it is stored in the form of fat. Protein is not stored in the body in appreciable amounts. If an excess of protein is fed, the unused nitrogen portion is discarded as urea in the urine and the carbon fraction is used as a source of energy. From an economic standpoint, it is unprofitable to feed more protein than needed to meet the nutritional requirements of the pig.

The use of crystalline amino acids, particularly lysine, now makes it possible to feed diets containing less protein than is commonly recommended for swine.

Symptoms of protein (amino acid) deficiency are reduced feed intake, stunted growth, poor hair and skin condition, and lowered reproduction.

MINERALS AND VITAMINS OF SWINE AFFECTED BY MANAGEMENT CHANGES

Mineral and vitamin additions to the diets of swine have increased since the 1980s along with changes in housing, feeding, and management. Among the more important mineral and vitamin changes are the following:

1. Confinement production has denied swine access to pasture crops and soils, which provided minerals and vitamins.

2. Slotted floors have prevented coprophagy (recycling of feces), which may be high in B-vitamins and vitamin K, synthesized by microorganisms in the large intestine.

3. Reduced use of multiple protein sources in diets, which often compliment each other in providing minerals and vitamins, has lessened proteins as a source of these nutrients.

4. Reduced daily feed intake of sows during gestation, calls for increased mineral and vitamin concentration in the ration.

5. The trend to earlier weaning, at 3 to 4 weeks of age, calls for a higher quality of baby pig diet in all nutrients.

6. The bioavailability of nutrients in heat dried grains and feed ingredients varies widely. Also, the presence of inhibitors and molds in feeds may result in reduced absorption, increasing the requirements for certain vitamins.

MINERALS

Of all common farm animals, the pig is most likely to suffer from mineral deficiencies. This is due to the following peculiarities of swine husbandry:

1. Hogs are fed principally upon cereal grains and

Fig. 8-6. Amino acid deficiency in littermate pigs. Pig B was fed an adequate diet containing Opaque-2 corn (high-lysine corn). Pig C was fed inadequate amounts of lysine and tryptophan. (Courtesy, Cornell University)

their byproducts, all of which are relatively low in mineral matter, particularly in calcium.

2. The skeleton of the pig supports greater weight in proportion to its size than that of any other farm animal.

3. Hogs are fed to grow at a maximum rate for an early market, before they are mature.

4. Hogs reproduce at a younger age than other classes of livestock.

5. Increased confinement rearing, without access to soil or forage, which would tend to balance the mineral deficiencies of the grains.

The functions of minerals are extremely diverse. They range from structural functions in some tissues to a wide variety of regulatory functions in other tissues.

Swine require at least 13 known inorganic elements, including calcium, chlorine, copper, iodine, iron, magnesium, manganese, phosphorus, potassium, selenium, sodium, sulfur, and zinc. Also, cobalt is required in the synthesis of vitamin B-12. Pigs may also require other trace elements, such as arsenic, boron, bromine, cadmium chromium, fluorine, lead, lithium, molybdenum, nickel, silicon, tin, and vanadium, which

TABLE
SWINE MINERAL

Minerals Which May Be Deficient Under Normal Conditions	Conditions Usually Prevailing Where Deficiencies Are Reported	Functions of Mineral	Deficiency Symptoms/Toxicity
Major or Macrominerals			
Salt (NaCl)	Salt deficiencies may exist when the protein supplement is all or chiefly of plant origin, although herbivorous animals require more salt than swine.	Sodium and chlorine are the principal extracellular cation and anion, respectively, in the body. Chlorine is the chief anion in gastric juice. Improves appetite, promotes growth, helps regulate body pH, and is essential for hydrochloric acid formation in the stomach.	**Deficiency symptoms**—Poor and depraved appetite, unthrifty condition, and failure to grow. **Toxicity**—Nervousness, weakness, staggering, epileptic seizures, paralysis, and death.
Calcium (Ca)	When the protein supplements are chiefly of plant origin and little forage is used. When swine are raised in confinement without vitamin D added to the ration. When feed intake is restricted during gestation. When there is a poor calcium-phosphorus ration. Retention of calcium is affected by source of dietary protein (or phytic acid content) and the level of magnesium.	Bone and teeth formation; nerve function; muscle contraction; blood coagulation; cell permeability. Essential for milk production.	**Deficiency symptoms**—Loss of appetite and poor growth, lack of thrift, lameness and stiffness, weakened bone structure, and impaired reproduction. Severe cases may show reduced serum calcium and tetany. Rickets may develop in young pigs, or osteomalacia in older animals. Paralysis of the hind legs. **Toxicity**—An excess level of calcium tends to reduce the performance of pigs and increase the pig's zinc requirement.
Phosphorus (P)	Rations containing only plant ingredients; late gestation; lactation; high-calcium rations; swine in confinement without vitamin D added to the ration; poor calcium to phosphorus ratio. Retention of phosphorus is affected by source of dietary protein (or phytic acid content) and the level of magnesium.	Bone and teeth formation; a component of phospholipids which are important in lipid transport and metabolism and cell-membrane structure. In energy metabolism. A component of RNA and DNA, the vital cellular constituents required for protein synthesis. A constituent of several enzyme systems.	**Deficiency symptoms**—Loss of appetite and poor growth, lameness and stiffness, weakened bone structure, reduced inorganic blood phosphorus, depraved appetite, breeding difficulties, and rickets in young pigs, or osteomalacia in older animals. Paralysis of the hind legs, which is called posterior paralysis. **Toxicity**—An excess of phosphorus tends to reduce the performance of pigs, but is not toxic as such.
Magnesium (Mg)	Some research suggests that the magnesium in natural ingredients is only 50–60% available to the pig.	Essential for normal skeletal development, as a constituent of bone, cofactor in many enzyme systems, primarily in the glycolytic system.	**Deficiency symptoms**—Hyperirritability, muscular twitching, reluctance to stand, weak pasterns, loss of equilibrium, and tetany, followed by death. **Toxicity**—The toxicity level of magnesium is not known.

have been shown to have a physiological role in one or more species. These elements are required at such low levels, however, that their dietary essentiality for the pig has not been proven. Most of them are believed to be present in adequate quantities in natural feed ingredients. However, the confinement of swine and the use of simpler swine diets with fewer ingredients may necessitate consideration of their importance in the future.

SWINE MINERAL CHART

Table 8-3, Swine Mineral Chart, gives in summary form the following pertinent information relative to each mineral listed: (1) conditions usually prevailing where deficiencies are reported, (2) functions, (3) deficiency symptoms/toxicity, (4) mineral requirements, (5) recommended allowances, and (6) practical sources. Further elucidation of minerals is contained in the narrative that follows Table 8-3.

8-3
CHART

Mineral Requirements[1]		Recommended Allowances[1]	Practical Sources of the Mineral	Comments
Minerals/ Animal/Day	Mineral Content of Diet			
*Na, variable according to class, age, and weight of swine (see Tables 9-1 and 9-2).	*Na, variable according to class, age, and weight of swine (see Tables 9-5 and 9-6).	Salt variable according to class, age, and weight of swine (see Tables 9-8 and 9-9).	Salt in loose form.	In iodine-deficient areas, stabilized iodized salt should be used. When pigs are salt starved, precaution should be taken to prevent over consumption of salt.
*Variable according to class, age, and weight of swine (see Tables 9-1 and 9-2).	*Variable according to class, age, and weight of swine (see Tables 9-5, 9-6, and 9-7).	Variable according to class, age, and weight of swine (see Tables 9-8 and 9-9).	Ground limestone, gypsum, or oystershell flour. Where both Ca and P are needed, use monocalcium phosphate, dicalcium phosphate, tricalcium phosphate, defluorinated phosphate, or bone meal.	Because cereal grains (which largely form the ration of swine) are low in Ca, swine are more apt to suffer from Ca deficiencies than from any of the other minerals except salt. *Most favorable Ca:P ratio is between 1:1 and 1.5:1. Sow's milk contains a Ca:P ratio of 1.3:1.
*Variable according to class, age, and weight of swine (see Tables 9-1 and 9-2).	*Variable according to class, age, and weight of swine (see Table 9-5, 9-6, and 9-7).	Variable according to class, age, and weight of swine (see Tables 9-8 and 9-9).	Where both CA and P are needed, use monocalcium phosphate, dicalcium phosphate, tricalcium phosphate, defluorinated phosphate, or bone meal.	*About 60–75% of the P in cereal grains and their byproducts, and in oilseed meals, is organically bound in the form of phytate and poorly available to the pig. *Most favorable Ca:P ratio is between 1:1 and 1.5:1. Sow's milk contains a Ca:P ratio of 1.3:1. Excess levels of P reduce performance of pigs.
*Variable according to class, age, and weight of swine (see Tables 9-1 and 9-2).	*0.04% of as-fed ration.	Practical rations are adequate in magnesium.	Magnesium oxide, magnesium sulfate, or magnesium carbonate. Dolomitic limestone.	Milk contains adequate magnesium for suckling pigs.

(Continued)

TABLE 8-3

Minerals Which May Be Deficient Under Normal Conditions	Conditions Usually Prevailing Where Deficiencies Are Reported	Functions of Mineral	Deficiency Symptoms/Toxicity
Major or Macrominerals (continued)			
Potassium (K)		Major cation of intracellular fluid where it is involved in osmotic pressure and acid-base balance. Electrolyte balance and neuromuscular function. Required in enzyme reaction involving phosphorylation of creatinine. Influences carbohydrate metabolism.	**Deficiency symptoms**—Loss of appetite, slow growth, poor hair and skin condition, decreased feed efficiency, inactivity, lack of coordination, and cardiac impairment. **Toxicity**—The toxic level of potassium is not well established. Pigs can tolerate up to 10 times the requirement if plenty of drinking water is provided.
Sulfur (S)		For synthesis of sulfur-containing compounds, such as glutathione, taurocholic acid, and chondroitin sulfate.	
Trace or microminerals			
Cobalt (Co)	If vitamin B-12 is limited.	An essential component of vitamin B-12.	**Deficiency symptoms**—No deficiency symptoms reported in swine. However, supplemental cobalt prevents lesions associated with zinc deficiency. **Toxicity**—A level of 400 ppm of cobalt is toxic to the young pig and may cause loss of appetite, stiff-leggedness, humped back, incoordination, muscle tremors, and anemia.
Copper (Cu)	Suckling pigs kept off soil.	Essential element in a number of enzyme systems and necessary for synthesizing hemoglobin and preventing nutritional anemia. Hemoglobin serves as a carrier of oxygen throughout the body. When fed at 100 to 250 ppm, copper stimulates growth in pigs.	**Deficiency symptoms**—Slow growth, poor hair and skin condition, lameness and stiffness, weakened bone structure, weak and crooked legs, anemia, and cardiac and vascular disorders. **Toxicity**—Depressed hemoglobin levels and jaundice.
Iodine (I)	Iodine-deficient areas/soils (in northwestern U.S. and in the Great Lakes region) when iodized salt is not fed. Where feeds come from iodine-deficient areas.	Needed by the thyroid gland for making thyroxin, an iodine-containing hormone which controls the rate of body metabolism or heat production.	**Deficiency symptoms**—Loss of appetite, slow growth, poor hair and skin condition, impaired breeding or gestation, offspring dead or weak at birth, pigs hairless at birth, and/or enlarged thyroid. **Toxicity**—An 800-ppm iodine level in the ration depresses growth, hemoglobin level, and liver iron concentration in growing pigs. During the last 30 days of gestation and during lactation, 1,500 to 2,500 ppm of iodine was found harmful to sows.
Iron (Fe)	Suckling pigs kept off soil.	*Iron is required as a component of hemoglobin in red blood cells. Iron also is found in muscle as myoglobin, in serum as transferrin, in the placenta as interoferrin, in milk as lactoferrin, and in the liver as ferritin and hemosiderin. Iron also plays an important role in the body as a constituent of a number of metabolic enzymes.	**Deficiency symptoms**—Loss of appetite, slow growth, poor hair and skin condition, paleness of mucous membranes, high mortality in young pigs, susceptibility to disease, thumps (characterized by labored breathing), and anemia. The number of grams of hemoglobin per 100 ml of blood is a rapid, reliable indicator of the iron status of the pig. A hemoglobin level of 8 g/100 ml indicates borderline anemia; a level of 7 g or less/100 ml indicates anemia. **Toxicity**—In 3- to 10-day-old pigs, the toxic oral dose of iron from ferrous sulfate is approximately 273 mcg/lb *(601 mcg/kg)* body weight.

(Continued)

Mineral Requirements[1]		Recommended Allowances[1]	Practical Sources of the Mineral	Comments
Minerals/ Animal/Day	Mineral Content of Diet			
*Variable according to class, age, and weight of swine (see Tables 9-1 and 9-2).	*Variable for young pigs (see Tables 9-5 and 9-6). *No estimates available for finishing and breeding swine.	Practical rations are adequate in potassium.	Corn contains 0.33% potassium, and other cereals contain 0.42–0.49% potassium.	Potassium is the third most abundant mineral in the body of the pig, exceeded only by calcium and phosphorus.
		The addition of inorganic sulfate to low-protein rations has not been beneficial.		
No requirements for cobalt have been established.		Practical diets are adequate in cobalt.	Cobalt chloride, cobalt sulfate, cobalt oxide, or cobalt carbonate. Also, several good commercial minerals containing cobalt are on the market.	
*Variable according to class, age, and weight of swine (see Tables 9-1 and 9-2).	*Variable according to class, age, and weight of swine (see Tables 9-5 and 9-6).	Variable according to class, age, and weight of swine (see Tables 9-8 and 9-9).	Copper sulfate, copper carbonate, and copper chloride are about equally effective. The copper in copper sulfide and copper oxide is poorly available to the pig.	Beyond the suckling period, natural feedstuffs usually contain enough copper. *When fed at a level of 100 to 250 ppm, copper will increase rate and efficiency of gains of pigs to breeding age.
*Variable according to class, age, and weight of swine (see Tables 9-1 and 9-2).	*0.06 mg/lb (0.14 mg/kg) as-fed diet.	Variable according to class, age, and weight of swine (see Table 9-8).	Stabilized iodized salt containing 0.007% iodine. Calcium iodate, potassium iodate, and pentacalcium orthoperiodate.	The majority of the iodine in the bodies of swine is present in the thyroid.
*Variable according to class, age, and weight of swine (see Tables 9-1 and 9-2). *Newborn pigs require 7 to 16 mg of absorbed iron daily for normal growth.	*Variable according to class, age, and weight of swine (see Tables 9-5 and 9-6). *The iron requirement of young pigs fed milk or purified liquid diets is 23 to 68 mg/lb (50 to 150 mg/kg) of milk solids. The iron requirement of pigs fed a dry, casein-based diet is about 50% higher/unit of dry matter than for those fed a similar diet in liquid form.	Variable according to class, age, and weight of swine (see Tables 9-8 and 9-9).	A single intramuscular injection of 200 mg of iron, in the form of iron dextran, given the first 3 days of life; or Oral administration of iron from iron chelates within the first few hours of life. Ferrous sulfate, ferric chloride, ferric citrate, ferric choline citrate, and ferric ammonium citrate are effective in preventing iron deficiency anemia. The iron in ferric oxide is largely unavailable.	*Pigs are born with about 50 mg of iron. *Iron has a detoxifying effect when added to gossypol-containing diets. Add iron from soluble source to free gossypol at a weight ratio of 1:1. *Milk is deficient in iron (sow's milk contains an average of 1 mg of iron/liter). Pigs should be encouraged to eat grain diet as soon as old enough. *Natural feed ingredients usually supply enough iron to meet post-weaning requirements.

(Continued)

Minerals Which May Be Deficient Under Normal Conditions	Conditions Usually Prevailing Where Deficiencies Are Reported	Functions of Mineral	Deficiency Symptoms/Toxicity
Trace or Microminerals (continued)			
Manganese (Mn)		Functions as a component of several enzymes involved in carbohydrate, lipid, and protein metabolism. A component in the organic matrix of bone.	**Deficiency symptoms**—Abnormal skeletal growth, increased fat depositioin, irregular or absent estrus cycles, resorbed fetuses, small and weak pigs at birth, and reduced milk production. **Toxicity**—The toxic level of manganese is not clearly defined. But high levels result in depressed feed intake, reduced growth, and limb stiffness.
Selenium (Se)	When diets consist almost exclusively of ingredients grown on selenium-deficient soils.	Functions as a part of glutathione peroxidase, an enzyme which enables the tripeptide glutathione to perform its role as a biological antioxidant in the body. The mutual sparing effect of selenium and vitamin E stems from their shared antiperoxidant roles. But high levels of vitamin E do not completely eliminate the level for selenium.	Deficiency symptoms—Sudden death, impaired reproduction, reduced milk production, and impaired immune response. **Toxicity**—Loss of appetite, loss of hair, fatty infiltration of the liver, degenerative changes in the liver and kidney, edema, and occasional separation of the hoof and skin at the coronary band. Dietary arsenicals help to alleviate selenium toxicity.
Zinc (Zn)	High levels of calcium in relation to zinc levels impair zinc utilization and increase the requirements.	Zinc is a component of many metalloenzymes and the hormone insulin. So, it plays an important role in protein, carbohydrate, and lipid metabolism.	Deficiency symptoms—Parakeratosis or swine dermatitis, pigs have a mangy appearance, reduced appetite, unthriftiness, poor growth rate, and diarrhea, and there may be vomiting. It affects swine of all ages. Zinc deficiency results in gilts producing fewer and smaller pigs; in boars with retarded testicular development; and in young pigs with retarded thymic development. **Toxicity**—Growth depression, arthritis, hemorrhage in axillary spaces, gastritis, and enteritis. High dietary calcium reduces the severity of zinc toxicity.

[1]As used herein, the distinction between "mineral requirements" and "recommended allowances" is as follows: In mineral requirements, no margins of safety are included intentionally; whereas in recommended allowances, margins of safety are provided in order to compensate for variations in feed compositions, environment, and possible losses during storage or processing.

Where preceded by an asterisk, the mineral requirements, recommended allowances, and other facts presented herein were taken from *Nutrient Requirements of Swine*, 9th rev. ed., NRC-National Academy of Sciences, 1988.

MAJOR OR MACROMINERALS

Salt (sodium and chlorine), calcium, and phosphorus, magnesium, potassium, and sulfur are the major or macrominerals.

SALT (NaCl)

Salt contains both sodium and chlorine, vital elements found in the fluids and soft tissues of the body. It improves the appetite, promotes growth, helps regulate body pH, and is essential for hydrochloric acid formation in the stomach.

Although swine require less salt than other classes of farm animals, it is generally advantageous to supply them with some of it, particularly if the protein supplement is not derived from animal or marine sources. A lack of salt is marked by a poor and depraved appetite, unthrifty condition, and failure to grow.

CALCIUM (Ca)/PHOSPHORUS (P)

Calcium and phosphorus are important in skeleton development. Also, they aid in blood clotting, muscle contraction, and energy metabolism.

About 99% of the calcium and 80% of the phosphorus in the body are found in the skeleton and teeth.

Replacement gilts need greater amounts of calcium and phosphorus than barrows. Thus, with split-sex feeding, replacement gilts can be fed higher levels of calcium and phosphorus (0.75 and 0.65% respectively) for maximizing bone development.

A deficiency of either calcium or phosphorus in the diet of the pig can result in poor and inefficient

(Continued)

Mineral Requirements[1]		Recommended Allowances[1]	Practical Sources of the Mineral	Comments
Minerals/ Animal/Day	Mineral Content of Diets			
*Variable according to class, age, and weight of swine (see Tables 9-1 and 9-2).	*Variable according to class, age, and weight of swine (see Tables 9-5 and 9-6).	Variable according to class, age, and weight of swine (see Tables 9-8 and 9-9).	Manganous oxide.	Manganese is usually present in adequate amounts in most swine diets, but it may not be adequate for the optimum reproductive performance of sows.
*Variable according to class, age, and weight of swine (see Tables 9-1 and 9-2).	*The dietary requirement for selenium is between 0.1 and 0.3 ppm (see Tables 9-5 and 9-6). *In 1987, the FDA approved up to 0.3 ppm selenium in the diet of all pigs.	Variable according to class, age, and weight of swine (see Tables 9-8 and 9-9).	Sodium selenite or sodium selenate.	Environmental stress may increase the incidence and degree of selenium deficiency. *Caution:* Toxic level of selenium is in range of 2.27–3.63 mg/lb *(5–8 mg/kg)* selenium in the feed.
*Variable according to class, age, and weight of swine (see Tables 9-1 and 9-2).	*Variable according to class, age, and weight of swine (see Tables 9-5 and 9-6). The zinc requirement is increased when excessive levels of calcium are fed.	Variable according to class, age, and weight of swine (see Tables 9-8 and 9-9).	Zinc carbonate or zinc sulfate.	It has been shown that parakeratosis is caused by zinc and calcium forming an unavailable complex.

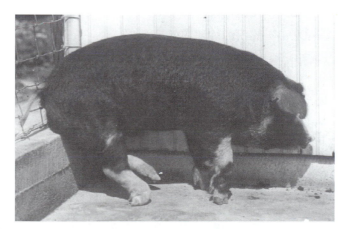

Fig. 8-7. Calcium deficiency. Note abnormal bone development and rachitic condition in advanced stage of deficiency. Lack of calcium retards normal skeletal development, but it does not usually depress total gain. (Courtesy, USDA)

gains, rickets or osteomalacia, broken bones, and posterior paralysis.

A large excess of either calcium or phosphorus interferes with the absorption of the other. Thus, it is important to have a suitable ratio between the two minerals. The most favorable calcium to phosphorus ratio is 1.2:1 to 1.5:1. Also, vitamin D is necessary for the proper utilization of these two minerals.

An excess of calcium interferes with zinc absorption and results in parakeratosis. A combination of a high level of calcium (over 0.9%) and a marginal zinc level can result in this condition.

It is important to supplement swine diets with both calcium and phosphorus. Cereal grains, which make up the bulk of swine diets, are quite low in calcium and are only fair sources of phosphorus. Moreover, much of the phosphorus in cereal grains is present in the form of phytin, a form of phosphorus which is poorly

Fig. 8-8. Phosphorus deficiency. *Left:* Typical phosphorus-deficient pig in advanced stage of deficiency. Leg bones are weak and crooked. *Right:* This pig received the same ration as the one on the left, except that the ration was adequate in available phosphorus. (Courtesy, Purdue University, West Lafayette, IN)

utilized. A range of 8 to 60% of phosphorus availability has been reported in cereal grains, but, for practical purposes, an availability of 30% is a reasonable estimate.

Feeds of animal origin, such as meat meal and fish meal, are high in calcium and in available phosphorus.

Experimentally, supplementing swine diets with phytase has been effective in improving the availability of phosphorus in corn and soybean meal. This results in less inorganic phosphate in the diet and lessens the phosphorus in the manure, thereby lessening the phosphorus pollution of the land. But, as is the case with all additives and ingredients in the diet, swine producers will need to determine the cost effectiveness of (1) adding phytase versus (2) adding higher levels of phosphorus, much of which is unavailable.

When used, phytase should be added according to the suppliers directions.

■ **Phytase in swine feeds**—In Europe, areas of intensive swine production have received much of the blame for phosphorus pollution of the land. *The North Sea Agreement* signed by the Paris Commission in 1989 committed participating countries to reducing phosphate outputs to 50% by 1995. This spurred research. The initial work on the use of the microbially derived phytase enzyme in pig diets was conducted in the Netherlands (Simon, *et al.*, 1990). Since the early work, further experiments have indicated that if phytase is used properly up to 50% reduction of the feed phosphate output may be feasible. It follows that if most of the phosphorus present in the cereal grains fed to pigs were made available during digestion, it would be sufficient to satisfy the requirements of growing pigs without any inorganic phosphorus mineral supplement being added to the diet, resulting in less

phosphorus being excreted in the feces and polluting the land. More experimental work is needed, with focus on improved technology in producing and adding phytase to the diet, and on lowering the cost of phytase.[1]

MAGNESIUM (Mg)

Magnesium is a cofactor in many enzyme systems and a constituent of bone. Apparently, the magnesium requirement is met by grain-soybean meal diets, or by diets containing grain and protein supplements.

POTASSIUM (K)

Potassium is the most abundant mineral in muscle tissue.

Grain-soybean meal diets normally contain enough potassium to meet the requirements for all classes of swine.

The dietary potassium requirement for the pig is increased by high levels of dietary chloride, sulfate, and other anions, as it is for the chick.

SULFUR (S)

Sulfur is an essential element. However, the sulfur-containing amino acids (cystine and methionine) appear adequate to meet the pig's need for synthesis of sulfur-containing compounds. The addition of inorganic sulfate to low-protein swine rations has not been beneficial.

TRACE OR MICROMINERALS

Minerals that are required in small amounts are known as trace or microminerals. These include cobalt, copper, iodine, iron, manganese, selenium, and zinc.

COBALT (Co)

Cobalt is a component of vitamin B-12. The intestinal microflora of the pig are capable of synthesizing vitamin B-12 provided sufficient cobalt is present. But only a minimum level of dietary cobalt is necessary for this process. Intestinal synthesis is of greater importance if preformed vitamin B-12 is limiting.

---

[1]Khan, N., *Feed Mix*, Vol. 3, Nov. 5, 1995, pp. 14–18.

There is no evidence that pigs have a requirement for cobalt, other than for vitamin B-12 synthesis.

Cobalt can partially substitute for zinc. Also, supplemental cobalt will prevent lesions associated with zinc deficiency.

A level of 400 ppm of cobalt is toxic to the young pig. Selenium and vitamin E provide some protection against toxicity from excessive levels of dietary cobalt.

COPPER (Cu)

The pig requires copper for the synthesis of hemoglobin and for the synthesis and activation of several oxidative enzymes necessary for normal metabolism.

A deficiency of copper leads to poor iron mobilization. A level of 6 ppm in the ration is adequate for baby pigs.

When fed at 100 to 250 ppm, copper stimulates growth in pigs, apparently due to the antibacterial action of these high levels of copper.

IODINE (I)

The dietary iodine requirement is not well established. Moreover, it is increased by goitrogens in certain feedstuffs, including rapeseed, linseed, lentils, peanuts, and soybeans. A level of 0.14 ppm of iodine in a corn-soybean meal diet will prevent goiter in growing pigs; and a level of 0.35 ppm of added iodine will prevent iodine deficiency in sows.

The incorporation of iodized salt (0.007% iodine) at a level of 0.2% of the diet provides sufficient iodine (0.14% ppm) to meet the needs of growing pigs fed grain-soybean meal diets.

Calcium iodate, potassium iodate, and pentacalcium orthoperiodate are nutritionally available forms and more stable in salt mixtures than sodium iodide or potassium iodide.

IRON (Fe)

Iron is necessary for the formation of hemoglobin in the red blood cells and the prevention of nutritional anemia. Hemoglobin serves as a carrier of oxygen throughout the body.

As the unborn pig develops, a supply of iron is stored in its body. The amount stored varies greatly between pigs of the same litter. But in no case is the amount of iron adequate to keep the pig growing at its maximum for more than 10 days or 2 weeks after birth unless some supplemental source of iron is available during the suckling period.

Sow's milk is very low in iron; and, to date, research has not uncovered any way of increasing its iron content. Thus, if suckling pigs are confined with no access to soil or feed, serious losses from anemia are likely. Once a pig begins to consume natural feedstuffs, the danger of anemia is practically nil because most feeds contain sufficient amounts of iron to meet the pig's requirement.

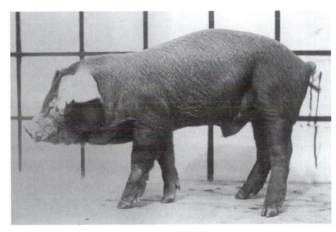

Fig. 8-10. Suckling pig with nutritional anemia, caused by a lack of iron, characterized by swollen condition about the head and paleness of the mucous membranes. (Courtesy, College of Veterinary Medicine, University of Illinois, Urbana)

Anemic pigs lose their appetite and become weak and inactive. In more advanced stages of the deficiency, the pig's breathing becomes labored, a condition that is sometimes called *thumps*. In this condition they are more susceptible to other diseases and parasites. Death may occur in severe cases.

The most commonly used sources of iron to prevent anemia in newborn pigs are injectable and oral products, with the injectable iron preferred. An intramuscular injection of 200 mg of iron dextran given at 1 to 3 days of age will prevent the anemia problem.

Fig. 8-9. Hairlessness in pigs caused by a deficiency of iodine. In iodine-deficient areas, farm animals should receive iodized salt throughout the year. (Courtesy, Department of Veterinary Pathology and Hygiene, College of Veterinary Medicine, University of Illinois, Urbana)

Need for a second injection depends on how much iron was given in the first injection and how much iron is available to the baby pigs during the lactation period. Baby pigs can receive iron orally from consuming creep feed or sow feed. Need for a second injection also depends on the blood hemoglobin concentration, which is a rapid and reliable indicator of the iron status of the pig. Blood hemoglobin levels of 10 mg/100 ml, or above, indicate adequate iron status.

Formerly, it was recommended that iron injections be given in the ham. However, when iron injections are given in the ham, permanent staining of the meat may occur. Because ham is one of the highest value cuts, the current recommendation is that the injection be given in the neck.

The postweaning dietary iron requirement is about 80 ppm.

MANGANESE (Mn)

Manganese functions with many enzymes in soft tissue metabolism and also in bone development. Deficiency symptoms are lameness, weakened bone structure, irregular estrus, offspring born dead or weak, and increased backfat. While manganese is usually present in adequate amounts without supplementation in most swine rations, it may not be adequate for the optimum reproductive performance of sows.

SELENIUM (Se)

Selenium functions as a part of glutathione peroxidase, an enzyme which enables the tripeptide glutathione to perform its role as a biological antioxidant in the body. This explains why deficiencies of selenium and vitamin E result in similar signs. However, high levels of vitamin E do not completely eliminate the need for selenium.

Currently, the U.S. Food and Drug Administration (FDA) allows the addition of up to 0.3 ppm of selenium in the diets of all pigs.

In some cases, selenium in the diet at a level of 5 ppm will produce toxic symptoms.

ZINC (Zn)

The requirement for zinc in swine diets is very low, but when high levels of calcium are fed, zinc utilization is impaired and the requirements are increased. A zinc deficiency results in a mangelike skin condition called *parakeratosis*. Other symptoms are poor growth, inefficient feed conversion, gilts producing fewer and smaller pigs, boars with retarded testicular development, and young pigs with retarded thymic development.

Boars have a higher zinc requirement than gilts, and gilts have a higher requirement than barrows.

CHELATED MINERALS

A chelated mineral is bound to a compound such as protein or an amino acid that helps to stabilize the mineral. Recent research has shown that chelated minerals are 0 to 15% more available. However, their cost may be two to three times greater than those of nonchelated minerals.

FEEDS AS A SOURCE OF MINERALS

The most satisfactory source of minerals for hogs is in the feed consumed. Thus, it is important to know whether the minerals in the ration are of the right kind and sufficient in amount. Certain general characteristics of feeds in regard to calcium and phosphorus (the two predominating mineral elements of the body) are worth noting:

1. The cereal grains and their byproducts and protein supplements of plant origin are low in calcium but fairly good in phosphorus. However, as mentioned earlier, the phosphorus in plants is not fully utilizable by swine.

2. The protein supplements of animal origin (skim milk, buttermilk, tankage, meat scraps, fish meal),

Fig. 8-11. Zinc deficiency. *Left:* Pig received 17 ppm of zinc and gained only 3 lb in 74 days. Note severe dermatosis ("mangy look"), or parakeratosis. *Right:* Pig received the same ration as the pig on the left, except that the diet contained 67 ppm of zinc. This pig gained 111 lb in 74 days. (Courtesy, Purdue University, West Lafayette, IN)

legume forage (pasturage and hay), and rape, are all rich in calcium.

3. Most protein-rich supplements are high in phosphorus.

ELECTROLYTES

Electrolytes (minerals) are essential for maintaining water balance in pigs. The major elements involved in electrolyte balance are sodium, chloride, potassium, magnesium, and calcium, with sodium, chloride, and potassium predominating. It is not recommended that electrolytes be included in swine diets at levels exceeding those given in the tables in Chapters 8 and 9 of this book even in times of stress such as those associated with weaning and with feeder pig sales and transfers.

Electrolyte balance is particularly important for starting pigs, because they are more susceptible to diarrhea, which can cause severe dehydration.

MINERAL SOURCES AND BIOAVAILABILITY

The major sources of the minerals commonly added to swine diets are listed in Table 8-4. In addition, Table 8-4 gives the bioavailability of minerals from several sources. Decisions on which source of mineral to use should be based primarily on price per unit of available element.

TABLE 8-4
MINERAL SOURCES AND BIOAVAILABILITY[1, 2]

Mineral Element	Source	Formula	Content of Element	RB[3]	Comments
			(%)	(%)	
Calcium	Calcium carbonate	$CaCO_3$	38	100	Limestone, oyster shell
	Curacao phosphate		34 to 36	Unk[4]	
	Defluorinated rock phosphate		30 to 34	92 to 95	< 1 part F to 100 parts P
	Dicalcium phosphate	$CaHPO_4 \bullet 2H_2O$ and $CaHPO_4$	20 to 24	Unk	Grey granules
	Monocalcium phosphate	$CaH_4(PO_4)_2 \bullet H_2O$	16 to 22	Unk	
	Soft rock phosphate		17 to 20	70	Colloidal phosphate
	Steamed bone meal		24 to 30	Unk	
Copper	Cupric acetate	$Cu(C_2H_3O_2)_2$	100		
	Cupric carbonate	$CuCO_3 \bullet Cu(OH)_2$	55.00	100	Dark-green crystals
	Cupric chloride	$CuCl_2 \bullet 2H_2O$	36.90	100	Green crystals
	Cupric oxide	CuO	75.0	0	Black powder or granules; not recommended as a copper supplement
	Cupric sulfate	$CuSO_4 \bullet 5H_2O$	25.2	100	Blue or ultramarine crystals
Iodine	Ethylenediamine dihydroiodide (EDDI)	$NH_2CH_2CH_2NH_2 \bullet 2HI$	79.5	Unk	White
	Calcium iodate	$Ca(IO_3)_2$	64.0	90	Stable source
	Calcium periodate	$Ca_5(IO_5)_2$	39.28	90	
	Potassium iodide	KI	68.17	100	Used in iodized salt (0.01%)
Iron	Ferric chloride	$FeCl_3 \bullet 6H_2O$	20.66	44 to 95	
	Ferric oxide	Fe_2O_3	57.00	0	Red—used as a coloring pigment; not recommended as an iron supplement
	Ferrous carbonate	$FeCO_3$	40	0 to 74	Beige
	Ferrous fumarate	$FeC_4H_2O_4$	32.54	95	Reddish-brown
	Ferrous oxide	FeO	75.37	Unk	Black powder
	Ferrous sulfate (1 H_2O)	$FeSO_4 \bullet H_2O$	30	100	Green to brown crystals
	Ferrous sulfate (7 H_2O)	$FeSO_4 \bullet 7H_2O$	21.4	100	Greenish crystals

(Continued)

TABLE 8-4 (Continued)

Mineral Element	Source	Formula	Content of Element	RB[3]	Comments
			(%)	(%)	
Manganese	Manganese dioxide	MnO_2		30	Black powder
	Manganous carbonate	$MnCO_3$	46.40	40	Rose-colored crystals
	Manganous chloride	$MnCl_2 \cdot 4H_2O$	27.48	100	Rose-colored crystals
	Manganous oxide	*MnO*	60	70	Green to brown powder
	Manganous sulfate	$MnSO_4 \cdot H_2O$	29.5	100	White to cream powder
Phosphorus	Curacao phosphate		12 to 15	100	
	Defluorinated rock phosphate		16 to 18	87	< 1 part F to 100 parts P
	Dicalcium phosphate	*$CaHPO_4 \cdot 2H_2O$ and $CaHPO_4$*	18.5	100 to 105	Grey granules
	Monocalcium phosphate	*$CaH_4(PO_4)_2 \cdot H_2O$*	21	110	
	Monosodium phosphate	$NaH_2PO_4 \cdot H_2O$	22 to 26	100	Large white crystals
	Soft rock phosphate		9	100	
	Steamed bone meal		12 to 14	82	
Selenium	Sodium selenate	Na_2SeO_4	41.8	100	White crystals
	Sodium selenite	*Na_2SeO_3*	45	100	White to light pink crystals
Zinc	Zinc carbonate	$ZnCO_3$	51.63	100	White crystals
	Zinc oxide	*ZnO*	72	68	Greyish powder
	Zinc sulfate (1 H_2O)	*$ZnSO_4 \cdot H_2O$*	35.5	100	White crystals
	Zinc sulfate (7 H_2O)	$ZnSO_4 \cdot 7H_2O$	22.25	100	

[1]Adapted by the authors from: Reese, D. E., *et al.*, *Swine Nutrition Guide*, pub. by University of Nebraska and South Dakota State University, 1995.

[2]Most common sources are in italic.

[3]RB = relative bioavailability.

[4]Unk = unknown.

VITAMINS

Vitamins are complex organic compounds that are required in minute amounts, which are essential for health and normal body functions. Like amino acids, each vitamin has a specific function to perform. Vitamins are classified into two groups—fat-soluble and water-soluble. The body can store reserves of the fat-soluble vitamins for a considerable period of time. But stores of the water-soluble vitamins are depleted rapidly.

Some vitamins are present in feed ingredients in adequate amounts. Others are produced in the pig's body in adequate amounts. However, for optimal performance, several vitamins need to be added to swine diets.

Because of the greater prevalence of confinement feeding, swine are more likely to suffer from vitamin deficiencies than any other class of four-footed animals.

SWINE VITAMIN CHART

Table 8-5, Swine Vitamin Chart, gives in summary form the following pertinent information relative to each vitamin listed: (1) conditions usually prevailing where deficiencies are reported, (2) functions, (3) deficiency symptoms/toxicity, (4) vitamin requirements, (5) recommended allowances, and (6) practical sources. Further elucidation of vitamins is contained in the narrative.

FAT-SOLUBLE VITAMINS

The primary fat-soluble vitamins of practical importance for swine are vitamins A, D, and E. Vitamin K may be of concern under some circumstances.

VITAMIN A

The vitamin A needs of swine can be met by either vitamin A or carotene. Vitamin A as such does not occur in plants. However, green plants and yellow corn contain a yellow pigment called *carotene* which can be converted to vitamin A by the animal body. The combination of vitamin A and carotene present in the diet is referred to as its vitamin A activity.

Carotene is easily destroyed by the ultraviolet rays of the sun and by heat. The carotene content of corn and legume hay usually deteriorates quite rapidly in storage. Therefore, a synthetic concentrate is a more practical and reliable source of vitamin A than natural sources. Most commercial feed companies fortify their swine feeds with a stabilized form of vitamin A, which is active over a considerable period of time.

Vitamin A is essential for vision, reproduction, growth, and the maintenance of differentiated epithelia and mucous secretions of swine.

Swine are able to store Vitamin A in the liver, and to draw from this storage during periods of low intake.

Vitamin A deficiency signs in growing pigs are incoordination of movement, loss of control of the hind legs, weakness of the back, and night blindness. Sows may fail to come into heat, they may resorb their fetuses, or they may have young born dead with various deformities and defects. Vitamin A is also needed for normal vision and growth of new cells which line the respiratory, digestive, and reproductive tracts.

Scientists from around the world have evidence that vitamin A may increase litter size by 0.5 to 0.8 pigs per litter. But more research is needed to determine proper dose and time of injection, and to make sure there are no side effects.

VITAMIN D

Vitamin D is sometimes referred to as the "sunshine" vitamin, since the action of sunlight on a compound in the skin will produce it. So long as hogs are exposed to the sun, there is no danger of a deficiency.

Living plants do not contain vitamin D. Plants that mature or are cut and cured in the sun contain some vitamin D as a result of radiation by sunlight. Pigs can utilize equally well either vitamin D_2 (from plant products) or vitamin D_3 (from animal products). Irradiated yeast is a good source of vitamin D_2.

Vitamin D is needed for the efficient assimilation of calcium and phosphorus; hence, it is required for the growth of strong bones. A lack of vitamin D will result in stiffness and lameness, broken or deformed bones, enlargement of joints, and general unthriftiness; known as rickets in young pigs and osteomalacia in mature hogs.

It is noteworthy that the vitamin D requirement is less when a proper balance of calcium and phosphorus exists in the diet.

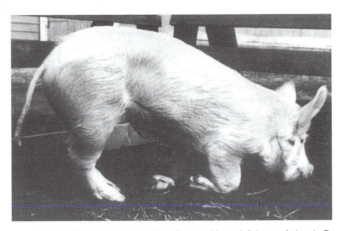

Fig. 8-12. Rickets (advanced case) caused by a deficiency of vitamin D. The pig was fed indoors, without exposure to sunlight and without adequate vitamin D. Because of leg abnormalities, it was unable to walk. Later, the pig responded to vitamin D. (Courtesy, University of Saskatchewan, Saskatoon, Canada)

VITAMIN E

Vitamin E is a biological antioxidant which protects unsaturated fat against oxidation. Eight naturally occurring compounds called *tocopherols* have vitamin E activity, the most active of which is alpha-tocopherol. Also, alpha-tocopherol is very stable during storage and/or in mixed feeds. Since cell membranes in the animal body contain unsaturated fat, a vitamin E deficiency may result in oxidative damage to the cell. This is manifested in the pig by liver necrosis, pale muscle, mulberry heart, edema, and sudden death.

For many years, the primary source of vitamin E in feed was the natural form of vitamin E (d-a-tocopherol) found in green plants and seeds. However, oxidation rapidly destroys natural vitamin E. For example, vitamin E losses of 50 to 70% can occur in alfalfa stored at 90°F for 12 weeks; and losses of 5 to 30% can occur during dehydration of alfalfa.

Also, storage of high-moisture grain or its treatment with organic acids greatly reduces its vitamin E content. Therefore, predicting the amount of vitamin E activity in feed ingredients is difficult.

The trace element selenium also functions with vitamin E in protecting the body against oxidative damage. The need for vitamin E is more acute when swine feeds are low in selenium. Thus, in areas where feed ingredients are low in selenium and where a majority of swine are raised in confinement without access to forages, supplemental vitamin E or selenium, or both, are important.

Vitamin E toxicity has not been reported in swine. Levels as high as 45 IU/lb of ration have been fed to growing pigs without toxic effects.

Vitamins Which May Be Deficient Under Normal Conditions	Conditions Usually Prevailing Where Deficiencies Are Reported	Functions of Vitamin	Deficiency Symptoms/Toxicity
Fat-soluble vitamins			
A	Absence of green forages, either pasture or green hay—especially under confined conditions. Where the ration consists chiefly of white corn, milo, barley, wheat, oats, or rye; or byproducts of these grains; or yellow corn that has been stored more than one year.	Essential for normal maintenance and functioning of the epithelial tissues, particularly of the eye and the respiratory, digestive, reproductive, nerve, and urinary systems.	Deficiency symptoms—Night and day blindness, very irritable, poor appetite and slow growth, lameness, incoordination of movement, loss of control of the hind legs, and weakness of the back. Low resistance to respiratory infections. Sows may fail to come in heat, may resorb their fetuses, and may have young born dead with various deformities and defects. **Toxicity**—A roughened hair coat, scaly skin, hyperirritability and sensitivity to touch, bleeding from the cracks which appear in the skin above the hooves, blood in the urine and feces, loss of control of the legs accompanied by inability to rise, periodic tremors, and death.
D	Limited sunlight and/or limited quantities of sun-cured hay in drylot rations.	Aids in assimilation and utilization of calcium and phosphorus, and necessary in the normal bone development of animals—including the bones of the fetus.	Deficiency symptoms—Rickets in young pigs, or osteomalacia in mature hogs. Both conditions result in large joints and weak bones. In severe vitamin D deficiency, pigs may exhibit signs of calcium and magnesium deficiency, including tetany. **Toxicity**—Reduced feed intake and growth rate; and death. Vitamin D_3 is more toxic than D_2 in swine.
E	Rations containing excessive amounts of highly unsaturated fatty acids or oxidized fats. Swine feeds low in selenium, especially where swine are raised in confinement without access to forages.	Antioxidant. Muscle structure. Reproduction. High levels of vitamin E in the diet may increase the immune response.	Deficiency symptoms—Loss of appetite and slow growth. Increased embryonic mortality and muscular incoordination in suckling pigs from sows fed vitamin E-deficient rations during gestation and lactation. A wide variety of pathological conditions. **Toxicity**—Vitamin E toxicity in swine has not been demonstrated.
K	Moldy feed. High antibiotic levels, which may make for inadequate intestinal synthesis of vitamin K.	Essential for prothrombin formation and blood clotting.	Deficiency symptoms—Bleeding condition in young pigs, which responds to injection or oral administration of vitamin K. Slow growth and hyperirritability. **Toxicity**—Concentrations of 50 mg of menadione pyrimidinol bisulfite (MPB)/lb *(110 mg of MPB/kg)* of diet were not toxic to weanling pigs.
Water-soluble vitamins			
Biotin	When pigs are fed dried, raw egg white or given sulfa drugs. Marginal deficiency may exist when hogs are fed cereal grain diets, housed in individual stalls or on slotted floors (which lessens coprophagy), and/or have no access to green forage.	Biotin is important metabolically as a cofactor for several enzymes. It may improve sow and litter performance when added to gestation-lactation diets.	Deficiency symptoms—Excessive hair loss, skin ulcerations and dermatitis, exudate around the eyes, cracking of the hooves, and cracking and bleeding of the footpads.

8-5
CHART

Vitamin Requirements[1]		Recommended Allowances[1]	Practical Sources of the Vitamin	Comments
Vitamins/ Animal/Day	Vitamin Content of Diet			
*Variable according to class, age, and weight of swine (see Tables 9-1 and 9-2).	*Variable according to class, age, and weight of swine (see Tables 9-5 and 9-6).	Variable according to class, age, and weight of swine (see Tables 9-8 and 9-9).	Either vitamin A or various pro-vitamins.	Based upon liver storage, the biopotency of 1 mg of carotene in corn fed to weanling pigs is 261 IU of vitamin A. Meals from artificially dehydrated forages are much higher in carotene than sun-cured products. Taken together, liver storage, levels of plasma vitamin A, and pressure of cerebrospinal fluid give reliable estimates of the vitamin A status of the pig.
*Variable according to class, age, and weight of swine (see Tables 9-1 and 9-2).	*Variable according to class, age, and weight of swine (see Tables 9-5 and 9-6).	Variable according to class, age, and weight of swine (see Tables 9-8 and 9-9).	Vitamin D_2 (ergocalciferol) and vitamin D_3 (cholecalciferol) are similar in biological activity for swine. The action of ultraviolet light on the ergosterol that is present in plants forms ergocalciferol; and the photochemical conversion of 7-dehydro-cholesterol in the skin of animals forms colecalciferol. Irradiated yeast. Exposure to sunlight. Sun-cured hay (10% alfalfa in the total ration will normally supply sufficient vitamin D).	Grains, grain byproducts, and high-protein feedstuffs are practically devoid of vitamin D; therefore, unless swine are exposed daily to the ultraviolet rays of the sun, the diet should be fortified with vitamin D. The vitamin D requirement is less when a proper balance of calcium and phosphorus exists in the ration. One IU vitamin D is defined as the biological activity of 0.025 mcg of cholecalciferol.
*Variable according to class, age, and weight of swine (see Tables 9-1 and 9-2). Many dietary factors affect the vitamin E requirement, including the selenium level, unsaturated fatty acids, sulfur amino acids, retinol, copper, iron, and synthetic antioxidants.	*Variable according to class, age, and weight of swine (see Tables 9-5 and 9-6).	Variable according to class, age, and weight of swine (see Tables 9-8 and 9-9).	Alpha tocopherol. Predicting the amount of vitamin E activity in feed is difficult.	The 8 naturally occuring tocopherols differ in their biological activities, with d-alpha-tocopherol being the most active. One IU of vitamin E is the equivalent in biopotency of 1 mg dl-alpha-tocopherol acetate.
*Variable according to class, age, and weight of swine (see Tables 9-1 and 9-2).	*Supplement the as-fed diet with menadione at level of 0.2 mg/lb (0.5 mg/kg).	Under practical conditions, the vitamin K requirement is met by vitamin K in feedstuffs and by intestinal synthesis.	The following water-soluble forms of menadione are commonly used to supplement swine diets: menadione sodium bisulfite complex (MSB), menadione sodium bisulfate complex (MSBC), and menadione pyrimidinol bisulfite (MPB).	Vitamin K exists in 3 forms: phylloquinone (K_1), menaquinone (K_2), and menadione (K_3).
*Variable according to class, age, and weight of swine (see Tables 9-1 and 9-2).	*Variable according to class, age, and weight of swine (see Tables 9-5 and 9-6).	Variable according to class, age, and weight of swine (see Tables 9-8 and 9-9).	Biotin.	The protein avidin in raw egg white makes biotin unavailable to pigs. Heat treatment inactivates avidin and makes egg white safe for feeding to pigs.

(Continued)

TABLE 8-5

Vitamins Which May Be Deficient Under Normal Conditions	Conditions Usually Prevailing Where Deficiencies Are Reported	Functions of Vitamin	Deficiency Symptoms/Toxicity
Water-soluble vitamins (continued)			
Choline	Baby pigs fed a synthetic milk diet containing not more than 0.8% methionine.	Involved in nerve impulses; a component of phospholipids; donor of methyl groups; and involved in the mobilization and oxidation of fatty acids in the liver.	Deficiency symptoms—Unthriftiness, lack of coordination, fatty infiltration of the liver, poor reproduction, poor lactation, and decreased survival of the young. **Toxicity**—No signs of choline toxicity have been reported in swine.
Folacin (Folic Acid)		Metabolic reactions involving incorporation of single carbon units into larger molecules. Folacin is involved in the conversion of serine to glycine and homocystine to methionine.	**Deficiency symptoms**—Poor growth, fading hair color, and anemia. Folic acid may increase the number of pigs born alive by approximately one pig.
Niacin (Nicotinic Acid, Nicotinamide)		Niacin is a component of the coenzymes which are essential for the metabolism of carbohydrates, proteins, and lipids.	**Deficiency symptoms**—Loss of appetite and decreased gain, followed by diarrhea, occasional vomiting, dermatitis, and loss of hair.
Pantothenic Acid (B-3)	Long periods of inadequate pantothenic acid intake.	As a component of coenzyme A, pantothenic acid is important in the catabolism and synthesis of 2-carbon units evolved during carbohydrate and fat metabolism.	**Deficiency symptoms**—A goose-stepping gait, loss of appetite, poor growth, diarrhea, loss of hair, reduced fertility, and breeding failure.
Riboflavin (B-2)		A component of 2 coenzymes. It is important in the metabolism of proteins, fats, and carbohydrates.	**Deficiency symptoms**—Loss of appetite, poor growth, rough hair coat, diarrhea, cataracts, vomiting, reproductive failure in the sow, pigs dead or weak at birth, and crooked legs and incoordination.
Thiamin (B-1)	Thiamin is heat labile. So, excess heat can reduce the thaimin content of the diet ingredients.	As a coenzyme in energy and protein metabolism. Promotes appetite and growth, required for normal carbohydrate metabolism, and aids reproduction.	**Deficiency symptoms**—Loss of appetite and poor growth, diarrhea, dead or weak offspring, slow pulse, low body temperature, and flabby heart.
Vitamin B-6 (Pyridoxine, Pyridoxal, Pyridoxamine)		An important cofactor for many amino acid enzyme systems. Vitamin B-6 also plays a crucial role in central nervous system function.	**Deficiency symptoms**—Loss of appetite and poor growth, unsteady gait, anemia, exudate around the eyes, and epileptic-like fits (convulsions).
Vitamin B-12 (Cobalamins)	Pigs fed ingredients of plant origin and housed on slotted floors.	Vitamin B-12 contains the trace element cobalt in its molecule. As a coenzyme, B-12 is involved in the synthesis of methyl groups derived from formate, glycine, or serene, and their transfer to homocystine to reform methonine. It is also important in the methylation of uracil to form thymine, which is converted to thymidine and used for the synthesis of DNA.	**Deficiency symptoms**—Loss of appetite, reduced weight gain, rough skin and hair coat, irritability, hypersensitivity, hind leg incoordination, and anemia.

(Continued)

Vitamin Requirements[1]		Recommended Allowances[1]	Practical Sources of the Vitamin	Comments
Vitamins/ Animal/Day	Vitamin Content of Diet			
*Variable according to class, age, and weight of swine (see Tables 9-1 and 9-2).	*Variable according to class, age, and weight of swine (see Tables 9-5 and 9-6).	Practical diets are adequate in choline.	Choline chlorides or choline dihydrogen.	Choline does not qualify as a true vitamin, because it is required at far greater levels than true vitamins and is not known to participate in any enzyme system. Studies have shown that more live pigs are born and weaned when sows receive supplemental choline throughout gestation.
*Variable according to class, age, and weight of swine (see Tables 9-1 and 9-2).	*Variable according to class, age, and weight of swine (see Tables 9-5 and 9-6).	Practical diets plus intestinal synthesis are adequate.	Synthetic folacin.	Folacin includes a group of compounds with folic acid activity.
*Variable according to class, age, and weight of swine (see Tables 9-1 and 9-2).	*Variable according to class, age, and weight of swine (see Tables 9-5 and 9-6).	Variable according to class, age, and weight of swine (see Tables 9-8 and 9-9).	Nicotinamide. Nicotinic acid.	Niacin occurs in corn, wheat, and milo in bound form; hence, it may be unavailable to the pig. Also, the dietary tryptophan level affects the niacin requirement because of the conversion of tryptophan to niacin.
*Variable according to class, age, and weight of swine (see Tables 9-1 and 9-2).	*Variable according to class, age, and weight of swine (see Tables 9-5 and 9-6).	Variable according to class, age, and weight of swine (see Tables 9-8 and 9-9).	Calcium pantothenate (only the D isomer has vitamin activity). Dried milk products, condensed fish solubles, and alfalfa meal.	Widely distributed and occurs in practically all feedstuffs. However, the quantity present may not always be sufficient to meet the needs of the pig.
*Variable according to class, age, and weight of swine (see Tables 9-1 and 9-2).	*Variable according to class, age, and weight of swine (see Tables 9-5 and 9-6).	Variable according to class, age, and weight of swine (see Tables 9-8 and 9-9).	Synthetic riboflavin. Yeast. Milk and milk products. Meat scraps and fish meal.	Riboflavin is apt to be lacking in swine diets that do not contain animal product sources.
*Variable according to class, age, and weight of swine (see Tables 9-1 and 9-2).	*Variable according to class, age, and weight of swine (see Tables 9-5 and 9-6).	Practical diets are adequate in thiamin.	Thiamin hydrochloride. Thiamin mononitrate. Rice polish. Wheat germ meal. Yeast. The oilseed meals. Distillers' solubles.	
*Variable according to class, age, and weight of swine (see Tables 9-1 and 9-2).	*Variable according to class, age, and weight of swine (see Tables 9-5 and 9-6).	Practical diets are adequate in vitamin B-6.	Pyridoxine hydrochloride. Rich supplemental sources include rice polish, wheat germ, and yeast.	The vitamin B-6 content of normal feed is usually sufficient.
*Variable according to class, age, and weight of swine (see Tables 9-1 and 9-2).	*Variable according to class, age, and weight of swine (see Tables 9-5 and 9-6).	Variable according to class, age, and weight of swine (see Tables 9-8 and 9-9).	Synthetic B-12, which is produced commercially by microbial fermentation. Protein supplements of animal origin. Fermentation products.	Vitamin B-12 is apt to be lacking in swine diets. Synthesis of vitamin B-12 by intestinal flora may supplement dietary sources. B-12 contains the trace element cobalt; hence, the synthesis of B-12 in the intestines is dependent on the presence of cobalt in the feed. This may be the major, if not the only, function of cobalt as an essential nutrient.

(Continued)

TABLE 8-5

Vitamins Which May Be Deficient Under Normal Conditions	Conditions Usually Prevailing Where Deficiencies Are Reported	Functions of Vitamin	Deficiency Symptoms/Toxicity
Water-soluble vitamins (continued)			
Vitamin C (Ascorbic Acid, Dehydroascorbic Acid)		Vitamin C is a water-soluble antioxidant that is involved in the formation and maintenance of collagen, absorption and movement of iron, metabolism of fats and lipids, cholesterol control, sound teeth and bones, strong capillary walls and healthy blood vessels, metabolism of folic acid, and as a general antioxidant.	**Deficiency symptoms**—No specific symptoms noted when there is a deficiency.

[1]As used herein, the distinction between *vitamin requirements* and *recommended allowances* is as follows: In vitamin requirements, no margins of safety are included intentionally; whereas in recommended allowances, margins of safety are provided in order to compensate for variations in feed composition, environment, and possible losses during storage or processing.

VITAMIN K

Vitamin K exists in three forms: Phylloquinone (K$_1$), Menaquinone (K$_2$), and menadione (K$_3$). Menadione is the synthetic form of vitamin K which has the same cyclic structure as vitamin K$_1$ and K$_2$. Vitamin K$_1$ occurs naturally in green plants. Vitamin K$_2$ is present in microorganisms and is formed by intestinal bacteria.

Vitamin K is one of the essential factors necessary for proper blood clotting. Generally, sufficient amounts of vitamin K$_2$ are synthesized by bacteria in the digestive tract to meet the needs of swine, with the vitamin K$_2$ absorbed directly from the gut or obtained by coprophagy. However, this synthesis may be inadequate in situations where high antibiotic levels are used, where clotting inhibitors (dicoumarol) may be present from molds in the feed, or where there is excess calcium.

The deficiency symptoms of vitamin K are slow growth, hemorrhage, prolonged blood-clotting time, and hyperirritability.

WATER-SOLUBLE VITAMINS

Biotin, niacin, pantothenic acid, riboflavin, and vitamin B-12 are the water-soluble vitamins most likely to be deficient in swine diet. However, occasionally the other water-soluble vitamins are deficient. So, choline, folacin, thiamin, vitamin B-6, and vitamin C are also discussed in the sections that follow.

BIOTIN

Biotin is important metabolically as a cofactor for several enzymes.

Biotin is present in adequate amounts in most common feedstuffs, but its bioavailability varies greatly among ingredients.

Biotin deficiency has been associated with foot lesions and toe cracks in sows.

In general, biotin supplementation has not improved the performance of baby pigs or of growing-finishing hogs fed a variety of feedstuffs. However, biotin supplementation of sow rations has significantly improved reproductive performance, including the number of pigs farrowed and weaned, litter weaning weight, and number of days from weaning to estrus. Also, biotin supplementation improved hoof hardness of sows.

CHOLINE

Choline is included with the vitamins, but does not qualify as a true vitamin because it is required at far greater levels than true vitamins and is not known to participate in any enzyme system.

Choline functions as a "methyl donor" in metabolism and can lower the requirement of methionine to the extent that methionine is used for this function. It is also a constituent of some important phospholipids in the body, and it is involved in the mobilization and oxidation of the fatty acids in the liver. While choline is probably present in adequate amounts in most practical swine diets, studies have shown that more live pigs per litter are born and weaned when sows receive supplemental choline throughout gestation. Also, since proteins supply most of the choline in swine diets, a reduction in ration protein when synthetic lysine is incorporated in the diet may at the same time create a choline-deficient situation.

Choline-deficiency symptoms are lowered reproduction, pigs weak at birth, and lack of coordination.

Until recently, spraddle legs in baby pigs were attributed to choline deficiency. Although the cause of

(Continued)

Vitamin Requirements[1]		Recommended Allowances[1]	Practical Sources of the Vitamin	Comments
Vitamins/ Animal/Day	Vitamin Content of Diet			
		Practical diets plus intestinal synthesis are adequate.	Vitamin C (ascorbic acid).	Normally, pigs are able to synthesize vitamin C in amounts sufficient to meet their requirements. However, there is limited evidence that dietary ascorbic acid is beneficial under some conditions.

Where preceded by an asterisk, the vitamin requirements, recommended allowances, and other facts presented herein were taken from *Nutrient Requirements of Swine*, 9th rev. ed., NRC-National Academy of Sciences, 1988.

spraddle legs is not fully understood, many factors may contribute to the condition, including genetics, management, slick floors, mycotoxins, and/or a virus.

FOLACIN (Folic Acid/Folate)

Folacin includes a group of compounds with folic acid activity. Folic acid participates in many enzymatic reactions that appear to be essential in assuring embryo survival.

A deficiency of folacin causes a disturbance in the metabolism of single-carbon compounds, including the synthesis of methyl groups, serine, purines, and thymine. Folacin is involved in the metabolic conversion of amino acids: of serine to glycine, and of homocysteine to methionine.

Folacin deficiency in pigs leads to slow weight gain, fading hair color, and anemia.

The folacin in feedstuffs commonly fed to swine, along with bacterial synthesis within the intestinal tract, usually adequately meets the requirement for all classes of swine.

Studies at Kansas State University indicate that supplementation of the sow gestation diet with folic acid increased the number of pigs born alive by approximately one pig per litter. Further experiments are needed to determine the optimum level of folic acid supplementation.

NIACIN (Nicotinic Acid, Nicotinamide)

This vitamin plays an important role in body metabolism as a constituent of two coenzymes, nicotinamide-adenine dinucleotide (NAD) and nicotinamide-adenine dinucleotide phosphate (NADP). These coenzymes in the pig are essential for the metabolism of carbohydrates, proteins, and lipids.

A niacin deficiency results in *pig pellagra*, which is characterized by diarrhea, rough skin, and retarded growth.

Recent research has shown that the niacin in cereal grains is in a bound form, and that the niacin of corn may be almost completely unavailable to swine. It should be assumed that all niacin in cereal grains and their byproducts is completely unavailable. The protein source and tryptophan content of the diet can also affect the niacin requirement since tryptophan, an amino acid, can be converted to niacin.

PANTOTHENIC ACID (B-3)

Pantothenic acid is a constituent of coenzyme A which plays a key role in energy metabolism.

The biological availability of pantothenic acid is high from barley, wheat, and soybean meal, but low from corn and grain sorghum.

A lack of this vitamin may result in poor growth,

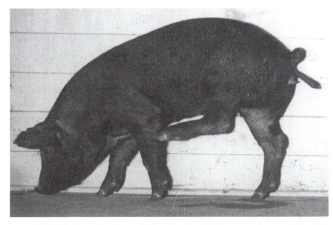

Fig. 8-13. Pig showing pantothenic acid deficiency symptoms. Note high goose-stepping gait. (Courtesy, University of California)

diarrhea, loss of hair, and a high-stepping gait of the hind legs—often called *goose stepping*.

Dried milk products, condensed fish solubles, and alfalfa meal are good natural sources of pantothenic acid. It is also available in synthetic form as calcium pantothenate.

RIBOFLAVIN (B-2)

Riboflavin is sometimes referred to as Vitamin B-2. It functions in the body as a constituent of two coenzymes, flavin mononucleotide and flavin adenine dinucleotide. Riboflavin is important in the metabolism of proteins, fats, and carbohydrates.

In growing swine, a deficiency may cause loss of appetite, stiffness, dermatitis, and eye problems. Poor conception and reproduction have been noted in gilts fed riboflavin-deficient rations. Pigs may be born prematurely, dead, or too weak to survive.

Milk products and other animal proteins, alfalfa meal, and distillers' solubles are good natural sources of riboflavin. Also, synthetic riboflavin is available.

THIAMIN (B-1)

Thiamin is essential for carbohydrate and protein metabolism. The coenzyme thiamin pyrophosphate is essential for the oxidative decarboxylation of alpha-keto acids.

Pigs must have a dietary source of thiamin, because, unlike ruminants, they cannot synthesize sufficient of it in the digestive tract. Normally, the thiamin content of feeds is sufficient to meet the needs of swine since cereal grains, which are major swine feeds, are good sources. However, deficiencies may result from: (1) heating feed ingredients excessively in processing, because thiamin is heat labile; (2) treating feedstuffs with sulfur dioxide, which inactivates the thiamin; or (3) feeding unprocessed fish or fish scraps of certain types of fish that contain the antithiamin factor known as *thiaminase*.

Fig. 8-14. Thiamin deficiency in littermate pigs. *Right:* Pig received no thiamin. *Left:* Pig received the equivalent of 2 mg thiamin/100 lb liveweight. Otherwise, their diets were the same. (Courtesy, USDA)

VITAMIN B-6 (Pyridoxine, Pyridoxal, Pyridoxamine)

Vitamin B-6 occurs in feedstuffs as pyridoxine, pyridoxal, pyridoxamine, and pyridoxal phosphate. Pyridoxal phosphate is an important cofactor for many amino acid enzyme systems. Vitamin B-6 plays a key role in central nervous system function.

Supplementation of grain—soybean meal rations with vitamin B-6 is generally unnecessary, because the concentration and availability of vitamin B-6 in the feed ingredients will meet the pig's requirement.

A deficiency of vitamin B-6 will reduce appetite and growth rate. Advanced deficiency will result in exudate around the eyes, convulsions, lack of coordination, coma, and death.

VITAMIN B-12 (Cobalamins)

Vitamin B-12 was discovered in 1948. Originally, it was known as the *animal protein factor* because of its association with ingredients of animal origin.

Vitamin B-12 stimulates the appetite, increases rate of growth, improves feed efficiency, and is necessary for normal reproduction.

Pigs require B-12 but responses to supplementation have been variable, primarily because of the synthesis of vitamin B-12 by microorganisms in the environment and within the intestinal tract, along with the pig's inclination toward coprophagy (feces eating).

VITAMIN C (Ascorbic Acid, Dehydroascorbic Acid)

Vitamin C (ascorbic acid) is an antioxidant that is involved in a variety of metabolic processes. It is also essential for hydroxylation of proline and lysine, which are integral constituents of collagen. Collagen is essential for growth of cartilage and bone. Vitamin C enhances the formation of intercellular material, the formation of bone matrix, and the formation of tooth dentin.

A dietary source of vitamin C is essential for primates and guinea pigs, but domestic swine can synthesize this vitamin. However, some claims have been made that, under certain conditions—as during periods of excessive stress—pigs may not be able to synthesize vitamin C fast enough to meet their requirement. Yet, the conditions under which supplemental vitamin C may be beneficial are not well defined, so no recommendation for vitamin C is given for the pig.

UNIDENTIFIED FACTORS

Some unidentified factor or factors may, under certain circumstances, be involved in securing optimum results during the critical periods (early growth and gestation-lactation). Sources of the unknown factor or factors are distillers' dried solubles, fish solubles, dried whey, grass juice concentrate, alfalfa meal, brewers' dried yeast, liver, and pasture.

NATURAL AND SYNTHETIC SOURCES OF VITAMINS

Green leafy plants, grasses, and alfalfa are excellent sources of vitamins for swine. However, with confinement rearing very little plant material is available. Additionally, fewer ingredients are being used in diet formulations. So, when formulating swine diets, it is recommended that all vitamin and mineral levels be *added* levels. Further, it is recommended that synthetic vitamins be added.

VITAMIN SOURCES AND BIOAVAILABILITY

The major sources of vitamins for swine are listed in Table 8-6. In addition, Table 8-6 gives the bioavailability of vitamins from several sources.

Although vitamins are present in grains and protein supplements, it is usually better to rely on vitamins supplied by the sources listed in Table 8-6. *The reason:* Vitamins in grains and protein sources may be lost during drying, storage, and processing. *Note:* For this reason, all the vitamin recommendations in Chapters 8 and 9 of this book are *added* levels.

TABLE 8-6
VITAMIN SOURCES AND BIOAVAILABILITIES[1,2]

Vitamin	1 IU Equals	Sources	RB[3]	Comments
			(%)	
Vitamin A	*3 mcg retinol or 344 mcg vitamin A acetate or 1 USP unit*	Vitamin A acetate (all-transretinyl acetate)	Unk[4]	Use coated form
	0.55 mcg vitamin A palmitate	Vitamin A palmitate	Unk	Used primarily in food
	0.36 mcg vitamin A propionate	Vitamin A propionate	Unk	Used primarily in injectibles
Vitamin D	0.025 mcg cholecalciferol or 1 USP unit or 1 ICU	Vitamin D_3 (cholecalciferol)	Unk	Coated form more stable
Vitamin E	*1 mg dl-a-tocopheryl acetate*	dl-a-tocopheryl acetate (all rac)	100	
	0.735 mg d-a-tocopheryl acetate	d-a-tocopheryl acetate (RRR)	136	
	0.909 mg d-a-tocopherol	dl-a-tocopherol (all rac)	110	Very unstable
	0.671 mg d-a-tocopherol	d-a-tocopherol (RRR)	244	Very unstable
Vitamin K	1 Ansbacher unit = 20 Dam units = 0.0008 mg menadione	Menadione sodium bisulfite (MSB)	100	Coated form more stable
		Menadione sodium bisulfite complex (MSBC)	100	Legal for poultry only
		Menadione dimethylphrimidinol bisulfite (MPB)	100	
Riboflavin	No IU—use mcg or mg	Cyrstalline riboflavin	100	
Niacin	No IU—use mcg or mg	Niacinamide	100	
		Nicotinic acid	100	
Pantothenic acid	*No IU—use mcg or mg*	d-calcium pantothenate	100	
		dl-calcium pantothenate	50	
		dl-calcium pantothenate + calcium chloride complex	50	

(Continued)

TABLE 8-6 (Continued)

Vitamin	1 IU Equals	Sources	RB[3]	Comments
			(%)	
Vitamin B-12	1 mcg cyanocobalamin or 1 USP unit or 11,000 LLD (*L. lactis* Dorner) units	Cyanocabalamin	Unk	
Choline	No IU—use mcg or mg	Choline chloride	Unk	Hygroscopic
Biotin	No IU—use mcg or mg			
	d-biotin	Unk		
Folic acid	No IU—use mcg or mg	Folic acid	Unk	Coated form more stable

[1]Adapted by the authors from: Reece, D. E., *et al.*, *Swine Nutritioin Guide*, published by University of Nebraska and South Dakota State University, 1995.

[2]Most common sources are in italic.

[3]RB = relative bioavailability.

[4]Unk = unknown.

WATER

Water is so common that it is seldom thought of as a nutrient. However, it is the largest single part of nearly all living things. Water accounts for as much as 80% of the body weight of a pig at birth and declines to approximately 50% in a finished market animal.

Water performs many tasks in the body. It makes up most of the blood which carries nutrients to the cells and carries waste products away; it is necessary in most of the body's chemical reactions; it is the body's built-in cooling system—it regulates the temperature, and it serves as a lubricant.

Life on earth would not be possible without water. An animal can live longer without feed than without water.

The water requirement of hogs is related to feed intake and body weight. Under normal conditions, a pig will consume ¼ to ⅓ gal of water for every pound of dry feed.

Table 8-7 gives the estimated water consumption of various classes of swine.

The need for water is increased by diarrhea, high

Fig. 8-15. Pigs drinking from nipple waterers. (Courtesy, National Pork Producers Council, Des Moines, IA)

salt intake, high ambient temperature, fever, or lactation.

For all classes of swine confined to pens, it is recommended that at least one nipple drinker device be provided for every 15 pigs in the group, with a minimum of two devices per group. One nipple drinker should be provided for every 10 pigs in the nursery.

Water can serve as a carrier for dewormers, medicinals, oral vaccines, or water-soluble nutrients when administered through a properly controlled dispensing system.

FEEDS FOR SWINE

Throughout the world, swine are raised on a great variety of feeds, including numerous byproducts. Except when on pasture or when ground dry forages are

TABLE 8-7
WATER CONSUMPTION OF SWINE

Class of Swine	Water Consumption
	(gal/head/day)
Gestating sows	2–3
Lactating sows	4–5
Starting pigs (13–45 lb)	0.5–1
Growing pigs (45–130 lb)	1
Finishing pigs (130–250 lb)	1.5–2

incorporated in the ration, they eat relatively little roughage. Only about 4% of the total feed consumed by swine in the United States is derived from roughage.

Corn and swine production have always been closely associated. Yet, the agriculture of the 50 states is very diverse, and the diet of the pig is readily adapted to the feeds produced locally. A similar adaptation in feeding practices is found in other countries. Thus, in most sections of the world, swine are fed predominantly on homegrown feeds. Ireland depends largely upon potatoes and dairy byproducts; the swine industry of Denmark has been built up to augment the dairy industry, with milk and whey supplementing homegrown and imported cereals (mostly barley); in Germany, the pig is fed on such crops as potatoes, sugar beets, and green forage; and in China, the pig is primarily a scavenger, competing very little for grains suitable for human consumption.

Thus, a swine producer generally has a wide variety of ingredients from which to choose in formulating a diet. Each ingredient may contain several nutrients in varying amounts. Because ingredients vary in price, and in amount and quality of nutrients contained, judgment must be exercised in the choice made.

The nutrient present in the largest amount determines how an ingredient is classified. Thus, corn is classed as an energy feed because it is high in calories, and soybean meal is classified as a protein supplement because it is high in protein content.

CONCENTRATES

Because of their simple monogastric stomach, swine consume more concentrates and less roughages than any other class of large farm animals. This characteristic gives pigs limited opportunity to consume large quantities of calcium and of vitamin-rich and better quality protein roughages, with the result that they suffer from more nutritional deficiencies than any other species except poultry.

Although most concentrate feeds are not suitable as the sole ration for hogs, it must be realized that swine can utilize a larger variety of feeds to greater advantage than other farm animals. In general, the grain crops—corn, barley, wheat, oats, rye, and the sorghums—constitute the major component of the swine diet. However, sweet potatoes and peanuts are successfully and extensively used in the South, soybeans in the central states, and peas in the Northwest. In those districts where they are grown, potatoes (cull) also are utilized in considerable quantities in feeding hogs. In addition, in almost every section of the country byproduct feeds are fed to hogs—including the byproducts of the fishing, meat-packing, milling, and dairy industries. Human food wastes, such as refuse or garbage (commercial garbage must be cooked), are also fed extensively.

The protein and vitamin requirements of the monogastric pig differ very greatly from those of the ruminant, for the latter improves the quality of proteins and creates certain vitamins through bacterial synthesis.

Despite all this, it is possible to meet the nutritive needs of the pig on concentrated feeds by keeping in mind the following facts when balancing the diet:

1. The cereal grains and their byproducts are relatively good sources of phosphorus, but low in calcium and the other minerals.
2. Except for the carotene content of yellow corn and green peas, the grains are very poor sources of the vitamins.
3. Most cereal grains supply proteins of poor quality.
4. Protein supplements of animal origin and soybean meal generally supply proteins of high quality, whereas proteins of plant origin other than soybean meal generally supply proteins of low quality.
5. Because of the inadequacies of most concentrates, it is usually necessary to rely on fortifications with minerals and vitamins.

ENERGY FEEDS

Carbohydrates and fats may be classed together as energy feeds for swine. Ration ingredients commonly used as sources of energy are corn, barley, sorghum, and wheat.

In this country, corn and swine production have always gone hand in hand. Normally, about one-fourth of the U.S. corn crop is fed to hogs. It is usually the cheapest source of energy, but price fluctuations frequently justify consideration of other energy feeds.

Although barley, sorghum, and wheat are higher in protein than corn, it is noteworthy that their protein is generally of the same poor quality as corn. It is noteworthy, too, that all of the cereal grains have about the same vitamin and mineral deficiencies.

Further discussion of each of the energy feeds commonly fed to swine is found in Chapter 10, Grains and Other High Energy Feeds for Swine.

PROTEIN FEEDS

Protein is made up of nitrogenous compounds called amino acids. Protein feeds vary in the kind and amount of amino acids they contain. During the digestion process, the protein in feed is broken down into the various amino acids and the pig recombines them into the kind of protein needed for muscle development, repair of worn-out tissue, etc. Thus, the real need of the pig is for amino acids, not protein as such.

The pig can synthesize some of the amino acids, with the result that they are not required in the diet. However, 10 of the amino acids are termed *essential*, because the body cannot manufacture them in sufficient quantity to permit maximum growth and performance. It is important that ingredients rich in the essential amino acids be used in formulating the diet.

Although it is a common practice to refer to "percent protein" in a ration, this term has little meaning in swine feeds unless there is knowledge concerning the amino acids present. A protein feed is considered to be of good quality when it contains all the essential amino acids in the proportions and amounts needed by the pig.

Soybean meal is by far the leading high-protein supplement of the United States. About half of the total high-protein feeds fed to animals (including both oilseed meals and animal proteins) consists of soybean meal. Although soybean meal is marginal in methionine, it is otherwise very well balanced in amino acids. It must be supplemented with minerals and vitamins. Usually, it is not fed free-choice because of its high palatability, which results in pigs eating more than is needed to meet their protein needs.

Although soybean meal is the most widely used protein supplement, many other protein feeds are suitable, and are used; among them, cottonseed meal, dried skim milk, fish meal, linseed meal, meat scraps, meat and bone scraps, peanut meal, rapeseed meal, and tankage.

Further discussion of each of these protein feeds is found in Chapter 11, Protein Feeds for Swine.

PASTURES

In modern U.S. swine production, pastures have been relegated to a minor role, especially in big industrialized hog operations. Nevertheless, good legume pasture Is an excellent feed for hogs.

With excellent pasture that possesses a heavy legume content, sows can be fed 2 lb less grain and 0.5 lb less protein supplement per head per day during gestation.

For growing and finishing hogs, high-quality legume pasture can furnish part of the protein supplement. The protein level in complete ground rations can be reduced by 2% for growing-finishing hogs fed on legume pasture as compared to confinement feeding.

Pastures and their use are discussed in Chapter 12, Pastures and Forages for Swine, and pastures versus confinement hog production is discussed in Chapter 14, Swine Management.

DRY FORAGES

Alfalfa is practically the only dry forage fed to swine in the United States.

If the price of alfalfa meal is right—if it is a cheaper source of protein and energy than corn and other grains—15 to 35% (or even higher levels) alfalfa meal may, to advantage, be incorporated in gestating-lactating sow diets and 2 to 5% may be used in growing-finishing diets. These levels serve as a safety factor to ensure the presence of certain vitamins, minerals, and unidentified factors.

Further discussion of dry forages is found in Chapter 12, Pastures and Forages for Swine.

SILAGES

Good-quality silage is an excellent feed for brood sows; hence, it may be used to advantage where it is available on dairy and beef farms. Unless the sow herd is very large, however, it will not likely pay to construct a silo especially for hogs.

Further discussion of silages is found in Chapter 12, Pastures and Forages for Swine

HOGGING DOWN CROPS

Pigs are sometimes allowed to do their own harvesting. Small grain crops that have been badly lodged or otherwise damaged, or that cannot be harvested because of weather, may be harvested by hogs. In the South, when the price and quality are down, such crops as peanuts, sweet potatoes, chufas, and other root and tuber crops are sometimes harvested by hogs.

Further discussion of hogging down crops is found in Chapter 10, Grains and Other High Energy Feeds for Swine.

GARBAGE

From the remote day of domestication forward, swine have been considered scavengers—often fed on table scraps and other wastes. Even today, in most of the developing countries, pigs, and, to a lesser extent other livestock, consume precious few products that are suitable for human food.

Today, all 50 states of the U.S. have laws requiring that municipal garbage be cooked.

Further discussion of garbage is found in Chapter 10, Grains and Other High Energy Feeds for Swine.

FEED ADDITIVES

Certain feed additives have become somewhat standard ingredients of swine diets, especially for pigs from birth to market weight. They are not nutrients as such; hence, they should not be considered as dietary essentials.

Feed additives are covered in more detail in Chapter 9, Swine Feeding Standards, Ration Formulation, and Feeding Programs.

FEED EVALUATION

Fig. 8-16. Evaluating feeds by feeding trials.

Profit is the ultimate criterion of success in any livestock operation, and the cost of nutrients is an important factor in determining success. The feed composition values presented in this and other books merely represent an averaging of an accumulation of data concerning the nutritive value of feeds. Considerable variation is inherent in the nutrient content of different samples of feeds. Thus, the successful producer must recognize the value of a well-planned feed analysis program.

An example of the variability of feedstuffs is presented in Table 8-8. Investigators at North Carolina

TABLE 8-8
VARIATION IN THE COMPOSITION OF 48.5% PROTEIN SOYBEAN MEAL IN NORTH CAROLINA[1]

Nutrient	Number of Samples	Average Composition	Low	High
		------ (%) ------		
Moisture	1,423	11.86	1.22	15.17
Protein	1,425	48.69	41.88	54.86
Fat	128	2.08	0.68	4.70
Fiber	1,424	3.20	2.10	33.00

[1]Adapted by the authors from "Ingredient Quality Analysis and Reporting," by D. W. Murphy and J. B. Ward, *Feedstuffs*, June 7, 1976.

State University summarized the quality of various feed ingredients used within the state. Table 8-8 shows the variation that they found in the composition of samples of 48.5% soybean meal that were monitored.

Feeds are analyzed by physical, chemical, or biological procedures. Although physical evaluation may be the least accurate, it provides a quick and easy means of obtaining considerable information about the overall quality of a feed. The chemical procedure is more accurate than a physical evaluation, but it takes time. The biological method necessitates considerable time and expense, and the results are often variable.

PHYSICAL EVALUATION

In order to produce or buy superior feeds, swine producers need to know what constitutes feed quality, and how to recognize it. They need to be familiar with those recognizable characteristics of feeds which indicate high palatability and nutrient content. If in doubt, observation of the pigs consuming the feed will tell them, for pigs prefer and thrive on high-quality feed.

The easily recognizable characteristics of good grains and other concentrates are:

1. Seeds are not split or cracked.
2. Seeds are of low-moisture content—generally containing about 88% dry matter.
3. Seeds have a good color.
4. Concentrates and seeds are free from mold.
5. Concentrates and seeds are free from rodent and insect damage.
6. Concentrates and seeds are free from foreign material, such as iron filings.
7. Concentrates and seeds are free from rancid odor.

CHEMICAL ANALYSIS

Today, feeds are being analyzed routinely through highly sophisticated chemical procedures. Many agricultural experiment stations, as well as most large feed companies, have facilities to analyze feeds for both the prevention and diagnosis of nutritional problems.

A chemical analysis gives a solid foundation on which to start in the evaluation of feeds. Thus, feed composition tables serve as a basis for ration formulation and for feed purchasing and merchandising. Commercially prepared feeds are required by state law to be labeled with a list of ingredients and a guaranteed analysis. Although state laws vary slightly, most of them require that the feed label (tag) show in percent the minimum crude protein and fat; and maximum crude fiber and ash. Some feed labels also include maximum salt, and/or minimum calcium and phosphorus. These figures are the buyer's assurance that the feed contains the minimal amounts of the higher cost items—protein and fat; and not more than the stipulated amounts of the lower cost, and less valuable, items—the crude fiber and ash.

PROXIMATE ANALYSIS (Weende Procedure)

For more than 100 years, feeds have been analyzed by a method developed by two scientists, Henneberg and Stohmann, at the Weende Experiment Station in Germany. This method is called the proximate analysis, or the Weende system, of feed analysis. Feeds are broken down into six components: (1) moisture, (2) ash, (3) crude protein, (4) ether extract, (5) crude fiber, and (6) nitrogen-free extract.

BOMB CALORIMETRY

When compounds are burned completely in the presence of oxygen, the resulting heat is referred to as gross energy or the heat of combustion. The bomb calorimeter is used to determine the gross energy of feed, waste products from feed (for example, feces and urine), and tissues.

The calorie is defined as the amount of heat required to raise the temperature of 1 g of water 1°C (precisely from 14.5° to 15.5°C). With this fact in mind, we can readily see how the bomb calorimeter works.

Briefly stated, the procedure is as follows: An electric wire is attached to the material being tested, so that it can be ignited by remote control; 2,000 g of water are poured around the bomb; 25 to 30 atmospheres of oxygen are added to the bomb; the material is ignited; the heat given off from the burned material warms the water; and a thermometer registers the change in the temperature of the water. For example,

Fig. 8-17. Bomb calorimeter for the determination of gross energy (caloric content). (Courtesy, Parr Instrument Company, Moline, IL)

if 1 g of feed is burned and the temperature of the water increases 1°C, 2,000 calories/gram are given off.

It is noteworthy that the determination of the heat of combustion with a bomb calorimeter is not as difficult or time-consuming as the chemical analyses used in arriving at TDN values.

Note: Although the *bomb calorimeter* is useful in obtaining gross energy (GE), in order to determine animal utilization of a feed, animal trials of one type or another must be carried out.

BIOLOGICAL ANALYSIS

Quite often, biological assays are used in the analysis of micronutrients in feeds. There are two basic types of biological assays—(1) microbiological assays, and (2) the use of nutrient-deficient animals.

Biological assays tend to be laborious and time-consuming. Large numbers of samples are needed to produce statistically reliable results, and quite often data obtained from these assays are highly variable. The assay utilizing nutrient-deficient animals is particularly cumbersome because (1) the animals should be of approximately the same age, sex, and weight; and (2) time is required to induce deficient conditions in these animals.

QUESTIONS FOR STUDY AND DISCUSSION

1. Several factors affect the nutritive requirements of swine. Name them.

2. Why is knowledge of swine body composition important?

3. Table 8-1 gives the range in body composition of pigs from birth to maturity. Detail the changes in body composition from birth to maturity of water, fat, protein, and ash.

4. Describe the digestive system of the pig. Compare the digestive system of the pig (a nonruminant) and the digestive system of a ruminant.

5. Define nutrients. Name the nutrients required by the pig.

6. Discuss the nutrient needs for each of the following body functions:
   a. Maintenance
   b. Growth
   c. Reproduction
   d. Lactation
   e. Fattening

7. Table 8-2 shows a dramatic change in the feed conversion of pigs from pre-nursery to finisher. How do you explain this?

8. What is a calorie and how is it determined?

9. What is the difference between the gross energy of a feed and the digestible energy of a feed?

10. How is the total digestible nutrients (TDN) of a feed calculated? Why is this system of measuring energy in feedstuffs being used less?

11. What is an essential amino acid? Which amino acids are essential for swine?

12. Why is protein most frequently the limiting factor in the diet of swine, from the standpoint of both quality and quantity? What is a limiting amino acid?

13. Explain how minerals and vitamins of swine have been affected by management changes since the 1980s.

14. What peculiarities of swine husbandry are conducive to swine suffering from mineral deficiencies?

15. Distinguish between macrominerals and microminerals.

16. Give (a) the function, and (b) the deficiency symptoms resulting from a lack of each of the following minerals in the pig: salt, calcium, phosphorus, iodine, and iron.

17. Discuss the relationship of calcium, phosphorus, and vitamin D in swine nutrition.

18. Why is the availability of phosphorus estimated at 30% when formulating diets?

19. What are chelated minerals? Should they be recommended?

20. Discuss the sources and bioavailability of minerals.

21. Give (a) the function, and (b) the deficiency symptoms resulting from a lack of each of the following vitamins in the pig: A, D, E, choline, pantothenic acid, and riboflavin.

22. What vitamins are most apt to be deficient in ordinary swine diets?

23. Before the advent of modern feed formulations and vitamins, how did producers ensure that swine received all the necessary vitamins?

24. Discuss the sources and bioavailability of vitamins.

25. Discuss the importance of water for the pig.

26. Classify feeds used for swine and indicate the role of pastures, dry forages, silages, hogging down crops, and garbage.

27. List and describe the ways feeds may be analyzed. Why are feed analyses so important?

SELECTED REFERENCES

Title of Publication	Author(s)	Publisher
Animal Feeding & Nutrition	M. H. Jorgens	Kendall/Hunt Publishing Co., Dubuque, IA, 1982
Animal Science, Ninth Edition	M. E. Ensminger	Interstate Publishers Inc., Danville, IL, 1991
Bioenergetics and Growth	S. Brody	Reinhold Publishing Corp., New York, NY, 1945
Body Composition in Animals and Man	National Research Council	National Academy of Sciences Washington, DC, 1968
Effect of Environment on Nutrient Requirements of Domestic Animals	National Research Council	National Academy Press, Washington, DC, 1981
Feeds & Nutrition Digest	M. E. Ensminger J. E. Oldfield W. W. Heinemann	The Ensminger Publishing Company, Clovis, CA, 1990
Feeds & Nutrition, Second Edition	M. E. Ensminger J. E. Oldfield W. W. Heinemann	The Ensminger Publishing Company, Clovis, CA, 1990
Fire of Life, The	M. Kleiber	John Wiley & Sons, Inc., New York, NY, 1961
Fundamentals of Nutrition	L. E. Lloyd B. E. McDonald E. W. Crampton	W. H. Freeman and Co., San Francisco, CA, 1978
Livestock Feeds and Feeding	D. C. Church	O&B Books, Inc., Corvallis, OR, 1984
Nutrient Requirements of Swine, Ninth Revised Edition	National Research Council	National Academy of Sciences, Washington, DC, 1988
Pork Industry Handbook		Cooperative Extension Service, University of Illinois, Urbana-Champaign
Science of Animals That Serve Mankind, The	J. R. Campbell F. J. Lasley	McGraw-Hill Book Company, New York, NY, 1985
Stockman's Handbook, The, Seventh Edition	M. E. Ensminger	Interstate Publishers Inc., Danville, IL, 1992
Swine Feeding and Nutrition	T. J. Cunha	Academic Press, Inc., New York, NY, 1977
Swine Nutrition	E. R. Miller D. E. Ullrey A. J. Lewis	Butterworth-Heinemann, Boston, MA, 1991
Swine Nutrition Guide	J. F. Patience P. A. Thacker	Prairie Swine Center, University of Saskatchewan, Saskatoon, Saskatchewan, Canada, 1989
Swine Production and Nutrition	W. G. Pond J. H. Maner	AVI Publishing Company, Inc., Westport, CT, 1984

9

SWINE FEEDING STANDARDS, DIET FORMULATION, AND FEEDING PROGRAMS[1]

The largest swine-feed production facility in the U.S., with a capacity of 18,000 tons per week, operated by Murphy Farms, Rose Hill, North Carolina. (Courtesy, Dr. Garth W. Boyd and Ms. Rhonda Campbell)

[1]In the preparation of this chapter, the authors adapted applicable material from the following two publications: *Swine Nutrition Guide*, published by the University of Nebraska and South Dakota State University, 1995; and *Kansas Swine Nutrition Guide*, published by Kansas State University, updated 1995.

Many factors affect swine feeding standards, diet formulation, and feeding programs; primarily, feedstuffs, ingredient quality, feed intake, nutrient interactions, nutrient bioavailability, feed processing, safety margin, nonnutrient additives, genotype and productivity, health, economics, and water. These factors interact with each other. Thus, their net output determines the level of swine production and profitability. As a result of research and management, these factors are constantly changing. So, to maximize performance and profits, the modification of the nutrition and feeding programs of today's swine herds never ends.

FACTORS INVOLVED IN FORMULATING SWINE DIETS

Before anyone can intelligently formulate a swine diet, it is necessary to know (1) the nutrient requirements of the particular pigs to be fed, which calls for feeding standards; (2) the availability, nutrient content, and cost of feedstuffs; (3) the acceptability and physical condition of feedstuffs; (4) the average daily consumption of the pigs to be fed; and (5) the presence of substances which may be harmful to product quality.

NUTRITIVE NEEDS OF SWINE

The nutrient needs of swine are influenced by age, function, disease level, nutrient interaction, environment, etc. It has been established that the pig has a requirement for over 40 different nutrients. Fortunately, not all of them are of practical concern.

FEEDING STANDARDS

Feeding standards are tables listing the amounts of one or more nutrients required by different species of animals for specific productive functions, such as growth, finishing, gestating, and lactating. Most feeding standards are expressed in either (1) quantities of nutrients required per day, and/or (2) concentration in the diet. The first type is used where animals are provided a given amount of a feed during a 24-hour period, and the second is used where animals are provided a diet without limitation on the time in which it is consumed.

Today, the most up-to-date feeding standards in the United States are those published by the National Research Council (NRC) of the National Academy of Sciences. Periodically, a specific committee, composed of outstanding researchers who have worked extensively with the class of animal whose requirements are being reviewed, revises the nutrient requirements of

each species for different functions. Thus, the nutritive needs of each type of livestock are dealt with separately and in depth.

Feeding standards have been established for swine, and, through the use of these standards, the producer can formulate diets tailored to meet the nutrient requirements of the pigs.

Although feeding standards are excellent and needed guides, there are still many situations where nutrient needs cannot be specified with great accuracy. For example, maximal carcass leanness may require greater concentrations of certain nutrients than maximal rate of gain. Lean pigs deposit more protein in their gain, which increases their requirements for protein and individual amino acids. Leanness is associated positively with feed efficiency; thus, requirements for maximal feed efficiency are generally greater than those for maximal weight gain. Maximal blood hemoglobin concentration may necessitate higher levels of iron than those needed for maximal rate or efficiency of gain, and maximal bone ash generally requires higher levels of calcium and phosphorus than those needed for maximal weight gain. Moreover, feeding standards tell nothing about the palatability, physical nature, or possible digestive disturbances of a diet. Neither do they give consideration to individual differences, management differences, and the effects of such stresses as weather, disease, and parasitism. Thus, there are many variables that alter the nutrient needs and utilization of pigs—variables that are difficult to include quantitatively in feeding standards, even when feed quality is well known. Ultimately, each swine manager must select a set of standards that will permit the greatest economic return in the particular environment. Optimum standards will relate to the genetic potential of the herd and to the price and availability of feed ingredients in the region.

NATIONAL RESEARCH COUNCIL (NRC) REQUIREMENTS

The NRC nutritional requirements, or standards, are presented in Tables 9-1, 9-2, 9-3, 9-4, 9-5, 9-6, and 9-7. It is not intended that these standards impart the impression that such figures are absolute, final, and unchangeable. Rather, they should be used as guides based on research. Also, these figures are, for the most part, requirements (rather than allowances); hence, they do not provide for margins of safety to compensate for variations in feed composition, environment, and possible losses of nutrients during storage or processing.

In using Tables 9-1, 9-2, 9-3, 9-4, 9-5, 9-6, and 9-7, the following additional points should be recognized:

TABLE 9-1
DAILY NUTRIENT INTAKES AND REQUIREMENTS OF SWINE ALLOWED FEED *AD LIBITUM*[1]

Intake and Performance Levels	Lb / Kg	2.2–11 / 1–5		11–22 / 5–10		22–44 / 10–20		44–110 / 20–50		110–242 / 50–110	
		(lb)	*(g)*	*(lb)*	*(g)*	*(lb)*	*(g)*	*(lb)*	*(g)*	*(lb)*	*(g)*
Expected weight gain per day		0.4	*200*	0.6	*250*	1.0	*450*	1.5	*700*	1.8	*820*
Expected feed intake per day		0.6	*250*	1.0	*460*	2.1	*950*	4.2	*1,900*	6.9	*3,110*
Expected efficiency (gain/feed)		0.800		0.543		0.474		0.368		0.264	
Expected efficiency (feed/gain)		1.25		1.84		2.11		2.71		3.79	
Digestible energy intake (kcal per day)		850		1,560		3,230		6,460		10,570	
Metabolizable energy intake (kcal per day)		805		1,490		3,090		6,200		10,185	
Energy concentration (kcal ME per lb diet)		1,461		1,470		1,474		1,479		1,486	
Energy concentration *(kcal ME per kg diet)*		*3,220*		*3,240*		*3,250*		*3,260*		*3,275*	
Protein per day		0.1	*60*	0.2	*92*	0.4	*171*	0.6	*285*	0.9	*404*

Nutrient	Requirement (Amount Per Day)				
Indispensable amino acids:					
Arginine (g)	1.5	2.3	3.8	4.8	3.1
Histidine (g)	0.9	1.4	2.4	4.2	5.6
Isoleucine (g)	1.9	3.0	5.0	8.7	11.8
Leucine (g)	2.5	3.9	6.6	11.4	15.6
Lysine (g)	3.5	5.3	9.0	14.3	18.7
Methionine + cystine (g)	1.7	2.7	4.6	7.8	10.6
Phenylalanine + tyrosine (g)	2.8	4.3	7.3	12.5	17.1
Threonine (g)	2.0	3.1	5.3	9.1	12.4
Tryptophan (g)	0.5	0.8	1.3	2.3	3.1
Valine (g)	2.0	3.1	5.3	9.1	12.4
Linoleic acid (g)	0.3	0.5	1.0	1.9	3.1
Major or macrominerals:					
Calcium (g)	2.2	3.7	6.6	11.4	15.6
Chlorine (g)	0.2	0.4	0.8	1.5	2.5
Magnesium (g)	0.1	0.2	0.4	0.8	1.2
Phosphorus, total (g)	1.8	3.0	5.7	9.5	12.4
Phosphorus, available (g)	1.4	1.8	3.0	4.4	4.7
Potassium (g)	0.8	1.3	2.5	4.4	5.3
Sodium (g)	0.2	0.5	1.0	1.9	3.1
Trace or microminerals:					
Copper (mg)	1.50	2.76	4.75	7.60	9.33
Iodine (mg)	0.04	0.06	0.13	0.27	0.44
Iron (mg)	25	46	76	114	124
Manganese (mg)	1.00	1.84	2.85	3.80	6.22
Selenium (mg)	0.08	0.14	0.24	0.28	0.31
Zinc (mg)	25	46	76	114	155
Fat-soluble vitamins:					
Vitamin A (IU)	550	1,012	1,662	2,470	4,043
Vitamin D (IU)	55	101	190	285	466
Vitamin E (IU)	4	7	10	21	34
Vitamin K (menadione) (mg)	0.02	0.02	0.05	0.10	0.16
Water-soluble vitamins:					
Biotin (mg)	0.02	0.02	0.05	0.10	0.16
Choline (g)	0.15	0.23	0.38	0.57	0.93
Folacin (Folic acid) (mg)	0.08	0.14	0.28	0.57	0.93
Niacin (Nicotinic acid, Nicotinamide), available (mg)	5.00	6.90	11.88	19.00	21.77
Pantothenic acid (Vitamin B-3) (mg)	3.00	4.60	8.55	15.20	21.77
Riboflavin (Vitamin B-2) (mg)	1.00	1.61	2.85	4.75	6.22
Thiamin (Vitamin B-1) (mg)	0.38	0.46	0.95	1.90	3.11
Vitamin B-6 (Pyridoxine, Pyridoxal, Pyridoxamine) (mg)	0.50	0.69	1.42	1.90	3.11
Vitamin B-12 (Cobalamins) (mcg)	5.00	8.05	14.25	19.00	15.55

[1]Adapted by the authors from *Nutrient Requirements of Swine*, 9th rev. ed., National Research Council, National Academy Press, 1988, p. 51, Table 5-2.

TABLE 9-2
DAILY NUTRIENT INTAKES AND REQUIREMENTS
OF INTERMEDIATE-WEIGHT BREEDING ANIMALS[1]

		Mean Gestation or Farrowing Weight of:	
		Bred Gilts, Sows, and Adult Boars	Lactating Gilts and Sows
Intake and Performance Levels	Lb *Kg*	358.3 *162.5*	363.8 *165.0*
		(lb) *(kg)*	*(lb)* *(kg)*
Daily feed intake		4.2 *1.9*	11.7 *5.3*
Digestible energy . . . . (Mcal per day)		6.3	17.7
Metabolizable energy . . . . (Mcal per day)		6.1	17.0
		(g)	*(g)*
Crude protein per day		0.5 *228*	1.5 *689*

Nutrient	Requirement (Amount Per Day)	
Indispensable amino acids:		
Arginine . . . . (g)	0.0	21.2
Histidine . . . . (g)	2.8	13.2
Isoleucine . . . . (g)	5.7	20.7
Leucine . . . . (g)	5.7	25.4
Lysine . . . . (g)	8.2	31.8
Methionine + cystine . . . . (g)	4.4	19.1
Phenylalanine + tyrosine . . . . (g)	8.6	37.1
Threonine . . . . (g)	5.7	22.8
Tryptophan . . . . (g)	1.7	6.4
Valine . . . . (g)	6.1	31.8
Linoleic acid . . . . (g)	1.9	5.3
Major or macrominerals:		
Calcium . . . . (g)	14.2	39.8
Chlorine . . . . (g)	2.3	8.5
Magnesium . . . . (g)	0.8	2.1
Phosphorus, total . . . . (g)	11.4	31.8
Phosphorus, available . . . . (g)	6.6	18.6
Potassium . . . . (g)	3.8	10.6
Sodium . . . . (g)	2.8	10.6
Trace or microminerals:		
Copper . . . . (mg)	9.5	26.5
Iodine . . . . (mg)	0.3	0.7
Iron . . . . (mg)	152	424
Manganese . . . . (mg)	19	53
Selenium . . . . (mg)	0.3	0.8
Zinc . . . . (mg)	95	265
Fat-soluble vitamins:		
Vitamin A . . . . (IU)	7,600	10,600
Vitamin D . . . . (IU)	380	1,060
Vitamin E . . . . (IU)	42	117
Vitamin K (menadione) . . . . (mg)	1.0	2.6
Water-soluble vitamins:		
Biotin . . . . (mg)	0.4	1.1
Choline . . . . (g)	2.4	5.3
Folacin (Folic acid) . . . . (mg)	0.6	1.6
Niacin (Nicotinic acid, Nicotinamide), available . . . . (mg)	19.0	53.0
Pantothenic acid (Vitamin B-3) . . . . (mg)	22.8	63.6
Riboflavin (Vitamin B-2) . . . . (mg)	7.1	19.9
Thiamin (Vitamin B-1) . . . . (mg)	1.9	5.3
Vitamin B-6 (Pyridoxine, Pyridoxal, Pyridoxamine) . . . . (mg)	1.9	5.3
Vitamin B-12 (Cobalamins) . . . . . (mcg)	28.5	79.5

[1]Adapted by the authors from *Nutrient Requirements of Swine*, 9th rev. ed., National Research Council, National Academy Press, 1988, p. 52, Table 5-4.

1. Feedstuffs produced in various parts of the country vary in nutritive value.

2. The environment in which pigs are produced can modify the requirements.

3. Animals bred for high performance have nutritional needs that are quite different from average performers.

TABLE 9-3
DAILY ENERGY AND FEED REQUIREMENTS
OF PREGNANT GILTS AND SOWS[1]

		Weight of Bred Gilts and Sows at Mating[2]		
Intake and Performance Levels	Lb *Kg*	265 *120*	309 *140*	353 *160*
		(lb) *(kg)*	*(lb)* *(kg)*	*(lb)* *(kg)*
Mean gestation weight[3]		314.2 *142.5*	358.3 *162.5*	402.3 *182.5*
Energy required:				
Maintenance[4] . (Mcal DE per day)		4.53	5.00	5.47
Gestation weight gain[5] . . . . (Mcal DE per day)		1.29	1.29	1.29
Total . . . . . . (Mcal DE per day)		5.82	6.29	6.76
Feed required per day[6]		4.0 *1.8*	4.2 *1.9*	4.4 *2.0*

[1]Adapted by the authors from *Nutrient Requirements of Swine*, 9th rev. ed., National Research Council, National Academy Press, 1988, p. 53, Table 5-6.

[2]Requirements are based on 55-lb *(25-kg)* maternal weight gain plus 44-lb *(20-kg)* increase in weight due to the products of conception; the total weight gain is 99 lb *(45 kg)*.

[3]Mean gestation weight is weight at mating + (total weight gain/2).

[4]The animal's daily maintenance requirement is 110 kcal of DE/kg$^{0.75}$.

[5]The gestation weight gain is 1.10 Mcal of DE/day for maternal weight gain plus 0.19 Mcal of DE/day for conceptus gain.

[6]The feed required/day is based on a corn-soybean meal ration containing 3.34 Mcal of DE/kg.

TABLE 9-4
DAILY ENERGY AND FEED REQUIREMENTS
OF LACTATING GILTS AND SOWS[1]

		Weight of Lactating Gilts and Sows at Postfarrowing		
Intake and Performance Levels	Lb *Kg*	320 *145*	364 *165*	408 *185*
		(lb) *(kg)*	*(lb)* *(kg)*	*(lb)* *(kg)*
Milk yield		11.0 *5.0*	13.78 *6.25*	16.5 *7.5*
Energy required:				
Maintenance[2] (Mcal DE per day)		4.5	5.0	5.5
Milk production[3] (Mcal DE per day)		10.0	12.5	15.0
Total . . . . . . (Mcal DE per day)		14.5	17.5	20.5
Feed required per day[4]		9.7 *4.4*	11.7 *5.3*	13.5 *6.1*

[1]Adapted by the authors from *Nutrient Requirements of Swine*, 9th rev. ed., National Research Council, National Academy Press, 1988, p. 53, Table 5-6.

[2]The animal's daily maintenance requirement is 110 kcal of DE/kg$^{0.75}$.

[3]Milk production requires 2.0 Mcal of DE/kg of milk.

[4]The feed required/day is based on a corn-soybean meal ration containing 3.34 Mcal of DE/kg.

TABLE 9-5
NUTRIENT REQUIREMENTS IN THE DIET OF SWINE ALLOWED FEED *AD LIBITUM* (90% DRY MATTER)[1]

Intake and Performance Levels	Lb Kg	2.2–11 1–5		11–22 5–10		22–44 10–20		44–110 20–50		110–242 50–110	
		(lb)	*(g)*	*(lb)*	*(g)*	*(lb)*	*(g)*	*(lb)*	*(g)*	*(lb)*	*(g)*
Expected weight gain per day		0.4	*200*	0.6	*250*	1.0	*450*	1.5	*700*	1.8	*820*
Expected feed intake per day		0.6	*250*	1.0	*460*	2.1	*950*	4.2	*1,900*	6.9	*3,110*
Expected efficiency (gain/feed)		0.800		0.543		0.474		0.368		0.264	
Expected efficiency (feed/gain)		1.25		1.84		2.11		2.71		3.79	
Digestible energy intake (kcal per day)		850		1,560		3,230		6,460		10,570	
Metabolizable energy intake (kcal per day)		805		1,490		3,090		6,200		10,185	
Energy concentration (kcal ME per lb diet)		1,461		1,470		1,474		1,479		1,486	
Energy concentration *(kcal ME per kg diet)*		*3,220*		*3,240*		*3,250*		*3,260*		*3,275*	
Protein (%)		24		20		18		15		13	

Nutrient	Requirement (Percent or Amount/Lb [Kg] Diet)[2]									
	(lb)	*(kg)*	*(lb)*	*(kg)*	*(lb)*	*(kg)*	*(lb)*	*(kg)*	*(lb)*	*(kg)*
Indispensable amino acids:										
Arginine (g)	0.60		0.50		0.40		0.25		0.10	
Histidine (g)	0.36		0.31		0.25		0.22		0.18	
Isoleucine (g)	0.76		0.65		0.53		0.46		0.38	
Leucine (g)	1.00		0.85		0.70		0.60		0.50	
Lysine (g)	1.40		1.15		0.95		0.75		0.60	
Methionine + cystine (g)	0.68		0.58		0.48		0.41		0.34	
Phenylalanine + tyrosine (g)	1.10		0.94		0.77		0.66		0.55	
Threonine (g)	0.80		0.68		0.56		0.48		0.40	
Tryptophan (g)	0.20		0.17		0.14		0.12		0.10	
Valine (g)	0.80		0.68		0.56		0.48		0.40	
Linoleic acid (g)	0.1		0.1		0.1		0.1		0.1	
Major or macrominerals:										
Calcium (g)	0.90		0.80		0.70		0.60		0.50	
Chlorine (g)	0.08		0.08		0.08		0.08		0.08	
Magnesium (g)	0.04		0.04		0.04		0.04		0.04	
Phosphorus, total (g)	0.70		0.65		0.60		0.50		0.40	
Phosphorus, available (g)	0.55		0.40		0.32		0.23		0.15	
Potassium (g)	0.30		0.28		0.26		0.23		0.17	
Sodium (g)	0.10		0.10		0.10		0.10		0.10	
Trace or microminerals:										
Copper (mg)	2.7	*6.0*	2.7	*6.0*	2.3	*5.0*	1.8	*4.0*	1.4	*3.0*
Iodine (mg)	0.06	*0.14*	0.06	*0.14*	0.06	*0.14*	0.06	*0.14*	0.06	*0.14*
Iron (mg)	45	*100*	45	*100*	36	*80*	27	*60*	18	*40*
Manganese (mg)	1.8	*4.0*	1.8	*4.0*	1.4	*3.0*	0.9	*2.0*	0.9	*2.0*
Selenium (mg)	0.14	*0.30*	0.14	*0.30*	0.11	*0.25*	0.07	*0.15*	0.05	*0.10*
Zinc (mg)	45	*100*	45	*100*	36	*80*	27	*60*	23	*50*
Fat-soluble vitamins:										
Vitamin A (IU)	2,200		2,200		1,750		1,300		1,300	
Vitamin D (IU)	220		220		200		150		150	
Vitamin E (IU)	16		16		11		11		11	
Vitamin K (menadione) (mg)	0.2	*0.5*	0.2	*0.5*	0.2	*0.5*	0.2	*0.5*	0.2	*0.5*
Water-soluble vitamins:										
Biotin (mg)	0.04	*0.08*	0.02	*0.05*	0.02	*0.05*	0.02	*0.05*	0.02	*0.05*
Choline (g)	0.27	*0.6*	0.23	*0.5*	0.18	*0.4*	0.14	*0.3*	0.14	*0.3*
Folacin (Folic acid) (mg)	0.1	*0.3*	0.1	*0.3*	0.1	*0.3*	0.1	*0.3*	0.1	*0.3*
Niacin (Nicotinic acid, Nicotinamide), available (mg)	9.1	*20.0*	6.8	*15.0*	5.7	*12.5*	4.5	*10.0*	3.2	*7.0*
Pantothenic acid (Vitamin B-3) (mg)	5.4	*12.0*	4.5	*10.0*	4.1	*9.0*	3.6	*8.0*	3.2	*7.0*
Riboflavin (Vitamin B-2) (mg)	1.8	*4.0*	1.6	*3.5*	1.4	*3.0*	1.1	*2.5*	0.9	*2.0*
Thiamin (Vitamin B-1) (mg)	0.7	*1.5*	0.5	*1.0*	0.5	*1.0*	0.5	*1.0*	0.5	*1.0*
Vitamin B-6 (Pyridoxine, Pyridoxal, Pyridoxamine) (mg)	0.9	*2.0*	0.7	*1.5*	0.7	*1.5*	0.5	*1.0*	0.5	*1.0*
Vitamin B-12 (Cobalamins) (mcg)	9.1	*20.0*	7.9	*17.5*	6.8	*15.0*	4.5	*10.0*	2.3	*5.0*

[1]Adapted by the authors from *Nutrient Requirements of Swine*, 9th rev. ed., National Research Council, National Academy Press, 1988, p. 50, Table 5-1.

[2]These requirements are based upon the following types of pigs and diets: 2.2- to 11-lb *(1- to 5-kg)* pigs, a diet that includes 25 to 75% milk products; 11- to 22-lb *(5- to 10-kg)* pigs, a corn-soybean meal diet that includes 5 to 25% milk products; 22- to 242-lb *(10- to 110-kg)* pigs, a corn-soybean meal diet. In the corn-soybean meal diets, the corn contains 8.5% protein; the soybean meal contains 44%.

TABLE 9-6
NUTRIENT REQUIREMENTS IN THE DIET OF BREEDING SWINE[1]

Intake Levels	Bred Gilts, Sows, and Adult Boars	Lactating Gilts and Sows
Digestible energy . . . . . (kcal per lb diet)	1,515	1,515
Digestible energy . . . . . (kcal per kg diet)	3,340	3,340
Metabolizable energy . . . (kcal per lb diet)	1,456	1,456
Metabolizable energy . . . (kcal per kg diet)	3,210	3,210
Crude protein . . . . (%)	12	13

Nutrient	Requirement (Percent or Amount/Lb [Kg] Diet)[2]			
	(lb)	(kg)	(lb)	(kg)
Indispensable amino acids:				
Arginine . . . . (g)	0.00		0.40	
Histidine . . . . (g)	0.15		0.25	
Isoleucine . . . . (g)	0.30		0.39	
Leucine . . . . (g)	0.30		0.48	
Lysine . . . . (g)	0.43		0.60	
Methionine + cystine . . . . (g)	0.23		0.36	
Phenylalanine + tyrosine . . . . (g)	0.45		0.70	
Threonine . . . . (g)	0.30		0.43	
Tryptophan . . . . (g)	0.09		0.12	
Valine . . . . (g)	0.32		0.60	
Linoleic acid . . . . (g)	0.1		0.1	
Major or macrominerals:				
Calcium . . . . (g)	0.75		0.75	
Chlorine . . . . (g)	0.12		0.16	
Magnesium . . . . (g)	0.04		0.04	
Phosphorus, total . . . . (g)	0.60		0.60	
Phosphorus, available . . . . (g)	0.35		0.35	
Potassium . . . . (g)	0.20		0.20	
Sodium . . . . (g)	0.15		0.20	
Trace or microminerals:				
Copper . . . . (mg)	2.27	5.00	2.27	5.00
Iodine . . . . (mg)	0.06	0.14	0.06	0.14
Iron . . . . (mg)	36.29	80.00	36.29	80.00
Manganese . . . . (mg)	4.54	10.00	4.54	10.00
Selenium . . . . (mg)	0.07	0.15	0.07	0.15
Zinc . . . . (mg)	22.68	50.00	22.68	50.00
Fat-soluble vitamins:				
Vitamin A . . . . (IU)	4,000		2,000	
Vitamin D . . . . (IU)	200		200	
Vitamin E . . . . (IU)	22		22	
Vitamin K (menadione) . . . . (mg)	0.23	0.50	0.23	0.50
Water-soluble vitamins:				
Biotin . . . . (mg)	0.09	0.20	0.09	0.20
Choline . . . . (g)	0.57	1.25	0.45	1.00
Folacin (Folic acid) . . . . (mg)	0.14	0.30	0.14	0.30
Niacin (Nicotinic acid, Nicotinamide), available . . . . (mg)	4.54	10.00	4.54	10.00
Pantothenic acid (Vitamin B-3) . . . . (mg)	5.44	12.00	5.44	12.00
Riboflavin (Vitamin B-2) . . . . (mg)	1.70	3.75	1.70	3.75
Thiamin (Vitamin B-1) . . . . (mg)	0.45	1.00	0.45	1.00
Vitamin B-6 (Pyridoxine, Pyridoxal, Pyridoxamine) . . . . (mg)	0.45	1.00	0.45	1.00
Vitamin B-12 (Cobalamins) . . . . . (mcg)	6.80	15.00	6.80	15.00

[1]Adapted by the authors from *Nutrient Requirements of Swine*, 9th rev. ed., National Research Council, National Academy Press, 1988, p. 52, Table 5-3.

[2]These requirements are based upon corn-soybean meal diets, feed intakes, and performance levels listed in Tables 9-2, 9-3, and 9-4. In the corn-soybean meal rations, the corn contains 8.5% protein; the soybean meal contains 44%.

TABLE 9-7
REQUIREMENTS FOR SEVERAL NUTRIENTS
OF BREEDING HERD REPLACEMENTS ALLOWED FEED *AD LIBITUM*[1]

Intake Levels		Weight Of:			
		Developing Gilts		Developing Boars	
	Lb	44–110	110–242	44–110	110–242
	Kg	20–50	50–110	20–50	50–110
Energy concentration (kcal ME per lb ration)		1,476	1,479	1,470	1,476
Energy concentration (kcal ME per kg ration)		3,255	3,260	3,240	3,255
Crude protein . . . . . . (%)		16	15	18	16
Nutrient:[2]					
Lysine . . . . (%)		0.80	0.70	0.90	0.75
Calcium . . . . (%)		0.65	0.55	0.70	0.60
Phosphorus, total . . . (%)		0.55	0.45	0.60	0.50
Phosphorus, avail. . . . (%)		0.28	0.20	0.33	0.25

[1]Adapted by the authors from *Nutrient Requirements of Swine*, 9th rev. ed., National Research Council, National Academy Press, 1988, p. 53, Table 5-5.

[2]Sufficient data are not available to indicate that requirements for other nutrients are different from those in Table 9-5 for animals of these weights.

RECOMMENDED NUTRIENT ALLOWANCES

Nutrient requirements generally represent the minimum quantity of the nutrients that should be incorporated in a diet while nutrient allowances take into consideration a margin of safety. Nutrient allowances make provisions for variation in feed composition; possible losses during storage and processing; day-to-day and period-to-period differences in needs of animals; age and size of animal; stage of gestation and lactation; the kind and degree of activity; the amount of stress; the system of management; the health, condition, and temperament of the animal; and the kind, quality, and amount of feed—all of which exert a powerful influence in determining nutritive needs.

Since the modern swine producer is interested in maximum performance, it follows that the input of nutrients must be ample to bring this about. For this reason, the recommended allowances of the various important nutrients that should be included in diet formulation for optimum performance are given in Tables 9-8 and 9-9.

TABLE 9-8
RECOMMENDED MINIMUM NUTRIENT ALLOWANCES FOR SWINE (AS-FED BASIS)[1, 2]

	Type of Diet/Body Weight, Lb							
	Starter 1 < 15	Starter 2 15 to 25	Starter 3 25 to 50	Grower I 50 to 80	Grower II 80 to 120	Finisher 120 to Market Wt.	Gestation	Lactation
Total Level (%)								
Protein	20 to 22	18 to 20	18	17	16	14	14	15
Lysine	1.50	1.25	1.15	0.95	0.80	0.65	0.60	0.70
Tryptophan	0.21	0.18	0.17	0.15	0.13	0.11	0.13	0.14
Threonine	0.86	0.71	0.69	0.59	0.50	0.43	0.42	0.50
Calcium	0.90	0.90	0.80	0.75	0.75	0.65	0.90	0.90
Phosphorus	0.80	0.80	0.70	0.65	0.65	0.55	0.80	0.80
Additions/ton[3]								
Minerals								
Salt[4] . . . . (lb)	5	5	7	7	7	7	10	10
Copper[5] . . . . . . (g)	15	205	205	15	15	10	15	15
Iodine . . . . (g)	0.27	0.27	0.27	0.27	0.27	0.18	0.27	0.27
Iron . . . . (g)	150	150	150	150	150	100	150	150
Manganese . . . . . (g)	36	36	36	36	36	24	36	36
Selenium[6] . . . . . . (g)	0.27	0.27	0.27	0.09	0.09	0.09	0.09	0.09
Zinc . . . . (g)	3,000	150	150	150	150	100	150	150
Vitamins								
Vitamin A . (million IU)	10	10	10	8	8	6	10	10
Vitamin D_3 (thousand IU)	1,500	1,500	1,500	1,200	1,200	900	1,500	1,500
Vitamin E (thousand IU)	40	40	40	32	32	24	40	40
Vitamin K[7] . . . . . . (g)	4	4	4	3.2	3.2	2.4	4	4
Riboflavin . . . . . . (g)	7.5	7.5	7.5	6	6	4.5	7.5	7.5
Niacin . . . . (g)	45	45	45	36	36	27	45	45
Pantothenic acid . . . (g)	26	26	26	20.8	20.8	15.6	26	26
Choline . . . . (g)	150	150	150	120	120	90	500	500
Biotin . . . . (mg)							200	200
Folic acid . . . . . . (mg)							1,500	1,500
Vitamin B-12 . . . . (mg)	30	30	30	24	24	18	30	30

[1]Adapted by the authors from: Goodband, R. D., *et al.*, *Kansas Swine Nutrition Guide*, published by Kansas State University, 1994, updated 1995.

[2]For corn or milo-soybean meal based diets. All diets (except the gestation diet) are provided *ad libitum* (full-fed).

[3]For convenience, the mineral and vitamin additions/ton for starter 1 can be included in all diets to market weight.

[4]For pigs < 25 lb, added salt levels may be lowered or eliminated if the diet contains > 10% dried whey.

[5]To convert from g/ton to ppm, multiply by 1.1.

[6]Maximum legal addition 0.3 ppm for pigs < 50 lb and 0.1 ppm for pigs > 50 lb.

[7]Menadione activity.

TABLE 9-9
RECOMMENDED NUTRIENT ALLOWANCES FOR REPLACEMENT GILTS AND BOARS (AS-FED BASIS)[1, 2]

	Developing Gilts (Body Weight/Lb)		Developing Boars (Body Weight/Lb)	
	80 to 120	120 to 240	80 to 120	120 to 240
Total Level (%)				
Protein	20	16	20	17
Lysine	1.15	0.85	1.15	0.90
Tryptophan	0.18	0.14	0.18	0.15
Threonine	0.72	0.57	0.72	0.60
Calcium	0.90	0.80	0.90	0.80
Phosphorus	0.80	0.70	0.80	0.70
Additions/ton				
Minerals				
Salt . . . . (lb)	7	7	7	7
Copper[3] . . . . (g)	15	15	15	15
Iodine . . . . (g)	0.27	0.27	0.27	0.27
Iron . . . . (g)	150	150	150	150
Manganese . . . . (g)	36	36	36	36
Selenium[4] . . . . (g)	0.9	0.09	0.09	0.09
Zinc . . . . (g)	150	150	150	150
Vitamins				
Vitamin A . . . . (million IU)	10	10	10	10
Vitamin D$_3$ . . . . (thousand IU)	1,500	1,500	1,500	1,500
Vitamin E . . . . (thousand IU)	40	40	40	40
Vitamin K[5] . . . . (g)	4	4	4	4
Riboflavin . . . . (g)	7.5	7.5	7.5	7.5
Niacin . . . . (g)	45	45	45	45
Pantothenic acid . . . . (g)	26	26	26	26
Choline . . . . (g)	500	500	500	500
Biotin . . . . (mg)	200	200	200	200
Folic acid . . . . (mg)	1,500	1,500	1,500	1,500
Vitamin B-12 . . . . (mg)	30	30	30	30

[1]Adapted by the authors from: Goodband, R. D., *et al.*, *Kansas Swine Nutrition Guide*, published by Kansas State University, 1994, updated 1995.

[2]For corn or milo-soybean meal based diets. All diets are provided *ad libitum* (full-fed).

[3]To convert from g/ton to ppm, multiply by 1.1.

[4]Maximum legal addition (0.1 ppm).

[5]Menadione activity.

FEED SUBSTITUTION TABLE

Successful swine producers are keen students of values. They recognize that feeds of similar nutritive properties can and should be interchanged in the diet as price relationships warrant, thus making it possible at all times to obtain a balanced diet at the lowest cost.

Table 9-10, Feed Substitution Table for Swine, is a summary of the comparative values of the most common U.S. and Canadian feeds. In arriving at these values, two primary factors besides chemical composition and feeding value were considered; namely, palatability and carcass quality.

In using this feed substitution table, the following facts should be recognized:

1. That, for best results, different ages and groups of animals within classes should be fed differently.

2. That individual feeds differ widely in feeding value. Barley and oats, for example, vary widely in feeding value according to the hull content and the test weight per bushel.

3. That, based primarily on available supply and price, certain feeds—especially those of medium protein content, such as peanuts and peas (dried)—are used interchangeably as (a) grains and byproduct feeds, and/or (b) protein supplements.

4. That the feeding value of certain feeds is materially affected by preparation; thus, potatoes and beans should always be cooked for hogs. The values herein reported are based on proper feed preparation in each case.

For these reasons, the comparative values of feeds shown in the feed substitution table which follows are not absolute. Rather, they are reasonably accurate approximations based on averaged-quality feeds.

TABLE 9-10
FEED SUBSTITUTION TABLE FOR SWINE (AS-FED BASIS)

Feedstuff	Relative Feeding Value (lb for lb) In Comparison With the Designated (underlined) Base Feed Which = 100	Maximum Percentage of Base Feed (or comparable feed or feeds) Which It Can Replace for Best Results	Remarks
ENERGY FEEDS: GRAINS, BYPRODUCT FEEDS, ROOTS AND TUBERS:[1]			
Corn, No. 2	*100*	*100*	Corn is the leading U.S. swine feed, about 25% of the total production being fed to hogs. Corn is high in energy, but low in lysine and tryptophan.
Alfalfa hay, early bloom, or Alfalfa meal, dehy	75–85	10–50	Low energy, good source of carotene and B vitamins, unpalatable to baby pigs. None in starter, 0–5% grower-finishing, 0–50% gestation, 0–10% lactation.
Bakery waste	95–110	20–40	Bakery wastes average about 10% protein and 13% fat. Variable salt content. May constitute up to 20% of starter diets and up to 40% of grower-finisher, gestation, and lactation diets.
Barley	90–95	100	Of variable feeding value due to wide spread in test weight per bushel. Should be ground or rolled. Low lysine.
Beans (cull)	90	33–66	Cook thoroughly; on a dry weight basis, limit to one-half the grain diet for pigs under 100 lb *(45.4 kg)* and two-thirds of the grain diet for pigs above 100 lb *(45.4 kg)*; supplement with animal protein.
Beet pulp	70–80	10	Bulky, high fiber, laxative. May constitute up to 10% of gestation and lactation diets.
Carrots (or beets, mangels, or turnips)	12–20	25	
Cassava, dried meal	85	33.33	Low in methionine. Available as a swine feed in the tropics.
Corn and cob meal	80–90	0–70	Bulky, low energy. May constitute up to 70% of gestation diets.
Corn gluten feed (23% protein)	100	5–40	Corn gluten feed is bulky, low in lysine, and not too palatable. It may constitute up to 5% of starter diets, 10% of grower-finisher and lactation diets, and 40% of gestation diets.
Corn, high lysine	100–105	100	Superior to corn in lysine.
Corn meal	100	20	
Corn silage (25–30% D.M.)	20–30	0–90	Bulky, low energy, feed to sows only.
Emmer	80–90	80	Emmer may be used about like barley.
Fat (stabilized)	185–210	5	High energy, reduces dust.
Hominy feed	95	50	Hominy feed will produce soft pork if it constitutes more than one-half the grain diet. Hominy feed is subject to rancidity.
Millet (Proso)	85–90	50	Low lysine.
Molasses, beet	70	5	Used in pelleting. Laxative at high levels.

(Continued)

TABLE 9-10 (Continued)

Feedstuff	Relative Feeding Value (lb for lb) In Comparison With the Designated (underlined) Base Feed Which = 100	Maximum Percentage of Base Feed (or comparable feed or feeds) Which It Can Replace for Best Results	Remarks
ENERGY FEEDS: GRAINS, BYPRODUCT FEEDS, ROOTS AND TUBERS:[1] (Continued)			
Molasses, cane (74% D.M.)	70	5	Used in pelleting. High levels (above 15% of the diet) cause soft, watery feces.
Molasses, citrus	70	5	It takes pigs 5–7 days to get used to the bitter taste of citrus molasses.
Oats	70–80	15–70	Grind for swine. Feeding value varies according to test weight per bushel. Oats may constitute up to 15–20% of grower-finisher diets and up to 70% of gestation diets.
Oats, groats	110–115	20	Palatable, but expensive. Primarily used in starter diets in which it may constitute up to 20%.
Peanuts	120–125	100	Peanuts are usually fed by hogging off.
Peas, dried	90–100	100	Normally peas should be fed to swine as a protein supplement. Two tons of peas equal 1 ton of grain plus 1 ton of soybean meal. Peas are low in methionine.
Potatoes, Irish (24% D.M.)	25–28	25–50	Not palatable in the raw state; must be cooked. When cooked and fed in a ratio of 3 lb *(1.4 kg)* of potatoes to 1 lb *(0.45 kg)* of grain, they are worth 25–28% as much as corn. May constitute 25% of grower-finisher diets and 50% of gestation diets.
Potatoes (Irish), dehy	100	33.33	
Potatoes (sweet)	20–25	33.33–50	Cooking also improves the feeding value of sweet potatoes.
Potatoes (sweet), dehy	90	33.33	
Rice (rough rice)	80–85	50	Low energy, low lysine. Rice should be ground.
Rice bran	100	33.33	If more than one-third of the grain consists of rice bran, soft pork will result.
Rice polishings	100–120	33.33	Limited because feed becomes rancid in storage and soft pork will be produced.
Rice screenings	95	50	
Rye	90	20–30	Should be limited because it is unpalatable. Grind for swine. Watch for ergot.
Sorghum, grain	95	100	Check protein content and add supplement as necessary. Both very dry grain and bird-resistant varieties should be ground. Grain sorghum is low in lysine.
Spelt	65–80	25	Low energy, low lysine. Value varies according to the amount of hulls.
Sugar	70–80	0–5	High palatability, no protein. Normally, used only in starter diets.
Sunflower seed	100	50	High energy, high fiber.
Triticale	90–95	50	Higher levels not palatable. Watch for ergot.
Wheat	100–105	100	Feed whole if self-fed. Otherwise, grind coarsely; fine grinding makes it pasty and unpalatable. Low in lysine. Wheat-corn mixtures are more efficient than wheat alone.
Wheat bran	65–75	15–25	Bulky, high fiber, laxative. Bran is particularly valuable at farrowing time.
Wheat middlings	103	20	May be used as a partial substitute for grain.
Wheat standard middlings	85–100	10–30	Use as a partial grain substitute.
Wheat red dog and wheat white shorts	115–120	25	
Whey, dry	50	5–20	High lactose, very palatable. Use for baby pigs.
Whey, liquid	15	5–20	High lactose, very palatable. Use for baby pigs.

TABLE 9-10 (Continued)

Feedstuff	Relative Feeding Value (lb for lb) In Comparison With the Designated (underlined) Base Feed Which = 100	Maximum Percentage of Base Feed (or comparable feed or feeds) Which It Can Replace for Best Results	Remarks
PROTEIN FEEDS:			
Soybean meal (41–50%)	*100*	*100*	Well balanced in amino acids. Best quality of all plant protein supplements. Very palatable.
Blood meal (80%)	120–130	20	High in protein (above 80%), high in lysine, but low in isoleucine. Not very palatable.
Buttermilk, dry	90–105	100	Good amino acid balance.
Buttermilk, liquid	15	100	Pound for pound, worth one-tenth as much as dried buttermilk.
Buttermilk, semisolid	33.33–50	100	Pound for pound, worth one-tenth as much as dried buttermilk.
Canola meal (32–44%)	90	75	Low in goitrogenic compounds.
Copra meal (coconut meal) (21%)	50	25	
Corn, distillers' dried grains w/solubles	65–75	5	Used primarily as B-vitamin and unidentified sources, usually at about 5% of the diet.
Corn gluten meal (60% protein) . .	90	50	Bulky, low in lysine, and not too palatable.
Cottonseed meal (36–48%)	85	33.33	Except when new glandless cottonseed meal is used, high levels may produce gossypol poisoning; hence, the level of cottonseed meal in swine diets should not exceed 8–9% of the total diet. Cottonseed meal is low in lysine.
Fish meal (60%)	115	100	Excellent balance of amino acids, and good source of calcium and phosphorus.
Linseed meal (35%)	80	25–50	Low in lysine; slightly laxative.
Malt sprouts	100	10	Malt sprouts contain a growth factor(s). They result in increased feed intake and gain.
Meat and bone meal (50%)	100	100	Low in tryptophan; good source of calcium and phosphorus.
Meat scraps (50–55%)	100	100	
Peanut meal (45%)	95	50	Becomes rancid when stored too long. Low in lysine; very palatable.
Peanuts	60–70	50	Peanuts are usually fed by hogging off. High levels will produce soft pork.
Peas, dried	50	50	
Rapeseed meal (32–44%)	85–90	33.33	Rather unpalatable. Contains goitrogenic compounds that can be hazardous. But is usually detoxified.
Shrimp meal	90–100	50	
Skim milk, dried	90–120	100	Excellent-quality protein; very palatable; expensive. Especially good in prestarter and starter diets, of which it may constitute up to 10%.
Skim milk, liquid			Pound for pound, worth one-tenth as much as dried skim milk.
Soybeans, full fat, cooked	90–100	25–40	High energy. At high levels, will produce soft pork.
Sunflower meal (36–45%)	90–95	50	For swine, it should be combined with high-lysine supplements such as meat scraps or fish meal.
Tankage (60%)	110	100	Good source of calcium and phosphorus. Low in tryptophan. Not palatable.
PASTURES AND DRY LEGUMES:			
Pasture, good		5–20% of grain, and 20–50% of protein supplement.	Pasture and dry legumes are sources of good-quality proteins, of minerals, and of vitamins.
Alfalfa meal		It can replace all of pasture, in drylot rations.	Low energy, good source of carotene and B vitamins, unpalatable to baby pigs. For drylot diets, include 5–10% alfalfa in diet of grower-finishing pigs, up to 50% in diet of gestating sows, and up to 10% for lactating sows.

[1]Roots and tubers are of lower value than the grain and byproduct feeds due to their higher moisture content.

FEED ADDITIVES

The use of feed additives in swine diets has been extensive in the United States for more than 40 years. Most swine producers use additives because of their demonstrated ability to increase growth rate, improve feed utilization, and reduce mortality and morbidity from clinical and subclinical infections.

Most, but not all, additives used for swine fall into the following five classifications:

1. Antibiotics
2. Chembiotics or chemotherapeutics
3. Anthelmintics or dewormers
4. Copper compounds
5. Probiotics

An antibiotic is a compound synthesized by living organisms, such as bacteria or molds, which inhibits the growth of another.

Chembiotics are compounds similar to antibiotics but they are produced chemically rather than microbiologically.

Anthelmintics or dewormers are compounds added to swine diets to help control worms.

Copper compounds, such as copper sulfate, have growth-stimulating value similar to antibiotics. They are also effective as therapeutic treatment for certain intestinal disorders that do not respond satisfactorily to antibiotics or chembiotics.

Probiotics, which means "in favor of life," have an opposite effect to antibiotics on the microorganisms of the digestive tract. It is theorized that probiotics increase the population of desirable microorganisms rather than kill or inhibit undesirable organisms.

Table 9-11 is a partial list of approved additives for use in swine diets. Note that Table 9-11 presents for each additive (1) level of use (g/ton), (2) indications for use, and (3) withdrawal period.

TABLE 9-11
SWINE FEED ADDITIVES[1, 2]

Feed Additive	Level	Indications for Use	Withdrawal Period
	(g/ton)		(days)
Apralan (Apramycin)	150	Feed continuously as the only diet for 14 days.	28
Arsanilic acid	45–90	Growth promotion, control of bacterial enteritis (scours).	5
Auerozol	250	Reduce cervical abscesses; treatment of bacterial enteritis; atrophic rhinitis.	15
100g CTC			
100 g sulfamethazine			
50 g penicillin			
Aureomycin	10–50	Growth and feed efficiency.	None
(Chlortetracycline; CTC)	50–100	Bacterial enteritis and rhinitis.	None
	100–200	Bacterial enteritis (scours).	None
	200–400	Reduce leptospirosis shedding.	None
Bacitracin[3]	10–30	Growth and feed efficiency.	None
	250	Treatment and prevention of bacterial enteritis (scours).	None
CSP 250	250 (total)	Reduce abscesses; growth promotion; aids in rhinitis control; treatment of salmonellosis.	7
100 g CTC			
100 g sulfathiazole			
50 g penicillin			
Denagard (Tiamulin)	10	Growth performance (weaning to 125).	None
	35	Swine dysentery.	2
Flavomycin (Bambermycin)	2–4	Growth and feed efficiency.	None
Lincomycin	20	Growth performance.	None
	40	Control of dysentery.	None
	100	Treatment of bacterial enteritis (scours).	6
	200	Reduce severity of mycoplasmal pneumonia.	6

(Continued)

TABLE 9-11 (Continued)

Feed Additive	Level	Indications for Use	Withdrawal Period
	(g/ton)		*(days)*
Mecadox (Carbadox)	10–25	Growth and feed efficiency.	70
	50	Control of dysentery and enteritis; growth promotion in pigs up to 75 lb.	70
Neo-Terramycin	120–290 (total)	Control of bacterial enteritis (scours).	
70–140 g Neomycin base (100–200 g)		Neomycin at 140 g.	10
50–150 g Oxytetracycline		Neomycin < 140 g.	5
Penicillin	10–50	Growth and feed efficiency.	None
Terramycin	25–50	Growth and feed efficiency in starter pigs (10–30 lb).	None
(Oxytetracycline; OTC)	7.5–10	Growth and feed efficiency in growing-finishing pigs (30–200 lb).	None
	50	Prevention of scours and bacterial enteritis (scours).	None
	100	Treatment of scours and bacterial enteritis (scours).	None
	50–150	Aids in rhinits control.	None
	500	Treatment and control of leptospirosis fed prior to farrowing.	5
Tylan (Tylosin)	10–20	Growth and feed efficiency in finishing pigs.	None
	20–40	Growth and feed efficiency in growing pigs.	None
	20–100	Growth and feed efficiency in starter pigs.	None
	40–100	Control of bacterial enteritis (scours).	None
	100	Aids in rhinitis control.	None
Tylan-Sulfa	200 (total)	Maintain weight and efficiency with rhinitis; prevention of bacterial enteritis (scours); control of pneumonias.	15
100 g Tylosin			
100 g Sulfamethazine			
Stafac (Virginiamycin)	5–10	Growth and feed efficiency.	None
	25	Control of bacterial enteritis (scours) up to 120 lb.	None
	100	Treatment and control of bacterial enteritis (scours).	None
Anthelmintics			
Atgard (Dichlorous)	348	Control internal parasites (pigs).	None
	479	Control internal parasites in bred and open sows and boars.	None
	334–500	Aid in improving litter production.	None
Banminth (Pyrantel tartrate)	96	Internal parasite control.	1
Hygromycin B	12	Control ascarids, nodular, and whipworms.	15
Ivomec (Ivermectin)	2 ppm for 7 days	Internal and external parasites.	5
Safegard (Fenbendazole)	9 mg/kg body weight over 3–12 day's time	Internal parasite control.	None
Tramisol (Levamisol)	720	Internal parasite control.	3

[1]Adapted by the authors from: Goodband, R. D., *et al., Kansas Swine Nutrition Guide*, published by Kansas State University, 1994, updated 1995.

[2]This is only a partial list of feed additives and combinations. Information was adapted from 1994 Feed Additive Compendium. Feed additives and their combinations may not be used in any way other than specified on the label.

[3]Bacitracin is available in several forms including the manganese, zinc, and methylene disalicylate (MD form) derivatives.

In addition to the additives listed in Table 9-11, there are numerous other additives that are used to increase acceptability of the diet of pigs, preserve quality of the diet, or improve digestion and utilization of the feed; among such additives are antioxidants, mold inhibitors, flavors, sweeteners, pellet binders, clays, enzymes, organic acids, yucca extract, and electrolytes.

Regardless of the feed additive used, the following safety precautions should be observed:

1. Use only additives approved for use by swine. Regulations change frequently, so check them often.
2. Follow label directions carefully.
3. Use minimal effective amounts.
4. Add or apply precise quantities.
5. Certain feed additives must be withdrawn from the feed before slaughter to insure residue-free carcasses. Consult the feed label for withdrawal time for the specific feed additive that is being fed.

HOW TO BALANCE DIETS

When in confinement, animals have access only to the feed provided by the caretaker. Therefore, it is important to provide balanced diets.

Suggested diets are given in subsequent sections of this chapter. Generally these diets will suffice, but it is recognized that diets should vary with conditions, and that many times they should be formulated to meet the conditions of a specific farm or the practices common to an area.

Good producers should know how to balance diets. They should be able to select and buy feeds with informed appraisal; to check on how well manufacturers, dealers, or consultants are meeting their needs; and to evaluate the results.

Fig. 9-1. Pigs in confinement rely totally on the feed supplied by the producer, hence, any errors in diet formulation may result in lowered or uneconomical production. (Courtesy, Iowa State University, Ames, IA)

Diet formulation consists of combining feeds that will be eaten in the amount needed to supply the daily nutrient requirements of the animal. This may be accomplished by the methods presented later in this chapter, but first the following pointers are necessary:

1. In computing diets, more than simple arithmetic should be considered, for no set of figures can substitute for experience and swine intuition. Formulating diets is both an art and a science—the art comes from animal know-how, experience, and keen observation; the science is largely founded on mathematics, chemistry, physiology, and bacteriology. Both are essential for success.

2. Before attempting to balance a diet, the following major points (factors) should be considered:

a. **Availability and cost of the different feed ingredients.** Preferably, cost of ingredients should be based on delivery after processing—because delivery and processing costs are quite variable.

b. **Moisture content.** When considering costs and balancing diets, feed should be placed on a comparable moisture basis; usually, either "as-fed" or "moisture-free." This is especially important in the case of high-moisture grain.

c. **Composition of the feeds under consideration.** Feed composition tables, or average analysis, should be considered as guides, because of wide variations in the composition of feeds. For example, the protein and moisture content of sorghum is quite variable. Wherever possible, especially with large operations, it is best to take a representative sample of each major feed ingredient and have a chemical analysis made of it for the more common constituents—protein, fat, fiber, nitrogen-free extract, and moisture; and often calcium, phosphorus, and carotene. Such ingredients as oil meals and prepared supplements, which must meet specific standards, need not be analyzed so often, except as quality control measures.

Despite the recognized value of a chemical analysis, it is not the total answer. It does not provide information on the availability of nutrients to the animal; it varies from sample to sample, because feeds vary and a representative sample is not always easily obtained, and it does not tell anything about the associated effect of feedstuffs. Nor does a chemical analysis tell anything about taste, palatability, texture, undesirable physiological effects such as laxativeness, and amino acid content. Nevertheless, a chemical analysis does give a solid foundation on which to start in evaluating feeds. Also, with chemical analysis at hand, and bearing in mind that it is the composition of the total feed (the finished diet) that counts, the person formulating the diet can determine more intelligently the quantity of protein to buy, and the kinds and amounts of minerals and vitamins to add.

d. **Quality of feed.** Numerous factors determine the quality of feed, including—

■ Stage of harvesting—For example, early cut forages tend to be of higher quality than those that are mature.

■ Freedom from contamination—Contamination from foreign substances such as dirt, sticks, and rocks can reduce feed quality, as can aflatoxins, pesticide residues, and a variety of chemicals.

■ Uniformity—Does the feed come from one particular area or does it represent a conglomerate of several sources?

■ Length of storage—When feed is stored for extended periods, some of its quality is lost due to its exposure to the elements. This is particularly true with forages.

e. **Degree of processing of the feed.** Often, the value of feed can be either increased or decreased by processing. For example, heating some types of grains makes them more readily digestible to swine and increases their feeding value.

f. **Soil analysis.** If the origin of a given feed ingredient is known, a soil analysis or knowledge of the soils of the area can be very helpful; for example, (1) the phosphorus content of soils affects plant composition, (2) soils high in molybdenum and selenium affect the composition of the feeds produced, (3) iodine- and cobalt-deficient areas are important in animal nutrition, and (4) other similar soil-plant-animal relationships exist.

g. **The nutrient requirements and allowances.** These should be known for the particular class of swine for which a diet is to be formulated; and, preferably, they should be based on controlled feeding experiments. Also, it must be recognized that nutrient requirements and allowances must be changed from time to time, as a result of new experimental findings.

3. In addition to providing the proper quantity of feed and to meeting the protein and energy requirements, a well balanced and satisfactory diet should be:

a. Palatable and digestible.

b. Economical. Generally speaking, this calls for the maximum use of feeds available in the area.

c. Adequate in protein content, but not higher than is actually needed, for, generally speaking, medium and high protein feeds are in scarcer supply and higher in price than high energy feeds.

In this connection, it is noteworthy that the newer findings in nutrition indicate that (1) much of the value formerly attributed to proteins, as such, was due to the amino acids, vitamins, and minerals which they furnished, and (2) lower protein content diets may be used successfully provided they are of good quality and fortified properly with the needed vitamins and minerals.

d. Well fortified with the needed minerals, but mineral imbalances should be avoided.

e. Well fortified with the needed vitamins.

f. Fortified with antibiotics, or other antimicrobial agents, as justified.

g. One that will enhance, rather than impair, the quality of pork produced.

4. In addition to considering changes in availability of feeds and feed prices, diet formulation should be altered at stages to correspond to changes in weight and productivity of animals.

STEPS IN DIET FORMULATION

The ideal diet is one that will maximize production at the lowest cost. A costly diet may produce phenomenal gains in livestock, but the cost per unit of production may make the diet economically infeasible. Likewise, the cheapest diet is not always the best since it may not allow for maximum production.

Therefore, the cost per unit of production is the ultimate determinant of what constitutes the best diet. Awareness of this fact separates successful producers from marginal or unsuccessful ones.

The following four steps should be taken in an orderly fashion in order to formulate an economical diet:

1. **Find and list the nutrient requirements and/or allowances for the specific animal to be fed.** It should be remembered that nutrient requirements generally represent the minimum quantity of the nutrients that should be incorporated while allowances take into consideration a margin of safety. Factors to be considered are:

a. Age.

b. Sex.

c. Body size.

d. Type of production. Is the animal being fed for maintenance, growth, fattening, reproduction, or lactation?

e. Intensity of production. Is the growing animal gaining 0, 1, 2, or 3 lb per day? Is the lactating animal at the peak of milk production?

2. **Determine what feeds are available and list their respective nutrient compositions.** In diets for swine, protein, energy, vitamins A, D, E, and B-complex vitamins, several macrominerals, and several mi-

crominerals are generally considered in diet formulation. Furthermore, adequate amounts of the essential amino acids must be supplied in the diet. Because of these considerations, it is easy to see why many large producers are using the computer for diet formulations.

3. **Determine the cost of the feed ingredients under consideration.** Not only should the cost of the feed be considered, but also the cost of mixing, transportation, and storage. Some feeds require antioxidants and/or refrigeration to prevent spoilage. Others lose nutritive value when stored for extended periods.

4. **Consider the limitations of the various feed ingredients and formulate the most economical diet.** Remember that the ultimate goal is to formulate a diet that minimizes the cost per unit of production.

ADJUSTING MOISTURE CONTENT

The majority of feed composition tables are listed on an "as-fed" basis, while most of the National Research Council nutrient requirement tables are on either an "approximate 90% dry matter" or a "moisture-free basis." Since feeds contain varying amounts of dry matter, it would be much simpler, and more accurate, if both feed composition and nutrient requirement tables were on a dry basis. In order to facilitate diet formulation, the authors list both the "as-fed" and "moisture-free" contents of feeds in Chapter 21, Feed Composition Tables.

The significance of water content of feeds becomes obvious in the examples given in Table 9-12.

TABLE 9-12
COMPARATIVE PROTEIN CONTENT OF THREE FEEDS ON AS-FED AND MOISTURE-FREE BASES

Feed	Water	Dry Matter	Protein	
			As-fed	Moisture-free
	(%)	(%)	(%)	(%)
Corn, grain . . .	12	88	9.6	10.9
Milk	87	13	3.6	27.4
Potatoes	76	24	2.2	9.2

When comparing the protein content of milk and potatoes to corn, milk has a much higher protein content on a moisture-free basis and potatoes are similar to corn. The same principle applies to other nutrients, also.

■ **To convert as-fed diets to a moisture-free basis**—This may be done by using the following formulas:

Formula 1

When the diet is listed on an as-fed basis, and the producer wishes to compare the content of the various ingredients with the requirements on a moisture-free basis, the equation is—

$$\% \text{ nutrient in dry diet (total)} =$$
$$\frac{\% \text{ nutrient in wet diet (total)}}{\% \text{ dry matter in diet (total)}} \times 100$$

For example, a diet containing 34% dry matter and 7% protein on a wet basis becomes a 20.6% protein diet on a moisture-free basis.

Formula 2

If the dry matter content of the ingredient, the percentage of the ingredient in the wet diet, and the percent dry matter wanted in the diet are known, it is possible to calculate the amount of that ingredient in the diet on a moisture-free basis.

$$\text{Amount of ingredient in dry diet} =$$
$$\frac{\% \text{ of ingredient in wet diet}}{\% \text{ dry matter wanted in diet}} \times \begin{array}{c} \% \text{ dry matter} \\ \text{of ingredient} \end{array}$$

Therefore, if a 34% dry matter diet is desired, and if an ingredient containing 25% dry matter is incorporated at a level of 30% of the wet diet, the ingredient constitutes 22% of the moisture-free weight in the diet.

Formula 3

If the producer wants to change the amounts of the ingredients from an as-fed basis to a moisture-free basis, the equation below should be used.

Parts on a wet basis = % ingredient in wet diet × % dry matter of the ingredient

This calculation should be done for each ingredient and the products added. Each product should then be divided by the sum of the products.

■ **To convert a moisture-free basis to an as-fed basis**—To convert the components of a dry diet to that of a wet diet having a given percent of dry matter, one can use the following equation:

$$\text{Parts of ingredient in wet diet} =$$
$$\frac{\% \text{ ingredient in dry diet} \times}{\% \text{ dry matter in ingredient}} \begin{array}{c} \% \text{ dry matter desired in diet} \end{array}$$

The total number of parts should be summed and water added to make 100 parts.

METHODS OF FORMULATING DIETS

In the sections that follow, five different methods of diet formulation are presented: (1) the computer method, (2) the square method, (3) the simultaneous equation method, (4) the 2 × 2 matrix method, and (5) the trial-and-error method. Despite the sometimes confusing mechanics of each system, if done properly, the end result of all five methods is the same—a diet that provides the desired allowance of nutrients in correct proportions economically (or at least cost), but, more important, so as to achieve the greatest net returns—for it is net profit, rather than cost, that counts. Since feed represents by far the greatest cost item in swine production, the importance of balanced diets is evident.

An exercise in diet formulation follows for purposes of illustrating the application of each of these five methods.

COMPUTER METHOD OF FORMULATING DIETS

Fig. 9-2. Computer diet formulation. (Courtesy,California State University, Fresno)

Today, practically all large swine operations use computers for diet formulations.

Despite their sophistication, there is nothing magical or mysterious about balancing diets by computer. Although they can alleviate many human errors in calculations, the data which come out of a computer are no better than those which go into it. The people back of the computer—the producer and the nutritionist who prepare the data that go into it, and who evaluate

and apply the results that come out of it—become more important than ever. This is so because an electronic computer doesn't know anything about (1) feed palatability; (2) calcium-phosphorus ratio; (3) limitations that must be imposed on certain feeds to obtain maximum utilization; (4) the goals in the feeding program—such as growing or finishing; (5) byproduct feeds for which there may not be a suitable market; (6) feed processing and storage facilities; (7) the health, environment, and stress of the animals; and (8) those responsible for actual feed preparation and feeding. The computer also demands much more of the nutritionist in terms of precise information relative to nutrient composition, availability, requirements, and cost.

A balanced swine diet contains 40 or more nutrients. When using manual calculation methods, it is impractical to consider more than three or four nutrients at a time. With computers, all nutrients can be considered simultaneously.

■ **Computer software and hardware**—Computer type and size of memory and disk drive must meet the criteria of the software developer; otherwise, the software may not be usable. So, the selection of the software should precede the selection of the hardware.

Numerous companies market computer software for diet formulation. The software varies from the very simple and straightforward to the very complex packages intended for large feed manufacturers. University personnel and nutrition consultants are good sources of information on software and hardware.

■ **Linear programming**—With the advent of the microcomputer and electronic spreadsheets, linear programming evolved.

Linear programming is a mathematical technique in which a large number of simultaneous equations are solved in such a way as to meet the minimum and maximum levels of nutrients and levels of feedstuffs specified by the user at the lowest possible cost.

Computer ration formulation programs use a linear equation such as—

$$\text{Requirements} = ax_1 + bx_2 + cx_3 + dx_4$$

Where a, b, c, and d represent the amounts of each of the four ingredients in the diet, and x_1, x_2, x_3, and x_4 represent the amount of a specific nutrient that is in each of the four ingredients. With a computer, a large number of ingredients and nutrients can be considered simultaneously. It is much more efficient than hand calculations.

■ **Setup of the computer formulation**—*A program is a precise series of directives given to a computer which enables it to solve problems.* In least-cost formulation, all the computer does is solve a series of simultaneous equations through a sophisticated system of matrix algebra.

Nutrient requirements and certain restrictions are listed in what are called *rows*. The ingredients of the individual feeds are listed under the term *columns*. The values called for in the solution of the formulation are listed under what is called the *right-hand side*.

For example, the various components of the diet that are to be looked at in the diet, such as cost, energy, protein, and minerals, would constitute the rows. The various feeds to be reviewed for the diet would have their respective cost, energy, protein, and mineral values listed under columns. If we wanted the diet to equal 1 ton, with 400 lb protein, and certain mineral and vitamin specifications, we would list the specifications under the section called *right-hand side*.

In addition to setting up the rows, columns, and right-hand side, any limitations concerning minimum, maximum, or fixed amounts of any particular feedstuff must be included in the program. For example, the producer may want fish meal in the diet at a level of at least 5% but not to exceed 10%. These restrictions would have to be included in the information fed into the computer.

■ **Information provided by a computer-based feed formulation program**—Tables 9-13 and 9-14 outline the type of information provided by a computer feed formulation program.[1]

In Table 9-13, the ingredient summary itemizes

---

[1]This section and Tables 9-13 and 9-14 were adapted by the authors from Patience, J. F. and P. A. Thacker, *Swine Nutrition Guide*, published by Prairie Swine Centre, University of Saskatchewan, Saskatoon, pp. 135–136.

the ingredients selected and the amount of each required in the diet to meet pig requirements. Limits (maximums and minimums) that were set up in the original feed specifications are also shown. For example, an upper limit of 40% was placed on wheat and 15% on canola meal. It can be seen in the example that wheat was priced competitively, because it went to its upper limit. The premixes are at their lower limits because they are expensive.

The output from the computer shows the competitiveness of the price of each ingredient. For example, canola meal would have to drop to $247 per ton in order for more canola to come into the formula. If the price of canola meal rose to $310 per ton, the computer would select less of it. In this example, the price of canola could fluctuate between $247 and $310 per ton with little change in use rate. These constraints on canola result from the high energy content requested for this diet. There is no lower price limit for wheat because the maximum amount is already being used. Barley appears to be competitively priced, since it is very close to its lower price limit and is much cheaper than its upper price limit. These programs are very useful for determining the value of certain ingredients of various diets used.

Table 9-14 provides a somewhat similar summary for nutrients. Nutrients that are at their lower limit are forcing the cost of the diet up. For example, digestible energy, lysine, sodium, calcium, and phosphorus are all at their lower limit which means that if any of these could be lowered, the cost of the diet would be reduced. If a reduction in these ingredients causes animal performance to suffer though, reducing the diet cost would result in lower performance.

TABLE 9-13
TYPICAL INGREDIENT OUTPUT FROM A FEED FORMULATION PROGRAM

Ingredients	Formula			Price		
	Actual	Minimum	Maximum	Low	Actual	High
	(%)	*(%)*	*(%)*	*($)*	*($)*	*($)*
Barley	43.101	—	—	1.00	1.12	1.94
Wheat	40.000	—	40.000	—	1.28	1.40
Soybean meal — 46.5%	10.991	—	—	4.12	4.37	4.85
Canola meal	1.683	—	15.000	2.47	2.85	3.10
L-lysine HCl	0.179	—	—	22.95	54.50	67.35
Dicalcium phosphate	1.708	—	—	—	4.00	20.72
Limestone	1.357	—	—	—	0.60	11.27
Salt	0.380	—	—	—	0.80	—
Mineral premix	0.300	0.300	—	—	3.24	—
Vitamin premix	0.300	0.300	—	—	13.41	—
Total	100.000				175.04	

TABLE 9-14
TYPICAL NUTRIENT OUTPUT FROM A FEED FORMULATION PROGRAM

Nutrient	Requirements			Constraint		
	Actual	Minimum	Maximum	Unit Cost	Increment	Decrement
				($)	*($)*	*($)*
Digestible energy . . . . . . (kcal/kg)	3,200.00	3,200.00	—	0.001	11.844	58.518
Protein . . . . (%)	15.73	—	—	—	—	—
Lysine . . . . (%)	0.85	0.85	—	0.700	33.770	0.720
Methionine . . . . (%)	0.26	—	—	—	—	—
T.S.A.A.[1] . . . . (%)	0.62	0.50	—	—	—	—
Tryptophan . . . . (%)	0.21	0.13	—	—	—	—
Threonine . . . . (%)	0.58	0.50	—	—	—	—
Leucine . . . . (%)	1.20	—	—	—	—	—
Isoleucine . . . . (%)	0.65	—	—	—	—	—
Valine . . . . (%)	0.80	—	—	—	—	—
Phenylalanine . . . . (%)	0.81	—	—	—	—	—
Tyrosine . . . . (%)	0.39	—	—	—	—	—
Arginine . . . . (%)	0.97	—	—	—	—	—
Glycine . . . . (%)	0.70	—	—	—	—	—
Histidine . . . . (%)	0.28	—	—	—	—	—
Sodium . . . . (%)	0.15	0.15	0.25	0.050	16.780	15.000
Chloride . . . . (%)	0.27	0.15	—	—	—	—
Calcium . . . . (%)	0.90	0.90	1.00	0.050	1.000	5.170
Phosphorus . . . . (%)	0.75	0.75	0.90	0.210	15.000	33.740
Available phosphorus . . . . . (%)	0.36	—	—	—	—	—

[1]T.S.A.A. refers to "total sulfur amino acids." It is calculated as the sum of methionine plus cysteine; thus, it reflects the supportive role of cysteine in the diet.

SQUARE (OR PEARSON SQUARE) METHOD

The square method is simple, direct, and easy. Also, it permits quick substitution of feed ingredients in keeping with market fluctuations, without disturbing the protein content.

In balancing diets by the square method, it is recognized that one specific nutrient alone receives major consideration. Correctly speaking, therefore, it is a method of balancing one nutrient requirement, with no consideration given to the vitamin, mineral, and other nutritive requirements.

To compute diets by the square method, or by any other method, it is necessary to have available (1) nutrient requirements (see Tables 9-1 to 9-7), (2) nutrient allowances (see Tables 9-8 and 9-9), and (3) feed composition tables (see Chapter 21).

The following example will show how to use the square method in formulating a swine diet:

Example. *A swine producer has 40-lb pigs to* which he/she desires to feed a 16% protein diet until they reach 120-lb weight. The corn on hand contains 9.5% protein. The producer can buy a 36% protein supplement, which is reinforced with minerals and vitamins. What percent of the diet should consist of each corn and the 36% protein supplement?

Step by step, the procedure in balancing this diet is as follows:

1. Draw a square, and place the number 16 (desired protein level) in the center thereof.

2. At the upper left-hand corner of the square, write *protein supplement* and its protein content (36); at the lower left-hand corner, write *corn* and its protein content (9.5).

3. Subtract diagonally across the square (the smaller number from the larger number), and record the difference at the corners on the right-hand side (36 − 16 = 20; 16 − 9.5 = 6.5). The number at the upper right-hand corner gives the parts of concentrate by weight, and the number at the lower right-hand corner gives the parts of corn by weight to make a diet with 16% protein.

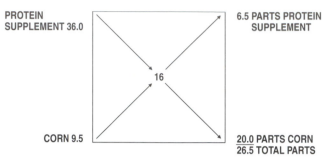

Fig. 9-3. The square method for balancing grain and protein supplementation.

4. To determine what percent of the diet would be corn, divide the parts of corn by the total parts: 20 ÷ 26.5 = 75% corn. The remainder, 25%, would be supplement.

The square method may also be used to balance diets on an amino acid basis rather than on a crude protein basis. This provides a more precise indication of the adequacy of protein in the diet. Lysine is the most critical amino acid in swine diets.

Example. *A swine producer wants a corn-soybean meal diet to furnish 0.76% lysine to grower pigs (40 to 120 lb). What percent of the diet should consist of each corn and soybean meal?*

Briefly, the following steps are involved:

1. Corn contains about .20% lysine and soybean meal contains about 2.27% lysine.
2. Set up square as shown in Fig. 9-4.
3. To determine what percent of the diet would be corn, divide the parts of corn by the total parts: 1.51 ÷ 2.07 × 100 = 73% corn. The remainder, 27%, would be soybean meal.

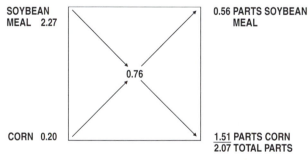

Fig. 9-4. The square method for balancing amino acid content of a diet.

SIMULTANEOUS EQUATION METHOD

In addition to the square method, it is possible to formulate diets involving two sources and one nutrient quickly through the solving of simultaneous equations:

Example. *A producer has on hand corn containing about 9% protein, and can buy a 40% protein supplement, which is reinforced with minerals and vitamins. A diet for 30-lb pigs, containing 18% protein, is desired.*

Step by step, the procedure in balancing this diet is as follows:

1. Let X = amount of corn to be used in 100 lb of mixed feed, and Y = amount of 40% protein supplement to be used in 100 lb of mixed feed. We know that the corn contains 9% protein and the protein supplement 40% and that the diet should be 18% protein. Therefore, the equation we must solve is as follows:

$$0.09X + 0.40Y = 18 \text{ (lb of protein in 100 lb of feed)}$$

2. In order to solve for two unknowns (X and Y), we must create a "dummy equation." This can be done in the following manner:

$$X + Y = 100 \text{ lb of feed}$$

3. We must now multiply our dummy equation by 0.09 in order that our X term will cancel out with the original equation. Therefore:

$$X + Y = 100 \text{ becomes } 0.09X + 0.09Y = 9$$

4. Subtracting our new dummy equation from the original equation, we can solve for Y as shown below:

Original equation:	$0.09X + 0.40Y = 18$
Dummy equation:	$-0.09X - 0.09Y = -9.0$
	$0.00X + 0.31Y = 9.0$

$$Y = \frac{9}{0.31} \text{ or 29.03 lb of 40\% protein supplement per 100 lb of feed or 29.03\%}$$

5. We can now substitute our newly acquired value for Y in the original equation and solve for X, as follows:

$$X = 100 - 29.03$$
$$= 70.97 \text{ lb of corn per l00 lb of feed or 70.97\%}$$

TRIAL-AND-ERROR METHOD

In the trial-and-error method, consideration is given to meeting whatever allowances are decided upon for each of the nutrients that one cares to list and consider.

A producer decides to use a diet containing 14% protein, 1,500 kcal of digestible energy per lb, 0.75% calcium, and 0.60% phosphorus.

1. Considering available feeds and common feeding practices, the next step is arbitrarily to set down a diet, and see how well it measures up to the desired

allowances. The approximate composition of the available feeds may be arrived at from the feed composition tables (Chapter 21) if an actual chemical analysis is not available. Where commercial supplements are used, the guarantee on the feed tag may be used.

Try the following diet:

Per Ton

(lb)

Corn	1,790
Soybean meal	100
Meat with bone meal	100
Vitamin and mineral premix	10
	2,000

2. Compare the desired allowances to that supplied by the proposed diet. Calculations show that this diet supplies 13.1% protein, 1,534 kcal of digestible energy per pound, 0.55% calcium, and 0.52% phosphorus. Therefore, it falls somewhat short and the producer decides to increase the meat with bone meal by 50 lb and reduce the corn by 50 lb giving the following new diet:

Per Ton

(lb)

Corn	1,740
Soybean meal	100
Meat with bone meal	150
Vitamin and mineral premix	10
	2,000

Through calculations based on the composition of these feeds, the producer finds that this diet will now supply 14% protein, 1,518 kcal of digestible energy per pound, 0.82% calcium, and 0.63% phosphorus. So, the producer decides it approximates the desired allowances and may be considered satisfactory.

FORMULATION WORKSHEET

When formulating diets, it is advisable to record the diet on a worksheet similar to that in Fig. 9-5. This worksheet is merely an example of the format that should be used. A similar sheet can be worked up for micronutrient composition of premixes for vitamins and

MACRONUTRIENT WORKSHEET

Ration Number: Date:

Ingredient	✓ if Mixed	Amount	Crude Fiber	Ether Extract (Fat)	N-Free Extract	Crude Protein	Energy	Calcium (Ca)	Phosphorus (P)	Vitamin A
		(lb)	*(lb)*	*(lb)*	*(lb)*	*(lb)*	TDN = lb NE = Mcal ME = Mcal	*(lb)*	*(lb)*	*(IU)*
TOTAL										
NUTRIENT REQUIREMENTS										
NUTRIENT BALANCE (Total—Nutrient Requirements)										

Fig. 9-5. Formulation worksheet for macronutrients.

minerals as well as for amino acids. The worksheet serves three purposes:

1. It provides a means of reviewing and double checking the calculations used to formulate the diet. If there is a gross error, it should become obvious when listed on the worksheet.

2. It can be used to organize mixing procedures. It is vital that the person mixing feed be able to refer to a worksheet on which can be recorded what has been mixed and what mixing order should follow.

3. The worksheet can be filed for future reference. If any questions should arise when the feed is being used, the worksheet provides an orderly record of the content of the feed and its mixing.

Each type of diet should be assigned a number for future reference, and the date of formulation and/or mixing should be recorded. In addition to listing the feed ingredients and their respective amounts, the nutrient requirements to be fulfilled by the diet should be listed on the worksheet immediately below the totals of various components of the feed. By subtracting the totals contained in the feed from the nutrient requirements, the person formulating the diet can then determine if there are any severe excesses or deficiencies in the diet.

POINTERS IN FORMULATING DIETS AND FEEDING SWINE

In formulating diets and in feeding swine, the following points are noteworthy:

1. Feeds of similar nutritive properties can be interchanged in the diet as price relationships warrant.

2. If wheat, barley, oats, or grain sorghum is used instead of corn as the grain in a diet, the protein supplement may be slightly reduced, because these grains have a higher protein content than corn.

3. Pacific Coast grains are generally lower in protein content than grains produced elsewhere.

4. When proteins of animal origin predominate, less calcium supplementation is necessary.

5. Where there is insufficient sunlight or where dehydrated alfalfa meal is fed, vitamin D should be added.

6. Where the diet consists chiefly of white corn, barley, wheat, oats, rye, kafir, or byproducts of these grains, there may be a deficiency of vitamin A.

7. Except for gestating sows and boars of breeding age, hogs are generally self-fed. All of the ingredients may be mixed together and placed in the same self-feeder.

8. Full-fed finishing hogs will consume 4 to 5 lb of feed daily per 100 lb liveweight until they weigh 100 lb. They will eat 3 to 4 lb daily per 100 lb weight from this stage until marketing.

HOME MIXED VS COMMERCIAL FEEDS

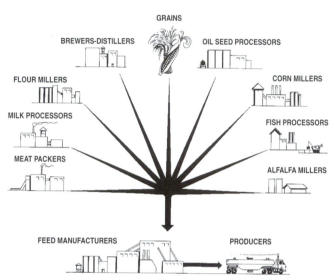

Fig. 9-6. Commercial feed companies obtain their raw materials for feeds from many sources. Over 100 different ingredients are processed into various (1) complete feeds, or (2) concentrates that are fed with homegrown grains.

The swine producer has the following options from which to choose for home mixing feeds:

1. Purchase of a commercially prepared protein supplement (likely reinforced with vitamins and minerals), which may be blended with local or homegrown grain.

2. Purchase a commercially prepared vitamin-trace mineral premix which may be mixed with an oil meal, and then blended with local or homegrown grain.

A suggested vitamin premix is given in Table 9-15. *Note:* This vitamin premix is designed to be fed to all ages of swine by adjusting its inclusion rate, and by using an add pack (Table 9-17) for gestating and lactating sow diets.

A suggested trace mineral premix is given in Table 9-16. *Note:* This mineral premix is designed to be fed to all ages of swine by adjusting its inclusion rate.

3. Use a base mix. Base mixes contain all needed ingredients except grain and protein and usually account for 2.5 to 5.0% of the diet by weight.

Because feed processing systems differ from farm to farm, several base mix specifications have been included in Table 9-18 for producers who do not choose or do not have the milling capabilities to handle the small inclusion rates associated with a premix program. These base mixes contain approximately the

TABLE 9-15
VITAMIN PREMIX[1,2]

Vitamin	Guaranteed per Pound of Premix	Source
Vitamin A	2,000,000 USP units	A 650
Vitamin D$_3$	300,000 USP units	D$_3$ 400
Vitamin E	8,000 Int units	E 50%
Vitamin K (Menadione)	800 mg	MPB[3] 100%
Vitamin B-12	6 mg	B-12 600
Riboflavin	1,500 mg	Riboflavin 95%
Pantothenic acid	5,200 mg	Cal pan 100%
Niacin	9,000 mg	Niacin 99.5%
Choline	30,000 mg	Choline Cl 60%

[1]Adapted by the authors from: Goodband, R. D., *et al.*, *Kansas Swine Nutrition Guide*, published by Kansas State University, 1994, updated 1995.

[2]Feeding rates for this premix: Sow and starter diets—5 lb/ton; grower diets—4 lb/ton; finisher diets—3 lb/ton.

[3]Instead of MPB, several suppliers use MSBC (another excellent source of menadione).

TABLE 9-16
TRACE MINERAL PREMIX[1,2]

Mineral	Guaranteed per Pound of Premix	Source
Zinc	50 g	Zn oxide or sulfate
Iron	50 g	Ferrous sulfate
Manganese	12 g	Mn Oxide or sulfate
Copper	5 g	Cu sulfate
Iodine	90 mg	Ca iodate
Selenium	90 mg	Na selenite

[1]Adapted by the authors from: Goodband, R. D., *et al.*, *Kansas Swine Nutrition Guide*, published by Kansas State University, 1994, updated 1995.

[2]Feeding rates for this premix: Sow and starter diets—3 lb/ton; finisher diets—2 lb/ton.

TABLE 9-17
SOW ADD PACK[1,2]

Vitamin	Guaranteed per Pound of Premix	Source
Choline	70,000 mg	Choline Cl 60%
Biotin	40 mg	Biotin
Folic acid	300 mg	Folic acid

[1]Adapted by the authors from: Goodband, R. D., *et al.*, *Kansas Swine Nutrition Guide*, published by Kansas State University, 1994, updated 1995.

[2]Feeding rates for this premix: Sow diets—5 lb/ton.

same calcium, phosphorus, vitamin, and trace mineral levels as recommended in Table 9-8. Furthermore, these base mixes can be substituted for the individual ingredients (monocalcium phosphate, limestone, salt, vitamin and trace mineral premixes) in the suggested diet formulations listed in Tables 9-23 to 9-34 and provide similar nutrient content.

Commercial feeds are just what the term implies—instead of being home mixed, these feeds are mixed by commercial feed manufacturers who specialize in the business.

The commercial feed manufacturer has the distinct advantages of (1) purchasing feed in quantity lots, making possible price advantages; (2) economical and controlled mixing; (3) the hiring of scientifically trained personnel for use in determining the diets; and (4) quality control. Many producers have neither the know-how nor the quantity of business to provide these services on their own. Because of these several advantages, commercial feeds are of special interest to smaller swine producers.

Numerous types of commercial feeds, ranging from additives to complete diets, are on the market, with most of them designed for a specific species, age, or need. Among them, are complete diets, concentrates, pelleted or cubed forages, protein supplements (with or without reinforcements of vitamins and/or minerals), vitamin and/or mineral supplements, additives, milk replacers, starters, growers, finishers, fitting diets, diets for different levels of production, diets for gestation and lactation, and medicated feeds.

In summary, it may be said that there exist two good alternative sources of most feeds and diets— home mixed or commercial—and the able manager will choose wisely between them.

■ **State commercial feed laws**—Nearly all states have laws regulating the sale of commercial feeds. These benefit both producers and reputable feed manufacturers. In most states the laws require that every brand of commercial feed sold in the state be licensed, and that the chemical composition be guaranteed.

Samples of each commercial feed are taken each year, and analyzed chemically in the state's laboratory to determine if manufacturers lived up to their guarantees. Additionally, skilled microscopists examine the sample to ascertain that the ingredients present are the same as those guaranteed. Flagrant violations on the latter point may be prosecuted.

Results of these examinations are generally published, annually, by the state department in charge of such regulatory work. Usually, the publication of the guarantee alongside any "short-changing" is sufficient to cause the manufacturer to rectify the situation promptly, for such public information soon becomes known to both users and competitors.

TABLE 9-18
BASE MIXES FOR SWINE[1, 2]

| Item | Guaranteed Nutrient Levels per Pound of Base Mix | | | | | Source |
	Starter[3]	Grower	Finishing	Grower-Finisher[4]	Sow	
Vitamin A .... (USP)	110,000	100,000	100,000	100,000	100,000	A 650
Vitamin D₃ ...... (USP)	16,600	15,000	15,000	15,000	15,000	D₃ 400
Vitamin E .... (IU)	440	400	400	400	400	E 50%
Vitamin K (Menadione) (mg)	44	40	40	40	40	MPB 100%[5]
Vitamin B-12 ...... (mg)	0.33	0.30	0.30	0.30	0.30	B-12 600
Riboflavin .... (mg)	80	75	75	75	75	Riboflavin 95%
Pantothenic acid .... (mg)	285	260	260	260	260	Cal Pan 100%
Niacin .... (mg)	500	450	450	450	450	Niacin 99.5%
Choline .... (mg)	1,650	1,500	1,500	1,500	5,000	Choline Cl 60%
Biotin .... (mg)	—	—	—	—	2	Biotin
Folic acid .... (mg)	—	—	—	—	15	Folic acid
Zinc .... (mg)	1,650	1,875	1,650	1,875	1,500	Zn sulfate or oxide
Iron .... (mg)	1,650	1,875	1,650	1,875	1,500	Ferrous sulfate
Manganese .... (mg)	400	450	400	450	360	Mn oxide or sulfate
Copper .... (mg)	2,050	185	165	185	150	Cu sulfate
Iodine .... (mg)	3	3.3	3	3.3	2.7	Ca iodate or EDDI
Selenium .... (mg)	3	3.3	3	3.3	2.7	Na selenite
Calcium .... (%)	16.2	16.5	20	18.5	16.5	Limestone, Dical or
Phosphorus .... (%)	9.3	8.65	7.5	8.1	9.0	Dical or Monocal Phos
Lysine .... (%)	2.66	2.93	3.75	2.93	—	Lysine-HCl
Methionine .... (%)	1.66	—	—	—	—	DL-Methionine
Salt .... (%)	5.5	10.0	10.0	10.0	10.0	NaCl
Feeding rate ..... (lb/ton)	90	80	60	80/60	100	
Bag size[6] ...... (lb/bag)	45	40	60	60	50	

[1]Adapted by the authors from: Goodband, R. D., *et al.*, *Kansas Swine Nutrition Guide*, published by Kansas State University, 1994, updated 1995.

[2]These base mixes, when used at their recommended inclusion rate, will provide approximately the same vitamin and mineral levels as the vitamin and trace mineral premixes in Tables 9-15, 9-16, and 9-17.

[3]This base mix is to be included in Phase II starter diet (15 to 25 lb) with 200 lb/ton edible grade, spray dried whey and 50 lb/ton spray-dried blood meal. An antibiotic for growth promotion may be added to this base mix.

[4]This base mix is a compromise option for producers that can only handle one grower-finisher base mix.

[5]Instead of MPB, several suppliers use MSBC (another excellent source of menadione).

[6]These base mixes can be provided in bulk or bagged form.

SELECTING COMMERCIAL FEEDS

There is a difference in commercial feeds! That is, there is a difference from the standpoint of what a swine producer can purchase with available feed dollars. The smart operator will know how to determine what constitutes the best in commercial feeds for specific needs. The producer will not rely solely on the appearance or aroma of the feed, nor on the feed salesperson. The most important factors to consider or look for in buying a commercial feed are:

1. **Reputation of the manufacturer.** This should be determined by (a) conferring with other producers who have used the particular products, and (b) checking on whether or not the commercial feed under consideration has consistently met its guarantees. The

latter can be determined by reading the bulletins or reports published by the respective state departments in charge of enforcing feed laws.

Quite often, a feed that costs a little more will be of a higher quality than its competition. Therefore, the increased costs can be justified by the increased performance of the animals to which the diet is fed.

Many of the larger manufacturers offer services other than selling feed, including the consulting services of well-trained and experienced staff and excellent publications.

2. **Specific needs.** Feed needs vary according to (a) the class, age, and productivity of the animals, and (b) whether the animals are fed primarily for maintenance, growth, finishing (or show-ring fitting), reproduction, or lactation. The wise operator will buy different formula feeds for different needs.

Feeding swine has become a sophisticated and complicated process. Feed manufacturers have extensive resources with which to formulate and test diets for different needs. As a result, most manufacturers have a large selection of feeds—one of which should be applicable to the needs of the producer. It is essential that the producer make clear to the feed salesperson the needs of the animals to be fed.

3. **Labeling.** Most states require that mixed feeds carry labels guaranteeing the ingredients and the chemical makeup of the feed. The feed tag should contain the following information:

■ **Net weight of the feed**—Usually in pounds.

■ **Brand name and product name**—The brand name refers to any word, name, symbol, or logo which identifies the feed of a distributor and distinguishes it from the feeds of other manufacturers. The product name identifies the specific use for the feed.

■ **Guaranteed analysis**—Most feed labels give minimum and/or maximum guarantees of certain nutrients within the feed. Laws vary from state to state as to what analyses must be guaranteed, but most, if not all, of the following analyses are generally listed: (a) dry matter, (b) crude protein, (c) crude fat, (d) crude fiber, (e) ash, (f) nitrogen-free extract, (g) calcium and phosphorus, and (h) other nutrients. Generally, feeds with more protein and fat and less fiber indicate quality. Most labels list minimum values of crude protein and crude fat and maximum values of crude fiber. Both maximum and minimum guarantees are listed for calcium, phosphorus, and salt on most feed tags.

A high-fiber content often indicates a low feeding value. Feeds for swine generally contain negligible amounts of fiber. But they are higher in fats and micronutrients than ruminant feeds.

■ **Listing of ingredients**—The feed tag lists the ingredients of the feed in descending order of quantity used. While the exact quantities of the ingredients are not generally given, the buyer can obtain a rough idea as to the composition of the feed.

■ **Directions for use**—If the feed is to be used for a specific purpose, directions may be given on the label. They should specify what kind of animal and for what particular purpose the feed was formulated.

■ **Name and mailing address of the manufacturer**—The manufacturer is responsible for the quality of the feed. Any failure to meet guarantees or any contamination problems incurred with the feed makes the manufacturer liable for penalties and damages incurred from the feed. For this reason, the manufacturer should be identified clearly on the product.

■ **Warnings**—When drugs have been added to commercial feed, the feed label must clearly indicate that it is medicated. Many states require that the name of the drug with a listing of the quantity of active ingredients added be stated on the feed tag. Also, the purpose of medication, any restrictions of use, and withdrawal period for the medicated feed should be so stated.

4. **Quality control.** A good commercial feed manufacturer will follow a sound quality control program. Producers should look for and evaluate this program.

5. **Flexible formulas.** Feeds with flexible formulas are usually the best buy. This is because the price of feed ingredients in different source feeds varies considerably from time to time. Thus, a good feed manufacturer, having access to least-cost computer programs, will shift formulas as prices change, in order to give the producer the most for the money. This is as it should be, for (a) there is no one best diet, and (b) if substitutions are made wisely, the price of the feed can be kept down and the feeder will continue to get equally good results.

FEED PROCESSING

There are many different methods of processing feeds for swine. Grinding with a hammermill or roller mill is the most common method. Although there is not full agreement relative to particle size, most swine producers favor an average particle size of 700 to 900 microns for all grains except wheat. Finely ground wheat creates palatability problems.

If there are whole kernels in the feed, in all probability the feed is not ground finely enough and you may be losing 5% to 8% in efficiency.

If properly designed, either a hammermill or a roller mill is satisfactory for processing. However, hammermills can change from grinding one grain to an-

Fig. 9-7. Diets provided nursery pigs in phase feeding program at Farmers Hybrid, Des Moines, Iowa. Left to right the nursery rations are:

Phase	% Protein	% Lysine	Pig Wt Lb
1	26	1.60	9–12
2	25	1.53	12–16
3	24	1.40	14–20
4	22	1.24	20–50

(Courtesy, Farmers Hybrid, Des Moines, IA)

other by changing screens. But a hammermill requires more energy than a roller mill and will produce a higher percentage of fines and dust.

In addition to grinding, the most common forms of processing feeds for swine are: (1) cooking, (2) extruding, (3) fermenting, (4) full-fat heat pressed soybeans, (5) liquid, (6) micronizing, (7) paste, (8) roasting, (9) soaking, and (10) steam flaking.

MIXING/FEEDING SYSTEMS

Basically, there are four systems of preparing diets for a swine operation; namely—

■ **Complete feed**—Complete feeds, which may be mixed by a commercial feed company or farm mixed, are complete and ready to feed. The use of complete diets affords an accurate means of controlling the intake of minerals, vitamins, and feed additives. Although convenient, complete feeds are usually the most expensive. Additionally, flexibility is limited when diet changes are needed.

■ **Grain and supplement**—Mixing producer-raised grain and a supplement (most commonly a 40% protein supplement) has been popular for a long time. This system may be more expensive than the base mix system.

■ **Premix system**—The premix consists of minerals and/or vitamins and additives. It is mixed with the main ingredients of the diet. Normally, the premix is added at a level of 2 to 10 lb per ton of the otherwise complete feed. (See Tables 9-15, 9-16, and 9-17 for suggested premixes.)

■ **Base mix system**—A base mix completes protein quality (amino acids) and provides minerals, vitamins, and additives. It is mixed with grain and high protein

sources available locally. Generally, the base mix system is cost effective for farm mixing and fits well into many portable feed systems. (See Table 9-18 for a suggested base mix.)

QUALITY CONTROL

Quality control is an essential component of the manufacture of swine feeds. It is particularly important in the processing and mixing of baby pig diets. A sound quality control program assures that the feed consumed by swine contains the desired concentration of nutrients.

Since feed usually represents 65 to 70% of the total cost of producing swine, it makes good business sense for pork producers to ensure that pigs receive feed that has been properly formulated using good quality ingredients, and manufactured properly.

FEEDING PROGRAMS AND DIETS

As is obvious from the previous discussion, the nutritive requirements of swine vary according to age, weight, and stage of production—gestation or lactation. Furthermore, diets must be reformulated from time to time in keeping with new developments and changing prices. In the sections which follow, recommendations for feeding and suggested diets for the various categories of swine are given. These categories include (1) feeding baby pigs (segregated early weaning, phase feeding, and creep feeding), (2) feeding growing pigs, (3) feeding finishing pigs, (4) feeding replacement gilts and boars, (5) feeding brood sows (limit feeding and flushing), (6) feeding gestating sows, (7) feeding lactating sows, and (8) feeding show and sale animals. Suggested diets are intended to serve as useful guides—not the final answer. Wise producers will adapt suggested diets to meet their needs.

FEEDING BABY PIGS

It is important that newborn pigs receive colostrum during the first 24 hours post-farrowing. Colostrum contains the antibodies necessary for building the baby pig's disease resistance.

Equalizing litters within 24 to 48 hours and transferring pigs so that litters contain pigs of similar weight can improve pig survival.

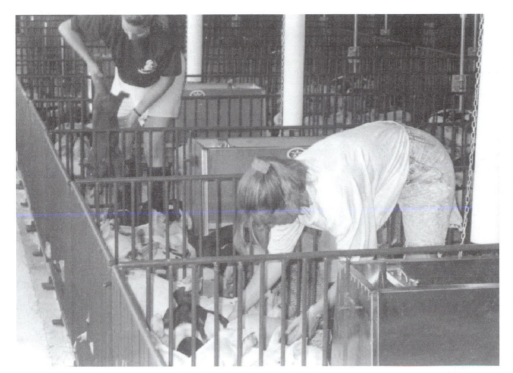

Fig. 9-8. Pigs in modern nursery facility at Murphy Farms. (Courtesy, Dr. Garth W. Boyd and Ms. Rhonda Campbell, Rose Hill, NC)

SEGREGATED EARLY WEANING (SEW)

Segregated early weaning (SEW) is an infectious disease control procedure the primary objective of which is to improve productivity of the growing/finishing phase by preventing the transfer of disease. To accomplish this, the piglets are usually weaned at 10 to 17 days of age, rather than at the conventional 21- to 28-day weaning. The pigs may be segregated in either of two ways: (1) by transporting the piglets from the farrowing site to another site, or (2) by removing the sows to a centralized breeding/gestating facility and leaving the piglets in the site in which they were born.

The segregation of weaned pigs prevents the vertical transmission (from sow to her piglets) of infectious disease. The weaned pigs are placed in groups and each group remains segregated by the use of all-in/all-out production. Segregation of age groups of pigs through the use of all-in/all-out production also prevents the horizontal transmission of disease from group to group.

The transfer of colostral immunity from the sow protects the piglets from infection until they are weaned. Infectious organisms circulating in the herd and vaccination of sows stimulate colostral immunity. The shedding of infectious organisms may be decreased by medicating sows. Piglets may also be medicated near the time of weaning to eliminate bacterial pathogens.

Weaning age is critical to the success of SEW because the maternal immunity provided by colostrum decreases at different rates for different diseases. For example, vertical transmission of the pseudorabies virus has been prevented by using a 21-day weaning age, whereas the organism causing rhinitis requires a weaning age of 10 days because colostral immunity does not prevent infection.

The major **advantage** of SEW production is the increased performance of the growing and finishing hogs. It is conjectured that this advantage is due to the decreased burden of infectious diseases, especially respiratory disease, stimulating the pigs' immune system. If the piglets have less need to partition nutrients to the immune system, they can use the nutrients for growth. The decreased pathogenic burden also results in pigs with greater appetites. Moreover, the decreased infectious disease load results in the decreased use of drugs and vaccines, which lowers production costs. Still another advantage of SEW production is that it makes it possible to maintain the genetic base of the sow herd. For example, if the herd becomes infected with an organism, disease-free offspring can be raised on another site without disposing of the sow herd.

Potential **disadvantages** of the SEW program are: (1) the short lactation period is associated with increased weaning-to-service intervals, reduced farrowing rates, and decreased subsequent litter size; (2) the moving of animals from site to site increases expenses and requires coordination; (3) the transportation of very young pigs provides special challenges; and (4) the weaning of piglets at 5 to 10 days of age requires a modification of nursery equipment and management.

■ **Diets for SEW and other early weaned pigs**—Because of the dramatic changes in the digestive system of early weaned pigs, a specialized nutritional program should be utilized, compared to pigs weaned at older ages.

Diets and feeding programs for SEW and other early weaned pigs are presented in Tables 9-19, 9-20, 9-21, and 9-22.

TABLE 9-19
DIETS FOR SEW AND OTHER EARLY WEANED PIGS
7 TO 17 DAYS OF AGE[1, 2]

Ingredient	Day 7 SEW Diet		Transition Diet	
	(%)	*(lb/ton)*	*(%)*	*(lb/ton)*
Edible-grade spray-dried whey	25.0	500	20.0	400
Edible-grade lactose	5.0	100	—	—
Spray-dried animal plasma	6.7	134	2.5	50
Spray-dried blood meal	1.75	35	2.5	50
Select menhaden fish meal	6.0	120	2.5	50
Corn	33.35	667	39.85	797
Soybean meal, 46.5% protein	12.7	254	23.25	465
Soybean oil or Choice white grease	6.0	120	5.0	100
Monocalcium phosphate (21% P)	0.75	15	1.3	26
Limestone (38% calcium)	0.45	9	0.725	14.5
L-lysine HCl	0.15	3	0.15	3
DL-methionine	0.15	3	0.125	2.5
Salt	0.1	2	0.2	4
Medication[3]	1.0	20	1.0	20
Remainder will be vitamins and trace minerals[4]	0.9	18	0.9	18
Total	100.0	2,000	100.0	2,000

[1]Adapted by the authors from: Goodband, R. D., *et al., Kansas Swine Nutrition Guide*, published by Kansas State University, 1994, updated 1995.

[2]After vitamins and trace minerals are added, any remaining portion of the diet should be filled with corn to achieve 100%.

[3]Producer should specify which medication is desired in the Phase I diet.

[4]All vitamins and trace mineral levels are listed in Table 9-20.

Instructions:

The SEW diet should be pelleted in 1/8 or 3/32" pellets.
The Transition diet should be pelleted in 1/8, 3/32, or 5/32" pellets.

The SEW diet should be fed to early weaned pigs (<17 days) from weaning to 11 lb. The Transition diet should be fed from 11 to 15 lb for pigs that were fed the SEW diet from weaning to 11 lb.

Diet must be provided in 50 lb bags.

TABLE 9-20
VITAMINS AND TRACE MINERALS FOR SEW DIETS
LISTED IN TABLE 9-19[1, 2]

Nutrient	Amount Added per Ton	Source
Vitamin A	10,000,000 USP units	A 650
Vitamin D_3	1,500,000 USP units	D_3 400
Vitamin E	40,000 Int units	E 50%
Vitamin K (Menadione)	4,000 mg	MPB[3] 100%
Vitamin B-12	30 mg	B-12 600
Riboflavin	7,500 mg	Riboflavin 95%
Pantothenic acid	26,000 mg	Cal Pan 100%
Niacin	45,000 mg	Niacin 99.5%
Choline	150,000 mg	Choline Cl 60%
Zinc	2,700 g	Zn oxide[4]
Iron	150 g	Ferrous sulfate
Manganese	36 g	Mn oxide or sulfate
Copper	15 g	Cu sulfate
Iodine	270 mg	Ca iodate
Selenium	270 mg	Na selenite

[1]Adapted by the authors from: Goodband, R. D., *et al., Kansas Swine Nutrition Guide*, published by Kansas State University, 1994, updated 1995.

[2]Total levels and sources of vitamin and trace minerals that must be added per ton of complete diets listed in Table 9-20.

[3]Instead of MPB, several suppliers use MSBC (another excellent source of menadione).

[4]Zinc source MUST be zinc oxide.

Instructions:

Ingredient quality is imperative for this diet. Therefore, the following guidelines should be followed as sources for the major ingredients listed for the Day 7 SEW and Transition diets.

Ingredient	Source
Extra-grade, edible-grade, spray-dried whey	Land O Lakes or equivalent
Select menhaden fish meal	Zapata Proteins or equivalent
Spray-dried porcine plasma	Merrick's, American Proteins or equivalent
Spray-dried blood meal	California Spray Dry, American Proteins or equivalent

TABLE 9-21
PHASE FEEDING PROGRAMS FOR SEW AND OTHER EARLY WEANED PIGS

Early Weaning (5 to 17 Day Weaning)		Conventional Weaning (17 to 21 Day Weaning)	
5–11 lb	SEW		
11–15 lb	Transition	11–15 lb	Phase I
15–25 lb	Phase II	15–25 lb	Phase II
25–50 lb	Phase III	25–50 lb	Phase III

TABLE 9-22
FEED ALLOWANCE PER PIG (WEANING TO 50 LB)
FOR VARIOUS PHASE FEEDING PROGRAMS

	Weaning Age, d:	7	14	21	24
Diet, lb	Initial Weight, lb:	6	9	11.5	13.5
SEW		5	2	—	—
Transition		5	5	—	—
Phase I		—	—	4	1.5
Phase II		15	15	15	15
Phase III		50	50	50	50

PHASE FEEDING

Phase feeding is the feeding of each of several diets for a relatively short period of time in order to meet the young pig's nutrient requirements. Phase feeding is practical in big operations.

When one diet is fed to young pigs for a long period of time, it is usually under the pig's nutrient requirements to begin with, and over fortified for the older pig. By phase feeding, the producer can minimize this over- and under-feeding and provide a more economical feeding program for the pig.

Because the baby pig undergoes dramatic changes in digestive development, the most common application of phase feeding is for starter pigs. Table 9-22 lists the expected feed intakes of the appropriate starter diets based on various ages at weaning. Adhering to these expected feed use guidelines will help minimize over-feeding expensive starter diets. By phase feeding, the producer can match the baby pig's nutrient requirements and digestive capabilities with the most economical diet possible, yet get maximum performance in the nursery. Although these diets are very expensive, the small amount of feed consumed and the excellent feed efficiency justify their cost.

Diets for phase feeding nursery pigs are given in Table 9-23.

TABLE 9-23
DIETS FOR PHASE FEEDING NURSERY PIGS (WEANING TO 50 LB)[1, 2]

Ingredient	Phase I (< 15 lb) 1	Phase II (15 to 25 lb) 1	2	3	4	Phase III (25 to 50 lb) 1
Corn or milo	897	1,204	1,138	1,198	1,133	1,337
Dried whey	400	200	200	200	200	—
Soybean meal (46.5%)	321	454	460	430	435	571
Spray-dried porcine plasma	150	—	—	—	—	—
Fat	100	—	60	—	60	—
Spray-dried blood meal	35	50	50	—	—	—
Select menhaden fish meal	—	—	—	100	100	—
Monocalcium phosphate	38	38	39	26	27	31
Limestone	14	17	16	10	10	18
Medication	20	20	20	20	20	20
Salt	2	5	5	5	5	7
Vitamin premix[3]	5	5	5	5	5	5
Trace mineral premix[4]	3	3	3	3	3	3
Selenium premix	3	3	3	3	3	3
L-Lysine HCl	3	3	3	3	3	3
DL-Methionine	3	1	1	—	—	—
Zinc oxide	7.6	5	5	5	5	—
Copper sulfate	—	—	—	—	—	1.5
	2,000	2,000	2,000	2,000	2,000	2,000
Calculated analysis, %						
Lysine	1.50	1.25	1.25	1.25	1.25	1.15
Ca	0.90	0.90	0.90	0.90	0.90	0.80
P	0.80	0.80	0.80	0.80	0.80	0.70

[1]Adapted by the authors from: Goodband, R. D., *et al., Kansas Swine Nutrition Guide*, published by Kansas State University, 1994, updated 1995.

[2]All diets fed *ad libitum.*

[3]Suggested vitamin premix Table 9-15.

[4]Suggested trace mineral premix Table 9-16.

CREEP FEEDING

The practice of self-feeding concentrates to young pigs in a separate area away from their dams is known as creep feeding.

Research has shown that very little creep feed will be consumed before three weeks of age. Often, more creep feed is wasted than consumed before three

weeks of age. It is recommended that creep feed be fed on a daily basis in order that the diet be fresh.

FEEDING GROWING PIGS

Pigs ranging in weight from about 50 to 120 lb are known as growing pigs. In a farrow-to-finish operation, grower diets represent approximately 20 to 25% of the feed usage.

In Table 9-24, the grower stage is broken down into two phases: 50 to 80 lb and 80 to 120 lb, better to meet the pig's nutrient requirements. Note, too, that several grower diets are listed in Table 9-24.

The growing pig deposits lean tissue at a fast rate. This calls for high levels of lysine and other amino acids. Because gilts consume about ½ lb less feed per day than barrows, they may not eat enough to fully meet their requirements.

■ **Split-sex feeding**—*Split-sex feeding refers to sorting gilts from barrows and feeding each group separate diets.* It may be practical in large operations. Because gilts consume less feed than barrows, their diets can be fortified with extra amino acids for growth rate and feed efficiency, along with additional calcium and phos-

Fig. 9-9. Creep feeding young pigs in a light, warm, dry, and draft-free area. (Photo by J. C. Allen and Son, West Lafayette, IN)

TABLE 9-24
GROWER DIETS (50 TO 120 LB)[1,2]

Ingredient	50 to 80 Lb				80 to 120 Lb			
	1	2	3	4	1	2	3	4
Milo or corn	1,360	1,395	1,430	1,465	1,499	1,534	1,569	1,604
Soybean meal (46.5%)	578	542	507	471	435	400	364	328
Monocalcium phosphate	26	27	27	28	29	29	30	31
Limestone	18	18	18	18	19	19	19	19
Salt	7	7	7	7	7	7	7	7
Lysine HCl	3	3	3	3	3	3	3	3
Vitamin premix[3]	4	4	4	4	4	4	4	4
Trace mineral premix[4]	3	3	3	3	3	3	3	3
Selenium premix	1	1	1	1	1	1	1	1
	2,000	2,000	2,000	2,000	2,000	2,000	2,000	2,000
Calculated analysis, %								
Lysine	1.15	1.10	1.05	1.00	0.95	0.90	0.85	0.80
Ca	0.75	0.75	0.75	0.75	0.75	0.75	0.75	0.75
P	0.65	0.65	0.65	0.65	0.65	0.65	0.65	0.65

[1]Adapted by the authors from: Goodband, R. D., *et al., Kansas Swine Nutrition Guide*, published by Kansas State University, 1994, updated 1995.

[2]All diets fed *ad libitum.*

[3]Suggested vitamin premix Table 9-15.

[4]Suggested trace mineral premix Table 9-16.

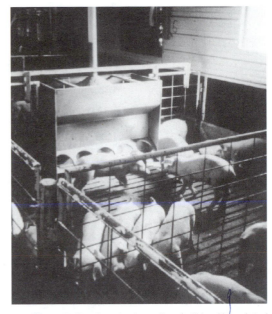

Fig. 9-10. Growing pigs in modern facilities. Note slatted floors. Note, too, that feed is delivered to the self-feeders from exterior bulk bins via augers mounted on ceiling. (Courtesy, Kansas State University, Manhattan)

TABLE 9-25
SUGGESTED LYSINE LEVELS FOR HIGH-LEAN GROWTH PIGS FOR WINTER-SUMMER AND SPLIT-SEX FEEDING[1]

Phase (Wt, Lb)	Winter		Summer	
	Gilts	Barrows	Gilts	Barrows
50 to 80	1.10	1.05	1.15	1.10
80 to 120	1.00	0.95	1.05	1.00
120 to 160	0.90	0.80	0.95	0.85
160 to 200	0.80	0.70	0.85	0.75
200 to market	0.70	0.60	0.75	0.65

[1]Adapted by the authors from: Goodband, R. D., *et al., Kansas Swine Nutrition Guide,* published by Kansas State University, 1994, updated 1995.

TABLE 9-26
SUGGESTED LYSINE LEVELS FOR MEDIUM-LEAN GROWTH PIGS FOR WINTER-SUMMER AND SPLIT-SEX FEEDING[1]

Phase (Wt, Lb)	Winter		Summer	
	Gilts	Barrows	Gilts	Barrows
50 to 80	1.00	0.95	1.05	1.00
80 to 120	0.90	0.85	0.95	0.90
120 to 160	0.80	0.70	0.85	0.75
160 to 200	0.70	0.60	0.75	0.65
200 to market	0.60	0.55	0.65	0.55

[1]Adapted by the authors from: Goodband, R. D., *et al., Kansas Swine Nutrition Guide,* published by Kansas State University, 1994, updated 1995.

phorus for bone development if they are going to be retained in the breeding herd. With split-sex feeding, gilts are generally fed 10% more lysine in the diet than barrows.

■ **Lysine requirement of lean, rapidly growing pigs, winter vs summer, and split-sex groups—** Lean, rapidly growing pigs have a higher amino acid requirement than slow-growing pigs. Additionally, the winter and summer requirements for lysine differ. The dietary levels of lysine (1) for growing pigs (50 to 160 lb) of high- and average-lean growth potential, (2) for winter and summer, and (3) for split-sex feeding are presented in Tables 9-25 and 9-26.

FEEDING FINISHING PIGS

Pigs ranging in weight from approximately 120 lb to market weight are known as finishing pigs. The feed consumed by finishing pigs represents approximately 50 to 55% of the feed usage on a farrow-to-finish operation. So, decisions to change or modify finishing diets must be based on economics. It follows that summer vs winter diets, and/or split-sex feeding can be economically justified for finishing hogs. Table 9-27

Fig. 9-11. Finishing pigs in modern facilities. The floors are totally slatted with concrete slats. Feed is delivered to the self-feeders from exterior bulk bins via augers mounted on the ceiling. (Courtesy, Iowa State University, Ames)

gives suggested finishing diets for 120 lb to market. The dietary levels of lysine (1) for finishing pigs (160 lb to market) of high- and average-lean growth potential, (2) for winter and summer, and (3) for split-sex feeding are presented in Tables 9-25 and 9-26.

TABLE 9-27
FINISHING DIETS (120 LB TO MARKET)[1, 2]

Ingredient	1	2	3	4	5	6
	-- (lb) --					
Milo or corn	1,583	1,618	1,652	1,687	1,721	1,757
Soybean meal (46.5%)	363	327	292	256	221	185
Monocalcium phosphate	20	21	22	22	23	23
Limestone	18	18	18	19	19	19
Salt	7	7	7	7	7	7
L-lysine HCl	3	3	3	3	3	3
Vitamin premix[3]	3	3	3	3	3	3
Trace mineral premix[4]	2	2	2	2	2	2
Selenium premix	1	1	1	1	1	1
	2,000	2,000	2,000	2,000	2,000	2,000
Calculated analysis, %						
Lysine	0.85	0.80	0.75	0.70	0.65	0.60
Ca	0.65	0.65	0.65	0.65	0.65	0.65
P	0.55	0.55	0.55	0.55	0.55	0.55

[1]Adapted by the authors from: Goodband, R. D., *et al., Kansas Swine Nutrition Guide*, published by Kansas State University, 1994, updated 1995.

[2]All diets fed *ad libitum.*

[3]Suggested vitamin premix Table 9-15.

[4]Suggested trace mineral premix Table 9-16.

FEEDING REPLACEMENT GILTS AND BOARS

Prospective breeding gilts should be kept from getting too fat. Meat-type animals can usually be left on a high-energy diet until they reach 175 to 200 lb without becoming too fat. It is neither necessary nor desirable that females intended for breeding purposes carry the same degree of finish as market animals. After selecting replacement gilts, they should be fed as follows:

Fig. 9-12. Replacement gilts, performance tested and indexed. (Courtesy, Babcock Swine, Inc., Plainview, MN)

1. Give about 5 lb per head per day through their second heat period.

2. Flush—full feed—after the second heat period until breeding on the third heat period.

3. After breeding, limit the feed intake to 3 to 5 lb per day. Overfeeding during gestation can cause embryonic death and thus decrease litter size.

The feed allowance of young boars should vary according to the condition of the animals, the climatic condition, and the individuality. If the animals are inclined to get too fat, which is likely to happen in self-feeding, the diet may well contain a considerable amount of bulky feeds; otherwise, limited feeding may be necessary.

The feed requirements of herd boars are about the same as those of females of equal weight. They should always be kept in thrifty, vigorous condition and virile. In no case should boars be overfat, nor should they be in a thin, run-down condition. Normally, the following feed allowances will suffice: for boars weighing 120 to 150 lb, 6 to 9 lb of feed daily; for mature boars, 5 to 7 lb of feed daily. A more liberal diet must be provided in the wintertime and when the sire is in heavy service. The feed allowance should be varied with the age, development, temperament, breeding demands, and roughage consumed.

Table 9-9 gives the *Recommended Nutrient Allowances for Replacement Gilts and Boars.*

Replacement gilt and boar diets are given in Tables 9-28, 9-29, and 9-30.

TABLE 9-28
DIETS FOR REPLACEMENT GILTS AND BOARS BEING FED 3 LB *(1.4 KG)* PER DAY

Ingredient	Protein Concentration	Ration Number							
		1		2		3		4	
	(%)	*(lb)*	*(kg)*	*(lb)*	*(kg)*	*(lb)*	*(kg)*	*(lb)*	*(kg)*
Ground yellow corn[1]	8.9	1,637	*742.0*	1,542	*699.0*	1,647	*747.0*	1,582	*717.0*
Soybean meal, solvent	44	250	*113.4*	250	*113.4*	170	*77.1*	140	*63.5*
Dehydrated alfalfa meal	17	—	—	100	*45.4*	—	—	100	*45.4*
Meat and bone meal	50	—	—	—	—	100	*45.4*	100	*45.4*
Calcium carbonate (39% Ca)	—	15	*6.8*	15	*6.8*	10	*4.5*	5	*2.3*
Dicalcium phosphate (22% Ca, 18.5% P) . .	—	60	*27.2*	55	*24.9*	35	*15.9*	35	*15.9*
Iodized salt	—	15	*6.8*	15	*6.8*	15	*6.8*	15	*6.8*
Vitamin premix[2]	—	20	*9.1*	20	*9.1*	20	*9.1*	20	*9.1*
Trace mineral premix[2]	—	3	*1.4*	3	*1.4*	3	*1.4*	3	*1.4*
Total		2,000	*907.0*	2,000	*907.0*	2,000	*907.0*	2,000	*907.0*
Calculated analysis:									
Protein . . . .(%)		12.8		13.2		13.5		13.5	
Metabolizable energy (ME) . . . . (kcal/lb)		1,410		1,392		1,417		1,399	
. . . . (kcal/kg)		*3,109*		*3,069*		*3,124*		*3,085*	

[1]Ground oats can replace corn up to 20% of the total diet. If more than 20% oats is used in the diet, the level of feeding should be increased because of the low energy content of the oats. Ground milo, wheat, or barley can replace the corn.

[2]Suggested vitamin and trace mineral premixes are given in Tabels 9-15 and 9-16, respectively.

TABLE 9-29
DIETS FOR REPLACEMENT GILTS AND BOARS BEING FED 4 LB *(1.8 KG)* PER DAY

Ingredient	Protein Concentration	Diet Number							
		1		2		3		4	
	(%)	*(lb)*	*(kg)*	*(lb)*	*(kg)*	*(lb)*	*(kg)*	*(lb)*	*(kg)*
Ground yellow corn[1]	8.9	1,742.5	*790.2*	1,647.5	*747.2*	1,757.5	*797.1*	1,677.5	*760.8*
Soybean meal, solvent	44	175	*79.4*	170	*77.1*	90	*40.8*	70	*31.7*
Dehydrated alfalfa meal	17	—	—	100	*45.4*	—	—	100	*45.4*
Meat and bone meal	50	—	—	—	—	100	*45.4*	100	*45.4*
Calcium carbonate (39% Ca)	—	20	*9.1*	15	*6.8*	10	*4.5*	10	*4.5*
Dicalcium phosphate (22% Ca, 18.5% P) . .	—	35	*15.9*	40	*18.1*	15	*6.8*	15	*6.8*
Iodized salt	—	10	*4.5*	10	*4.5*	10	*4.5*	10	*4.5*
Vitamin premix[2]	—	15	*6.8*	15	*6.8*	15	*6.8*	15	*6.8*
Trace mineral premix[2]	—	2.5	*1.1*	2.5	*1.1*	2.5	*1.1*	2.5	*1.1*
Total		2,000	*907.0*	2,000	*907.0*	2,000	*907.0*	2,000	*907.0*
Calculated analysis:									
Protein . . . .(%)		11.6		11.9		12.3		12.4	
Metabolizable energy (ME) . . . . (kcal/lb)		1,435		1,413		1,441		1,419	
. . . . (kcal/kg)		*3,164*		*3,116*		*3,177*		*3,129*	

[1]Ground oats can replace corn up to 20% of the total diet. If more than 20% oats is used in the diet, the level of feeding should be increased because of the low energy content of the oats. Ground milo, wheat, or barley can replace the corn.

[2]Suggested vitamin and trace mineral premixes are given in Tabels 9-15 and 9-16, respectively.

TABLE 9-30
DIETS FOR REPLACEMENT GILTS AND BOARS BEING FED 5 LB *(2.3 KG)* PER DAY

Ingredient	Protein Concentration	Ration Number							
		1		2		3		4	
	(%)	*(lb)*	*(kg)*	*(lb)*	*(kg)*	*(lb)*	*(kg)*	*(lb)*	*(kg)*
Ground yellow corn[1]	8.9	1,838	*833.6*	1,753	*795.0*	1,853	*840.0*	1,800	*816.3*
Soybean meal, solvent	44	100	*45.4*	85	*38.5*	—	—	—	—
Dehydrated alfalfa meal	17	—	—	100	*45.4*	—	—	50	*22.7*
Meat and bone meal	50	—	—	—	—	115	*52.2*	50	*22.7*
Calcium carbonate (39% Ca)	—	15	*6.8*	10	*4.5*	5	*2.3*	8	*3.6*
Dicalcium phosphate (22% Ca, 18.5% P) . .	—	25	*11.3*	30	*13.6*	5	*2.3*	20	*9.1*
Iodized salt	—	10	*4.5*	10	*4.5*	10	*4.5*	10	*4.5*
Vitamin premix[2]	—	10	*4.5*	10	*4.5*	10	*4.5*	10	*4.5*
Trace mineral premix[2]	—	2	*0.9*	2	*0.9*	2	*0.9*	2	*0.9*
Total		2,000	*907.0*	2,000	*907.0*	2,000	*907.0*	2,000	*907.0*
Calculated analysis:									
Protein . . . .(%)			10.4		10.5		11.1		10.8
Metabolizable energy (ME) . . . . (kcal/lb)			1,452		1,430		1,456		1,441
. . . . (kcal/kg)			*3,202*		*3,153*		*3,210*		*3,177*

[1]Ground oats can replace corn up to 20% of the total diet. If more than 20% oats is used in the diet, the level of feeding should be increased because of the low energy content of the oats. Ground milo, wheat, or barley can replace the corn.

[2]Suggested vitamin and trace mineral premixes are given in Tabels 9-15 and 9-16, respectively.

FEEDING BROOD SOWS

Fig. 9-13. Modern gestating facility, with each sow in an individual stall. (Courtesy, Kansas State University)

The nutrition of brood sows is critical, for it may materially affect conception, reproduction, and lactation. Proper feeding of sows should begin with replacement gilts and continue through each stage of the breeding cycle—flushing, gestation, farrowing, and lactation.

FLUSHING SOWS AND GILTS

The practice of feeding sows and gilts more liberally so that they gain weight from 1.0 to 1.5 lb daily from 1 to 2 weeks before the opening of the breeding season until they are safely in pig is known as flushing.

Experiments and experiences indicate that flushing is effective with gilts, but not with second and third litter sows. (See Chapter 7, section headed "Flushing Sows and Gilts.")

FEEDING GESTATING SOWS

The nutrients fed the pregnant gilt or sow must first take care of the usual maintenance needs. If the gilt is not fully mature, nutrients are required for both maternal growth and growth of the fetus. Quality and quantity of proteins, minerals, and vitamins become particularly important in the diet of young, pregnant gilts, for their requirements are much greater and more exacting than those of the mature sow.

Approximately two-thirds of the growth of the fetus is made during the last month of the gestation period. It may be said, therefore, that the demands resulting from pregnancy are particularly accelerated during the latter third of the gestation period. Again, the increased needs are primarily for proteins, vitamins, and minerals.

During gestation, it is also necessary that body reserves be stored for subsequent use during lactation. With a large litter and a sow that is a heavy milker, the demands for milk production are generally greater than can be supplied by the diet fed at the time of lactation. Although desired gains will vary somewhat according to the initial condition, mature sows are generally fed to gain about 70 lb during the pregnancy period, and first litter pregnant gilts are fed to gain about 90 lb. This means that from the day of mating until farrowing time gilts should be fed to gain about 0.9 lb per day and mature sows about 0.7 lb per day. This calls for a daily feed allowance of approximately 4 to 4.5 lb per head, with variation according to environmental conditions.

It is important that the condition of dry sows should be regulated so that they are neither too fat nor too thin at farrowing time. Overly fat sows may have difficulty in farrowing and give birth to weak or dead pigs. Sows that are too thin at farrowing tend to become suckled down during lactation. Thus, one way or another, limited feeding is a must for gestating gilts and sows.

■ **Limit feeding**—The success of limit-fed gilts and sows depends upon controlling the intake of each female. Care must be taken to see that each female gets her share. This may be accomplished by any one of the following feeding systems (these feeding systems are detailed later in this chapter in the section headed "Feeding Systems"):

1. By adding sufficient bulk.
2. By interval feeding.
3. By group hand-feeding.
4. By individual feeding.

Daily nutrient allowances for females during gestation are given in Table 9-31. Suggested gestation diets are given in Table 9-32.

TABLE 9-31
RECOMMENDED DAILY NUTRIENT LEVELS DURING GESTATION[1]

Nutrient	Amount/Head/Day	
Crude protein	250	g
Lysine	11	g
Tyrptophan	2.5	g
Threonine	8	g
Minerals		
Calcium	16	g
Phosphorus	14.5	g
Salt	9	g
Copper	30	mg
Iodine	0.54	mg
Iron	300	mg
Manganese	72	mg
Selenium[2]	0.18	mg
Zinc	300	mg
Vitamins		
Vitamin A	20,000	IU
Vitamin D	3,000	IU
Vitamin E	80	IU
Vitamin K[3]	8	mg
Riboflavin	15	mg
Niacin	90	mg
d-Pantothenic acid	52	mg
Vitamin B-12	0.06	mg
Folic acid	3	mg
Biotin	0.4	mg
Choline	1,000	mg

[1]Adapted by the authors from: Goodband, R. D., *et al., Kansas Swine Nutrition Guide*, published by Kansas State University, 1994, updated 1995.

[2]Legal addition if fed at 4 lb per head per day.

[3]Methadione sodium bisulfite (MSB) or equivalent.

TABLE 9-32
SUGGESTED GESTATION DIETS[1, 2]

Ingredient	1	2	3	4
	--------------- (lb) ---------------			
Milo or corn	1,684	1,638	1,603	1,533
Soybean meal (46.5%)	235	272	308	379
Monocalcium phosphate	37	46	45	44
Limestone	20	20	20	20
Salt	10	10	10	10
Vitamin premix[3]	5	5	5	5
Trace mineral premix[4]	3	3	3	3
Sow add pack[5]	5	5	5	5
Selenium premix	1	1	1	1
	2,000	2,000	2,000	2,000
Calculated analysis, %				
Lysine	0.55	0.60	0.65	0.75
Ca	0.80	0.90	0.90	0.90
P	0.70	0.80	0.80	0.80

[1]Adapted by the authors from: Goodband, R. D., *et al., Kansas Swine Nutrition Guide*, published by Kansas State University, 1994, updated 1995.

[2]Suggested feeding rates during gestation: Diet 1, > 5 lb/day; Diets 2 and 3, 4 to 5 lb/day; Diet 4, gilt gestation, 4 lb/day.

[3]Suggested vitamin premix Table 9-15.

[4]Suggested trace mineral premix Table 9-16.

[5]Suggested sow add pack Table 9-17.

Note: Laxative feeds should not be fed to sows beginning 4 to 5 days before farrowing (formerly a common practice) unless there is a constipation problem.

FEEDING LACTATING SOWS

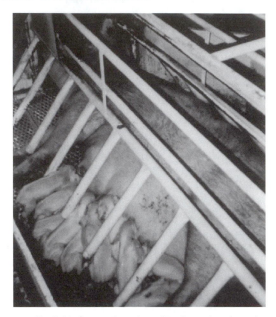

Fig. 9-14. Sow and newborn litter in modern farrowing facility. At peak production, a good sow will produce 25 lb of milk per day. So, lactating sows need to be fed liberally. (Courtesy, Kansas State University, Manhattan)

Many sows perform best when they are allowed to consume all the feed they can beginning the day they farrow. However, some people find it easier to detect lactational problems in sows if they are limit-fed during the first three days after farrowing. If limited feeding is practiced after farrowing, provide at least 3 lb of feed the day after farrowing and increase the offering 3 lb/day thereafter. By day four postfarrowing, the sow should be given *ad libitum* (free choice) access to feed.

The nutritive requirements of a lactating sow are more rigorous than those during gestation. They are very similar to those of a milk cow, except they are more exacting relative to quality proteins and the B vitamins because of the absence of rumen synthesis in the pig.

At peak production, a good sow will produce 25 lb of milk per day. A sow's milk is also richer than cow's milk in all nutrients, especially in fat. Thus, sows suckling litters need a liberal allowance of concentrates rich in essential amino acids, minerals, and vitamins.

It is essential that suckling pigs receive a generous supply of milk, for at no other stage in life will they make such economical gains. The gains made by pigs from birth to weaning are largely determined by the milk production of the sows; and this in turn is dependent upon the diet fed and the sow's inherent ability to produce milk. The lactating sow should be provided with a liberal feed allowance—ranging from 2.5 to 4.5 lb daily for each 100 lb weight. Generous feeding during lactation, with a small shrinkage in weight, is more economical than a stingy allowance of feed, for the nutrients in milk must come either from the feed or from the sow's body. Lactating sows are commonly self-fed, because even when hand-fed they are practically on full feed.

Daily nutrient allowances of females during lactation are given in Table 9-33. Suggested lactation diets are given in Table 9-34.

TABLE 9-33
RECOMMENDED NUTRIENT ALLOWANCES FOR
LACTATING GILTS AND SOWS[1,2]

Nutrient	Amount/Head/Day	
Crude protein	899	g
Lysine	44	g
Tyrptophan	11	g
Threonine	32	g
Minerals		
Calcium	48	g
Phosphorus	43	g
Salt	27	g
Copper	90	mg
Iodine	1.6	mg
Iron	900	mg
Manganese	216	mg
Selenium	0.54	mg
Zinc	900	mg
Vitamins		
Vitamin A	60,000	IU
Vitamin D	9,000	IU
Vitamin E	240	IU
Vitamin K[3]	24	mg
Riboflavin	45	mg
Niacin	270	mg
d-Pantothenic acid	156	mg
Vitamin B-12	0.18	mg
Folic acid	9	mg
Biotin	1.2	mg
Choline	3,000	mg

[1]Adapted by the authors from: Goodband, R. D., *et al., Kansas Swine Nutrition Guide,* published by Kansas State University, 1994, updated 1995.

[2]Assumes at least 12 lb/d feed intake of a diet containing 0.80% lysine.

[3]Menadione sodium bisulfite (MSB) or equivalent.

TABLE 9-34
SUGGESTED LACTATION DIETS[1, 2]

Ingredient	1	2	3	4
	---------- (lb) ----------			
Milo or corn	1,568	1,497	1,429	1,359
Soybean meal (46.5%)	343	415	486	557
Monocalcium phosphate	45	44	42	41
Limestone	20	20	19	19
Salt	10	10	10	10
Vitamin premix[3]	5	5	5	5
Trace mineral premix[4]	3	3	3	3
Sow add pack[5]	5	5	5	5
Selenium premix	1	1	1	1
	2,000	2,000	2,000	2,000
Calculated analysis, %				
Lysine	0.70	0.80	0.90	1.00
Ca	0.90	0.90	0.90	0.90
P	0.80	0.80	0.80	0.80

[1]Adapted by the authors from: Goodband, R. D., *et al., Kansas Swine Nutrition Guide,* published by Kansas State University, 1994, updated 1995.

[2]All diets fed *ad libitum.*

[3]Suggested vitamin premix Table 9-15.

[4]Suggested trace mineral premix Table 9-16.

[5]Suggested sow add pack Table 9-17.

FEEDING SHOW AND SALE SWINE

Any of the diets listed in the preceding sections for the respective classes and ages of swine are suitable for use in fitting show animals of similar classification. Because of the high cost of labor, the recent trend has been toward self-feeding both young breeding animals and market barrows and gilts that are being fitted for show. Many of them are left on self-feeders right up to show time, others are hand-fed during the last month or two of the fitting period. However, most experienced exhibitors feel that they can get superior bloom and condition by either (1) hand-feeding, or (2) using a combination of hand-feeding and self-feeding (hand-feeding twice daily and allowing free access to a self-feeder). When hand-feeding, they also prefer mixing and feeding the entire diet in the form of a slop.

Also, most experienced caretakers prefer using high protein content diets for fitting purposes. They feel they get more bloom that way.

In fitting show barrows, it may be necessary to decrease or discontinue slop feeding 2 to 4 weeks before the show to avoid paunchiness and lowering of the dressing percentage.

When oatmeal (oat groats, rolled hulled oats) is not too high priced, many successful hog fitters replace up to 50% of the grain (corn, wheat, barley, oats, and/or sorghum) in the diet with oatmeal. They do this especially when fitting hogs—both breeding animals and barrows—in the younger age groups. Oatmeal is highly palatable, lighter, and less fattening than corn.

Additional details regarding fitting and showing swine are given in Chapter 5, Selecting, Fitting, and Showing Swine.

FEEDING SYSTEMS

The choice of the feeding system(s) and the choice of the diet(s) must go hand in hand. For example, if the grain and the protein supplements are to be self-fed in separate feeders or compartments, it is important that they be of equal palatability; otherwise, pigs will consume too much of one and too little of the other. A listing and discussion of each of the common feeding systems follows.

■ **Complete mixed rations**—Most large swine producers favor the use of complete mixed feeds because they (1) require less labor and management than free-choice feeding, (2) lend themselves better to automation, and (3) provide better control of nutrient intake.

Where producers do their own mixing, they favor simplified diets. Fortunately, a simple diet of corn and soybean meal, fortified with minerals and vitamins, will generally give as good results as a more complex ration consisting of many ingredients. Of course, with large volume buying and computerized formulations, large swine operations and commercial feed companies can use more complex diets (with many ingredients) advantageously, especially from the standpoints of enhancing palatability and balancing amino acids, minerals, and vitamins.

Fig. 9-15. A well-fitted Yorkshire boar in the show-ring. (Courtesy, American Yorkshire Club, Inc., West Lafayette, IN)

■ **On-floor feeding**—On-floor feeding is particularly suited to the controlled feeding of growing-finishing swine or the breeding herd. Feeding in the sleeping area encourages cleanliness, since pigs are less inclined to defecate where they eat. Feed wastage is reduced to a minimum when the animals do not have more feed available than they will consume at one eating. Even though automated, restricted feeding requires close attention, because the daily feed intake of pigs is affected by weather.

■ **Free-choice feeding**—In free-choice feeding, the different components of the diet, such as (1) a grain and (2) a protein supplement fortified with minerals and vitamins, are fed in separate compartments of a self-feeder. Pigs are allowed to eat as much of each component as they desire.

Generally, pigs fed free-choice diets in separate feeders or compartments will not make as uniform or as fast gains as pigs fed a complete mixed diet. The free-choice system requires more supervision, as the palatability of the grain or the protein supplement may vary and the pigs will then overeat or undereat the supplement or the grain. There is very little, if any, difference in economy of gain between feeding a free-choice or a complete ground mixed diet.

■ **Liquid feeding**—Liquid feeding usually involves mixing predetermined amounts of feed and water prior to, or at the time of, feeding. When properly used, this method can practically eliminate feed dust in the feeding area and minimize wastage. Ratios of feed and water can be varied to produce a free-flowing liquid or a thick paste. In some cases, feed is automatically dropped into the water in the feed trough. Research has shown no difference in rate of gain or feed per pound of gain in pigs full fed on liquid or dry feeds. Neither does liquid feeding have any effect on dressing percentage, carcass measurements, or carcass quality.

■ **Limit feeding**—With gestating sows, limit feeding to 4 to 6 lb per head daily is a must in order to keep them from getting too fat. Overly fat sows have difficulty in farrowing and give birth to weak or dead pigs. With growing-finishing pigs, it is a way in which to increase slightly the proportion of lean to fat in the carcass. A discussion of limit feeding of (1) gilts and sows, and (2) growing-finishing pigs follows.

1. **Gilts and sows.** Replacement gilts should be started on a limited feeding program at 180 to 200 lb; and all gestating sows and gilts should be limit fed. Limit feeding may be accomplished by any one of the following methods:

a. **By feeding bulky, fibrous feed,** like silage, haylage, or alfalfa, with such feed constituting at least one-third of the diet. Actually, this is a way in which to lower the energy content of the

Fig. 9-16. Tie stalls for sows on the farm of Stephen Viggers, Washington County, Iowa. In this system, each sow is fitted with a collar and individually fed like a stanchioned dairy cow. In such individual confinement, a sow eats just what you want her to have—no more. (*Farm Journal* photo—Dick Seim)

diet. Although bulky feeds will hold the weight down, they usually do not lower feed cost.

b. **By interval feeding,** in which gilts and sows are turned to self-feeders for 2 to 8 hours every second or third day. Under this system, gilts will usually eat around 12 lb of feed at a time (or an average of 4 lb per day) and older sows will consume around 15 lb (or an average of 5 lb per day). The amount of feed consumed in interval feeding may be controlled either (1) by varying the interval, from every other day to twice a week, (2) by varying the length of time that the gilts and

Fig. 9-17. Swine feeding is modern and automatic. Feed is stored in outside bins and then augered into a controlled environment. (Courtesy, *National Hog Farmer*, St. Paul, MN)

Fig. 9-18. Feed augered in from outside storage bins fills self-feeders in an environmentally controlled building. Complete self-fed diets for growing and finishing pigs complement automation. (Photo by J. C. Allen and Son, West Lafayette, IN)

sows are left on the self-feeders (from 2 to 8 hours), or (3) by hand-feeding.

Experiments and experiences show that reproductive performance is the same with either interval feeding or daily hand-feeding a limited amount. Turning sows to self-feeders at intervals requires less labor than daily hand-feeding, but under some conditions it results in greater stress on fences and equipment.

c. **By group hand-feeding** a limited diet to several sows. This is apt to result in the "bossy" sows getting too much and the "timid" sows getting too little. This problem can be partially alleviated by feeding over a large area.

d. **By individual feeding** in either individual pens or in tie stalls, tethered by a neck collar or belt (see Fig. 9-16).

2. **Growing-finishing pigs.** Sometimes growing-finishing pigs are limit fed in order to produce leaner carcasses. Usually, it is started when pigs weigh around 100 lb and feed is limited to about 85 to 95% of what pigs of comparable age consume when self-fed. Limit feeding of market hogs results in slower gains, increased labor, and more mechanization. Thus, unless sufficient premium is paid for the modestly leaner carcasses, it cannot be justified.

■ **Pelleted diets**—The use of a complete pelleted diet for growing-finishing hogs will increase the average daily gain by 2 to 5% and improve the feed efficiency by approximately 5 to 10%. Thus, when a complete diet is purchased, buying a pelleted feed may be more economical than buying a meal. But the advantage of pelleting will usually not be sufficient to offset the cost of hauling grain to the mill and having a pelleted diet made. Also, pellet machines are costly; hence, the purchase of such equipment cannot be justified with the volume of hogs handled by most swine producers.

OTHER FEED AND MANAGEMENT ASPECTS

In addition to the subject matter covered in Chapter 8 and thus far in this chapter, there are other feed and management aspects of great importance in swine production. Some of these are detailed in the sections that follow.

NUTRITIONAL ANEMIA

Anemia is a blood condition in which there is a deficiency of hemoglobin which transports oxygen to various parts of the body. It is caused by a deficiency of iron and copper in the diet, and it is most likely to occur in nursing pigs that do not have access to soil.

The four basic reasons for anemia in nursing pigs raised in confinement are:

1. **Low body storage of iron in the newborn pig.** The accompanying table shows that the pig has a comparatively low tissue level of iron at birth, but a high level of iron at maturity. Thus, the iron requirement of the pig is greater than for other species.

| | Iron Concentration | |
Species	Newborn	Adult
	(ppm)	
Pig	29	90
Human	94	74
Rabbit	135	60
Rat	59	60

2. **Low iron content of sow's colostrum and milk.** Only approximately 10% of the pig's iron requirements are met by colostrum, and this drops to about 5% about the third day.

3. **Elimination of contact with soil iron.** Pigs farrowed and raised in confinement have no contact with the soil.

4. **Tremendous growth rate of the nursing pig.** During the first weeks of life, the pig makes much greater increase of birth weight than other species. As a result, anemia develops more rapidly in pigs than in other species.

Anemic pigs show listlessness, rough hair coat, wrinkled skins, drooping ears and tails, pale membranes around the mouth and eyes, and labored breathing.

The following nutritional anemia preventive measures may be used:

1. Give each baby pig an intramuscular injection (in the neck) of 200 mg of iron dextran at 1 to 3 days

of age. If less than 200 mg of iron is given in the first injection, a second iron shot may be needed.

2. Give the pigs iron tablets or paste at 2 to 3 days of age. Repeat the treatment every 7 to 10 days until the pigs are eating creep diet adequately. If pills are given, it is important to see that the pigs swallow them and not spit them out.

SCOURING

One of the major problems facing the swine producer is scouring in baby pigs. It is estimated that about 20% of pig losses between farrowing and weaning is caused by scouring. Scouring in pigs may be due to feeding practices, management practices, environment, or disease. More information regarding scouring is given in Chapter 15, Swine Health, Disease Prevention, and Parasite Control.

WEANING PIGS

Today, because quality prestarters and starter diets are available, and because of improved feeding and management practices, early weaning at 7 to 14 days of age is feasible. However, weaning at 3 to 4 weeks of age is generally considered more practical for most swine producers than earlier.

The earlier pigs are weaned, the better the required feeding and management practices. Advantages of early weaning (3 to 4 weeks of age) are:

1. Heavier pigs at 8 weeks of age, with fewer runts.
2. Lower sow feed costs.
3. More litters per year.
4. Less weight loss of the sow.
5. Greater flexibility in rebreeding or selling sows.
6. Greater turnover of sows through the farrowing unit; hence, less total farrowing area space required and lower facility cost per sow.

Further advantages, disadvantages, and recommendations of early weaning are given in Chapter 14, Swine Management.

FEED REQUIRED TO PRODUCE A POUND OF MARKET PIG

Nationally, it has been estimated that it requires 4.0 lb of feed to produce 1 lb of market hog, exclusive of the feed required by sows and boars to produce pigs. This appears high—and it is high. But remember that 10 to 20% of all pigs farrowed die before weaning. Remember, too, that many swine producers are inef-

ficient, just as many operators in all walks of life are inefficient.

Table 9-35 figures are realistic *goals for well-managed swine operations.*

TABLE 9-35
ESTIMATED FEED REQUIRED TO PRODUCE 240-LB MARKET PIG[1]

Stage of Production	Feed Required per 240-Lb Market Pig
	(lb)
Sow gestation diet (includes pregestation and breeding)	110
Boar diet	8
Lactation diet	45
Starter diet (creep to 40 lb)	54
Grower-finisher diet (40 to 240 lb)	700
Total, lb	917
Per 100 lb of pork produced $\frac{917}{240} \times 100 = 382$ lb	

[1]See Appendix for conversion of U.S. customary to metric.

The values given in Table 9-35 are estimates based on realistic standards for apportioning the quantities of sow and boar feed to each pig and the feed conversion normally attained during the starter and grower-finisher periods. Although these data do not provide for pig deaths after weaning, normal milling losses, and feed wastage, it is assumed that these losses would likely be offset by sow weight gains which are not considered in the pounds of hog produced. Data obtained in commercial herds where accurate records have been kept indicate that 382 lb of feed per 200 lb of live hog produced is a realistic goal for a practical swine operation. However, to achieve this level of efficiency, a sound feeding and management program must be followed, including limit feeding of pregnant sows, high conception rates, large litters weaned, early weaning and rebreeding, low death losses, minimal disease problems, balanced diets, and minimal feed wastage.

EFFECT OF SEX ON PERFORMANCE OF GROWING-FINISHING PIGS

When full fed, boars consume 10 to 15% less feed daily than barrows or gilts and are 10 to 15% more efficient in feed conversion. Also, boars gain faster than barrows or gilts. Barrows gain approximately 0.10 lb faster per day than gilts, which reduces their age at slaughter by 10 days. Feed per pound of gain is similar for barrows and gilts. Gilts yield carcasses having 0.11 in. less backfat, 0.52 square in. larger loin eye area, and 1.8% more lean cuts than barrows. Dressing percentage usually favors barrows, which is consistent with their greater depth of backfat.

SOFT PORK

Feed fats are laid down in the body without undergoing much change. Thus, when finishing hogs are liberally fed on high-fat content feeds in which the fat is liquid at ordinary temperatures, soft pork results. This condition prevails when hogs are liberally fed such feeds as soybeans, peanuts, mast, or garbage. The fat of the cereal grains is also liquid at ordinary temperatures, but fortunately the fat content in these feeds is relatively low. When such feeds are liberally fed to swine, most of the pork fat is actually formed from the more abundant carbohydrates in these feeds.

Soft pork is undesirable from the standpoint of both the processor and the consumer. It remains flabby and oily even under refrigeration. In soft pork, there is a higher shrinkage in processing; the cuts do not stand up and are unattractive in the showcase; it is difficult to slice the bacon; and the cooking losses are higher through loss of fat. For these reasons, hogs that are liberally fed on those feeds known to produce soft pork are heavily discounted on the market.

The firmness of pork carcasses may be judged by (1) grasping the flank below the ham, (2) lifting one end of the cut while permitting the other end to rest on the table (a firm pork cut will not bend readily), or (3) applying a slight pressure of the thumb (not gouging) on a cut surface. Experimentally, either the iodine number or the refractive index is used in determining the degree of softness; this is a measure of the degree of unsaturation.

Unless the producer is willing to take the normal reduction in price (about $1 per cwt), it is recommended that feeds which normally produce soft pork be fed liberally only to pigs under 85 lb in weight and to the breeding herd. For growing-finishing pigs over 85 lb in weight, soybeans and peanuts should not constitute more than 10% of the diet if a serious soft

pork problem is to be averted.

Experimental evidence and practical observation have shown, however, that when a diet producing hard fat is given following a period of feeds rich in unsaturated fats, the body fat gradually becomes harder. It has also been found that this process takes place more rapidly if the animals are first fasted for a period before the change in diet is made. This practice is called *hardening off*. Thus, many hogs that are, for practical reasons, finished primarily on such feeds as soybeans, peanuts, or garbage, are hardened off with a diet of corn or some other suitable grain.

HOG-CORN RATIO

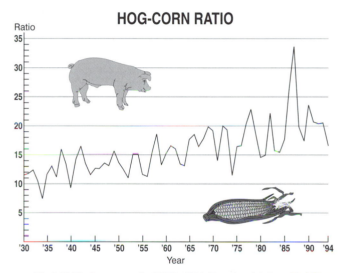

Fig. 9-20. The hog-corn ratio, 1930 to 1994. (Based on data from *Pork Facts*, 1995–1996, National Pork Producers Council, p. 13)

The hog-corn ratio is determined simply by dividing the price received for live hogs per hundredweight by the current price for corn per bushel at that time. For example if:

$$\left.\begin{array}{l}\text{Price of Hogs @ \$45/cwt}\\\text{Price of Corn} = \$2.25/\text{bu}\end{array}\right\} \text{ hog-corn ratio} = 20$$

Corn is the primary ingredient of the pig's diet in the majority of feeding areas in the U.S. and feed constitutes 60 to 65% of the total costs of hog production.

Therefore, when corn is cheap relative to hog prices, many farmers opt to add value to it by feeding it to hogs rather than selling it outright. When the corn price is high, relative to hog prices, farmers may choose to sell the corn. Therefore, above a certain ratio, farmers will more often expand production of hogs, whereas below that level, they will usually cut

Fig. 9-19. Soft pork. Feed fats do affect body fats. The bacon belly on the left came from a hog liberally fed on soybeans. (Courtesy, University of Illinois)

back production. In this way, the ratio serves as a general indicator of future trends.

The hog-corn ratio is less important as a guide for mega swine operations than for family farms. The reason: A pork producer who has a large fixed investment in facilities for only one use—producing hogs—is not going to use the size of the nation's corn crop as a major factor in expansion decisions.

POISONS AND TOXINS

Swine are susceptible to a number of poisons, any one of which may be disastrous in a herd. Among them are the following: moldy feed, including three species of mycotoxins—aflatoxins, ergot poisoning, and estrogenic syndrome—pitch poisoning, lead poisoning, mercury poisoning, pesticides, plant poisoning, involving

TABLE 9-36

Poison	Source	Symptoms and Signs
Arsenic (As)	Arsenic used to control insects and weeds, and to defoliate crops. Overdosing of phylarsonic compounds as feed additives to swine for growth and disease control.	The onset is sudden; characterized by groaning, restlessness, rapid breathing, muscular incoordination, blindness, and photosensitization. Death in 3–4 hours, or, if less material is consumed, in a few weeks. Necropsy reveals severe hemorrhagic inflammation of stomach and intestines, with perhaps areas of erosion on mucous membranes.
Copper sulfate ($CuSO_4 \cdot 5H_2O$)	Overdosing with copper sulfate (bluestone), or the use of too concentrated solution, in treating for parasites.	Reduced growth, lower hemoglobin, and death. Abdominal pains, vomiting, and diarrhea. Necropsy shows stomach and intestines inflamed and stomach lining coated with copper sulfate.
Ergot (a parasitic fungus)	It replaces the seed in the heads of grasses and cereal grains, in which it appears as a purplish-black, hard banana-shaped dense mass from ¼–¾ in. *(5.6 to 16.9 mm)* long. Most common in rye, wild rye, bromegrass, and dallisgrass.	Acute ergot poisoning, caused by large quantities eaten at one time, may produce paralysis of the limbs and tongue, disturbance of the gastrointestinal tract, and abortion. It is a cumulative poison; hence, poisoning may develop from lesser quantities eaten over a long period of time. Chronic poisoning produces gangrene of the extremities, with subsequent sloughing off of hooves, ears, and tail. Delirium, spasms, and paralysis may occur before death.
Fluorine (F)	Ingesting excessive quantities of fluorine through either the feed or water.	Abnormal teeth (especially mottled enamel) and bones, stiffness of joints, loss of appetite, emaciation, reduction in milk flow, diarrhea, and salt hunger.
Lead (Pb)	Lead is discharged into the air from auto exhaust fumes and other sources. Lead pollution of feed and food crops as a result of lead being deposited on the leaves and other edible portions of the plant by direct fallout. Inhaling airborne lead. Lead may get into feed or food and water from contact with lead pipes, utensils, or discharged storage batteries.	Symptoms develop rapidly in young animals, but slowly in mature animals. Loss of appetite and evidence of gastroenteritis. Feces may become very dark gray and be tinged with blood. Salivation, champing of the jaws, frenzy, blindness, convulsions, coma, and death. Mature animals usually have diarrhea, show incoordination, especially in the hind limbs, and prostration.
Mercury (Hg)	Mercury is discharged into air and water from industrial operations and is used in herbicide and fungicide treatments. Consumption of seed grains treated with fungicides that contain mercury, for the control of fungus diseases of oats, wheat, barley and flax. Mercury poisoning has occurred where mercury from industrial plants has been discharged	Gastrointestinal, renal, and nervous disturbances; but impossible, on basis of symptoms, to differentiate mercury from other poisons. Case history of animals consuming mercury-treated grains should be considered strong circumstantial evidence.

several plants that are toxic to swine, and blue-green algae. The source, symptoms and signs, distribution and losses, prevention, and treatment of some potential poisons are covered in Table 9-36 which follows.

Fig 9-21. Selenium poisoning (alkali disease). An "alkalied" pig. Notice the general rundown condition, thinness of hair, and diseased feet. (Courtesy, South Dakota Agricultural Experiment Station)

SOME POTENTIAL POISONS

Distribution and Losses Caused By	Prevention	Treatment	Remarks
Arsenic has long been a leading cause of chemical poisoning. Much of the vomitoxin problem is associated with head scab in small grains.	Keep animals away from arsenic.	Handled by the veterinarian. If caught in time, first remove the material from the animal. Sodium thiosulfate may be used, and supportive treatment may be indicated. British Anti-Lewisite (Dimercaprol) is a specific antidote for some forms of arsenic poisoning.	Accumulation of arsenic in soils may sharply decrease crop growth and yields, but it is not a hazard to animals or humans that eat plants grown in these fields, provided they do not eat the foliage of sprayed plants.
	Never give copper sulfate in a concentration greater than 2% or in a dose of more than 4 fl oz of a 1% solution.		Dietary Cu toxicosis is closely related to molybdenum concentration. Cu:Mo ratio of 10:1 may be toxic.
Ergot is found throughout the world.	Never feed heavily ergot-infested hay or grain.	If noticed in time, stricken animals may recover if put on good feed. Tannin used as a drench is an antidote, and sedations, such as chloral hydrate, may be given to nervous animals.	Six different alkaloids are involved in ergot poisoning.
The water in parts of AK, CA, SC, and TX has been reported to contain excess fluorine. Occasionally, throughout the U.S. high-fluorine phosphates are used in mineral mixtures.	Avoid the use of feeds, water, or mineral supplements containing excessive fluorine.	Any damage may be permanent, but animals which have not developed severe symptoms may be helped to some extent, if the source of excess fluorine is eliminated.	Fluorine is a cumulative poison.
At one time, it was a component of sprays used to control insects and plant diseases and of paints. But leaded paint is now banned by law.	Avoid sources of lead.	If damage to tissue has been extensive, treatment is of little value; in any event it should be handled by a veterinarian. Protein (milk, eggs, blood serum) may reduce G.I. absorption. Calcium disodium EDTA.	It is a cumulative poison. When incorporated in the soil, nearly all the lead is converted into forms that are not available to plants. Any lead taken up by the plant roots tends to stay in the roots, rather than move up to the top of the plant. Lead poisoning can be diagnosed positively by analyzing the blood tissue for lead content.
When, through ignorance or negligence, mercury-treated grain is fed to animals.	Do not feed livestock seed grains treated with a mercury-containing fungicide. Surplus of treated grain should be burned and the ash buried deep in the ground.	Treatment is not too satisfactory. Protein (milk, eggs, blood serum) may reduce G.I. absorption. Sodium thiosulfate and BAL may be used.	Ultimate diagnosis depends upon demonstrating the presence of mercury in the tissues, especially in the kidneys and liver. Food and Drug Administration prohibits use of mercury-treated grain for feed or food. Mercury is a cumulative poison. Grain crops produced from mercury-treated seed and crops produced on soils treated with

(Continued)

TABLE 9-36

Poison	Source	Symptoms and Signs
Mercury (Hg) (Continued)	into water and then accumulated in fish and shellfish.	
Mycotoxins (toxin-producing molds; *e.g., Aspergillus flavus, Penicillium cyclopium, P. islandicum,* and *P. palitans*)	Aflatoxin (most studied of the group) associated with peanuts, Brazil nuts, silage, corn and most other cereals, hay, and grasses. The mold can produce toxic compounds on virtually any food (even synthetic) that will support growth. Another mycotoxin which is very troublesome in certain areas is vomitoxin, which is produced by *fusarium* molds.	Molds affect animals in a variety of ways, from decreased production to sudden death. Usually, the first sign is loss of appetite and weight. A few animals will abort, and an occasional animal will die. With high intakes of mycotoxin, or with the several types of molds, any one or a combination of the following symptoms may develop; liver damage, hyperkeratosis, a typical interstitial pneumonia, bloody slimy scours, arched back, dry gangrene at end or tail or top of hoof, hemorrhagic hepatitis, renal damage, lameness, and/or swollen legs. Vomitoxin supresses the immune system.
Nitrate-nitrite poisoning	Consuming high-nitrate feeds—feeds with a high-nitrate content due to nitrate fertilization, drought, etc. Eating nitrate or nitrite fertilizer, or drinking pond water containing same.	Accelerated respiration and pulse rate; diarrhea; frequent urination; loss of appetite; general weakness, trembling, and a staggering gait; frothing from the mouth; abortion; blue color of mucous membrane, muzzle and udder due to lack of oxygen; dark brown blood; and death in 90–150 minutes after eating lethal doses of nitrite.
Pitch (clay pigeon poisoning; coal tar poisoning)	Expended clay pigeons; roofing material, certain types of tar paper, and plumbers' pitch.	An acute, highly fatal disease, characterized, clinically, by depression, and pathologically, by striking liver lesions. Anemia and jaundice.
Plants	Pigweed *(Amaranthus retroflexus)*, cocklebur *(Xanthium strumarium)*, and nightshade *(Solanum nigrum)*.	**Pigweed:** Weakness, trembling and incoordination; paralysis of rear legs; coma and death within 48 hours of onset of signs. **Cocklebur:** Within 8–24 hours after ingestion, depression, nausea, weakness, spasms, labored breathing, and death. **Nightshade:** Anorexia, constipation, depression, incoordination, trembling spasms, coma, and death.
Protein poisoning	Heavy feeding of high protein feeds.	None, providing animal does not have a kidney ailment.
Salt poisoning (NaCl—sodium chloride)	Brine from cured meats; wet salt. Where large amounts of brine or salt have been mixed in hog slop. When excess salt is fed following salt starvation. When salt is improperly used to govern self-feeding of concentrate.	Sudden onset—1–2 hours after ingesting salt; extreme nervousness; muscle twitching and fine tremors; much weaving, wobbling, staggering, and circling; blindness; weakness; normal temperature, rapid but weak pulse, and very rapid and shallow breathing; diarrhea; convulsions; death from within a few hours up to 48 hours.
Selenium poisoning	Consumption of plants grown on soils containing selenium.	A general loss of hair in swine. In severe cases, the hooves slough off, lameness occurs, food consumption decreases, and death may occur by starvation.

(Continued)

Distribution and Losses Caused By	Prevention	Treatment	Remarks
			mercury herbicides have not been found to contain harmful concentrations of the element.
Widely distributed throughout the world. In addition to the effect of mycotoxins on the animal's health, milk is contaminated by the residues or mycotoxins, or their metabolic products.	The prime cause of aflatoxin is moisture; hence, proper harvesting, drying, and storage are important factors in lessening contamination and toxin production. Propionic and acetic acids, and sodium propionate, will inhibit mold growth; hence, their use in preserving high-moisture grains is encouraged.	Remove the source of the mold. Animals suffering from molds frequently respond to vitamin B injections. Iron therapy may be helpful, since hemorrhaging is a frequent problem.	Certain molds produce toxins, or mycotoxins. Aflatoxin has been clearly shown to be a carcinogen (tumor producing). Ultraviolet irradiation and anhydrous ammonia under pressure will reduce the toxicity of aflatoxins and, if continued long enough, will deactivate them entirely.
Excessive nitrate content of feeds is an increasingly important cause of poisoning in farm animals, due primarily to more and more high nitrogen fertilization. Nitrates and nitrites are water-soluble and may be leached from the soil into groundwater.	Check potential sources such as water and feed. When in doubt, have the feed analyzed. Properly store nitrogen fertilizer.	A 4% solution of methylene blue (in a 5% glucose or a 1.8% sodium sulfate solution) administered by a veterinarian intravenously at the rate of 10 mg/kg liveweight.	Nitrate does not appear to cause the actual toxicity, and may only cause gastrointestinal irritation in pigs. Nitrite is far more toxic. Nitrite converts hemoglobin to methemoglobin, which cannot transport oxygen.
Wherever there is pitch.	Do not allow hogs access to pitch-containing or coal tar-containing products.	No known treatment.	Pastures containing clay pigeons are dangerous for years; deaths having been reported 35 years after area was used for trap-shooting.
Primarily, pigs on pastures.	Keep weeds down. Be sure pasture contains adequate good forage.	Remove animals from the source. Oral administration of mineral oil and possibly injection of physostigmine for cocklebur poisoning.	Producers should recognize these plants at all stages of their growth.
None.			Despite occasional diagnosis of protein poisoning, there is no proof that heavy feeding of high protein feeds is harmful. Generally, feeds of high protein content are more costly and there is a temptation to feed too little of them.
Salt poisoning is relatively rare except in pigs.	If animals have not had salt for a long time, they should first be hand-fed salt, gradually increasing daily allowance until they leave a little in the mineral box; then self-feed.	Provide large quantities of fresh water to affected animals. Those unable to drink should be given water via stomach tube, by the veterinarian.	Indians and pioneers handed down many legendary stories about huge numbers of wild animals that killed themselves simply by gorging at a newly found salt lick after having been salt-starved for long periods of time. Salt poisoning in pigs does not occur unless there is water deprivation. Even normal salt concentration may be toxic if water intake is low.
In certain regions of western U.S.—especially certain areas in SD, MT, WY, NE, KS, and perhaps areas in other states in the Great Plains and Rocky Mountains. Also, in Canada.	Abandon areas where soils contain selenium, because crops produced on such soils constitute a menace to both farm animals and humans.	Although arsenic has been shown to counteract the effects of selenium toxicity, there appears to be no practical method of treating other than removal of animals from affected areas.	Chronic cases occur when animals consume feeds containing 8.5 ppm of selenium over an extended period; acute cases occur on 25–125 ppm. The toxic level of selenium is in the range of 2.27–4.54 mg/lb of feed.

(Continued)

TABLE 9-36

Poison	Source	Symptoms and Signs
Zearalenone *(estrogenic syndrome)*	An estrogenic mycotoxin produced by *Fusarium graminearum* in corn.	Swelling and edema of the vagina and possibly mammary development in prepubertal gilts. Preputial enlargement in boars. Abortions or reduction in live pigs born.

QUESTIONS FOR STUDY AND DISCUSSION

1. List the primary factors which affect swine feeding standards, diet formulation, and feeding programs.

2. Why aren't nutritional requirements, or standards, like NRC Tables 9-1 to 9-7, absolute, final, and unchangeable?

3. Why should there be a swine diet for every need?

4. Explain the difference between nutrient requirements and nutrient allowances.

5. Of what value are feed substitution tables? What factors are usually considered in arriving at the comparative values of feeds listed in a feed substitution table?

6. Why are antibiotics used as feed additives? How do antibiotics and probiotics differ in their respective functions?

7. List some factors (pointers) which must be considered before attempting to balance a diet by any method.

8. List the four methods of balancing diets, and give the advantages and disadvantages of each.

9. What precautions should be taken when using diets formulated by the computer?

10. Select a specific class of swine and formulate a balanced diet using those feeds that are available at the lowest cost. List the diet in terms of both as-fed and moisture-free bases.

11. Will the "least-cost diet" always make for the greatest net returns?

12. Under what circumstances would you (a) home mix a hog feed, or (b) buy a commercial hog feed?

13. What factors should you consider or look for in buying a commercial hog feed?

14. List and discuss the forms of processing feeds for swine.

15. List and discuss the four basic systems of preparing diets for swine.

16. Why is quality control essential in manufacturing swine feeds?

17. What is segregated early weaning (SEW)? When and how should it be used?

18. What is phase feeding? When and how should it be used?

19. Define and discuss (a) creep feeding, (b) growing pigs, and (c) split-sex feeding.

20. Discuss the lysine requirements of (a) lean, rapidly growing pigs, (b) winter vs summer, and (c) split-sex groups.

21. What are finishing pigs? Why should decisions to change or modify finishing diets be based on economics?

22. At what stage and age should replacement gilts be sorted out from finishing pigs? How should they be fed thereafter?

23. Would you recommend flushing (a) second and third litter sows, (b) gilts? If so, how would you do it?

24. One way or another, limited feeding is necessary for gestating gilts and sows. List and detail four systems of limit feeding.

25. Would you recommend *ad libitum* (free choice) or limited feeding of sows immediately after farrowing?

26. How would you formulate a diet and feed show and sale swine?

27. List and discuss the federal feeding systems.

(Continued)

Distribution and Losses Caused By	Prevention	Treatment	Remarks
Wherever corn is fed.	Prevent favorable conditions when storing corn.	Withdrawal of offending feed.	Associated with vulvovaginitis and rectal prolapses. Wet corn with high temperatures followed by low temperatures favor zearalenone production. Suspected diets or corn should be analyzed for the presence of zearalenone.

28. Discuss the cause, symptoms, and prevention of nutritional anemia in pigs.

29. List and discuss the advantages and disadvantages of early weaning.

30. Estimate the total feed required to produce a 240-lb market pig, then break it down into the amount required for each of the following stages of production: (a) sow gestation diet (including pregestation and breeding), (b) boar diet, (c) lactation diet, (d) starter diet (creep to 40 lb), and (e) grower-finisher diet (40 to 240 lb).

31. Since boars gain faster and are more efficient in feed conversion than barrows or gilts, why don't we feed out more boars?

32. Hams derived from peanut-fed hogs are soft pork. Yet, when they are highly advertised they may be sold at a premium. So, why be concerned about soft pork?

33. Of what significance is the hog-corn ratio?

34. Name two mycotoxins and indicate some of the symptoms/signs and problems caused by each.

SELECTED REFERENCES

Title of Publication	Author(s)	Publisher
Animal Science	M. E. Ensminger	Interstate Publishers, Inc., Danville, IL, 1991
Effect of Environment on Nutrient Requirements of Domestic Animals		National Academy Press, Washington, D.C. 1981
Feed Formulations	T. W. Perry	The Interstate Printers & Publishers, Inc., Danville, IL, 1982
Feeds & Nutrition, Second Edition	M. E. Ensminger J. E. Oldfield W. W. Heinemann	The Ensminger Publishing Company, Clovis, CA, 1990
Nutrient Requirements of Swine	National Research Council	National Academy of Sciences, Washington, DC, 1988
Pork Facts	Staff	National Pork Producers Council, Des Moines, IA, 1995–1996
Stockman's Handbook, The	M. E. Ensminger	Interstate Publishers, Inc., Danville, IL, 1992
Swine Feeding and Nutrition	T. J. Cunha	Academic Press, Inc., New York, NY, 1977
Swine Nutrition	E. R. Miller D. E. Ullrey A. J. Lewis	Butterworth-Heinemann, Boston, MA, 1991
Swine Nutrition Guide	J. F. Patience P. A. Thacker	Prairie Swine Centre, University of Saskatchewan, Saskatoon, Canada, 1989

Swine Production and Nutrition	W. G. Pond J. H. Maner	AVI Publishing Co., Inc., Westport, CT, 1984
Swine Production in Temperate and Tropical Environments	W. G. Pond J. H. Maner	W. H. Freeman & Co., San Francisco, CA, 1974

In addition to the above selected references, valuable publications on feeding and feeds for swine can be obtained from:

1. Division of Publications
   Office of Information
   U.S. Department of Agriculture
   Washington, DC 20250

2. Your state agricultural college

3. Feed manufacturers and pharmaceutical houses.

Plate 1. Native Chinese breed. This breed is very prolific; litters of 20 or more are not uncommon. (Courtesy, Iowa State University, Ames, IA)

Plate 4. Modern Duroc barrow. (Courtesy, *Seedstock*, W. Lafayette, IN)

Plate 2. Modern lean Hampshire barrow. (Courtesy, *Seedstock*, W. Lafayette, IN)

Plate 5. Muscular Chester White barrow. (Courtesy, *Chester White Record*, Peoria, IL)

Plate 3. Landrace sow. (Courtesy, American Landrace Assn., W. Lafayette, IN)

Plate 6. Hereford gilts. (Courtesy, National Hereford Hog Assn., Flandreau, SD)

Plate 7. Evaluating. (Courtesy, Farmland Industries, Inc., Kansas City, MO)

Plate 10. Junior exhibitors with their World Pork Expo Reserve Champion Poland China Barrow. (Courtesy, *Poland China Record*, Peoria, IL)

Plate 8. Measuring back fat on a sow. (Courtesy, Dekalb Swine Breeders, Inc., Dekalb, IL)

Plate 11. Junior exhibitor showing Hampshire hog in World Pork Expo. (Courtesy, National Pork Producers Council, Des Moines, IA)

Plate 9. A Yorkshire boar class. (Courtesy, American Yorkshire Club, W. Lafayette, IN)

Plate 12. Large White sow. Note the row of 8 teats (nipples) on a side. (Courtesy, American Diamond Swine Genetics, Prairie City, IA)

Plate 13. Yorkshire sow and litter. (Courtesy, Land O Lakes, FT. Dodge, IA)

Plate 16. Ultrasound equipment used in pregnancy determination of a gestating sow. (Courtesy, National Pork Producers, Council, Des Moines, IA)

Plate 14. Transgenics. (Courtesy, Dekalb Swine Breeders, Inc., Dekalb, IL)

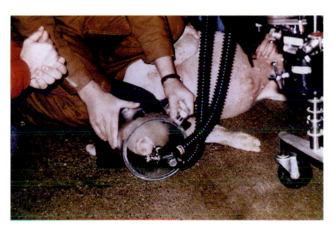

Plate 17. Halothane testing of a pig for determination of stress susceptibility. (Courtesy, National Pork Producers Council, Des Moines, IA)

Plate 15. Dispensing semen. (Courtesy, Lone Willow Farm, Roanoke, IL)

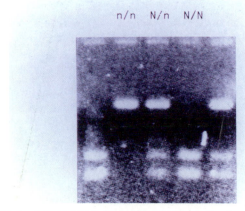

Plate 18. DNA test result showing the bands for a stress negative (nn), stress carrier (Nn), and stress susceptible (NN), pig. (Courtesy, National Pork Producers Council, Des Moines, IA)

Plate 19. Feed truck delivering bulk feed to feed bins. (Courtesy, Oklahoma State University, Stillwater, OK)

Plate 22. Individual units each with its feed bin. (Courtesy, Farmers Hybrid Co., Des Moines, IA)

Plate 20. Self-feeder and partially slotted floor for growing pigs. (Courtesy, Iowa State University, Ames, IA)

Plate 23. Weaned nursery pigs being started on feed. (Courtesy, Farmers Hybrid Co., Des Moines, IA)

Plate 21. Finishing pigs self-fed in environmentally controlled building. (Courtesy, Land O Lakes, Ft. Dodge, IA)

Plate 24. Finishing hogs fed a complete diet. (Courtesy, Compart's Boar Store, Nicollet, MN)

Plate 25. Office and lab at Lone Willow Farm. (Courtesy, Lone Willow Farm, Roanoke, IL)

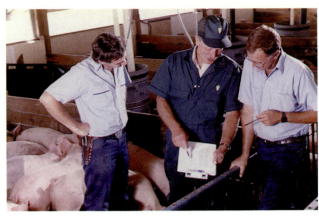

Plate 28. Regular consultation with a veterinarian is important to maintain herd health. (Courtesy, National Pork Producers Council, Des Moines, IA)

Plate 26. Computer used in keeping complete and current production and financial records on a modern hog farm. (Courtesy, National Pork Producers Council, Des Moines, IA)

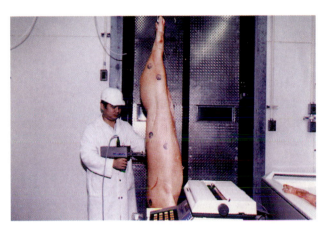

Plate 29. Successful managers produce high quality pork. This shows a carcass being probed. (Courtesy, National Pork Producers Council, Des Moines, IA)

Plate 27. Good management and good records go hand in hand. (Courtesy, Land O Lakes, Ft. Dodge, IA)

Plate 30. "Honey wagon" applying liquid manure to the land—our land. (Courtesy, Oklahoma State University, Stillwater, OK)

Plate 31. Environmentally controlled and automated unit. (Courtesy, Land O Lakes, Ft. Dodge, IA)

Plate 34. Exterior of modern nursery owned by Keith Rhoda, Prairie City, IA. It has 8 seperate rooms with 24 pens in each room. Each room has it's own flush system and bulk bin. (Courtesy, American Diamond Swine Genetics, Prairie City, IA)

Plate 32. Nebraska-style finishing barn. The curtain in front is on temperature control, and will rise and fall as set. (Courtesy, Compart's Boar Store, Nicollet, MN)

Plate 35. Interior of modern nursery owned by Keith Rhoda, Prairie City, IA. (Courtesy, American Diamond Swine Genetics, Prairie City, IA)

Plate 33. Individual farrowing houses. (Courtesy, National Hereford Hog Assn., Flandreau, SD)

Plate 36. Lagoon, a popular manure management option. (Courtesy, National Pork Producers Council, Des Moines, IA)

Plate 37. Duroc boars ready for export to South Korea. (Courtesy, Lone Willow Farm, Roanoke, IL)

Plate 40. Auction selling. (Courtesy, Farmland Industries, Inc., Kansas City, MO)

Plate 38. Evaluating a Large White boar. (Courtesy, Lieske Genetics, Henderson, MN)

Plate 41. Carcass measurements on live animal. (Courtesy, Dekalb Swine Breeders, Inc., Dekalb, IL)

Plate 39. Breeding stock representative explaining genetics program to producer. (Courtesy, Dekalb Swine Breeders, Inc., Dekalb, IL)

Plate 42. Carcasses on the rail. (Courtesy, National Pork Producers Council, Des Moines, IA)

Plate 43. Glazed ham with vegetable strata. (Courtesy, National Live Stock and Meat Board, Chicago, IL)

Plate 46. Grilled frankfurters with assorted toppings. (Courtesy, National Live Stock and Meat Board, Chicago, IL)

Plate 44. Pork chops—the other white meat. (Courtesy, Dr. Garth Boyd and Ms. Rhonda Campbell, Murphy Farms, Rose Hill, NC)

Plate 47. Mixed sausage grill. (Courtesy, National Live Stock and Meat Board, Chicago, IL)

Plate 45. Spicy sausage burritos. (Courtesy, National Live Stock and Meat Board, Chicago, IL)

Plate 48. Ham with saucy apple wedges. (Courtesy, National Live Stock and Meat Board, Chicago, IL)

Modern swine confinement building. Note outside feed bins, from which feed is augered into self-feeders in each pen. (Courtesy, DeKalb Swine Breeders, DeKalb, IL)

10

GRAINS AND OTHER HIGH ENERGY FEEDS FOR SWINE[1]

---

[1]In the preparation of this chapter, the authors adapted applicable material from the following two publications: *Swine Nutrition Guide*, published by the University of Nebraska and South Dakota State University, 1995; and *Kansas Swine Nutrition Guide*, published by Kansas State University, updated 1995.

The feed ingredient definitions used in Chapter 10 include those used in the *1996 Official Publication* of the *Association of American Feed Control Officials (AAFCO)*.

Pigs need energy for maintenance, growth, reproduction, and lactation. The bulk of the pig's energy is met by carbohydrates and fats. Fats and oils are dense sources of energy, containing about 2.25 times more calories than carbohydrates. Although cereal grains and fats/oils will meet the pig's energy needs, they must be supplemented with amino acids (protein), minerals, and vitamins.

The energy content of feedstuffs and the energy requirements of pigs are generally expressed as total digestible nutrients (TDN), digestible energy (DE), and/or metabolizable energy (ME).

The primary energy sources for swine are the cereal grains: corn, milo, wheat, barley, and their by-products. In addition, fat is often used to increase the energy density of swine diets. Most common cereal grains and fats are quite palatable and digestible.

The relative feeding value of common swine feed energy sources is shown in Table 10-1. Corn—the most extensively used swine feed in the United States—is

TABLE 10-1
RELATIVE FEEDING VALUES AND MAXIMUM USAGE RATES OF ENERGY SOURCES
(A * Denotes No Nutritional Limitations in a Balanced Diet)[1, 2]

Ingredient (As-Fed)	Feeding Value Relative to Corn[4]	Maximum Recommended Percent of Complete Diets[3]			
		Starter	Growing-Finishing	Gestation	Lactation
	(%)	(%)	(%)	(%)	(%)
Alfalfa meal, dehy	70 to 80	0	10	25	0
Alfalfa hay, early bloom	65 to 75	0	10	60	0
Bakery waste, dehy	110 to 120	*	*	*	*
Barley (48 lb/bu)	90 to 100	25	*5	*	*6
Beet pulp	80 to 90	0	10	50	10
Corn distillers' grains w/solubles, dehy	110 to 120	5	15	40	10
Corn gluten meal	95 to 105	5	10	*	10
Corn, high lysine	100 to 110	*	*	*	*
Corn, high oil	100 to 110	*	*	*	*
Corn, hominy feed	95 to 105	0	60	60	60
Corn, yellow (> 40 lb/bu)	100	*	*	*	*
Fats/oils (stabilized)	190 to 200	5	5	5	5
Millet, proso	85 to 95	40	*	*	40
Milo, grain sorghum (> 48 lb/bu)	95 to 97	*	*	*	40
Molasses (77% DM)	55 to 65	5	5	5	5
Oats (38 lb/bu)	85 to 95	15	30	*	10
Oats, high lysine	85 to 95	30	60	*	10
Oats groats	110 to 120	*	*	*	*
Rye[7]	85 to 95	0	25	20	10
Triticale[8]	95 to 105	20	40	40	40
Wheat bran	80 to 90	0	10	30	10
Wheat, hard (> 55 lb/bu)[9]	100 to 110	30	*	*	40
Wheat middlings	110 to 120	5	25	*	10

[1]Adapted by the authors from: Reese, D. E., et al., Swine Nutrition Guide, published by the University of Nebraska and South Dakota State University, 1995.

[2]Assumes diets are balanced for essential amino acids, minerals, and vitamins.

[3]Higher levels may be fed although performance may decrease. Economic considerations should influence actual inclusion rates.

[4]Corn = 100%. Values apply when ingredients are fed at no more than the maximum recommended percentage of complete diet. A range is presented to compensate for quality variation.

[5]For maximum performance, limit barley to two-thirds of the grain for 45- to 130-lb pigs. No limitation for pigs > 130 lb.

[6]Increased fiber in barley will reduce the ME/lb of feed. Thus, less should be used when feed intake is low.

[7]Ergot free.

[8]Low trypsin inhibitor varieties. Feed value tends to be highly variable.

[9]Coarsely ground.

assumed to have a feeding value of 100%; and milo (grain sorghum) is assumed to have a feeding value of 95 to 97% that of corn. Thus, milo can replace corn in the diet when the price of milo is less than 95% of the price of the same weight of corn. For example, if the price of milo is less than $0.038/lb it is a better buy than corn that costs $0.04/lb. *Note:* The feeding value of milo is slightly less than that of corn because it has less ME and available lysine.

The relative feeding values of the energy sources given in Table 10-1 apply when ingredients are included in diets in quantities no greater than those shown. When ingredients are included in diets at lower levels than indicated in Table 10-1, their feeding value may increase slightly. Average daily gain and reproductive performance will not normally be reduced by replacing corn with any of the energy sources shown in Table 10-1. A range in feeding value is presented because of the variation in ingredient quality and individual producer goals. *Note:* In addition to the relative feeding value of energy sources, the producer should consider such factors as storage costs, availability, palatability, and carcass quality.

Backfat thickness may decrease up to 0.1 in. when oats, barley, or other lower energy ingredients replace all of the corn in the diet.

Table 10-1 may be augmented with the composition of some of the more commonly used grains and other high energy feeds listed in Chapter 21 of this book.

GRAINS

Grains are seeds from cereal plants—members of the grass family, Gramineae. They contain large quantities of carbohydrates. Corn, grain sorghums, barley, and oats are the primary grains fed to swine. Rice and

Fig. 10-1. Finishing pigs self-fed on a slatted floor. Note vertical pipe which delivers feed to the self-feeder. (Courtesy, Land O Lakes, Ft. Dodge, IA)

wheat are consumed primarily by humans, but many of their milling byproducts are fed to pigs. Rye, millet, emmer, spelt, and triticale are fed to hogs in limited amounts.

Grains provide the swine producer with an excellent source of highly digestible energy, but it is important that the characteristics of each available grain be thoroughly understood, thus making it possible to formulate economical diets which take full advantage of their nutritive properties and compensate for their deficiencies. A brief discussion of each of the more important grains utilized by swine will follow.

CORN (Maize)

Corn has long been the most important U.S. cereal grain and swine feed. The annual yield, ranging from 6 to 9 billion bushels, is equal to that of any other three cereals combined. About one-fourth to one-half of this production is fed to hogs. For this reason, the feeding value of corn is usually accepted as the standard with which other cereals are compared.

Corn is an excellent energy feed for all classes of swine. It is an ideal finishing feed because it is high in digestible carbohydrate (starch), low in fiber, and very palatable.

In spite of its virtues, corn alone will not keep pigs alive. It contains 7 to 9% protein, but the protein is deficient in practically all of the essential amino acids required by the weanling pig, especially lysine and tryptophan. It is also so deficient in calcium and other minerals, and so inadequate in vitamin content, that pigs will die if they are limited to a diet containing only corn. So, corn must be supplemented with a protein that makes up its amino acid deficiencies. Equally important are the needed minerals and vitamins. When properly supplemented, corn is an excellent energy feed for all classes of swine. *Note:* Several researchers have attempted to predict the lysine content of corn based on crude protein concentration. However, the poor relationship between lysine and protein concentration makes this method impractical.

White corn is low in carotene (the precursor of vitamin A). However, when pigs have access to a good pasture, when the diet is supplemented adequately with alfalfa meal, or when vitamin A is added, white and yellow corn appear to be equal in feeding value. It is noteworthy that the carotene content of yellow corn decreases with storage; about 25% of its original vitamin A value may be lost after one year of storage and 50% after two years.

Corn may be fed ground and combined in definite proportions with other feeds.

Contrary to the opinion prevailing in some quarters, experimental studies have not revealed any dif-

Fig. 10-2. Corn—as high as an elephant's eye and climbing clear up to the sky—has long been the most important U.S. cereal grain and swine feed. (Photo by J. C. Allen and Son, West Lafayette, IN)

feeds; among them, the following: bakery waste, barley, beans (cull), fat (tallow, greases), lard, molasses (cane or beet), oats, potatoes (cull, Irish), rye, sorghum (milo), spelt, triticale, and wheat.

When rating corn—the chief swine feed—as 100%, Table 10-1 shows how some other grain and high energy feeds, properly fed, compare with it, pound for pound.

■ **High-moisture corn**—High-moisture corn may be substituted for dry corn on a dry matter basis with little effect on overall performance, in either feed conversion or rate of gain.

Fig. 10-3. Confinement farrow-to-finish in Pennsylvania using high-moisture corn. (Courtesy A. O. Smith, Harvestore Products, Arlington Heights, IL)

ference between hybrid and open-pollinated yellow dent corns in feeding value for growing-finishing hogs.

From a market standpoint, the swine producer should be familiar with the following federal grades of corn:

No. 1, which must not exceed 14.0% moisture.
No. 2, which must not exceed 15.5% moisture.
No. 3, which must not exceed 17.5% moisture.
No. 4, which must not exceed 20.0% moisture.
No. 5, which must not exceed 23.0% moisture.
Sample grade, which exceeds 23.0% moisture.

Although moisture content is the chief criterion in determining federal grades of corn, the percentage of unsound kernels and amount of foreign materials are also factors. From a swine feeding standpoint, the value of the several grades of corn is about equal, providing they are placed on a comparable dry matter basis. The same situation applies to soft corn (corn frosted before maturity). Of course, handling and storage problems are increased with higher moisture content. Corn with more moisture than in No. 2 corn will not keep well in storage.

Yellow corn is usually the cheapest source of energy over much of the United States. But price fluctuations frequently justify consideration of other

■ **High-lysine corn (Opaque-2)**—Corn is now being bred that is much higher than normal corn in lysine and tryptophan; hence, it has a better balance of the amino acids for swine. Also, the high-lysine corn (called Opaque-2) is higher in total protein than normal corn. However, Opaque-2 is lower in leucine than regular corn.

Currently, little high-lysine corn is available commercially. However, corn breeders are working ceaselessly away to perfect Opaque-2, for it would make for improved nutrition of both animals and people throughout the world, wherever corn (maize) is grown.

Based on experiments and experiences, the following recommendations are made relative to high-lysine corn:

1. Check the moisture carefully when mechanically drying high-lysine corn, because it dries more rapidly than normal corn.

2. Grind high-lysine corn more coarsely than regular corn. A ½-in. screen works best. Some producers prefer a roller mill when processing high-lysine corn.

3. If high-lysine corn contains (a) 0.38% lysine or higher on an 86% dry matter basis, or (b) 0.44% lysine

or higher on a 100% dry matter basis, the following recommendations are applicable:

    a. Reduce growing-finishing diets 2% in crude protein below the protein level being fed in diets containing normal corn.

    b. Feed a pregestation and gestation diet containing 12% crude protein.

    c. Feed a lactation diet containing 14% crude protein.

If swine producers consider growing high-lysine corn, they must evaluate such economic factors as the lower yield of high-lysine corn vs the price of normal corn plus supplemental lysine.

■ **Waxy corn**—*Waxy corn is the name given to a variety of corn which is sometimes grown for industrial use because of its special type of starch. There appears to be no advantage or disadvantage associated with incorporating waxy corn in the diet of swine. Thus, the decision to produce and feed waxy corn should be based on agronomic and economic factors, and not on the effects of waxy corn on the performance of swine.*

BARLEY

Barley, the world's most ancient and most widely grown cereal, ranks third in importance among U.S. grain crops. It is grown extensively in sections too cool for corn to thrive, such as in the northern part of the United States. In Canada and northern Europe, it is the most common grain for swine. It is an excellent feed, producing firm pork of high quality.

Compared with corn, barley contains somewhat more protein (lysine) and fiber (due to the hulls) and somewhat less carbohydrates and fats (see Chapter 21). Like oats, the feeding value of barley is quite variable, due to the wide spread in test weight per bushel. When properly supplemented, ground barley is worth about 90 to 100% as much as corn for hogs. Fine grinding (600 to 700 microns) barley improves its feeding value. Barley is not well adapted to self-feeding, free choice, because pigs usually eat more of the accompanying protein supplements than is required to balance the diet. This is because barley is less palatable and higher in protein content than corn. For this reason, it is best to feed the barley and supplement together as a mixed diet.

Barley that is moderately infested with scab (Gibberella saubinetti) may be fed to growing-finishing pigs, provided the vomitoxin concentration in the final diet is less than 1 ppm. In no case should scab-infested barley be fed to gestating-lactating sows or young pigs.

EMMER AND SPELT

Emmer and spelt resemble barley somewhat in appearance and feeding value for hogs, for the hulls are not usually removed from the kernels in threshing. Three-fourths of the acreage of these grains (mostly emmer) is produced in North and South Dakota.

When ground and limited to not more than one-third of the diet, they are equal to barley in feeding value for hogs.

MILLET (Hog Millet, Proso Millet, or Broomcorn Millet)

Millet has been raised since prehistoric times in the Old World as an important bread grain crop. In this country, hog millet, often called broomcorn millet, is sometimes grown for livestock feeding purposes in the northern part of the Great Plains, where the growing season is too short for the grain sorghums. The yields usually range from 10 to 30 bushels per acre. Ground millet is worth 85 to 95% as much as corn for hogs.

OATS

Fig. 10-4. Oats, a favorite feed for brood sows and boars. (Photo by J. C. Allen and Son, West Lafayette, IN)

In total annual U.S. tonnage, oats rank fourth among the cereals.

Oats contain more lysine than corn or milo, but they are high in fiber.

Because of their high fiber content (1) oats are best utilized in sow gestation diets and (2) they should be limited to 30% of the diet of growing/finishing pigs. In larger quantities, oats retard the rate of gain, becoming about 80% as valuable as corn. For mature breeding animals the oats content of the diet may well be higher than one-third.

Grinding increases by 25 to 30% the feeding value of oats for swine. Oatmeal and rolled oats, such as are used for human consumption, are excellent feeds in fitting young pigs for the show. But they are usually expensive.

Leftover seed oats that have been treated with mercuric compounds should not be fed to swine.

Hulled oats are particularly valuable in diets for very young pigs; 100 lb of them being equal to 140 lb of corn. However, it takes 155 to 165 lb of whole oats to produce 100 lb of hulled oats.

RICE

Rice is one of the most important cereal grains of the world and provides most of the food for half the human population of the earth. It forms a major part of the diet of Oriental peoples, with whose chopsticks every young American associates rice. In this country, it is grown extensively for human food in Louisiana, Arkansas, Texas, and California. When low priced or offgrade, it is frequently fed to swine. Also, a considerable part of the byproducts of rice processing is profitably utilized as swine feed.

The whole rice fed to swine is known as ground rough rice (or ground paddy rice). It is finely ground, threshed rice (or rough rice) from which the hull has not been removed. From the standpoint of chemical composition and feeding value for swine, ground rough rice compares favorably with oats. For growing-finishing pigs, it is worth 80 to 85% as much as corn when it is finely ground and limited to less than one-half of the grains in the diet. Ground rough rice produces firm pork of good quality.

RYE

Rye thrives on poor, sandy soils. In such areas, it is sometimes marketed through hogs, though it generally sells at a premium for bread-making or brewing.

Rye is similar to wheat in chemical composition (see Chapter 21), but it is less palatable to pigs. For the latter reason, it should be fed in combination with

Fig. 10-5. A head of rye infested with ergot (a common fungus disease).

more relished feeds and limited to no more than one-half of the swine diet. When properly supplemented, ground rye has a feeding value approximately 85 to 95% that of corn. Because of the small, hard kernel, rye should be ground for hog feed.

Rye that is mildly infested with ergot (a common fungus disease in rye) may be fed to growing-finishing hogs, providing it does not constitute more than 10% of the total diet. Since ergot may cause abortion, ergot-infested rye should never be fed to pregnant sows. Likewise, ergot should not be permitted in the diet of lactating sows or suckling pigs.

SORGHUM (Kafir, Milo)

The grain sorghums—which include a number of varieties of earless plants, bearing heads of seeds—are assuming an increasingly important role in American agriculture. New and higher-yielding varieties have been developed and become popular. As a result, more and more grain sorghums are being fed to hogs.

Kafirs and milos are among the more important sorghums grown for grain. Other less widely produced types include sorgo, feterita, durra, hegari, kaoliang,

Fig 10-6. A Texas kafir crop. Kafir is usually threshed for swine, but grinding is not necessary when self-feeding. Sometimes kafir is harvested by hogging down. (Courtesy, USDA, Soil Conservation Service)

and shallu. These grains are generally grown in regions where climatic and soil conditions are unfavorable for corn, primarily in the central and south plains states.

Like white corn, the grain sorghums are low in carotene (vitamin A value). Also, they are deficient in other vitamins, and in proteins and minerals (see Chapter 21). Because the energy content of corn is slightly higher than that of milo, the feed efficiency of pigs fed corn diets will be slightly better than that of pigs fed milo, but average daily gains will be the same. Feeding trials show that the grain sorghums are worth about 95 to 97% as much as shelled corn for growing-finishing pigs.

One disadvantage of milo is that it is more variable in nutrient content than corn. In addition, because a milo kernel is smaller and harder than a corn kernel, fine grinding (1/8- or 5/32-in. screen) or rolling is recommended for best utilization.

Sorghum-fed hogs yield firm carcasses of good quality.

TRITICALE

This is a hybrid cereal derived from a cross between wheat (Triticum) and rye (Secale), followed by doubling the chromosomes in the hybrid. The objective of the cross: to combine the grain quality, productivity, and disease resistance of wheat with the vigor and hardiness of rye. The first such crosses were made in 1875 when the Scottish botanist, Stephen Wilson, dusted pollen from a rye plant onto the stigma of a wheat plant. Only a few seeds developed and germinated; and the hybrids were found to be sterile. In 1937, the French botanist, Pierre Givaudon, produced some fertile wheat-rye hybrids with the ability to re-

produce. Intensive experimental work to improve triticale was undertaken at the University of Manitoba, beginning in 1954.

In comparison with wheat, triticale (1) has a larger grain, but there are fewer of them in each head (spike), (2) has a higher protein content, with a slightly better balanced amino acid composition and more lysine, and (3) is more winter hardy.

So far, the main use for triticale has been as a feed grain, pasture, green chop, and silage crop for animals.

WHEAT

Through the ages, wheat has been made into precious bread for humankind. It was utilized for this purpose by the aboriginal Swiss lake dwellers; by the ancient Egyptians, Assyrians, Greeks, and Romans; and by the Hebrews, who, according to the Book of Revelations, exchanged "a measure of wheat for a penny." Even today, most U.S. wheat is utilized as a bread grain.

The total annual U.S. wheat tonnage is second only to corn. However, because it is produced mainly for the manufacture of flour and other human foods, it is generally too high in price to feed to hogs. When the price is favorable or when the grain has been damaged by insects, frost, fire, or disease, it may be more profitable to market it through hogs than for human consumption.

Compared with corn, wheat is higher in protein (and lysine) and carbohydrates, lower in fat, and slightly higher in total digestible nutrients. Wheat, like white corn, is deficient in carotene (see Chapter 21). Pound for pound, it is up to 10% more valuable than corn in producing gains on hogs, thus having the unique distinction of being the only cereal grain that excels corn as a swine feed (see Table 10-1).

Since the kernels are small and hard, wheat should be ground. Because wheat tends to flour when processed, it should be coarsely ground (3/16-in. screen) or rolled. If ground too finely, feed intake may be decreased and performance lowered.

Wheat produces good quality pork.

BYPRODUCT FEEDS FROM GRAINS

Most of the grains are milled in some manner for the preparation of foods for human consumption. In these milling processes a number of byproducts are produced which are generally considered to be of better value to humans but they can be, and are, used extensively as swine feeds.

CORN MILLING BYPRODUCTS

In the modern manufacture of cornstarch, sugar, syrup, and oil for human food, the following common byproduct feeds are obtained: (1) germ meal (dry), (2) germ meal (wet), (3) gluten feed, (4) gluten meal, and (5) hominy feed (grits).

GERM MEAL (DRY)

This is a byproduct of the dry milling of corn for corn meal, corn grits, hominy feed, and other corn products.

For best results in swine feeding, it is recommended that corn germ meal (dry milled) be utilized in the same manner as recommended for corn germ meal (wet milled).

GERM MEAL (WET)

This is a byproduct of the wet milling of cornstarch, corn syrup, and other corn products. The corn germs obtained in these processing operations are dried, crushed, and the oil extracted. Then the remaining residue or cake is ground, producing corn germ meal.

Corn germ meal contains about 22% protein and 10% fiber. When used in combination with higher quality proteins, corn germ meal is a satisfactory feed for swine. As a protein feed for pigs, it should not constitute more than half of the supplement, with the other half consisting of a protein source which supplies the amino acids lacking in germ meal.

When price relationships are favorable, corn germ meal can be used as a partial grain substitute in the diet, but it should not constitute more than 20% of the total feed. When so limited, corn germ meal is equal in value to corn, pound for pound.

GLUTEN FEED

Corn gluten feed is the corn byproduct remaining after the extraction of most of the starch and germ in the wet milling manufacture of cornstarch or syrup, with or without fermented corn extractives or corn germ meal. This feed contains approximately 25% protein and 7.5% fiber. Because of its bulkiness, unpalatability, and poor quality proteins, it is recommended that corn gluten feed be incorporated in the diet of ruminants rather than fed to swine.

GLUTEN MEAL

Corn gluten meal is similar to corn gluten feed without the bran. It averages about 43% protein and 3.0% fiber.

Because the proteins are of low quality, corn gluten meal should never be fed as the sole protein supplement to swine. When price relationships are favorable, it may be satisfactorily incorporated in confinement diets as part (25 to 50%) of the protein supplement in combination with supplements which supply the amino acids that gluten meal lacks. For pasture feeding, it may constitute 50% of the protein supplement—provided the other half is of animal or marine origin. When used in this manner, corn gluten meal is of about equal value to any of the common protein-rich plant supplements.

HOMINY FEED (Grits)

Hominy feed, a byproduct of corn milling, is a mixture of corn bran, corn germ, and a part of the starchy portion of either white or yellow corn kernels or mixtures thereof, as produced in the manufacture of pearl hominy, hominy grits, or corn meal. Hominy feed must contain not less than 4% crude fat. If prefixed with the words *white* or *yellow*, the product must correspond thereto.

Hominy feed is a satisfactory substitute for corn in swine feeding, but it is somewhat higher in fiber and less palatable than the grain from which it is derived. Yellow hominy feed supplies carotene (vitamin A), while white hominy feed lacks this factor. From the standpoint of gains and feed efficiency, properly supplemented hominy feed is worth 95 to 105% as much as corn, but it will produce soft pork if it constitutes more than half the grain diet.

In order to minimize the chance of hominy feed becoming rancid, it should be used fresh and stored in a cool, well ventilated building.

RICE MILLING BYPRODUCTS

The common byproducts resulting from processing rice are: (1) brewers' rice, (2) ground brown rice, (3) polishings, and (4) rice bran. In swine feeding, these products are used as grain replacements or energy feeds. Their compositions are given in Chapter 21.

BREWERS' RICE
(Chipped or Broken Rice)

Chipped rice consists of the small broken kernels of rice resulting from the milling operations. It is used chiefly in the brewing industry. Chipped rice is similar to corn in composition and feeding value. For best results in swine feeding, it should be finely ground, mixed with more palatable feeds, and included in a properly balanced diet. The carcasses of hogs fed on chipped rice are hard and firm.

GROUND BROWN RICE

Ground brown rice is the product obtained in grinding the rice kernels after the hull has been removed. It is nearly equal to corn as a swine feed and produces hard pork.

POLISHINGS

Rice polishings are the finely powdered byproduct obtained in polishing the rice kernels after the hulls and bran have been removed. They average about 12.8% protein, 13.4% fat, and 2.7% fiber, and are high in thiamin. Because of the high fat content, rice polishings become rancid in storage and produce soft pork unless fed in amounts restricted to less than one-third of the diet. When incorporated in a properly balanced diet, they are equal or superior in feeding value to corn for all classes of swine, including young pigs and gestating-lactating sows.

RICE BRAN

Rice bran is the pericarp, or bran layer, obtained in milling rice for human food, together with such quantity of hull fragments as is unavoidable in the regular milling operations. It must contain not more than 13% crude fiber. When the calcium carbonate exceeds 3% (Ca 1.2%), the percentage must be declared in the brand name. When limited to one-third of the swine diet, rice bran is equal in value to corn. Higher proportions have a lower feeding value and produce soft pork.

WHEAT MILLING BYPRODUCTS

In the average wheat milling operations, about 75% of the weight of the wheat kernel is converted into flour, with the remaining 25% going into byproduct feeds. Among the latter are the following hog feeds: (1) middlings, (2) mill run, (3) red dog, (4) shorts, and (5) wheat bran.

MIDDLINGS

Wheat middlings are a byproduct from spring wheat. They consist mostly of fine particles of bran and germ, with very little of the red dog flour. According to the definition of the Association of American Feed Control Officials, wheat flour middlings must not contain more than 9.5% fiber.

In general, this feed is characterized by fair amounts of proteins (about 17%) which are somewhat lacking in quality. It is high in phosphorus, low in calcium, and deficient in both carotene and vitamin D. Because of these deficiencies, middlings are best used in combination with protein supplements of animal origin, and then limited to about 1 lb per head daily. When fed to hogs in this manner, they are worth 110 to 120% as much as corn, pound for pound.

MILL RUN

Wheat mill run consists of the coarse wheat bran, fine particles of wheat bran, wheat shorts, wheat germ, wheat flour, and the offal from the "tail of the mill."

According to the Association of American Feed Control Officials, this product must contain not more than 9.5% crude fiber. Because of its bulkiness, it is recommended that wheat mill run be fed to the older hogs.

The terms used to designate the common wheat byproducts resulting from flour milling differ according to (1) the section of the country, and (2) whether they are obtained from winter or spring wheat. The byproducts feeds resulting from the milling of spring wheat are commonly known as bran, standard middlings, flour middlings, and wheat red dog. The byproducts resulting from the milling of winter wheat are known as bran, brown shorts, gray shorts, and white middlings. Wheat flour byproducts are distinguished by their fiber content.

Although the wheat milling products are higher in protein content than the cereal grains (and this fact is taken advantage of in formulating diets), in actual swine feeding practice they function more as grain replacements than as protein supplements.

RED DOG

Wheat red dog (red dog flour, or wheat red dog flour), which is a byproduct from milling spring wheat, consists chiefly of the aleurone layer together with small quantities of flour and fine bran particles. It must contain not more than 4% crude fiber. Red dog may be used in the diets of young pigs because of its lower fiber and high digestibility. When limited to less than one-sixth of the diet, it is slightly higher in feeding value than corn.

SHORTS

Wheat shorts consist of fine particles of wheat bran, wheat germ, wheat flour, and the offal from the "tail of the mill." The Association of American Feed Control Officials states that wheat shorts shall contain not more than 7.0% crude fiber. When fed at less than one-fifth of the diet of growing-finishing pigs, they have about 87% of the value of corn.

WHEAT BRAN

Wheat bran is the coarse outer covering of the wheat kernel. It contains a fair amount of protein (averaging about 16%), a good amount of phosphorus, and is laxative in action. Because of its bulk, wheat bran is not used extensively in diets for growing-finishing pigs. It does have a very definite place in the diets of pregnant and suckling sows. It is a good plan to include wheat bran in the diet to the amount of one-third by weight immediately before and after farrowing.

OTHER MILLING BYPRODUCTS

The milling of other grains such as barley, sorghum, rye, and oats consists of generally similar procedures and comparable byproducts. The usefulness of these byproducts for swine feeding must be considered basically in terms of their contributions as sources of energy. Generally, the byproduct feeds are higher in protein and fiber than the seeds from which they are derived but, because of the poor amino acid balance of the protein, they have their greatest usefulness as energy substitutes when economics warrant it.

BYPRODUCTS FROM THE BREWING AND DISTILLING INDUSTRIES

Considerable quantities of grains are used in the brewing of beers and ales and in the distilling of liquors. After processing, the remaining byproduct can be readily adapted to primarily ruminant feeding programs, but a small portion is used in swine diets. In addition to those feeds, solubles and yeast products from these industries are used in livestock feeds, though to a much lesser extent.

BREWERS' BYPRODUCTS

Barley is the primary grain used for the brewing of beers and ales. The initial step of brewing involves the malting of the barley. The *malt sprouts* and *malt hulls* are then separated from the malted barley, forming two byproducts which can be used as feed.

The clean malted barley is mixed with other grains (generally corn or rice) and a flavoring agent (hops), to form a mash. This mash is then cooked in water to enhance further enzymatic activity, and following cooking, it is separated into liquid and solid fractions. The liquid—called *wort*—undergoes yeast fermentation to form the alcoholic beverage of either beer or ale.

■ **Brewers' condensed solubles**—This byproduct is obtained by condensing the liquids resulting as byproducts of the production of beer or wort. It must contain not less than 20% total solids, and on a dry matter basis it must contain at least 70% carbohydrates. The guaranteed analysis shall include maximum moisture.

■ **Brewers' dried grains**—This is the dried extracted residue of barley malt, alone or in mixture with other cereal grain or grain products resulting from the manufacture of wort or beer. It may contain pulverized dried spent hops in an amount not to exceed 3% evenly distributed.

■ **Brewers' dried yeast**—Brewers' dried yeast is the dried nonfermentative, nonextracted yeast of the botanical classification, *Saccharomyces*, resulting from tile brewing of beer and ale. It must contain 35% crude protein. It must be labeled according to its crude protein content.

Brewers' dried yeast is an excellent source of highly digestible protein of good quality. It can replace up to 80% of the animal protein portion of swine diets when properly supplemented with calcium. However, in swine feeds it is used primarily as a source of B vitamins and unidentified growth factors.

■ **Brewers' wet grains**—This is the extracted residue resulting from the manufacture of wort from barley malt alone or in mixture with other cereal grains or grain products. The guaranteed analysis shall include the maximum moisture.

■ **Dried spent hops**—This byproduct is obtained by drying the material filtered from hopped wort.

■ **Malt cleanings**—Malt cleanings are byproducts produced in the cleaning of malted barley or from the recleaning of malt which does not meet the minimum standards for crude protein in malt sprouts.

It must be designated and sold according to its crude protein content.

■ **Malt hulls**—Barley grain is covered by a hull. During the cleaning of malted barley, the hulls are removed and subsequently used as feed. They contain about 22% crude fiber.

■ **Malt sprouts (malt culms)**—Malt sprouts are obtained by removal of the sprouts of malted barley. These sprouts may include some of the hulls and other parts of the malt, but they must contain not less than 24% crude protein. Since only about one-half of the crude protein is true protein, malt sprouts can be used most effectively in ruminant diets.

DISTILLERS' BYPRODUCTS

A large number of distilled spirits are produced throughout the world, each characterized by (1) the area of origin, (2) type of material used, (3) preparation of those materials, (4) proportions of materials, (5) fermentation conditions, (6) distillation processes, (7) maturation processes, and (8) mixture techniques.

Distillers' grains can be fed fresh, dried, or ensiled. The dried product is by far the easiest to handle and store. This product is less palatable to livestock than brewers' grains, but it contains more crude protein and less crude fiber. It is fed primarily to ruminants as a source of protein, although some is fed to swine.

■ **Condensed distillers' solubles**—The byproduct feed officially known as condensed distillers' solubles is a product obtained following the removal of ethyl alcohol by distillation from the yeast fermentation of a grain or a grain mixture by condensing the thin stillage fraction to a semisolid. As with the other distillers' product, the predominating grain is listed before the feed term.

Condensed distillers' solubles have long been noted as a source of B vitamins and unidentified growth factors for swine and poultry.

■ **Distillers' dried grains**—As officially defined by AAFCO, this is the product obtained after the removal of ethyl alcohol by distillation from the yeast fermentation of a grain or a grain mixture by separating the resultant coarse grain fraction of the whole stillage and drying it by methods employed in the grain distilling industry. Since a variety of grains are used in this process, the predominating grain is declared as the first word in the name—for example, corn distillers' dried grains.

■ **Distillers' dried grains with solubles**—This is the product obtained after the removal of ethyl alcohol by distillation from the yeast fermentation of a grain or a grain mixture by condensing and drying at least three-fourths of the solids of the resultant whole stillage by methods employed in the grain distilling industry. The predominating grain precedes the term distillers' dried grains with solubles.

■ **Distillers' dried solubles**—This byproduct is obtained after the removal of ethyl alcohol by distillation from the yeast fermentation of a grain mixture by condensing the thin stillage fraction and drying by methods employed in the grain distilling industry. The predominating grain must be declared in the first word of the name.

■ **Distillers' wet grains**—This is the product obtained after the removal of ethyl alcohol by distillation from the yeast fermentation of a grain mixture. The guaranteed analysis shall include the maximum moisture.

■ **Grain distillers' dried yeast**—Grain distillers' dried yeast is the dried nonfermentative yeast of the botanical classification, *Saccharomyces*, resulting from the fermentation of grains and yeasts which is separated from the mash either before or after distillation It must contain at least 40% crude protein. This byproduct is extremely rich in all of the B complex vitamins except vitamin B-12.

DRIED BAKERY PRODUCT

Bakery wastes consisting of bread, cookies, cake, crackers, flours, and doughs can be dried and sold as livestock feed under the name *dried bakery product*. Since it is high in digestible fat and carbohydrate, it is often used to replace grain when economically feasible. However, like grains, dried bakery product is low in vitamin A, protein, and minerals. It contains rather large amounts of salt; and if the salt content exceeds 3.5%, the maximum amount of salt must be so labeled in the name of the product. It can replace all of the grain in swine diets without affecting palatability.

Bakery wastes are worth 110 to 120% as much as corn for hogs.

OTHER HIGH ENERGY FEEDS

Although feed grains and their milling byproducts comprise the vast majority of the energy feeds, numerous other feeds are routinely used to supply energy to livestock, including swine. Seeds from plants other than *Gramineae* can be used effectively (for example, beans). Fats provide an extremely concentrated source of energy. Molasses is a liquid energy feed that is highly palatable and digestible. When the price and availability are advantageous, roots, tubers, and certain other byproduct feeds are fed to swine.

BEANS (Cull Beans)

Throughout the United States, several varieties of beans—including navy, Lima, kidney, pinto, and tepary beans—are raised for human food. Occasionally, low prices result in some of these beans being marketed through livestock channels. Also, cull beans— consisting of the discolored, shrunken, and cracked seeds, together with some foreign materials sorted out from the first-quality dry beans—are frequently cooked and fed to swine.

Chemically, beans of all varieties closely resemble

peas, but their feeding value is much lower. For swine, beans should be cooked thoroughly, limited to two-thirds of the grain diet (on a dry weight basis), and supplemented with a good quality protein source. When prepared and fed to swine in this manner, beans are 90% as valuable as corn.

BUCKWHEAT

Although not a cereal, buckwheat has the same general nutritive characteristics as the cereal grains. It is used chiefly for flour manufacture for making pancakes, a breakfast delicacy. However, when low in price or offgrade, buckwheat is often utilized for stock feeding. Because it is high in fiber—containing about 10%—and somewhat unpalatable, buckwheat should not constitute more than one-third of the swine diet. It should be ground for hogs. Some pigs, especially white ones, suffer from photosensitization (sensitivity to light) when fed buckwheat. Also, limited amounts of buckwheat flour byproduct without hulls, and buckwheat middlings are fed to pigs.

CITRUS FRUITS

Citrus fruits—drop and offgrade oranges, tangerines, grapefruit, and lemons—are sometimes used as hog feed in those areas where they are produced. They should be fed along with some grain and a protein supplement.

■ **Citrus meal and dried citrus pulp**—Citrus meal consists of the finer particles of dried citrus, whereas citrus pulp is the bulkier portions. Both are byproducts of citrus-canning factories which make citrus fruit juices, canned fruit, and other products. They consist of the dried and ground peel, residue of the inside portions, and occasional cull fruits of the citrus family, with or without extraction of part of the oil of the peel.

These products somewhat resemble dried beet pulp in chemical composition, but they are less palatable. They contain about 6% protein, 2.5 to 5% fat, 12 to 20% fiber, and 50 to 65% nitrogen-free extract. Citrus meal and citrus pulp are best suited as cattle feeds, but they may be satisfactorily used for growing-finishing pigs if limited to 5% of the diet. Higher levels result in smaller rate and efficiency of gains, digestive disturbances, and lower dressing percentage and carcass grade.

FATS AND OILS

Fats and oils such as white grease, beef tallow, corn oil, and soybean oil are excellent high energy feeds; they contain 2.25 times as much metabolizable

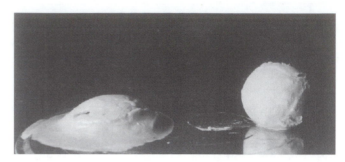

Fig 10-7. Feed fats affect body fats. The soft lard sample (left) came from hogs fed a high soybean diet. Both samples of lard had been exposed to room temperature, 70°F for two hours prior to photographing. (Courtesy, University of Illinois)

energy as most of the cereal grains. Additionally, a small amount of fat in the diet is desirable because fats are the carriers of the fat-soluble vitamins, and there is evidence that some species (humans, swine, rats, and dogs) require certain fatty acids.

Research indicates—

1. That feeding sows at the rate of 5 lb per day of a 5% fat diet for 10 days before farrowing has the potential to improve pig survivability if preweaning survival is below 80%, due to increases in milk yield and milk fat content.

2. That the addition of 3 to 5% fat to growing-finishing diets will improve feed efficiency and often daily gain. A rule of thumb is that feed efficiency is usually improved 2% for each 1% increment of added fat in growing-finishing diets.

3. That adding fats to *ad libitum* diets generally tends to increase backfat thickness.

4. That fats and oils in diets assist in dust control in confinement operations.

New commercial products that contain *dried fat* may reduce part of the commercial problems of adding liquid fat on the farm, but the economic feasibility of using these products must be evaluated.

From the above facts it may be concluded that adding fat to swine diets should be evaluated in terms of economic considerations. For example, if adding fat will increase diet cost by 5%, there must be more than 5% improvement in feed efficiency before it is practical.

GARBAGE

Municipal garbage has long been fed to hogs. But beginning about 1940, the practice declined because of a gradual lowering in the feeding value of garbage, along with other competition for it. Now less than 1% of the U.S. hogs are garbage fed. However, the recent development of garbage recycling processes, along with high-priced grains, has created renewed interest in garbage as a hog feed.

Prior to 1940, the garbage feeder figured that a ton of city garbage would produce 60 to 100 lb of pork; whereas, at the present time, it is estimated that a similar quantity will not produce more than 30 lb of pork. The change in feeding value may be largely attributed to improved refrigeration, the effective use of leftovers, and the change of the general eating habits of humans; for example, the increasing use of frozen and highly processed foods. Institutional, hotel, and restaurant garbage is superior to household garbage.

Garbage may be utilized either as a feed for a sow and pig enterprise or for finishing feeder pigs that are obtained from other sources. Usually, the venture seems most successful when a combination of grain and garbage feeding is practiced.

It is also observed that the most successful garbage feeders use concrete feeding floors, practice rigid sanitation, and take every precaution to prevent diseases and parasites. Unless considerable grain is fed to market hogs, especially after weights are over 100 lb, soft pork and paunchiness will result in garbage-fed hogs.

Swine are more likely to become infested with trichinella, vesicular exanthema, and certain other diseases when fed on raw garbage. For this reason, all states now have laws requiring that commercial garbage be cooked.

As a rule of thumb, about 4 lb of heavy garbage may be considered as equivalent to 1 lb of concentrate.

MOLASSES

Three kinds of molasses are fed to swine—cane molasses, beet molasses, and citrus molasses. The first two are byproducts of sugar manufacture; the cane molasses being derived from sugar cane and the beet molasses from sugar beets. As the name implies, citrus molasses is a byproduct of the citrus industry. The different types of molasses are available in both liquid and dehydrated forms.

Molasses is used in the following ways: (1) as an appetizer, (2) to reduce the dustiness of a diet, (3) as a binder for pelleting, (4) to supply unidentified factors, (5) in the case of cane molasses, to provide trace minerals, and (6) to provide a carrier for vitamins in liquid supplements. But there are better sources of energy for pigs than molasses.

CANE MOLASSES (Blackstrap)

A large quantity of this carbohydrate feed is produced in the southern states, and an additional tonnage is imported. Molasses must contain not less than 43% total sugars expressed as invert. If its moisture content exceeds 27%, its density determined by double dilution must not be less than 79.5 Brix. Molasses is

very low in phosphorus (see Chapter 21), thus indicating the need for feeding this mineral supplement when molasses constitutes much of the diet.

Molasses may cause scours in pigs unless they are started on it gradually and then limited in quantities fed. Molasses has its highest feeding value when it is used at a level of 3 to 10% of the diet. Also, heavier pigs can use molasses more effectively than lighter pigs. Diets of hogs weighing 150 to 240 lb may contain up to 40% molasses. Loose feces caused by the molasses are not a problem, providing the animals do well.

BEET MOLASSES

Beet molasses is a byproduct of beet sugar factories. It must contain not less than 48% total sugars expressed as invert and its density must not be less than 79.5 Brix. Beet molasses has a laxative effect, thus necessitating that pigs be started on it gradually and then fed limited amounts, 5 to 10 lb daily. It may constitute up to 40% of the diet for growing-finishing pigs weighing over 100 lb.

Beet molasses had best not be fed to pigs under 100 lb or to gestating sows. In fact, it is generally recommended that it be fed to farm animals other than swine. When properly used, beet molasses is equal to cane molasses as a swine feed, which means that it is 55 to 65% as valuable as corn.

CITRUS MOLASSES

Citrus molasses must contain not less than 45% total sugars expressed as invert and its density must not be less than 17.0 Brix.

Citrus molasses is unpalatable to hogs, due to its bitter taste. Since cattle do not object to the taste, most citrus molasses is fed to them. When mixed with more palatable feeds, citrus molasses can be fed to swine at about the same levels as cane molasses; and the two products have about the same feeding value.

PEANUTS

Peanuts are an important cash and feed crop of the South. They are grown for human consumption as peanuts, peanut butter, or peanut oil.

In the past, about one-third was harvested by hogging down (turning pigs into the field to root out the nuts). Generally, Spanish peanuts were grown for hogging down early in the fall, whereas runner peanuts were grown for winter use. Most practical swine producers used the brood sows and the young pigs for gleaning the fields, saving new fields for finishing hogs. Peanuts were frequently planted with corn when the crop was to be hogged down. Even where the peanuts

Fig. 10-8. Peanuts and corn in Georgia. This was a common planting in the South when both crops were hogged down. (Courtesy, USDA)

were harvested for sale through human channels, it was common practice to allow hogs to glean the fields. Today, the practice of harvesting peanuts by hogging down is limited except when the quality of the crop is poor or prices are down, or when gleaning the fields.

With the hulls on, peanuts contain about 25% protein and 36% fat. They are deficient in carotene, vitamin D, and calcium, and only fair in phosphorus.

Pigs in confinement will make satisfactory gains on a diet with harvested peanuts as the only concentrate, plus salt and a calcium supplement such as ground limestone or oyster shell flour. Even so, a high quality alfalfa hay, meal, or vitamin supplement should be provided in addition, thus alleviating any vitamin deficiency.

Pigs should have free access to salt and a calcium supplement when hogging down peanuts.

Unfortunately, the feeding of large quantities of peanuts will produce soft pork, unless they are restricted to pigs weighing less than 85 lb. Under most conditions, such a limitation is impractical. Furthermore, there is some evidence that peanut feeding increases backfat. More research is needed on the feeding of peanuts.

ROOT AND TUBER CROPS

Several root and tuber crops are fed to hogs, including sweet potatoes, white or Irish potatoes, chufas, cassavas, and Jerusalem artichokes. Roots are palatable, succulent, and laxative, but they are low in protein, calcium, and vitamin D, and except for carrots and sweet potatoes, they have little or no carotene (vitamin A value).

SWEET POTATOES

Cull or unmarketable sweet potatoes are a palatable swine feed. Sometimes sweet potatoes are fed to hogs in the South, usually by allowing the pigs to glean the fields after digging. They are high in carbohydrates in proportion to their protein and mineral content. Best

results are usually obtained when pigs grazing sweet potatoes are given one-third to one-half the usual grain allowance plus a protein supplement, a suitable mineral supplement, and a vitamin supplement. Sweet potatoes are too bulky for young pigs. It requires 400 to 500 lb potatoes to equal 100 lb of cereal grain when fed to hogs, the higher values being obtained in confinement feeding. Cooking improves the feeding value of sweet potatoes. They produce firm pork of good quality, although market animals fed on sweet potatoes are paunchy and have a low dressing percentage.

Dehydrated sweet potatoes have approximately 90% the feeding value of corn when used at levels of one-third to one-half in a well balanced diet.

POTATOES (Irish Potatoes)

Potatoes are sometimes used as hog feed when they have little value on the market, especially if grain prices are high. Since potatoes are not palatable or as readily digested in the raw state, they must be steamed or boiled (preferably in salt water, with any cooking water discarded) to increase palatability. When properly cooked, and when not replacing more than 50% of the grain diet, 350 to 400 lb of potatoes will be equivalent in feeding value to 100 lb of the common cereal grains. Potatoes should not be fed to sows during the latter part of gestation or immediately after farrowing.

When limited to 10 to 20% of the diet of growing-finishing pigs, dehydrated Irish potatoes (known as potato meal or potato flakes) are about equal to corn, pound for pound. Of course, the cost of dehydrating is too great to consider the process practical, but surplus dehydrated potatoes are sometimes available at a nominal price.

CHUFAS

Chufas, a southern tuber crop, were formerly planted from April to June for hogging down. The chufas sedge, frequently a weed in damp fields on southern farms, will remain over winter in the ground without killing. Chufas should be supplemented properly with proteins and minerals. They will produce 300 to 600 lb of pork per acre, after making allowance for the supplemental feeds consumed. Chufas produce soft pork.

CASSAVAS

These are a fleshy root crop that grows in Florida and along the Gulf Coast, yielding 5 to 6 tons per acre. The roots, containing 25 to 30% starch, are used in the production of the tapioca of commerce and as a swine feed.

When not constituting over one-third of the dry

matter in the diet, cassavas are a satisfactory feed for hogs. Dried cassava meal is worth about 85% as much as corn for growing-finishing hogs.

JERUSALEM ARTICHOKES

Jerusalem artichokes, a hardy perennial vegetable, produce tubers resembling the potato in composition, except that the chief carbohydrate is inulin instead of starch. They yield 6 to 15 tons of tubers per acre. The tubers live over the winter in the ground, and, even when hogged down or dug in the fall, usually enough tubers will remain to make the next crop. When hogging down artichokes, pigs should be fed grain and supplement, because they will make but little gain on the tubers alone.

OTHER ROOT CROPS

Such root crops as beets, mangels, carrots, and turnips are never produced in this country specifically for hog feed. However, when they are not salable for human food, they are sometimes fed to hogs. Because of their high water content (usually 80 to 90%), about 9 lb of roots are required to equal the feeding value of 1 lb of grain. Best results are secured when the roots are cut in small pieces, fed raw, and limited to a replacement of not more than one-fourth of the grain diet.

SOYBEANS

Soybeans, the precious beans from the Orient, have advanced from a minor to a major U.S. grain crop since World War II. For the most part, soybeans are processed, with the oil used for human food and the byproducts therefrom used for livestock feeds. However, a widespread interest in their use as a home-grown feed for swine has been created.

The protein of raw soybeans is poorly utilized by pigs due to the presence of a factor which inhibits protein digestion. However, cooking or roasting at the proper temperature destroys this factor and makes soybeans a satisfactory feed for hogs.

Fig. 10-9. Soybeans, the precious beans from the Orient; now a major U.S. grain crop. (Courtesy, American Soybean Assn., Hudson, IA)

Research has shown that whole cooked soybeans can be used to replace soybean meal or other forms of protein supplement in swine diets. Whole cooked beans have little effect on rate of gain, but they usually improve feed efficiency by 5 to 10% due to the higher fat content of the whole soybeans, which makes for a higher energy diet. However, this improvement in feed efficiency may be offset by the lower protein content of whole soybeans; whole cooked beans average about 37% crude protein, whereas soybean meal usually runs 44%. Also, due to the higher energy of the beans, the protein content of the diet must be 1 to 2% higher than in a soybean meal diet.

In order to process whole soybeans, a large investment in cooking or roasting equipment is required. Such an investment commits the operator to long-term use even though lower prices of soybean meal, relative to the price of whole soybeans, may not justify the practice.

It is noteworthy that hogs fed cooked whole soybeans have a softer carcass than those fed a soybean meal diet. Whether this condition will influence the price packers are willing to pay for live hogs is yet to be determined.

VELVET BEANS

Velvet beans are grown chiefly for forage, but occasionally the beans are harvested for livestock feed. When hulled and ground, they become velvet bean meal. When the beans and pod are ground together, they become ground velvet bean and pod. Regardless of the type of preparation, velvet beans should not constitute more than one-fourth of the diet for any class of swine; otherwise, toxicity (due to dihydroxyphenylalanine, which is closely related to epinephrine)—with accompanying severe diarrhea and vomiting—will result. Cooking decreases the toxicity and increases the palatability and digestibility, but it does not generally make them entirely satisfactory. If used, it is recommended that velvet beans be fed only to heavier pigs and not to brood sows, and that a high quality protein supplement be fed.

In the Gulf Coast region, velvet beans are sometimes grown with corn or with corn and peanuts for hogging down. Velvet bean pasture is not recommended for swine.

HOGGING DOWN CROPS

Today, hogging down is limited to crops (1) that cannot be harvested for one reason or another, (2) that are of low quality, or (3) that are not profitable to harvest in the conventional manner.

QUESTIONS FOR STUDY AND DISCUSSION

1. How are the energy content of feedstuffs and the energy requirements of pigs usually expressed?

2. For your area, what type of high energy feeds would you expect to be readily available?

3. Define the term *grain*. What grains are grown primarily for food for humans? What grains are produced chiefly for feed for livestock?

4. What are the advantages of feeding high lysine corn? List the disadvantages of feeding or producing this kind of corn. Discuss the factors that should be considered relative to waxy corn.

5. The following energy feeds are available at a low cost:

      a. oats
      b. corn germ meal
      c. rice polishings
      d. wheat bran
      e. dried bakery products
      f. cull beans
      g. citrus fruits
      h. cull potatoes

What recommendations would you make for the use of each of these in swine diets?

6. List the milling byproducts of (a) corn, (b) rice, and (c) wheat which may be used in swine diets.

7. Why are so few of the byproducts from the brewing and distilling industries used in swine diets?

8. High energy feeds possess certain nutritive deficiencies. Using the information from this chapter and Chapter 21, list some of these deficiencies for the major high energy feeds—corn, oats, barley, and milling byproducts.

9. What are the advantages of adding fat to a swine diet? What class of swine, and at what levels, would you recommend feeding fats and oils?

10. Why has garbage feeding declined in the United States?

11. For what reasons would molasses be added to the diet? List the sources of molasses.

12. Soybean meal is a common protein source for swine diets. What is the advantage of feeding cooked or roasted soybeans? (Check Chapter 21 to compare their composition.)

13. List some high energy feeds that may cause soft pork.

14. Why are crops hogged down? What crops are sometimes hogged down?

SELECTED REFERENCES

Title of Publication	Author(s)	Publisher
Association of American Feed Control Officials Incorporated, Official Publication	Association of American Feed Control Officials, Inc.	Association of American Feed Control Officials, Inc., Annual
Feeds & Nutrition Digest	M. E. Ensminger J. E. Oldfield W. W. Heinemann	The Ensminger Publishing Company, Clovis, CA, 1990
Feeds & Nutrition, Second Edition	M. E. Ensminger J. E. Oldfield W. W. Heinemann	The Ensminger Publishing Company, Clovis, CA, 1990
Nontraditional Feed Sources for Use in Swine Production	P. A. Thacker R. N. Kirkwood	Butterworth Publishers, Stoneham, MA, 1990
Swine Feeding and Nutrition	T. J. Cunha	Academic Press, Inc., New York, NY, 1977
Swine Nutrition Guide	J. F. Patience P. A. Thacker	Prairie Swine Centre, University of Saskatchewan, Saskatoon, Canada, 1989
Swine Production and Nutrition	W. G. Pond J. H. Maner	AVI Publishing Co., Inc., Westport, CT, 1984
Swine Production in Temperate and Tropical Environments	W. G. Pond J. H. Maner	W. H. Freeman and Co., San Francisco, CA, 1974

Soybeans, source of soybean meal, America's leading protein supplement. (Courtesy, American Soybean Assn., St. Louis, MO)

11

PROTEIN AND AMINO ACIDS FOR SWINE[1]

[1]In the preparation of this chapter, the authors adapted applicable material from the following two publications: *Swine Nutrition Guide*, published by the University of Nebraska and South Dakota State University, 1995; and *Kansas Swine Nutrition Guide*, published by Kansas State University, updated 1995.

The feed ingredient definitions used in Chapter 11 include those used in the 1996 Official Publication of the *Association of American Feed Control Officials (AAFCO)*.

Prior to 1890, no one was concerned about adding proteins to livestock diets, and vitamins were unknown. The flour mills in Minneapolis dumped wheat bran into the Mississippi River, because nobody wanted to buy it. Most of the linseed meal was shipped to Europe. Soybeans were little known outside the Orient, and tankage had not been processed.

Before 1890, swine were fed whatever home-grown grains—chiefly corn—and farm wastes as were available. They were marketed at 10 to 16 months of age. Sows farrowed once a year. But, what was time to a hog!

Fig. 11-1. Prior to 1890 many swine were fed chiefly corn. (Courtesy, SCS, USDA)

PROTEINS

Beginning about 1900, scientists discovered that the kind or quality of protein in livestock feeds was of tremendous importance, thus ushering in the golden era in nutrition. Soon the race was on for protein-rich feeds. Many of the byproducts that once polluted the streams of the nation were in unprecedented demand.

Protein in plants is largely concentrated in the actively growing portions, especially the leaves and seeds. Plants possess the ability to synthesize their own proteins from such relatively simple soil and air compounds as carbon dioxide, water, nitrates, and sulfates. Thus, plants, together with some bacteria which are able to synthesize these products, are the original sources of all proteins.

Proteins in animals are much more widely distributed than in plants. The proteins of the animal body are primary constituents of many structural and protective tissues—such as bones, ligaments, hair, hoofs, skin, and the soft tissues which include the organs and muscles. The total protein content of the bodies of pigs ranges from 8.3 to 19.6% (see Chapter 8, Table 8-1). By way of further contrast, it is noteworthy that, except for the bacterial action in the rumen, animals lack the ability of plants to synthesize proteins from inorganic materials. Hence, they must depend upon plants or other animals as a source of dietary protein.

Proteins are found in most of the feeds commonly fed to animals. The amount of protein, its digestibility, and the balance of essential amino acids are important factors that must be considered in balancing diets. In general, animal proteins are superior to plant proteins for swine and other monogastric animals. For example, zein (a corn protein) is a low-quality or unbalanced protein, being deficient in the essential amino acids lysine and tryptophan. On the other hand, animal proteins are excellent sources of lysine, and many (especially milk and eggs) are abundant in tryptophan.

Since protein feeds are usually among the more costly components of swine diets, it is important to provide enough protein for the animal to perform its assigned function, but to avoid feeding more than is necessary.

In the past, it was common practice to use several sources of protein in swine diets so that their respective amino acid profiles would complement each other. Today, the use of the computer in formulating diets and the increased availability of amino acid supplements, such as methionine hydroxy analog (MHA), allow nutritionists to formulate complete diets with a minimum number of protein feeds. The computer can rapidly determine what specific amino acids must be added and at what levels. Thus, the trend is toward fewer protein feed sources, properly supplemented with specific amino acids.

Ingredients that contain more than 20% of their total weight in crude (total) protein are generally classified as protein feeds. Protein supplements may be further categorized according to source of origin as (1) oilseed cake and meal proteins, (2) animal and marine proteins, and (3) mill products. Table 11-1 shows the U.S. tonnage of each of these three sources fed to U.S. livestock during the 10-year period 1984–1993. Oilseed and animal/marine proteins are the chief protein supplements used in swine diets. It is noteworthy, too, that of the total tonnage of oilseed and animal/marine proteins fed to U.S. livestock during the period 1984–1993, 85% came from oilseed and only 15% from animal/marine products.

Fig. 11-2. Amino acid balance is especially important in starter diets. (Courtesy, DeKalb swine Breeders, Inc., DeKalb, IL)

TABLE 11-1
U.S. TONNAGE OF PROTEIN FEEDS FOR LIVESTOCK[1]

| Year | Oilseed Cake and Meal | | Animal/ Marine Total[3] | Mill Products[4] |
	Soybeans	Other Oilseed Meals[2]		
	(1,000 tons)	*(1,000 tons)*	*(1,000 tons)*	*(1,000 tons)*
1984	19,480	2,339	3,855	40,504
1985	19,090	2,150	3,723	36,545
1986	20,387	1,669	3,599	37,559
1987	21,293	2,270	3,551	39,232
1988	19,657	2,220	3,305	36,902
1989	22,263	1,928	3,375	38,813
1990	22,934	2,194	3,260	38,537
1991	23,008	2,540	3,269	39,315
1992	24,251	2,146	3,321	40,362
1993	25,000	2,031	3,510	42,344

[1]From: *Agricultural Statistics 1994*, USDA, p. 45, Table 69.

[2]Other oilseed meals includes cottonseed, linseed, peanut, and sunflower cake and meal.

[3]Animal and marine totals includes tankage and meat meal, fish meal, and dried skim milk and whey for feed.

[4]Mill products includes wheat millfeeds, gluten feed and meal, rice millfeeds, brewers' dried grains, distillers' dried grains, dried molasses beet pulp, and alfalfa meal.

AMINO ACIDS

Pigs do not have a specific requirement for crude protein. Rather, pigs of all ages and stages of the life cycle require amino acids to enable them to grow and reproduce. *Amino acids are the structural units of proteins.* During digestion, proteins are broken down into amino acids. Then, the amino acids are absorbed into the bloodstream and used to build new proteins, such as muscles.

Diets that are *balanced* with respect to amino acids contain a desirable level and ratio of the 10 essential amino acids required by pigs for maintenance, growth, reproduction, and lactation. The 10 essential amino acids that must be provided in swine diets are: arginine, histidine, isoleucine, leucine, lysine, methionine, phenylalanine, threonine, tryptophan, and valine.

The proteins of corn and other cereal grains are deficient in certain essential amino acids, especially lysine, tryptophan, and threonine. Protein supplements are used to correct amino acid deficiencies in grains. For example, the correct combination of grain and soybean meal provides a good balance of amino acids except in starter diets.

■ **Alternative amino acid sources**—Table 11-2 lists alternate amino acid sources that can be used to replace soybean meal. Note that protein sources are classified into two major categories: plant proteins and animal proteins.

Soybean meal is the only plant protein that compares favorably with animal protein in terms of quality of amino acid content and ratio, and that can be used as the only protein source in most swine diets. Soybean meal may serve as the only protein source, with the exception of starter diets, which should contain some animal products.

In order to determine the relative feeding value of alternative protein sources, it is important to compare the lysine level in the new protein source to soybean meal. The relative feeding values of some alternative protein sources are listed in Table 11-2. This can be

TABLE 11-2
ALTERNATIVE AMINO ACID SOURCES[1]

Source	Protein	Relative Value as a Lysine Source		
		Lysine	Percent	Pounds[2]
	(%)	(%)	(%)	
Plant proteins:				
Soybean meal	44	2.90	100	100
Soy protein isolate	92	5.20	179	56
Soy protein concentrate	66	4.20	145	69
Yeast, brewers' dried	45	3.23	111	90
Soybean meal	46.5	3.01	104	96
Canola meal	38	2.27	78	127
Sunflower meal	45.5	1.68	58	173
Cottonseed meal	41	1.51	52	192
Wheat gluten, spray-dried	74	1.30	44	223
Alfalfa meal	17	0.80	28	363
Corn gluten meal	42.1	0.78	27	372
Wheat middlings	16	0.68	24	417
Wheat bran	15	0.56	19	518
Animal proteins:				
Blood meal, spray-dried	86	8.02	277	36
Porcine plasma, spray-dried	70	6.10	210	48
Fish meal	60	4.75	164	61
Egg protein, spray-dried	48	3.30	114	88
Meat and bone meal	50	2.80	97	104
Skim milk, dried	33	2.54	87	114
Fish solubles, dried	54	1.73	60	167
Whey, dried	12	0.97	33	299

[1]Adapted by the authors from: Goodband, R. D., *et al.*, *Kansas Swine Nutrition Guide*, published by Kansas State University, 1994, updated 1995.

[2]Pounds required to equal 100 lb of 44% soybean meal when various feeds replace up to 50% of the soybean meal.

utilized to determine the comparative economic value of the protein source as a partial or complete replacement of 44% soybean meal. These feeding values were calculated by dividing the lysine content of the feed ingredient by that of 44% soybean meal (2.90% lysine) and multiplying by 100 to put them on a percentage basis.

Example: Assuming that 44% soybean meal can be purchased at $250 per ton, what would a ton of 46.5% soybean meal be worth? Because the lysine content of 46.5% soybean meal is 3.01% and 44% soybean meal has 2.90% lysine, 46.5% soybean meal has 104% the feeding value of 44% soybean meal (3.01 ÷ 2.90 × 100 = 104%). Therefore, if 104% is multiplied by the cost of 44% soybean meal (104% ×

$250), 46.5% soybean meal is of greater value than 44% soybean meal if it costs less than $260 per ton.

■ **"Hidden" costs of alternative amino acid sources**—When selecting alternative amino acid sources, in addition to ingredient cost the swine producer should consider storage costs, anti-nutritional factors, product variability, fiber content, spoilage, and under- or over-processing. These factors are especially problematic in byproduct protein sources. Because byproduct feed ingredients tend to vary in composition, proper information regarding chemical composition is necessary to ensure optimum pig performance.

■ **Maximum level at which some feed ingredients can replace soybean meal**—When substituting other protein sources for soybean meal, it is important to know the maximum level at which the new feed ingredient can replace soybean meal without seriously affecting performance. Table 11-3 presents a suggested level (or range) of commonly used amino acid sources that can be used in starter, growing-finishing, gestation, and lactation diets to replace all or a part of the soybean meal. For example, corn gluten meal can replace 25% of the soybean meal protein in the diet of growing-finishing, gestation, and lactation diets.

TABLE 11-3
SUGGESTED RANGE OF COMMONLY USED AMINO ACID SOURCES[1]

Feed	Percent of Protein Portion of the Diet[2]			
	Starter	Growing-Finishing	Gestation	Lactation
Alfalfa meal, dehydrated	0	0–25	0–50	0–10
Blood meal, spray-dried	0–10	0–25	0–25	0–25
Fish meal	0–50	0–25	0–25	0–25
Meat and bone meal	0–25	0–25	0–25	0–25
Soybean meal	0–50	0–100	0–100	0–100
Canola meal	0	0–50	0–50	0–50
Sunflower meal	0	0–25	0–25	0–25
Corn gluten meal	0	0–25	0–25	0–25
Cottonseed meal	0	0–75	0–25	0–25
Tankage	0–25	0–25	0–25	0–25
Yeast, brewers dried	0–10	0–10	0–10	0–10
Egg protein, spray-dried	0–10	0	0	0
Porcine plasma, spray-dried	0–30	0	0	0
Soy protein concentrate	0–30	0	0	0
Soy protein isolate	0–30	0	0	0
Wheat gluten, spray-dried	0–10	0	0	0

[1]Adapted by the authors from: Goodband, R. D., *et al.*, *Kansas Swine Nutrition Guide*, published by Kansas State University, 1994, updated 1995.

[2]Percentages suggest maximum allowable inclusion rates for protein sources. Economics and pig performance standards must be considered for actual inclusion rates.

■ **Soybean meal as the sole source of supplemental protein in the diet**—Soybean meal can serve as the sole supplemental protein in the diet for pigs heavier than 25 lb. Younger and lighter pigs have a reduced ability to utilize the complex proteins found in soybean meal. Additionally, starting pigs may develop allergic reactions to certain proteins in soybean meal. For this reason, in the diet of starting pigs it is desirable to use less allergenic, highly digestible amino acid sources in diets for starter pigs such as spray-dried plasma proteins and blood meal, menhaden fish meal, and/or soybean protein concentrate.

■ **Ideal protein or amino acid balance**—*The ideal protein or ideal amino acid balance is a protein that provides a perfect pattern of essential and nonessential amino acids in the diet without any excesses or deficiencies.* This pattern is supposed to reflect the exact amino acid requirements of the pig. Thus, an ideal protein would provide exactly 100% of the recommended level of each amino acid. But there is little evidence to indicate that the performance of pigs fed diets containing an ideal balance of amino acids is better or worse than that of pigs fed practical corn or soybean-meal–based diets. However, if excess amino acids are reduced, nitrogen excreted through the urine and feces will be reduced, thereby lessening the nitrogen in the manure. Unless there is reason to reduce the nitrogen in the manure, producers should choose sources of amino acids that will produce lowest cost gains.

■ **Limiting amino acid**—If a diet is inadequate in one essential amino acid, protein synthesis cannot proceed beyond the rate at which that amino acid is available. This is called the *limiting amino acid*. Standard swine diets are formulated to meet the pig's requirements for lysine—the most limiting amino acid. But when formulating diets to meet the lysine requirements, excesses of many other amino acids usually exist.

■ **Amino acid availability**—Excessive heat in drying feed ingredients will reduce the availability of amino acids, especially lysine. If soybean meal or dried whey look darker than usual or have a burnt smell, likely the protein quality has been reduced. Also, the presence of an anti-nutritional factor may affect availability of amino acids. If a high percentage of byproduct feed ingredients or feed ingredients that have been overprocessed are being used, consideration should be given to balancing the diet on an available amino acid basis.

■ **Formulating diets on an available lysine basis**—Formulating diets on an available lysine basis is more precise than formulating on a protein basis. When formulated on a protein basis, the diet could be defi-

cient in lysine, resulting in reduced pig performance. Lysine is often the most limiting amino acid in a grain-soybean meal based diet. Generally, if the lysine recommendation is met the other amino acids will be adequate, also. An exception to this rule occurs when plasma proteins, blood meal, distillers' dried grains with solubles, and other byproducts are used in the diet.

■ **Crystalline amino acids in swine diets**—Whether or not it is economical to use crystalline amino acids in swine diets depends on the price of amino acids and the prices of grain and supplemental protein sources. The use of L-lysine HCl as a source of crystalline lysine is often economically sound. Crystalline methionine is commercially available and inexpensive. Crystalline tryptophan and threonine can be purchased in feed-grade forms, but currently they are rather expensive. Crystalline lysine and tryptophan together in the same source are now commercially available. It is anticipated that other sources combining these crystalline amino acids, as well as others, will be developed in the future.

Three pounds of L-lysine HCl (containing 78% pure lysine) plus 97 lb of corn contribute the same amount of digestible lysine as 100 lb of 44% crude protein soybean meal.

The level of crystalline amino acids supplemented will depend on the feed utilized in the formulation; it is usually dependent on the second limiting amino acid. In most swine diets, lysine is the first limiting amino acid and either tryptophan or threonine is the second limiting amino acid. However, starter pig diets containing large amounts of plasma proteins and blood meal may benefit from supplementation with crystalline methionine.

Caution: Research indicates that limit-fed pigs (for example, fed once per day) use crystalline amino acids less efficiently than pigs fed free-choice (eating several times per day). Under these circumstances, replace-

Fig. 11-3. Amino acid balancing requires good mixing. (Courtesy, Iowa State University, Ames, IA)

ment of intact protein (such as soybean meal) with lysine may lead to a deficiency of other amino acids. An amino acid deficiency causes reduced litter gain and sow lactation feed intake.

In order to assume proper distribution of crystalline amino acids in a complete feed, amino acids must first be combined with a carrier to achieve a minimum volume before they are added to the mixer.

PLANT PROTEINS

Even though they are not especially high in protein by comparison with other feedstuffs, the vegetative portions of many plants supply an extremely large portion of the protein in the total diet of livestock, simply because these portions of feeds are consumed in large quantities. Needed protein not provided in these feeds is commonly obtained from one or more of the oilseed byproducts—soybean meal, canola meal, cottonseed meal, linseed meal, peanut meal, safflower meal, sunflower seed meal, or coconut (copra) meal. The protein content and feeding value of these products vary according to the seed from which they are produced, the geographical area in which they are grown, the amount of hull and/or seed coat included, and the method of oil extraction used. Sometimes, the unprocessed seed is used to provide both a source of protein and a concentrated source of energy. The oil-bearing seeds are especially high in energy because of the oil that they contain.

Additional plant proteins are obtained as byproducts from grain milling, brewing and distilling, and starch production. Most of these industries use the starch in grains and seeds, then dispose of the residue, which contains a large portion of the protein of the original plant seed.

OILSEED MEALS

Several rich oil-bearing seeds are produced for vegetable oils for human food (margarine, shortening, and salad oil), and for paints and other industrial purposes. In the processing of these seeds, protein-rich products of value in swine feedings are obtained. Among such high-protein feeds are soybean meal, canola meal (rapeseed meal), coconut (copra) meal, cottonseed meal, linseed meal, peanut meal, safflower meal, sesame meal, and sunflower seed meal.

SOYBEAN MEAL

The United States produces about one-half of the world's soybeans; and soybean meal is now the most widely used protein supplement in the United States (see Table 11-1).

Soybean meal is the ground residue (soybean oil cake or soybean oil chips) remaining after the removal of most of the oil from soybeans. The oil is extracted by any one of three processes: (1) the expeller process, (2) the hydraulic process, or (3) the solvent process. Well-cooked soybean meal produced by each of the extraction processes is of approximately the same feeding value. Regardless of the method of extraction employed, reputable manufacturers now use the proper heat treatment so that the protein quality is good.

Soybean meal normally contains 41, 44, 48, or 50% protein, according to the amount of hull removed. It must contain not more than 7.0% crude fiber. It may contain calcium carbonate or an anticaking agent not to exceed 0.5%.

Although soybean meal is marginal in methionine, it is otherwise very well balanced in amino acids; and the protein of soybean meal is of better quality than the other protein-rich supplements of plant origin. It is, however, low in calcium, phosphorus, carotene, and vitamin D.

Since soybean meal is extremely palatable to pigs, when self-fed, free-choice, as the supplement to grain, they will often eat much more of it than is required to balance the diet. This is uneconomical because of the higher cost of the supplement in comparison with cereal grains and other high energy feeds. This wasteful practice may be alleviated by mixing the ground grain and soybean meal together and self-feeding as a complete diet.

Soybean meal is excellent as the only protein supplement to corn for swine of all ages above 25 lb weight providing the proper proportions are used to yield a good balance of the amino acids. It must be supplemented with minerals and vitamins.

CANOLA MEAL (Rapeseed Meal)

World canola (rapeseed) production currently ranks third among oilseeds. Canola is grown mainly in Canada, but it is increasing in the United States.

Canola was created from specially selected rapeseed by Canadian plant scientists in the 1970s. The old rapeseed was high in glucosinolate compounds, which, when fed to animals at high levels, made for palatability problems and lowered performance to goitrogenic action. Canola changed this. The new canola is low in glucosinolates in the meal, and low in erucic acid (a long-chain fatty acid) in the oil. The feed ingredient definition of the AAFCO specifies that the oil component of the seed contain less than 2% erucic acid, that the solid component of the seed contain less than 30 micromoles of glucosinolates per gram of air dry, oil free solid, and that canola meal must not exceed a maximum of 12% crude fiber.

Canola meal averages between 35 and 40% crude protein and its amino acids compare favorably with soybean meal. The diets of starter and growing-finishing pigs may contain up to 15% ground canola seed. However, sows should not receive more than 10% canola seed in their diets.

Caution: The lowering of the erucic acid and glucosinolate in rapeseed (canola) has proceeded at different rates in different countries. But the change is nearly complete in Canada.

COCONUT MEAL (Copra Meal)

This is the byproduct from the production of oil from the dried meats of coconuts. The oil is generally extracted by either (1) the mechanical process, or (2) the solvent process. Coconut meal averages about 21% protein content. Since the proteins are of not too high quality, it is generally recommended that coconut meal not constitute more than one-fourth of the protein supplement of swine diets.

COTTONSEED MEAL

Today, the U.S. cotton crop ranks fifth in value, being exceeded only by corn, soybeans, wheat, and hay. Among the oilseed meals, cottonseed meal ranks second in tonnage to soybean meal and cake. The processing steps in making cottonseed meal are as follows: (1) cleaning the seeds; (2) dehulling the seeds; (3) crushing the kernels; (4) extracting the oil by (a) mechanical (or screw pressed), (b) solvent, or (c) partially mechanically extracted and then solvent extracted process; and (5) grinding the remaining residue or cake, thus forming cottonseed meal.

The protein content of cottonseed meal can vary from about 22% in meal made from undecorticated seed to 60% in flour made from seed from which the hulls have been removed completely. Thus, by screening out the residual hulls, which are low in protein and high in fiber, the processor is able to make a cottonseed meal of the protein content desired— usually 36, 41, 45, 48, or 50%.

For the monogastric pig, cottonseed meal is low in lysine and tryptophan and deficient in vitamin D, carotene (vitamin A value), and calcium. Also, it contains a toxic substance known as gossypol, varying in amounts with the seed and the processing. But, it is rich in phosphorus.

It is recommended that the cottonseed meal content of practical swine diets (especially diets for growing swine) be limited to one-half the protein supplement of the diet. At this level, it is unlikely that the total diet will contain more than 0.01% free gossypol.

Fig. 11-4. A victim of gossypol poisoning, resulting from feeding too much cottonseed meal high in gossypol. This pig died soon after the picture was taken. (Courtesy, University of Arkansas)

Pig performance begins to be reduced at gossypol concentrations of 0.04% of the diet.

Solvent extracted, gossypol-free cottonseed meal can be used to replace 75% of the protein source in growing-finishing diets when balanced on a lysine basis.

More and more glandless cottonseed, free of gossypol, is being produced. It alleviates (1) any restrictions as to levels of meal, and (2) the need to add iron.

LINSEED MEAL

Linseed meal is a byproduct of flax, a fiber plant which antedates recorded history. In this country, most of the flax is produced as a cash crop for oil from the seed and the resulting byproduct, linseed meal. Practically none of the U.S. flax crop is grown for fiber, for it is more economical to import it from those countries where cheaper labor is available.

Most of the nation's flax is produced in North Dakota, South Dakota, Minnesota, and Texas. Normally, an additional quantity of seed is imported and processed in our plants.

The oil may be extracted from the seed by either of two processes: (1) the mechanical process (or so-called old process), or (2) the solvent process (or so-called new process). If solvent extracted, it must be so designated. Producers prefer the commonly used mechanical process, for the remaining meal is more palatable.

Linseed meal is the finely ground residue (known as cake, chips, or flakes) remaining after the oil has been extracted. It averages about 35% protein content, about half as much as tankage. The AAFCO stipulates that linseed meal must contain no more than 10% fiber.

For swine, the proteins of linseed meal do not

effectively make good the deficiencies of the cereal grains; linseed meal is low in the amino acids, lysine and tryptophan. Also, linseed meal is lacking in carotene and vitamin D, and is only fair in calcium and the B vitamins. It is laxative and produces a glossy coat on animals to which it is fed. Because of its deficiencies, linseed meal should not be fed to swine as the sole protein supplement.

Where good pasture is available for swine, linseed meal may be mixed with a protein-rich animal or marine product, with the linseed meal limited to 50% of the protein supplement.

Because of its laxative nature, linseed meal in limited quantities is a valuable addition to the diet of brood sows (about ¼ lb daily for mature sows) at farrowing time or for animals that are being forced for the show. Also, it imparts a desirable "bloom" to the hair of show animals.

PEANUT MEAL

Peanut meal, a byproduct of the peanut industry, is ground peanut cake, the product which remains after the extraction of part of the oil of peanuts by pressure or solvents. It is a palatable, high-quality vegetable protein supplement used extensively in livestock and poultry feeds. Peanut meal ranges from 41 to 50% protein and from 4.5 to 8% fat. It is low in methionine, lysine, and tryptophan; and low in calcium, carotene, and vitamin D.

The AAFCO stipulates that peanut meal must contain no more than 7% fiber.

Since peanut meal tends to become rancid when held too long—especially in warm, moist climates—it should not be stored longer than 6 weeks in the summer or 2 to 3 months in the winter.

If a calcium supplement is provided—such as ground limestone or oystershell flour—peanut meal is satisfactory as the only protein supplement for pigs or brood sows on good pasture or for well-grown pigs in confinement. For young pigs and gestating-lactating sows in confinement, peanut meal—like other proteins of vegetable origin—should not constitute more than one-half the protein supplement, the other half being composed of a high-quality protein source such as soybean meal or an animal or marine protein.

SAFFLOWER MEAL

A large proportion of the safflower seed is composed of hull—about 40%. Once the oil is removed from the seeds, the resulting product contains about 60% hulls and 18 to 22% protein. Various means have been tried to reduce this high-hull content. Most meals contain seeds with part of the hull removed, thereby yielding a product of about 15% fiber and 40% protein.

Safflower meal protein is low in lysine and methionine. Its use in swine diets is limited.

SESAME MEAL

Little sesame is grown in the United States despite the fact that it is one of the oldest cultivated oilseeds. The oil meal is produced from the entire seed. Solvent extraction yields higher protein (45%) but lower fat levels (1%) than either the screw-press or hydraulic method which produces meals containing about 38% protein and 5 to 11% oil. Sesame meal is a poor source of lysine but an excellent source of methionine.

SUNFLOWER SEED MEAL

In 1993, 2,486,000 acres of sunflowers were harvested in the United States, and 2.6 billion pounds of sunflower seed was produced.

Sunflower oil meal varies considerably depending on the extraction process and whether the seeds are dehulled. Meal from prepressed solvent extraction of dehulled seeds contains about 44% protein, as opposed to 28% for whole seeds. Screw-pressed sunflower seed meal ranges from 28 to 45% protein. It should not be used as the only protein supplement for swine, because it is low in lysine.

Because of its high fiber content (22 to 24%), sunflower meal should be limited in swine diets. It may replace up to 25% of the protein in the diet of growing-finishing pigs.

LEAF PROTEIN CONCENTRATES

In recent years, considerable attention has been focused on the extraction of protein from green, leafy plants. Green leaves are among the best sources of protein in feeds. To date, most of the research on leaf protein concentrate has centered around the processing of alfalfa; but there is a great potential for LPC development from other sources—for example, vegetable packinghouse byproducts and field wastes.

The advantages of using alfalfa for a protein resource are numerous; among them, (1) it is a perennial crop that can be harvested several times in one growing season; and (2) being a legume, it has the ability to convert nitrogen into protein. Hence, alfalfa yields more protein per acre than any other crop.

It would appear that the most efficient means of preserving the nutrients in alfalfa would be to process the alfalfa immediately after cutting to reduce field losses. Leaves could then be separated from the stems, thereby yielding a high-protein leaf meal. The stems could then be fed as green chop, ensiled or dehydrated to form a medium-protein dry roughage.

Unfortunately, this is not economically feasible on a large-scale basis. Until new developments lower the cost of processing, dehydrated alfalfa meal will remain as a protein supplement of second importance.

■ **Alfalfa meal**—Alfalfa meal is the product obtained from the grinding of the entire alfalfa hay. It may be either sun-cured or dehydrated. The following guarantees are recommended by the AAFCO for the various grades of alfalfa meal and ground alfalfa hay:

Crude Protein	Crude Fiber Not More Than
(%)	(%)
15	30
17	27
18	25
20	22
22	20

Alfalfa meal contains proteins of the right quality to balance out the amino acid deficiencies of the cereal grains; it is a rich source of minerals, especially calcium; and most important, it is an excellent source of vitamins needed by the pig. If leafy and green and not over a year old, alfalfa is high in carotene (provitamin A). Sun-cured hay is also a good source of vitamin D, the anti-rachitic vitamin. Alfalfa is an excellent source of the long list of B vitamins.

Despite alfalfa's virtues, its high fiber (25 to 30%), low energy density, and low palatability, make it unsuited in the diets of nursing pigs and lactating sows.

PULSE PROTEINS

Pulses are the seeds of leguminous plants. They are used primarily for human consumption, but they can be fed to swine effectively when the price is right. Although there are over 13,000 species within the family *Leguminosae*, only about 20 species are used for food and/or feed. These various pulses include: beans (common beans, dry beans, snap beans, kidney beans, navy beans, mung beans, and lima beans), chickpeas, cowpeas, field beans (horse or broad beans), field peas, and pigeonpeas. Their crude protein content ranges from 19 to 31% (see Table 21-1). Soybeans and peanuts are pulses, but they are used almost entirely as oilseed meals in livestock diets.

All of the pulses contain components which possess antinutritional properties. Fortunately, processing procedures, such as cooking, germination, and fermentation, can reduce the risks of feeding pulses to livestock. Among the chemical factors that can create problems in feeding pulses are protease inhibitors, goitrogens, cyanogens, antivitamins, metal-binding factors, lathyrogens, and phytohemagglutinins. When considering pulses for feeding swine, producers should be certain of the necessary processing procedures.

SOYBEANS, RAW

Raw soybeans, especially weather damaged or low test-weight beans, are often attractive alternatives to add to diets for gestating swine. As the pig becomes older, its susceptibility to trypsin inhibitors decreases.

Raw soybeans may be used in gestation diets, but not in lactation diets.

SOYBEANS, FULL-FAT (Heat Processed Soybeans)

The protein of raw soybeans is poorly utilized by young, growing pigs due to the presence of antitrypsin, a powerful growth inhibitor that affects young pigs.

Cooking or roasting at the proper temperature (250°F for 2½ to 3½ minutes in a roaster) destroys this factor and makes soybeans a satisfactory feed for young pigs. However, cooking whole soybeans for gestating brood sows is not necessary.

On-farm roasting or extruding and feeding full-fat soybeans may be more economical than selling the beans and buying back soybean meal and oil. Because whole or full-fat soybeans have less protein and lysine than soybean meal (32 to 37% protein and 2.1 to 2.4% lysine, respectively), it is necessary to add 20 to 25% more whole soybeans than soybean meal to have a similar protein level in the diet. But this will supply approximately 3% added fat to the diet, which will improve feed efficiency by 3 to 5%.

Soybeans, full fat (heat processed beans) must be sold according to its crude protein, crude fat, and crude fiber content.

SOY PROTEIN CONCENTRATE

Soy protein concentrate is produced by removing the water, soluble sugars, ash, and other minor constituents from defatted soy flour by either alcohol, dilute acid, or warm water extraction.

Soy protein concentrate must contain not less than 65% protein on a moisture-free basis. It contains approximately 4.2% lysine. Research indicates that soy protein concentrate can effectively replace dried skim milk in starter pig diets.

SOY PROTEIN ISOLATE

Soy protein isolate, which is the highest soy protein source (it must contain not less than 90% protein on a moisture-free basis), is produced as follows: Defatted soy flakes are insolubilized by reducing the pH to 4.5 (isoelectric point). At this point, the isoelectric proteins are separated from the insoluble materials. Then, the removal of insoluble fibrous material by either decantation or centrifugation completes the protein isolation procedure. The final product can be spray-dried to give an isoelectric protein, or neutralized to pH 7.0 and dried to give the common soy protein isolate.

Soy protein isolate is an effective replacement for dried skim milk in starter pig diets.

WHEAT GLUTEN, SPRAY-DRIED

Spray-dried wheat gluten is the protein fraction of wheat remaining after the starch has been extracted for use in human food products.

Wheat gluten contains approximately the same crude protein content as spray-dried porcine plasma (75 vs 68%, respectively), but it is very low in lysine (1.3%).

Amino acid supplemented starter diets containing spray-dried wheat gluten provide growth performance similar to diets containing dried skim milk.

ANIMAL PROTEINS

Protein supplements of animal origin are derived from (1) meat packing and rendering operations, (2) poultry and poultry processing, (3) milk and milk processing, and (4) fish and fish processing. Before the discovery of vitamin B-12, it was generally considered necessary to include one or more of these protein supplements in the diets of hogs and chickens. With the discovery and increased availability of synthetic vitamin B-12, high-protein feeds of animal origin have become less essential, though they are still included to some extent in diets for most monogastric animals.

MEAT PACKING BYPRODUCTS

Although the meat or flesh of animals is the primary object of slaughtering, modern plants process numerous and valuable byproducts, including protein-rich livestock feeds. The Association of American Feed Control Officials (AAFCO) description of selected meat packing byproducts follows.

MEAT MEAL

Meat meal is produced by the newer dry-rendering method, whereas tankage is produced by the old wet-rendering method. Meat meal is lighter colored than, and does not have the strong odor of, tankage.

Meat meal is the rendered product from mammal tissues, exclusive of any added blood, hair, hoof, horn, hide trimmings, manure, stomach and rumen contents except in such amounts as may occur unavoidably in good processing practices. It shall not contain added extraneous materials not provided for by this definition. The calcium level shall not exceed the actual level of phosphorus by more than 2.2 times. It shall not contain more than 12% pepsin indigestible residue and not more than 9% of the crude protein in the product shall be pepsin indigestible. The label shall include guarantees for minimum crude protein, minimum crude fat, maximum crude fiber, minimum phosphorus, and minimum and maximum calcium.

MEAT AND BONE MEAL

When bone is added, the word *bone* must be inserted in the name of both *meat meal* and *meat meal tankage*; and they are designated as *meat and bone meal* and *meat and bone meal tankage*, respectively.

Meat and bone meal is the rendered product from mammal tissues, including bone, exclusive of any added blood, hair, hoof, horn, hide trimmings, manure, stomach and rumen contents, except in such amounts as may occur unavoidably in good processing practices. It shall not contain added extraneous materials not provided for in this definition. It shall contain a minimum of 4.0% phosphorus and the calcium level shall not be more than 2.2 times the actual phosphorus level. It shall not contain more than 12% pepsin indigestible residue and not more than 9% of the crude protein in the product shall be pepsin indigestible. The label shall include guarantees for minimum crude protein, minimum crude fat, maximum crude fiber, minimum phosphorus, and minimum and maximum calcium.

MEAT MEAL TANKAGE

Generally, meat meal tankage contains about 60% crude protein whereas meat meal runs about 50%.

Meat meal tankage is the rendered product from mammal tissues, exclusive of any added hair, hoof, horn, hide trimmings, manure, stomach and rumen contents, except in such amounts as may occur unavoidably in processing factory practices. It may contain added blood or blood meal; however, it shall not contain any other added extraneous materials not provided for by this definition. The calcium level shall not exceed the actual level of phosphorus by more than 2.2 times. It shall not contain more than 12% pepsin indigestible residue and not more than 9% of the crude protein in the product shall be pepsin indigestible. The label shall include guarantees for minimum crude protein, minimum crude fat, maximum crude fiber, minimum phosphorus, and minimum and maximum calcium.

MEAT AND BONE MEAL TANKAGE

This is the rendered product from mammal tissues, including bone, exclusive of any added hair, hoof, horn, hide trimmings, manure, stomach and rumen contents except in such amounts as may occur unavoidably in good processing practices. It may contain added blood or blood meal; however, it shall not contain any added extraneous materials not provided for in this definition. It shall contain a minimum of 4.0% phosphorus and the calcium level shall not be more than 2.2 times the actual phosphorus level. It shall not contain more than 12% pepsin indigestible residue and not more than 9% of the crude protein in the product shall be pepsin indigestible. The label shall include guarantees for minimum crude protein, minimum crude fat, maximum crude fiber, minimum phosphorus, and minimum and maximum calcium.

HYDROLYZED HAIR

This product is prepared from clean, undecomposed hair, by heat and pressure to produce a product suitable for animal feeding. Not less than 80% of its crude protein must be digestible by the pepsin digestibility method.

GLANDULAR MEAL AND EXTRACTED GLANDULAR MEAL

This is obtained by drying liver and other glandular tissues from slaughtered mammals. When a significant portion of the water soluble material has been removed, it may be called *extracted glandular meal*.

MEAT PROTEIN ISOLATE

This is produced by separating meat protein from fresh, clean, unadulterated bones by heat processing followed by low temperature drying to preserve function and nutrition. This product is characterized by a fresh meaty aroma, a 90% minimum protein level, 1% maximum fat, and 2% maximum ash.

BLOOD PRODUCTS

Historically, blood has been cooked and dried as blood meal for use in livestock feeds using the traditional vat cooking process. Although such processing results in a high protein ingredient, it has found limited use in pig diets because of lack of palatability and low availability of lysine. Spray drying and newer flash drying procedures have significantly improved both the palatability and lysine availability of blood meal.

Spray-dried porcine plasma and spray-dried blood meal, byproducts from pig slaughter, have revolutionized nutritional programs for early-weaned pigs.

■ **Spray-dried porcine plasma**—This product is made up of the albumin, globin, and globulin fractions of blood and contains 68% protein and 6.1% lysine. The blood is collected in refrigerated tanks, and prevented from coagulating by adding sodium citrate. The plasma fraction is separated from the blood cells by centrifugation and stored at 25°F until the product is spray dried.

■ **Spray-dried blood meal**—This product is processed similar to porcine plasma, except it contains both the plasma and red-blood cell fractions.

Synthetic methionine should be added to starter diets containing either spray-dried porcine plasma or spray-dried blood meal.

Both spray-dried blood products are effective protein sources in starter pig diets.

POULTRY WASTES

Byproduct feedstuffs are derived from all segments of the poultry industry—from hatching all the way through processing for market; and they come from the broiler and turkey segments of the industry

Fig. 11-5. Proper balance of amino acids is less critical for finishing hogs than for young pigs or breeding swine. (Courtesy, Land O Lakes, Ft. Dodge, IA)

as well as from egg production. Centralization of these industries into large units with enough volume of wastes to make it feasible to process the potential feeds has opened new markets for what was previously a disposal problem. Certain precautions have had to be included to make the products most useful and safe, but considerable amounts of poultry products are currently being used in diets of various animals, both ruminants and monogastrics.

POULTRY

This is the clean combination of flesh and skin with or without accompanying bone, derived from the parts or whole carcasses of poultry or a combination thereof, exclusive of feathers, heads, feet, and entrails. It shall be suitable for use in animal food. If it bears a name descriptive of its kind, it must correspond thereto. If the bone has been removed, the process may be so designated by use of the appropriate feed term.

HYDROLYZED WHOLE POULTRY

This is the product resulting from the hydrolyzation of whole carcasses of culled or dead, undecomposed, poultry including feathers, heads, feet, entrails, undeveloped eggs, blood, and any other specific portions of the carcass. The product must be consistent with the actual proportions of whole poultry and must be free of added parts; including, but not limited to entrails, blood, or feathers. The poultry may be fermented as a part of the manufacturing process. The product shall be processed in such a fashion as to make it suitable for animal food, including heating (boiling at 212°F, or 100°C at sea level, for 30 minutes, or its equivalent, and agitated, except in steam cooking equipment). The product may, if acid treated, be subsequently neutralized. If the product bears a name descriptive of its kind, the name must correspond thereto.

HYDROLYZED POULTRY BYPRODUCTS AGGREGATE

This is the product resulting from hydrolyzation, heat treatment, or a combination thereof, of all byproducts of slaughter poultry, clean and undecomposed, including such parts as heads, feet, undeveloped eggs, intestines, feathers, and blood. The parts may be fermented as a part of the manufacturing process. The product shall be processed in such a fashion as to make it suitable for animal food, including heating (boiling at 212°F, or 100°C at sea level, for 30 minutes, or its equivalent, and agitated, except in steam cooking equipment). It may, if acid treated, be subsequently neutralized. If the product bears a name descriptive of its kind, the name must correspond thereto.

POULTRY BYPRODUCT MEAL

This consists of the ground, rendered, clean parts of the carcass of slaughtered poultry, such as necks, feet, undeveloped eggs, and intestines, exclusive of feathers, except in such amounts as might occur unavoidably in good processing practices. The label shall include guarantees for minimum crude protein, minimum crude fat, maximum crude fiber, minimum phosphorus, and minimum and maximum calcium. The calcium level shall not exceed the actual level of phosphorus by more than 2.2 times.

Because of the heads and feet, poultry byproducts are lower in nutritional value than the flesh of animals, including poultry. Thus, the biological value of the proteins is lower than the other animal proteins. They may be successfully used in pig diets, however, provided they are not the sole source of proteins.

POULTRY BYPRODUCTS

This must consist of non-rendered clean parts of carcasses of slaughtered poultry such as heads, feet, viscera, free from fecal content and foreign matter except in such trace amounts as might occur unavoidably in good factory practice. Because of the heads and feet, poultry byproducts are lower in nutritional value than the flesh of animals, including poultry.

HYDROLYZED POULTRY FEATHERS

This is the product resulting from the treatment under pressure of clean, undecomposed feathers from slaughtered poultry, free of additives, and/or accelerators. Not less than 75% of its crude protein content must be digestible by the pepsin digestibility method.

Although hydrolyzed feather meal is high in protein, it is rather low in nutritional value, being low in the amino acids histidine, lysine, methionine, and tryptophan. The amino acids which are present are readily available.

Because of the deficiencies of several amino acids, care must be exercised when incorporating feather meal into swine feed. The addition of fish meal or meat meal tends to complement feather meal and facilitates its use. In practice, feather meal rarely exceeds 5% of the diet of swine.

POULTRY HATCHERY BYPRODUCT

This is a mixture of egg shells, infertile and unhatched eggs, and culled chicks which have been cooked, dried, and ground, with or without removal of part of the fat.

Hatchery byproducts are the most valuable of the

feed byproducts from poultry. However, these products deteriorate quite rapidly if not cooled promptly.

EGG PRODUCT

This product is obtained from egg graders, egg breakers, and/or hatchery operations that is dehydrated, handled as liquid, or frozen. These shall be labeled as per USDA regulations governing eggs and egg products (7CFR, Part 59). This product shall be free of shells or other non-egg materials except in such amounts which might occur unavoidably in good processing practices, and contain a maximum ash content of 6% on a dry matter basis.

DAIRY PRODUCTS

Skimmed milk, buttermilk, and whey have long been used on or in close proximity to the farms where they are produced. However, in their liquid form it is impossible to ship them long distances or to store them and it is difficult to maintain sanitary feeding conditions in using them.

Along about 1910, processes were developed for drying buttermilk. Soon thereafter, special plants were built for dehydrating buttermilk, and the process was extended to skimmed milk and whey. Beginning about 1915, dried milk byproducts were incorporated in commercial poultry feeds. Subsequently, they have been used for swine feeding. But even today poultry consume a larger quantity of dried milk byproducts than swine.

But, milk byproducts furnish a relatively small percentage of the total high-protein feeds consumed by livestock.

The superior nutritive values of milk byproducts are due to their high-quality proteins, vitamins, a good mineral balance, and the beneficial effect of the milk sugar, lactose. In addition, these products are palatable and highly digestible. They are an ideal feed for young pigs and for balancing out the deficiencies of the cereal grains. The chief limitation to their wider use is price, together with the perishability and bulkiness of the liquid products. Liquid products coming from cows that have not been tested and found free from tuberculosis should be pasteurized.

Although whole milk is an excellent feed—worth about twice as much as skimmed milk—it is usually too expensive to feed to swine. For this reason, only the milk byproduct feeds will be discussed. In general, liquid milk products contain 10% dry matter; semisolid milk products, 30% dry matter; and dried milk products, 90% dry matter.

SKIMMED MILK

Because of the removal of the fat, skimmed milk supplies but little vitamin A value. However, in comparison with whole milk, it is higher in content of protein, milk sugar, and minerals. Like all milk products, skimmed milk is low in vitamin D and iron.

Skimmed milk is the best single protein supplement for swine. It is especially valuable for young pigs prior to and immediately after weaning. The addition of either pasture or a choice legume hay will supplement skimmed milk with the needed vitamins A and D.

Skimmed milk should be fed consistently sweet or sour, because abrupt changes are apt to produce digestive disturbances. Where a choice is possible, fresh skimmed milk is recommended.

The amount of skimmed milk to feed will vary according to (1) the available supply, (2) the relative price of feeds, (3) the kind of grain diet fed, and (4) whether or not pasture is available.

Under confinement conditions and with corn constituting all or most of the grain diet, a daily allowance of 1 to 1.5 gal of skimmed milk per pig will be about right to balance the diet. With the same grain diet and good pasture, the skimmed milk allowance may be cut in half. With barley or wheat (feeds of higher protein content) replacing the corn, about two-thirds of these amounts of skimmed milk will suffice. Since the protein requirement of the swine diet decreases with the age of the animal while total daily feed consumption increases, a constant daily intake of skimmed milk on this basis will about balance the swine diet throughout the life of the pig.

The feeding value of skimmed milk varies with the age of the pigs, the type of diet, the price of other feeds, and the amount of milk fed. Naturally, it has a higher value per pound when fed in limited amounts than when an excess is used. On the other hand, when the supply of milk is abundant and cheap, a larger proportion may be fed profitably—especially when grain is scarce and high in price. Roughly, it can be figured that (1) 15 lb of skimmed milk will replace 1 lb of tankage or meat meal, or (2) 6 lb of skimmed milk will replace 1 lb of complete feed.

CONDENSED SKIMMED MILK

This is the residue obtained by evaporating defatted milk. It contains 27% minimum total solids.

DRIED WHOLE MILK
(Dried Milk, Feed Grade)

Dried whole milk is the residue left following the drying of milk. It contains at least 26% milk fat. The label must contain a guarantee for minimum crude protein and for minimum crude fat.

BUTTERMILK

Buttermilk and skimmed milk have approximately the same composition and feeding value for swine, providing buttermilk has not been diluted by the addition of churn washings. Accordingly, the discussion of skimmed milk is also applicable to buttermilk; and it may be considered that (1) 15 lb of buttermilk will replace 1 lb of tankage or meat meal, or (2) 6 lb of buttermilk will replace 1 lb of complete feed.

CONDENSED BUTTERMILK

Condensed buttermilk is made by evaporating buttermilk to about one-third of its original weight. It contains 27% minimum total solids, .055% minimum milk fat for each percent of total solids, and 0.14% maximum ash for each percent of total solids.

One lb of condensed buttermilk is worth about 3 lb of liquid buttermilk. But it requires about 3 lb of condensed buttermilk to equal 1 lb of dried milk. Another basis of evaluating condensed buttermilk is that, pound for pound, it is worth approximately one-half as much as tankage.

Like the other milk byproducts, condensed buttermilk is (1) very palatable to pigs, (2) more valuable for young pigs than for older animals, and (3) more valuable for confinement feeding than for pigs on pasture. Perhaps the greatest value of condensed buttermilk is in the capacity of an appetizer. Swine exhibitors make extensive use of it in their fitting diets.

DRIED SKIMMED MILK AND DRIED BUTTERMILK

As the names indicate, these products are dehydrated skimmed milk and buttermilk, respectively. They contain less than 8% moisture, and average 32 to 35% protein. One lb of dried skimmed milk or dried buttermilk has about the same composition and feeding value as 10 lb of its liquid form. Though dried skimmed milk and dried buttermilk are excellent swine feeds, they are generally too high priced to be economical for this purpose except in pig starter diets.

Because of the palatability, pigs will consume excessive amounts of dried skimmed milk and dried buttermilk when self-fed, free-choice.

Prior to World War II, dried skimmed milk was the most widely used dried milk product included in feeds. During and since the war, however, much of it has been marketed as a human food.

DRIED CULTURED SKIMMED MILK AND CONDENSED CULTURED SKIMMED MILK

These two milk products are obtained following the culturing of skimmed milk with lactic acid bacteria. The dried product contains a maximum of 8% moisture, and the condensed product contains at least 27% total solids.

CHEESE RIND (Cheese Meal)

This is a byproduct from the manufacture of processed cheese, consisting of the cheese trimmings from which most of the fat has been removed. It contains around 60% protein and 9% fat. It is slightly more valuable than tankage as a swine feed.

MILK PROTEIN PRODUCTS

The four milk protein products currently available are:

1. **Dried milk protein.** This feed is obtained by drying the coagulated protein residue resulting from the controlled coprecipitation of casein, lactalbumin, and minor milk proteins from defatted milk.

2. **Dried milk albumin.** This product is produced by drying the coagulated protein residue separated from whey. It contains a minimum of 80% crude protein on a moisture-free basis.

3. **Casein.** Casein is the solid residue that remains after the acid or rennet coagulation of defatted milk. It contains a minimum of 80% crude protein.

4. **Dried hydrolyzed casein.** This is the residue obtained by drying the water-soluble product resulting from the enzymatic digestion of casein. It contains at least 74% crude protein.

Although the quality and quantity of protein from milk protein products are excellent, these sources of protein are generally too expensive to be used routinely as swine feeds except for baby pigs.

WHEY

Whey is a byproduct of cheese making. Practically all of the casein and most of the fat go into the cheese, leaving the whey which is high in lactose (milk sugar), but low in protein (0.6 to 0.9%) and fat (0.1 to 0.3%). However, its protein is of high quality. In general, whey has a feeding value equal to about one-half that of skimmed milk or buttermilk; thus, (1) 30 lb of whey will replace 1 lb of tankage, or (2) 12 lb of whey will replace 1 lb of complete feed. Pigs should be gradually accustomed to whey, after which it may be fed free-choice. Because of its low protein content, growing pigs should receive a protein-rich supplement in addition to grain and whey. In order to prevent the spread of disease, whey should be pasteurized at the factory and fed under sanitary conditions.

Numerous whey products are commercially available. Those recognized by the Association of American Feed Control Officials follow:

1. **Condensed cultured whey.** This is the residue obtained by evaporating cultured whey. The label of this product must contain the minimum percent of total cultured whey solids.

2. **Condensed hydrolyzed whey.** This residue, obtained by evaporating lactase enzyme hydrolyzed whey, contains at least 50% total solids and 0.3% total glucose and galactose for each percent total solids.

3. **Condensed whey.** This is the product resulting from the removal of a considerable portion of water of either cheese whey or casein whey. The minimum percent total whey solids must be prominently declared on the label.

Based on their respective nutrient contents, 1 lb of condensed whey should be worth about 9 lb of liquid whey for hogs.

4. **Condensed whey-product.** This is the residue obtained by evaporating whey from which a portion of lactose has been removed. The minimum percent of total whey-product solids, crude protein, and lactose, and the maximum percent ash, must be prominently shown on the label.

5. **Condensed whey solubles.** This is the product obtained by concentrating the whey residue after the removal of whey protein, with or without partial removal of lactose. Minimum percent of solids, crude protein, and lactose, and maximum percent ash must be prominently declared on the label.

6. **Dried hydrolyzed whey.** This product is the residue obtained by drying lactose hydrolyzed whey. It contains at least 30% glucose and galactose.

7. **Dried whey.** This is derived from drying whey from cheese manufacture. It contains not less than 11% protein nor less than 61% lactose; and it is rich in riboflavin, pantothenic acid, and some of the important unidentified factors. One lb of dried whey contains about the same nutrients as 13 to 14 lb of liquid whey.

8. **Dried whey-product.** When a portion of the lactose (milk sugar) which normally occurs in whey is removed, the resulting dried residue is called dried whey-product. According to the Association of American Feed Control Officials, the minimum percent of crude protein and lactose, and the maximum percent ash must be prominently declared on the label of the dried whey-product. Dried whey-product is a rich source of the water-soluble vitamins. When price relationships warrant, dried whey-product may replace alfalfa meal in confinement diets for swine on a comparable protein content basis.

9. **Dried whey solubles.** This product is the prod-

uct obtained by drying the whey residue from the manufacture of lactose following the removal of milk albumin and partial removal of lactose. The minimum percent of crude protein and lactose, and the maximum percent ash must be prominently declared on the label.

OTHER DAIRY PRODUCTS

Other dairy products available in certain areas are: (1) dairy food byproducts, (2) condensed modified whey solubles, (3) dried (dry) whey protein concentrate, (4) dried cultured whey product, (5) dried cultured whey, (6) dried chocolate milk, (7) dried cheese, and (8) dried cheese product.

MARINE BYPRODUCTS

In the beginning, marine wastes were dumped into the sea. Later, some of them were dried at high temperatures—usually in an open flame dryer—and used in fertilizers. About 1910, it was discovered that waste materials from fish canning plants and inedible fish were a desirable protein source for livestock feeding. Again experiment stations led the way in determining the feeding value of these new products which are now incorporated in swine and poultry diets from coast to coast.

FISH MEAL

Fish meal—a byproduct of the fisheries industry—consists of dried, ground whole fish or fish cuttings—either or both—with or without the extraction of part of the oil. If it contains more than 3% salt, the

Fig. 11-6. A typical menhaden fishing boat. Fishing is big business, of which fish meal is one of the byproducts. (Courtesy, National Fisheries Institute, Inc., Washington, DC)

salt content must be a part of the brand name. In no case shall the salt content exceed 7%. It must not contain more than 10% moisture.

The feeding value of fish meal varies somewhat, according to:

1. **The method of drying.** It may be vacuum, steam, or flame dried. The older flame drying method exposes the product to a higher temperature. This makes the proteins less digestible and destroys some of the vitamins.

2. **The type of raw material used.** It may be made from the offal produced in fish packing or canning factories, or from the whole fish with or without extraction of part of the oil.

Fish meal made from offal containing a large proportion of heads is less desirable because of the lower quality and digestibility of the proteins. Although few feeding comparisons have been made between the different kinds of fish meals, it is apparent that all of them are satisfactory when properly processed raw materials of good quality and moderate fat content are used. A high fat content may impart a fishy taste to eggs, meat, and milk.[2] Also, such meal is apt to become rancid in storage.

It is of interest to the swine producer to know the sources of the commonly used fish meals. These are:

1. **Menhaden fish meal.** This is the most common kind of fish meal used in the eastern states. It is made from menhaden herring (a very fat fish not suited for human food) caught primarily for its body oil. The meal is the dried residue after most of the oil has been extracted.

2. **Sardine meal or pilchard meal.** This is made from sardine canning waste and from the whole fish, principally on the West Coast.

3. **Herring meal.** This is a high-grade product produced in the Pacific Northwest and in Alaska.

4. **Salmon meal.** This is a byproduct of the salmon canning industry in the Pacific Northwest and in Alaska.

5. **White fish meal.** This is a byproduct from fisheries making cod and haddock products for human food. Its proteins are of very high quality.

Fish meal should be purchased from a reputable company on the basis of protein content. It varies in protein content from 57 to 77%, depending on the kind of fish from which it is made. When of comparable quality, fish meal is even superior to tankage or meat meal as a protein supplement for swine. The protein

---

[2]The Indiana Station (Vestal, *et al.*, *Jr. An. Sci.*, Vol. 4, No. 1) found that the addition to the diet of 0.5 and 1.5% fish oil produced a fishy flavor in pork, which was more pronounced in the roasts and bacon than in the chops.

Fig. 11-7. A close-up view of fish cuttings, which will be used in making fish meal. The meal is the dried residue after most of the oil has been extracted. (Courtesy, U.S. Department of Interior, Stuttgart, AR)

of a good-quality fish meal is 92 to 95% digestible. If fish meal is poorly processed or improperly stored, the digestibility of protein decreases dramatically. Since fish meals are cooked, there is danger that certain amino acids—notably lysine, cystine, tryptophan, and histidine—will be denatured, but these losses are minimized when proper processing techniques are used.

Fish meals containing high levels of fat are considered to be low quality. If they are incorporated into feeds, they tend to impart a fishy flavor to the animal products. Also, problems of rancidity are greater in high-fat fish meals.

Fish meal is an excellent source of minerals. Calcium and phosphorus are especially abundant, being present in the amounts of 3 to 6% and 1.5 to 3.0%, respectively. Many of the trace minerals, especially iodine, required by swine can be supplied in part by fish meal.

Fish meal is not a particularly good source of vitamins. Most of the fat-soluble vitamins are lost during the extraction of oil, but a fair amount of the B vitamins remain. However, fish meal is one of the richest sources of vitamin B-12 and unidentified growth factors.

The difference between fish meal and tankage or meat meal is not so marked for pigs on pasture as for pigs in confinement. When fed as the only protein supplement to corn for pigs on pasture, fish meal is worth approximately 10% more, pound for pound, than tankage or meat meal. Because of the generally higher cost of fish meal, it is commonly recommended that it be mixed with one or more protein-rich feeds of plant origin in providing supplements for pigs on pasture.

FISH RESIDUE MEAL

This is the dried residue from the manufacture of glue from nonoily fish. The provisions relative to salt content are the same as those for fish meal.

CONDENSED FISH SOLUBLES

This is a semisolid byproduct obtained by evaporating the liquid remaining from the steam rendering of fish, chiefly sardines, menhaden, and redfish. The gluewater that comes from processing contains about 5% total solids. After this liquid is evaporated or condensed, it contains about 50% total solids. Condensed fish solubles, containing approximately 30% crude protein, are a rich source of the B vitamins and unknown factors. They are particularly rich in pantothenic acid, niacin, and vitamin B-12.

Unfortunately, fish solubles cannot be stored successfully in dried form but must be left in a sticky, semisolid form. For this reason, their use is largely limited to commercially mixed diets or supplements for swine in confinement.

DRIED FISH SOLUBLES

Dried fish solubles are obtained by dehydrating the gluewater of fish processing. They contain at least 60% crude protein.

SHRIMP MEAL AND CRAB MEAL

Along the coastal regions of the United States, large quantities of potential feedstuffs are produced by the shellfish industry. Byproducts from the shrimp and crab industries are presently being converted to high-protein feeds.

Shrimp meal is the ground, dried waste of the shrimp industry, which may consist of the head, hull (or shell), and/or whole shrimp. The provisions relative to salt content are the same as those for fish meal. It is either steam dried or sun dried, with the former method being preferable. On the average, it contains 32% protein and 18% minerals.

Crab meal, the byproduct of the crab industry, is composed of the shell, viscera, and flesh. Mineral content is exceedingly high, about 40%. It must contain not less than 25% crude protein. A rule of thumb for feeding crab meal is that 1.6 lb of crab meal can replace 1 lb of fish meal.

FISH LIVER AND GLANDULAR MEAL

This marine byproduct is obtained by drying the entire viscera of fish. The AAFCO specifies that it must contain at least 18 mg of riboflavin per pound and that at least 50% of dry weight must consist of livers.

FISH PROTEIN CONCENTRATE

Fish protein concentrate is prepared through the solvent extraction processing of clean, undecomposed whole fish or fish cuttings. The product cannot contain more than 10% moisture and must contain at least 70% protein, according to the AAFCO. The solvent residues must conform to the rules as set forth by the Food Additive Regulations.

FISH BYPRODUCTS

This must consist of non-rendered, clean, undecomposed portions of fish (such as, but not limited to, heads, fins, tails, ends, skin, bone, and viscera) which result from the fish processing industry.

DRIED FISH PROTEIN DIGEST

This is the dried enzymatic digest of clean, undecomposed whole fish or fish cuttings using the enzyme hydrolysis process. The product must be free of bones, scales, and undigested solids, with or without the extraction of part of the oil. It must contain not less than 80% protein and not more than 10% moisture.

CONDENSED FISH PROTEIN DIGEST

This is the condensed enzymatic digest of clean, undecomposed whole fish or cuttings using the enzyme hydrolysis process. The product must be free of bones, scales, and undigested solids, with or without the extraction of part of the oil. It must contain not less than 30% protein.

FISH DIGEST RESIDUE

This is the clean, dried, undecomposed residue (bones, scales, undigested solids) of the enzymatic digest resulting from the enzyme hydrolysis process of producing fish protein digest. It must be designated according to its protein, calcium, and phosphorus content.

CONDENSED FISH SOLUBLES

This is obtained by evaporating excess moisture from the stickwater, aqueous liquids, resulting from the wet rendering of fish into fish meal, with or without removal of part of the oil. Minimum percent of solids, minimum percent of crude protein, and minimum percent of crude fat must be guaranteed.

SINGLE-CELL PROTEIN (SCP)

This refers to protein obtained from single-cell organisms, such as yeast, bacteria, and algae, that have been grown on specially prepared media. Production of this type of protein can be attained through the fermentation of petroleum derivatives or organic waste, or through the culturing of photosynthetic organisms in special illuminated ponds.

Historically, algae was a staple food of the Aztecs of Mexico. The high-protein algae was carried as rations by warriors. For centuries, the natives of Lake Chad in Africa have dried and eaten algae from the lake. Of course, yeast and bacteria have long been used in the baking, brewing, and distilling industries, in making cheese and other fermented foods, and in storing and preserving foods. Dried brewers' yeast, a residue from the brewing industry, and torula yeast resulting from the fermentation of wood residue and other cellulose sources, are currently marketed.

A wide variety of materials can be used as substrates for the growth of these organisms. Current research deals with the use of industrial byproducts which otherwise would have little or no economic value. Byproducts from the chemical, wood and paper, and food industries have shown considerable promise as sources of nutrients for single-cell organisms; among them, (1) crude and refined petroleum products, (2) methane, (3) alcohols, (4) sulfite waste liquor, (5) starch, (6) molasses, (7) cellulose, and (8) animal wastes.

The potential of single-cell protein as a high-protein source for both humans and livestock is enormous, but many obstacles must be overcome before it becomes widely used. It has been calculated that a single-cell protein fermenter covering one-third square mile could yield enough protein to supply 10% of the world's needs. There are, however, serious problems involving palatability, gastrointestinal disturbances, uric acid accumulation, and simple economics must be solved before widescale production of single-cell protein becomes a reality. There is a limited amount of single-cell protein on the market in the form of brewers' yeast and torula yeast, but these products are generally too expensive to use as a major protein source.

TYPES OF SINGLE-CELL PROTEIN

Single-cell protein can be produced by nonphotosynthetic and photosynthetic organisms. Of the nonphotosynthetic organisms, yeasts are the most popular sources, but bacteria and fungi are also currently being investigated as potential sources. The photosynthetic organisms—algae—are grown in ponds that are illuminated and fortified with simple salts such as carbonates, nitrates, and phosphates.

■ **Nonphotosynthetic organisms**—Traditionally, yeasts have been used as sources of vitamins and unidentified factors. For this purpose yeasts are classified into two groups:

1. **Yeast products used chiefly as a source of B vitamins and protein.** This group includes brewers' dried yeast, torula dried yeast, grain distillers' dried yeast, and molasses distillers' dried yeast.

2. **Irradiated dried yeast.** This is yeast that has been exposed to ultraviolet light, and which may be used as a source of vitamin D for four-footed animals.

Yeast contains about 45% protein, but one disadvantage to the utilization of yeasts as protein sources is that they are deficient in the sulfur amino acids. Fortunately, methionine hydroxy analog (MHA) is a cheap commercial product that can make up for this deficiency.

Bacteria grow very rapidly and tend to have a more balanced amino acid profile than yeasts do. Additionally, they generally contain more protein. Since humans will probably avoid this type of protein for esthetic reasons, this type of single-cell protein, in all likelihood, will be developed as livestock feed. Currently, there are two serious problems that exist with the culture of bacterial protein: (l) susceptibility to phages (bacteria-destroying agents), and (2) high ribonucleic acid content.

Until recently, the fungi have received little attention as single-cell protein. Since fungi can be harvested with relative ease and possess considerable enzymatic activity, they will likely be intensively studied in the near future.

■ **Photosynthetic organisms**—Photosynthetic organisms, such as algae, provide nutrients in relatively large quantities in a limited area. They convert sun energy directly into food. Depending on the type of organism, temperature, and latitude, yields of harvested protein can be attained on the order of 4 to 16 tons per acre yearly.

These organisms contain 5 to 15% dry matter. Nutrient composition (moisture-free) tends to be rather variable, with protein ranging from 8 to 75%, carbohydrate from 4 to 40%, lipids from l to 86%, and ash from 4 to 45%. The protein obtained from this type of single-cell protein is deficient in the sulfur amino acids, methionine and cystine. The biological value of algae protein has been estimated to range from 50 to 70. Additionally, algae provide a means of recycling animal wastes and renovating waste water.

PROBLEMS ASSOCIATED WITH SINGLE-CELL PROTEIN

Although single-cell protein appears to be an excellent alternative or supplemental source of protein, several problems must be overcome before it becomes a widely used feedstuff; among them, palatability, digestibility, nucleic acid content, toxicities, protein quality, and economics.

■ **Palatability**—Microbial cells must be processed to some extent if they are to be palatable. Otherwise, animals will not eat them. Yeasts are bitter tasting, and algae and bacteria have characteristically unpleasant tastes which can depress intake. At the present time, processes making single-cell protein tasteless are being developed so that this source of protein can be used as a supplement to more conventional feeds.

■ **Digestibility**—Ways of making single-cell protein more digestible must be developed if it is to be competitive with traditional protein feeds. Digestibility among single-cell product sources tends to be extremely variable. When eaten alone, the digestibility of algae is low, while mixing algae with other feeds improves digestibility. If the organisms are not killed prior to being used as a feed, digestibility is dramatically reduced. Certain forms of processing can improve the digestibility of some single-cell protein products. For example, if algae are cooked prior to use, digestibility can be doubled in many cases. However, processing beyond the killing of yeasts does little to improve digestibility.

■ **Nucleic acid content**—Much of the nitrogen found in single-cell organisms is in the form of nucleic acids. When the purines of the nucleic acids are metabolized, uric acid is formed. In humans, uric acid is relatively insoluble with the result that uric acid deposits accumulate and lead to kidney stones or gout. Some of the current research is aimed at reducing the nucleic acid concentration.

Although the feeding of SCP to livestock may not produce physiological problems, it will be necessary to demonstrate to the consuming public that problems will not arise from the consumption of animal products produced from the feeding of SCP.

■ **Toxicities**—Toxicities arising from the use of single-cell protein can result from two sources: (1) from toxins produced by the microorganisms themselves, and (2) from contaminated microorganisms. The second type of risk is probably the most likely to occur. Much of the single-cell protein will be derived from processes whereby byproducts of industry are used as substrates. Thus, the microorganisms will in some ways reflect the chemical composition of the byproduct. For example, if chemical residues, such as pesticides, or large amounts of trace minerals are present in the byproduct substrate,

the microorganism will, in all likelihood, absorb these chemicals. When livestock consume these contaminated microorganisms, toxicities may result. Hence, before widespread use can be made of single-cell proteins, there is the problem of convincing people and agencies of their wholesomeness.

■ **Protein quality**—Research to date has indicated that SCP is deficient in the sulfur-containing amino acids, and possibly in lysine and isoleucine. While the amino acid profile of SCP is more balanced than that of the cereal grains, it is clearly inferior to the traditional protein supplements. However, this inferiority can often be nullified by adding commercially available methionine, or by combining with other protein sources. Also, the advances in genetic engineering offer hope of developing new microorganisms capable of producing ideal amino acid profiles.

■ **Economic problems**—As long as the more traditional sources of protein—such as the oilseed meals, meat, fish, eggs, and milk—are readily available at reasonable prices, single-cell protein will remain a relatively obscure alternative.

As the world's human population increases, there will be an increasing demand for cheap protein. This demand could dry up the sources of traditional protein, thereby opening the way for intensive development of alternative sources of protein.

We are also in an era of concern for the maintenance of the quality of our environment. This means that greater emphasis will be placed on the transformation of industrial byproducts into commodities that can be used by both livestock and humans. Single-cell protein is one way that this challenge can be met.

It is noteworthy that full-scale production of high-protein animal supplement, made up of dried cells from a microorganism that lives on methanol, is now underway at Bellingham, in northern England. An annual production of 50,000 tons has been projected.[3]

MIXED PROTEIN SUPPLEMENTS

With the knowledge of the protein sources available, protein supplements can be mixed which will fill the needs of various ages of swine at a favorable price. Protein supplements containing minerals and vitamins can be purchased commercially or mixed on the farm. In turn, these may be combined with different cereal grains to obtain a diet of the desired level of protein. Tables 11-4 and 11-5 give suggested formulas for 35% and 40% protein supplements, while Table 11-6 shows how these protein supplements (40% and 35%, re-

---

[3]Sherwood, M., *Science News*, Vol. 11, February 14, 1981, p. 106.

Fig. 11-8. Mixed protein supplements combined with farm (local) grain have been replaced by complete mixed diets on big swine operations. But mixed protein supplements are often economical and practical on smaller operations. (Courtesy, Land O Lakes, Ft. Dodge, IA)

spectively) may be combined with different cereal grains to obtain diets with varying levels of protein. (For specific uses of these diets, refer to Chapter 9.)

Many protein supplements, particularly those of vegetable origin, are lacking in certain of the essential amino acids. Further, most protein feeds are deficient in certain vitamins and minerals, and perhaps in unknown factors. Therefore, pigs in confinement that are fed a diet consisting only of cereal grain and any of the common protein supplements are apt to suffer deficiency trouble, with unthrifty and runty pigs resulting. Before the time of vitamin and mineral premixes and amino acid balancing, proteins from different sources were combined so that the deficiencies in amino acids, in vitamins, and/or in minerals of one would be offset by surpluses from another. As a result, mixed protein supplements proved more efficient than single supplements.

TABLE 11-4
THIRTY-FIVE PERCENT PROTEIN SUPPLEMENTS

Ingredient	Protein Concentration	Supplement Number							
		1		2		3		4	
	(%)	(lb)	(kg)	(lb)	(kg)	(lb)	(kg)	(lb)	(kg)
Solvent soybean meal	44	1,670	757.4	1,575	714.3	1,100	498.9	1,050	476.2
Dehydrated alfalfa meal	17	—	—	100	45.4	—	—	200	90.7
Meat and bone meal	50	—	—	—	—	400	181.4	400	181.4
Corn	8.9	—	—	—	—	290	131.5	145	65.8
Calcium carbonate (39% Ca)	—	80	36.3	75	34.0	50	22.7	45	20.4
Dicalcium phosphate (22% Ca, 18.5% P)	—	140	63.5	140	63.5	60	27.2	60	27.2
Iodized salt	—	50	22.7	50	22.7	40	18.1	40	18.1
Trace mineral premix	—	10	4.5	10	4.5	10	4.5	10	4.5
Vitamin Premix	—	50	22.7	50	22.7	50	22.7	50	22.7
Total		2,000	907.0	2,000	907.0	2,000	907.0	2,000	907.0

Calculated analysis:

Protein ...(%)		36.74	35.50	35.49	35.45
Calcium ...(%)		3.31	3.26	3.39	3.42
Phosphorus ...(%)		1.80	1.78	1.74	1.73
Salt added ...(%)		2.50	2.50	2.00	2.00
Lysine ...(%)		2.51	2.40	2.21	2.19
Methionine ...(%)		0.53	0.51	0.50	0.49
Cystine ...(%)		0.56	0.54	0.51	0.52
Tryptophan ...(%)		0.53	0.52	0.41	0.42
Metabolizable energy (ME) ...(kcal/lb)		1,223	1,206	1,253	1,212
...(kcal/kg)		2,697	2,659	2,763	2,672

Feeding Directions

These supplements can be used to make growing-finishing, gestation, or lactation diets. Table 11-6 shows how these protein supplements may be combined with different cereal grains to obtain a diet of the desired level of protein.

The meat and bone meal was considered to have 8.1% calcium and 4.1% phosphorus. If meat and bone meal with a higher concentration of calcium and phosphorus is used, the amount of dicalcium phosphate should be reduced accordingly.

The corn can be replaced by wheat midds, corn distillers' grains with solubles, or other grain byproducts.

The level of feed additives will depend on the type of diet in which the supplement is going to be used, but should be 3 to 5 times higher than desired in the complete diet.

TABLE 11-5
FORTY PERCENT PROTEIN SUPPLEMENTS

Ingredient	Protein Concentration	Supplement Number 1		2		3		4	
	(%)	(lb)	(kg)	(lb)	(kg)	(lb)	(kg)	(lb)	(kg)
Solvent soybean meal	48.5	1,628	738.3	—	—	1,238	561.5	1,458	661.2
Solvent soybean meal	44	—	—	1,223	554.6	—	—	—	—
Dehydrated alfalfa meal	17	—	—	—	—	100	45.4	—	—
Meat and bone meal	50	—	—	550	249.4	400	181.4	—	—
Fish meal, menhaden	61	—	—	—	—	—	—	200	90.7
Calcium carbonate (39% Ca)	—	90	40.8	45	20.4	55	24.9	80	36.3
Dicalcium phosphate (22% Ca, 18.5% P) . .	—	160	72.6	60	27.2	85	38.5	140	63.5
Iodized salt	—	50	22.7	50	22.7	50	22.7	50	22.7
Trace mineral premix	—	12	5.4	12	5.4	12	5.4	12	5.4
Vitamin Premix	—	60	27.2	60	27.2	60	27.2	60	27.2
Total		2,000	907.0	2,000	907.0	2,000	907.0	2,000	907.0

Calculated analysis:

		1	2	3	4
Protein . . . .(%)		39.48	40.66	40.87	41.46
Calcium . . . .(%)		3.67	3.92	3.76	3.74
Phosphorus . . . .(%)		2.01	2.05	2.02	2.05
Salt added . . . .(%)		2.50	2.50	2.50	2.50
Lysine . . . .(%)		2.69	2.55	2.60	2.86
Methionine . . . .(%)		0.55	0.56	0.56	0.66
Cystine . . . .(%)		0.59	0.57	0.59	0.58
Tryptophan . . . .(%)		0.55	0.46	0.49	0.56
Metabolizable energy (ME) . . . . (kcal/lb)		1,299	1,212	1,280	1,319
. . . . (kcal/kg)		2,864	2,672	2,822	2,908

Feeding Directions

These supplements can be used to make growing-finishing, gestation, or lactation diets. Table 11-6 shows how these protein supplements may be combined with different cereal grains to obtain a diet of the desired level of protein.

The meat and bone meal was considered to have 8.1% calcium and 4.1% phosphorus. If meat and bone meal with a higher concentration of calcium and phosphorus is used, the amount of dicalcium phosphate should be reduced accordingly.

The level of feed additives will depend on the type of diet in which the supplement is going to be used, but should be 4 to 6 times higher than desired in the complete diet.

TABLE 11-6
RATIO OF GRAIN TO PROTEIN SUPPLEMENTS NEEDED TO OBTAIN THE DESIRED LEVEL
OF PROTEIN IN A DIET, WITH AND WITHOUT ALFALFA MEAL[1] (As-Fed Basis)

% Protein Desired	40% Supplement		35% Supplement		Corn (8.9% Crude Protein)		Ground Barley (11.6% Crude Protein)		Ground Oats (11.7% Crude Protein)		Ground Wheat (12.7% Crude Protein)		Alfalfa Meal (15.3% Crude Protein)	
	(lb)	(kg)	(lb)	(kg)	(lb)	(kg)	(lb)	(kg)	(lb)	(kg)	(lb)	(kg)	(lb)	(kg)
18	575	261	—	—	1,325	602	—	—	—	—	—	—	100	45
18	500	227	—	—	750	341	650	295	—	—	—	—	100	45
18	525	238	—	—	1,000	454	—	—	375	170	—	—	100	45
18	450	204	—	—	—	—	925	420	—	—	525	238	100	45
18	375	170	—	—	—	—	—	—	550	250	975	443	100	45
18	—	—	675	307	1,225	556	—	—	—	—	—	—	100	45
18	—	—	575	261	675	307	650	295	—	—	—	—	100	45

(Continued)

TABLE 11-6 (Continued)

% Protein Desired	40% Supplement		35% Supplement		Corn (8.9% Crude Protein)		Ground Barley (11.6% Crude Protein)		Ground Oats (11.7% Crude Protein)		Ground Wheat (12.7% Crude Protein)		Alfalfa Meal (15.3% Crude Protein)	
	(lb)	(kg)	(lb)	(kg)	(lb)	(kg)	(lb)	(kg)	(lb)	(kg)	(lb)	(kg)	(lb)	(kg)
18	—	—	650	295	1,000	454	—	—	—	—	250	114	100	45
18	—	—	500	227	—	—	1,000	454	—	—	400	182	100	45
18	—	—	500	227	—	—	—	—	500	227	900	409	100	45
18	600	272	—	—	1,400	636	—	—	—	—	—	—	—	—
18	525	238	—	—	875	397	600	272	—	—	—	—	—	—
18	550	250	—	—	1,000	454	—	—	450	204	—	—	—	—
18	425	193	—	—	—	—	925	420	—	—	650	295	—	—
18	400	182	—	—	—	—	—	—	700	317	900	409	—	—
18	—	—	725	329	1,275	579	—	—	—	—	—	—	—	—
18	—	—	650	295	750	341	600	272	—	—	—	—	—	—
18	—	—	675	307	1,000	454	—	—	325	148	—	—	—	—
18	—	—	500	227	—	—	850	386	—	—	650	295	—	—
18	—	—	500	227	—	—	—	—	550	250	950	431	—	—
16	450	204	—	—	1,450	658	—	—	—	—	—	—	100	45
16	350	159	—	—	850	386	700	317	—	—	—	—	100	45
16	300	136	—	—	—	—	950	431	—	—	650	295	100	45
16	400	182	—	—	1,150	522	—	—	350	159	—	—	100	45
16	275	125	—	—	—	—	—	—	625	284	1,000	454	100	45
16	—	—	550	250	1,350	613	—	—	—	—	—	—	100	45
16	—	—	425	193	775	352	700	317	—	—	—	—	100	45
16	—	—	375	170	—	—	875	397	—	—	650	295	100	45
16	—	—	500	227	1,050	477	—	—	350	159	—	—	100	45
16	—	—	325	148	—	—	—	—	575	261	1,000	454	100	45
16	375	170	—	—	925	420	700	317	—	—	—	—	—	—
16	475	216	—	—	1,525	692	—	—	—	—	—	—	—	—
16	425	193	—	—	1,225	556	—	—	350	159	—	—	—	—
16	250	114	—	—	1,150	522	—	—	—	—	700	318	—	—
16	250	114	—	—	—	—	—	—	750	341	1,000	454	—	—
16	—	—	450	204	850	386	700	317	—	—	—	—	—	—
16	—	—	575	261	1,425	647	—	—	—	—	—	—	—	—
16	—	—	525	238	1,125	511	—	—	350	159	—	—	—	—
16	—	—	200	91	1,100	499	—	—	—	—	700	318	—	—
16	—	—	175	80	—	—	—	—	725	329	1,100	499	—	—
14	325	148	—	—	1,575	715	—	—	—	—	—	—	100	45
14	225	102	—	—	875	397	800	363	—	—	—	—	100	45
14	275	125	—	—	1,325	602	—	—	300	136	—	—	100	45
14	—	—	375	170	1,525	692	—	—	—	—	—	—	100	45
14	—	—	350	114	850	386	800	363	—	—	—	—	100	45
14	—	—	350	159	1,250	568	—	—	300	136	—	—	100	45
14	350	159	—	—	1,650	749	—	—	—	—	—	—	—	—
14	250	114	—	—	950	431	800	363	—	—	—	—	—	—
14	300	136	—	—	1,300	590	—	—	400	182	—	—	—	—
14	—	—	425	193	1,575	715	—	—	—	—	—	—	—	—

(Continued)

TABLE 11-6 (Continued)

% Protein Desired	40% Supplement		35% Supplement		Corn (8.9% Crude Protein)		Ground Barley (11.6% Crude Protein)		Ground Oats (11.7% Crude Protein)		Ground Wheat (12.7% Crude Protein)		Alfalfa Meal (15.3% Crude Protein)	
	(lb)	*(kg)*	*(lb)*	*(kg)*	*(lb)*	*(kg)*	*(lb)*	*(kg)*	*(lb)*	*(kg)*	*(lb)*	*(kg)*	*(lb)*	*(kg)*
14	—	—	300	*136*	900	*409*	800	*363*	—	—	—	—	—	—
14	—	—	350	*159*	1,250	*568*	—	—	400	*182*	—	—	—	—
12	200	*91*	—	—	1,700	*772*	—	—	—	—	—	—	100	*45*
12	150	*68*	—	—	1,350	*613*	400	*182*	—	—	—	—	100	*45*
12	150	*68*	—	—	1,450	*658*	—	—	300	*136*	—	—	100	*45*
12	—	—	250	*114*	1,650	*749*	—	—	—	—	—	—	100	*45*
12	—	—	175	*80*	1,325	*602*	400	*182*	—	—	—	—	100	*45*
12	—	—	200	*91*	1,400	*636*	—	—	300	*136*	—	—	100	*45*
12	225	*102*	—	—	1,775	*806*	—	—	—	—	—	—	—	—
12	150	*68*	—	—	1,250	*568*	600	*272*	—	—	—	—	—	—
12	200	*91*	—	—	1,500	*681*	—	—	300	*136*	—	—	—	—
12	—	—	275	*125*	1,725	*783*	—	—	—	—	—	—	—	—
12	—	—	200	*91*	1,300	*590*	500	*227*	—	—	—	—	—	—
12	—	—	225	*102*	1,475	*670*	—	—	300	*136*	—	—	—	—

[1]In order to obtain an 18% protein feed, one could mix 575 lb of 40% supplement, 1,325 lb of corn, and 100 lb of alfalfa meal. Likewise, a 14% supplement without alfalfa meal (for use on pasture) could be obtained by mixing 300 lb of 35% supplement, 900 lb of corn, and 800 lb of barley.

QUESTIONS FOR STUDY AND DISCUSSION

1. What parts of plants contain the highest concentration of protein?

2. What is "protein quality"?

3. Discuss the relative importance of plant and animal protein sources (see Table 11-2). Which sources have become increasingly more important in recent years?

4. Explain why pigs do not have a specific requirement for crude protein.

5. Name six alternate amino acid sources from each (a) plants and (b) animals.

6. Discuss each of the following: (a) soybean meal as the sole source of supplemental protein in swine diets, (b) an ideal protein, (c) limiting amino acid, (d) amino acid availability, (e) formulating swine diets on an available lysine basis, (f) the use of crystalline amino acids in swine diets.

7. List the commonly used oilseed meals. Why are they called oilseed meals?

8. Discuss the feeding value of oilseeds in comparison to soybean meal.

9. Define *pulses*. List five types of pulses and indicate what precautions should be taken when pulses are to be used as either feed or food.

10. Briefly discuss the concept of leaf protein concentrates as applicable to swine.

11. What is alfalfa meal, and what is its value in swine diets?

12. What are the following ingredients and how should they be used: soybeans, full-fat; soybeans, raw; soy protein concentrate; soy protein isolate; and wheat gluten, spray-dried?

13. Compare tankage and meat meal to meat and bone meal.

14. What are spray-dried porcine plasma, and spray-dried blood meal? How are these products being used?

15. What byproducts and residues are included in hatchery byproducts?

16. What are the limitations to the use of dairy products?

17. Differentiate between skimmed milk and buttermilk.

18. What is whey and what nutritional value does it have? List the whey products.

19. List some common fish meals, and describe the feeding value of fish meals.

20. What are condensed fish solubles?

21. What is a single-cell protein? Is any currently used as animal feed?

22. Briefly discuss the advantages and disadvantages inherent with single-cell protein.

20. Describe 35 and 40% protein supplements and their use in formulating diets of 14% protein when barley and corn are the grain source.

SELECTED REFERENCES

Title of Publication	Author(s)	Publisher
Alternative Sources of Protein for Animal Production	National Research Council	National Academy of Sciences, Washington, DC, 1973
Association of American Feed Control Officials Incorporated, Official Publication	Association of American Feed Control Officials, Inc.	Association of American Feed Control Officials, Inc., annual
Feeds & Nutrition Digest	M. E. Ensminger J. E. Oldfield W. W. Heinemann	The Ensminger Publishing Company, Clovis, CA, 1990
Feeds & Nutrition, Second Edition	M. E. Ensminger J. E. Oldfield W. W. Heinemann	The Ensminger Publishing Company, Clovis, CA, 1990
Nonnutritional Feed Sources For Use in Swine Production	P. A. Thacker R. N. Kirkwood	Butterworth Publishers, Stoneham, MA, 1990
Protein Resources and Technology Status and Research Needs		U.S. Government Printing Office, Washington, DC, 1975
Swine Feeding and Nutrition	T. J. Cunha	Academic Press, Inc., New York, NY, 1977
Swine Nutrition Guide	J. F. Patience P. A. Thacker	Prairie Swine Centre, University of Saskatchewan, Saskatoon, Canada, 1989
Swine Production and Nutrition	W. G. Pond J. H. Maner	AVI Publishing Co., Inc., Westport, CT, 1984
Swine Production in Temperate and Tropical Environments	W. Pond J. Maner	W. H. Freeman and Co., San Francisco, CA, 1974

Gestating sow on pasture. (Courtesy, McClean County Hog Service, Inc., LeRoy, IL)

12

FORAGES FOR SWINE

Pigs, pastures, and profits are in vogue in Britain, where it is known as "intensive outdoor production." Not only that, they're environmentally and animal welfare friendly; there are no manure hauling and no utility bills; and the cost of outdoor farrowing is much less than for confinement farrowing.

Experimentally, and by some swine producers, it is being tried in the United States. For the mega swine operations, the back 40 would likely be too distant; so, it is expected that they will continue to favor confinement production.

The outdoor system of the 1990s, successfully pioneered in Britain and being tried in the United States, is very different from the earlier outdoor system in the United States. The pasture plots of the modern system being promulgated in the '90s are of radial design, modified on the outside to accommodate a rectangular area (see Fig. 12-1).

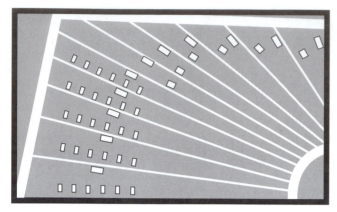

Fig. 12-1. Radial design pastures with individual huts. Some divide the pens with an electric fence.

GESTATING SOWS AND OUTDOOR FARROWING

There are no gestation crates, and no farrowing crates!

The modern outdoor farrowing system can be established and maintained with a minimum of capital—perhaps less than half the per sow cost of a modern confinement facility. Typically, an outdoor farrowing unit consists of the following:

1. **Hub area.** The hub area provides (a) access to each pen and (b) boar pens.

2. **Gestation pens.** Each gestation pen is equipped with a metal or A-hut (wooden), shade, self-feeder, and water.

3. **Farrowing pens.** Each equipped with a steel or wood hut, shade, self-feeder, creep-feeder, and water.

4. **Nursery unit.** Each equipped with huts, creep-feeder, self-feeder, and water.

5. **Growing-finishing unit.** Each equipped with a shelter, a large self-feeder, and water.

Sows are generally pen bred to farrow every week of the year. Pigs are weaned at 21 to 28 days of age. A climate that ranges from 25 to 85°F is ideal for outdoor production.

■ **Radial design vs traditional rectangular swine pastures**—Presently, only a relatively few U.S. swine pastures are of radial design. It is anticipated that they will increase. However, traditional rectangular swine pastures will continue, also. Of whatever design, swine pastures will be used primarily (1) by small- and medium-sized producers, and (2) for breeding animals. Mega producers will continue to favor complete confinement.

Chapter 12, Forages for Swine, is applicable to all swine pastures regardless of shape or design.

■ **Feed savings on pasture**—Research reports on feed savings by swine on pasture vary considerably, depending on type of pasture, class and age of hogs, and management system. On the average, however, bred sows and gilts on good legume pasture require about half as much grain and much less supplemental protein than those in drylots; and good pasture for growing-finishing pigs will effect a savings of 3 to 10% of the grain and of as much as 33% of the protein supplement in comparison to confinement production. Thus, the decision on whether to raise swine on pasture or in drylot should be based primarily on (1) net returns, and (2) whether the land can be put to a more profitable alternative use.

FARROW-TO-MARKET ON PASTURE

Farrow-to-market on pasture requires more labor, but less capital, than confinement systems. Also, farrow-to-market on pasture provides an opportunity to phase into hog production gradually, allowing the producer to learn and develop the necessary skills before expanding into a confinement system that requires more capital and intensive management. Young or inexperienced swine producers often choose a pasture production system.

The primary disadvantage with pasture systems is that some producers use the flexibility of the system to move "in and out" of swine production at the wrong time. They are prone to increase when market prices and profits are high, and to decrease or stop production when prices and profits are low.

Growing hogs (40 to 125 lb) will consume 0.75 lb pasture dry matter plus 3.75 lb of the diet per day.

Finishing hogs (125 to 240 lb) will consume 1.0 lb pasture dry matter plus 5.0 lb of the diet per day.

ADVANTAGES OF PASTURES

Many practical swine producers feel that pasture production does have certain real advantages over confinement production. It should be acknowledged, however, that in much of the earlier work in which comparisons were made between confinement and pasture diets, inadequate confinement diets were used. As a result, actually such trials were comparisons between the advantages of adequate pasture diets over inadequate confinement diets. The truth is that before swine nutritionists improved swine diets, profitable confinement rearing was nearly impossible.

In general, the following advantages may be cited in favor of pasture production over confinement production of swine:

1. **Saves feed.** Pastures make for a saving in feed costs in both grain and protein supplements. With properly balanced diets used in each case, the feed saving effected through the utilization of swine pastures is about as follows:

   a. Good pastures will reduce (1) the grain required in producing 100 lb of pork by 15 to 20%, and (2) the protein supplement required in producing 100 lb of pork by 20 to 50%.

   b. An acre of good pasture will result in a saving of 500 to 1,000 lb of grain and 300 to 500 lb of protein supplement.

   c. With mature brood sows, good pastures may lower feed costs by 50%. With pastures that possess a heavy legume content, sows can be fed 2 lb less grain and .5 lb less protein supplement per head per day during gestation. In fact, sows may get the major portion of their feed from good pastures up to 6 to 8 weeks before farrowing. The condition of sows is the best guide as to the amount of concentrate feeding necessary.

   d. When growing-finishing hogs are on high-quality legume pasture, the protein level of the diet can be reduced by 2% as compared to confinement feeding. When finishing pigs are on a limited-feeding program, only one-half to three-fourths as many animals per acre can be carried as when pigs are full-fed.

   e. Good pastures furnish a convenient and economical way to compensate for the protein, mineral, and vitamin deficiencies of grain and other high-energy feeds. This does not infer that protein, mineral, and vitamin supplements need not be added to the diet. Rather, the problem is simplified, and it may be solved at lower cost.

   f. Good pastures make for a slight saving in minerals—about 3 lb per acre.

   g. As a result of a 3-year study conducted by the Alabama Agricultural Experiment Station, researchers reported that they reduced feed costs by as much as $40.00 per sow per year by placing them on a rotational grazing system. The rotation pasture system used in the study consisted of a perennial pasture of orchardgrass and crimson clover; Tifleaf millet, a summer annual; and ryegrass and clover, a winter annual.

2. **Lessens nutritional deficiencies.** Pastures lessen nutritional deficiencies in swine, chiefly because of (a) their high-quality proteins, (b) their vitamins (an abundance of carotene and the water-soluble vitamins, plus the vitamin D value of the sunlight to which animals on pasture are exposed), (c) their unknown factors (most, if not all, of which they contain), and (d) their minerals (especially calcium). These nutrients are not present in adequate amounts in all confinement diets consisting of grains, oil meals, and mill feeds. Of course, with present knowledge of nutrition and available supplements, many confinement diets are as well balanced as most pasture diets.

3. **Lessens communicable diseases.** Hogs on pasture come in contact with each other less than hogs in confinement, with the result that fewer communicable disease problems are encountered than where hogs are in close confinement. For example, it is easier to control the spread of diseases such as TGE or other types of baby pig scours when pigs are farrowed in individual houses rather than in a central farrowing house.

4. **Lessens capital required for buildings and equipment.** Less expensive buildings (including movable houses) and equipment can be used in a pasture system than in confinement, with the result that it requires less capital investment on a per hog basis.

5. **Makes for greater flexibility.** Pasture operations are more flexible than confinement programs—an important consideration where renters are involved or where other uncertainties exist about a long-range program.

6. **Does not require as high levels of skill and management.** Pasture rearing does not require as high levels of skill and management as are necessary to make confinement production work. Also, it requires more competent labor to operate a fully automated, highly mechanized, confinement complex than a pasture operation.

7. **Makes for approved soil conservation practices on rolling land.** Where there is rolling land not suitable for cropping and/or need for organic matter, a pasture system may be preferred to confinement. Certainly, pastures conserve the maximum fertility value of the manure and lessen erosion. When animals are on pasture, 80% of the plant nutrients may be returned to the soil, and pigs will spread their own manure.

8. **Improves reproduction.** Pastures provide a desirable way of life for breeding animals, chiefly because of improved nutrition and valuable exercise. Thus, they result in more satisfactory litters being farrowed and in a more abundant milk flow. Also, boars on pasture are more vigorous and surer breeders.

DISADVANTAGES OF PASTURES

Among the disadvantages sometimes attributed to pasture systems are:

1. **It requires more labor.** Running hogs on pas-

ture does not lend itself to automation and laborsaving devices to the extent of confinement production, primarily in feeding and watering. Also, it may make for operator discomfort and inconvenience.

2. **Lower rate of gain.** Growing-finishing hogs on pasture usually make lower rates of gain.

3. **It may prevent more remunerative uses of land.** On many hog farms, operators can make more money from growing corn, soybeans, and other crops than they can from pastures. Often, the feed saved by the pasture will not pay for the cost of the pasture.

3. **It does not facilitate manure handling.** Although less manure has to be handled in a pasture system than in confinement, it is more difficult to automate and handle manure where hogs are scattered over a large area.

4. **It prevents enlarging hog production without enlarging the farm.** With high priced land, this fact must be weighed when it is desired to increase the size of the hog operation.

5. **It cannot be used year round in cold regions.** In many areas the weather is too adverse for pasture facilities to be used during the winter months. Furthermore, there may be more death loss of young pigs during cool or cold weather.

DESIRABLE CHARACTERISTICS OF A GOOD PASTURE

Although it is recognized that no one forage excels all others in all the desired qualities, the following desirable characteristics may serve as criteria in the choice of pasture crops for swine:

1. Adapted to local soil and climatic conditions. Although this is a prime requisite, practical swine producers cannot afford to disregard the grazing qualities.

2. Palatable and succulent.

3. Ability to endure tramping and grazing.

4. Easy to grow, and grown at a nominal cost.

5. Provide tender and succulent growth for a short period or consistent growth over a long period.

6. Highly nutritious; rich in proteins. vitamins, and minerals, and low in fiber.

7. High carrying capacity.

8. Fit satisfactorily into the crop rotation.

9. Uncontaminated with diseases or parasites.

Over most of the United States, one or more adapted legumes possess most of these qualities. Fortunately, they can be selected without fear of bloat, for swine—unlike cattle and sheep—are not susceptible to this ailment.

MANAGEMENT OF SWINE AND PASTURES

The principles of good swine management on pastures are the same, regardless of the kind of pasture or its location. Further, good management of both the pastured animals and the pastures go hand in hand; they are inseparable. They both require attention in order to get the highest returns. In brief, good producers, good pastures, and good hogs go together.

In general, consciously or unconsciously, the practical operator gives attention to the following swine pasture management factors:

1. **Self-feeding and limit-feeding.** Gestating sows on pasture are usually self-fed and limit-fed. This is generally accomplished by interval feeding in which sows consume 2 or 3 days of feed in one day, and then wait 2 or 3 days before being provided access to feed again. Adjustment in daily feed intake is made by altering either the time on the feeder (2 to 12 hours) or time off the feeder (2 or 3 days). Third day feeding is not recommended for gilts because they gain less weight and farrow smaller pigs. So, gilts should be limit-fed daily.

2. **Quantity of protein.** In arriving at the quantity of protein supplement to feed swine on pasture, consideration should be given to the following:

a. In general, pigs full-fed on a good legume pasture require about 50% as much protein supplement as when full-fed in confinement. Pigs limit-fed on good pasture will eat more forage and require only 30 to 40% as much protein supplement as confinement-fed pigs.

b. Less protein supplement is required for older hogs than for young pigs, because mature animals consume more pasture—finally reaching the stage where pastures supply ample proteins to balance their needs.

c. Less protein supplement is required with pastures that are higher in protein content and more palatable.

d. Less protein supplement will result in slower gains. Regardless of the weight of the hogs—whether weaner pigs or 150-lb shoates—any lowering of the protein supplement below that required to balance the diet will result in less rapid gains, with younger pigs being relatively more affected. Thus, where an earlier market is apt to mean higher prices, it may, in the end, be more economical to put more money into the purchase of high priced protein supplements.

e. Less protein supplement will require more pasture acreage, because of higher forage consumption per animal.

f. Less protein supplement will result in more

rooting, as the pigs attempt to meet their requirements from the soil.

3. **Providing young, succulent pastures.** As pasture crops mature, they become less palatable and less desirable because of higher fiber content and lower amounts of proteins, minerals, and vitamins. Clipping at intervals during the growing season is the best method of keeping swine pastures young and succulent. Mowing is also effective in both weed and parasite control (the latter because of exposing the feces to drying).

4. **Fencing.** Hogs on pastures should be fenced properly, with consideration given to kind of fence and size of area. With smaller areas and higher concentration of hogs, it is recommended that the woven wire used meet the following specifications: 32- to 36-in. height, 6-in. mesh, number 9 top and bottom wires, and number 11 staves. With larger areas (10 acres or more) and less concentration of swine, a 26-in. woven wire is satisfactory, with the other specifications remaining the same as for the higher fence. With all hog fences, one strand of barbed wire—either 2 or 4 point—should be stretched tightly between the woven wire and the ground. Of course, it goes without saying that hog fences—like any farm fence—should be stretched properly and stapled to durable posts that are properly set and spaced, and have suitable corner posts that are well braced.

When pasturing swine, many producers install a permanent perimeter fence, but use temporary divider fences along with low voltage electricity.

5. **Shelter.** Swine on pastures should be provided with proper shelter, including both huts and shades. For flexibility in rotating pastures, portable equipment is preferable. Suitable movable houses and shades are illustrated in Chapter 16.

6. **Clean, fresh water.** Swine should have access to clean, fresh water at all times. Under a system of pasture rotation, it is often difficult and expensive to pipe water to the area. But usually the added expense will be justified. Where possible, automatic waterers should be provided. Hand-watering on pasture is inconvenient, labor-consuming, and otherwise unsatisfactory.

7. **Ringing.** When rooting is encountered—a somewhat normal habit with any pig, but a condition which is accentuated with deficient or limited diets—the animals should be rung. Any one of several types of rings may be used with success if properly inserted in the snout.

8. **Minerals.** The cereal grains and their byproducts, and most all protein-rich supplements, are high in phosphorus, whereas legume and canola pastures are a fair source of calcium. Other needed minerals, including iodized salt in iodine-deficient areas, should be incorporated in the diet.

9. **Early spring and late fall grazing.** Early spring and late fall grazing are desirable for swine, but undesirable for pastures. In the fall, the plants need protection to build up root reserves. In recognition of this situation, a compromise is usually necessary—pasturing early and late, but not enough to do too great harm to the pastures. Where plenty of pasture areas are available, good swine and good pasture management in this regard are usually achieved through the rotation of pastures.

10. **Overgrazing.** Continuous overgrazing is to be avoided. For this reason, it is recommended that, except for annuals, the forage be left to grow up so that one to two crops of hay may be harvested each season in addition to the pasture. As a rule of thumb, the grazing should be regulated so as to allow the top growth to maintain a height of 3 to 6 in. To accomplish this, the swine producer should always have more than just enough pasture.

An acre of good pasture plus full-feeding of grain and supplement should, without overgrazing, provide enough forage for 15 to 20 growing-finishing pigs from weaning to the end of the pasture season, or for 6 to 8 mature bred sows, or for 4 to 6 sows with litters. When limit-feeding is practiced, pastures will carry only one-half to two-thirds as many hogs per acre as when full-feeding.

11. **Combination grazing with other classes of animals.** More than one class of animals can be grazed on most pastures—the practice of so-called combination grazing—with benefit to both the animals and the pastures. Although the carrying capacity (or animal units per acre) is slightly higher when combination grazing is practiced, there is a tendency to overstock, through trying to add what would be a normal number of animals per acre of each class when pastured separately.

Swine and cattle or sheep can be advantageously grazed on the same area. But avoid overgrazing. And do not include cows near calving because of the hazard of hogs inflicting injury on the vulva of cows or harming newborn calves. The swine-cattle combination is particularly desirable because (a) cattle will eat the coarser vegetation, leaving young, succulent growth for swine; and (b) most diseases and parasites afflict only one or the other class of animals.

12. **Scattering droppings and handling manure.** Proper scattering of the droppings on pasture is desirable for two reasons: (a) it conserves the maximum fertility value of the manure; and (b) it lessens the disease and parasite hazard, through exposing the droppings to the action of the sun and the air. The droppings may be scattered easily by harrowing the area at intervals, especially in the early spring and late fall.

For best parasite and disease control, swine manure should never be hauled from swine barns or lots and scattered on pastures utilized by this particular class of livestock.

13. **Rotating pastures.** Modern swine producers realize the importance of rotating pastures in order to control parasites and lessen the disease hazard. Pigs should never be allowed to follow pigs on a particular pasture without plowing the field in the interim. In the more heavily parasitized areas, it is recommended that swine be kept off permanent pastures for 2 or 3 years. Also, in rotating pastures, avoid cross-drainage from one field to another.

14. **Year-round grazing.** Because of the many advantages of pastures (as listed in an earlier section in this chapter), good swine management involves as nearly year-round grazing as is possible. From this standpoint, the southern states have a very real advantage over other areas. Through planting different crops at different seasons, progressive southern swine producers are achieving year-round grazing.

Some Corn Belt swine producers are obtaining a 12-month hog pasture by using two crops only, ladino clover and rye. In the North, where year-round grazing cannot be secured, it is merely recommended that swine producers attain as long a grazing season as is possible, especially through arranging for early spring and late fall pastures.

15. **Permanent vs temporary pastures.** It is advisable to use temporary rather than permanent pastures in small lots or where a high concentration of hogs are kept during most of the pasture season. Also, temporary pastures are better suited to a system requiring frequent plowing—once or twice each year—from the standpoint of parasite and disease control. One or more such pastures are usually included in the plan in order to secure the longest possible grazing season.

16. **Puddling.** Pastures should be protected during periods of heavy rainfall or when irrigating by removing swine until the soil has dried sufficiently to prevent puddling.

17. **Pasture cooling of gestating sows.** Records of the Ft. Reno, Oklahoma Research Station show that, on average, sows farrowing in August and September farrow two less pigs and wean about 1 less pig than sows farrowing in spring. Also, weaning weights are lighter for August and September farrowed pigs.

Pasture cooling of bred sows can be accomplished several ways: Adequate shade over sand that is wet down several times on hot days can be used. A fogging nozzle over sand or concrete is more desirable and less time consuming. However, the best method of providing summer cooling on pasture is a permanent structure which provides shade, fogging, and a concrete wallow.

18. **Pollution.** Little pollution potential exists from pasture systems with low animal densities or numbers, or where pastures are rotated. However, in high-den-

Fig. 12-2. For disease control, pastures should be well-drained, rotated, and not overgrazed. Also, fresh, clean drinking water should be provided rather than relying on puddles. (Courtesy, The Upjohn Company, Kalamazoo, MI)

sity pasture systems involving a large number of hogs, a pollution potential does exist. Frequently, there is little vegetation in these pasture lots, and rainwater falling on the lot drains away, carrying some solids with it. Where the lots are steeply sloping or where a watercourse runs adjacent to or through the lots, the problem can be serious.

Where a pollution problem exists from pasture lots, the site should be abandoned or all runoff should be caught in a retention reservoir. After each rain, liquid must be removed from the reservoir by pumping and/or evaporation to ensure sufficient volume to capture all runoff from the next rain.

TYPES AND COMPOSITIONS OF PASTURES

Pastures selected for swine should be succulent and capable of high production, very palatable, high in protein and vitamins, and produce over a reasonably long growth period.

In general, temporary pastures are preferable to permanent pastures for swine, especially from the standpoint of disease and parasite control.

The legumes as a group have a higher protein, calcium, and carotene content than grasses. They can furnish an adequate supply of most vitamins with the exception of vitamins D and B-12.

The following types of pastures should be considered.

LEGUMES

Legumes are plants that have the ability to work symbiotically with bacteria to fix nitrogen from the air.

The most commonly used legumes for swine are: alfalfa, alsike clover, birdsfoot trefoil, crimson clover, ladino clover, lespedeza, red clover, soybean forage, sweetclover, and white clover.

■ **Alfalfa**—Most of the research on the use of forage in swine diets has been with alfalfa. It appears to be the most practical forage crop for the pig because it can be used for pasture, silage, and as an ingredient in the diet. Because of their maturity, sows make better use of alfalfa (and other forages) than do growing-finishing hogs. Potential benefits to accrue from feeding alfalfa during gestation include improved survival of the baby pigs during the nursing period and a reduced culling rate in the sow herd. Some research has shown that as much as 97% alfalfa can be included in gestation diets without impairing reproductive performance. For commercial operations, however, no more than 65% alfalfa should be used in gestation diets. Gestation diets containing approximately 60% alfalfa can be self-fed.

Fig. 12-3. Pigs in an alfalfa pasture in Nebraska. (Courtesy, Burlington Northern, Inc., St. Paul, MN)

Growing pigs show satisfactory performance on diets containing alfalfa, provided the level does not exceed 20% of the diet. Even at this level, there will be a 5 to 15% depression in feed efficiency compared to a grain-soybean meal diet. Receiving diets containing 10% alfalfa may improve gain and feed intake in newly arrived feeder pigs.

■ **Alsike clover**—This clover provides a leafy crop with fine stems, and it grows well in soils that are too acidic or too wet for red clover. It is often used in pasture mixtures.

■ **Birdsfoot trefoil**—Birdsfoot trefoil is palatable and similar in nutrient content to alfalfa. Unlike alfalfa, it grows well on poorly drained soils. While it is not as productive as alfalfa on good soils, birdsfoot trefoil yields have exceeded yields of alfalfa on the wetter soils. Most varieties perform better in cooler climates. It will normally outlive red clover by several years.

■ **Crimson clover**—Crimson clover provides a good spring forage and sometimes winter forage in warm climates.

■ **Ladino clover**—Under optimum conditions, ladino clover will not produce as much forage per acre as alfalfa. But the protein content of ladino clover is superior to that of alfalfa. Ladino clover works best as an all-summer pasture crop in the northeastern and north central states.

■ **Lespedeza**—Also called *Japanese clover*, this species is less palatable to pigs than all clovers except sweetclover. It cannot be grazed until mid-summer, but it does grow reasonably well without lime and fertilizer and will adapt to soils that cannot be used for red clover.

■ **Red clover**—Red clover is a short-lived, relatively easy-to-establish perennial legume that will grow on soils too acidic or too wet for alfalfa. Red clover does not yield as much forage early in the spring as alfalfa and it is not as drought resistant as alfalfa. It is useful for pasture or silage. It will provide good forage through most of the grazing season if it is not overgrazed nor allowed to become too mature. Several studies have shown that pigs on red clover forage gain as rapidly as those on alfalfa.

■ **Soybean forage**—Soybeans as a green forage are less valuable than alfalfa, red and ladino clovers, and canola (rape). The soybeans should be grown in rows to reduce damage from trampling. Unlike most forages, soybeans cannot regenerate new growth from the crowns. Soybeans are less sensitive to nutrient levels in the soil than are alfalfa and clovers. In hot climates, soybeans may out-yield other legumes during the same period of time.

Pigs grazing mature soybeans should also have access to grain fortified with vitamins and minerals. However, inhibitors in the raw soybeans will prevent the pig from utilizing the dietary protein efficiently. In addition, the oil contained in the beans tends to make the carcass soft.

■ **Sweetclover**—Since pigs find sweetclover unpalatable, it may be more suitable for soil improvement. Sweetclover may be planted on soils not adapted for alfalfa or other clovers. If biennial sweetclover is sown in the spring, the first season's growth is more succulent and palatable than that harvested during the second summer.

■ **White clover**—White clover is a practical perennial legume to use with permanent pasture, especially those containing bluegrass. White clover makes a high-quality pasture and it does well in years of frequent rain. Note that ladino clover is a large-type white clover. Dutch or common white clover is a small type.

BRASSICAS

■ **Canola (rape)**—Rape is a high-yielding, fast-growing annual forage that belongs to the brassica family. Related species include kale and swede. Canola provides an excellent pasture for swine. When overgrazing is avoided, it provides abundant, palatable forage for a long-growing season. Canola can lead to photosensitization (sun-burning) when grazed wet. Pigs with white skin are most sensitive.

Fig. 12-4. Berkshire pigs on canola pasture. (Courtesy, J. C. Allen and Son, West Lafayette, IN)

GRASSES

■ **Bluegrass**—Bluegrass may serve as a permanent pasture for swine. The pasture can be grazed early, but it contains less protein than do legumes and is usually dormant during the warmest part of the summer.

■ **Smooth bromegrass**—Bromegrass is a palatable crop that withstands heavy grazing. Its early spring growth enables it to be pastured for longer periods than many legumes. Studies show that pigs on bromegrass pastures require more grain and supplement than pigs grazing alfalfa. Bromegrass can be successfully mixed with legumes.

■ **Orchardgrass**—Orchardgrass, a perennial, is a hardy species that can tolerate trampling. It quickly loses its palatability if not grazed down to prevent the grass from becoming tall and mature.

■ **Sudangrass**—Sudangrass, an annual, is palatable to pigs, and when seeded thickly, it provides ample

forage during the hottest part of the summer when other species are dormant. The early growth of Sudangrass contains a cyanogen, which may be converted to prussic acid (extremely toxic to pigs) under certain conditions such as wilting, trampling, chewing, frost, and drought. Because of the near-neutral pH in the rumen, ruminants are more sensitive to cyanogens than are nonruminants. Poisoning can be avoided if the grass is grazed only after it reaches a height of at least 18 to 24 in. Because Sudangrass is low in protein, it is better adapted for sows and older market hogs.

■ **Timothy**—Timothy withstands heavy use, but it should only be included as a minor part of a pasture mixture since it is less desirable than most other pasture crops.

■ **Winter rye**—Winter rye seeded during late summer will provide a useful forage crop for winter or early spring grazing. Optimal planting time should provide just enough growth so that seed stems are starting to shoot when the plant enters winter dormancy. When pigs are allowed to graze rye during the winter and spring months, stock the pasture with no more than 8 growing pigs, or 3 to 5 sows, per acre.

■ **Winter wheat and barley**—These two cereal grains are at least as palatable and nutritious as rye, but they do not provide as much fall production as rye and they cannot be grazed as heavily. Note that fresh wheat forage contains significantly more crude protein than is contained in barley forage.

FORAGE ANALYSIS

Table 12-1 lists the average nutrient composition for various forages. Forage analysis should form the basis for diet formulation whenever practical. Crude protein, calcium, and phosphorus can be assayed at most analytical laboratories at a relatively low cost. In addition, most laboratories offer neutral detergent fiber (NDF) analysis. NDF provides an estimate of the percentage of cell walls contained in the forage. The cell walls contain all of the fiber, which is the least digestible component of the forage. The fibrous components of the cell wall primarily include cellulose, hemicellulose, and lignin. None of the lignin is digestible and only 30 to 40% of the hemicellulose and cellulose is digestible. The percent of cellulose in the forage is estimated by subtracting acid detergent fiber (ADF), another fiber analysis, from NDF. Percent lignin is estimated with an acid detergent lignin (ADL) analysis. Percent hemicellulose is calculated by subtracting ADL from ADF. If levels of hemicellulose and lignin are significantly higher than the average values shown in

Table 12-1, metabolizable energy and crude protein concentrations in the forage will likely be lower.

For further and more specific information on swine pasture crops for a particular farm, consult the local county extension agent or write to the state agricultural college.

No one plant embodies all the desirable characteristics of a good swine pasture. None of them will grow the year round, or during extremely cold or dry weather. Further, each of them has a period of peak growth which must be conserved for periods of low growth. Consequently, the progressive swine producer will find it desirable (1) to grow more than one species, (2) to plan for each month of the year, and (3) to secure year-round grazing, or as early spring and late fall grazing as is possible in the particular area. In general,

TABLE 12-1
NUTRIENT COMPOSITION OF SOME FORAGE CROPS[1,2]

Forage Crop[3]	Dry Matter Basis									
	Dry Matter	Crude Protein	Metab. Energy	Ether Extract	Crude Fiber	Hemi-cellulose	Cellulose	Lignin	Calcium	Phos-phorus
	(%)	(%)	(Kcal/lb)	(%)	(%)	(%)	(%)	(%)	(%)	
Alfalfa, fresh, er blm	23	19.0	986	3.1	25.0	8	23	7	2.33	0.31
Alfalfa, fresh, fl blm	25	14.0	905	2.8	31.0	13	27	10	1.53	0.27
Alfalfa meal, dehyd	92	18.9	1,005	3.0	26.2	—	24	11	1.52	0.25
Alfalfa hay, sun-cured, er blm	90	18.0	986	3.0	23.0	9	24	8	1.41	0.22
Alfalfa hay, sun-cured, lt blm	90	14.0	854	1.8	32.0	12	26	12	1.43	0.25
Alfalfa haylage, wltd, er blm	35	17.0	986	3.2	28.0	9	23	10	—	—
Alfalfa haylage, wltd, fl blm	45	14.0	905	2.7	33.2	12	25	12	—	—
Barley hay, sun-cured	87	8.7	923	2.1	27.5	—	—	—	0.23	0.26
Bluegrass, Kentucky, fresh, er vegetative	31	17.4	1,182	3.6	25.3	—	26	3	0.50	0.44
Bluegrass, Kentucky, hay, sun-cured	89	13.0	923	3.5	31.0	—	—	—	0.33	0.25
Brome, smooth, hay, sun-cured, midblm	90	14.6	923	2.6	31.8	22	31	4	0.29	0.28
Clover, alsike, hay, sun-cured	88	14.9	955	3.0	30.1	13	—	—	1.29	0.26
Clover, crimson, hay, sun-cured	87	18.4	936	2.4	30.1	—	—	—	1.40	0.22
Clover, ladino, hay, sun-cured	90	22.0	986	2.7	21.2	—	—	7	1.35	0.31
Clover, red, fresh, er blm	20	19.4	1,136	5.0	23.2	—	—	—	2.26	0.38
Clover, red, hay, sun-cured	89	16.0	905	2.8	28.8	9	26	10	1.53	0.25
Corn, dent yellow, silage, well-eared	33	8.1	1,150	3.1	23.7	—	—	—	0.23	0.22
Lespedeza, common-lespedeza, Korean, hay, sun-cured, er blm	93	15.5	905	—	28.0	—	—	—	1.23	0.25
Orchardgrass, fresh, midblm	31	11.0	932	3.5	33.5	27	33	6	0.23	0.23
Rape (canola), fresh, er blm	11	23.5	1,232	3.8	15.8	—	—	—	—	—
Rye, fresh	24	15.9	1,136	3.7	28.5	—	—	—	0.39	0.33
Ryegrass, perennial, hay, sun-cured	86	8.6	986	2.2	30.3	—	—	2	0.65	0.32
Sorghum, Sudangrass, fresh, midblm	23	8.8	1,036	1.8	30.0	25	34	5	0.43	0.36
Sweetclover, yellow, hay, sun-cured	87	15.7	886	2.0	33.4	—	—	—	1.27	0.25
Timothy hay, sun-cured, er blm	90	15.0	968	2.9	28.0	29	31	4	0.53	0.25
Timothy hay, sun-cured, fl blm	89	8.1	922	3.1	32.0	30	34	6	0.43	0.20
Trefoil, birdsfoot, hay, sun-cured	92	16.3	968	2.5	30.7	—	24	9	1.70	0.27
Wheat, fresh, er vegetative	22	28.6	1,200	4.4	17.4	18	24	4	0.42	0.40

[1]From: *Pork Industry Handbook*, Purdue University Cooperative Extension Service, West Lafayette, IN.

[2]Adapted from Nutrient Requirements of Beef, National Academy Press. Metabolizable energy values shown are for beef, since comparable values are unavailable for swine. Similarly, amino acid levels are unknown for most forages.

[3]er=early; blm=bloom; fl=full; dehyd-dehydrated; lt=late; wltd=wilted; midblm=midbloom.

a combination of permanent and temporary pastures will best achieve these ends.

In the South, year-round grazing is a reality on many a successful swine farm. By careful planning, other areas can approach this desired goal. Fig. 12-5 illustrates in graphic form the growth period of each of the common swine pasture plants of the Corn Belt and North Central states. As noted, by selecting the proper combination of crops, swine pastures for each month of the year are assured. A similar graph for the South, the Atlantic Coast, and the West can be developed.

Hogging down crops furnishes a profitable way of salvaging lodged or damaged crops. The practice of hogging down crops is discussed in Chapter 10.

DIETS FOR GESTATING SOWS ON PASTURE

Suggested diets for gestating sows on pasture are presented in Table 12-2.

SILAGES AND HAYLAGES

Good-quality silage is an excellent feed for brood sows; however, unless the sow herd is very large, it will probably not pay to construct a silo especially for hogs.

Silage should be fed fresh each day in amounts the sows will clean up in 2 to 3 hours. Sows will usually eat 10 to 15 lb, and gilts will usually eat 8 to 12 lb. Some wastage can be expected. It is important that the silage be supplemented with 1.0 to 1.5 lb of a good protein supplement per head daily. Additional grain should be fed if necessary to keep the sows in proper condition.

When feeding silage to hogs, the following precautions should be observed:

1. Never feed silage alone, as a poor pig crop will result.

2. Silage causes digestive upsets (diarrhea) in baby pigs. Do not feed it to lactating sows.

3. Avoid feeding moldy or frozen silage. It can cause sows to abort.

Even though considerable silage may be fed to brood sows to advantage, it is important that it be of good quality and that it be properly supplemented from a nutritional standpoint.

When feeding legume haylage, offer all the haylage sows will clean up (usually 6 to 8 lb per head per day), plus 2.5 lb of complete feed.

A swine producer will seldom build a silo just to feed brood sows, but if silage or haylage are available

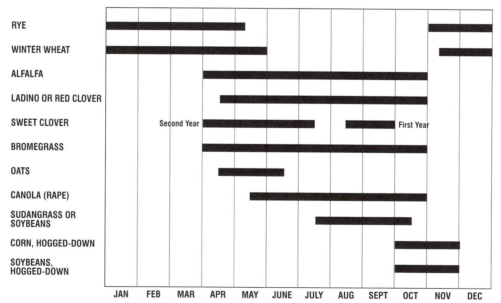

APPROXIMATE GRAZING PERIOD OF COMMON PASTURE CROPS FOR SWINE IN THE CORN BELT AND NORTH CENTRAL STATES

Fig. 12-5. As shown, by selecting the proper combination of crops, year-round grazing can be achieved. For example, a 12-month hog pasture may be obtained by using two crops only, rye and ladino clover.

TABLE 12-2
SUGGESTED DIETS TO SUPPLEMENT PASTURE FOR GESTATING SOWS[1, 2]

	Type of Pasture							
	Legume		Grass		Legume-Grass Mix		Rape (Canola)	
Ingredients								
Corn . . . . (lb)	1,756	1,744	1,371	1,335	1,521	1,495	1,858	1,857
Soybean meal, 48% . . . . (lb)	173	—	525	—	372	—	11	—
Soybean meal, 44% . . . . (lb)	—	186	—	562	—	398	—	12
Ground limestone . . . . . (lb)	—	—	16	16	24	24	18	18
Dicalcium phosphate . . . (lb)	—	—	64	63	59	59	89	89
Monosodium phosphate . . (lb)	47	46	—	—	—	—	—	—
Salt . . . . (lb)	12	12	12	12	12	12	12	12
Vitamin premix[3] . . . . . . (lb)	6	6	6	6	6	6	6	6
Trace mineral premix[4] . . . (lb)	6	6	6	6	6	6	6	6
Totals	2,000	2,000	2,000	2,000	2,000	2,000	2,000	2,000
Calculated Composition								
Metab. energy . . . . (kcal/lb)	1,499	1,492	1,470	1,449	1,469	1,442	1,453	1,452
Crude protein . . . . (%)	11.7	11.5	18.6	18.0	15.5	15.1	8.2	8.2
Lysine (estimated) . . . . . (%)	0.49	0.49	0.99	0.99	0.77	0.77	0.25	0.25
Calcium . . . . (%)	0.05	0.05	1.16	1.16	1.23	1.23	1.44	1.44
Phosphorus . . . . (%)	0.81	0.81	0.95	0.95	0.88	0.88	1.09	1.09

[1]From: *Pork Industry Handbook*, Purdue University Cooperative Extension Service, West Lafayette, IN.

[2]Assumptions: Sows will consume 3.5 lb pasture dry matter and will be fed 2.5 lb of the diet per day. Pasture and feed together will provide a minimum of 0.75 lb protein, 17 g lysine, 18 g calcium, and 14 g phosphorus per day, respectively.

[3]Should provide the following amounts per ton of complete feed: 8,000,000 IU vitamin A; 800,000 IU vitamin D; 64,000 IU vitamin E; 3.2 g vitamin K; 8 g riboflavin; 48 g niacin; 29 g pantothenic acid; 24 mg vitamin B-12; 1.6 g choline; 1.76 g folic acid; 320 mg biotin. (Concentrations higher than normal to compensate for reduced dry feed intake.)

[4]Should provide the following amounts per ton of complete feed: 15 g copper; 144 g iron; 144 g zinc; 40 g manganese; 320 mg iodine; 435 mg selenium. (Concentrations higher than normal to compensate for reduced dry feed intake.)

for cattle, using it for sows will produce excellent results.

DRY FORAGES

The favorable results obtained from feeding a corn-soybean meal diet, properly fortified with minerals and vitamins, in an era of relatively cheap grain, caused many researchers and swine producers to question the wisdom of using alfalfa meal in the diet. There was never any doubt about alfalfa meal being an excellent source of quality protein, carotene (vitamin A), B vitamins (riboflavin, pantothenic acid, and niacin), vitamin D if sun-cured, calcium, and unidentified factors. But these nutrients could be purchased more cheaply from other sources. Besides, its high

fiber (25 to 30%), low energy density, and low palatability, make it unsuited in diets for nursery pigs or lactating sows. As a result, the swing was away from the use of alfalfa meal in swine diets.

But scarce and high-priced grains have caused alfalfa to return in favor as a feedstuff, particularly in diets for gestating sows. If the price of alfalfa meal is right—if it is a cheaper source of protein and energy than corn and other grains—15 to 35% (or even higher levels) alfalfa meal may, to advantage, be incorporated in gestating sow diets and 2 to 5% may be used in growing-finishing diets. These levels serve as a safety factor to ensure the presence of certain vitamins, minerals, and unidentified factors.

Alfalfa is practically the only dry forage fed to swine in the United States.

QUESTIONS FOR STUDY AND DISCUSSION

1. Describe and evaluate gestating sows on pasture and outdoor farrowing.

2. Describe and evaluate farrow-to-market on pasture.

3. List and discuss the advantages and disadvantages of producing hogs on pasture.

4. How do you account for the current trend to confinement rearing of swine by large, highly specialized swine producers?

5. Compute the monetary return that might reasonably be expected from an acre of pasture grazed by hogs versus the return from alternate cropping programs.

6. Describe a good hog pasture.

7. Outline the management—feeding, watering, fencing, shelter, manure handling, grazing, and pollution control—of hogs on pasture.

8. What are the primary differences between good management of hogs on pasture versus good management of hogs in confinement?

9. What pasture(s) would you recommend for use for swine in (a) the West, (b) the South, (c) the Atlantic Coast states, and (d) the Corn Belt and North Central states? Justify your decision and prepare a chart showing how you could advantageously extend the grazing season in one of the areas.

10. Suppose you have access to good-quality silage and you wish to feed it to your hogs. What recommendations and precautions would you make?

11. How may alfalfa be used advantageously in feeding swine?

SELECTED REFERENCES

Title of Publication	Author(s)	Publisher
Crop Production	R. J. Delorit H. L. Ahlgren	Prentice-Hall, Inc., Englewood Cliffs, NJ, 1984
Feeds & Nutrition Digest	M. E. Ensminger J. E. Oldfield W. W. Heinemann	The Ensminger Publishing Company, Clovis, CA 1990
Feeds & Nutrition, Second Edition	M. E. Ensminger J. E. Oldfield W. W. Heinemann	The Ensminger Publishing Company, Clovis, CA 1990
Forages: The Science of Grassland Agriculture	M. E. Heath R. F. Barnes D. S. Metcalfe	The Iowa State University Press, Ames,. IA, 1985
Grass, Yearbook of Agriculture, 1948	U.S. Department of Agriculture	Superintendent of Documents, Washington, DC
Nontraditional Feed Sources for Use in Swine Production	P. A. Thacker R. N. Kirkwood	Butterworth Publishers, Stoneham, MA, 1990
Plant Science	J. Janick, et al.	W. H. Freeman and Co., San Francisco, CA, 1981
Producing Farm Crops	L. V. Boone	Interstate Publishers, Inc., Danville, IL, 1991
Stockman's Handbook, The, Seventh Edition	M. E. Ensminger	Interstate Publishers, Inc., Danville, IL, 1992
Swine Feeding and Nutrition	T. J. Cunha	Academic Press, Inc., New York, NY, 1977

Tail biting is an abnormal behavior in swine. (Photo by J. C. Allen and Son, West Lafayette, IN)

13

SWINE BEHAVIOR AND ENVIRONMENT

Successful hog producers are "students" of animal behavior. For example, they recognize when females are in heat (estrus), or when parturition is imminent. They know the squeal of a piglet in trouble, or the low-pitched rhythmical grunting sound a sow uses to call piglets to nurse. A producer should be able to walk through the facilities and spot sick animals by their behavior, thereby beginning early treatment or taking steps to correct some environmental problem. Moreover, confinement methods of rearing hogs have renewed the interest in and increased the need for understanding animal behavior.

Confinement must, however, embrace far more than ventilation and heating, along with ample feed and water. The producer needs to be concerned more with the natural habitat of animals. Nature ordained that they do more than eat, sleep, and reproduce. For example, studies on the behavior of swine show that they spend much of their day in active investigative behavior, primarily rooting and manipulating their movement. When free ranging, pigs may spend 40% of their day resting, 35% investigating novel surroundings, 15% eating, and 10% in other activities. What happens when pigs are confined in a building on slotted floors? How is the nervous energy dissipated that would normally be used to satisfy the drives for investigating and rooting?

Evidently, environmental deficiencies are manifested by tail biting, gastric ulcers, poor maternal care and loss of young, or other physiological functions resulting in a sudden death syndrome or tissue degeneration.

What can be done about it? Preventing disorders by merely cutting off the tails of pigs to prevent tail biting is not unlike trying to control malaria fever in humans by the use of drugs without getting rid of mosquitoes. Rather, we need to recognize such a disorder for what it is—a warning signal that conditions are not right. Correcting the cause of the disorder is the best solution. Unfortunately, this is not usually the easiest. Correcting the cause may involve trying to emulate the natural conditions of swine, such as altering space per animal and group size, promoting exercise, and gradually changing diets. Over the long pull, selection provides a major answer to correcting confinement and other behavioral problems; we need to breed swine adapted to producer-made environments.

This chapter presents some of the principles and applications of animal behavior. Those who have grown up around farm animals and dealt with them in practical ways have already accumulated substantial workaday knowledge about animal behavior. Those with urban backgrounds may need to familiarize themselves with the behavior of animals, better to feed and care for them and to recognize the early signs of illness. To all, the principles and applications of animal behavior

Fig. 13-1. Producers have drastically altered the environment of swine. The only time a pig produced in total confinement leaves a building is on the way to market. (Photo by J. C. Allen and Son, West Lafayette, IN)

depend on understanding, which is the intent of this chapter.

ANIMAL BEHAVIOR

Animal behavior is the reaction of animals to certain stimuli, or the manner in which they react to their environment. The individual and comparative study of animal behavior is known as *ethology.* Through the years, behavior has received less attention than the quantity and quality of the pork produced by swine. But modern breeding, feeding, and management have brought renewed interest in behavior, especially as a factor in obtaining maximum production and efficiency. With the restriction, or confinement, of herds, many abnormal behaviors evolved to plague those who raise them, including cannibalism, loss of appetite, stereotyped movements, poor parental care, overaggressiveness, dullness, degenerate sexual behavior, tail biting, and a host of other behavior disorders. Not only has confinement limited space, but it has interfered with the habitat and social organization to which, through thousands of years of evolution, the species became adapted and best suited. This is due to a genetic time lag. Swine producers altered the environment faster than they altered the genetic makeup

HOW SWINE BEHAVE

Animals behave differently, according to species. Also, some behavioral systems or patterns are better developed in certain species than in others. Moreover, ingestive and sexual behavioral systems have been most extensively studied because of their importance commercially. Nevertheless, most swine exhibit the following eight general functions or behavioral systems, each of which will be discussed:

1. Ingestive behavior (eating and drinking)

2. Eliminative behavior
3. Sexual behavior
4. Maternal behavior (mothering)
5. Agonistic behavior (combat)
6. Gregarious behavior
7. Investigative behavior
8. Shelter-seeking behavior

1. **Ingestive behavior (eating and drinking).** This type of behavior is characteristic of all animals of all species and all ages. Animals cannot live without feed and water. Moreover, for high production, animals must have aggressive eating habits. They must consume large quantities of feed.

Fig. 13-2. Within minutes of birth, piglets begin sucking and by 48 hours old, have established a "teat order." (Courtesy, Lynn Smith, Troy, NH; reprinted with permission from *Small Farmers' Journal*, Eugene, OR)

The first ingestive behavior trait common to all young mammals is sucking. Within minutes after birth, piglets find the udder and begin to nurse. When nursing occurs after parturition is complete, the sow calls the piglets using a low-pitched rhythmical grunt and the aroused piglets approach the sow, squealing in response. However, by its calls, a single hungry piglet may arouse the sow and the rest of the litter to nurse. Within 48 hours of birth, the piglets establish a "teat order"—each piglet suckles a particular teat. This is a form of dominance hierarchy. Anterior teats are suckled by the most dominant piglets. Once the teat order is established there is very little aggression. The sow will nurse her piglets about every 40 to 60 minutes, though the frequency is lower at night and declines as the piglets grow older.

Mature pigs possess teeth in the upper and lower jaws; hence, they bite, chew, and swallow feed. By nature, pigs love to root. If given the opportunity, they will stick their noses into the ground and lift forward and upward, moving soil out of the way and exposing earthworms, grubs, and roots. In the Perigord region

of France, pigs are trained to use this natural desire and search for truffles—a type of underground mushroom.

2. **Eliminative behavior**. If given an opportunity, pigs have very tidy toilet habits. They like to keep their bedding area clean and dry. Hence, they usually deposit their feces in a corner of the pen, away from the sleeping quarters. Modern methods of raising pigs in restricted quarters, which are often overcrowded, have disturbed their natural eliminative patterns. Some producers, however, have taken advantage of the social and ingestive behavior of swine to train them to eliminate in specific areas.

3. **Sexual behavior.** Reproduction is the first and most important requisite of swine breeding. Without young being born and born alive, the other economic traits are of academic interest only. Thus, it is important that all those who breed swine should have a working knowledge of sexual behavior.

Sexual behavior involves courtship and mating. It is largely controlled by hormones. Sows in heat (estrus) are nervous and active, and their vulvas are swollen. Natural estrous detection involves important nose-to-nose contact between the boar and the sow. The sow in heat will stand to be mounted by a boar. When sows do this they assume a typical mating stance—rigid limbs and cocked ears. This stance is assumed when pressure is applied to the rump; hence, producers use this as a means of detecting estrus. However, in some sows, the assumption of the mating stance may depend upon their hearing, smelling, or seeing the boar. Courtship of the boar involves eliciting the immobility stance. The boar often nudges the sow or gilt around the head or in the flanks with his head and nose and emits a courting song—a regular series of soft guttural grunts. Sows also respond to the smell of a boar—his pheromones. Chemically, these pheromones are androstenols, which are metabolites of testosterone. Receptive sows assume the mating stance when the boar attempts to mount. Some boars will mount several times before successfully mating, which takes 3 to 20 minutes.

Abnormal sexual behavior is most often noted in males. For example, boars reared in all male groups form stable homosexual relationships. Also some boars will mount inanimate objects.

Fig. 13-3. When an estrous female is touched on the back by a boar she assumes a stationary position—the mating stance or immobility response.

4. **Maternal behavior (mothering).** Nest building activities occur 1 to 3 days before farrowing. In a field or pasture, sows will choose a depression in the ground and line it with grass, straw, and other material, forming a nest. Farrowing crates and concrete floors prevent nest building, but sows show increased activity the last 24 hours or more before farrowing. They may grind their teeth, bite and root at the rails, and frequently stand up and lie down. Just before parturition, however, the sow quiets down and lies on one side. Sows usually remain on this side, but some sows will get up and down between the birth of piglets, thereby increasing the chances of crushing a piglet.

Many domestic animals lick their newborn, but swine do not. Sows seldom pay attention to their piglets until the last one is born. Then the sow is very protective of her pigs, especially if they squeal. She will go toward an intruder with mouth open and will emit a series of sharp, barking grunts in rapid succession. She continues to mother her pigs until they are weaned, but after 2 to 3 days of separation she loses interest in them. If pigs are left with the sow for 3 or 4 months, she will usually wean them herself. Sows will readily accept pigs from another litter, provided the transfer is made the first day or two following farrowing. Exchanging of pigs among sows, in order to even out the size litters is a common practice in herds where many sows are farrowing about the same time.

Some nervous sows eat their pigs during or immediately after farrowing. If this trait is observed, all pigs, both live and dead, along with the placental membranes should be removed as soon as possible, before the sow has an opportunity to eat them. Once the sow has acquired a taste for flesh, she may develop a permanent pig-eating habit. Usually, such nervous sows calm down following farrowing, after which their pigs may be returned to them and they will express normal protective behavior.

5. **Agonistic behavior (combat).** This type of behavior includes fighting, flight, and other related reactions associated with conflict. Among all species of farm mammals, males are more likely to fight than females. Nevertheless, females may exhibit fighting behavior under certain conditions. Castrated males are usually quite passive, which indicates that hormones, (especially testosterone) are involved in this type of behavior. Thus, farmers have for centuries used castration as a means of producing docile males, particularly swine, cattle, and horses.

Boars that are run together from a very young age seldom fight. Perhaps they have already settled their social rank. On the other hand, bringing together sexually mature strange boars almost always results in a fight. Sows and barrows will also fight, but they do not exhibit the jaw-clicking and saliva-producing (champing) characteristic of fighting boars. A sow will try to

Fig. 13-4. Agonistic behavior (combat). (Courtesy, Land O Lakes, Ft. Dodge, IA)

bite, whereas a boar will slash his opponent with his tusks.

When strange boars are first penned together, they smell one another and begin to circle as they "size up" each other. They frequently strut shoulder to shoulder with the hair on their crest bristled, ears cocked, and head raised in an alert, threatening position. In a serious encounter, the combatants utter deep-throated barking grunts and champ throughout the fight. As the fighting becomes intense, each boar repeatedly thrusts his head and neck sideways and upward, with his jaws open and his teeth bared. If the boars have tusks, slashes are usually inflicted on the shoulders of each other.

Fighting boars await the opportunity to discontinue shoulder contact and to nip at the ears or the neck and front legs. Sometimes, they even charge the side of the opponent with their mouth wide open.

Fighting may continue for as long as an hour, or it may end very quickly. In any event, it will continue until the dominant boar is satisfied and the loser retreats, with the winner biting and slashing him as he scampers away.

Other types of agonistic behavior among pigs includes (1) piglets fighting to establish "teat order," (2) aggression actions around the feeder, and (3) tail biting. The last two forms of agonistic behavior are likely related to high stocking rates.

6. **Gregarious behavior.** Gregarious behavior refers to the herding instinct. In the wild state, swine roved through the forest in herds. Usually these wild groups consisted of 5 to 10 sows, under the leadership of a boar. The wild boar, with his large and long head and strong tusks, was a formidable match for most any enemy.

Under domestication, swine retain their gregarious nature. However, humans have altered it a great deal. Today, hogs are usually confined to a very limited area.

Feral pigs, however, return to living in herds of less than 10 individuals, but groups of up to 80 animals sometimes form.

7. **Investigative behavior.** All animals are curious and have a tendency to explore their environment. Investigation takes place through seeing, hearing, smelling, tasting, and touching. Whenever an animal is introduced into a new area, its first reaction is to explore it. Experienced producers recognize that it is important to allow animals time for investigation before attempting to work them, either when they are placed in new quarters or when new animals are introduced into the herd.

Pigs are curious, too. When a strange person approaches a herd of hogs, an alarm, or "woof," is sounded and the animals scatter—scampering as fast as they can for a short distance. In the meantime, if the intruder remains stationary, either standing or sitting, the pigs invariably return to investigate by smelling, rooting, and nibbling. Of course, when pigs are placed in confinement, they have little area to investigate.

Fig. 13-5. Pigs are curious too. In this case the photographer was the curiosity. (Photo by R. O. Parker)

8. **Shelter-seeking behavior.** Hogs are very sensitive to extremes of heat and cold; hence, shelter seeking is a very important trait with them. It is particularly important that swine be provided with shade during hot weather so that they may avoid the direct rays of the sun, because they do not possess an adequate cooling mechanism. In hot weather, hogs will wallow in water if given the opportunity.

Hogs pant rapidly when they are hot; sleep stretched out full length when they are hot, so as to expose the maximum body surface to the air; and sleep curled up and huddled together during cold weather, thereby exposing minimal body surface to the air. Baby pigs are particularly sensitive to cold and they seek and need warmth.

SOCIAL RELATIONSHIPS

Social behavior may be defined as any behavior caused by or affecting another animal, usually of the same species, but also, in some cases, of another species.

Social organization may be defined as an aggregation of individuals into a fairly well integrated and self-consistent group in which the unity is based upon the interdependence of the separate organisms and upon their responses to one another.

The social structure and infrastructure in the herd are of great practical importance. Some of the ideas on peck order have had to undergo changes as a result of increased understanding of the social organization within the herd.

It is now obvious that there is no simple hierarchical descent—stepladder or peck order in numerical progression. The social structure is much more complex. As a general rule, the older an animal is at the time it is introduced into a group, the less integrated it becomes in the group.

DOMINANCE

When put together, unacquainted pigs fight and establish a linear type of dominance. The top ranking pig has precedence in access to feed and is the winner in an agonistic encounter. Also, studies have demonstrated that pigs of a high rank are heavier. However, there may be situations where two pigs occupy the same social rank or where a triangular type social rank exists. Changes in social rank are common in the middle and lowest ranking pigs, but it is uncommon for the top animal to change its position. In fact some studies have shown that the most dominant pig can be removed for as long as a month and still resume its position upon returning to the group.

Once the social rank order is established, it results in a peaceful coexistence. Thereafter, when the dominant one merely threatens, the subordinate animal submits and avoids conflict. Of course, there are some pairs that fight every time they chance to meet. Also, if strange animals are introduced into such a group, social disorganization results in the outbreak of new fighting, as a new social rank order is established.

Several factors influence social rank; among them, (1) age—both young animals and those that are senile rank toward the bottom; (2) early experience—once a subordinate in a particular herd, usually always a subordinate; (3) weight and size; and (4) aggressiveness or timidity.

Social rank becomes of importance when a group of animals is fed in confinement; and it becomes doubly important if limit-feeding is practiced. Under such circumstances, the dominant individuals crowd the subordinate ones away from the trough or feeder, with the result that the latter may go hungry. Dominants should be sorted out, and, if possible, grouped together. Of course, they will fight it out until a new social order is established. In the meantime, both feed efficiency and gains will suffer. But, as a result of removing the dominants, the feed intake of the rest of the animals will be improved, followed by greater feed efficiency and profit. Among the more settled animals, social facilitation will become more evident. After the dominants have been removed, the rest of the animals will settle down into a new hierarchy, but within the limits of their dominance. Their interaction or social facilitation will be far more likely to have a calming effect on this group, to both the economic and practical advantages of the operator.

Dominance and subordination are not inherited as such, for these relations are developed by experience. Rather, the capacity to fight (agonistic behavior) is inherited, and, in turn, this determines dominance and subordination. Hence, when combat has been bred into the herd, such herds never have the same settled appearance and docility that is desired of high-production and intensive animals.

LEADERSHIP

Leader-follower relationships are important in hogs. The young follow their mothers; hence, they continue to follow their elders.

Right off, it is important to distinguish leader–follower relationships from dominance; in the latter, the herd is driven, rather than led. After the dominants have been removed from the herd, the leader–follower phenomenon usually becomes more evident. It is well known that the dominant animal is not necessarily the leader; in fact, it is very rarely the leader. It pays too much attention to other matters of dominance in its relationship within the herd, with the result that it does not have the quality of leadership.

Taking a page from history, it is noteworthy that some of the world's greatest leaders, for example, Napoleon, have been small in stature. So it is with animals. The leader may be small, but intelligent. Sometimes the smallest pig in the pen emerges as the leader of his mates.

INTERSPECIES RELATIONSHIPS

Social relationships are normally formed between members of the same species. However, they can be developed between two different species. In domestication this tendency is important (1) because it permits several species to be kept together in the same pasture or corral, and (2) because of the close relationship between caretakers and animals. Such interspecies relationships can be produced artificially, generally by taking advantage of the maternal instinct of females and using them as foster mothers. For example, cows and bitches (dogs) have raised pigs.

COMMUNICATION AMONG SWINE

Communication involves a signal by one pig which upon being received by another pig influences its behavior. Communication between pigs may be via sound, smell, and/or visual displays.

■ **Sound**—Sound communication is of special interest because it forms the fundamental basis of human language. The gift of language alone sets humans apart from the rest of the animals and gives them enormous advantages in their adaptation to their environment and in their social organization.

Sound is also an important means of communication among pigs. They use sounds in many ways; among them, (1) feeding, in sounds of hunger by young, or food finding; (2) distress calls, which announce the approach or presence of an enemy; (3) sexual behavior, courting songs, and related fighting; and (4) mother-young interrelations to establish contact and evoke care behavior. These have already been discussed.

■ **Smell**—The sense of smell is well developed in pigs. It is their sense of smell that allows pigs to find food when they root. Sows or gilts in heat search out a boar via olfactory cues or pheromones.

■ **Visual displays**—Visual signs play some role in pig communication. Intended agonistic behavior and sexual behavior are in part recognized by visual displays. For example, boars will make the hair on top of their necks rise up (bristle). This serves to make them look larger and more formidable.

NORMAL SWINE BEHAVIOR

The producer needs to be familiar with behavioral norms of animals in order to detect and treat abnormal situations—especially illness. Many sicknesses are first suspected because of some change in behavior—loss of appetite (anorexia); listlessness; labored breathing; posture; reluctance or unusual movement; persistent rubbing; and altered social behavior, such as one animal leaving the herd and going off by itself—these are among the useful diagnostic tools.

Some of the signs of good health are:

1. Contentment.
2. Alertness.
3. Eating with relish.
4. Sleek coat with pliable and elastic skin.
5. Bright eyes and pink membranes.
6. Normal feces and urine.
7. Normal temperature (102.0 to 103.6°F), pulse rate (60 to 80 beats per minute), and breathing rate (8 to 13 breaths per minute).

Vision and sleep are also important aspects of swine behavior. The eyes of pigs, like many animals, are on the side of the head. This gives them an orbital or panoramic view—to the front, to the side, and to the back—virtually at the same time. Sometimes this type of vision is referred to as rounded or globular. The field of vision directly in front of a pig is binocular, but on the sides and toward the back it is monocular. Vision of this type leads pigs to interpretations different from the binocular type vision of humans. The resting position of swine varies according to temperature—in the summer, they sleep stretched full length; in cold weather, they sleep curled up. In any event, pigs sleep soundly—they even snore.

LEARNED BEHAVIOR

Given the proper reinforcement—usually food—pigs can be easily trained or more specifically conditioned. Conditioning is the type of learning in which the pig responds to a certain stimulus. The type that is used in swine production is operant conditioning—the pig learns to operate some aspect of the environment. For example, pigs can be operantly conditioned to push a panel to turn on a heat lamp or obtain a food reward. In practical application, an example is the lifting of the cover of a self-feeder, or drinking from a watering device.

ABNORMAL SWINE BEHAVIOR

Abnormal behavior of domestic animals is not fully understood. Like human behavior disorders, more study is needed. However, studies of captured wild animals have demonstrated that when the amount and quality, including variability, of the surroundings of an animal are reduced, there is an increased probability that abnormal behavior will develop. Also, it is recognized that confinement of animals creates a lack of space which often leads to unfavorable changes in habitat and social interactions for which the species have become adapted and best suited over thousands of years of evolution. Abnormal behavior may take many forms—ingestive, eliminative, sexual, maternal, agonistic, or investigative. Abnormal sexual behavior is particularly distressing because the whole of production depends on the animal's ability to reproduce. Tail biting is a common form of abnormal agonistic behavior. It accompanies close confinement, resulting when pigs are prevented from rooting, nibbling, and chewing—normal behavior. Thus far, docking is the best way to stop pigs from tail biting. Swine producers have tried all sorts of things to prevent tail biting. Some have substituted other materials for the pigs to bite, such as rubber tires or chains hung near the pigpen. Others have tried spraying the tails with distasteful chemicals, but none of these methods work very well.

APPLIED SWINE BEHAVIOR

The presentation to this point has been for the purpose of understanding. However, knowledge and understanding must be put into practice to be of value. Hence, the producer must make practical barnyard applications of swine behavior.

BREEDING FOR ADAPTATION

The wide variety of livestock in different parts of the world reflects a continuous process of natural and artificial selection which has resulted in the survival of animals well adapted to climate and other environmental factors. Adaptations relate to survival of the animals, but they do not necessarily entail maximum productivity of food for humans.

Selection should be from among animals kept under an environment similar to that which it is expected that their offspring shall perform. Moreover, these animals should be those that demonstrate high productivity in their environment. Therefore, pigs should be bred and selected that adapt quickly to artificial environments—animals that not only survive, but that thrive, under the conditions that are imposed upon them. Properly combined heredity and environment complement each other, but when one or the other is disregarded they may oppose each other.

TRAINING AND ADAPTATION

Early training and experience are extremely important. In general, young animals learn more quickly and easily than adults; hence, advance preparation for adult life will pay handsome dividends. Furthermore, stress can be reduced or avoided entirely if animals proceed through a graduated sequence of events leading to an otherwise noxious experience.

ANIMAL ENVIRONMENT

Environment may be defined as all the conditions, circumstances, and influences surrounding and affecting the growth, development, and production of animals. The most important influences in the environment are the feed and quarters (space and shelter).

The branch of science concerned with the relation of living things to their environment and to each other is known as ecology.

People achieve environmental control through clothing, vacationing in resort areas, and air-conditioned homes and cars. In swine, environmental control involves space requirements, light, air temperature, relative humidity, air velocity, wet bedding, ammonia build-up, dust, odors, and manure disposal, along with proper feed and water. Control or modification of these factors offers possibilities for improving animal performance. Although there is still much to be learned about environmental control, the gap between awareness and application is becoming smaller. Research on animal environment has lagged, primarily because it requires a melding of several disciplines—nutrition, physiology, genetics, engineering, and climatology. Those engaged in such studies are known as ecologists.

In the present era, pollution control is the first and most important requisite in locating a new swine unit, or in continuing an old one. The location should be such as to avoid (1) the neighbors complaining about odors, insects, and dust; and (2) pollution of surface and underground water. Without knowledge of animal behavior, or without pollution control, no amount of capital, native intelligence, and sweat will make for a successful swine enterprise.

Pollution of the environment is of worldwide concern. In 1984, the government of the Netherlands began a serious attack on pollution. Attention is being directed to the following three key areas:

1. The over application of minerals in manure to the land.
2. The reduction in ammonia emissions from slurry during storage and spreading.
3. The implementation of production techniques that reduce both the initial amount of manure and its content of nitrates and phosphates.

The following factors are of special importance in any discussion of animal environment:

1. Feed and nutrition.
2. Water.
3. Weather.
4. Facilities.
5. Health.
6. Stress.

FEED AND NUTRITION

The most important influence in the environment is the feed. Swine may be affected by (1) too little feed, (2) diets that are too low in one or more nutrients, (3) an imbalance between certain nutrients, or (4) objection to the physical form of the diet—for example, it may be ground too finely.

Forced production and the feeding of forages and grains which are often produced on leached and depleted soils have created many problems in nutrition. These conditions have been further aggravated through the increased confinement of animals, many animals being confined to pens all or a large part of the year. Under these unnatural conditions, nutritional diseases and ailments have become increasingly common.

Also, nutritional reproductive failures plague swine operations. Generally speaking, energy is more important than protein in reproduction. After giving birth, feed requirements increase tremendously because of milk production; hence, a sow suckling young needs approximately 50% greater feed allowance than during the pregnancy period. Otherwise, she will suffer a serious loss in weight, and she may fail to come in heat and conceive.

The following additional feed-environmental factors are pertinent:

1. **Regularity of feeding.** Animals are creatures of habit; hence, they should be fed at regular times each day, by the clock.

2. **Underfeeding.** Too little feed results in slow and stunted growth of pigs; in loss of weight, poor condition, and excessive fatigue of mature animals; and in poor reproduction, failure of some females to show heat, more services per conception, lowered young crop, and light birth weights.

3. **Overfeeding.** Too much feed is wasteful. Besides, it creates a health hazard; there is usually lowered reproduction in breeding animals, and a higher incidence of digestive disturbances. Animals that suffer from mild digestive disturbances are commonly referred to as "off feed."

4. **Deficiency of nutrient(s).** A deficiency of any essential nutrient will lower production and feed efficiency.

5. **Some feed ingredients and diets influence milk composition.** The diet fed during gestation and lactation affects some of the nutrients in the milk. Minerals in milk such as zinc and manganese can be increased by feeding higher levels, but milk levels of other minerals such as calcium, phosphorus, and iron are not increased by dietary means. Experiments have demonstrated that the fat content of sow's milk can be increased by feeding a high fat diet. This may benefit the piglets.

The results of a 10-year swine experiment conducted by the senior author and his colleagues at Washington State University, designed to study the effect of plane of nutrition on meat animal improvement, support the contention that selection of breeding animals should be carried on under the same environmental conditions as those under which commercial animals are produced (see Chapter 6, section headed "Heredity and Environment").

WATER

Animals can survive for a longer period without feed than without water. Water is one of the largest constituents in the animal body, ranging from 40% in very fat, mature hogs to 80% in newborn pigs. Deficits or excesses of more than a few percent of the total body water are incompatible with health, and large deficits of about 20% of the body weight lead to death.

The total water requirement of swine varies primarily with the weather (temperature and humidity); feed (kind and amount); the age and weight of animal; and the physiological state. The need for water increases with increased intakes of protein and salt, and with increased milk production of lactating sows. Water quality is also important, especially with respect to the content of salts and toxic compounds.

The water content of feeds ranges from about 10% in air-dry feeds to more than 80% in fresh, green forage. Feeds containing more than 20% water are known as *wet feeds*.

Home range behavior exists among most wild and feral animals, with each home range including at least one watering hole—generally a stream, spring, lake, or pond. The frequency of watering wild and feral swine, as well as domestic swine, is determined primarily by temperature and humidity—the higher the temperature and humidity, the more frequent the watering.

When hand-fed, pigs generally eat all their feed, then drink; when self-fed, they alternate between eating and drinking.

Under practical conditions, the frequency of watering is best determined by the animals, by allowing them access to clean, fresh water at all times.

WEATHER

Weather affects the maintenance requirements of swine. Extreme weather can cause wide fluctuations in animal performance. The difference in weather impact from one year to the next, and between areas of the country, causes difficulty in making a realistic analysis of buildings and management techniques used to reduce weather stress. Weather may be modified by facilities. The research data clearly show that winter quarters and summer shades almost always improve production and feed efficiency. The issue is clouded only because the additional costs incurred by shelters have frequently exceeded the benefits gained by the improved performance, particularly in those areas with less severe weather and climate.

The maintenance requirements of animals increase as temperature, humidity, and air movements depart from the comfort zone. Likewise, the heat loss from animals is affected by these three items.

Environmentally controlled buildings overcome the problems imposed by weather. With the shift to confinement structures and high-density production operations, building design and environmental control became more critical. Research has shown that animals are more efficient—that they produce and perform better, and require less feed—if raised under ideal conditions of temperature, humidity, and ventilation. The cost per head is, however, much higher for environmentally controlled facilities. Thus, the decision on whether or not confinement and environmental control can be justified should be determined by economics. Manure disposal and pollution control are also considerations.

Environmentally controlled buildings are costly to construct, but they make for the ultimate in animal comfort, health, and efficiency of feed utilization. Also, like any confinement building, they lend themselves to automation, which results in a saving in labor; and, because of minimizing space requirements, they effect a saving in land cost. If they malfunction, however, they can cause animals to suffocate and result in large economic losses. Today, environmental control is rather common in swine housing.

Before an environmental system can be designed for animals, it is important to know their (1) heat production, (2) vapor production, and (3) space requirements. This information is as pertinent to designing livestock buildings as nutrient requirements are to balancing diets. This information is presented in Chapter 16, Buildings and Equipment for Swine.

THERMONEUTRAL ZONE
(Comfort Zone)

Fig. 13-6 and the definitions that follow are pertinent to an understanding of thermal zones.

In Fig. 13-6 *heat production (metabolism)* is plotted against *ambient temperature* to depict the relationship between chemical and physical heat regulation. Note, too, the broad range of accommodation to low (cool) temperatures in contrast to the restricted range of accommodation to high (warm) temperatures. Definitions of terms pertaining to Fig. 13-6 follow.

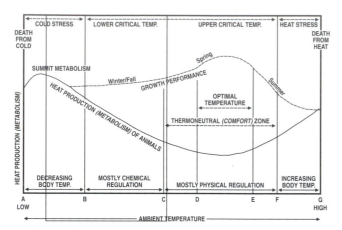

Fig. 13-6. Diagram showing the influence of thermal zones and temperature on homeotherms (warm-blooded animals).

■ *Thermoneutral (comfort) zone (C to F) is the range in temperature within which the animal may perform with little discomfort, and in which physical temperature regulation is employed.*

■ *Optimum temperature (D to E) is the temperature at which the animal responds most favorably, as determined by maximum production (gains, milk, wool, work, eggs) and feed efficiency.*

■ *Lower critical temperature (B) is the low point of the cold temperature beyond which the animal cannot maintain normal body temperature.* The chemical temperature regulation is employed in the zone below C. When the environmental temperature reaches below point B, the chemical-regulation mechanism is no longer able to cope with cold, and the body temperature drops, followed by death. The French physiologist, Giaja, used the term *summit metabolism* (maximum sustained heat production) to indicate the point beyond which a decrease in ambient temperature causes the homeothermic mechanisms to break down, resulting in a decline in both heat production and body temperature and eventually death of the animal.

■ *Upper critical temperature (F) is the high point on the range of the comfort zone, beyond which animals are heat stressed and physical regulation comes into play to cool them.*

The comfort zone, optimum temperature, and both upper and lower critical temperatures vary with different species, breeds, ages, body sizes, physiological and production status, acclimatizations, feed consumed (kind and amount), the activity of the animal, and the opportunity for evaporative cooling.

The temperature varies according to age, too. For example, the comfort zone of newborn pigs is 90 to 95°F for the first few days, whereas the comfort zone of sows after farrowing is 55 to 75°F.

Swine that consume large quantities of roughage or high-protein feeds produce more heat during digestion; hence, they have a different critical temperature than the same animals fed a high-concentrate, moderate-protein diet.

Stresses of both high and low temperatures are increased with high humidity. The cooling effect of evaporating sweat is minimized and the respired air has less of a cooling effect. As humidity of the air increases, discomfort at any temperature, and nutrient utilization, decrease proportionately.

Air movement (wind) results in body heat being removed at a more rapid rate than when there is no wind. In warm weather, air movement may make the animal more comfortable, but in cold weather it adds to the stress temperature. At low temperatures, the nutrients required to maintain the body temperature are increased as the wind velocity increases. In addition to the wind, a drafty condition where the wind passes through small openings directly onto some portion or all of the animal body will usually be more detrimental to comfort and nutrient utilization than the wind itself.

ADAPTATION, ACCLIMATION, ACCLIMATIZATION, AND HABITUATION OF SWINE TO THE ENVIRONMENT

Every discipline has developed its own vocabulary. The study of adaptation/environment is no exception. So, the following definitions are pertinent to a discussion of this subject:

Adaptation refers to the adjustment of animals to changes in their environment.

Acclimation refers to the short-term (over days or weeks) response of animals to their immediate environment.

Acclimatization refers to evolutionary changes of a species to a changed environment which may be passed on to succeeding generations.

Habituation is the act or process of making animals familiar with, or accustomed to, a new environment through use or experience.

Species differences in response to environmental factors result primarily from the kind of thermoregulatory mechanism provided by nature, such as type of coat (hair, wool, feathers), and sweat glands. Thus, hogs, which have a light coat of hair, are very sensitive to extremes of heat and cold. On the other hand, nature gave cattle an assist through growing more hair for winter and shedding hair for summer, with the result that they can withstand higher and lower temperatures than hogs. The long-haired, shaggy yak of Tibet and the woolly Scotch Highland cattle of Scotland are as cold tolerant as the arctic-dwelling caribou and reindeer.

FACILITIES

Optimum facility environments can only provide the means for animals to express their full genetic potential of production, but they do not compensate for poor management, health problems, or improper diets.

Research has shown that swine are more productive and feed-efficient when raised in an ideal environment. The primary reason for having facilities, therefore, is to modify the environment. Proper barns and other shelters, shades, wallows, sprinklers, insulation, ventilation, heating, air conditioning, and lighting can be used to approach the desired environment. Also, increasing attention needs to be given to other stress sources such as space requirements, and the grouping of animals as affected by class, age, size, and sex.

The principal scientific and practical criteria for decision making relative to the facilities for swine in modern, intensive operations is the productivity and cost of production of animals, which can be achieved only by healthy animals under minimal stress. So, the investment in environmental control facilities is usually balanced against the expected increased returns.

The capital investment in modern confinement swine facilities is very high. Dr. S. E. Curtis, Pennsylvania State University, presented the following figures.

> Nowadays, the capital investment in intensive swine production facilities for the gestation, farrowing, and weaning phases is at least $2,000 per sow, whereas that in extensive production (outdoor facilities) is between $300 and $500.[1]

As a result of a 10-year study, the Scottish Agricultural College, Aberdeen, Scotland, designed a "freedom farrowing crate" to improve sow and piglet comfort, maintain easy management for farmers, and satisfy animal welfare concerns. The sows chose (1) cramped farrowing sites over spacious ones; and (1) either a 3- or 4-walled structure (favored by 78% of the sows) over a 1- or 2-wall structure; narrow space when farrowing (44%) over intermediate-sized farrowing areas (29%), wide-sized (12%), while the rest (15%) laid down somewhere else in the pen. The conclusion: *Sows have a significant preference for confined farrowing sites, perhaps it satisfies their need for security and isolation.* As a result of this study, the Scottish scientists designed a *freedom farrowing crate* that is only a little bigger than the space needed for the sow to stand up and lie down comfortably. Also, they designed a creep box for the baby pigs, with two

heating elements (one on the floor and the other around the base of the wall at snout level).

HEALTH

Health is the state of complete well-being, and not merely the absence of disease.

Disease is any departure from the state of health.

Parasites are organisms living in, on, or at the expense of another living organism.

Diseases and parasites (external and internal) are ever present swine environmental factors. Death takes a tremendous toll. Even greater economic losses result from retarded growth, poor feed efficiency, carcass condemnations, decreases in meat quality, and increases in labor and drug costs.

Any departure from the signs of good health constitutes a warning of trouble. Most sicknesses are ushered in by one or more signs of poor health—by indicators that tell expert caretakers that all is not well—that tells them that their animals will go off feed tomorrow, and that prompts them to do something about it today.

Fig. 13-7. Producers should be able to walk through their facilities and spot sick animals by their behavior. "Poor doers" like this one are often the victims of disease. (Courtesy, The Upjohn Company, Kalamazoo, MI)

Among the signs of ill health are lack of appetite; listlessness; droopy ears; sunken eyes; humped-up appearance; abnormal feces—either very hard or watery feces suggests an upset in the water balance or some intestinal disturbance following infection; abnormal urine (repeated attempts to urinate without success or off-colored urine should be cause for suspicion); abnormal discharges from the nose, mouth, and eyes; unusual posture—such as standing with the head down or extreme nervousness; persistent rubbing; hairless spots, dull hair coat, and dry, scruffy hidebound

---

[1]*Livestock Environment IV*, Fourth International Symposium, University of Warwick, Coventry, England, July 6–9, 1993, Supplement, American Society of Agricultural Engineers, pp. 1246–1254.

skin; pale, red, or purple mucous membranes lining the eyes and gums; reluctance to move or unusual movements; higher than normal temperature; labored breathing—increased rate and depth; altered social behavior such as leaving the herd and going off alone; and sudden drop in weight gains.

It is estimated that diseases and parasites in the United States decrease swine production by 15 to 20%. In the developing countries, diseases and parasites take an even greater toll—they decrease swine production by 30 to 40%.

STRESS

Stress is the nonspecific response of the body to any demand. As used herein, stress indicates an environmental condition that is adverse to an animal's well-being, either external or internal.

Swine experience many periods of stress— physical or psychological tension or strain. Stress of any kind affects animals. Among the external forces which may stress swine are previous nutrition, abrupt diet changes, change of water, space, level of production, number of animals together, changing quarters or mates, irregular care, transporting, excitement, presence of strangers, fatigue, illness, management, weaning, temperature, and abrupt weather changes.

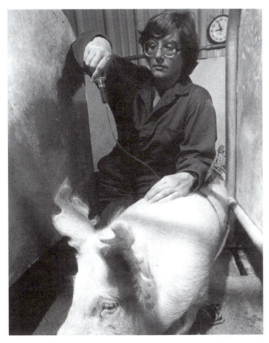

Fig. 13-8. Pigs are stressed, too. This shows a U.S. Department of Agriculture Environmental Physiologist taking a blood sample from a confined hog in an animal stress study at the U.S. Animal Research Center, Clay Center, Nebraska. (Courtesy, USDA)

In the life of a pig, some stresses are normal, and they may even be beneficial—they can stimulate favorable action on the part of an individual. Thus, we need to differentiate between stress and distress. Distress—not being able to adapt—is responsible for harmful effects. The trick is to manage stress so that it doesn't become distress and cause damage and to recognize the warning signals of distress.

The principal criteria used to evaluate, or measure, the well-being or stress of people are: increased blood pressure, increased muscle tension, body temperature, rapid heart rate, rapid breathing, and altered endocrine gland function. In the whole scheme, the nervous system and the endocrine system are intimately involved in the response to stress and the effects of stress.

The principal criteria used to evaluate, or measure, the well-being or stress of swine are: growth rate or production, efficiency of feed use, efficiency of reproduction, body temperature, pulse rate, breathing rate, mortality, and morbidity. Other signs of swine well-being, any departure from which constitutes a warning signal, are: contentment, alertness, eating with relish, sleek coat and pliable and elastic skin, bright eyes and pink eye membranes, and normal feces and urine.

Stress is unavoidable. Wild animals were often subjected to great stress; there were no caretakers to modify their weather, often their range was overgrazed, and sometimes malnutrition, predators, diseases, and parasites took a tremendous toll.

Domestic animals are subjected to different stresses than their wild ancestors, especially to more restricted areas and greater animal density. However, in order to be profitable, their stresses must be minimal.

■ **Porcine stress syndrome**—Stress is a special problem in pigs. Some pigs subjected to the stress of management procedures of handling, crowded transportation or sudden environmental change, adversely react and may even succumb. In the 1960s, Dr. David Topel of Iowa State University coined the term *porcine stress syndrome* (PSS). Pigs suffering from PSS demonstrate the following sequence of events when subjected to stress:

1. Difficulty in moving.
2. Signs of trembling or muscle tremors.
3. Irregular blotching of the skin.
4. Labored breathing and increased body temperature.
5. Total collapse and a shocklike death.

Moreover, pigs which demonstrate the PSS are usually associated with low-quality or pale, soft, and exudative (PSE) pork.

POLLUTION

Pollution is the issue of the decade. Anything that defiles, desecrates, or makes impure or unclean the surroundings pollutes the environment and can have a detrimental effect on swine health and performance. Thus, gases, odorous vapors, and dust particles from swine wastes (feces and urine) in buildings directly affect the quality of the environment. Muddy lots may also pollute the environment. For healthy and productive pigs, each of these pollutants must be maintained at an acceptable level.

Pollution potential, affecting the environment of both people and swine, increased as the U.S. swine industry moved toward specialization, mechanization, high animal density, and confinement.

To operate compatibly within the community, to provide maximum self-protection, and to avoid neighbor complaints and legal actions seeking either monetary damages or court injunctions, swine producers must be aware of some basic information and strategy concerning pollution and apply pollution control measures appropriate to the location.

The most troublesome swine pollutants are odors, dust, manure, muddy lots, pesticides, and nitrates in water. A discussion of each of these follows.

ODOR

The hog odor issue heats up in swine producing areas where neighbors and communities complain of offensive odors. Today, neighbors point to confinement hog operations. Their thoughts collide. One neighbor comments: "Those pigs have given a lot of people work." Another neighbor declares: "It smells bad out there." Jobs and smelly pig manure are at the center of a debate that has moved from farms and towns to state capitals.

To protect farmers from people moving next door then complaining about odors, right-to-farm laws have been enacted in 42 states.

In 1995, the Iowa legislature adopted a provision for the nation's No. 1 hog-producing state specifying that a livestock operation is not a nuisance unless it is proved by clear and convincing evidence that (1) the operation unreasonably and continuously interferes with a person's enjoyment of their life or property; and (2) the injury was caused by the negligent operation of the facility. Producers are required to have manure management plans which includes having adequate land for applying livestock waste. The 1995 Iowa law also imposes siting distances from residences, businesses, churches, schools, and public areas.

Swine producers can no longer thumb their noses at those who criticize the offensive odors associated with hog farms. More and more people are holding their noses and complaining about the smell of neighbors' swine buildings, pits, and lagoons.

Researchers are exploring odor control solutions including: (1) composting manure, (2) the microbial flora and odors that make up manure and the part each plays in producing odors, (3) slowing down or stopping microbial activity, (4) the effect of swine diets on odors, (5) the evaluation of different products for treating odors in manure or mixing in the feed to prevent odors, and methods and rates of spreading manure on the land.

Instead of spending so much money on lawsuits, hog farmers are spending more and more dollars on solutions to the problem.

DUST

Dust may be defined as a mixture of small particles of different sizes of dry matter. Dust is a contributing factor to both swine and human health, especially with respect to respiratory diseases.

High levels of dust particles resulting from automated dry feed handling systems, dander and hair from hogs, and dried manure particles can occur inside swine buildings. Manure gases can cling to those dust particles in such a way that inhaling these gas-laden particles is uncomfortable and objectionable. Particulate matter also includes viral, bacterial, and fungal agents from the building environment and carries them into the respiratory system of people and hogs.

MANURE

If you have hogs, you have manure. Today, that means you have a problem.

Of course, there is no one best manure management system for all situations. But, one way or another, science and technology must evolve with ways of disposing of manure, and this must be accomplished without polluting streams or the atmosphere or being offensive to the neighbors.

If not managed properly, animals may produce the following pollutants in troublesome quantities: manure, gases/odors, dust, and flies/other insects. Also, they may pollute water supplies.

When manure and urine are stored and undergo anaerobic digestion, dangerous and disagreeable gases are produced. The ones of primary concern are: hydrogen sulfide, ammonia, carbon dioxide, and methane.

FOUR METHODS OF HANDLING MANURE

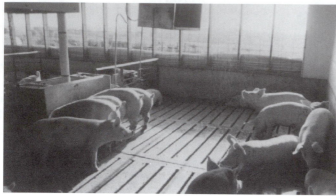

Fig. 13-9. Slotted floors through which the feces and urine pass to a storage area below or nearby. (Courtesy, Land O Lakes, Ft. Dodge, IA)

Fig. 13-10. Portable reel offers convenience of movement when irrigating from a swine lagoon. (Courtesy, Dr. Garth Body and Ms. Rhonda Campbell, Murphy Farms, Rose Hill, NC)

Fig. 13-11. A liquid manure knife injection applicator. Note the hose attached which is an umbilical supply line from the manure storage. (Courtesy, Iowa State University, Ames, IA)

Fig. 13-12. Irrigating a pasture from a swine lagoon. Note the cattle grazing the pasture. (Courtesy, Dr. Garth Boyd and Ms. Rhonda Campbell, Murphy Farms, Rose Hill, NC)

MUDDY LOTS

Muddy lots often plague livestock producers, especially during the winter months. Mud increases scours and other diseases in newborn animals and reduces production and feed efficiency in older animals.

Fig. 13-13. Muddy lots like this one are a source of pollution. (Courtesy, *Wallaces' Farmers & Iowa Homestead*)

PESTICIDES

Although science and technology have been the great multipliers in increasing our food supply, potential food supplies are still destroyed by the ravages of pests.

A pesticide is a substance that is used to control pests. Pesticides are an integral part of modern agricultural production and contribute greatly to the quality of food, clothing, and forest products we enjoy. Also, they protect our health from disease and vermin. Pesticides have been condemned, however, for polluting the environment, and in some cases for posing human health hazards. Unfortunately, opinions relative to pesticides tend to become polarized. A report by the National Research Council summarized the situation as follows:

> Users of pesticides fear that they will be regulated to the point where pests cannot be effectively controlled, with concomitant losses of food while opponents of the use of pesticides fear that people are being poisoned and that irreversible damage is being done to the environment.

No pest control system is perfect; and new pests keep evolving. So, research and development on a wide variety of fronts should be continued. We need to develop safer and more effective pesticides, both chemical and nonchemical. In the meantime, there is need for prudence and patience.

NITRATES IN WATER

The 10 ppm drinking water standard for nitrate levels (NO_3–N) was established by the U.S. Environmental Protection Agency. It was set at this level to protect the most nitrate-sensitive segment of the human population—infants under three months of age.

Nitrogen (N) enters surface water primarily by run-off. Nitrates reach groundwater primarily by leaching. Organic nitrogen (N) or ammoniated (NH_4+) sources are converted to (NO_3–N) by a process called nitrification. Primary agricultural sources of nitrogen include commercial fertilizer, animal waste, and breakdown of residual nitrogen in soils and crop residues.

ZONING

In recent years, some of the concern over the problem of pollution of the environment (air, water, and soil) and its effect on human health has stemmed from animals kept in the suburbs, with the neighbors protesting on the basis that they make for more flies, dust, and odors. This has resulted in zoning.

Thus, persons who desire to keep an animal(s) in their backyard, on a plot of land within city limits, should first study the zoning ordinance to determine if there are any restrictions against it.

Most zoning regulations are local ordinances covering a municipality, a county, or certain zoning districts. Also, there are rapid and basic changes in zoning ordinances; many cities and counties are rewriting and making changes in them. Thus, a prospective backyard animal keeper should check with local zoning authorities before purchasing an animal or animals.

POLLUTION LAWS AND REGULATIONS

Invoking an old law (the Refuse Act of 1899, which gave the Corps of Engineers control over runoff or seepage into any stream which flows into navigable waters), the U.S. Environmental Protection Agency (EPA) launched a program to control water pollution by requiring that all cattle feedlots which had 1,000 head or more the previous year must apply for a permit by July 1, 1971. The states followed suit; although differing their regulations, all of them increased legal pressures for clean water and air. Then followed the Federal Water Pollution Control Act Amendments, enacted by Congress in 1971 charging the EPA with developing a broad national program to eliminate water pollution.

Owners/operators of animal feeding facilities with more than 1,000 animal units must apply. Animal units are computed as follows: multiply number of slaughter and feeder cattle by 1.0; multiply number of mature dairy cattle by 1.4; multiply number of swine weighing over 55 lb by 0.4; multiply the number of sheep by 0.1; and multiply the number of horses by 2.0. (See Table 13-1, footnote 1, for what constitutes 1,000 animal units.)

TABLE 13-1
SUMMARY OF REGULATIONS

Feedlots with 1,000 or More Animal Units[1]	Feedlots with Less than 1,000 but with 300 or More Animal Units[2]	Feedlots with Less than 300 Animal Units
Permit required for all feedlots with discharges[3] of pollutants.	Permit required if feedlot— 1. Discharges[3] pollutants through an unnatural conveyance, or 2. Discharges[3] pollutants into waters passing through or coming into direct contact with animals in the confined area. Feedlots subject to case-by-case designation requiring an individual permit only after on-site inspection and notice to the owner or operator.	No permit required unless— 1. Feedlot discharges pollutants through an unnatural conveyance, or 2. Feedlot discharges pollutants into waters passing through or coming into direct contact with the animals in the confined area, and 3. After on-site inspection, written notice is transmitted to the owner or operator.

[1]More than 1,000 feeder or slaughter cattle, 700 mature diary cows (milked or dry), 2,500 swine weighing over 55 lb (24.9 kg), 500 horses, 10,000 sheep or lambs, 55,000 turkeys, 100,000 laying hens or broilers with continuous overflow watering, 30,000 laying hens or broilers with liquid manure handling, 5,000 ducks, or any combination of these animals adding up to 1,000 animal units.

[2]More than 300 feeder or slaughter cattle, 200 mature diary cows (milked or dry), 750 swine weighing over 55 lb (24.9 kg), 150 horses, 3,000 sheep, 16,500 turkeys, 30,000 laying hens or broilers with continuous overflow watering, 9,000 laying hens or broilers with liquid manure handling, 1,500 ducks, or any combination of these animals adding up to 300 animal units.

[3]Feedlot not subject to requirement to obtain permit if discharge occurs only in the event of a 25-year, 24-hour storm event.

FOOD SAFETY AND DIET/HEALTH CONCERNS

Many food safety and diet/health concerns are unwarranted. American consumers are prone to over-react to rumors relative to their food. They care little about what they put on their backs, but they are greatly concerned about what goes into their stomachs.

America's food supply is the safest in the world! Nevertheless, there is need for constant vigilance and improvement, especially in animal products which are subject to all the hazards of other foods (spoilage, pesticides, toxicities), plus being capable of transmitting, or serving as passive carriers, of certain diseases to humans.

In colonial times, the livestock producer slaughtered animals and processed meats, milked the cows, and gathered the eggs; then, delivered the products door-to-door to urban customers. If the products were not acceptable (spoiled meat, sour milk, cracked eggs), the matter was resolved quickly and on the spot, or the producer lost a customer. Today, the public expects the livestock team—farmers, processors, and retailers—to provide wholesome and safe products free from disease agents, toxic substances, and pesticide and drug residues.

Uptake of pesticides by animals, leading to residues in animal products, can result either from direct application of pesticides to animals or from animals ingesting feeds carrying pesticide residues. Drug residues are caused by (1) producers failing to withdraw drugs from livestock far enough in advance of marketing products; (2) contaminated feed storage, mixing, and handling equipment; and/or (3) the wastes (feces and urine) of treated animals coming in contact with untreated animals. *Reading and following the directions on the label is the key to safe pesticide and drug use.*

Dr. C. Everett Koop, former U.S. Surgeon General, is the authority for the following statement:

> People who are worried about pesticides fail to recognize that cancer rates in the United States have dropped remarkably over the past 40 years. During this period of time, stomach cancer has dropped more than 75%, and rectal cancer has dropped more than 65%. The only cancer going up today is cigarette-induced lung cancer.

Dr. Koop continued:

> The same hysteria that sometimes accompanies food safety issues also has affected how Americans look at diet and health. However, one major issue—cholesterol—is waning, as consumers realize how much it has been oversold. The cholesterol bubble has been pricked and is slowly deflating. While cholesterol is a risk factor in coronary heart disease, scientists consider other risk factors, such as smoking, hypertension, and heredity, to be much more significant than cholesterol. Because cholesterol is manufactured in the body naturally, diet does not have the direct relationship to blood levels of cholesterol that many misled laymen assume.[2]

Because the welfare of the nation is dependent

---

[2]Koop, Dr. C. Everett, *Gulf Coast Cattleman*, Vol. 55, No. 9, Dec. 1989, p. 9.

upon the health of its people, animal (and other) products are carefully monitored by various government agencies to assure consumers that they are wholesome and safe; and because of recognizing the importance of consumers in the safety of their food, the private sector may do additional testing. The agencies most responsible for this important work are:

1. The U.S. Department of Health and Human Services, including the following agencies: The Center for Disease Control, the Food and Drug Administration (FDA), and the National Institute of Health.

2. The U.S. Department of Agriculture, including the following agencies: the Agricultural Research Service, the Animal and Plant Health Inspection Service, the Cooperative State Research Service, the Federal Extension Service, the Labeling and Registration Section, and the Veterinary Service Division.

3. State and local government agencies.

4. International organizations engaged in health and/or nutrition activities, including the World Health Organization (WHO), and Food and Agriculture Organization (FAO).

5. Private industry groups such as the National Pork Producers Council and the National Dairy Council.

6. Professional organizations, including dentists, dietitians, doctors, health educators, nurses, and public health workers.

7. Food processors and retailers.

CONSERVE ENERGY

Fossil fuels—the stored photosynthesis of previous millennia—are like a bank account. There is nothing wrong with drawing upon either of them, but neither is inexhaustible. It is highly imprudent not to be aware of big withdrawals and not to cover them. Within a short span of a few years, the world made the transition from a positive energy balance based upon the capture of the energy of the sun via green plants, crops, and forests to an imbalance, or even a negative balance, by resorting primarily to the bank of trapped sun energy of fossil fuels that had accumulated over millions of years. Currently, the global use of energy resources is increasing at an alarming rate.

Everyone knows that cars and airplanes are powered by fuel. But how many people realize that modern, mechanized food production in the developed countries requires an extra input of fuel, which is mostly of fossil origin? How many people know that the developed nations have used fossil fuels to supplement the natural energy that comes directly from the sun on a day-to-day basis, and that they have grown dependent upon them?

Of course, the direct input of fuel into food pro-

duction is of rather recent origin. It all began in a very small way about 1840, when fuel-powered ships transported fertilizer (guano, and later bone meal) from South America to Europe. Then, after 1910, transportation vehicles relied almost exclusively on fossil fuels. But the direct use of fossil fuels in agriculture started with the manufacturing of chemical fertilizer beginning in 1922. Following closely in period of time, the tractor was substituted for horses, mules, and oxen—eventually almost completely replacing them.

In addition to food production on the farm, there are two other important steps in the food line as it moves from the producer to the consumer; namely, processing and marketing, both of which require bigger energy inputs than to produce the food on the farm (see Table 13-2).

Table 13-2 points up the increasing drain that modern food production is putting on the energy supply. In 1996, U.S. farms put in 2.8 calories of fuel per calorie of food grown, 3.1 times more than the on-farm energy input in 1940.

TABLE 13-2
MODERN FOOD PRODUCTION IS INEFFICIENT IN ENERGY UTILIZATION—THE STORY FROM PRODUCER TO CONSUMER[1]

Year	On the Farm	Food Processing	Marketing and Home Cooking	Total/ Person/ Year
1940[2]				
Million kcal	0.9	2.2	2.1	5.2
Percent	18.0	42.0	40.0	100.0
1996[3]				
Million kcal	2.8	5.7	4.6	13.1[4]
Percent	21.4	43.5	35.1	100.0
Increase, times 1940–1996	3.1	2.6	2.2	2.5

[1]Energy in million kcal used per capita to produce 1 million kcal of food in the U.S.

[2]Values from Borgstrom, G., "The Price of a Tractor," *Ceres*, FAO of the U.N., Rome, Italy, Nov.–Dec. 1974, p. 18, Table 3

[3]Authors' estimate based on several reports detailing trends in energy usage.

[4]This means that in 1996 it required 13.1 million kcal to produce 1 million kcal of food for each person, a daily consumption of 2,740 kcal (1,000,000 ÷ 365 = 2,740).

Table 13-2 also shows that, in the United States in 1996, a total of 13.1 calories were used in the production, food processing, and marketing-cooking for every calorie of food consumed, with a percentage distribution of the total cost of energy at each step from producer to consumer as follows: on the farm 21.4%; food processing, 43.5%; and marketing and home cooking, 35.1%. In 1940, it took only 5.2 calories—somewhat less than half the 1996 figure—to get 1 calorie of food on the table. It's noteworthy, too, that more energy is required for food processing and marketing-home cooking than for growing the product; and that, from 1940 to 1996, the on-the-farm energy requirement increased by 3.1 times, in comparison with an increase of 2.6 and 2.2 times for each of the other steps—processing and marketing/home cooking.

Prior to the advent of machines and fuel in crop production, 1 calorie of energy input on the farm produced about 16 calories of food energy. Today, on the average, U.S. farms put in about 2.8 calories of fuel per calorie of food grown; hence, to produce a daily intake of 3,000 calories of edible food from cultivated crops may require 8,400 calories of energy from fossil fuels—an exhaustible source. It's more surprising yet—and thought-provoking—to know that, even today in the poorer or developing countries, it takes only 1 calorie to produce each 10 calories of food consumed. The Oriental wet rice peasant uses only 1 unit of energy to produce 50 units of food energy. This gives the Orientals a favorable position among the major powers as the energy crisis worsens.

SUSTAINABLE AGRICULTURE

Fig. 13-14. Sustainable agriculture involves such ages old, and 21st century new, practices as terracing. (Courtesy, USDA, Soil Conservation Service)

Endangered species—and more! Today, it is endangered planet, endangered people and animals, and endangered agriculture. Among the deluge of warnings of environmental catastrophes are:

■ Pollution-caused warming of the atmosphere, known as the *greenhouse effect*, threatening weather changes that could render large areas of the planet unproductive and uninhabitable.

■ Toxic and radioactive wastes and dumped garbage that could poison drinking water and despoil the land.

■ Chemical pollution that is depleting the atmosphere's protective ozone layer.

■ Slashing and burning of tropical rain forests, driving thousands of species to extinction, increasing the amount of carbon dioxide in the atmosphere, and contributing to the greenhouse effect that warms the earth.

Awareness of these and other similar environmental catastrophes are being observed too little and too late. Remember that it took nature thousands of years to form the rain forest, but it took a mere 25 years for people to destroy much of it. And when a rain forest is gone, it is gone forever!

Although less dramatic, the Amazon rain forest story has been, or is being, repeated all over the world in the form of the greenhouse effect, toxicities, polluted streams, and/or other harbingers of threats to our environment. Too long we have managed our nonrenewable resources as though there is no tomorrow! Now, the situation is being righted. Worldwide, environmental quality and economic efficiency are in vogue. In the United States, this movement is called *sustainable agriculture*.

Sustainable agriculture is often described as farming that is ecologically sound and economically viable. It may be high or low input, large scale or small scale, a single crop or diversified farm, and use either organic or conventional inputs and practices. Obviously, the actual practices will differ from farm to farm. A definition follows.

Sustainable agriculture is farming with reduced off-farm purchased inputs of pesticides, herbicides, and fertilizers, along with reduced negative impact on natural resources and improved environmental quality and economic efficiency, while producing and distributing abundant, nutritious, affordable, high-quality foods and fibers for American and world markets.

The development of improved crops, cropping systems, irrigation, farm management, and marketing will be needed to make farms more profitable and sustainable. Typically, such farms will rely more on biological resources and management than on nonrenewable inputs of energy and chemicals. The foundation of a sustainable farm system is a comprehensive understanding of the land, the farm resources and operations, and potential short- and long-term markets.

Many of the practices advocated under sustainable agriculture are not new; they involve such timeless agricultural practices as soil erosion control, the protection of groundwater, the use of legumes as a source of nitrogen, biological insect and weed control, and the use of pastures as a primary feed source.

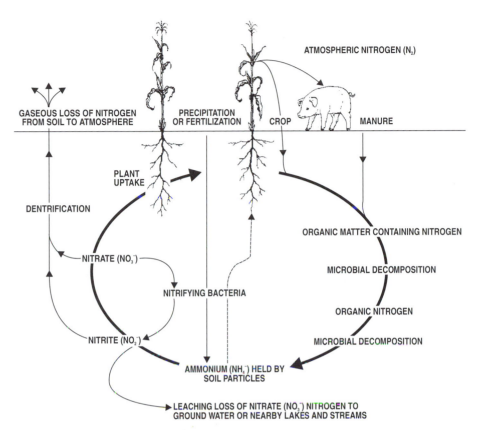

Fig. 13-15. At its best, a sustainable agriculture enhances the *cycle of nature*; manure is applied to soils and decomposed by microbes; complex protein in manure is broken down to release nitrogen as ammonia (NH_4); aerobic microbes convert ammonium nitrogen to nitrite (NO_2), thence to nitrate (NO_3) nitrogen; nitrate is (1) taken up by plants and built back into protein compounds; (2) leached downward when the soil is saturated—contaminating surface and groundwater if excessive nitrogen has been applied to the soil; and/or (3) released into the atmosphere when soils are wet for extended periods of time and the absence of air causes anaerobic microbes to convert the nitrate nitrogen to gaseous form.

ANIMAL WELFARE/ ANIMAL RIGHTS

In recent years, the behavior and environment of animals in confinement have come under increased scrutiny of animal welfare/animal rights groups all over the world. For example, in 1987 Sweden passed legislation designed (1) to phase out layer cages as soon as a viable alternative can be found; (2) to discontinue the use of sow stalls and farrowing crates; (3) to provide more space and straw bedding for slaughter hogs; and (4) to forbid the use of genetic engineering, growth hormones, and other drugs for farm animals except for veterinary therapy. Also, the law provides for fining and imprisoning violators.

The United Kingdom Farm Animal Welfare Council deems *Five Freedoms* as essential guarantees of ani-

mal welfare in any system of livestock husbandry; namely, freedom from: (1) hunger and thirst; (2) discomfort; (3) pain, injury, and disease; (4) fear and distress; and freedom to (5) display normal behavior.[3]

Animal welfarists see many modern practices as unnatural, and not conducive to the welfare of animals. In general, they construe animal welfare as the well-being, health, and happiness of animals; and they believe that certain intensive production systems are cruel and should be outlawed. The animal rightists go further; they maintain that humans are animals, too, and that all animals should be accorded the same moral protection. They contend that animals have es-

---

[3]*Livestock Environment IV*, Fourth International Symposium, University of Warwick, Coventry, England, July 6–9, 1993, Supplement, American Society of Agricultural Engineers, pp. 1255–1266.

sential physical and behavioral requirements, which, if denied, lead to privation, stress, and suffering; and they conclude that all animals have the right to live.

Livestock producers know that the abuse of animals in intensive/confinement systems leads to lowered production and income—a case in which decency and profits are on the same side of the ledger. They recognize that husbandry that reduces labor and housing costs often results in physical and social conditions that increase animal problems. Nevertheless, means of reducing behavioral and environmental stress are needed so that decreased labor and housing costs are not offset by losses in productivity. The welfarist/rightists counter with the claim that the evaluation of animal welfare must be based on more than productivity; they believe that there should be behavioral, physiological, and environmental evidence of well being, too. And so the arguments go!

But wild animals are often more severely stressed than domesticated animals. They didn't have caretakers to store feed for winter or to irrigate during droughts; to provide protection against storms, extreme temperatures, and predators; and to control diseases and parasites. Often survival was grim business. In America, the entire horse population died out during the Pleistocene Epoch. Fossil remains prove that members of the horse family roamed the plains of America (especially the area that is now known as the Great Plains of the United States) during most of tertiary period, beginning about 58 million years ago. Yet no horses were present on this continent when Columbus discovered America in 1492. Why they perished is still one of the unexplained mysteries. As the disappearance was so complete and so sudden, many scientists believe that it must have been caused by some contagious disease or some fatal parasite. Others feel that perhaps it was due to multiple causes, including (1) climatic changes, (2) competition, and/or failure to adapt. Regardless of why horses disappeared, it is known that conditions were favorable to them at the time of their reestablishment by the Spanish conquistadors about 500 years ago.

Animal welfare issues tend to increase with urbanization. Moreover, fewer and fewer urbanites have farm backgrounds. As a result, the animal welfare gap between town and country widens. Also, both the news media and the legislators are increasingly from urban centers. It follows that the urban views that are propounded will have greater and greater impact in the years ahead.

SOME NEW ANIMAL/ ENVIRONMENTALLY-FRIENDLY INNOVATIONS

Scientists and technologists are working ceaselessly away in developing new facilities and equipment designed to be animal/environmentally-friendly; among them, are the following:

■ **Schematic cross-section of a sow feeder.**

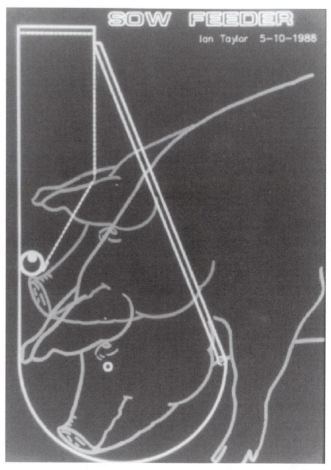

Fig. 13-16. This feeder is designed to accommodate both the pigs' physical and behavioral attributes. Note that this feeder provides sufficient space to accommodate the natural range of eating motions, as depicted in this illustration of a sow's eating movements occurring within the spatial limitations of an appropriately proportioned sow feeder. (Courtesy, Ian A. Taylor, Animal Environment Specialists, Inc., Columbus, OH)

■ MoorComfort® Gestation System, patented.

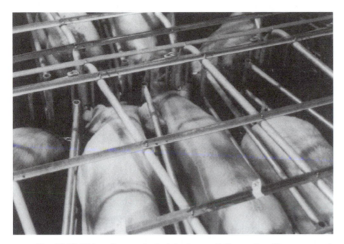

Fig. 13-17. This unique patented design gestation crate utilizes appropriately proportioned pivoting sides to allow sows (or boars) greater freedom of movement, including the ability to turn around, within the same space as conventional fixed-sided stalls (patent holder—QDM Inc.). As a consequence, this system also increases the individual sow's choice of social and thermal factors while maintaining the benefits of individual feeding, health monitoring, and protection from aggression. It is critical that the correct dimensions and floor plan options be employed to ensure optimum system performance. (Courtesy, Ian A. Taylor, Animal Environment Specialists Inc., Columbus, OH)

Fig. 13-18. Increasing concerns over effluent management issues, including odor and nutrient overloading of soil and ground water, prompt interest in new treatment systems that will provide increased control. A very simple but effective system employing a patented application of tangential flow separation methodology combined with simple proprietary chemistry has been developed to provide, economically, increased control. These TFS-based systems, already successfully used across other industries, selectively remove phosphorus and other effluent constituents of concern from the effluent flow, together with solids. This process increases the disposal options available, including cost recovery via the recycling of water and the use of the manure concentrate as a high-quality fertilizer or as a nutrient source in refeeding options. This very space efficient system can be matched with optional secondary treatment steps to allow "lagoonless" swine production. The small mobile demonstration unit shown will treat effluent flows up to 20 gallons per minute. (Courtesy, Ian A. Taylor, Animal Environment Specialists Inc., Columbus, OH)

QUESTIONS FOR STUDY AND DISCUSSION

1. Why has increased confinement of animals made for great interest in the subject of animal behavior?

2. Define *animal behavior* and *ethology*.

3. Why has there been a genetic time lag; why have swine producers altered the environment faster than the genetic makeup of animals?

4. List the eight general functions of behavioral systems that swine exhibit.

5. Discuss the following behavioral systems as they pertain to swine:

    a. Ingestive behavior

    b. Eliminative behavior

    c. Maternal behavior (mothering)

    d. Sexual behavior

6. How is the dominance hierarchy established when several swine are brought together to form a herd? How is it maintained?

7. Explain the difference between dominance and leader-follower.

8. Discuss how swine communicate with each other, and the importance of each method.

9. Those who care for swine need to be familiar with behavioral norms in order to detect and treat abnormal situations—especially illness. Describe a normal pig.

10. Discuss each of the following abnormal behaviors of swine: homosexuality, and tail biting.

11. List and discuss the significance of one example of the practical application of swine behavior in breeding for adaptation.

12. How may environment affect swine?

13. Discuss how each of the following environmental factors affects swine:

    a. Feed and nutrition

    b. Water

    c. Weather

    d. Facilities

    e. Health

    f. Stress

14. Define *pollution*. Why has pollution become such a great issue?

15. Detail why the following pollutants are especially troublesome in swine operations:
   - a. Odor
   - b. Dust
   - c. Manure
   - d. Muddy lots
   - e. Pesticides
   - f. Nitrates in water

16. Of what value is zoning?

17. Assume that you plan to establish a herd of 1,000 swine. With what pollution laws and regulations must you comply? How would you go about applying?

18. Is America's food supply sufficiently safe?

19. Why should a swine producer conserve energy?

20. What is sustainable agriculture?

21. Describe the cycle of nature.

22. Define (a) animal welfare, and (b) animal rights.

23. List and discuss the United Kingdom's Farm Animal Welfare Council's *Five Freedoms*.

SELECTED REFERENCES

Title of Publication	Author(s)	Publisher
Agricultural Waste Utilization and Management	F. J. Humenik, Chairman, Ex. Com.	American Society of Agricultural Engineers, St. Joseph, MI, 1985
Agriculture and Groundwater Quality	P. F. Pratt, Chairman	Council for Agricultural Sciences and Technology, Ames, IA, 1985
Animal Behavior	J. P. Scott	The University of Chicago Press, Chicago, IL, 1972
Animal Behavior, Third Edition	V. G. Dethier E. Stellar	Prentice-Hall, Inc., Englewood Cliffs, NJ, 1970
Animal Behavior and its Application	D. V. Ellis	Lewis Publishers, Inc., Chelsea, MI, 1986
Animal Behavior in Laboratory and Field	Ed. by A. W. Stokes	W. H. Freeman and Co., San Francisco, CA, 1968
Animal Behaviour: A Synthesis of Ethology and Comparative Psychology	R. A. Hinde	McGraw-Hill Book Co., New York, NY, 1970
Animal Cytology and Evolution	M. J. White	Cambridge University Press, Cambridge, England, 1973
Animal Science, Ninth Edition	M. E. Ensminger	Interstate Publishers, Inc., Danville, IL, 1991
Applied Animal Ethiology	W. R. Stricklin, Guest Editor	Elsevier Science Publishers, Amsterdam, The Netherlands, Vol. 11, No. 4, Feb. 1984
Behavior of Domestic Animals, The	Ed. by E. S. E. Hafez	The Williams & Wilkins Company, Baltimore, MD, 1975
Bibliography of Livestock Waste Management	J. R. Miner D. Bundy G. Christenbury	Office of Research and Monitoring U.S. Environmental Protection Agency, Washington, DC, 1972
Biology of the Pig, The	W. G. Pond K. A. Houpt	Cornell University Press, Ithaca, NY, 1978
Development and Evolution of Behavior	Ed. by L. R. Aronson, *et al.*	W. H. Freeman and Co., San Francisco, CA, 1970

Dictionary of Farm Animal Behavior, Second Edition	J. F. Hurnik A. B. Webster P. B. Siegel	Iowa State University Press, Ames, IA, 1995
Domestic Animal Behavior	J. V. Craig	Prentice-Hall, Inc., Englewood Cliffs, NJ, 1981
Environmental Management of Animal Agriculture	S. E. Curtis	Animal Environment Services, Mahomet, IL, 1981
Ethology: The Biology of Behavior	I. Eibl-Eibesfeldt	Holt, Rinehart and Winston, New York, NY, 1975
Farm Animal Welfare	B. E. Rollin	Iowa State University Press, Ames, IA, 1995
Impact of Stress, The Proceedings	Ed. by R. E. Moreng J. R. Herbertson	Colorado State University, Ft. Collins, CO, 1986
Integrating Sustainable Agriculture, Ecology, and Environmental Policy	Ed. by R. K. Olson	Food Products Press, An Imprint of The Haworth Press, Inc., New York, NY, 1992
Introduction to Animal Behavior, An: Ethology's First Century, Second Edition	P. H. Klopfer	Prentice-Hall, Inc., Englewood Cliffs, NJ, 1974
Kinships of Animals and Man	A. H. Morgan	McGraw-Hill Book Co., New York, NY, 1955
Livestock Behavior, a Practical Guide	R. Kilgour C. Dalton	Westview Press, Boulder, CO, 1984
Livestock Environment IV	Ed. by E. Collins C. Boon	American Society of Agricultural Engineers, 1993
Our Friendly Animals and Whence They Came	K. P. Schmidt	M. A. Donahue & Co., Chicago, IL, 1938
Principles of Animal Behavior	W. N. Tavolga	Harper & Row, New York, NY, 1969
Principles of Animal Environment	M. L. Esmay	Avi Publishing Co., Westport, CT, 1978
Proceedings '95, International Livestock Odor Conference, 1995	Papers by Program Staff	Iowa State University, Ames, IA, 1995
Readings in Animal Behavior	Ed. by T. E. McGill	Holt, Rinehart and Winston, New York, NY, 1973
Science of Animals That Serve Mankind, The	J. R. Campbell J. F. Lasley	McGraw-Hill Book Co., New York, NY, 1975
Scientific Aspects of the Welfare of Food Animals	F. H. Baker, Chairman	Council for Agricultural Science and Technology (CAST), Ames, IA, 1981
Social Hierarchy and Dominance	Ed. by M. W. Schein	Dowden, Hutchinson & Ross, Inc., Stroudsburg, PA, 1975
Stockman's Handbook, The, Seventh Edition	M. E. Ensminger	Interstate Publishers, Inc., Danville, IL, 1992
Stress Physiology in Livestock: Vol. 1, Basic Principles; Vol. 2, Ungulates; Vol. 3, Poultry	M. K. Yousef	CRC Press, Inc., Boca Raton, FL, 1985
Sustainable Agriculture and the 1995 Farm Bill	N. P. Clarke, Cochair P. B. Ford, Cochair	CAST, Ames, IA, 1995
Waste Management and Utilization in Food Production and Processing	L. L. Boersma, Cochair I. P. Murarka, Cochair	CAST, Ames, IA, 1995

Why worry?

Managers must fill many roles. This shows a manager checking the temperature in a nursery. Note the two thermometers; one above the manager's hand, and the other at floor level. (Courtesy, Land O Lakes, Ft. Dodge, IA)

14

SWINE MANAGEMENT

Swine management is the art of caring for and handling swine. It gives point and purpose to everything else. It can make or break a swine operation.

MANAGER'S ROLE

Managers of a big and turbulent swine operation must fulfill many rules in order to be successful; among them, planner, organizer, leader, controller, and facilitator of changes. These primary functions are shown in Fig. 14-1.

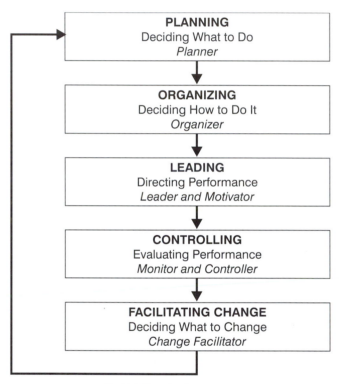

PLANNING
Deciding What to Do
Planner

ORGANIZING
Deciding How to Do It
Organizer

LEADING
Directing Performance
Leader and Motivator

CONTROLLING
Evaluating Performance
Monitor and Controller

FACILITATING CHANGE
Deciding What to Change
Change Facilitator

Fig. 14-1. The role of a manager.

The development of sound management ability is a gradual process which incorporates both formal training and practical experience. By itself, formal management education is insufficient in developing sound management abilities. Similarly, practical management experience in the absence of formal training in the use of management skills can hamper success.

SWINE MANAGEMENT SYSTEMS AND PRACTICES

Swine management systems and practices vary widely between areas and individual swine producers.

They are influenced by size of operations; available land and its value for alternate uses; markets; available capital, labor, and feed; and the individual whims of producers.

INDOOR VS OUTDOOR INTENSIVE SWINE PRODUCTION

About 1955, the confinement system started evolving, although for sometime it had been recognized that hogs are better adapted to confinement rearing than other classes of four-footed farm animals. By 1960, it was estimated that 3 to 5% of the market hogs in the United States were produced in confinement. Then, in 1980 a survey revealed that 81% of U.S. swine producers confined sows at farrowing time, 66.2% provided confinement for nurseries, 70.5% confined growing-finishing pigs, but only 24.3% confined the sow herd.[1] This shift to confinement went hand in hand with (1) greater knowledge of nutrition, (2) improved disease prevention and parasite control, (3) improvements in buildings and equipment, and (4) the growth of large-scale, highly specialized operations. Contrary to the opinion held by some, this does not mean that the use of pastures for hog production is antiquated. Rather, there now exists two alternatives for the management of swine, instead of just one; and the able manager will choose wisely between them or use the best of each system.

In the mid-1990s, intensive outdoor production was in vogue in Britain, and Denmark was test-marketing "Free-Range Pork."

■ **Texas Tech University, Lubbock, Texas, researchers compared the two systems—indoor vs outdoor intensive pork production**—They reported that "Outdoors yields fewer but heavier pigs than indoors." Details follow:[2]

A total of 358 litters were evaluated in two environments.

The indoor facility featured totally confined breeding, farrowing, and nursery units.

The outdoor unit was located on a 34-acre site. Gilts were bred in radial-designed paddocks (shaped like the spokes of a wagon wheel with a central hub and triangle-shaped paddocks). Then gilts were penned together in the breeding-gestation paddocks. Bred gilts remained together until they were moved into a farrowing hut, each equipped with a fender (guard rail) 12 in. high. The fender remained up for three days after farrowing. Wheat straw was used for

[1]Survey of randomly selected swine producers from the *Hog Farm Management* circulation list and published as "1980 Market Profile," Miller Publishing Company, Minneapolis, MN, p. 42.

[2]*National Hog Farmer*, January 15, 1995.

bedding. Farrowing paddocks were operated on an all-in/all-out basis.

Gilts and sows were bred to farrow every three weeks, year around. Pigs were weaned at 28 days.

Diets of both the indoor and outdoor groups were identical, except the outdoor diet was cubed.

Pertinent results follow:

1. **Pigs weaned per litter.** The outdoor gilts weaned 7.5 pigs per litter, compared with 8.5 pigs per litter for the indoor gilts. Outdoor litters had a 27.5% death loss vs 12.3% indoor death loss. Crushing was the primary cause of preweaning piglet death in the outdoor farrowing huts.

2. **The outdoor pigs were heavier.** The outdoor pigs weighed more than the indoor pigs at weaning (4 weeks of age) and at the end of the nursery phase (9 weeks of age).

3. **The two systems were similar economically.** If the pork producer sold feeder pigs, the total dollar receipts for each unit would have been similar because of the heavier weights of the outdoor pigs and the lower overhead cost.

4. **Outdoor system was harder on labor.** Sometimes, the caretakers had to work longer and do more heavy lifting (the huts) with the outdoor group. Also, they were exposed to all kinds of weather.

5. **Data in Table 14-1 tell the story.** Table 14-1 summarizes the results. It shows how environment affected sow and litter productivity.

TABLE 14-1
EFFECTS OF FARROWING ENVIRONMENT
ON SOW AND LITTER PERFORMANCE[1]

Measure	Indoor	Outdoor	SE	P-Value[2]
Number litters	148	210	—	—
Piglets born . . . . . . no./litter	10.99	11.01	0.27	0.97
Born live . . . . no./litter	9.91	10.59	0.30	0.08
Found dead . . . . . . no./litter	1.47	0.38	0.15	0.0001
Birth weight alive . . lb/piglet	4.14	4.25	0.06	0.17
Piglet birth weight . . . lb/litter	43.0	44.9	1.20	0.22
Pigs weaned . . . . . no./litter	8.6	7.5	0.28	0.007
Pre-weaning mortality . . . %	12.3	27.5	2.17	0.0001
Weaning weight . . . . lb/pig	14.3	15.3	0.48	0.106
Total weaning weight . lb/pig	120.5	114.6	4.45	0.30
Piglet gain . . . . . lb/day/pig	0.36	0.39	0.012	0.11

[1]Data in table are least squares means averaged over the first and second parities.

[2]P-value for environment effect.

In the mid-1990s, outdoor swine units were common in the United Kingdom. The warm climate of the southern part of the United States is suited to outdoor intensive production. In addition to a dry climate, for success of outdoor, intensive production it is important that animal density be kept low so that disease incidence is low and waste buildup is not a problem.

Many changes have occurred since the early days of outdoor pork production in the United States. The pork industry has moved from the low investment, low-intensity system of the 1950s to the '90s modern high-investment, high-intensity confinement systems. Many improvements in production accompanied the move indoors. Some of these changes can be adapted and applied to improve intensive outdoor production. (Also, see Chapter 12, Forages for Swine, chapter introduction and section headed "Gestating Sows and Outdoor Farrowing.")

GILTS VS OLDER SOWS

Controlled experiments and practical observations, in which gilts have been compared with older (tried) sows, bear out the following facts: (1) gilts have fewer pigs in their litters than do older sows, and (2) the pigs from gilts average slightly smaller in weight at birth and also tend to make somewhat slower gains.

Despite these disadvantages, gilts have certain advantages, especially for the commercial pork producer. Their chief superiority lies in the fact that they continue to grow and increase in value while in reproduction. Although overweight young sows bring less on the market than prime barrows, from this standpoint alone they generally return a handsome profit when the price of pork is sufficiently favorable. Many practical commercial producers, who probably are less close to the herd at farrowing time than purebred breeders, are of the opinion that the smaller gilts crush fewer pigs than do older and heavier sows.

In no case is it recommended that the purebred breeder rely only on gilts. Tried sows that are regular producers of large litters with a heavy weaning weight and that are good mothers and producers of progeny of the right type should be retained in the herd as long as they are fertile.

MULTIPLE FARROWING

Multiple farrowing refers to that type of program in which there is a scheduling of breeding so that the litters arrive in a greater number of farrowing periods throughout the year. Big operations and environmentally controlled buildings have made for year-round production.

There is nothing mysterious or complicated about multiple farrowing. It does, however, entail some planning and close attention to management details.

Among the factors **favorable** to multiple farrowing are:

1. It makes for a more stable hog market, with fewer high and low market receipts and price fluctuations.

2. It distributes the work load for the swine producer.

3. It makes for better use of existing buildings and equipment; for example, the farrowing house can accommodate more litters because of the longer farrowing season.

4. It provides a more sustained flow of hogs to market, which, from the standpoint of the packer, is desirable because—(a) it makes for more complete use of labor and plant capacity, and (b) it enables the processor more nearly to meet the demands of the retailer and consumer. Also, the producer's income is distributed throughout the year.

5. It provides retailers with a steady supply of pork for their trade.

6. It avoids sharp price rises, which the consumer dislikes.

Among the factors **unfavorable** to multiple farrowing are:

1. The swine enterprise is more confining for a longer period of time, for the reason that competent help must be available over a more prolonged farrowing season.

2. The possibility of a disease outbreak may be increased because of the possible build-up of pathogenic organisms. That is, with multiple farrowing, it is not possible to clean out and air out for a long period of time.

3. Farrowing in seasons of extreme cold or heat increases building and equipment costs.

4. It requires more management know-how.

No one expects the seasonal pattern of hog production to be completely eliminated, but, because of the several recognized advantages of multiple farrowing to both the processor and the producer, it will continue to lessen some of the market gluts of the past. (See the subsequent section in this chapter headed "Farrowing Management.")

MANAGING HERD BOARS

Herd boars influence the swine breeding program in two important ways: (1) They provide a source of genetic improvement and (2) they affect the farrowing rate and litter size. In addition, replacement boars are a potential source for the production of disease in the herd.

Boars should be purchased 45 to 60 days before they are needed for breeding. This allows ample time

Fig. 14-2. A mature herd boar, 50% inbred Yorkshire and 50% inbred Lacombe. (Courtesy, McLean County Hog Service, Inc., LeRoy, IL)

to locate superior animals, and following their purchase, to check their health, acclimate them, and test-mate and/or evaluate them for breeding soundness. Boars should be ready for use when they are about eight months of age.

■ **Transporting newly purchased boars**—Many seedstock suppliers offer a delivery service to their customers. Proper care in transporting boars insures maximum animal performance by minimizing stresses, injuries, and disease.

■ **Quarantine of new boars**—Newly purchased boars should be isolated in clean, comfortable facilities for a minimum of 30 days; preferably 60 days. Following isolation, new additions should be tested for brucellosis, pseudorabies, transmissible gastroenteritis (TGE), and such other diseases as the herd veterinarian recommends. In the meantime, the isolation period should be continued until all health tests are received and evaluated.

■ **Feeding boars in isolation**—During the isolation period, boars should be fed a diet similar to that fed by the seller. This reduces stress associated with relocation. Gradually, the diet should be changed to match that of the buyer. Depending upon the diet, age, condition, and housing, boars should be fed 5 to 6.5 lb of a balanced 14% crude protein diet per day—5 to 5.5 lb for younger boars; 5 to 6.5 lb for mature boars. A well-balanced sow gestation diet is satisfactory. Mega hog operations may have enough boars to provide a special boar diet.

■ **Test mating and semen evaluation of boars**—Boars should be test mated at 7 to 8 months of age. Records indicate that about 1 in 10 untried boars has a fertility problem. Use an in-estrus (heat) gilt for test mating, and assist the boar if necessary. If possible, also collect and evaluate a semen sample.

■ **Age and service of boar**—The recommended maximum number of services per boar, by age, is given in Table 14-2.

TABLE 14-2
RECOMMENDED MAXIMUM NUMBER OF SERVICES
PER BOAR, BY AGE

Boar	Individual Mating System Maximum Matings		Pen Mating System[1] Boar to Sow Ratio
	Daily	Weekly	7- to 10-Day Breeding Period
Young (8 to 12 months)	1	5	1.2 to 4
Mature (more than 12 months)	2	7	1.3 to 5

[1]Assumes that all sows are weaned on the same day.

Table 14-2 may be used as a guide in determining boar power.

Individual mating systems provide an opportunity to control the use of boars and the opportunity to observe and record matings, but more labor and higher cost facilities are usually required. Individual mating systems are essential for proper management for a weekly farrowing or for all-in/all-out production.

Pen mating requires less labor and lower cost facilities, but the resulting farrowing rate may be lower than for sows which are individually mated. When pen mating is used, consider dividing weaned sows in two or more groups and rotating boars on a 12- to 24-hour interval.

■ **Boar culling and replacement**—Farrowing rate and litter size are generally best when boars 9 to 20 months of age are used. However, as long as boars remain structurally sound and are aggressive breeders, fertility is generally maintained until they are 3 years of age or more. Most boars are culled for reasons of lameness, lack of aggressiveness, and size, rather than because of poor semen quality. Most well-managed herds replace 25 to 50% of their boars annually.

■ **Breeding area**—Provide a good breeding area with good footing. Avoid wet, slippery floors.

■ **Boar housing**—When using an individual mating system, boars are generally penned separately in individual crates about 28 in. wide and 7 ft long, or in pens 6 ft × 8 ft. Individual housing of boars eliminates fighting, riding, and competition for feed.

Groups of boars used in a pen mating system should be penned together when they are removed from a sow group.

■ **Keep boars cool during summer**—Boars subjected to temperature over 85°F may have reduced semen quality, resulting in lowered farrowing rate and smaller litter size.

Heat stress may be determined by monitoring the respiration rate of boars. The normal respiration rate for a boar is 25 to 35 breaths per minute. During heat stress, respiration rate may increase to 75 to 100 breaths per minute. Respiration rate can be determined by watching the movement of the rib cage; it expands and contracts with each single breath.

A group of boars used in pen mating can be kept cool by the use of a sprinkler installed under a shade built over a sand or concrete floor. For boars housed in environmentally controlled buildings, consideration should be given to the use of thermostatically controlled drippers, evaporative coolers, or geothermal systems.

■ **Routine boar management**—Some routine good boar management procedures include (1) observing boars daily for abnormal behavior such as lack of appetite, listlessness, or lameness; (2) having a high-low thermometer in the breeding area and in the boar housing facility; (3) vaccinating boars for reproductive diseases such as erysipelas, leptospirosis, and parvovirus; (4) treating for mange and lice; and (5) cutting the tusks of boars and trimming preputial hairs around the sheath, every 6 to 8 months.

Note: The care, management, and feeding of boars is fully covered in Chapters 7 and 9 of this book; hence, the reader is referred thereto.

MANAGING SOWS AND GILTS

In order to achieve maximum reproductive efficiency, a high level of management must be expended on the breeding herd. Good management of sows and gilts in the herd will pay handsome dividends by increasing the number of live pigs farrowed and marketed.

Fig. 14-3. A prolific Hampshire sow, structurally sound and feminine. (Courtesy, Compart's Boar Store, Nicollet, MN)

GILT MANAGEMENT

Selection of superior replacement females is very important. Gilts should be selected from family lines that have superior productivity, including mothering ability, and superior carcass traits. A good indication of the gilt's ability to reproduce normally is cycling at an early age—at 4½ to 6 months of age. Table 14-3 presents average figures which may be used to evaluate the reproductive ability of swine.

TABLE 14-3
AVERAGE TIME FOR PUBERTY, ESTRUS, AND OVULATION IN SWINE

Reproductive Traits	Average Time
Age at puberty	4½ to 6 months
Weight at puberty	150 to 230 lb
Duration of estrus	2 to 3 days
Length of estrous cycle . . .	18 to 24 days (20–21 avg.)
Weaning to estrus	3 to 7 days (5 avg.)
Time of ovulation	35 to 40 hours after onset of estrus

The general recommendations regarding age at first breeding are (1) to breed gilts during the third heat period, or (2) to breed gilts when they reach 230 to 240 lb body weight. It is recommended that gilts not reaching puberty or the desired breeding weight by 7 months of age be culled from the breeding herd.

There are seasonal differences in age at first heat. In general, fall-born gilts reach puberty at earlier age and lighter weight than spring-born gilts.

Anestrous conditions (absence of standing heat) may be the result of one or more of the following conditions:

1. Faulty heat detection.
2. Hot weather stress.
3. Silent heat (ovulation with no visible sign of heat).
4. Sickness.
5. Nutritional deficiency(s) (especially lack of protein and/or energy).
6. Social stress.
7. Pregnancy.

SOW MANAGEMENT

Selection of sows that cycle within 7 days postweaning is very important in keeping management schedules running smoothly. If a sow fails to conceive within 28 days postweaning, she should be culled. With each 21-day delay in conception, the sow must produce one to two extra pigs just to pay for the additional labor and feed. Likewise, if cycling gilts do not conceive

after three estrous cycles, they should be culled so as not to increase the number of "hard breeders" in future generations.

When adequate nursery facilities are available, weaning at 3 to 4 weeks of age is recommended so that the sows can be returned to production as soon as possible. New research indicates that, when adequate facilities, diet, and management are available, piglets may be weaned at 10 to 14 days of age without jeopardizing their ability to grow efficiently.

Heavy-milking sows mobilize minerals from their skeletons to facilitate lactation. To prevent sow breakdown (especially of the pelvic arch), the lactation diet should be properly fortified with minerals. In addition, dry pens with plenty of space and good footing should be provided, so as to prevent slippage. Also, the use of individual stalls will reduce the incidence of injuries.

When postweaning scours are a problem, weaning should be postponed if possible.

Sows in thin condition should be on a high plane of nutrition and gaining weight before breeding. This will assure maximum ovulation rate.

Synchronization of heat in sows is a relatively simple matter. When litters are weaned from a group of sows at the same time, a high proportion of the sows will come into heat within 3 to 7 days postweaning. Adequate boar power is very essential in order to take advantage of synchronization of postweaning heat.

■ **Effect of high temperature on sows**—High temperature (above 85°F) will delay or prevent the occurrence of heat, reduce ovulation rate, and increase early embryonic death. A group of sows in a pen or pasture may be kept cool by the use of a sprinkler installed under a shade built over a sand or concrete floor. For sows housed in environmentally controlled buildings, consideration should be given to the use of thermostatically controlled drippers, evaporative cooling, or geothermal systems.

HEAT DETECTION / OVULATION / CONCEPTION

Heat is the time that the female accepts the male for mating. Heat detection is more effective in the presence of a sexually mature boar (by placing a boar in the pen with the female, or by nose-to-nose fenceline contact with females). The caretaker can confirm heat by applying back pressure to each female in the presence of a boar. Most sows and gilts in heat will respond by standing stiffly and attempting to stiffen their ears (making them erect [called "popping their ears"]). If they do not stand solidly and pop their ears, they are not in heat.

Achieving a high conception rate and good litter size necessitates getting sperm into the female's reproductive tract at the time when pregnancy rate and litter size will be maximized. Regardless of the method of breeding, (*i.e.*, pen mating, hand mating, or artificial insemination) an adequate number of live sperm must be in the reproductive tract a few hours before ovulation occurs or conception rate and litter size will be reduced. Fig. 14-4 shows the effect on conception rate of breeding at various times relative to the time of ovulation.

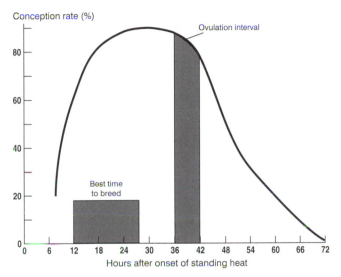

Fig. 14-4. Effect of time of insemination on conception rate in swine.

Note that when heat lasts 48 hours a female will ovulate 8 to 12 hours before the end of standing heat or 36 to 40 hours after its onset. When mating occurs too early or too late, conception rate and litter size drop dramatically.

With once-a-day heat detection, breed the females each day they will stand. With twice-a-day detection, breed at 12 to 24 hours after they are first detected in heat.

Producers using unobserved (pen) mating must have plenty of boar power. For every 10 sows, use one mature boar (over 1 year of age) per 21-day breeding period. Decrease that ratio to 4 to 6 sows for each young boar (less than 1 year old). A sow-to-boar ratio of 4 to 1 for mature boars and 2 to 1 for young boars is recommended when sows are weaned in groups. Reducing the breeding period in a pen mating system to no more than 7 to 10 days simplifies baby pig management at farrowing, but there is a high probability that only 85 to 90% of the sows will cycle within that period.

■ **Double mating**—A boost in conception rate and litter size can be obtained by using more than one boar on each female (double mating). Double mating

is easily accomplished when using hand mating or AI. When pen breeding, boars should be rotated from pen to pen at least once every day and rested periodically.

MANAGING THE GILT POOL

The gilt pool refers to a group of females selected as potential brood sows to fill vacancies in the sow herd. Empty farrowing crates indicate a lack of planning and poor use of the gilt pool.

Under normal conditions, about 15 to 30% of the weaned sows from each farrowing are culled. During problem breeding periods (hot weather, disease, or infertile boars), the number of replacement gilts increases. So, as many as three replacement gilts for each farrowing crate to be filled may be necessary.

PREGNANCY DETECTION

Electronic pregnancy diagnosis is a reality! With electronic detectors, pregnancy at 30 to 45 days after breeding can be determined with 90 to 95% accuracy. Accuracy drops off rapidly after 45 days. The Doppler instrument which detects fetal heartbeats may be used over a longer period of time, but it costs more. Some instruments can be used for both pregnancy detection and measuring backfat thickness and loin eye size.

FARROWING MANAGEMENT

Regardless of the frequency of farrowing groups entering the farrowing facilities and the breeding system being used, producers should strive to farrow all females in one facility within one week. The all-in/all-out management practice can be achieved with wise use of the gilt pool and group weaning.

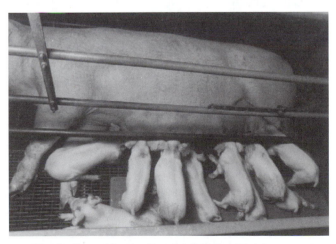

Fig. 14-5. Sow and litter in a farrowing crate. Crates help keep sows from accidentally injuring their offspring. (Courtesy, National Pork Producers Council, Des Moines, IA)

Being present when sows farrow will save one or more pigs per litter.

Brief periods of exercise and a bulky diet several days prior to farrowing will minimize constipation problems. All feed should be removed the day of farrowing, but make sure that the sows have plenty of fresh, cool water. (Also see Chapter 9, section headed "Feeding Brood Sows.")

SUMMARY

Commercial gilts should be selected from the largest, healthiest litters (based on farrowing and weaning data) as replacements on the basis of their ability to come into heat at an early age and conceive within three heat periods after their first exposure to a boar. Sows should be retained on the basis of their ability to conceive within seven days after weaning, or the earliest time that fits the management schedule of the producer, and their ability to wean a large, healthy litter. Use of the sow productivity index developed by the National Swine Improvement Federation is recommended as an added selection tool. Regardless of how sperm are placed in the female's reproductive tract, they must be there a few hours before ovulation to maximize the chance of getting the best pregnancy rate and litter size.

During gestation, gilts should be fed so they will gain about 90 lb and sows should gain about 70 lb. Depending on facilities, lactation may last 10 to 28 days to help ensure that baby pigs get a good start. At farrowing, additional pigs will be saved if an attendant is present to correct problems when they occur. Adequate records of individual performance during all phases of the reproductive cycle will be of benefit in upgrading the herd and making it more profitable.

Note: The care, management, and feeding of sows and gilts is fully covered in Chapters 7 and 9 of this book; hence, the reader is referred thereto.

TYPES OF PRODUCTION SYSTEMS AND THEIR MANAGEMENT

Commercial swine producers may specialize as follows:

1. Farrow-to-finish production.
2. Feeder pig production.
3. Growing-finishing production.

The type of production chosen will depend on the interest and experience of the producer, as well as the equipment, facilities, labor, feed supply, and the markets.

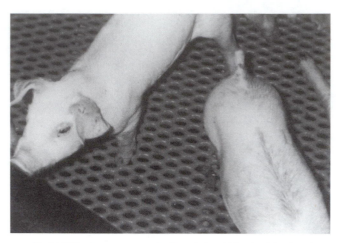

Fig. 14-6. Pigs in a nursery on plastic coated metal flooring. (Courtesy, Iowa State University, Ames)

FARROW-TO-FINISH PRODUCTION

Farrow-to-finish producers farrow and produce pigs to weaning, then grow and finish them for market. For success, farrow-to-finish operators must have expertise in all phases of swine production.

(Also see Chapter 9 of this book, section headed "Feeding Programs and Diets.")

FEEDER PIG PRODUCTION

Feeder pig production refers to the production and sale of immature pigs weighing 30 to 60 lb, usually throughout the year, for growing and finishing on other farms. Feeder pig production is best suited for those

Fig. 14-7. Feeder pigs on a slotted floor through which the feces and urine pass to a pit below. (Courtesy, Iowa State University, Ames)

who have a surplus of labor available and a limited feed supply. It makes for a two-phase system in swine production, similar to the two-phase system so well known in the cattle industry where some operators specialize in the cow-and-calf system (the production of feeder cattle) and others in finishing cattle. Until recent years, it was generally assumed that a two-phase system lent itself more logically to beef cattle than to swine because of the western range and of fewer disease problems. But several important scientific and technological developments which occurred in the swine industry in the 1950s and early 1960s ushered in considerable two-phase production of hogs. Among such developments were: (1) specific pathogen-free (SPF) herds and other improved disease control measures; (2) confined and continuous production—which increased specialization in breeding, in farrowing, and in finishing; and (3) increased mechanization.

Among the **advantages** of raising feeder pigs as compared to growing-finishing pigs are:

1. It provides an opportunity to use efficiently a maximum amount of labor, for about two-thirds of the labor of raising hogs occurs by the time pigs reach weaning age.

2. It requires less grain per dollar of product sold. To maintain a brood sow and raise a litter of pigs to 40 lb requires only 25 to 30% as much feed as is needed to feed them to market weight of 240 lb.

3. It requires less feed and manure handling.

4. It allows a rapid turnover in the volume of pigs that can be handled each year. The farrowing schedule can be planned to provide frequent sales for consistent income.

5. It can be started with relatively small capital inputs.

The main **disadvantage** to feeder pig production is that producers must depend on those engaged in growing-finishing operations for a market; hence, they are somewhat limited as to time of marketing and volume of sales.

Knowledge of the following is pertinent to successful feeder pig production:

1. **Basic requirements.** The two most important requirements of a good feeder pig production program are:

    a. **High-level of management.** Since 10 to 20% of the pigs born never reach weaning age, the importance of good management during this critical period is obvious.

    b. **Dependable market.** Most feeder pig producers are not equipped to feed hogs from weaning to market. Also, modern systems of multiple farrowing make it necessary to market pigs on schedule to make room for younger pigs to be farrowed and raised to weaning weights. Thus, if dependable markets are not available, feeder pig producers can find themselves in the unenviable position of having unsold pigs on their hands and more of them on the way.

2. **Variation in feeder pig prices.** Feeder pig prices are more erratic than slaughter hog prices—they move upward and downward more than slaughter hog prices, and they do so more rapidly. This is attributed to the fact that there is no well organized nationwide system of marketing feeder pigs where dependable price information is available.

3. **Method of marketing feeder pigs.** There are three methods of marketing feeder pigs:

    a. **Contract arrangements.** This refers to the sale of feeder pigs directly from producers to finishers on a contract basis.

    b. **Competitive organized markets.** The auction method of selling feeder pigs is good for smaller feeder pig producers. Well-conducted auctions provide the advantages of (1) uniform lots in type, color, and size; (2) selling a large number of pigs quickly and at reasonable cost; (3) buyers who are interested in various sizes, qualities, and numbers; and (4) open competition.

    c. **Private treaty.** This refers to the direct negotiation between producer and buyer. Often, newspaper ads bring the two parties together and then a per head price is established after visual inspection. The major disadvantage to this system is its uncertainty.

Note: The feeding of growing (feeder) pigs is fully covered in Chapter 9 of this book; hence, the reader is referred thereto.

GROWING-FINISHING PRODUCTION

Growing-finishing producers buy feeder pigs weighing 30 to 60 lb, then grow, finish, and market them.

Growing-finishing production is the easiest job in a complete swine operation. Death losses in the growing-finishing phase are generally low.

For those who have large grain supplies but limited labor and facilities, the purchase of feeder pigs, then growing, finishing, and marketing them, provides a good means of marketing their grains. Profits will depend on the price of feeder pigs, the price of feed, the efficiency of feed conversion, the price of slaughter hogs, and the labor and capital investment costs.

Note: The feeding of growing pigs and the feeding of finishing pigs are fully covered in Chapter 9 of this book; hence, the reader is referred thereto.

Fig. 14-8. Interior of a finishing unit with one curtain open. Such units usually house 1,000 to 1,200 pigs of about the same age. (Courtesy, Iowa State University, Ames)

SEPARATE SEX FEEDING OF BARROWS AND GILTS[3]

There are well known sex differences in the performance of growing-finishing pigs. (See Chapter 9 of this book, section headed "Effect of Sex on Performance of Growing-Finishing Pigs.") This prompts the question: Should barrows and gilts be fed separately?

Following a cooperative study involving nine research stations in the North Central Region of the United States, Cromwell, G. L., *et al.*, concluded that there is justification for feeding barrows and gilts separately based on the higher protein requirements of gilts than of barrows.

Note: Separate sex feeding of barrows and gilts will necessitate 2 or 3 additional grower diets and 2 or 3 additional finisher diets, along with separate feed storage facilities for each diet. So, separate sex feeding of barrows and gilts may not be practical except in very large swine operations.

SWINE SKILLS

The essential skills include those routine things which the producer must do by hand, whereas systems and practices are activities involving policies and programs. In order to be successful and to get satisfaction from producing swine, the operator should be able to perform the skills which follow.

---

[3]The results of this study appeared in the University of Kentucky, Lexington, *1989 Swine Research Report*.

CLIPPING THE BOAR'S TUSKS

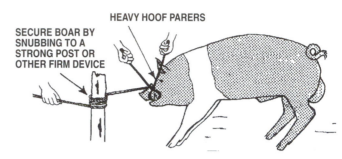

Fig. 14-9. Removing the boar's tusks.

It is never safe to allow the boar to have long tusks, for they may inflict injury upon other boars or even prove hazardous to the caretaker. Above all, such tusks should be removed well in advance of the breeding season, at which time it is necessary to handle the boar a great deal. The common procedure in preparation for removing the tusks consists of drawing a strong rope over the upper jaw and tying the other end securely to a post or other object. As the animal pulls back and the mouth opens, the tusks may be cut with a hoof parer.

REMOVING THE NEEDLE TEETH

Newborn pigs have eight small, tusklike teeth (so-called needle or black teeth), two on each side of both the upper and lower jaws. As these are of no benefit to the pig, most swine producers prefer to cut them off soon after birth. This operation may be done

Fig. 14-10. Clipping the needle teeth, using forceps made especially for this purpose. (Courtesy, DeKalb Swine Breeders, Inc., DeKalb, IL)

with a small pair of wire cutters or with forceps made especially for the purpose. In removing the teeth, care should be taken to avoid injury to the jaw or gums, for injuries may provide an opening for bacteria. For this reason only the tips of needle teeth should be clipped—about two-thirds of each tooth.

The needle teeth are very sharp and are often the cause of pain or injury to the sow, particularly if the udder is tender. Moreover, the pigs may bite or scratch each other, and infection may start and cause serious trouble.

MARKING OR IDENTIFYING SWINE

The common method of marking or identifying swine consists of ear notching the litters. Pigs are generally marked at the same time that the needle teeth are removed. Purebred breeders find it necessary to employ a system of marking so that they may determine the parentage of the individuals for purposes of registration and herd records. Even in the commercial herd, a system of identification is necessary if the gilts are to be selected from the larger and more efficient litters. The ear notches are usually made with a special V-notcher. Most of the breed associations are in position to recommend a satisfactory marking system, and many require the Universal Ear Notching System shown in Fig. 14-11. This is the most common notching system, though there are others.

Plastic ear tags, branding, and tattooing are also used for identifying swine.

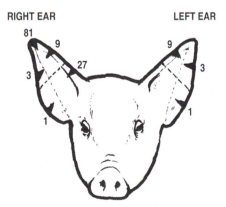

Fig. 14-11. Universal Ear Notching System (also known as the 1-3-9-27 system) used, or recommended, by most swine registry associations. The right ear is used for litter mark, and the left ear is used for individual pig number. Up to 161 litters can be marked with this system.

■ **Swine with microchips**—Electronic swine tracking systems are being tested in both Europe and the United States. The goal is to assign a number to each farm and each pig.

The system can be used to eradicate such swine diseases as brucellosis and pseudorabies.

Each animal will carry its own history—its diet, reproductive history, health, and medication.

Processors will be able to track microbiological contamination to its source—to the feed or whatever.

Ultimately, meat retailers will also be able to trace meat to its source.

Such chips are now being tested. But, presently, the receiver for one cannot "read" the signal sent out by a different system's transmitter.

CASTRATION

Castration is the removal of the testicles from the male and the ovaries from the female. In this country, female swine usually are not castrated (spayed).

Male pigs are castrated to maintain the quality of the meat, to prevent uncontrolled breeding, and to prevent the development of the boar odor or flavor that occurs in the cooked meat of a boar. As a result of castration, male pigs take on the conformation of a sow rather than a boar.

Boars that are no longer useful in a breeding program may be castrated to remove the boar odor before marketing; such an operation is known as *stagging.* By the time the castration wound has healed (in 3 to 4 weeks), the odor usually disappears enough to allow the stag to be marketed.

All male pigs that are not to be used for breeding purposes should be castrated at 1 to 10 days of age. Researchers at the University of Maryland's Eastern Shore Swine Research and Education Facility compared piglets castrated at 1 to 10 days of age with intact boar piglets. At 21 days of age, there was no significant difference in the weight and performance of the two groups. In addition, young pigs are easier to hold and restrain and they bleed very little. Also, the early castrates receive antibody protection from the sow's milk.

The best way to restrain, or hold, swine that are to be castrated depends on the age and size of the animal and the number of helpers available. A very young pig is held with one hand and castrated with the other. A heavier pig may be (1) suspended by its hind legs with the back toward the helper (whose knees are clamped against the pig's ribs, near the shoulders); or (2) held on its back on the top of a table (this requires either (a) a castration crate or rack, or (b) two helpers—one grasping the front legs and the other the rear legs). Large boars are usually snared around the upper jaw and behind the tusks, with the free end of the snare tied to a post; then further restraint is applied by either tying all four legs or by hoisting the hind legs,

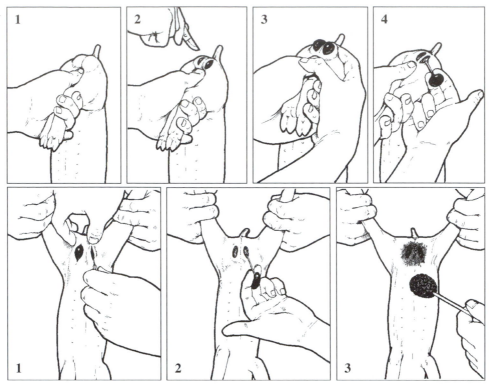

Fig. 14-12. Two methods of castrating young pigs. When pigs are castrated at 1 to 5 days of age, it is a one-person operation, as shown in the top series of drawings; (1) thumb used to push the testicles and tighten scrotal skin, (2) incision made over each testicle, (3) exposed testicles push through the incisions, and (4) each testicle grasped and pulled upward. Older pigs are castrated by holding them upside down, as shown in the bottom series of drawings, (1) testicles pushed down and incision made between the legs, (2) each testicle is grasped, gently pulled taut and the cord is cut, and (3) antiseptic is applied.

with the animal castrated in either a standing or a lying position.

The following method may be used for pigs less than 1 week old:

1. Use a surgical scalpel handle that accepts a hooked blade (Bard Parker No. 3 handle and No. 12 blade).

2. Clean the scrotal area with a mild disinfectant solution.

3. Hold pigs by the hind legs with one hand using the thumb or first finger to push up the testicles until the scrotal skin is tightened.

4. Push the tip of the hooked blade through the scrotum and direct it forward and upward toward the tail.

5. Repeat procedure over the other testicle. An alternative is to make the cut across the lower part of the scrotum to remove both testicles.

6. Push the exposed testicles out through the incisions.

7. Grasp each testicle separately and pull upward and remove with as much loose tissue as possible.

8. Properly performed, very little bleeding occurs and the incision remains clean.

9. Keep the knife in a disinfectant solution between pigs, and an antiseptic powder or spray may be applied to the incision.

Older pigs (over a week) are castrated by holding them upside down by the hind legs and then pushing the testicles down and cutting under the body between the legs. This method allows for good drainage but it requires two people or a pig holder.

Some swine producers routinely handle this management phase, others call upon the veterinarian; perhaps the most important thing is that it be done at the proper time. Pigs with undescended testicles or ruptures (scrotal hernias) should be operated on by a veterinarian.

RINGING

By instinct most hogs do some rooting, but it is likely to be especially damaging to pastures.

When rooting starts, the herd should be "ringed"; and this applies to all hogs past weaning age. Older animals can be restrained by a rope or snare placed around the snout, whereas young pigs can be held.

Many types of rings can be and are used, but the fish-hook type is most common. Rings (usually 1 to 3 rings) are placed in the snout, just back of the cartilage but away from the bone; although some producers prefer to use a ring that is placed through the septum (the partition of the nose). Others cut the cartilage on top of the snout, but this causes a rather severe setback and should be practiced with caution.

TAIL DOCKING

Trimming or "docking" baby pigs' tails is practical for farrow-to-finish operations, and it is mandatory in some graded feeder pig sales. Tail docking seems to be the best method of preventing, or at least reducing, tail biting.

The tail should be clipped to about 0.75 to 1.0 in. from the bone of the tail. Either sterilized wire cutters or an electric cauterizing blade can be used. Most times the procedure is performed when needle teeth are clipped and/or iron injections are given. A wound protectant spray or dip may be applied to the tail stump.

INJECTIONS

Shortly after birth a pig receives its first injection of iron dextran to prevent nutritional anemia. Proper injection involves the right size needle for the job and the best site for the injection. For piglets, a 0.5 to 1.0 in. 20-gauge needle works for thin liquids while an 18-gauge needle is best for thick liquids. Most sow injections should be given with 0.75 to 1.0 in. 16- to 14-gauge needles. Of course, needles should be clean and sharp. Common injection sites are the neck and the ham (with the neck preferred so as not to hazard damaging the more valuable ham). During injections, pigs should be adequately restrained.

WEANING PIGS

The optimum age to wean pigs varies considerably, depending on nutritional programs, facilities, environment, health, and management. Presently, the conventional age to wean pigs in the United States is 3 to 4 weeks, with a range of 10 days to 7 weeks. Segregated early weaning (SEW), which is an infectious disease control procedure, calls for weaning at 10 to 17 days of age (see Chapter 9, section headed "Segregated Early Weaning [SEW]").

The earlier pigs are weaned, the better. Advan-

tages to early weaning (10 days to 4 weeks of age) are:

1. Heavier pigs at 8 to 9 weeks of age, with fewer runts.
2. Lower sow feed costs.
3. More litters per year.
4. Less weight loss of the sow.
5. Greater flexibility in rebreeding or selling sows.
6. Greater turnover of sows through the farrowing unit; hence, less total farrowing area space required and lower facility cost per sow.

Successful early weaning encompasses the following:

1. A sound breeding and feeding program of gestating sows to ensure large, healthy pigs at birth; there is a high positive relationship between birth weight and weight at weaning.
2. Good milking sows that supply plenty of nutrients to get the pigs off to a good start.
3. Good baby pig and sow management during lactation to ensure strong, uniform, and healthy pigs at weaning.
4. A good feeding program, beginning with a good quality prestarter diet. Unless piglets are weaned at more than 21 days of age, the labor and expense associated with creep feeding is not economical. (Also see Chapter 9, section headed "Creep Feeding.")
5. Increase in costs for labor, facilities, and medication.

For best results, the guidelines given in Table 14-4 should be observed when planning for early weaning.

The age of pigs at weaning may vary from herd to herd, according to the facilities available, intensity

TABLE 14-4
GUIDELINES TO SUCCESSFUL EARLY WEANING[1]

Guideline	Age in Weeks				
	1	2	3	4	5
Minimum pig weight . . . . (lb)	5	9	12	15	21
Nursery temperature at pig level . . . . (°F)	85	85	83	81	79
Minimum floor space per pig . . . . . (sq ft)[2]	3	3	3	3	3
Maximum number of pigs per linear foot of feeding space	5	5	5	5	5
Maximum number of pigs per nipple waterer[3]	8	8	8	8	8
Maximum number of pigs per group	10	10	10	15	25

[1]See Appendix for conversion of U.S. customary to metric.

[2]The figures given herein are for solid floors. On slotted floors, this may be lowered 2 sq ft per pig from 1 to 5 weeks of age.

[3]Where bowls are used instead of nipples, there should be one bowl for each 12 pigs.

of operation, and managerial skills of the producer. Generally, pigs can be weaned over a wide age range; however, the younger the pigs, the more demanding the management required to do it successfully. Observance of the following guides will reduce the stress at weaning.

1. Wean only pigs weighing more than 12 lb.
2. Wean over a 2- to 3-day period, weaning the larger pigs in the litter first.
3. For 3-week-old pigs provide an environmental temperature of 80 to 85°F.
4. Group pigs according to size.
5. Limit numbers in a pen to 25.
6. Limit feed intake for 48 hours if post-weaning scours are a problem.
7. Provide 1 feeder hole for 4 to 5 pigs and 1 waterer for each 20 to 25 pigs.
8. Medicate drinking water if scours develop.

GROUPING HOGS

Grouping, along with separating hogs by sexes, ages, and size, is important. The following practices are generally advocated by successful producers:

1. **Gilts to be retained for breeding herd.** They should be separated from market hogs at four to five months of age.
2. **Pregnant gilts and sows.** They should be kept separate during the gestation period, unless they are self-fed a bulky diet.
3. **Boars of different ages.** Junior and mature boars should not be run together. Boars of the same age or size can be run together during the off-breeding season.
4. **Adjusting size of litter.** Where possible, the size of litters should be adjusted to the number of functioning teats and nursing ability of the sow. Transferring pigs from sow to sow should be done as early as possible; three to four days after farrowing is usually the maximum length of time that this can be done. If the odor of the pigs is masked, it may be possible to transfer at a later date.
5. **Running sows and litters together.** Pigs should be about 2 weeks old before placing sows and litters together, although small groups may be put together as early as 1 week. The age difference between such litters should not be more than 1 week in a central farrowing house or 2 weeks on pasture. Not more than 4 sows and litters should be grouped together in a central farrowing house; and not more than 6 on pasture.
6. **Creep feeding.** A maximum of 40 pigs per creep may be allowed.
7. **Early weaning.** In early weaning, not over 10 pigs should be placed together up to 3 weeks of age;

20 may be placed together at 3 to 4 weeks of age; and 25 at 5 weeks of age.
8. **Pigs of different weights.** Growing-finishing pigs of varying weights should not be run together. It is recommended that the range in weight should not exceed 20% above or below the average.

SANITATION AND HEALTH

Sanitation is the foundation of an effective herd health program. It is the employment of hygienic measures to promote health and prevent diseases, but sanitation is more than cleaning and disinfecting. It is an attitude. Sanitation begins with the producer who understands diseases and their transmission (see Chapter 15). Some of the factors or practices involved in sanitation are:

■ **Environmental effects**—Moisture and high humidity favor the survival and transmission of disease-causing organisms. Drying and proper ventilation are important to maintain herd health.

■ **Design and construction of facilities**—Buildings and equipment should be durable and easily and completely cleaned. All areas should have proper drainage so water will not stand and invite filth.

■ **Vacating facilities**—Combined with thorough cleaning and disinfecting, removing all animals from an area can help break disease cycles, since many disease-causing organisms cannot survive very long outside the pig's body. This includes pasture and dirt lot rotation. For best results, a facility should be empty 3 to 4 weeks or longer, but a few days will help.

■ **Cleaning and disinfecting**—Thorough cleaning followed by disinfecting complements vacated areas. Moreover, all equipment and surroundings should always be maintained in a clean and neat state.

■ **Footbaths**—To prevent the spread of disease from one production unit to another or from farm to farm, footbaths filled with a suitable disinfectant should be located at entry ways. Solutions in the footbaths should be kept fresh and/or replaced whenever organic material accumulates.

■ **Washing sows**—To minimize the exposure of newborn pigs to parasite eggs and other microorganisms, sows should be washed with warm soap or detergent and mild germicidal solutions before farrowing. This should be done immediately before entering the farrowing house. Of course, sows will be placed in cleaned and disinfected stalls.

■ **Carcass and afterbirth disposal**—Dead animals and afterbirth can spread disease, and should be removed immediately by a licensed rendering com-

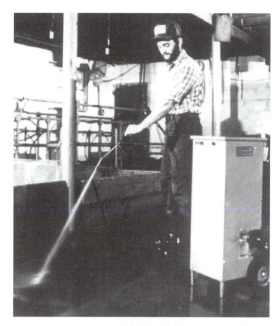

Fig. 14-13. Thorough cleaning, disinfecting, and drying out is an important aspect of disease control. (Courtesy, Hess & Clark, Inc., Ashland, OH)

pany, burned or properly buried 3 ft deep, down grade from water sources and covered with a generous amount of quicklime.

One of the early practices of sanitation was the McLean County System of Swine Sanitation developed by Drs. B. H. Ransom and H. B. Raffensperger. The system was developed in McLean County Illinois, after which it was named, as the result of a trial period, commencing in 1919. Today, the McLean County System of Swine Sanitation is of historic interest only. But it was effective in lessening swine parasites and diseases by application of the following four simple steps: (1) cleaning and disinfecting the farrowing quarters, (2) washing the sow before placing her in the farrowing quarters, (3) hauling the sow and pigs to clean pasture, and (4) keeping the pigs on clean pasture until they were at least four months old.

Further details regarding diseases, their spread and control, are given in Chapter 15.

ALL-IN/ALL-OUT (AIAO)

With all-in/all-out (AIAO) management, pigs are moved in groups through each of the following stages: (1) farrowing, (2) nursery, and (3) growing-finishing. Normally, a barn or room holds one week's farrowings, but in AIAO 2 or 3 weeks of production are housed together for a time. After each group is moved to the next production phase, the entire barn or room is washed and disinfected before a new group of pigs is moved in. AIAO can benefit almost any type of operation, including farrowing operations, feeder pig producers, and grower-finishers—large or small.

AIAO is not new. It was first used in farrowing and nursery units to combat scours *(E. coli)* and respiratory problems.

Now, AIAO is being added to such health sanitary/security procedures as footbaths, separate farm entrances, security fences, bird netting, shower-in/shower-out, controlling traffic, and buying disease-free breeding stock. AIAO is being used to control a host of swine diseases at farrowing, in the nursery, and in grower-finisher operations.

Note: There are a few diseases that cannot be controlled by AIAO, which call for repopulation, such as actinobacillus pleuropneumonia, swine dysentery, pseudorabies (PRV), interstitial pneumonia of young pigs, and SIRS complex (Mystery Swine Disease).

BEDDING SWINE

Bedding or litter is used primarily for the purpose of keeping animals clean and comfortable. But bedding has the following added values from the standpoint of the manure:

1. It soaks up the urine, which contains about one-half the total plant nutrients of manure.
2. It makes manure easier to handle.
3. It absorbs plant nutrients, fixing both ammonia and potash in relatively insoluble forms that protects them against losses by leaching. This characteristic of bedding is especially important in peat moss, but of little significance with sawdust and shavings.

Often bedding is eliminated, especially in the confinement-type operations, primarily because it reduces costs and makes cleaning easier. Slotted floors of confinement operations are responsible for the elimination of bedding.

KIND AND AMOUNT OF BEDDING

The kind of bedding material selected should be determined primarily by (1) availability and price, (2) absorptive capacity, (3) cleanness (this excludes dirt or dust which might cause odors or stain hogs), (4) ease of handling, (5) ease of cleanup and disposal, (6) nonirritability from dust or components causing allergies, (7) texture or size, and (8) fertility value or plant nutrient content. In addition, a desirable bedding should not be excessively coarse, and should remain well in place and not be too readily kicked aside.

Table 14-5 lists some common bedding materials and gives the average water absorptive capacity of each. In addition to these bedding materials, many other products can be and are successfully used for this purpose, including leaves of many kinds, tobacco stalks, buckwheat hulls, and shredded paper.

TABLE 14-5
WATER ABSORPTION OF BEDDING MATERIALS

Material	Water Absorbed by Air-Dry Bedding
	(lb/cwt)
Barley straw	210
Cocoa shells	270
Corn stover (shredded)	250
Corncobs (crushed or ground)	210
Cottonseed hulls	250
Flax straw	260
Hay (mature, chopped)	300
Leaves (broadleaf)	200
(pine needles)	100
Oat hulls	200
Oat straw (long)	280
(chopped)	375
Peanut hulls	250
Peat moss	1,000
Rye straw	210
Sand	25
Sawdust (top-quality pine)	250
(run-of-the-mill hardwood)	150
Sugarcane bagasse	220
Tree bark (dry, fine)	250
(from tanneries)	400
Vermiculite[1]	350
Wheat straw (long)	220
(chopped)	295
Wood chips (top-quality pine)	300
(run-of-the-mill hardwood)	150
Wood shavings (top-quality pine)	200
(run-of-the-mill hardwood)	150

[1]This is a micalike mineral mined chiefly in South Carolina and Montana.

Naturally, the availability and price per ton of various bedding materials varies from area to area, and from year to year. Thus, in the New England states shavings and sawdust are available, whereas other forms of bedding are scarce, and straws are more plentiful in the central and western states.

Table 14-5 shows that bedding materials differ considerably in their relative capacities to absorb liquid. Also, it is noteworthy that cut straw will absorb more liquid than long straw. But there are disadvantages to chopping; chopped straws do not stay in place, and they may be dusty.

From the standpoint of the value of plant nutrients per ton of air-dry material, peat moss is the most valuable bedding, and wood products (sawdust and shavings) the least valuable.

The suspicion that sawdust and shavings will hurt the land is rather widespread, but unfounded. It is true that these products decompose slowly, but this process can be expedited by the addition of nitrogen fertilizers. Also, when plowed under, they increase soil acidity, but the change is both small and temporary.

The minimum desirable amount of bedding to use is the amount necessary to absorb completely the liquids in manure. Some helpful guides to the end that this may be accomplished follow:

1. Per 24-hour confinement, the minimum bedding requirements of hogs, based on uncut wheat or oats straw, is 0.5 to 1 lb. With other bedding materials, the quantities will vary according to their respective absorptive capacities (see Table 14-5). Also, more than these minimum quantities of bedding may be desirable where cleanliness and comfort of the animal are important. Comfortable animals lie down more and utilize a higher proportion of the energy of the feed for productive purposes.

2. Under average conditions, about 500 lb of bedding are used for each ton of excrement.

3. Farrowing sows should be bedded lightly with chopped or short material that will not interfere with the movement of the piglets.

REDUCING BEDDING NEEDS

In most areas, bedding materials are becoming scarcer and higher in price, primarily because geneticists are breeding plants with shorter straws and stalks and because of more competitive and remunerative uses for some of the materials.

Producers may reduce needs and costs as follows:

1. **Collect liquid excrement separately.** Where the liquid excrement is collected separately in a cistern or tank, less bedding is required than where the liquid and solid excrement are kept together.

2. **Chop bedding.** Chopped straw, waste hay, fodder, or cobs will go further and do a better job of keeping animals dry than long materials.

3. **Ventilate quarters properly.** Proper ventilation lowers the humidity and helps keep the bedding dry.

4. **Feed and water away from sleeping quarters.** Animals should be fed and watered in areas removed from their sleeping quarters. With this type of arrangement, they defecate less in the sleeping area.

5. **Provide exercise area.** Where possible and practical, provide an exercise area, without confining animals to or near their sleeping quarters.

6. **Consider slotted floors.** Slotted or wire floors alleviate the need for bedding.

7. **Consider rubber mats.** Rubber (either solid or foam rubber) bedding replacers (or more correctly speaking, they are bedding-savers, for a limited amount of bedding is usually sprinkled over the top) are now available for use in hog houses. The life expectancy of solid rubber mats is 12 years; for foam rubber, about 4 years.

MANURE

The term manure refers to a mixture of animal excrement (consisting of undigested feeds plus certain body wastes) and bedding.

When the senior author was a boy on a Missouri farm, we fed livestock to produce manure, to grow more crops, to feed more livestock, to produce more manure. But time was! The use of chemical fertilizer expanded many fold; labor costs rose to the point where it was costly to conserve and spread manure on the land; more animals were raised in confinement; and a predominantly urban population didn't appreciate what they referred to as "foul-smelling, fly-breeding stuff." As a result, what to do with manure is a major problem on many swine establishments.

China has kept its soils productive for thousands of years, primarily through the use of night soil (human waste) and every other kind of manure, applied to the land in primitive, but effective, fashion. All over China, a familiar saying is: "The more pigs, the more manure; and the more manure, the more grain." Indeed, manure is very precious in China; it is carefully conserved and added to the land. Manure is used as a way in which to increase yields of farmland already under cultivation. China has long practiced sustainable agriculture.

Sustainable agriculture is farming with reduced off-farm purchases of fertilizer, pesticides, and herbicides, along with reduced negative impact on energy and other natural resources and improved environmental quality.

The challenge to American swine producers is (1) to practice more and more sustainable agriculture, and at the same time, (2) to improve environmental quality and the social impact of their operation.

Note well: All the methods of handling and using manure presented in this section have shortcomings.

In many cases, the cost of waste disposal will exceed the value to the user.

(Also see Chapter 13, all sections under pollution.)

AMOUNT, COMPOSITION, AND VALUE OF MANURE PRODUCED

The quantity, composition, and value of manure produced vary according to species, weight, kind and amount of feed, and kind and amount of bedding. Fig. 14-14 compares the manure production of swine with that of the other livestock species. Among the farm animals, swine produce the most manure per 1,000 lb of liveweight.

ANNUAL PRODUCTION OF MANURE

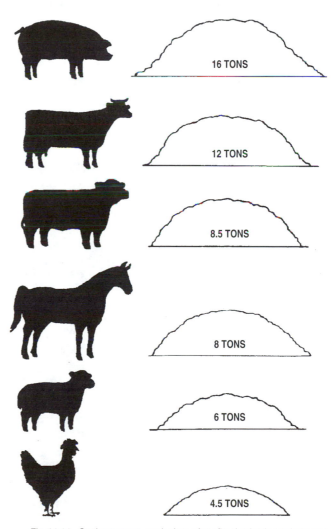

16 TONS

12 TONS

8.5 TONS

8 TONS

6 TONS

4.5 TONS

Fig. 14-14. On the average, each class of confined animals produces per year per 1,000 lb the tonnages of manure, free of bedding, shown above.

Swine manure, in an undiluted form, without bedding is produced in the quantities shown in Table 14-6.

TABLE 14-6
APPROXIMATE MANURE PRODUCTION, FREE OF BEDDING[1]

Animal Unit	Average Animal Weight		Manure Production[2]			
			Daily		Yearly	
	(lb)	(kg)	(cu in.)	(cm³)	(cu ft)	(m³)
Sow and litter	400	181	1,140	18,681	241	6.8
Prenursery pig	20	9	52	852	11	0.3
Nursery pig	55	25	121	1,983	26	0.7
Growing pig	115	52	242	3,966	52	1.5
Finishing pig	190	86	415	6,801	88	2.5
Gestating sow	325	147	346	5,670	73	2.1
Boar	400	181	432	7,079	91	2.6

[1]Adapted by the authors from *Swine Housing and Equipment Handbook*, Midwest Plan Service, Ames, Iowa, 1983, p. 56, Table 18.

[2]Solids and liquids, including 15% extra from waterers and washwater. Average density of manure is 60 lb/cu ft *(961 kg/m³)*.

Many factors affect the composition of manure; for example, diet, method of collection and storage, bedding, added water and the time and method of application. In general terms, about 75% of the nitrogen (N), 80% of the phosphorus (P), and 85% of the potassium (K) contained in swine feeds are returned as manure, as shown in Fig. 14-15. In addition, about 40% of the organic matter in feeds is excreted as manure. As a rule of thumb, it is commonly estimated that 80% of the total nutrients in feeds are excreted by animals as manure.

The urine comprises 40% of the total weight of the excrement of swine and 20% of that of horses. These figures represent the two extremes in farm animals. Yet the urine contains nearly 50% of the nitrogen, 4% of the phosphorus, and 55% of the potassium of average manure—roughly one-half the total nutrient content of manure (see Fig. 14-16). Also,

1,000 BUSHELS OF CORN CONTAIN:	ANIMALS RETAIN:	RETURNED IN MANURE:
1,000 LB N	250 LB N	750 LB N
170 LB P	34 LB P	136 LB P
190 LB K	19 LB K	171 LB K

Fig. 14-15. Animals retain about 20% of the nutrients in feed. The rest is excreted in manure.

the plant nutrients in the liquid portion of manure are more readily available to plants than those in the solid portion. It is, therefore, important to conserve the urine.

The actual monetary value of manure can and should be based on (1) increased crop yields, and (2) equivalent cost of a like amount of commercial fertilizer. Numerous experiments and practical observations have shown the measurable monetary value of manure in increased crop yields. Furthermore, the energy crisis of the 1970s prompted a re-evaluation of manure—the unwanted barnyard centerpiece for 40 years.

On the average, the fertilizer value of a ton of manure is about 10 lb of nitrogen, 3 lb phosphorus, and 8 lb potassium. On the assumption that nitrogen retails at 25¢, phosphorus at 20¢, and potassium at 10¢ per pound, then a ton of manure is worth about $4.00. Further guidelines relative to the fertilizer value of swine manure are given in Table 14-7.

Of course, the value of manure cannot be measured alone in terms of increased crop yields and equivalent cost of a like amount of commercial fertilizer. It has additional value for the organic matter which it contains, which almost all soils need, and which farmers and ranchers cannot buy in a sack or tank.

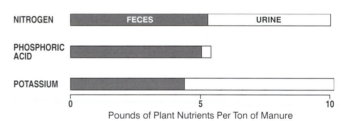

Fig. 14-16. Distribution of plant nutrients between liquid and solid portions of a ton of average farm manure. As noted, the urine contains about one-half the fertility value of manure.

Also, it is noteworthy that, due to the slower availability of its nitrogen and to its contribution to the soil humus, manure produces rather lasting benefits, which may continue for many years. Approximately one-half the plant nutrients in manure are available to and effective upon the crops in the immediate cycle of the rotation to which the application is made. Of the unused remainder, about one-half in turn, is taken up by the crops in the second cycle of the rotation; one-half the remainder in the third cycle, etc. Likewise, the continuous use of manure through several rounds of a rotation builds up a backlog which brings additional benefits, and a measurable climb in yield levels.

Swine producers sometimes fail to recognize the value of this barnyard crop because (1) it is produced whether or not it is wanted, and (2) it is available without cost.

TABLE 14-7
APPROXIMATE DRY MATTER AND FERTILIZER
NUTRIENT COMPOSITION OF SWINE MANURE
AT TIME APPLIED TO THE LAND[1, 2]

Manure Handling System	Dry Matter	Ammonium N[3]	P_2O_5[4]	K_2O[5]	Total N[6]
Solid	*(%)*	- - - - - - - - - - - - *(lb/ton)* - - - - - - - - - - - -			
Without bedding	18 (15–20)	7 (6–9)	9 (7–13)	8 (6–10)	10 (9–11)
With bedding	18 (17–20)	6 (5–8)	7 (5–10)	7 (6–9)	8 (7–10)
Liquid	*(%)*	- - - - - - - - - *(lb/1,000 gal.)* - - - - - - - - -			
Anaerobic storage	4 (2–7)	26 (21–31)	27 (13–20)	22 (12–30)	36 (28–55)
Lagoon[7]	1 (0.3–2)	4 (2–5)	2 (1–4)	4 (2–6)	4 (3–6)

[1]Adapted by the authors from *Pork Industry Handbook*, PIH-25, p. 4, Table 4.

[2]Application conversion factors: 1 bu = 40–60 lb solid manure; 1,000 gal. = about 4 tons; 27,154 gal. = 1 acre in.

[3]Ammonium N, which is available to the plant during the growing season.

[4]To convert to elemental P, multiply by 0.44.

[5]To convert to elemental K, multiply 0.83.

[6]Ammonium-N plus organic N, which is slow releasing.

[7]Includes feedlot runoff water and is sized as follows: single cell – 2 cu ft/lb animal weight; two-cell lagoon-cell 1, 1–2 cu ft/lb animal weight and cell 2, 1 cu ft/lb animal weight.

WAYS OF HANDLING MANURE

Basically, manure is handled as a solid or as a liquid. Solid manure results from catching and holding excrement in bedding or by allowing the liquids to run off leaving the solids behind. Liquids are generally wastes under slotted floors, or the runoff from lots and manure stacks.

Modern handling of manure involves maximum automation and a minimum loss of nutrients. Among the methods being used are scrapers, power loaders, conveyors, industrial-type vacuums, slotted floors with the manure stored underneath or emptying into irrigation systems, storage vats, spreaders (including those designed to handle liquids alone or liquids and solids together), dehydration, and lagoons. Actually, there is no one best manure management system for all situations; rather, it is a matter of designing and using that system which will be most practical for a particular set of conditions. But, to make an intelligent management selection, the following systems should be understood.

■ **Lagoon**—A lagoon is a waste treatment unit—a digester, a unit for biochemical breakdown of organic wastes (manure, straw). Except in a very dry climate with high evaporation rates, excess liquids must be field spread—not released to a watercourse.

Fig. 14-17. Open front growing-finishing units, showing the lagoon into which the waste drains. (Courtesy, Farmland Industries, Inc., Kansas City, MO)

■ **Oxidation ditch**—An oxidation ditch is a storage unit, oxygenated to promote aerobic bacterial digestion. The purpose is usually odor control, and the effluent is still a potential pollutant.

■ **Settling or debris basin**—A settling or debris basin is a separating and holding unit, usually part of a runoff control system. Liquids enter the basin and slow down. The undissolved solids settle out. The liquids are slowly drained off, leaving the solids to dry for removal and field spreading. A settling basin is usually smaller and much shallower than a lagoon. It is intended to dry out.

■ **Holding pond or basin**—Gutters, tanks, pits, and many lagoons are holding units, and any digesting that takes place is usually incidental.

When considering manure storage in a separate tank, earthen dam lagoon, or in a pit under slotted floors, the storage capacity can be computed as follows:

Fig. 14-18. Finishing pigs on slotted floor. (Courtesy, Land O Lakes, Ft. Dodge, IA)

Storage capacity = number of animals × daily manure production (see Table 14-6) × desired storage in days + extra water.

Water should be added to a pit (1) initially, and (2) subsequently to replace evaporation losses from wastes. Thus, if the manure is to be pumped, one-fifth to two-fifths of the storage volume may be needed for the extra water. For irrigation, there should be about 95% water and 5% manure. Water should be kept to a minimum if the manure is to be field spread with a tank wagon.

Generally, 3 to 6 months' storage capacity is desirable. Also, it should be noted that cleaning swine facilities with high-pressure water may double the volume of waste.

MANURE GASES

When stored inside a building, gases from liquid wastes create a hazard and undesirable odors. The common gases (95% or more) produced by manure decomposition are methane, ammonia, hydrogen sulfide, and carbon dioxide. Several have undesirable odors or possible animal toxicity, and some promote corrosion of equipment. Table 14-8 gives some properties of the more abundant gases.

Animals and people can be killed (asphyxiated) because methane and carbon dioxide displace oxygen.

Most gas problems occur when manure is agitated or when ventilation fans fail.

No one should enter a storage tank, unless (1) the space over the wastes is first ventilated with a fan, (2) another person is standing by to give assistance if needed, and (3) they are wearing self-contained breathing equipment—the kind used for fire fighting or scuba diving.

It is important that maximum building ventilation be provided when agitating or pumping wastes from a pit. Also, an alarm system (loud bell) to warn of power failures in tightly enclosed buildings is important, because there can be a rapid build-up of gases when forced ventilation ceases.

MANURE AS A FERTILIZER— PRECAUTIONS FOR USE

Historically, manure has been used as a fertilizer, and it is expected that manure will continue to be used primarily as a fertilizer for many years to come.

With today's heavy animal concentration in one location, the question is being asked: How many tons of manure can be applied per acre without depressing crop yield, creating salt problems in the soil, creating nitrate problems in feed, contributing excess nitrate to ground water or surface streams, or violating state regulations?

Based on earlier studies in the Midwest, before the rise of commercial fertilizers, it would appear that one can apply from 5 to 20 tons of manure per acre, year after year with benefit. (One in. of liquid manure containing 95% water and 5% solids spread over an acre of land will result in approximately 5.5 tons of solids per acre.)

Heavier than 20-ton applications can be made, but probably should not be repeated every year. With higher rates annually, there may be excess salt and nitrate build-up. Excess nitrate from manure can pollute streams and ground water and result in toxic levels of nitrate in crops. Without doubt the maximum rate at which manure can be applied to the land will vary widely according to soil type, rainfall, and temperature.

State regulations differ in limiting the rate of manure application. Missouri permits up to 30 tons per acre on pasture, and 40 tons per acre on cropland. Indiana limits manure application according to the amount of nitrogen applied, with the maximum limit set at 225 lb per acre per year. Nebraska requires only 0.5 acre of land for liquid manure disposal per acre of feedlot, which appears to be the least acreage for manure disposal required by any state.

Farmers with sufficient land should use rates of manure which supply only the nutrients needed by the crop rather than the maximum possible amounts suggested for pollution control.

■ **Precautions**—The following precautions should be observed when using manure as a fertilizer:

1. Avoid applying waste closer than 100 ft to waterways, streams, lakes, wells, springs, or ponds.
2. Do not apply where percolation of water down through the soil is not good, or where irrigation water is very salty or inadequate to move salts down.
3. Do not spread on frozen ground.

TABLE 14-8
PROPERTIES OF THE MORE ABUNDANT MANURE GASES

Gas	Weight (Air = 1)	Physiologic Effect	Other Properties	Comments
CH_4 Methane	0.50	Asphyxiant	Odorless, explosive	Explosive when present at only 5% the volume of air
NH_3 Ammonia	0.67	Irritant	Strong odor, corrosive	5,000 ppm dangerous level
H_2S Hydrogen sulfide	1+	Poison	Rotten-egg odor, corrosive	Fatal in 30 min. when present at 800–1,000 ppm
CO_2 Carbon dioxide	1.33	Asphyxiant	Odorless, mildly corrosive	Normally only about 0.03% of the volume of air

Fig. 14-19. A tank-type liquid swine manure knife injection applicator. By knifing liquid into the cropland, odor and nutrient losses are greatly reduced. (Courtesy, Iowa State University, Ames)

4. Distribute the waste as uniformly as possible on the area to be covered.

5. Incorporate (preferably by plowing, discing, or injecting) manure into the soil as quickly as possible after application. This will maximize nutrient conservation, reduce odors, and minimize runoff pollution.

6. Minimize odor problems by—

a. Spreading raw manure frequently, especially during the summer.

b. Spreading early in the day as the air is warming up, rather than late in the day when the air is cooling.

c. Spreading only on days when the wind is not blowing toward populated areas.

7. In irrigated areas, (a) irrigate thoroughly to leach excess salts below the root zone, and (b) allow about a month after irrigation before planting, to enable soil microorganisms to begin decomposition of manure.

OTHER USES OF MANURE

Manure will no doubt continue to be used as fertilizer, but in the years to come more may be used as a feed and converted into energy.

MANURE AS A FEED

Recycling manure as a livestock feed is the most promising of the nonfertilizer uses. Various processing methods are being employed. The most common method is the removal of coarse solids in raw manure with a liquid-solids separation device. These coarse materials are screened, stockpiled, and fed to ruminants after some disinfection stage. Another approach is to grow bacteria from waste through aerobic or oxidative treatment. Then, the resulting bacteria is used as a high-protein feed source (single-cell protein). A third method, and possibly the most practical, is the mixing of swine manure and high cellulose crop residues like cornstalks, to produce manure silage for feeding to ruminants.

It is noteworthy that the position of the Food and Drug Administration (FDA) relative to the feeding of animal waste (manure) is set forth by requiring submission of the following three basic categories of information: (1) establishing nutritive value (or efficacy); (2) determining safety to animals; and (3) determining that food from animals consuming such a product is safe for humans. Research on the feeding value of manure is not discouraged by the FDA but the commercial marketing of manure is approved only within the framework given above.

Fig. 14-20. Cattle grazing a pasture sprinkler irrigated from swine lagoon effluent. (Courtesy, Dr. Garth W. Boyd and Ms. Rhonda Campbell, Murphy Farms, Rose Hill, NC)

MANURE AS AN ENERGY SOURCE

Manure may also serve as a source of energy, which, of course, is not new. The pioneers burned dried bison dung, which they dubbed "buffalo chips," to heat their sod shanties. In this century, methane from manure has been used for power in European farm hamlets when natural gas was hard to get. While the costs of constructing plants to produce energy from manure on a large-scale basis may be high, some energy specialists feel that a prolonged fuel shortage will make such plants economical. India now has thousands of anaerobic digestion plants in operation.

Methane, of course, is usable like natural gas. There is nothing new or mysterious about this process. Sanitary engineers have long known that a family of bacteria produces methane when they ferment organic material under strictly anaerobic conditions. (Granddad called it swamp gas; his city cousin called it sewer gas.) However, it should be added that, due to capital and technical resources needed, for some time to come, the production of methane by anerobic digestion will likely be limited to municipal or corporate industries, but there is still the potential that in the future, with more research and trials, methane production will also be economically feasible for small livestock operations.

Some of the factors which determine the economical feasibility of methane gas from hog manure include the following:

1. Efficient on-site use of the methane gas and the digested slurry.
2. Efficient collection and management of high-quality (8 to 10% solids) manure.
3. Low-cost digester.
4. Large manure tonnage to achieve economics of scale.

It is estimated that the net output of methane from anaerobic digestion of swine manure could be about 2 cu ft per day per 150-lb hog.

WASTE MANAGEMENT WILL BE THE ISSUE IN THE DECADES TO COME

Manure spills from hog operations in North Carolina, Missouri, Iowa, and Minnesota in 1995 focused attention on animal waste for the decades to come. The two biggest concerns will be (1) nutrient management (primarily nitrogen, although heavy metals—copper and zinc, for example—may also limit manure application in the future), and (2) odor control. In preparation for more and more protest, swine producers need to do two things: (1) control their pollution,

and (2) keep informed of local, state, and national legislation.

■ **Tougher controls on hog manure are coming**—So, the swine industry should take the lead in minimizing the problem, because it is not going away.

MANAGEMENT PRACTICES AND ANIMAL WELFARE

To all animal caretakers, the principles and application of animal behavior and environment depend on understanding; and on recognizing that they should provide as comfortable an environment as feasible for their animals, for both humanitarian and economic reasons. This requires that attention be paid to environmental factors that influence the behavioral welfare of their animals as well as their physical comfort, with emphasis on the two most important influences of all in animal behavior and environment—feed and confinement.

Animal welfare issues tend to increase with urbanization. Moreover, fewer and fewer urbanites have farm backgrounds. As a result, the animal welfare gap between town and country widens. Also, both the news media and the legislators are increasingly from urban centers. It follows that the urban views that are propounded will have greater and greater impact in the future.

In the decades to come, swine producers must fit the farm to the hogs, and the corporate sow will need friends.

MANAGEMENT TO PREVENT DRUG RESIDUES IN PORK

Feed additives have been widely used in swine diets for more than 40 years. As a group, they are effective in improving the rate and efficiency of gains and in reducing mortality and morbidity. Certain feed additives require a withdrawal period prior to slaughter in order to insure that residues do not occur in carcasses. Chapter 9, Table 9-11, lists the common swine feed additives and gives the withdrawal times of those that require withdrawal.

The feed additives that have caused the greatest residue problem and received the most attention in recent years are the sulfonamides. The term, sulfonamide, includes sulfamethazine, sulfathiazole, and other sulfonamide drugs.

Drug residues in pork carcasses can be greatly reduced and even eliminated by adherence to the following practices:

1. Use only approved levels and combinations of drugs.

2. Follow good feed mixing practices (especially adequate mixing time) to insure that feed is mixed properly.

3. Maintain a record system to keep track of drug premixes and medicated feed usage.

4. Mix batch feeds in proper sequence to reduce the chance of carry-over of drugs into finishing feeds.

5. Clean out or flush feed mixing, conveying, and feeding equipment to reduce drug carryover into finishing feeds.

6. Adhere to proper withdrawal periods for drugs.

7. Prevent recycling of drugs via manure and urine.

8. Use on-farm testing programs to insure freedom from drug residues.

9. Read and follow the guidelines in the *Pork Quality Assurance Program* of the National Pork Producers Council.

QUESTIONS FOR STUDY AND DISCUSSION

1. Define management. List and discuss the five primary roles of a successful manager of a large swine operation. How can a young person become a manager?

2. What did the Texas Tech University researchers report as a result of their study of "indoor vs outdoor intensive pork production"? Will swine producers in the warmer areas of the United States return to outdoor production as modernized?

3. List the most important management aspects of herd boars, and discuss each of them.

4. List the most important management aspects of sows and gilts, and discuss each of them.

5. Describe each of the following production systems, and tell the place of each of them: (a) farrow-to-finish production, (b) feeder pig production, and (c) growing-finishing production.

6. Why are each of the following skills important: (a) clipping the boar's tusks, (b) removing the needle teeth, (c) marking or identifying, (d) castrating, (e) ringing, and (f) tail docking?

7. At what age, and how, would you wean pigs?

8. What is meant by all-in/all-out management? How is it done?

9. Is manure important in sustainable agriculture?

10. Is it possible to handle manure without it disturbing the neighbors?

11. List the pros and cons of each of the most frequently used ways of handling manure.

12. The authors of this book state that "waste management will be an issue in the decades to come." Do you agree or disagree with the authors?

13. List and discuss any common practices which you feel should be changed in order to improve animal welfare.

14. List and discuss management practices to prevent drug residue in pork.

SELECTED REFERENCES

Title of Publication	Author(s)	Publisher
Handbook of Livestock Management Techniques	R. A. Battaglia V. B. Mayrose	Burgess Publishing Company, Minneapolis, MN, 1981
Pork Industry Handbook	Selected Staff	Cooperative Extension Service, University of Illinois, Urbana-Champaign, updated frequently
Stockman's Handbook, The, Seventh Edition	M. E. Ensminger	Interstate Publishers, Inc., Danville, IL, 1992
Swine Production and Nutrition	W. G. Pond J. H. Maner	AVI Publishing Co., Westport, CT, 1984

Temperature-controlled nursery pens. (Courtesy, Iowa State University, Ames)

Exterior, showing curtains and a ventilation fan. (Courtesy, Iowa State University, Ames)

Picture of health. A painting of a Duroc boar by the noted artist Tom Phillips. (Courtesy, National Pork Producers Council, Des Moines, IA)

15

SWINE HEALTH, DISEASE PREVENTION, AND PARASITE CONTROL[1]

---

[1]The material presented in this chapter is based on factual information. However, when the instructions and precautions given herein are in disagreement with those of competent local authorities and/or reputable manufacturers, always follow the latter.

Contents	Page

Contents	Page

In the discussion that follows, an attempt is made to present a combination of scientific and practical information relative to swine health, disease prevention, and parasite control. It is intended that this should enhance the services of the veterinarian; and that producers can do a better job in controlling diseases when they have enlightened information at their disposal.

Keeping a herd healthy and disease-free can mean the difference between a profitable and an unprofitable operation.

Also, swine producers should be well informed relative to the relationship of swine diseases and parasites to other classes of animals and to human health, because many of them are transmissible between species. It is noteworthy, for example, that over 90 different types of infectious and parasitic diseases can be spread from animals to human beings.[2] Accordingly, other classes of animals and humans necessarily will be mentioned in the discussion which follows relative to swine diseases and parasites.

Some of the common swine diseases and parasites will be discussed in this chapter.

■ **Pharmaceutical compounds and terms**—*For regulatory purposes, pharmaceutical compounds are termed as follows, categories with which swine producers should be familiar:*

1. **Over-the-counter drugs.** These are drugs that are available to the general public. They are sold for general application by the purchaser in accordance with the application on the label.

2. **Prescription drugs.** These are drugs which are for use by, or on the order of, the veterinarian. These are always identified by the following statement on the label:

---

[2]Hull, T. G., *Diseases Transmitted from Animals to Man*, Charles C Thomas, Publisher, Springfield, IL.

Caution. Federal law restricts this drug to use by or on the order of a licensed veterinarian.

3. **Extra-label drugs.** This category is not a part of the law or regulations, but it is recognized in discretionary privileges granted to veterinarians. Use of over-the-counter drugs in therapies or dosages not approved by the labelling constitutes extra-label drug use. *Note:* No coccidiostats are approved for use in swine. So, the treatments given in this book for the control of coccidiosis are extra-label uses.

NORMAL TEMPERATURE, PULSE RATE, AND BREATHING RATE OF SWINE

Fig. 15-1 gives the normal temperature, heart rate (pulse rate), and respiration rate (breathing rate) of swine. In general, any marked and/or persistent deviations from these normals may be looked upon as a sign of animal ill health.

Swine producers should have an animal thermometer, which is heavier and more rugged than the ordinary human thermometer. Also, at the end opposite to the bulb, animal thermometers have an eye to which a 12-in. length of string with a clip is tied. The temperature is measured by inserting the thermometer full length in the rectum, where it should be left a minimum of three minutes. The clip on the string is attached to the hair of the animal.

In general, infectious diseases are ushered in with a rise in body temperature, but it must be remembered that body temperature is affected by barn or outside temperature, exercise, excitement, age, feed, etc. It is lower in cold weather, in older animals, and at night.

The pulse (heart rate) is felt at the front (anterior) rim of the ear or the underside of the tail. Also, the

SWINE VITAL SIGNS

RECTAL TEMPERATURE
102.5°F (101.6–103.6)
39.2°C (38.7–39.8)

HEART RATE
60 PER MINUTE (55–85)

RESPIRATION RATE
16 PER MINUTE (8–18)

Fig. 15-1. The normal temperature, heart rate, and breathing rate of swine with the ranges in parentheses.

heart beat can be felt by placing the palm of the hand between the left elbow and the side of the rib cage. It should be pointed out that the younger, the smaller, and the more nervous the animal, the higher the pulse

rate. Also, the pulse rate increases with exercise, excitement, digestion, and high outside temperature.

The breathing rate (respiration) can be determined by placing the hand on the flank, by observing the rise and fall of the flanks, or, in the winter, by watching the breath condensate in coming from the nostrils. Rapid breathing due to recent exercise, excitement, hot weather, or poorly ventilated buildings should not be confused with disease. Respiration is accelerated in pain and in febrile conditions.

LIFE CYCLE HERD HEALTH

Successful swine production necessitates the application of health-conserving, disease prevention, and parasite-control measures to the breeding, feeding, and management of the herd.

Although the basic principles and objectives of swine health remain the same, their application has changed with the increase in unit size, combined with greater intensification and sophistication. Two other relevant developments have evolved: The growth of large seedstock producers; and the production, marketing, and growing-finishing of feeder pigs.

Today, swine enterprises are so diverse and patterns of swine health are so varied that methods of disease prevention and parasite control must be adapted to each operation. So, Table 15-1, Life Cycle Herd Health, is presented for guide purposes in order to assist operators and their respective veterinarians in developing programs that are adapted to their individual enterprises. Special problems of each hog operation should be taken into consideration. Also, locale, type and size of operation, and government regulations will influence the herd health program.

TABLE 15-1
LIFE CYCLE HERD HEALTH

Time (age)	Disease and Parasite Control	Management and Breeeding
Gilts:		
4½–6 months		Select gilts at puberty (4½–6 months) and weighing 150–230 lb.
6½ months	Deworm; treat for lice and mange. Initiate fence-line (nose-to-nose) contact with boars. Vaccinate for erysipelas, leptospirosis (lepto bacterine are relatively short-lived. So, vaccinate gilts twice before first breeding.), parvovirus, and pseudorabies.	Choose gilts according to established genetic selection criteria and program. Isolate purchased gilts for 60 days. Blood sample purchased replacements for important disease(s) not already present in the herd.
7½ months	Repeat vaccinations.	
8 months		Breed gilts when they weigh 230–240 lb period (at least two matings per s

TABLE 15-1 (Continued)

Time (age)	Disease and Parasite Control	Management and Breeeding
Sows:		
3–7 days post weaning	Vaccinate sows for leptospirosis near each breeding.	Breed sows 3–7 days post weaning.
Gilts & Sows:		
Breeding		With once-a-day heat detection, breed females each day they will stand. With twice-a-day detection, breed 12–24 hours after they are first detected in heat. Double mate if possible.
30–45 days post-breeding		Pregnancy check (30–45 days post-breeding).
6 weeks prior to farrowing	*Clostridium* toxoid.	
4–6 weeks prior to farrowing	Atrophic rhinitis, *E. coli* bacterin, pseudorabies, rotavirus, TGE. Treat for lice and mange.	
2 weeks prior to farrowing	Atrophic rhinitis, *Clostridium*, *E. coli* bacterin, rotavirus, TGE.	
7–10 days prior to farrowing	Treat for lice and mange and deworm.	May include feed to prevent constipation. Wash sows thoroughly with detergent before entering farrowing house.
Farrowing		Record litter and sow information: Pig environment—90–95°F; Sow environment—65–70°F.
10 days to 4 weeks post-farrow	Erysipelas, leptospirosis, parvovirus, and pseudorabies for sows. Treat for lice and mange. All-in/all-out management, achieved by use of gilt pool and group weaning.	Wean pigs. Provide comfort, sanitation, and adequate diet.
Boars:		
4–6 months		Select and bring to farm at least 60 days prior to breeding. (Boars are ready for limited use at 8 months of age.) Isolate purchased boars for 60 days. Blood sample for important diseases not already in the herd.
First 30 days following purchase in isolation	Test for actinobacillus, brucellosis, leptospirosis, parvovirus, pseudorabies, and TGE. Treat for lice and mange and deworm.	Feed unmedicated feed, and observe for diarrhea, lameness, pneumonia, and ulcers.
Second 30 days following purchase in isolation	Vaccinate for erysipelas, leptospirosis, and parvovirus.	Commingle with cull gilts, and observe desire and ability to breed. Provide fenceline contact with gilts and sows to be bred.
7–8 months		Test mate and evaluate semen.
8 months		Boars should be ready for service. Individual mating is necessary for (1) weekly farrowing and (2) all-in/all-out production. When pen mating, rotate boars at 12- to 24-hour intervals.
Every 6 months	Revaccinate for erysipelas, leptospirosis, parvovirus, and pseudorabies; then, deworm and treat for lice and mange.	
Pigs:		
1 day		Clip needle teeth. Dock tails. Ear notch.
1–3 days	Iron injection.	Give each baby pig an injection in the neck of 200 mg of iron dextrin.
3–7 days	Vaccinate for atrophic rhinitis and TGE.	Start oral iron fed in tray.
1–10 days		Castrate.
10–14 days	Iron (injection or oral).	Start creep feed with oral iron mixed in.
3–4 weeks	Vaccinate for atrophic rhinitis.	Expose to pre-starter feed. Wean.
Weaning + 10 days	Treat for lice, mange, and then deworm.	
Weaning + 20 days	Vaccinate with *Actinobacillus pleuropneumoniae* and erysipelas bacterins.	Farrow-to-finish producers farrow the pigs then grow and finish them for market.
10–12 weeks	Vaccinate for pseudorabies, and revaccinate with *Actinobacillus pleuropneumoniae* and erysipelas bacterins.	Feeder pig production is a two-phase operation: Producers sell 30 to 60 lb pigs for growing and finishing on other farms.
5–6 months	Correctly follow all feed medication and vaccination withdrawal times prior to slaughter.	Health check 20% or 30 hogs from a market group.

PURCHASE OF FEEDER PIGS

Fig. 15-2. Feeder pigs in good condition, ready to market. (Courtesy, DeKalb Swine Breeders, Inc., DeKalb, IL)

Normally, death losses of feeder pigs from purchase to market should not exceed 2%. About half of these losses occur during the first week after arrival.

Feeder pigs that are in good condition and, preferably, purchased from a single reliable source, then handled as described below, will have the best chance of getting off to a good start with minimum death losses:

1. Abide by the legal health regulations of your state. These laws are for your protection. Vaccinations and other health certificates should be furnished by the seller and demanded by the buyer.

2. Avoid unnecessary handling, watch the pigs closely during this critical period, and work with your veterinarian to maintain optimum herd health.

3. Keep newly arrived feeder pigs isolated from other hogs for at least 3 weeks.

4. Avoid contact between feeder pigs and breeding stock, because feeder pigs can sometimes be carriers of diseases which can be spread to other hogs on the farm. Wear different outer clothing and boots when working with two sets of pigs to avoid spreading disease.

5. Provide warm, dry, draft-free, disinfected, well-bedded quarters.

6. Allow ample space for feeding, watering, and sleeping; provide one feeder space at the feeder for each 5 pigs, one waterer for each 30 pigs, and 4 to 6 sq ft of sleeping space per pig.

7. Feed a low protein diet containing (a) 12 to 14% protein, (b) extra fiber, and (c) a high level of vitamins and antibiotics. Additionally, 1 lb of ferrous sulfate and 0.5 lb of copper sulfate per 1,000 lb of feed may be helpful.

8. Use high antibiotic level for the first 3 days, preferably in the drinking water.

9. Dust bedding for control of lice and mange before pigs arrive. Do not spray pigs sooner than 5 days after arrival or when weather is unfavorable.

10. Do not worm or castrate pigs for at least 10 days after arrival.

11. Keep pigs of different sizes separated; otherwise, the bigger ones will crowd the smaller ones away from feed and water.

12. When signs of trouble appear, call your veterinarian.

SPECIFIC PATHOGEN-FREE (SPF) PIGS

Specific pathogen-free pigs are pigs that are free of disease at birth. Pathogen-free pigs are obtained from their dam 2 to 4 days prematurely by hysterectomy, caesarotomy, or hysterotomy. Also, pigs may be caught at natural birth in sterile canvas bags, in sterile basins, or on sterile canvas towels.

Hysterectomy means removal of the uterus. With this technique, the pigs are freed without transversing the birth canal, thereby eliminating any chance of the pigs becoming infected while passing through the birth canal. Although hysterotomy represented the ultimate in disease control, it had one major disadvantage. Many laboratories could not comply with inspection regulations; hence, they experienced difficulty in marketing the carcass of the sow. As a result, this forced commercial laboratories to obtain the pigs by Caesarean section (C-section), with methods developed to keep the newborn pigs separated from the contaminated environment of the sow.

This system, which is licensed, embraces the following provisions: (1) obtaining pigs by a surgical process (Caesarean section) 2 to 4 days before normal birth; (2) rearing pigs in individual isolation until 1 week old, and in groups of 8 to 12 until 4 weeks old; (3) rearing pigs in groups of 10 to 20 from 4 weeks old to maturity, on farms from which all other swine have been removed to which no new stock is introduced, and where the producer avoids contact with other swine; (4) resuming normal birth of SPF pigs on these clean farms; and (5) restocking other "clean" farms—farms that have no swine or only SPF swine, and on which the owner avoids contact with other swine.

Primary SPF herds are those originating from surgically derived stock and maintained in strict isolation. Any new blood lines added to the herd must also be obtained by surgical means. These primary herds are used to supply secondary multiplying herds, which in turn supply breeding stock to commercial swine producers.

The SPF method is drastic and costly. However, at the present time, it is the only means whereby atrophic rhinitis, mycoplasmal pneumonia (MP), and swine dysentery can be controlled and eradicated. In addition to these diseases SPF herds must be validated brucellosis-free, leptospirosis-free, and with no evidence of lice or mange.

The National Swine Repopulation Association, Lincoln, Nebraska, is responsible for supervising the SPF program and for issuing an Accreditation Certificate to those who qualify.

A diagrammatic outline of the SPF method is presented in Fig. 15-3.

OBTAIN ASEPTIC PIGS BY C-SECTION

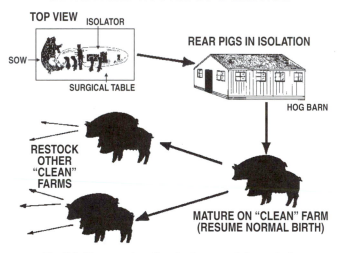

Fig. 15-3. Diagrammatic outline of swine repopulation method.

ALL-IN/ALL-OUT (AIAO)

All-in/all-out (AIAO) refers to the movement of pigs in groups through (1) farrowing, (2) nursery, and (3) growing-finishing. Normally, a barn or room holds one week's farrowing and is in continuous use. In AIAO, 2 or 3 weeks of production are bunched together for a time. After each group is moved to the next production phase, the entire barn or room is washed, disinfected, and repopulated.

Purdue University researchers compared the performance and health of growing-finishing pigs in an all-in/all-out system (AIAO) to the conventional (continuous use system) (CON) system. Based on four complete replications, they reported that the AIAO pigs gained faster (1.72 vs 1.52 lb/day), more efficiently

(3.04 vs 3.23 lb feed), and required fewer days to reach market weight of 231 lb (173 vs 185 days) than the CON pigs.[3]

(Also see Chapter 14, section headed "All-In/All-Out [AIAO].")

SEGREGATED EARLY WEANING (SEW)

Segregated early weaning (SEW) is an infectious disease control procedure the primary objective of which is to improve the productivity of the growing-finishing phase of swine production by preventing the transfer of diseases from sows to litters. This is accomplished by (1) weaning piglets at 10 to 17 days of age, rather than the more conventional 21- to 28-day weaning; and (2) segregating the piglets from their mothers either by transporting them from the farrowing site to another facility, or removing the sows to a centralized breeding/gestating facility and leaving the piglets in the site in which they were born. Early weaning (10 to 17 days) is critical to the success of SEW. Maternal immunity provided by colostrum decreases at different rates for different diseases. For example, the organism causing atrophic rhinitis requires a weaning age of 10 days, whereas the transmission by the pseudorabies virus has been prevented by using a 21-day weaning age. Thus, the segregation of early weaned pigs (SEW) prevents the vertical transmission of infectious diseases from sows to their piglets.

(Also see Chapter 9, section headed "Segregated Early Weaning [SEW].")

DISEASES OF SWINE

Adequate sanitation is the first and most important requirement that must be fulfilled if the swine enterprise is to be disease-free. Filthy quarters, feeding floors, and watering places favor the entrance of disease-producing germs into the body of the animal. In addition to maintaining a program of sanitation, the good caretaker is ever alert in observing any deviation from the normal in the functions of the animal—such as loss of appetite, lameness, digestive disturbances, etc.

A veterinarian should be called with the appearance of serious trouble, but often intelligent help may be given before a veterinarian can be reached. Moreover, the appearance of serious disorders can be recognized and control measures instituted before the

---

[3]Cline, T. R., *et al., Journal of Animal Science*, 70 (Supplement 1), 49, 1992.

spread of the disease has made much progress. The caretaker who is sufficiently familiar with the nature and causes of the common diseases is in a position to employ sound preventive measures.

This chapter is limited to nonnutritional diseases and ailments; the nutritional diseases and ailments of swine are covered in Chapter 8.

ANTHRAX (Splenic Fever, Charbon)

Anthrax, also referred to as splenic fever or charbon, is an acute infectious disease affecting all warm-blooded animals (and humans). It usually occurs as scattered outbreaks or cases, but hundreds of animals may be involved. Certain sections are known as anthrax districts because of the repeated appearance of the disease. Grazing animals are particularly subject to anthrax, especially when pasturing closely following a drought or on land that has been recently flooded.

Historically, anthrax is of great importance. It is one of the first scourges to be described in ancient and Biblical literature; it marks the beginning of modern bacteriology, being described by Koch in 1877; and it is the first disease in which immunization was effected by means of an attenuated culture, Pasteur having immunized animals against anthrax in 1881.

SYMPTOMS AND SIGNS[4]

The mortality is usually quite high. It runs a very short course and is characterized by a blood poisoning (septicemia). In swine, the disease is usually evidenced by swelling of the neck region (lymph nodes), which leads to death from suffocation and blood poisoning. It is accompanied by high temperature, loss of appetite, muscular weakness, difficult breathing, depression, and the passage of blood-stained feces.

CAUSE, PREVENTION, AND TREATMENT

The disease is identified by a microscopic examination of the blood or lymph nodes in which will be found the typical large, rod-shaped organisms causing anthrax, *Bacillus anthracis*. These organisms can survive for years in a spore stage, resisting all destructive agents. As a result, anthrax may remain in the soil for extremely long periods. Swine, however, rarely contact a sufficient dosage of anthrax spores from the soil to cause infection. Outbreaks in swine have been traced

---

[4]Currently, many veterinarians prefer the word *signs* rather than *symptoms*, but throughout this chapter the authors accede to the commonly accepted terminology among swine producers and include the word *symptoms*.

Fig. 15-4. Hog with anthrax. Note the swelling of the throat. This was rapidly followed by blood poisoning and death. (Courtesy, Department of Veterinary Pathology and Hygiene, College of Veterinary Medicine, University of Illinois)

to contaminated feed which in most cases contained products of animal origin.

Immunization is not generally practiced on a large scale since swine possess a level of natural resistance sufficient to prevent the disease unless there is heavy exposure to the bacillus. Herds that are infected should be quarantined and withheld from the market until the danger of disease transmission is past. The swine producer should never open the carcass of a dead animal suspected of having died from anthrax; instead, the veterinarian should be summoned at the first sign of an outbreak.

When the presence of anthrax is suspected or proved, all carcasses and contaminated material should be completely burned or, covered with lime and deeply buried, preferably on the spot. This precaution is important because the disease can be spread by dogs, coyotes, buzzards, and other flesh eaters and by flies and other insects.

When an outbreak of anthrax is discovered, all sick animals should be isolated promptly and treated. However, the treatment of affected animals is not too satisfactory. Penicillin or the tetracyclines are effective if given early. All exposed healthy animals should be vaccinated; pastures should be rotated; the premises should be quarantined; and a rigid program of sanitation should be initiated. These control measures should be carried out under the supervision of a veterinarian.

ATROPHIC RHINITIS

Atrophic rhinitis is quite widespread in the United States, and it has been reported in other countries.

Apparently, it affects swine only; for it does not seem to be related to atrophic rhinitis in humans. Nationwide, it is estimated that 80% of the swine herds are affected to some degree by turbinate atrophy. From 50 to 75% of market hogs have at least mild lesions of turbinate atrophy.

SYMPTOMS AND SIGNS

Atrophic rhinitis is a transmissible disease that is characterized by rhinitis—inflammation of the mucous membranes of the nose—and wasting away or lack of growth of the turbinate bones of the nose—small, scrollike structures that warm, moisten, and filter incoming air.

Fig. 15-5. Atrophic rhinitis may cause twisted noses in some animals of infected herd. (Photo by J. C. Allen and Son, West Lafayette, IN)

Persistent sneezing, which becomes more pronounced as the pigs grow older, is the first symptom. At 4 to 8 weeks of age, the snout begins to show wrinkles, and it may bulge and thicken. At 8 to 16 weeks of age, the snout and face may twist to one side. Nose bleeding is often seen. Affected pigs become rough all over, and make slow and inefficient gains. Actual death may be due to pneumonia. All ages of swine are susceptible, but the most severe cases develop when pigs are infected at only a few weeks of age. Swabs from the noses of breeding animals can be cultured for the presence of the causative organism.

CAUSE, PREVENTION, AND TREATMENT

In the United States, the primary causative agents of atrophic rhinitis are *Bordetella bronchiseptica* (Bb) and *Pasteurella multocida* (Pm). *Haemophilus parasuis* may cause mild turbinate atrophy. *Bordetella* is carried in the respiratory tract of numerous mammals including humans, rats, cats, and dogs, though not all strains have equal disease-causing capabilities. Its spread is via infected sow to her litter, the air in farrowing houses and nurseries, and nonswine sources such as cats. A calcium-phosphorus imbalance or a calcium deficiency in growing pigs has been shown to produce similar lesions.

Preventive measures—or at least measures to limit rhinitis—consist of some of the following:

1. Keep the percentage of replacements to a minimum—maintain older breeding stock.
2. Use SPF pigs.
3. Keep farrowing quarters and nursery clean. Use the all-in/all-out system if possible.
4. Isolate bred females in separate lots and never allow contact with any other swine except their offspring until they are culled. Keep individual litters separate until a month after removal of the sow at weaning time. Then, select and isolate new breeding stock from those litters which show no evidence of symptoms, and also react negatively to nasal swabbing tests for *Bordetella*.
5. On a continual basis, cull animals which are visibly affected.
6. Vaccinate, using Bb or Pm bacterins/toxoids, most of which are recommended to be given twice to the sow prefarrowing. Also, modified live intranasal Bp vaccines are available for piglets. Follow the directions of the manufacturer of the product used.

Treatment of rhinitis is based on the use of feed medications containing tetracyclines and/or sulfonamides.

The most effective management control of atrophic rhinitis is all-in/all-out farrowing and nursery with thorough sanitation between groups.

The most effective way to eradicate atrophic rhinitis from a herd is complete depopulation and repopulation with atrophic rhinitis-free swine. Atrophic rhinitis may be eradicated from herds that need to preserve their genetic base (*i.e.,* seedstock herds) by following the SPF program (Caesarean delivery, raised in isolation) or by segregated early weaning (SEW).

After a repopulation to eliminate the atrophic rhinitis, the following steps help to maintain a rhinitis-free herd:

1. Practice the closed herd concept.
2. Add boars that have passed three consecutive negative nasal swabbing tests, or use SPF boars.
3. Control the cat and rodent population.
4. Use nasal swabbing and culturing to monitor the herd.

BRUCELLOSIS

Brucellosis is an insidious (hidden) disease in which the lesions frequently are not evident. Although the medical term *brucellosis* is used in a collective way to designate the disease caused by the three different but closely related *Brucella* organisms, the species names further differentiate the germs as (1) *Brucella abortus*, (2) *B. suis*, and (3) *B. melitensis*.

Brucella abortus is known as Bang's disease (after Professor Bang, noted Danish research worker, who, in 1896, first discovered the organism responsible for bovine brucellosis), or contagious abortion in cattle; *B. suis* causes Traum's disease or infectious abortion in swine; and *B. melitensis* causes Malta fever, or abortion, in goats. The disease is known as Malta fever, Mediterranean fever, undulant fever, or brucellosis in humans. The causative organism is often associated with fistulous withers and poll-evil of horses.

Swine brucellosis control and eradication is important for two reasons: (1) the danger of human infection, and (2) the economic loss.

The most accurate, and possibly the most sensitive method, of diagnosis of swine brucellosis is isolation of *Brucella* organisms.

SYMPTOMS AND SIGNS

Unfortunately, the symptoms of brucellosis are often rather indefinite. It should be borne in mind that not all animals that abort are affected with brucellosis and that not all animals affected with brucellosis will abort. On the other hand, every case of abortion should be regarded with suspicion until proved noninfectious.

In swine, abortion and sterility are not as common as in cattle; infection may cause swollen joints and lameness, and swelling or atrophy of the testes, epididymis, and prostate in the male.

Fig. 15-6. Aborted swine fetuses, the result of *B. suis*. The act of abortion is the most readily observed symptom of this disease in sows. (Courtesy, Department of Veterinary Pathology and Hygiene, College of Veterinary Medicine, University of Illinois)

CAUSE, PREVENTION, AND TREATMENT

The disease is caused by bacteria called *Brucella suis* in swine, *B. abortus* in cattle, and *B. melitensis* in goats. The suis and melitensis types are seen in cattle, but the incidence is rare.

Humans are susceptible to all three types of brucellosis. The swine organism causes a more severe disease in human beings than the cattle organism, although not so severe as that induced by the goat type. Fortunately, far fewer people are exposed to the latter simply because of the limited number of goats and the rarity of the disease in goats in the United States. Producers are aware of the possibility that human beings may contract undulant fever from handling affected animals, especially at the time of parturition; from slaughtering operations or handling raw meats from affected animals; or from consuming raw milk or other raw by-products from goats or cows, and eating uncooked meats infected with brucellosis organisms. The simple precautions of pasteurizing milk and cooking meat, however, make these foods safe for human consumption.

The brucella organism is relatively resistant to drying, but is killed by the common disinfectants and by pasteurization. The organism is found in immense numbers in the various tissues of the aborted young and in the discharges and membranes from the aborted animal. It is harbored indefinitely in the udder and may also be found in the sex glands, spleen, liver, kidneys, bloodstream, joints, and lymph nodes.

Brucellosis appears to be commonly acquired through the mouth in feed and water contaminated with the bacteria, or by licking infected animals, contaminated feeders, or other objects to which the bacteria may adhere. There is also evidence that boars frequently transmit the disease through the act of service.

Freedom from disease should be the goal of all control programs. Strict sanitation; the recognition and removal of infected animals through testing programs; isolation at the time of parturition; and the control of animals, feed, and water brought into the premises is the key to the successful control and eradication of brucellosis.

Sound management practices, which include either buying replacement animals that are free of the disease or raising all females, are a necessary adjunct in prevention. Drainage from infected areas should be diverted or fenced off, and visitors (people and animals) should be kept away from animal barns and feedlots. Animals taken to livestock shows and fairs should be isolated on their return and tested 30 days later.

Vaccination procedures in swine have not been successful in brucellosis control programs. Instead, there is a joint state-federal and livestock industry program for the eradication of brucellosis. Guidelines for this program are set forth in the Uniform Methods and Rules for Brucellosis Eradication. Current information regarding the program is available from state veterinarians. In general, herds are established and maintained as validated brucellosis-free through surveillance blood testing. If a herd is infected or suspected of being infected with *B. suis*, three plans are recommended:

■ **Plan I**—Depopulation.

Market the entire herd for slaughter; clean and disinfect houses and equipment; restock premises with animals from Validated Brucellosis-free Herds.

■ **Plan II**—For use in salvaging irreplaceable bloodlines.

Retain weanling pigs for breeding and market infected adult animals for slaughter as soon as practical; isolate and test replacement gilts before breeding, saving only those that are negative and breeding them to negative boars; retest gilts after farrowing and before moving them from individual farrowing pens, segregate any reactors and retain only pigs from negative sows for breeding purposes; and repeat this procedure until the herd has two consecutive negative tests, at which time the herd is eligible for validation.

■ **Plan III**—For use in herds where only a few reactors are found and no clinical symptoms of brucellosis have been noted.

Market reactors for slaughter; retest herd at intervals, removing reactors for slaughter, until entire herd is negative for two successive tests. If herd is not readily freed of infection, abandon this plan in favor of Plan I or Plan II.

No known medicinal agent is effective in the treatment of brucellosis in any class of farm animals. Therefore, the producer should not waste valuable time and money on so-called "cures."

CHOLERA, HOG

This is a highly contagious viral disease affecting only swine.

A federal-state hog cholera eradication program was initiated in the United States in 1962. The last U.S. outbreak occurred in 1976. To eradicate hog cholera in the U.S. cost a total of approximately $140 million.

SYMPTOMS AND SIGNS

The symptoms appear after an incubation period of about a week. The disease is marked by a sudden onset, fever, loss of appetite, and weakness—although some pigs may die without showing any symptoms. Affected animals separate themselves from the herd, show a wobbly, scissor-like gait; and, although refusing feed, they may drink much water. The underside of the

Fig. 15-7. Hogs sick with hog cholera. Note the extreme depression and weakness. (Courtesy, USDA)

pigs may show a purplish red coloring (also seen in erysipelas and in other acute febrile diseases). Sometimes there may be chilling, causing the affected animals to bury themselves in the bedding or to pile up. There is constipation alternating with diarrhea, and coughing is often evident. There may be a discharge from the eyes. Often the disease is associated with pneumonia and/or enteritis. Since this disease is often confused with erysipelas, diagnosis should include a postmortem examination of one or more of the pigs that have recently died, a bacteriological check for the erysipelas bacteria, and perhaps a blood test for white cell count. The Fluorescent Antibody Test (FAT) is widely used by pathologists to diagnose this disease.

CAUSE, PREVENTION, AND TREATMENT

Hog cholera is caused by a virus.

The USDA first introduced the simultaneous serum-virus hog cholera vaccination in 1908. This type of immunization had the following effects:

1. It checked the widespread hog cholera epidemics of the old days, when so many hogs died that there was a stench in the air from burning the carcasses.

2. It kept the live, "hot" hog cholera virus (which it contained) spread all over the country.

3. It provided fairly good control; just good enough that American swine producers lived with the disease, rather than wipe it out as had been done in Canada where the use of the live hog cholera virus was forbidden.

In 1951, the USDA granted the first special licenses for the commercial sale of Modified Live Virus (MLV) preparations of rabbit, swine, or tissue culture origin. These preparations consist of modified (attenuated) live virus capable of producing long-lasting immunity. But, unfortunately, they were found capable of spreading the infection to susceptible animals. As a result, the USDA banned all interstate shipments of modified hog cholera vaccines after March 1, 1969.

The passage, in 1961, of Public Law 87-209 for hog cholera eradication in the United States, marked the beginning of the end of the disease in this country. But this drastic step was not without precedent. Both Canada and Great Britain had used the slaughter method to stamp out hog cholera.

Federal funds were made available in 1962, and the following four-phase program was initiated soon thereafter, with the objective of establishing a hog cholera-free swine population.

■ **Phase I: preparation**—Educating; surveys to determine incidence to improve and standardize diagnostic systems, and to promote disease reporting; and enforcement of garbage cooking.

■ **Phase II: reduction of incidence**—Quarantine of infected and exposed pigs, and stopping intrastate movement of infected pigs.

■ **Phase III: elimination of outbreaks**— Depopulation (with indemnity) of infected premises, and increasing the control of biologics used.

■ **Phase IV: protection against reinfection**—Prompt enforcement of all procedures in Phase III and a 21-day segregation of all swine imported from outside the state, with restrictions on the importation of pigs from states having endemic hog cholera.

The program was highly successful.

To prevent the reintroduction of hog cholera, countries free of the disease ban the import of live pigs, pork, and insufficiently heated pork products from countries where hog cholera is present.

CIRCLING DISEASE (Listerellosis, Encephalitis, Listeriosis)

Circling disease, also called listerellosis or listeriosis, is an infectious disease affecting mainly sheep, goats, and cattle; but it has been reported in swine, foals, and other animals and in humans. It occurs most often in winter and spring, with a mortality approaching 100%.

SYMPTOMS AND SIGNS

This disease principally affects the nervous system. Depression, staggering, circling, and strange awkward movements are noted. Larger pigs may show stilted movements in the front legs and may drag the hind legs. The course of the disease is very short, with paralysis and death the usual termination. Positive diagnosis can be made only by isolation and identification of the specific etiological agent.

CAUSE, PREVENTION, AND TREATMENT

Circling disease results from the invasion of the central nervous system by bacteria called *Listeria monocytogenes*. The method of transmission is unknown.

Presently, the occurrence of listeriosis in swine in the United States does not merit herd vaccination even if a suitable vaccine were available. Management practices paying careful attention to sanitation, nutrition, parasite control, and a comfortable environment offer the only means of prevention.

Listeria monocytogenes is susceptible to several antibiotics, but penicillin is the drug of choice.

DYSENTERY, SWINE (Bloody Scours, Hemorrhagic Dysentery, Black Scours, Vibrionic Dysentery)

Swine dysentery, an acute infectious disease, has been reported from coast to coast, but it is most common in the Corn Belt, where the swine population is densest. Outbreaks of the disease are usually associated with animals that pass through public auctions. In untreated herds, morbidity may be up to 90%, and mortality up to 30%.

SYMPTOMS AND SIGNS

The most characteristic symptom of swine dysentery is a profuse bloody diarrhea. Sometimes the feces are black instead of bloody and contain shreds of tissue. Most affected animals go off feed, and there is a moderate rise in temperature. Some pigs die suddenly after a couple days of illness, whereas others linger on for two weeks or longer. On autopsy or postmortem, the large intestine is found to be inflamed and bloody.

CAUSE, PREVENTION, AND TREATMENT

Swine dysentery appears to be caused by *S. hyodysenteriae*, which is ingested with contaminated fecal material. A diagnosis of swine dysentery is confirmed by the isolation and identification of *S. hyodysenteriae* from the lining of the colon or from feces. It can be carried from herd to herd on contaminated boots or vehicle tires.

Currently, one federally licensed vaccine is available, which may reduce the clinical signs. But it will not prevent infection or carriers.

Eradication of swine dysentery in infected herds requires (1) depopulation, clean-up, disinfection, and repopulation with *S. hyodysenteriae*-free stock; (2) medication (carbadox, lincomycin, and tiamulin); (3) sanitation; and (4) elimination of rodents.

EDEMA DISEASE (Enterotoxemia, Gut Edema, Gastric Edema, Edema of the Bowel)

This is an acute, usually fatal disease of young pigs, which usually occurs 1 to 3 weeks after weaning. Often it is found under conditions of excellent management and nutrition. Morbidity is usually low, but case mortality rate is very high—up to 100%. It appears to be increasing in this country, possibly due to earlier weaning.

SYMPTOMS AND SIGNS

The disease usually strikes the most thrifty, rapid-growing pigs. Often, the first indication of an outbreak is the sudden death of one or more pigs in a group. Constipation, inability to eat, swollen eyelids, and a staggering gait may be observed. Affected pigs may display nervous symptoms, such as fits or convulsions. As the disease progresses, the hog becomes completely paralyzed. Death usually follows in from a few hours to 2 to 3 days. The term *gut edema* is derived from the common postmortem findings of a jelly-like swelling (edema) of the stomach and portions of the intestines.

CAUSE, PREVENTION, AND TREATMENT

The disease is caused by certain serotypes of *E. coli* that produce a specific toxin.

The recommended prevention and control consists of antibiotics in the feed or water; restricted feeding, and small frequent feedings; creep feeding; and high fiber diets. Presently, no commercial vaccine is available.

If clinical signs have developed, treatment is ineffective. In general, treatment is aimed at reducing the factors believed to predispose an increase of *E. coli* and avoiding sudden changes in management. Overall, the edema disease treatment and management practice that has earned the best rating involves withholding (for 24 hours) or a sharp reduction in the amount of feed for a short but variable length of time, and avoiding stresses.

ENTERIC COLIBACILLOSIS (Baby Pig Scours, Colibacillosis, Scours)

Colibacillosis (scours) may infect piglets soon after birth, older nursing piglets, and weaned piglets.

Intestinal disorders can be caused by a variety of bacteria and viruses, and within a herd these infectious agents may cause disorders concurrently or sequentially. Therefore, it is often necessary for laboratory tests to identify the agent causing the intestinal disorder. In pigs less than seven days old, pathogenic (disease-causing) strains of *Escherichia coli* bacteria rapidly proliferate in the small intestine, producing toxins which produce diarrhea. The danger of diarrhea in baby pigs is the death of piglets through dehydration. This disease is often called baby pig scours, scours, or colibacillosis.

Baby pig scours is so acute that even with proper treatment death and setbacks in performance make it very costly to producers. It is far better to prevent this disease than to be continuously fighting it.

SYMPTOMS AND SIGNS

The symptoms and signs are: clear, watery to yellowish-brown pasty diarrhea; dehydration; depression, gauntness, inflamed perineum; and variable death loss—the highest death loss is in pigs from 0 to 4 days old.

Enteric colibacillosis infection in young unweaned pigs must be differentiated from other common infectious causes of diarrhea in pigs of this age group, including TGE virus, rotavirus, and coccidia. A diagnosis may be made by determining the fecal pH. The pH of the feces of pigs infected with enteric colibacillosis is alkaline, whereas the feces from diarrhea as a result of TGE virus or rotavirus infection, are acid.

CAUSE, PREVENTION, AND TREATMENT

Diagnosis requires laboratory tests to be certain that the cause is a pathogenic strain of *E. coli* since other bacteria and viruses can cause diarrhea and not all strains of *E. coli* produce disease.

Prevention of baby pig scours can be viewed as a three-pronged approach:

1. A good sanitation program reduces the number of *E. coli* bacteria in the environment. In an environment where the ventilation is poor, the humidity high, and it is dirty and wet, large numbers of *E. coli* are present. Moreover, in herds with baby pig scours, affected pigs should be treated and liquid stools covered to limit the spread of the bacteria.

2. An overall program of good management—

nutrition and herd health—ensures the delivery of vigorous piglets and satisfactory lactation. Moreover, proper management of the farrowing house to avoid stressful conditions for the piglets, primarily chilling, maintains the resistance of piglets to infections. Also, all-in/all-out farrowing rooms are effective in preventing and controlling baby pig scours.

3. Immunity passed from the sow to the piglet in the colostrum and milk provides protection until piglets start producing their own antibodies at about 10 days of age. Vaccination of the sow late in gestation with the pathogenic strain of *E. coli* promotes the formation of colostral antibodies which provides the piglets with passive immunity to the *E. coli* until their immune system starts producing antibodies. The veterinarian and the producer need to work together for this portion of the approach to control baby pig scours.

Prevention is better than treatment. However, once *E. coli* diarrhea is diagnosed, those pigs which are affected should be treated with an antibacterial drug which is known to be effective for the *E. coli* strain in the herd. It is known that *E. coli* is so adaptable that it can rapidly develop a resistance to antibacterial drugs. Besides being a treatment for the affected pigs, administering antibacterial agents lowers the number of pathogenic *E. coli* being shed in the watery feces.

ERYSIPELAS, SWINE

This is an acute or chronic infectious disease of swine, but it has also been reported in sheep, rabbits, and turkeys. When it attacks humans, it is called erysipeloid, which should not be confused with human erysipelas which is caused by a streptococcal bacteria. When the acute form of erysipelas occurs in swine, death losses are high, ranging from 50 to 75%.

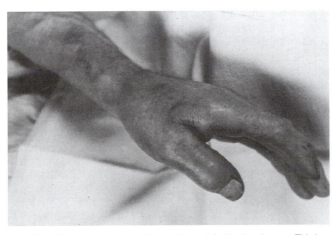

Fig. 15-8. Erysipeloid or swine erysipelas infection in a human. This is a troublesome and disabling wound infection. (Courtesy, Department of Veterinary Pathology and Hygiene, College of Veterinary Medicine, University of Illinois)

It is estimated that 30 to 50% of healthy swine are carriers.

SYMPTOMS AND SIGNS

The disease occurs in three forms: acute, subacute, and chronic. Additionally, a subclinical infection

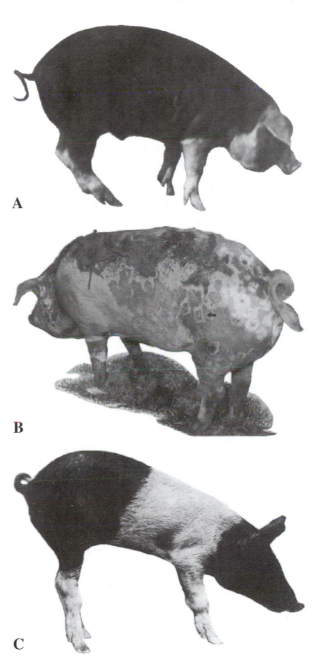

Fig. 15-9. Three forms of swine erysipelas: **A**, the acute form, showing hind feet "camped" up under the body giving evidence of pain; **B**, the "diamond-skin" form, showing typical skin lesions; and **C**, the chronic form, showing swelling on hind leg just below the hock. (Pictures **A** and **C**, courtesy, Department of Veterinary Pathology and Hygiene, College of Veterinary Medicine, University of Illinois; picture **B**, courtesy, Dr. L. M. Forland, Northwood, IA)

can occur with no visible signs and then develop into the chronic form.

The symptoms of the acute septicemia form resemble those seen in hog cholera. The affected animals show a high fever, and frequently have edema of the nose, ears, and limbs. Edema of the nose usually causes affected animals to breathe with a snoring sound. Also, there may be purplish patches under the belly similar to those described for hog cholera. The characteristic skin lesions called *diamond-skin* lesions appear 2 to 3 days after exposure. If severely affected animals do not die, areas of necrotic skin develop, which are dark, dry, and firm; and eventually these areas will slough, particularly on the ears and tail.

In the subacute form, the signs are less severe. Only a few skin lesions may appear, and they may be overlooked. The typical lesions are reddish, rectangular plaques in the skin. If the animal appears visibly sick, it will recover in a shorter time than the acutely affected animal.

The chronic form of erysipelas may follow the acute form, the subacute form or a subclinical infection with no visible signs. In the chronic form, the heart and/or the joints are usually the areas of localization. The joints of the knees and hocks are most commonly affected, showing enlargement and stiffness— arthritis. Those pigs chronically affected are usually very poor and unthrifty.

CAUSE, PREVENTION, AND TREATMENT

It has been demonstrated that the sole causative agent of erysipelas is *Erysipelothrix rhusiopathiae*, a bacterium. This organism is often found in the tonsils, gall bladders, and intestines of apparently normal pigs. Infection usually takes place by ingestion. Veterinary aid is necessary for an accurate diagnosis.

Swine erysipelas is prevented by a sound program of herd health management, including active immunization using either an attenuated (living avirulent) vaccine or bacterins (nonliving). Attenuated vaccines may be given orally in the drinking water or via injection. The immunization for erysipelas provides protection for only 3 to 5 months or slightly longer with a double treatment.

When an outbreak occurs, all sick animals should be isolated and the herd examined daily for new cases. Antiserum is available and it provides immediate passive immunity for about 2 weeks. During a herd outbreak, antiserum may be used to provide protection to suckling pigs not old enough to be vaccinated. Penicillin alone or in conjunction with antiserum provides a satisfactory treatment for acute cases. Tetracyclines and tylosin are also effective. Quarters which housed infected animals should be cleaned and disinfected. Lots or pastures which held infected animals

should remain vacant for several months, preferably including the summer season. Animals that are chronically affected should be eliminated from the herd, as they can be carriers of the disease.

FOOT-AND-MOUTH DISEASE

Foot-and-mouth disease is an old disease; it was first reported in 1514.

It is a highly contagious disease of cloven-footed animals (mainly swine, sheep, and cattle) characterized by the appearance of water blisters in the mouth (and in the snout in the case of hogs), on the skin between and around the claws of the hoof, and on the teats and udder. Fever is another symptom.

Humans are mildly susceptible but very rarely infected, whereas the horse is immune.

Fig. 15-10. Hog walking on its knees. Sore feet caused by foot-and-mouth disease. (Courtesy. USDA)

Unfortunately, one attack does not render the animal permanently immune, but the disease has a tendency to recur perhaps because of the multiplicity of the causative virus.

The disease is not present in the United States, but there have been at least 9 outbreaks (some authorities claim 10) in this country between 1870 and 1929, each of which was stamped out by the prompt slaughter of every affected and exposed animal. No United States outbreak has occurred since 1929, but the disease is greatly feared. Drastic measures are exercised in preventing the introduction of the disease into the United States, or, in the case of actual outbreak, in eradicating it.

Foot-and-mouth disease is constantly present in Europe, Asia, Africa, and South America. However, it has not been reported in New Zealand or Australia. Neither hogs nor uncooked pork products can be imported from any country in which it has been determined that foot-and-mouth disease exists.

In September 1946, an outbreak of foot-and-mouth disease appeared in Mexico. From that date until January 1, 1955, the Mexican-U.S. border was closed to imports of virtually all livestock and meat products, except for a 9-month period from September 1, 1952, to May 23, 1953. Likewise, the Canadian-United States border was closed, from February 1952, to March 1, 1953, due to an outbreak of the disease in Canada.

SYMPTOMS AND SIGNS

The disease is characterized by the formation of blisters (vesicles) and a moderate fever 3 to 6 days following exposure. These blisters are found on the mucous membranes of the tongue, lips, palate, cheeks, and on the skin around the claws of the feet, and on the teats and udder. In swine the blisters are usually found on and above the snout also.

Presence of these vesicles, especially in the mouth of cattle, stimulates a profuse flow of saliva that hangs from the lips in strings. Complicating or secondary factors are infected feet, caked udder, abortion, and great loss of weight. The mortality of adult animals is not ordinarily high, but the usefulness and productivity of affected animals is likely to be greatly damaged, thus causing great economic loss.

CAUSE, PREVENTION, AND TREATMENT

The infective agent of this disease is a small filterable virus, of which there are at least seven types, all of which are immunologically distinct from one another. Thus, infection with one does not protect against the other. The virus is present in the fluid and coverings of the blisters, in the blood, meat, milk, saliva, urine, and other secretions of the infected animal. The virus may be excreted in the urine for over 200 days following experimental inoculation. The virus can also be spread through infected biological products, such as small-pox vaccine and hog cholera virus and serum, and by the cattle fever tick.

Except for the nine outbreaks mentioned, the disease has been kept out of the United States by extreme precautions, such as quarantine at ports of entry and assistance with eradication in neighboring countries when introduction appears imminent. Neither live animals nor fresh or frozen meats can be imported from any country in which it has been determined that foot-and-mouth disease exists. Meat imports from these countries must be canned or fully cured.

Two methods have been applied in control: the slaughter method and the quarantine procedure. Then, if the existence of the disease is confirmed by diagnosis, the area is immediately placed under strict quarantine; infected and exposed animals are slaughtered and buried, with owners being paid indemnities based on their appraised value. Everything is cleaned and thoroughly disinfected.

No attempt is made to treat animals that are known to be infected.

Fortunately, the foot-and-mouth disease virus is quickly destroyed by a solution of the cheap and common chemical, sodium hydroxide (lye). Because quick control action is necessary, state or federal authorities should immediately be notified the very moment the presence of the disease is suspected.

Vaccines containing one or more immuno-types of the virus are manufactured in European countries and in Argentina. The immunity produced from such vaccines lasts for only 3 to 6 months; hence, animals must be vaccinated 3 to 4 times per year. Vaccines have not been used in the outbreaks in the United States because they have not been regarded as favorable to rapid, complete eradication of the infection.

A new vaccine against foot-and-mouth disease has been developed by Dr. H. Bachrach and other U.S. Department of Agriculture scientists who have used gene-splicing techniques that allow mass production of a safe and effective supply of the vaccine.

HEMOPHILUS PLEUROPNEUMONIA (HPP)

Hemophilus is a severe pneumonia striking young pigs (up to six months of age). While pigs affected with hemophilus will often die, the greatest economic losses result from pigs surviving the disease, since the chronic lung lesions from the disease create "poor doers." Average daily gains decrease and feed per pound of gain increases.

The distribution of hemophilus is worldwide, causing significant economic losses to the swine industries of such countries as Brazil, Canada, Denmark, Germany, Mexico, Switzerland, and, of course, the United States.

SYMPTOMS AND SIGNS

The incubation period is short, sometimes 8 to 12 hours, and it commonly strikes pigs from 40 lb to market weight. Often the first indication of hemophilus is the sudden death of apparently healthy pigs, though this sudden death follows a stressful period; for example, moving, mixing or rapid weather change. When observed, healthy pigs may develop labored breathing and die within minutes of a stress. Bleeding from the nose may be observed in some pigs, but not all. Pigs less severely affected, may demonstrate thumping, a fever of 104 to 107°F, depression, and reluctance to move. Prominent lesions include hemorrhagic, edema-filled lungs. Since the causative organism is spread via aerosol droplets, morbidity reaches 100% while mortality is 20 to 40%, when immediate and effective treatment is instituted.

CAUSE, PREVENTION, AND TREATMENT

Often a tentative diagnosis can be made based on the history, age of affected pigs, signs and symptoms, and lung lesions, but a definitive diagnosis may require the culture of the causative bacteria—*Haemophilus pleuropneumoniae*—from a lung lesion.

Presently, the best prevention seems to be prevention of its spread and introduction into a herd. The major source of infection is the carrier pig that has recovered from hemophilus. New animals should be isolated and tested to determine if they are carriers before being introduced into an uninfected herd. Vaccination with a killed vaccine is effective in protecting susceptible animals and animals exposed to the disease.

Treatment during an acute outbreak consists of maintaining high blood levels of antibiotics such as procaine penicillin and long-acting oxytetracycline (LA-200) by injection. Following an outbreak, antibiotic sensitivity tests should be used to determine the most effective drug. To reduce death losses, both healthy and sick swine should be treated. The treatment of animals showing clinical signs has not demonstrated uniform success due to the development of lung damage.

INFLUENZA, SWINE (Hog Flu)

This disease of swine resembles influenza in humans, but the effects may be more serious. Swine influenza is an acute respiratory disease. The onset is rapid, and in most cases all swine in an infected herd show symptoms at the same time. The disease is usually first observed in the fall when the weather becomes cold. Swine influenza occurs throughout the United States. However, the mortality seldom exceeds 2% and the principle loss from swine flu is due to the lingering debility which results in uneconomical gains.

SYMPTOMS AND SIGNS

The disease makes its appearance suddenly, following an incubation period of less than a week. High fever, loss of appetite, a cough, and discharge from the eyes and nose are seen. The animals seem reluctant to move but may sit up like dogs in an attempt to facilitate breathing which may in some cases be difficult.

CAUSE, PREVENTION, AND TREATMENT

The disease is caused by a type A influenza virus, which is similar to the virus that caused the worldwide flu epidemic of 1918 in humans. Indeed, there is substantial evidence that flu was first introduced into swine at that time—a classic example of a human disease transmitted to animals. Moreover, there is serologic and virologic evidence of cross infection of humans and swine with type A influenza virus. Swine flu is passed from pig to pig via direct contacts with infectious secretions or exudates, droplet infection, or small particle aerosols. Some investigators have hypothesized that the lungworm, whose intermediate host is the earthworm, serves as a host to the influenza virus; thereby perpetuating and transmitting the virus. The pig then eats earthworms which contain lungworm larvae which may in turn harbor the virus. However, this system is not clearly shown to be necessary for the survival of the virus, since some studies have demonstrated that hogs can become carriers.

Correcting any faulty feeding and management with the herd, and avoiding the introduction of new animals will help control swine flu. A killed vaccine is available. Vaccination is reported to decrease clinical signs and reduce virus shedding in challenged pigs. The only treatment seems to be the provision of warm, dry, clean quarters, minimum diets, and fresh, clean water accessible at all times. Some veterinarians believe that death losses from pneumonia due to secondary invading organisms can be lessened by the use of antibiotics or sulfonamides, administered hypodermically or in the drinking water. Expectorants are also often used if respiration is difficult.

LEPTOSPIROSIS

Leptospirosis was first observed in humans in 1915–1916, in dogs in 1931, and in cattle in 1934. It has been recognized as an important disease of swine in this country since about 1952, although it very probably has been present for a much longer period.

Humans may contract infections through skin abrasions, when handling or slaughtering infected animals, by swimming in contaminated water, through consuming uncooked foods that are contaminated, or through drinking unpasteurized milk.

SYMPTOMS AND SIGNS

Swine losses from leptospirosis are chiefly in baby pigs that are aborted or born dead or weak, and in the unthriftiness of market hogs. Affected animals may refuse to eat, become mildly depressed, lose weight, and often run a high fever for 2 or 3 days. Occasionally, there may be blood-stained urine. Very frequently, however, the symptoms in growing-finishing pigs are not apparent. Leptospirosis infection is usually confirmed on the basis of herd history, postmortem findings, and blood test. Leptospirosis should be suspect when reproductive failures are being considered.

CAUSE, PREVENTION, AND TREATMENT

The most common cause of leptospirosis in swine in the United States is *Leptospira pomona*, a corkscrew-shaped bacterium of the spirochete group. It can cause the disease in swine, cattle, sheep, horses, and humans, and it is readily transferred from one species to another.

Leptospirosis organisms enter the body via the skin or mucous membranes, and they are spread by infective urine—direct contact with urine or water contaminated by urine.

The following preventive measures are recommended:

1. Blood test all animals prior to purchase, isolate for 30 days, and then retest prior to adding them to the herd.

2. Vaccinate with leptospira bacterin, following the manufacturer's directions. Immunity from bacterins is relatively short-lived. So, gilts should be vaccinated twice before the first breeding, and sows should be vaccinated at or near each breeding—approximately every six months.

If leptospirosis strikes in a swine herd—and it is positively diagnosed—it may be brought under control by the following procedure, which should be carried out under the supervision of a veterinarian:

1. Spread animals out over a large area; avoid congestion in a pen or barn.

2. Fence off water holes, ponds, or slow-running streams.

3. Isolate sick animals and new additions to the herd.

4. Discard milk from diseased cows.

5. Clean and disinfect barns; exterminate rodents.

6. Administer leptospirosis vaccine to all sows on problem farms each year.

7. Keep different classes of livestock separated, because leptospirosis can be spread from one species to another.

Streptomycin and the tetracycline group of antibiotics are the treatment of choice. If afflicted animals are treated promptly, fairly good results can be expected. To help remove the carrier phase, feed high levels of the tetracycline drugs (400 to 500 g per ton in complete feeds for 14 days). Untreated, swine can remain carriers up to one year.

MASTITIS-METRITIS-AGALACTIA COMPLEX (MMA or Agalactia)

This is a disease of sows and gilts characterized by an inflammation of the mammary glands (mastitis), and inflammation of the uterus (metritis), and/or a failure to secrete milk (agalactia). Often, however, there is not total lack of milk but rather a reduction in the normal amount of milk (hypogalactia). The death rate in sows is low, but the loss of baby pigs may be high.

MMA appears to be a more serious problem in "multiple" and "confinement" farrowing than where portable houses on pasture are used.

SYMPTOMS AND SIGNS

The first signs of the disease usually appear within three days after farrowing, although symptoms usually can be seen before farrowing or before the pigs are weaned. A whitish or yellowish discharge of pus appears from the vagina of the infected animal and the temperature rises to 103 to 106°F or higher. The sow goes off feed and stops milking, and the pigs often develop diarrhea. Sometimes the problem is not recognized until the pigs starve to death.

CAUSE, PREVENTION, AND TREATMENT

The organism commonly thought to cause the MMA syndrome is *Escherichia coli*. Other organisms that have been associated with the syndrome include: *Actinobacillus, Actinomyces, Aerobacter, Citrobacter, Clostridia, Corynebacterium, Enterobacter, Klebsiella, Pseudomonas, Mycoplasma, Proteus, Streptococcus, Staphylococcus*, and *Chlamydia*. It is particularly noteworthy that *E. coli* is also the most common organism isolated as the cause of diarrhea in baby pigs during the first three weeks of life. However, the disease is complex, and proof of the cause is often difficult.

Prevention revolves around sound management, good nutrition, superior sanitation, and proper swine husbandry. When sows are kept in confinement, it is especially important to keep them from becoming overfat and constipated, and to reduce stress, particularly near the time of farrowing. The first condition can be controlled by limiting the diet, and the addition of 6 to 10% molasses or extra bran to the diet will prevent constipation. Also, underfeeding may contribute to the problem.

The list of treatments suggested by different people is long, confusing, and not always effective. Among them are:

1. Antibiotics to eliminate infections.

2. Oxytocin, a hormone, to cause milk let-down.

3. Relief of constipation by (a) using a warm,

soapy enema, (b) feeding a diet high in molasses or wheat bran, or (c) providing morning and evening exercise.

4. Supplemental milk and glucose to keep the baby pigs alive.

5. Oral antibiotic for the baby pigs if diarrhea occurs.

6. Vaccination of sows with a bacterin prior to farrowing; preferably using an autogenous bacterin prepared from the strains of bacteria involved in the herd, or using a stock bacterin.

MYCOPLASMAL PNEUMONIA
(Enzootic Pneumonia)

Mycoplasmal pneumonia of pigs is a chronic disease, and one of the world's most important swine diseases, affecting all ages. In the United States, most herds are affected to some degree; and the feed efficiency of infected pigs may be lowered by 20%.

SYMPTOMS AND SIGNS

Coughing is the first and most characteristic symptom manifested of mycoplasmal pneumonia. It generally begins 10 to 16 days following exposure to diseased animals, and it may persist indefinitely. Coughing is most marked when pigs come out to feed in the morning or following vigorous exercise. Diarrhea usually occurs when pigs first begin to cough, but it lasts only two or three days. Temperatures are only slightly elevated, and even when they reach 105°F the pigs do not look sick. In general, affected pigs eat well, but gains are slow and feed utilization poor.

If management is poor, many pigs develop serious pneumonia due to secondary bacterial infection of the lungs. Also, severe pneumonia occurs when mycoplasmal pneumonia is complicated by large numbers of roundworm larvae passing through the lungs. The age 14 to 26 weeks seems to be the most critical for such secondary lung complications.

CAUSE, PREVENTION, AND TREATMENT

The disease is caused by a small coccobacillary organism, *Mycoplasma hyopneumoniae*. Actually, three different diseases are caused by different mycoplasma species: mycoplasmal pneumonia, mycoplasmal polyserositis arthritis, and mycoplasmal arthritis.

Since mycoplasmal pneumonia is spread largely by direct contact from infected to normal animals, prevention consists in avoiding the purchase of animals from infected herds.

The disease is continued in infected herds by contact of young pigs with carrier sows, since about one in four sows remains a carrier. Two methods have been used for its eradication: (1) specific pathogen-free (SPF) pigs, and (2) farrowing old sows in isolation and ascertaining that the pigs are free of the disease before adding them to the herd.

Treatments with linconycin (Lincomix) (at 200 g/ton of feed for 21 days), tetracyclines, tylosin (Tylan), or timulin (Denagard) will reduce the severity of mycoplasmal pneumonia and improve performance. Also, good nursing will help.

PNEUMONIA

Pneumonia is a widespread disease of all animals, including swine. If untreated, 50 to 75% of affected animals die.

SYMPTOMS AND SIGNS

Pneumonia is an inflammation of the lungs. It is ushered in by a chill, followed by elevated temperature. There is quick, shallow respiration, discharge from the nostrils and perhaps the eyes, and a cough. Sick animals stand with their legs wide apart, and there is a drop in milk production, loss of appetite, constipation, crackling noises with breathing, and gasping for breath.

CAUSE, PREVENTION, AND TREATMENT

The causes of pneumonia are numerous, including (1) many bacteria, (2) inhalation of water on medicines given by untrained persons as a drench, (3) noxious gases, (4) mycoplasma, (5) viruses, and (6) lungworms. Changeable weather during the spring and fall, and damp or drafty buildings are conducive to pneumonia. When pneumonia is encountered, one should always try to establish what is causing it.

Preventive measures include providing good hygienic surroundings and practicing good, sound husbandry.

Sick animals should be segregated in quiet, clean quarters away from drafts, and given easily digested nutritious feeds. Treatment may include antibiotics or sulfonamides for acute pneumonias or for keeping down secondary bacterial pathogens.

PORCINE PARVOVIRUS (PPV)

Losses caused by this disease are usually unexpected since the sow or gilt typically appears healthy during gestation. The incidence of parvovirus is high and it has resulted in large economic losses. It seems that porcine parvovirus is one of the most important infectious agents indicated in SMEDI. Parvovirus occurs throughout the world.

SYMPTOMS AND SIGNS

Usually, the only clinical sign is maternal reproductive failure such as (1) return to estrus, (2) failure to farrow despite not showing signs of heat, (3) farrowing few pigs per litter, or (4) farrowing a large proportion of mummified fetuses. The observable signs depend upon the stage of pregnancy when the sow or gilt is infected. Gilts are especially vulnerable.

CAUSE, PREVENTION, AND TREATMENT

The disease is caused by a virus of the genus *Parvovirus*, though there is no indication that the porcine parvovirus is the same as the parvoviruses affecting species such as rats, mice, cats, dogs, cows, and sheep. They are antigenically different.

Diagnosis is made on the basis of differentiating parvovirus from other causes of reproductive failure in swine, and, if available, mummified fetuses can be submitted for immunofluorescence testing.

Prevention consists of naturally infecting or vaccinating gilts with porcine parvovirus before they are bred. Commonly, gilts are exposed to sows or moved to a potentially contaminated area to promote a natural infection and immunity. The effectiveness of this method is, however, uncertain. A vaccination can ensure that gilts develop immunity before breeding. Several inactivated vaccines are on the market.

There is no treatment for porcine parvovirus.

PORCINE REPRODUCTIVE AND RESPIRATORY SYNDROME (PRRS)

In 1987, this disease was first reported in North Carolina, Iowa, and Minnesota. Then, in 1988–1989, several outbreaks followed in Indiana. In 1990, the disease appeared in Canada; and in 1991, it was reported in several European countries.

When the disease first appeared in the United States, it was designated as the "Mystery Disease" because it could not be readily identified or treated. Because of the cyanosis (a bluish coloration of the skin [especially the ears, vulva, abdomen, and snout] due to deficient oxygenation of the blood) of afflicted animals, Europeans called it "Blue Ear Disease."

In the mid-1990s, PRRS was considered to be one of the most difficult and costly swine health problems.

SYMPTOMS AND SIGNS

As the name Porcine Reproductive and Respiratory Syndrome implies, the syndrome consists of two parts: (1) A reproductive component, characterized by late term abortions (107 to 112 days of pregnancy) accompanied by elevated numbers of stillbirths, mummies, and weak born pigs; and (2) a respiratory disease component, characterized by the piglets in the farrowing and nursery houses exhibiting labored breathing, loss of appetite, gauntness, and rough hair coats. As a result of PRRS infection, piglets are more susceptible to such other infections as bacterial pneumonia and *Streptococcus suis*. Cyanosis of the ears (Blue Ear Disease), vulva, abdomen, and snout may be observed; and afflicted hogs may have a fever (103 to 105°F). Affected herds gradually return to almost normal in 2 to 4 months.

CAUSE, PREVENTION, AND TREATMENT

Porcine Reproductive and Respiratory Syndrome (PRRS) is caused by a virus which was first isolated in 1991. Also, researchers have shown that there are different strains of the virus, and that the different strains cause diseases that vary in severity.

Prevention and control of PRRS involves (1) determining the status of the herd, (2) segregating age groups better to control viral shedding, (3) isolating incoming replacement boars and gilts a minimum of 45 days, (4) operating on an all-in/all-out basis, (5) depopulating nurseries if necessary, (6) changing feed if current feed is high in mycotoxins, and (7) vaccinating pigs under 16 weeks of age with the newly approved vaccine.

There is no specific treatment of PRRS-afflicted swine. Good nursing and good feed will help. The use of an antibiotic to reduce secondary infection is prudent.

PORCINE STRESS SYNDROME (PSS)

Some pigs are unable to adapt successfully to stressful management practices, and they exhibit a nonpathological disorder that is best described as a syndrome rather than a single abnormal characteristic. It commonly occurs in rapidly growing, heavily muscled pigs, resulting from intensive genetic selection in systems of partial or total confinement. It is often manifested as sudden unexplained death losses. Furthermore, the syndrome is related to low-quality pork products or so-called pale, soft, exudative (PSE) pork, when the stress susceptible pigs survive until slaughter.

SYMPTOMS AND SIGNS

A series of signs develop in PSS pigs when they are subjected to some stressful management practices such as loading and transport to market, overcrowding and fighting, vaccination, or a sudden change in the weather. The clinical signs are:

1. Muscle and tail tremors.
2. Dyspnea (short of breath).
3. Cyanosis (bluish or purplish discoloration of the skin and mucuous membranes due to deficient oxygenation).
4. Rapid rise in body temperature with signs of heat stress even in cool weather.
5. Total collapse, marked muscle rigidity, and hyperthermia.
6. Death in a shocklike state.

At slaughter, PSS pigs produce—

1. Pale, soft, watery musculature.
2. Pork showing protein denaturation and lowered quality.

While less than 1% of the swine in the United States actually die of PSS under normal marketing and management practices, the incidence of low-quality pork in ham and loin muscles runs about 8 to 10%.

CAUSE, PREVENTION, AND TREATMENT

Susceptibility to PSS is caused by a single autosomal recessive gene. The disease is manifested only in pigs that are homozygous recessive for this gene, which means that they inherited it from both the sire and the dam. Some producers have suggested that pigs that are heterozygous for the stress gene are the ideal market hog, i.e., lean, but not susceptible to PSS.

Prevention consists in using the DNA probe test to eliminate the gene from the population. The older tests involved exposing a pig to a stressor (e.g., halathane anesthesia, exercise, heat) then checking for clinical signs of muscle damage (elevation of serum creatine phosphokinase [CPK] or pyruvate kinase [PK] levels). These older tests will detect homozygous pigs only; and have largely given way to the newer DNA probe test. Also, the PSS syndrome can be lessened (but not entirely prevented) by avoiding excess stress.

Common treatments are: (1) removing stress, and (2) cooling pigs down.

POX, SWINE

Swine pox is widely distributed in the Midwest, where it is an important disease of young pigs.

SYMPTOMS AND SIGNS

Swine pox is characterized by small red spots, which appear over large parts of the body especially on the ears, neck, undersurface of the body, and inside the thighs. These spots grow rapidly and reach the size of a dime. A hard nodule develops in the center of each. Several days later, small, pea-sized vesicles (blisters) develop; at first these contain a clear fluid, but later the contents become puslike. Soon, these blisters dry up, leaving dark brown scabs, which fall off. Preceding these skin changes, some animals show fever, chills, and refusal to eat. Very few pigs die.

CAUSE, PREVENTION, AND TREATMENT

There are two types of swine pox, each caused by a different type of virus (swine pox virus and vaccinic virus). Swine which recover from the disease are immune to further attacks from the specific type of virus that caused the disease, but not to attacks from the other virus. One type of virus is related to that causing pox in various other species of animals, but the other is not.

The disease appears to be transmitted primarily by lice, and less often by other insects and by contact. Therefore, lice control and general sanitation are the best preventive measures. Vaccination is possible but not recommended for two reasons: (1) The disease has no great economic importance, and (2) the use of a live vaccine would introduce more virus in an infected environment.

No treatment for swine pox is known. Good management and nursing will help. Also, good, clean quarters prevent secondary infections of the skin lesions.

PSEUDORABIES (Aujeszky's Disease, Mad Itch)

First of all, it should be recognized that pseudorabies is not related to rabies, though some of the symptoms and signs are similar. Pseudorabies is an acute, infectious, and frequently fatal disease affecting most species of domestic and wild animals. Humans are resistant to the disease. It was first recognized as an infectious disease of cattle and dogs in Hungary by Aujeszky in 1902.

Pseudorabies is widespread and of considerable economic importance in the midwestern United States. The disease is also widespread in Europe, where it causes heavy losses. The continued increase and severity of the disease may be due to rearing more pigs in large confinement units or by the cessation of cholera vaccination.

SYMPTOMS AND SIGNS

Baby pigs may die with few, if any, clinical signs evidenced. However, death is usually preceded by fever which may exceed 105°F, dullness, loss of appetite, vomiting, weakness, incoordination, and convulsions. In pigs less than 3 weeks old, death losses frequently approach 100%. After 3 weeks of age, the signs and death losses decrease.

In adult pigs, the signs may be very mild and include fever, off feed, salivation, coughing, sneezing, vomiting, diarrhea, constipation, convulsions, itching, middle ear infections, and blindness. Sows infected in middle pregnancy may abort mummified fetuses. Sows infected late in pregnancy often abort or give birth to weak, *shaker*, or stillborn pigs. Piglets infected prior to birth usually die within two days.

In animals other than swine the disease is characterized by intense itching, self-mutilation, convulsions, and death.

Diagnosis should be confirmed by a laboratory test, usually performed on serum samples.

CAUSE, PREVENTION, AND TREATMENT

Pseudorabies is caused by a virus of the herpesvirus group. The main mode of spreading the disease is via direct contact with the nose serving as the primary point of entry for the virus. Nasal discharges and saliva containing the virus contaminate water, bedding, feed, and feeders, as well as the clothes worn by workers.

Vaccines for pseudorabies are available. Modified live virus (MLV), inactivated, and gene-deleted vaccines have been developed for pseudorabies control. The use of vaccines has been controversial, and opinions have varied based on the desire to control or eradicate the disease. Immunized animals generally survive the infection and shed the virus for awhile, thereby acting as carriers. Possibly eradication and preventive measures are better alternatives.

The following control measures are recommended for the protection of herds:

1. Dispose of dead pigs properly (burning or burying).
2. Buy tested breeding stock and isolate them for 30 days.
3. If you raise breeding stock, do not buy feeder pigs.
4. Get feeder pigs from a farm that has not had the disease.
5. Keep visitors away from swine premises.
6. Keep stray dogs, cats, and wildlife off the premises.
7. Keep swine and cattle separate.
8. Isolate show stock for 30 days after the fair is over.

Antibiotics may reduce the secondary pneumonic infections in older animals. When an outbreak occurs the premises should be quarantined and the movement of all people strictly controlled. If it is at all possible, sick animals should be separated from unaffected animals. *Note:* Regulatory control varies from state to state.

RABIES (Hydrophobia, "Madness")

Rabies (hydrophobia or "madness") is an acute infectious disease of all warm-blooded animals and humans. It is characterized by deranged consciousness and paralysis, and terminates fatally. This disease is one that is far too prevalent and, if present knowledge were applied, it could be controlled. Complete eradication would be difficult to achieve because of the reservoir of infection in wild animals.

When a human being is bitten by a dog that is suspected of being rabid, the first impulse is to kill the dog immediately. This is a mistake. Instead, it is important to confine the animal under the observation of a veterinarian until the disease, if it is present, has a chance to develop and run its course. If no recognizable symptoms appear in the animal within a period of two weeks after it inflicted the bite, it is usually safe to assume that there was no rabies at the time. Death occurs within a few days after the symptoms appear, and the dog's brain can be examined for specific evidence of rabies. With this procedure, unless the bite is in the region of the neck or head, there will usually be ample time in which to administer treatment to exposed human beings. As the virus has been found in the saliva of a dog at least five days before the appearance of the clinically recognizable symptoms, the bite of a dog should always be considered potentially dangerous until proved otherwise. In any event, when people are bitten or exposed to rabies, they should immediately report to the family doctor who usually administers Semple-type vaccine, though irradiated vaccines are used to some extent. With severe bites, especially those around the head, antiserum is particularly indicated.

SYMPTOMS AND SIGNS

Rabies in swine is not very common. The disease may manifest itself in two forms: the furious form, or the dumb form. It is often difficult to distinguish between the two forms, however. The furious type usually merges into the dumb form because paralysis always occurs just before death.

In the early stages, rabid swine generally exhibit irritability as manifested by aimless wandering, hoarse grunting, gnawing or rubbing the bitten spot, and sudden jumping if disturbed. In the later stages, the pigs become paralyzed and death follows.

CAUSE, PREVENTION, AND TREATMENT

Rabies is caused by a filterable virus which is usually carried into a bite wound by a rabid animal. It is generally transmitted to farm animals by dogs and certain wild animals, such as the bat, fox, skunk, etc.

Rabies can best be prevented by attacking it at its chief source, the dog. With the advent of an improved vaccine for the dog, it should be a requirement that all dogs be immunized. This should be supplemented by regulations governing the licensing, quarantine, and transportation of dogs. Complete eradication would be difficult to achieve because of the reservoir of infection in wild animals.

When swine are bitten or exposed to rabies, see your veterinarian. After the disease is fully developed, there is no known treatment.

ROTAVIRAL ENTERITIS

The importance of rotaviruses in pigs is a fairly recent development. In 1976, a rotavirus was first reported as being responsible for diarrhea in pigs. Now it is recognized that the infection is widespread and common in pigs. The mortality rate in piglets varies from 0 to 50%.

SYMPTOMS AND SIGNS

The disease causes diarrhea in piglets. Generally, it strikes 3 to 4 days following weaning. However, neonatal diarrhea due to rotavirus is occasionally reported. This diarrhea is characterized by a white or yellow stool which is liquid at first but later becomes creamy and then pasty before returning to normal. Piglets also become depressed, fail to eat, and resist moving. The severity of the disease is influenced by (1) the age of the piglet, (2) stresses such as chilling, (3) concurrent infections with pathogenic E. coli or the TGE virus, and (4) inadequate intake of immune milk. In general, the older the pig, the less severe the outcome, but the severity increases with weaning at any age. In many pigs, however, the infection may occur with no clinical signs.

CAUSE, PREVENTION, AND TREATMENT

The diarrhea is caused by a group of viruses known as rotaviruses—a name derived from the wheel-like appearance of the virus through the electron microscope. A rotavirus infection can be confused with transmissible gastroenteritis (TGE). A laboratory diagnosis is necessary to differentiate.

Most swine herds are infected with rotavirus. Piglets do, however, receive protection from the antibodies in the sow's milk and colostrum. The best methods for the prevention and control of rotaviruses in young pigs appears to be sure piglets get adequate colostrum and milk at an early age and to provide good sanitation and a comfortable (nonstressful) environment. Good sanitation prevents the build-up of rotavirus and the chance of concurrent E. coli infection. Orally administered modified live virus vaccines at 7 to 21 days or via drinking water at weaning have shown some benefit.

No known therapeutic agents are available for the specific treatment of porcine rotavirus infections. General supportive therapy, management procedures, and antibiotics are recommended to minimize mortality due to rotavirus and secondary bacterial infections. Electrolyte solutions containing glucose-glycine fed ad libitum minimize dehydration and weight loss induced by rotavirus infection.

SALMONELLOSIS

The concern with this disease arises because the causative bacteria are widespread and adaptable; they may cause diarrhea and septicemia in pigs; and they are a source of human salmonellosis. Most outbreaks of the disease have occurred in intensively reared, weaned pigs.

SYMPTOMS AND SIGNS

The signs may result primarily from two forms of the disease: septicemia or diarrhea (enterocolitis). Septicemic salmonellosis occurs mainly in weaned pigs less than 4 months old. It is characterized by restlessness, failure to eat, and a fever of 105 to 107°F. Generally, mortality is high but for some reason, morbidity is low, often less than 10%. The disease is spread via the fecal-oral route, and recovered pigs are carriers and fecal shedders for some time.

Between the ages of weaning and 4 months, outbreaks of enterocolitic salmonellosis may occur. This is characterized by a watery, yellow diarrhea, which rapidly spreads to most pigs in a pen within a few days. Pigs may experience repeated bouts with the disease. Affected pigs have an increased body temperature and decreased food intake. Mortality is low and is the result of several days of diarrhea which causes dehydration.

CAUSE, PREVENTION, AND TREATMENT

Most often the disease is caused by the bacteria *Salmonella choleraesuis* or *Salmonella typhimurium*. Other species of the *Salmonella* may also be involved in swine diseases. The septicemic form is usually caused by *S. choleraesuis*, while diarrhea is caused by *S. typhimurium*.

The recommended prevention and control of Salmonella infection consists of: (1) all-in/all-out pig flow and sanitation, and (2) vaccination with a killed bacterin or a modified live vaccine. *Note:* Vaccination against salmonellosis is controversial because of conflicting claims for safety and efficacy.

The major sources of *Salmonella* introduction into a herd are (1) the bringing in of new animals into a herd, and (2) contaminated feed. Quarantining new animals allows time for fecal shedding of *Salmonella* to stop. Wet feed and inadequately cooked feed support the growth of *Salmonella*. Other control measures when the disease is not evident include, primarily, keeping stresses to a minimum, cleanliness, and the use of disinfectants.

Often the *Salmonella* bacteria are resistant to a variety of drugs, hence, the choice of medication to control the disease should be based on the susceptibility of the *Salmonella* involved. The use of various antibiotics based on sensitivity is widely advocated. But antibiotics in the feed and water of critically ill animals have limited efficacy.

SMEDI (Stillbirth, Mummification, Embryonic Death, Infertility)

SMEDI refers to a syndrome rather than a disease caused by a specific pathogen. The factors involved result in stillbirths (S), mummified fetuses (M), embryonic deaths (ED), and infertility (I).

SYMPTOMS AND SIGNS

Signs of reproductive failure are: anestrus, failure to mate, bleeding at mating, repeat breedings, abortions, mummies, stillbirths, embryonic deaths, infertility, fewer pigs per litter, and pregnant sows fail to farrow.

1. *Stillborn piglets are piglets that are alive at the initiation of farrowing, but die intrapartum.*

2. *Mummies are the remains of fetal tissues after the maternal uterus has removed bodily fluids leaving*

only the nonabsorbable components of the fetuses, including the partially calcified skeletons.

3. *Embryonic deaths are deaths of developing young between the first and 114th day of gestation.*

4. *Infertility in swine refers to animals lacking the capability to reproduce.*

CAUSE, PREVENTION, AND TREATMENT

SMEDI (reproductive failure) occurs in all swine breeding operations, but it is regarded as significant only when reproductive levels fall below the expected norm. These norms vary from operation to operation and are based on such things as percentage of females cycling, conception and farrowing rates, average litter size, and number of pigs produced per sow per year.

Most reproductive problems involve management practices, nutrition, environmental effects, toxicoses, genetics, and/or disease conditions.

The first step in investigating swine reproductive failure is to identify the problem(s), followed by rectifying it.

The treatment directed toward reproductive problems often improves the reproductive management and productivity of the herd.

STREPTOCOCCIC INFECTION (Streptococcic Suis)

This type of infection appears to be fairly common in large, intensive hog farms. It is responsible for epidemic outbreaks of meningitis, septicemia, and arthritis.

SYMPTOMS AND SIGNS

In peracute cases, pigs may be found dead with no premonitory signs. Usually, however, signs of meningitis predominate. In most cases a progression of loss of appetite, depression, reddening of the skin, fever, incoordination, paralysis, paddling, opisthotonun, tremors, and convulsions develop.

CAUSE, PREVENTION, AND TREATMENT

The disease is caused by *Streptococci suis*.

Good management, with emphasis on sanitation, and minimizing stress from overcrowding, poor ventilation, and mixing/moving pigs, constitutes the best prevention and control. However, it occurs under both good and poor sanitary management conditions.

Early individual therapy with penicillin and supportive nursing prevent death and may result in complete recovery.

TETANUS (Lockjaw)

Tetanus is chiefly a wound infection disease that attacks the nervous system of almost all animals. Also, it is a human disease. In the central states, tetanus frequently affects pigs, lambs, and calves following castration or other wounds. It is generally referred to as *lockjaw*.

In the United States, the disease occurs most frequently in the South, where precautions against tetanus are an essential part of the routine treatment of wounds. The disease is worldwide in distribution.

SYMPTOMS AND SIGNS

The incubation period of tetanus varies from 1 to 2 weeks but may be from 1 day to many months. It is

Fig. 15-11. Pigs with tetanus or lockjaw. Such infection usually follows castration or other wounds. (Courtesy, Department of Veterinary Pathology and Hygiene, College of Veterinary Medicine, University of Illinois)

usually associated with a wound but may not directly follow an injury. The first noticeable sign of the disease is a stiffness first observed about the head. The animal often chews slowly and weakly and swallows awkwardly; hence, the common designation *lockjaw*. The animal then shows violent spasms or contractions of groups of muscles brought on by the slightest movement or noise, and usually attempts to remain standing throughout the course of the disease. If recovery occurs, it will take a month or more. In over 80% of the cases, however, death ensues, usually because of many factors, but in acute cases respiratory failure is likely the cause.

CAUSE, PREVENTION, AND TREATMENT

The disease is caused by an exceedingly powerful toxin (more than 100 times as toxic as strychnine) liberated by the tetanus organism, *Clostridium tetani*. This organism is an anaerobe (lives in absence of oxygen) which forms the most hardy spores known. It may be found in certain soils, horse feces, and sometimes in human excreta. The organism usually causes trouble when it gets into a wound that rapidly heals or closes over it. In the absence of oxygen, it then grows and liberates the toxin which apparently passes by diffusion out into the surrounding medium or environment and then spreads to the central nervous system where it seems to produce its most dire results.

The disease can largely be prevented by reducing the probability of wounds, by general cleanliness, proper wound treatment, and by administration of tetanus antitoxin in the so-called *hot* areas whenever surgery is performed or when an animal has received a wound from which tetanus may result. Tetanus antitoxin is of little or no value after symptoms have developed. All valuable animals should be protected with tetanus toxoid—a vaccination with neutralized tetanus toxin which stimulates antibodies.

All perceptible wounds should be properly treated, and the animals should be kept quiet and preferably should be placed in a dark, quiet corner free from flies. Supportive treatment is of great importance and will contribute towards a favorable course. Affected animals should be placed under the care of a veterinarian. The prognosis for swine is, however, poor.

TRANSMISSIBLE GASTROENTERITIS (TGE)

This is a rapidly spreading disease accompanied by symptoms of inflammation of the stomach and intestine. The incubation period may be only 18 hours. Once the disease strikes, it spreads rapidly; the entire herd may become noticeably affected in 2 to 3 days.

SYMPTOMS AND SIGNS

It is characterized by piglets vomiting, accompanied or rapidly followed by a watery and usually yellowish diarrhea, rapid loss in weight, dehydration, and high morbidity and mortality in pigs under two weeks of age. It affects swine of all ages, but the death loss in older swine is low. Death results from starvation, dehydration, and acidosis due to changes in the lining of the intestine.

CAUSE, PREVENTION, AND TREATMENT

The disease is caused by a virus that belongs to the group called *coronaviruses*.

Presently, commercial vaccines given to the sow before farrowing are of limited value in preventing the infection or diarrhea of TGE in piglets. The most effective preventive measure consists of exposing the sows to the virus before they farrow—a planned infection using minced intestines of young pigs known to have had TGE. They will then develop resistance or protection, and the pigs will get the antibodies from the colostrum and milk. However, this procedure should be used only where it is inevitable that sows will be exposed at farrowing time and where there is no danger of spreading the disease to neighboring herds. Isolation at farrowing time is also recommended, as is isolation of new additions to the herd. Also, continuous farrowing should be avoided because it tends to perpetuate the disease.

No effective treatment is known, but good feeding and management will help. Keeping fresh water available and providing a draft-free environment of about 90°F or higher can significantly reduce losses in piglets 3 to 4 days of age. Antibiotics or sulfonamides may minimize secondary bacterial complications. Electrolyte therapy will help.

TUBERCULOSIS

Tuberculosis is a chronic infectious disease of humans and animals. It is characterized by the development of nodules (tubercules) that may calcify or turn into abscesses. The disease spreads very slowly, and affects mainly the lymph nodes. There are three kinds of tubercle bacilli—the human, the bovine, and the avian (bird) types. As shown in Table 15-2, swine are subject to all three kinds, and practically every species of animal is subject to one or more of the three kinds.

Until the early 1900s, the incidence of tuberculosis in the United States, in animals and humans, steadily declined. But, presently, the incidence of this disease is increasing. In 1917, when a thorough nationwide eradication campaign was first initiated, 1 cow in 20 had the disease. In 1993, only 1 cow in 2,879 had the disease. Meanwhile, human mortality from tuberculosis has dropped from 150 per 100,000 in 1918 to 0.5 per 100,000 in 1992. Some decline in the incidence of the disease in poultry and swine has also been noted, but the reduction among these species has been far less marked.

SYMPTOMS AND SIGNS

Tuberculosis may take one or more of several forms. Humans get tuberculosis of the skin (lupus), of the lymph nodes (scrofula), of the bones and joints, of the lining of the brain (tuberculous meningitis), and of the lungs. For the most part, animals get tuberculosis of the lungs and lymph nodes; although in poultry the liver, spleen, and intestines are chiefly affected. In cows, the udder becomes infected in chronic cases. In swine, infection is most often contracted through ingestion of infected material; hence, the lesions often are in the abdominal cavity.

Many times an infected animal will show no outward physical signs of the disease. There may be a gradual loss of weight and condition and swelling of joints, especially in older animals. If the respiratory system is affected, there may be a chronic cough and labored breathing. Other seats of infection are lymph glands, udder, genitals, central nervous system, and

TABLE 15-2
RELATIVE SUSCEPTIBILITY OF HUMANS AND ANIMALS TO THREE DIFFERENT KINDS OF TUBERCULOSIS BACILLI

| Species | Susceptibility to Three Kinds of Tuberculosos Bacilli | | | Comments |
	Human Type	Bovine Type	Avian (Bird) Type	
Humans	Susceptible	Moderately susceptible	Questionable	Pathogenicity of avian type for humans is practically nil.
Swine	Moderately susceptible	Susceptible	Susceptible	Ninety percent of all swine cases are due to the avian type.
Cattle	Slightly susceptible	Susceptible	Slightly susceptible	
Chickens	Resistant	Resistant	Very susceptible	Chickens only have the avian type.
Horses and mules	Relatively resistant	Moderately susceptible	Relatively resistant	Rarely seen in these animals in the United States.
Sheep	Fairly resistant	Susceptible	Susceptible	Rarely seen in these animals.
Goats	Marked resistance	Highly susceptible	Susceptible	Rarely seen in these animals in the United States.
Dogs	Susceptible	Susceptible	Resistant	Highly resistant to the avian type.
Cats	Quite resistant	Susceptible	Quite resistant	Usually obtained from milk of tubercular cows.

Fig. 15-12. Boar dying with the bovine type of tuberculosis. Though swine are susceptible to all three types of tuberculosis, 90% of all swine cases are due to the avian type. For the most part, hogs get tuberculosis of the lungs and lymph nodes. (Courtesy, Department of Veterinary Pathology and Hygiene, College of Veterinary Medicine, University of Illinois)

the digestive system. The symptoms are similar, regardless of species.

CAUSE, PREVENTION, AND TREATMENT

The causative agent is a rod-shaped organism belonging to the acid-fast group of bacilli—*Mycobacterium tuberculosis*, *Mycobacterium bovis*, and *Mycobacterium avium*. The disease is usually contracted by eating food or drinking fluids contaminated with tuberculosis material. Hogs may also contract the disease by eating part of a tubercular chicken.

There are three principal methods of tuberculin testing—the intradermic, subcutaneous, and ophthalmic. The first of these is the method now principally used with swine. It consists of the injection of tuberculin into the dermis (the true skin), usually on the ear or vulva. Upon injection into an infected animal, tuberculin will set up in the body a reaction characterized by a swelling at the site of injection. In human beings, the X-ray is usually used for purposes of detecting the presence of the disease.

As a part of the federal-state tuberculosis eradication campaign of 1917, provision was made for indemnity payments on animals slaughtered.

An effective control program among swine embraces the following procedure: (1) disposing of tubercular swine, cattle, and poultry; (2) applying strict sanitation; and (3) rotating feedlots and pastures.

Preventive treatment for both humans and animals consists of pasteurization of milk and creamery by-products and the removal and supervised slaughter of reactor animals. Also, avoid housing or pasturing swine with chickens.

In human beings, tuberculosis can be arrested by hospitalization and complete rest, but in animals this method of treatment is neither effective nor practical. Mass screening programs and prompt treatment with such medications as isoniazid, streptomycin, and paraaminosalicylic acid have helped check the development of affected humans.

VESICULAR EXANTHEMA

Vesicular exanthema is almost exclusively a disease of swine, but horses are slightly susceptible. Humans are not affected. It first appeared in Buena Park, California, in 1932, where it was mistaken for and treated like foot-and-mouth disease. During the early 1950s, it reached epidemic proportions in the United States and was finally contained by 1956 after costing the federal government about $33 million.

The course of the disease is 1 to 2 weeks, and the mortality is low. Pregnant sows often abort and nursing sows fall off noticeably in milk production.

SYMPTOMS AND SIGNS

The symptoms are similar to foot-and-mouth disease. Small vesicles (like water blisters) appear around the head, particularly on the snout, nose, or lips. Also these blisters appear on the feet where the hair and the horny part of the hoof meet, on the ball of the feet, on the dewclaws, between the toes, and on the udder and teats of nursing sows. Affected animals go lame, have a high temperature, and usually go off feed for 3 to 4 days.

In order to distinguish vesicular exanthema from foot-and-mouth disease, tests may be made with horses and cattle. Also guinea pigs may be infected with foot-and-mouth disease, but are resistant to the virus of vesicular exanthema.

CAUSE, PREVENTION, AND TREATMENT

Vesicular exanthema is caused by a virus. Usually symptoms appear 24 to 48 hours after contact with the virus.

No immunizing agents are available. Preventive measures consist of (1) keeping hogs from infested areas, and (2) keeping them from consuming feed and water contaminated with the virus. Uncooked garbage that contains pork trimmings or animals from the sea, may be the source of the virus. Thus, cooking breaks the chain of infection from feed to susceptible swine, providing the principal means of control.

Good nursing will help, but no treatment is entirely successful. Rigid quarantine of affected areas (applicable to both hogs and pork products) and the destruction of affected herds appear to constitute the best control and eradication. The virus that causes vesicular exanthema is readily killed when exposed to a 2% sodium hydroxide (lye) solution or a 4% sodium carbonate (soda ash) solution.

Since vesicular exanthema is considered an exotic, reportable disease, federal and state authorities should be notified immediately of suspected cases.

PARASITES OF SWINE[5]

Hogs are probably more affected by parasites than any other class of livestock, with the possible exception of sheep. Infection with either internal or external parasites results in lack of thrift, poor development of young pigs, and increased susceptibility to diseases. Moreover, feed is always too costly to give to parasites. A total of more than 50 species of worm and protozoan parasites have been reported as being found in swine throughout the world. Fortunately, a number of these species occur only infrequently in the swine of this country; and other species, although widespread, are not of major importance under ordinary conditions.

Fig. 15-14. Animals that are heavily parasitized cannot utilize feed to the best advantage. Here are shown worms which were expelled by a single animal after medicinal treatment. (Courtesy, USDA)

Fig. 15-13. Parasites make the difference! Two pigs of the same age, but the smaller one was stunted by worms and other hog-lot infections. (Courtesy, USDA)

It is hoped that the following discussion on parasites may be helpful in (1) preventing their propagation, and (2) causing their destruction through the use of effective wormers (anthelmintics) or insecticides. Removal of parasites, however, does not ensure against reinfection; thus, if animals are kept on contaminated ground, the relief afforded by treatment may be only temporary. The prevention and control of parasites, therefore, is far more important than any treatment that

may be prescribed, no matter how successful the latter may be. In order to initiate and carry out control measures successfully, the swine producer should have a clear understanding of how each parasite develops and where it lives, stage by stage, from the egg to the adult worm. When the life history and habits of the parasite are definitely known, then and only then can plans be made to break the life cycle at some point that will destroy it.

Few dosages for the control of internal parasites are given; instead, users are admonished to follow the manufacturer's directions on the label. Also, only a limited number of pesticides are suggested for the control of external parasites because of (1) the diversity of environments and management practices under which they occur, (2) the varying restrictions on the use of pesticides from area to area, and (3) the fact that registered uses of pesticides change from time to time. Information about what is available and registered for use in a specific area can be obtained from the county agent, extension entomologist, or agricultural consultant.

At the outset, it should be emphasized that animals fed an adequate diet and kept under sanitary conditions are seldom heavily infected with parasites. A discussion of the most damaging internal parasites of hogs follows.

ASCARIDS (Large Roundworms; Ascaris suum)

This is the most common and one of the most injurious parasites of swine.

---

[5]The use of trade names of wormers and pesticides in this section does not imply endorsement, nor is any criticism implied of similar products not named; rather, it is recognized that producers, and those who counsel with them, are generally more familiar with trade names than with generic names.

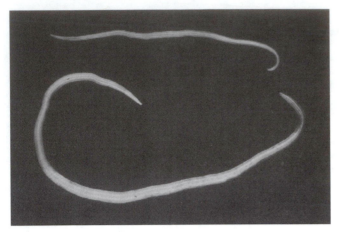

Fig. 15-15. Adult large roundworms (ascarids). They are usually yellowish or pinkish in color, 8 to 12 in. long, and almost the size of a lead pencil. (Courtesy, USDA)

DISTRIBUTION AND LOSSES CAUSED BY ASCARIDS

The roundworm is widespread throughout the world. At one time, it was reported that in the United States 20 to 70% or more of the hogs examined for this parasite were infected.[6] At times, especially in young pigs, the damage inflicted by roundworms may cause death of the animals. Lighter infections result in a stunting of growth and in uneconomical gains.

――――――

[6]Spindler, L. A., "Internal Parasites of Swine," *Keeping Livestock Healthy*, Yearbook of Agriculture, 1942, USDA, p. 766.

LIFE HISTORY AND HABITS

Technically, the parasite is known as *Ascaris suum*. The adult worm is usually yellowish or pinkish in color, 8 to 12 in. long, and almost the size of a lead pencil. The life history of the parasite may be briefly described as follows:

1. The female worms lay eggs in the small intestines and these are eliminated with the feces. These eggs are extremely resistant to the usual destructive influences.

2. A small larva develops in the egg and remains there until the egg is swallowed by the pig along with contaminated feed or water. Then it emerges from its shell, bores through the wall of the intestine, and thence gets into the bloodstream—by means of which, after a brief sojourn in the liver, it is carried to the lungs.

3. In the lungs, the larvae break out of the capillaries, enter the windpipe, and migrate to the throat. While in the throat, they are swallowed and swept to the small intestines where they develop into sexually mature worms, thus completing their life cycle.

DAMAGE INFLICTED; SYMPTOMS AND SIGNS OF AFFECTED ANIMALS

Principal damage is produced by the migrating larvae which produce liver and lung lesions. These lesions in the liver result in liver condemnations and in the lungs they create conditions favorable to the development of bacterial and viral pneumonias. The evident symptoms are exceedingly variable. Infected

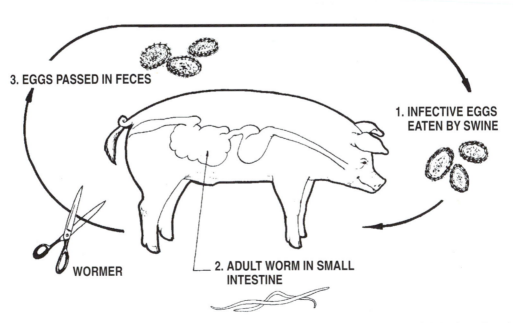

3. EGGS PASSED IN FECES

1. INFECTIVE EGGS EATEN BY SWINE

WORMER

2. ADULT WORM IN SMALL INTESTINE

Fig. 15-16. Diagram showing the life history and habits of the large roundworm. A proper wormer of choice (see scissors) and good sanitation are effective control and preventive measures.

young pigs become unthrifty in appearance and stunted in growth. Because of the presence of the larvae in the lungs, coughing and "thumpy" breathing are usually characteristic symptoms. There may be a yellow color to the mucous membrane, due to blockage of the bile ducts.

PREVENTION, CONTROL, AND TREATMENT

Prevention consists of keeping the young pigs away from infection. For pigs on pasture, the application of the McLean County System of Swine Sanitation has proved most effective in protecting pigs from the common roundworm infection. For pigs in confinement, thorough and frequent cleaning of buildings and floors is the best control program.

■ **McLean County System of Sanitation**—This system, devised by Drs. B. H. Ransom and H. B. Raffensperger of the U.S. Bureau of Animal Industry, was developed in McLean County, Illinois, as a result of a trial period of seven years, commencing in 1919. Though this program was worked out chiefly for the purpose of preventing infection of young pigs with the common roundworm, it is equally effective in lessening troubles from other parasites and in disease control, thus making possible cheaper and more profitable pork production. The system involves four simple steps:

1. Clean and disinfect the farrowing quarters.
2. Scrub sows with soap and water before moving into farrowing quarters.
3. Haul the sow and litter to clean pen or pasture.
4. Keep the pigs on clean quarters or pasture.

Although strict sanitation programs aid in the prevention of worm infections, sows and gilts should be routinely wormed 10 to 30 days before farrowing. Weanlings, feeder pigs, and boars should also be wormed on a routine basis.

A variety of wormers (anthelmintics), which are effective on ascarids, are available, and are constantly being improved and new ones are becoming available. Most are broad-spectrum wormers, which are effective on a variety of worms. It is recommended that several different wormers be used in rotation. Some of the common wormers that are effective on roundworms are: dichlorvos (Atgard), fenbendazole (Safe-Guard Type "A"), hygromycin B (Hygromix), ivermectin (Ivomec), levamisole (Tramisol) , piperazine (Wonder Wormer), and pyrantel tartrate (Banminth). Each should, of course, be used according to manufacturer's directions.

BLOWFLIES

The flies of the blowfly group include a number of species that find their principal breeding ground in dead and putrifying flesh, although they sometimes infest wounds or unhealthy tissues of live animals and fresh or cooked meat. All the important species of blowflies except the flesh flies, which are grayish and have three dark stripes on their backs, have a more or less metallic luster.

DISTRIBUTION AND LOSSES CAUSED BY BLOWFLIES

Although blowflies are widespread, they present the greatest problem in the Pacific Northwest and in the South and southwestern states. Death losses from blowflies are not excessive, but they cause much discomfort to affected animals and lower production.

LIFE HISTORY AND HABITS

With the exception of the group known as gray flesh flies, which deposit tiny living maggots instead of eggs, the blowflies have a similar life cycle to the screwworm, except that the cycle is completed in about one-half the time. (See discussion of the screwworm later in this chapter.)

DAMAGE INFLICTED; SYMPTOMS AND SIGNS OF AFFECTED ANIMALS

The blowfly causes its greatest damage by infesting wounds and the soiled hair or fleeces of living animals. Such damage, which is largely limited to the black blowfly (or wool-maggot fly), is similar to that caused by screwworms. The maggots spread over the body, feeding on the dead skin and exudates, where they produce a severe irritation and destroy the ability of the skin to function. Infested animals rapidly become weak and fevered; and although they recover, they may remain in an unthrifty condition for a long period.

Because blowflies infest fresh or cooked meat, they are often a problem of major importance around packinghouses or farm homes.

PREVENTION, CONTROL, AND TREATMENT

Prevention of blowfly damage consists of eliminating the pest and decreasing the susceptibility of animals to infestation.

As blowflies breed principally in dead carcasses, the most effective control is effected by promptly destroying all dead animals by burning or deep burial. The use of traps, poisoned baits, and electrified screens is also helpful in reducing trouble from blowflies. Suitable repellents, such as pine tar oil, help prevent the fly from depositing its eggs.

Wounds can be protected from blowflies by the application of Smear 62 or EQ335. After the wounds have been invaded by the larvae, they may be effectively treated with pressurized aerosols containing coumaphos, lindane, or ronnel.

COCCIDIOSIS

Coccidiosis—a parasitic disease affecting swine, cattle, sheep, goats, pet stock, and poultry—is caused by microscopic protozoan organisms known as coccidia, which live in the cells of the intestinal lining. Each class of domestic livestock harbors its own species of coccidia, thus there is no cross-infection between animals.

DISTRIBUTION AND LOSSES CAUSED BY COCCIDIOSIS

The distribution of the disease is worldwide. Except in very severe infections, or where a secondary bacterial invasion develops, infested farm animals usually recover. The chief economic loss is in lowered gains.

LIFE HISTORY AND HABITS

Infected animals may eliminate daily with their droppings thousands of coccidia organisms (in the resistant oocyst stage). Under favorable conditions of temperature and moisture, coccidia sporulate in 3 to 5 days, and each oocyst contains 2 to 4 infective sporocysts. The sporulated oocyst then gains entrance into an animal by being swallowed with contaminated feed or water. In the host's intestine, the outer membrane of the oocyst, acted on by the digestive juices, ruptures and liberates the eight sporozoites within. Each sporozoite then attacks and penetrates an epithelial cell, ultimately destroying it. While destroying the cell, however, the parasite undergoes sexual multiplication and fertilization with the formation of new oocysts. The parasite (oocyst) is then expelled with the feces and is again in a position to infect a new host.

The coccidia parasite abounds in wet, filthy surroundings; resists freezing and ordinary disinfectants; and can be carried long distances in streams.

DAMAGE INFLICTED; SYMPTOMS AND SIGNS OF AFFECTED ANIMALS

A severe infection with coccidia produces diarrhea, and the feces may be bloody. The bloody discharge is due to the destruction of the epithelial cells lining the intestines. Ensuing exposure and rupture of the blood vessels then produces hemorrhage in the intestinal lumen.

In addition to a bloody diarrhea, affected animals usually show pronounced unthriftiness and weakness. It generally occurs, as a recognizable disease, in pigs 7 to 21 days old. Morbidity is usually high while mortality varies, probably due to the number of oocysts ingested.

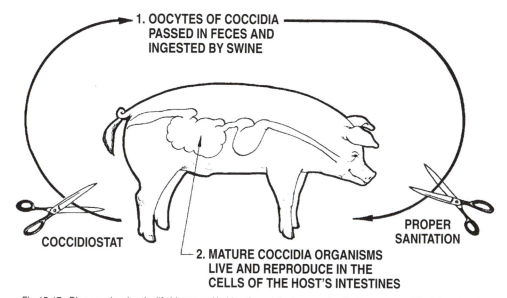

Fig. 15-17. Diagram showing the life history and habits of coccidia, the organism that causes coccidiosis in swine. As noted (see scissors) proper sanitation is the key to prevention, thus protecting animals from feed or water that is contaminated with the protozoan that causes the disease. Some of the coccidostats used for other species may prove useful, but no good approved treatment is available for pigs.

PREVENTION, CONTROL, AND TREATMENT

Coccidiosis can be prevented and controlled by (1) practicing all-in/all-out farrowing house management, (2) cleaning thoroughly and disinfecting with 5% Clorox, and (3) using raised farrowing crates with woven wire or expanded metal flooring.

No coccidiostats are approved for use in swine. All treatments presented herein are extra-label uses.

The addition of coccidiostats to sow feed is both useless and illegal—*it is not approved by the FDA*.

Pigs may be treated by (1) giving each pig 2 ml of 9.6% amprolium solution orally for three days, or (2) giving each pig trimethoprim/sulfa orally. Such pig treatments are labor-intensive and of limited efficacy.

KIDNEY WORMS (Stephanurus dentatus)

Kidney worms are one of the most damaging worm parasites affecting swine in the southern United States. The kidney worm, technically known as *Stephanurus dentatus*, is a thick-bodied black-and-white worm up to 2 in. long. Though especially harmful to swine, cattle may become infected when running with hogs.

DISTRIBUTION AND LOSSES CAUSED BY KIDNEY WORMS

The kidney worm is one of the most serious obstacles to profitable swine production in the South. Although it causes initial loss to the producer because of inefficient gains and lowered reproduction, the carcass also is affected, with damage to the liver, kidney, loin, leaf fat, and even the ham. This necessitates severe trimming or even condemnation of the carcass. All such carcass losses are ultimately borne by the swine producer in the form of lowered market prices. In the past, a blanket reduction in price was made on all market hogs coming from certain areas of the South where swine infections with kidney worms are notoriously heavy. This was done in anticipation of the usual damage to the carcass.

LIFE HISTORY AND HABITS

Adult kidney worms may be found around the kidneys and in cysts in the ureters (tubes leading from the kidneys to the bladder). The mature female worms lay numerous eggs that are discharged with the urine. It has been estimated that, in 1 day, as many as 1 million eggs may be passed in the urine of a moderately infected hog. When eggs fall on moist, shaded soil, a tiny larva hatches from each egg in 1 to 2 days, depending on temperature conditions. In another 3 to 5 days, the larva develops into the infective stage. Hogs then obtain kidney worms by swallowing the infective larvae with contaminated feed and through the skin on ground containing the larvae. Under warm, moist, shaded conditions, the larvae may survive for several months.

Kidney worm larvae can also enter the bodies of

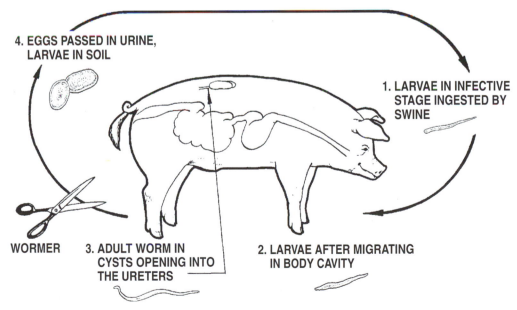

4. EGGS PASSED IN URINE, LARVAE IN SOIL

1. LARVAE IN INFECTIVE STAGE INGESTED BY SWINE

WORMER 3. ADULT WORM IN CYSTS OPENING INTO THE URETERS

2. LARVAE AFTER MIGRATING IN BODY CAVITY

Fig. 15-18. Diagram showing the life history and habits of the kidney worm. As noted (see scissors), a proper wormer of choice is effective in the control and prevention of kidney worms. Proper sanitation also complements this by limiting the contact with infective larvae.

pigs through the skin, though this is not considered a great source of infection. Regardless of the way in which the larvae enter the body, they get into the blood and migrate to the liver, lungs, and other organs—some of them finally reaching the kidneys. Upon reaching maturity, which may be 12 to 14 months after the pigs ingest the larvae, the adult female kidney worms begin producing eggs, thus completing the life cycle.

DAMAGE INFLICTED; SYMPTOMS AND SIGNS OF AFFECTED ANIMALS

There are no definite symptoms ascribable to kidney worm infection. The growth rate is markedly retarded, and health is impaired. Frequently, infected animals discharge pus in the urine. Most cases are diagnosed at necropsy. Diagnosis is only positively made by microscopically discovering the presence of eggs in the urine.

Primarily, kidney worms damage the liver, ureter, kidneys, and tissues surrounding the kidneys (loin).

PREVENTION, CONTROL, AND TREATMENT

The "gilt-only" method, first proved at the Coastal Plain Experiment Station, Tifton, Georgia, constitutes an effective prevention and control. With this method, gilts are bred; then, after farrowing and weaning off their first litter, they are sent to slaughter before mature kidney worms develop. This system is based on the fact that it may take the kidney worm as long as a year to reach the egg-laying stage. By following this system for 2 years, the swine producer may completely eliminate kidney worms. Also, several therapeutics are effective in controlling kidney worms when used according to manufacturer's directions; among them, fenbendazole, ivermectin, and levamisole hydrochloride.

LICE

The louse is a small, flattened, wingless insect parasite of which there are several species. Hog lice, which are the largest species, may range up to 0.25 in. in length. The two main types are sucking lice and biting lice. Of the two groups, the sucking lice are the most injurious.

Most species of lice are specific for a particular class of animals; thus, hog lice will not remain on the other farm animals, nor will lice from other animals usually infest hogs. Lice are always more abundant on weak, unthrifty animals and are more troublesome during the winter months than during the rest of the year.

Fig. 15-19. Male hog louse *(Haematopinus suis)*. This is a small, flattened, wingless insect parasite. Hog lice may range up to 0.25 in. in length. (Courtesy, USDA)

DISTRIBUTION AND LOSSES CAUSED BY LICE

The presence of lice upon animals is almost universal, but the degree of infestation depends largely upon the state of animal nutrition and the extent to which the owner will tolerate parasites. The irritation caused by the presence of lice retards growth, gains, and/or production of milk; and such diseases as swine pox may be transmitted by lice.

LIFE HISTORY AND HABITS

Lice spend their entire life cycle on the host's body. They attach their eggs or "nits" to the hair near the skin where they hatch in about 2 weeks. Two weeks later the young females begin laying eggs, and after reproduction they die on the host. Lice do not survive more than 2 to 3 days when separated from the host. They are found on all parts of the body, but, in particular, the areas such as the folds around the neck, jowl, flanks, and inside the legs and ears.

DAMAGE INFLICTED; SYMPTOMS AND SIGNS OF AFFECTED ANIMALS

Infestation shows up most commonly in winter in ill-nourished and neglected animals. There is intense irritation, restlessness, and loss of condition. As many lice are blood suckers, they devitalize their host. There may be severe itching and the animal may be seen scratching, rubbing, and gnawing at the skin. The hair may be rough, thin, and lack luster; and scabs may be evident. In pigs, the skin of infested animals becomes thick, cracked, tender, and sore. In some cases, the symptoms may resemble that of mange; and it must be kept in mind that the two may occur simultaneously.

With the coming of spring, when the hair sheds and when some animals may go to pasture, lousiness is greatly diminished.

PREVENTION, CONTROL, AND TREATMENT

Because of the close contact of domesticated animals, especially during the winter months, it is practically impossible to prevent herds from becoming slightly infested with the pests. Nevertheless, lice can be kept under control.

For more effective control, all members of the herd must be treated simultaneously, and this is especially necessary during the fall months about the time they are placed in winter quarters. Spraying or dipping during freezing weather should be avoided, however. It is also desirable to treat the housing and bedding.

With a power sprayer, from 100 to 200 lb/sq in. of pressure is adequate for spraying hogs.

Dusting is less effective than spraying or dipping, but may be preferable when few animals are to be treated or during the winter months.

The following insecticides are effective in the control of lice when used according to the manufacturer's label directions: amitraz (Takie), diazinon, fenvalerate, lindane, malathion, permethrin, and phosmet (Prolate-Star Bar). Also, ivermectin, which is an injectable, is very effective.

Where hog lice are encountered, it is suggested that the producer obtain from local authorities the current recommendations relative to the choice and concentration of the insecticide to use. Also, the producer should check with the state extension specialists relative to control measures.

LUNGWORMS (Metastrongylus elongates, M. pudendotectus, and M. salmi)

Lungworms are among the most widespread of the parasites that infect swine. Three species are common in hogs in this country; namely, *M. elongatus*, *M. pudendotectus*, and *M. salmi*. All lungworms are threadlike in diameter, 1 to 1.5 in. in length, and white or brownish in color. As the name would indicate, they are found in the bronchi, or air passages, of the lungs. Sheep and cattle also have lungworms.

DISTRIBUTION AND LOSSES CAUSED BY LUNGWORMS

The lungworm is found in all sections of the United States, but the heaviest infection of swine occurs in the southeastern states. In addition to the usual eco-

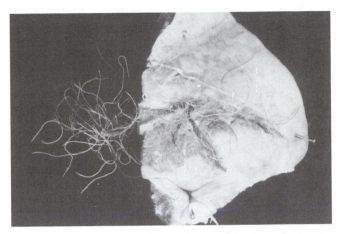

Fig. 15-20. Lower portion of a swine lung, partially cut open to show the nest of lungworms. Lungworms are threadlike in diameter, 1 to 1.5 in. in length, and white or brownish in color. (Courtesy, USDA)

nomic losses and lowered growth and feed efficiency caused by the lungworm, there is evidence to indicate that this parasite may be instrumental in the spread of swine influenza.

LIFE HISTORY AND HABITS

Female lungworms produce large numbers of thick-shelled eggs, each containing a tiny larva. The eggs are coughed up, swallowed, and eliminated with the feces of the pig. Earthworms, the intermediate hosts, feed on the feces, then swallow the eggs, which hatch in the earthworms' intestines. The larva then develops in the earthworm for about 10 days, after which it is capable of producing an infection. Infection of the pig results from the swallowing of the earthworm, which it usually acquires by rooting and feeding in places where earthworms abound—manure piles; rich, feces-contaminated soil; under trash; and in moist places. After ingestion by the pig, the lungworm larvae are liberated from the earthworm and migrate by way of the lymphatic and blood circulatory systems to the lungs. There they become localized, complete development, and begin to produce eggs within 24 days, thus completing the life cycle.

The female lungworm produces an incredibly large number of eggs. It is estimated that as many as 3 million eggs may be eliminated in the droppings of a heavily infected pig in a period of 24 hours. More than 2,000 larvae have been found in a single earthworm collected in a hog lot.

DAMAGE INFLICTED; SYMPTOMS AND SIGNS OF AFFECTED ANIMALS

Pigs heavily infected with lungworms become un-

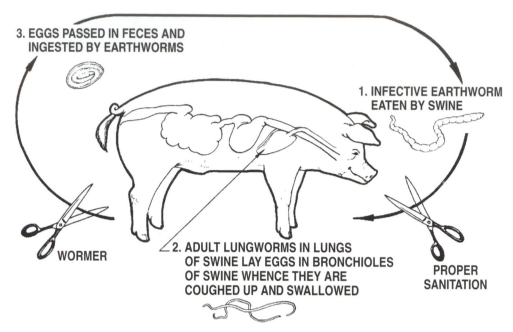

3. EGGS PASSED IN FECES AND
INGESTED BY EARTHWORMS

1. INFECTIVE EARTHWORM
EATEN BY SWINE

WORMER

2. ADULT LUNGWORMS IN LUNGS
OF SWINE LAY EGGS IN BRONCHIOLES
OF SWINE WHENCE THEY ARE
COUGHED UP AND SWALLOWED

PROPER
SANITATION

Fig. 15-21. Diagram showing the life history and habits of lungworms. As noted (see scissors), a proper wormer of choice is effective in the control and prevention of lungworms. Proper sanitation also complements this by limiting the contact with infective earthworms containing lungworm larvae.

thrifty, stunted, and are subject to spasmodic coughing. Positive diagnosis is made only by fecal examination or by postmortem examination. A cross section of the lungs exposes the white threadlike worms in the air tubes.

PREVENTION, CONTROL, AND TREATMENT

Prevention of lungworm infection consists of keeping hogs away from those areas where earthworms are likely to abound. Removal of manure piles and trash and the drainage of low places will help. In brief, the swine producer should provide clean, dry, well-drained lots—conditions which are not conducive for the intermediate host, the earthworm. Ringing the snout will also help in that it will prevent rooting.

The following products are effective in the control of lungworms when used according to the manufacturer's directions: fenbendazole, ivermectin, and levamisole hydrochloride.

MITES (Mange)

Mites produce a specific contagious disease known as mange (or scabies, scab, or itch). These small insect-like parasites, which are almost invisible to the naked eye, constitute a very large group. They attack members of both the plant and animal kingdoms.

Mites are responsible for the condition known as mange (scabies) in swine, sheep, cattle, and horses.

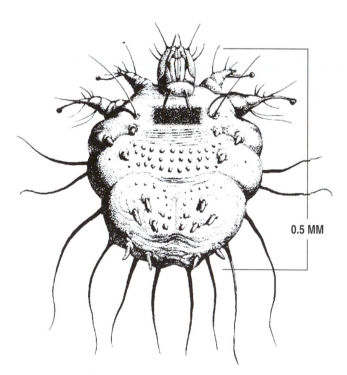

0.5 MM

Fig. 15-22. The sarcoptic mange mite *(Sarcoptes scabiei).* On the underneath side, it has four pairs of short, stumpy legs, some of which are provided with unjointed pedicles with suckerlike organs on their ends.

Each species of domesticated animals has its own species of mange mites. The mites from one species of animals cannot live normally and propagate permanently on different species. The disease appears to spread most rapidly among young and poorly nourished animals. Two types of mange affect swine—sarcoptic mange and demodectic mange. Of the two, sarcoptic mange is by far the most prevalent, showing worldwide distribution. Demodectic mange is uncommon in swine. Intense itching often accompanies the sarcoptic form since the mites form tunnels in the skin as part of their life cycle

DISTRIBUTION AND LOSSES CAUSED BY MITES

Injury from mites is caused by irritation and blood sucking and the formation of scabs and other skin affections. In a severe attack, the skin may be much less valuable for leather. Growth is retarded and animals are unthrifty.

LIFE HISTORY AND HABITS

The entire life cycle of the sarcoptic mange mite, *Sarcoptes scabiei* var. *suis*, occurs on the body of the host pig. After the female mite mates, she lays eggs at the rate of about 1 to 3 per day in tunnels which she carves into the upper two-thirds of the epidermis. In about 5 days the eggs hatch and then mature to adults within the tunnels in about 10 to 15 days. New females then mate near the surface and begin their tunnels. The cycle from egg to egg-laying female requires 10 to 15 days. Mature females die about 1 month after reaching maturity. Generally, mite infestations begin in the inner ear and then spread along the neck and across the body.

Mites are more prevalent during the winter months when animals are in close contact with each other and when treatment is more difficult. They spread among the young and among the poorly nourished animals.

DAMAGE INFLICTED; SYMPTOMS AND SIGNS OF AFFECTED ANIMALS

Infested animals do not eat properly and, as a result, they do not gain at normal rates.

Sarcoptic mites cause marked irritation, itching, and scratching, and crusting of the skin which is often accompanied or followed by the formation of a thick, tough, wrinkled skin. Often there are secondary skin infections. The only certain method of diagnosis is to demonstrate the presence of the mites.

Fig. 15-23. Pig with sarcoptic mange, caused by sarcoptic mites. Note the thickening, wrinkling, and harshness of the skin. (Courtesy, Department of Veterinary Pathology and Hygiene, College of Veterinary Medicine, University of Illinois)

PREVENTION, CONTROL, AND TREATMENT

Prevention consists of avoiding contact with diseased animals or infested premises. In case of an outbreak, the local veterinarian or livestock sanitary officials should be contacted.

Mites can be controlled by spraying animals with suitable insecticides and quarantine of infected herds. Producers should have a routine program for spraying for mange control. For example, gilts and sows should be sprayed 4 to 6 weeks before farrowing and then again 7 to 10 days prior to farrowing. A high-pressure

Fig. 15-24. Pigs are sprayed for external parasites before going on official test at the Minnesota swine evaluation station. (Courtesy, University of Minnesota)

spray (100–250 psi) is recommended and repeated treatments are important due to eggs in tunnels which would not be reached by the first spraying. The following insecticides are effective in the control of mites when used according to the manufacturer's directions: amitraz (Takie), diazinon, fenvalerate, lindane, malathion, permethrin, and phosmet (Prolate-Star Bar). Also, ivermectin, which is an injectable, is very effective. Because new insecticides are constantly being developed, and old ones banned or restricted, when hog mites are encountered, it is suggested that the producer obtain from local authorities the current recommendations relative to the choice and concentration of the insecticide to use. Generally, those products listed for the control of mange are also effective for louse control.

NODULAR WORMS
(Oesophagostomum spp.)

These parasites are most numerous in the southeastern United States. A parasitic survey conducted in North Carolina showed the nodular worm to be second to strongyloides in incidence.

They are called nodular worms because of the nodules or lumps they cause in the large intestine. Four species of *Oesophagostomum* occur in swine, but all of them are slender, whitish to grayish in color, and 0.3 to 0.5 in. in length.

DISTRIBUTION AND LOSSES CAUSED BY NODULAR WORMS

Nodular worms are widely distributed, but damage is heaviest in the southeastern states. In addition to the usual lack of thrift that accompanies parasite infections, the intestines of severely infected animals are not suited for either sausage casings or food (chitterlings).

LIFE HISTORY AND HABITS

The four species of nodular worms affecting swine are *O. dentatum*, *O. brevicaudum*, *O. georgianum*, and *O. quadrispinulatum*—all have similar life histories. The adult worms are localized in the large intestine of the host animal. The female worms deposit large numbers of partly developed eggs that become mixed with the intestinal contents and are eliminated with the feces. With favorable conditions of temperature and moisture, a larva emerges from each egg in 1 to 2 days. After another 3 to 6 days of development, the larvae are infective to swine. Pigs then become infected by swallowing the larvae while feeding on contaminated ground or grazing on contaminated pastures. In the digestive system of the host, the larvae travel to the large intestine where they penetrate into the wall and grow for the next 2 to 3 weeks, forming nodules. They then move into the lumen, or cavity, of the large intestine where they continue development. Within 3 to 7 weeks after ingestion by the pig, the worms are

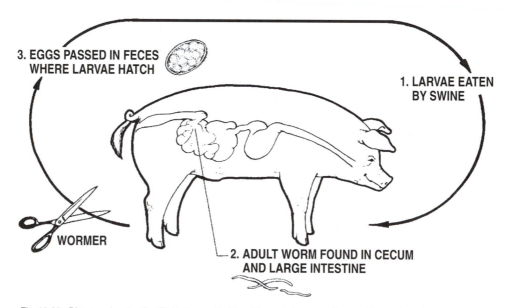

Fig. 15-25. Diagram showing the life history and habits of the nodular worm. As noted (see scissors), a proper wormer of choice may be used to control and prevent nodular worms. Rigid sanitation, accompanied by pasture rotation (when pastures are involved), complement a program for control and prevention of nodular worms.

fully grown and have mated and are producing eggs, thus starting a new cycle.

DAMAGE INFLICTED; SYMPTOMS AND SIGNS OF AFFECTED ANIMALS

No specific symptoms can be attributed to the presence of nodular worms. Weakness, anemia, emaciation, diarrhea, and general unthriftiness have been reported as due to infection with these parasites.

PREVENTION, CONTROL, AND TREATMENT

A strict program of swine sanitation, accompanied by a program of regular worming, constitutes a successful and practical preventive measure.

Pigs may be dewormed by administering dichlorvos, hygromycin B, levamisole hydrochloride, or pyrantel tartrate. Whichever drug is chosen, the manufacturer's directions should be read and carefully followed.

RINGWORM

Ringworm, or barn itch, is a contagious disease of the outer layers of skin. It is caused by certain microscopic molds or fungi (*Trichophyton, Achorion,* or *Microsporon*). All animals and humans are susceptible.

DISTRIBUTION AND LOSSES CAUSED BY RINGWORM

Ringworm is widespread throughout the United States. Though it may appear among animals on pasture, it is far more prevalent as a barn disease. It is unsightly, and affected animals may experience considerable discomfort; but the actual economic losses attributed to the disease are not too great.

LIFE HISTORY AND HABITS

The period of incubation for this disease is about one week. The fungi form spores that may live 18 months or longer in barns or elsewhere.

DAMAGE INFLICTED; SYMPTOMS AND SIGNS OF AFFECTED ANIMALS

Round, scaly areas almost devoid of hair appear mainly in the vicinity of the eyes, ears, side of the neck, or the root of the tail. Crusts may form, and the skin may have a gray, powdery, asbestoslike appearance. The infested patches, if not checked, gradually increase in size. Mild itching usually accompanies the disease.

PREVENTION, CONTROL, AND TREATMENT

The organisms are spread from animal to animal and through the medium of contaminated fence posts and brushes. Thus, prevention and control consists of disinfecting everything that has been in contact with infested animals. The affected animals should also be isolated. Strict sanitation is an essential in the control of ringworm.

The hair should be clipped, the scabs removed, and the area brushed and washed with mild soap. The diseased parts should be painted with tincture of iodine or salicylic acid and alcohol (1 part in 10) every 3 days until cleared up. Animals may be treated orally with nystatin or griseofulvin, or topically with iodine, captan, or salicylic acid.

SCREWWORMS
(Callitroga hominovora)

"Officially," the screwworm was eradicated from the southeastern U.S. in the late 1950s. Now many millions of sterile flies are being released to eradicate completely this pest (1) in Mexico, and (2) in the adjoining states of the United States when there is an outbreak. Since the screwworm fly may infest an animal through wounds and also through lesions caused by ticks, horse flies, or even horn flies, there is an added impetus to control these parasites.

They are not found in cold-blooded animals such as turtles, snakes, and lizards.

Wounds resulting from castrating, docking, dehorning, branding, and shearing afford a breeding ground for this parasite. Add to this the wounds from some types of vegetation, from fighting, and from blood-sucking insects, and ample places for propagation are provided.

DISTRIBUTION AND LOSSES CAUSED BY SCREWWORMS

Screwworm infestations occasionally occur in certain areas of Arizona, New Mexico, and Texas, and they often occur in many areas of Mexico and other Latin American countries. At one time, the screwworm was responsible for 50% of the normal annual livestock losses in the southern and southwestern states.

LIFE HISTORY AND HABITS

The primary screwworm fly is bluish green, with three dark stripes on its back and reddish or orange color below the eyes. The fly generally deposits its eggs in shinglelike masses on the edges or the dry portion of wounds. From 50 to 300 eggs are laid at one time, with a single female being capable of laying about 3,000 eggs in a lifetime. Hatching of the eggs occurs in 11 hours, and the young whitish worms (larvae or maggots) immediately burrow into the living flesh. There they feed and grow for a period of 4 to 7 days, shedding their skin twice during this period.

When these larvae have reached their full growth, they assume a pinkish color, leave the wound, and drop to the ground, where they dig beneath the surface of the soil and undergo a transformation to the hard-skinned, dark brown, motionless pupae. It is during the pupa stage that the maggot changes to the adult fly.

After the pupa has been in the soil from 7 to 60 days, the fly emerges from it, works its way to the surface of the ground, and crawls up on some nearby object (bush, weed, etc.) to allow its wings to unfold and otherwise mature. Under favorable conditions, the newly emerged female fly becomes sexually mature and will lay eggs 5 days later. During warm weather, the entire life cycle is usually completed in 21 days, but under cold, unfavorable conditions the cycle may take as many as 80 days or longer.

DAMAGE INFLICTED; SYMPTOMS AND SIGNS OF AFFECTED ANIMALS

The injury caused by this parasite is inflicted chiefly by the maggots. Early symptoms in affected animals are: loss of appetite and condition, and listlessness. Unless proper treatment is administered, the great destruction of many tissues kills the host in a few days.

PREVENTION, CONTROL, AND TREATMENT

Prevention in infested areas consists mainly of keeping animal wounds to a minimum and of protecting those that do materialize.

In 1958, the U.S. Department of Agriculture initiated an eradication program. Screwworm larvae were reared on artificial media. Two days before fly emergence, the pupae were exposed to gamma irradiation at a dosage which caused sexual sterility but no other deleterious effects. Sterile flies were distributed over the entire screwworm-infested region in sufficient quantity to outnumber the native flies, at an average rate of 400 males per square mile per week. The female mates only once and, therefore, when mated with a sterile male does not reproduce. There was a decline in the native population each generation until the native males were so outnumbered by sterile males that no fertile matings occurred and the native flies were eliminated. This program has virtually eliminated all the losses caused by screwworms in the United States. Unfortunately, the states bordering on Mexico

are periodically reinfested by mated female flies from Mexico. For this reason, permanent elimination of screwworms from the United States by the sterile-male technique cannot be hoped for until they are also eradicated in Mexico.

When maggots (larvae) are found in an animal, they should be removed and sent to the proper authorities for identification.

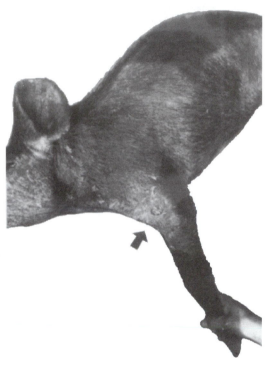

Fig. 15-26. Screwworm injury to the front leg of a year-old sow. (Courtesy, USDA)

Wounds can be protected from screwworm flies by the application of a wound dressing of Smear 62 (containing diphenylamine, benzol, turkey red oil, and lampblack) or EQ335 (containing 3% lindane and 35% pine oil). After wounds are invaded by larvae, they may be effectively treated with pressurized aerosols containing coumaphos, lindale, or ronnel.

Title 9 of the Code of Federal Regulations, Part 83, Screwworms, gives the screwworm control zone, the areas of recurring infestation, and the inspection and treatment requirements for movement from these areas.

(Also see later section in this chapter entitled, "Federal and State Regulations Relative to Disease Control.")

STOMACH WORMS
(Ascarops strongylina, Physoce phalus sexalatus, and Hyostrongylus rubidus)

Three species of small stomach worms infect swine. Two of the species, *A. strongylina* and *P. sexalatus*, are commonly known as "thick stomach worms." These parasites are reddish in color and nearly 1 in. long in the adult stage. The third species, *H. rubidus*, commonly known as the "red stomach worm," is a small, delicate, slender, reddish worm about 0.2 in. in length.

DISTRIBUTION AND LOSSES CAUSED BY STOMACH WORMS

The occurrence of the stomach worm is widespread in hogs throughout the United States. Early surveys in certain of the midwestern states demonstrated that as many as 90% of the animals examined harbored the worms, and, in the southeastern states, 50 to 80%.[7] However, it is noteworthy that the move from dirt lots to confinement farrowing and rearing has greatly reduced the prevalence of stomach worms in the United States.

---

[7]Spindler, L. A., "Internal Parasites of Swine, *Keeping Livestock Healthy*, Yearbook of Agriculture, 1942, USDA, p. 752.

LIFE HISTORY AND HABITS

It has been definitely established that the ordinary dung beetle serves as the intermediate host for the thick stomach worm. The female worm deposits eggs in the stomach of the pig, with each egg containing a tiny embryo. Passing with the feces to the outside of the body of the pig, the eggs are eaten by various species of dung beetles, which are the intermediate hosts.

After developing for about a month in the beetle, the larvae arrive at a stage infective to swine. Hogs feeding on contaminated ground then swallow the beetles. In the stomach of the pig, the parasites are liberated from the beetles and make their way into the mucous membrane of the stomach, where they grow to maturity.

The life history of the red stomach worm differs from that of the thick stomach worms in that no intermediate host is necessary, the infections being directly acquired. Eggs are shed in the feces and they hatch in 1 to 2 days. After 7 days the larvae are infective. Swine ingest them by feeding in wet areas where they survive, and then they enter the stomach, thus completing their cycle.

DAMAGE INFLICTED; SYMPTOMS AND SIGNS OF AFFECTED ANIMALS

Because of the burrowing tendency of this parasite, inflammation and gastric ulcers usually follow. In

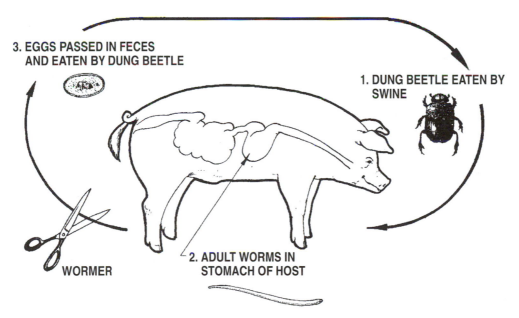

3. EGGS PASSED IN FECES AND EATEN BY DUNG BEETLE

1. DUNG BEETLE EATEN BY SWINE

WORMER

2. ADULT WORMS IN STOMACH OF HOST

Fig 15-27. Diagram showing the life history and habits of the thick stomach worm. As noted (see scissors), a proper wormer of choice is effective in the control and prevention of thick stomach worms. Good sanitation also helps by limiting the contact with dung beetles containing the encysted larvae of thick stomach worms.

external appearance, the affected animals usually become unthrifty and show a marked loss in appetite.

PREVENTION, CONTROL, AND TREATMENT

Preventive measures for the control of stomach worms are similar to those advocated for the control of ascarids; namely, sanitation and the use of wormers. Carbon disulfide, dichlorvos, fenbendazole, ivermectin, levamisole, and piperazene are effective in removing small stomach worms. They should, of course, be administered by following the manufacturer's directions. The McLean County System of Swine Sanitation should be applied to swine on pastures.

THORN-HEADED WORMS (Macracanthorhynchus hirudinaceus)

The thorn-headed worm is of considerable importance in the southern part of the United States, though its distribution is worldwide. It may be easily distinguished from the common intestinal roundworm by the presence of rows of hooks—a spiny proboscis (snout)—through which it attaches itself to the wall of the small intestine of the host. Thorn-headed worms are milk white to bluish in color, and cylindrical to flat in shape, ranging from 2 to 26 in. long. The female is longer than the male

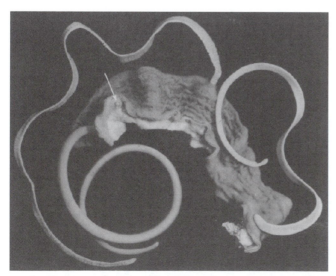

Fig. 15-28. Thorn-headed worms of swine. These worms are milk white to bluish in color and cylindrical to flat in shape. Females may be 26 in. long while males are only about 4 in. long. (Courtesy, USDA)

DISTRIBUTION AND LOSSES CAUSED BY THORN-HEADED WORMS

The thorn-headed worm is not a common parasite of hogs grown in the Corn Belt states, but it is of considerable economic importance in the Deep South. In addition to the usual slow growth, inefficient feed utilization, and death losses resulting from other parasites, thorn-headed worms weaken the intestine and make it unfit for sausage casings. This causes a financial loss to the packer that is passed along to the swine producer in the form of lowered meat prices.

LIFE HISTORY AND HABITS

Adult female thorn-headed worms produce numerous thick-shelled brownish eggs, each containing a fully developed larva. Each female may produce as many as 600,000 eggs per day at the peak of her egg-producing capacity. The eggs, which pass out with the feces, are very resistant to destruction.

Beetle grubs, the larvae of June beetles, or dung beetles, serve as the intermediate hosts. The grubs, feeding on infected manure or contaminated soil, swallow the parasite eggs. The eggs hatch in the bodies of the grubs and in 3 to 6 months develop to a stage that is infective to swine.

Pigs rooting in manure or trash piles, rich soil, or low-lying pastures swallow the grubs (of which they are very fond) or beetles. The young thorn-headed worms then escape from the bodies of the grubs or adult beetles through the process of digestion and develop to egg-laying maturity in 3 to 4 months.

DAMAGE INFLICTED; SYMPTOMS AND SIGNS OF AFFECTED ANIMALS

No special symptoms have been attributed to swine infected with thorn-headed worms, although these parasites are decidedly injurious. Infected swine exhibit the general unthriftiness that is commonly associated with parasites. A heavy infestation may even kill young pigs. Upon autopsy, a swelling or nodule may be evident at the point of attachment, and the intestinal wall may exhibit great weakness.

PREVENTION, CONTROL, AND TREATMENT

Prevention consists of keeping pigs from feeding in areas where they might obtain grubs of the June beetle or dung beetle. Thus, sanitation, clean ground, and nose ringing are effective preventive measures. Levamisole will remove thorn-headed worms.

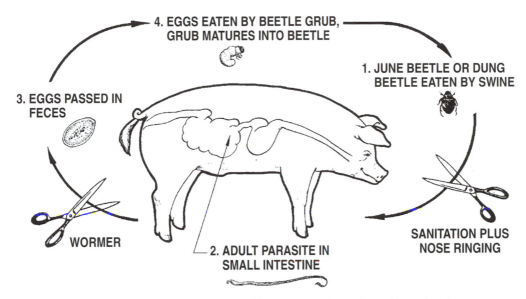

Fig. 15-29. Diagram showing the life history and habits of the thorn-headed worm. As noted (see scissors), a proper wormer of choice is effective in the control and prevention of thorn-headed worms. Nose ringing of pasture hogs and sanitation are complementary procedures.

THREADWORMS
(Strongyloides ransomi)

The pig is the only known host of *S. ransomi*. These worms are tiny. The parasitic females are 0.13 to 0.18 in. long, while the free-living forms are only about 0.04 in. long.

DISTRIBUTION AND LOSSES CAUSED BY THREADWORMS

It is the most important swine parasite in the southern and southeastern states, though it has worldwide distribution.

LIFE HISTORY AND HABITS

Small, embryonated eggs are passed in the feces. They hatch in 12 to 18 hours. These larvae develop into the infective form or free-living form 2 to 3 days after hatching; the infective larvae penetrate the skin and proceed to the lungs via the bloodstream, thence from the alveoli of the lungs to the bronchi, esophagus, stomach, and small intestine, where they become adult parthenogenetic females. Oral ingestion of the infective larvae can also produce infection. Furthermore, baby pigs may obtain the infective larvae from the colostrum and demonstrate an infection as early as 4 days of age. A free-living generation of adult males and females develop the infective parasitic larvae.

DAMAGE INFLICTED; SYMPTOMS AND SIGNS OF AFFECTED ANIMALS

Diagnosis is best made upon autopsy. However, in heavy infestations, it is a serious disease causing scours, anemia, and severe weight loss and death loss, particularly in young pigs. Light infections may show no symptoms.

PREVENTION, CONTROL, AND TREATMENT

Prevention is aided through (1) a program of strict sanitation, (2) selecting dry, unshaded areas for swine lots, (3) pasture rotation, and (4) a program of frequent worming, using an effective wormer of choice.

Levamisole (Tramisol), thiabendazole (Thibenzole), and ivermectin are effective for the treatment and control of threadworms. The manufacturer's directions for use and withdrawal time should be carefully read and followed.

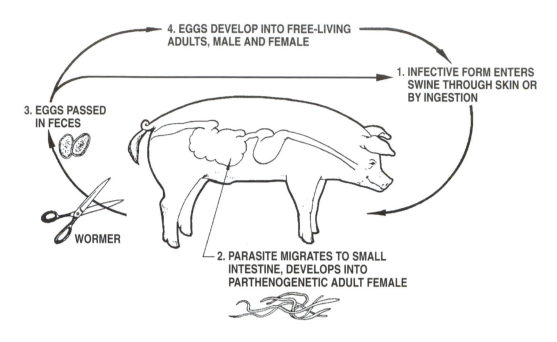

Fig. 15-30. Diagram showing the life history and habits of the threadworm. As noted (see scissors), a proper wormer of choice, complemented by sanitation, is effective in the control and prevention of threadworms.

TRICHINOSIS
(Trichinella spiralis)

This is a parasitic disease of human beings caused by *T. spiralis*. The main source of the disease is infected pork, eaten raw or improperly cooked. Although the parasite is present in the muscle tissue of swine, it does not induce symptoms in this species. Some infections have resulted from the consumption of bear meat in Canada, Alaska, and the northeastern and western states.

DISTRIBUTION AND LOSSES
CAUSED BY TRICHINOSIS

This disease appears to be worldwide, with the highest incidence occurring in areas where uncooked garbage is fed to hogs. It is estimated that less than 0.1% of pork from grain-fed hogs is infected with trichinosis and that there is less than 1% infection of pork from garbage-fed hogs.

LIFE HISTORY AND HABITS

The life cycle of trichina may be summarized as follows:

1. The adult parasite, which is a round worm from 0.06 to 0.16 in. in length, lives in the small intestine of humans, hogs, rats, and other animals. The female worms penetrate into the lining of the intestines where they produce numerous young or larvae. Males die soon after mating.

2. The larvae pass from the wall of the intestine into the lymph system, thence into the bloodstream, and finally into the muscle cells.

3. In the muscles, the larvae grow until they are about 0.04 in. long, then roll into a characteristic spiral shape, and become surrounded by a capsule. In this environment and stage of development (cysts), these larvae may live for years or until the raw or improperly cooked muscle tissue is eaten by humans or other species of meat eaters.

4. Upon gaining entrance to the intestines of another host, the worm starts a new life cycle.

DAMAGE INFLICTED; SYMPTOMS
AND SIGNS OF INFECTED ANIMALS

There are no specific symptoms in hogs, even when the parasite is present in the muscle tissue, its usual abode.

The symptoms in humans vary depending upon the degree of parasite infection. The disease is usually accompanied by a fever, digestive disturbances, swelling of infected muscles, and severe muscular pain (in

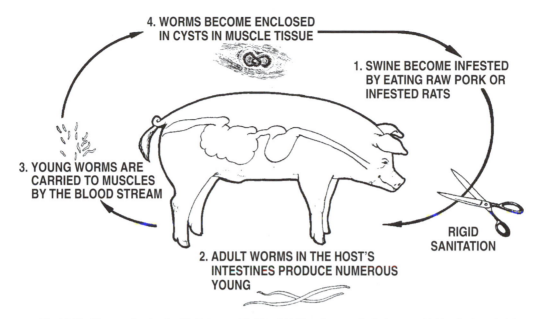

4. WORMS BECOME ENCLOSED IN CYSTS IN MUSCLE TISSUE

1. SWINE BECOME INFESTED BY EATING RAW PORK OR INFESTED RATS

3. YOUNG WORMS ARE CARRIED TO MUSCLES BY THE BLOOD STREAM

RIGID SANITATION

2. ADULT WORMS IN THE HOST'S INTESTINES PRODUCE NUMEROUS YOUNG

Fig. 15-31. Diagram showing the life history and habits of trichina, the parasite that causes trichinosis. As noted (see scissors) effective control of trichinosis in swine is obtained through rigid sanitation, including (1) destruction of rats, (2) proper carcass disposal of dead hogs, and (3) cooking all garbage and offal from slaughtering houses.

the breathing muscles as well as others). A specific skin test, serologic tests, or muscle biopsy, may also be applied to aid in the diagnosis in humans. Normally, the symptoms begin about 10 to 14 days after infected pork has been eaten. Infected humans should be under the care of a medical doctor.

PREVENTION, CONTROL, AND TREATMENT

Fortunately, a considerable amount of the trichinosis pork that is slaughtered is noninfective by the time it reaches the consumer. In processing, much of it is made safe through heating, freezing, etc. Nevertheless, a small amount of infected pork may be sold primarily as ready-to-eat pork from noninspected facilities, and the only certain protection to humans is to cook pork and pork products to 137°F internal temperature. Thus, prevention consists of the thorough cooking of all pork before it is eaten, either by humans or by swine. *Trichinella* is also destroyed by freezing for a continuous period of not less than 20 days at a temperature of not higher than 5°F.

Microscopic examination of pork is the only way in which to detect the presence of trichina. Such methods are regarded as impractical, and meat inspection in the United States does not include examination for *Trichinella*.

Trichinosis in swine may be prevented by (1) destruction of all rats on the farm, (2) proper carcass disposal of hogs that die on the farm, and (3) cooking

of all garbage and offal from slaughtering houses at a temperature of 212°F for 30 minutes. However, as all these preventive measures may not be feasible or practical or carried out with care, the surest and best protection currently available to pork consumers consists of cooking or freezing pork at the temperatures specified. It is noteworthy, however, that the state and federal governments regulate the cooking of garbage.

WHIPWORMS (Trichuris suis)

Whipworms are usually found attached to the walls of the cecum (blind gut) and large intestine of swine. They are 1 to 3 in. in length. The worms have a very slender anterior portion and a much enlarged posterior. The anterior resembles the lash of a whip and the posterior the handle; hence, the name whipworm.

DISTRIBUTION AND LOSSES CAUSED BY WHIPWORMS

This parasite is present in most all areas where swine are raised. It appears, however, that the heaviest infection of swine occurs in the southeastern states. Also, there is ample evidence that the whipworm is increasing in the Corn Belt and in the Southeast. One well-known practicing Iowa veterinarian is authority for the statement that "the whipworm is second only to

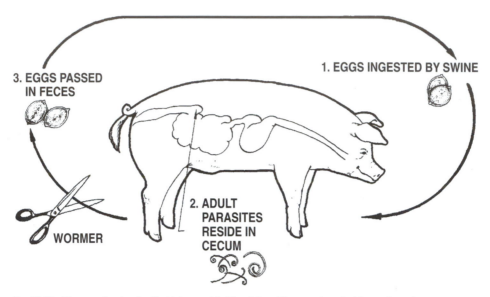

Fig. 15-32. Diagram showing the life history and habits of the whipworm. As noted (see scissors), a proper wormer of choice, complemented by dry, clean lots (or well-drained pastures), is the key to the control and prevention of whipworms.

the large intestinal roundworm (ascarid) in swine" in his particular area.

LIFE HISTORY AND HABITS

The life cycle of the whipworm is simple and direct; that is, no intermediate host is necessary to complete the life cycle. The eggs, which are produced in large numbers, are eliminated with the feces. An infective larva develops in the shell, and swine become infected by swallowing the eggs when feeding on soil where they are present. The eggs hatch in the stomach and intestine, and the larvae make their way to the blind gut, where they grow to maturity in about 10 weeks or longer.

DAMAGE INFLICTED; SYMPTOMS AND SIGNS OF AFFECTED ANIMALS

Infected animals may develop a bloody diarrhea, and in heavy infections the pigs become anemic and dehydrated. In massive infections, growth may be noticeably retarded, and the animals may become weak and finally die. Also, the parasites may cause sufficient damage to enable secondary infections to become established.

PREVENTION, CONTROL, AND TREATMENT

Clean, well-drained pastures, rotation grazing, and plenty of sunlight are an aid to the prevention and control of whipworms. Dichlorvos (Atgard), fenbenda-

zole, and hygromycin B (Hygromix) are effective in whipworm treatment and control programs. The manufacturer's directions for use and withdrawal should be carefully read and followed for the drug of choice.

CHOICE OF WORMER

Knowing what internal parasites are present within an animal is the first requisite to the choice of the proper drug, or anthelmintic. Since no one drug is appropriate or economical for all conditions, the next requisite is to select the right one; the one which, when used according to directions, will be most effective and produce a minimum of side effects on the animal treated. So, coupled with knowledge of the kind of parasites present, an individual assessment of each animal is necessary. Among the factors to consider are age, pregnancy, other illnesses and medications, and the method by which the drug is to be administered. Some drugs characteristically put animals off performance for several days after treatment, whereas others have less tendency to do so. Some drugs are unnecessarily harsh or expensive for the problem at hand, whereas a safe, inexpensive alternative would be equally suitable.

Each swine producer should, in cooperation with the local veterinarian and/or other advisor, evolve with a parasite control program and schedule. It is recommended that several different wormers be used in rotation. Also, a schedule of treatments should be prepared, based on knowledge of the life cycles of the various parasites. Possibly, before an active worming program is begun, the producer and veterinarian should have the feces of pigs in the herd examined for the types of worms present, and then choose the wormer. Often, in slotted floor systems, no worm eggs are found.

The recommendations made in this chapter as to the compounds to use for the control of parasites are intended as guidelines. Furthermore, it is recognized that wormers are constantly being improved, that new ones are becoming available, and that some old ones are banned. So, the producer should consult the local veterinarian relative to the choice of drug to use on the animals at that time.

FORMS OF PESTICIDES

Pesticides for use on animals may be purchased in several forms. The most common are emulsifiable concentrates, dusts, wettable powders, and oil solutions. When treating animals, be sure to use only pesticide formulations that are prepared specifically for livestock.

■ **Emulsifiable concentrates (EC)**—Emulsifiable concentrates, which are probably the most common type of formulation, are solutions of pesticides in petroleum oils or other solvents. An emulsifier has been added so that the solution will mix well with water. On occasion, usually after extended storage, an EC may separate into its various parts; in that case, it should be discarded. An emulsion may also separate if it is allowed to stand after the concentrate has been added to the water; periodic agitation will help prevent this.

■ **Dusts**—Dusts are applied directly to animals in the dry form and cannot be used as sprays.

■ **Wettable powders**—Wettable powders are also dry, but the addition of a dispersing and wetting agent allows them to be suspended in water for application. Continuous agitation of the mixture is important when treating with wettable powders.

■ **Oil solutions**—Oil solutions are pesticides dissolved in oil; no emulsifier is added. These materials are usually ready for use and should not be added to water.

PRECAUTIONS ON THE USE OF PESTICIDES

Agricultural chemicals are as vital to the health of animals as modern medicines are to the health of people. Producers depend on chemicals—insecticides, herbicides, fungicides, and similar materials—to control the pests that attack their animals or damage their feed crops; and, when necessary, they use biologics for the prevention or treatment of animal diseases. However, certain basic precautions must be observed when pesticides are to be used because, used improperly, they can be injurious to humans, domestic animals, wildlife, and beneficial insects. If used properly, pesticides should give satisfactory control and should not leave residues that exceed the tolerances established for any specific chemical. To avoid excessive residues in pork products, follow the label directions carefully with respect to dosage levels and chemical safety precautions. Remember that swine producers are responsible for residues in their own animals and products as well as for problems caused by drift from their property. Other pertinent information follows:

■ **Selecting pesticides**—Always select the formulation and the pesticide labeled for the purpose for which it is to be used.

■ **Storing pesticides**—Always store pesticides in original containers. Never transfer them to unlabeled containers or to feed or beverage containers. Store pesticides in a dry place out of reach of children, animals, or unauthorized persons.

■ **Disposing of empty containers and unused pesticides**—Properly and promptly dispose of all empty pesticide containers. Do not reuse. Break and bury glass containers. Chop holes in, crush, and bury metal containers. Bury containers and unused pesticides at least 18 in. deep in the soil in a sanitary landfill, or dump in a level, isolated place where water supplies will not be contaminated. Check with local authorities to determine specific procedures for your area.

■ **Mixing and handling**—Mix and prepare pesticides in the open or in a well-ventilated place. Wear rubber gloves and clean, dry clothing (respirator device may be necessary with some products). If any pesticide is spilled on you or your clothing, wash with soap and water immediately and change clothing. Avoid prolonged inhalation. Do not smoke, eat, or drink when mixing and handling pesticides.

■ **Applying**—Use only amounts recommended. Apply at the correct time to avoid unlawful residues. Avoid treating pigs younger than specified on the label. Avoid retreating more often than label restrictions. Avoid drift on nearby crops, pastures, livestock, or other nontarget areas. Avoid prolonged contact with all pesticides. Do not eat, drink, or smoke until all operations have ceased and hands and face are thoroughly washed. Change and launder clothing after each day's work.

■ **In case of an emergency**—If you accidentally swallow a pesticide, induce vomiting by taking 1 tablespoonful of salt in a glass of water. Repeat if necessary. Call a doctor.

■ **Withdrawal**—After spraying, dipping, or dusting your animals with pesticides, observe the prescribed number of days interval between the last treatment and slaughter. Refer to the container labels for this information.

IMMUNITY

No discussion of health and disease would be complete without a brief explanation of immunity. When an animal is immune to a certain disease, it simply means that it is not susceptible to that disease. There are two forms of immunity: natural and acquired.

When immunity to a disease is inherited, it is referred to as a natural immunity. For example, when sheep are exposed to hog cholera they never contract the disease because they have a type of natural immunity referred to as species immunity. Likewise, humans are naturally immune to Texas fever. Algerian sheep are said to be highly resistant to anthrax; this constitutes a type of natural immunity called racial immunity.

The body also has the ability, when properly stimulated by a given organism or toxin, to produce antibodies and/or antitoxins. When an animal has enough antibodies for overcoming particular (disease-producing) organisms, it is said to be immune to that disease. This type of immunity is referred to as acquired.

Acquired immunity or resistance is either active or passive. When the animal is stimulated in such manner (vaccination or actual disease) as to cause it to produce antibodies, it is said to have acquired active immunity. On the other hand, if an animal is injected with the antibodies (or immune bodies) produced by an actively immunized animal, it is referred to as an acquired passive immunity. Such immunity is usually conferred by the injection of blood serum from immunized animals, the serum carrying with it the substances by which the protection is conferred. Passive immunization confers immunity upon its injection, but the immunity disappears within 3 to 6 weeks. Young mammals secure passive immunity from the mother's colostrum the first few days following birth.

In active immunity, resistance is not developed until after 1 or 2 weeks; but it is far more lasting, for the animal apparently keeps on manufacturing antibodies. It can be said, therefore, that active immunity has a great advantage. There are exceptions, however—for example, the tetanus antitoxin.

DISINFECTANTS

A disinfectant is a bactericidal or microbicidal agent that frees from infection (usually a chemical agent which destroys disease germs or other micro organisms, or inactivates viruses).

The high concentration of swine and the continuous use of buildings often result in a condition referred to as disease build-up not limited to bacteria. As disease-producing organisms—viruses, bacteria, fungi, and parasite eggs—accumulate in the environment, disease problems can become more severe and be transmitted to each succeeding group of animals raised on the same premises. Under these circumstances, cleaning and disinfection become extremely important in breaking the life cycle. Also, in the case of a disease outbreak, the premises must be disinfected.

Under ordinary conditions, proper cleaning of buildings removes most of the microorganisms, along with the filth, thus eliminating the necessity of disinfection.

Effective disinfection depends on five things:

1. Thorough cleaning before application.
2. The phenol coefficient of the disinfectant, which indicates the killing strength of a disinfectant as compared to phenol (carbolic acid). It is determined by a standard laboratory test in which the typhoid fever germ is used as the test organism.
3. The dilution at which the disinfectant is used.
4. The temperature; most disinfectants are much more effective if applied hot.
5. Thoroughness of application and time of exposure.

Disinfection must in all cases be preceded by a very thorough cleaning, for organic matter serves to protect disease germs and otherwise interferes with the activity of the disinfecting agent.

Sunlight possesses disinfecting properties, but it is variable and superficial in its action. Heat and some of the chemical disinfectants are more effective.

The application of heat by steam, hot water, burning, or boiling is an effective method of disinfection. In many cases, however, it may not be practical.

A good disinfectant should (1) have the power to kill disease-producing organisms, (2) remain stable in the presence of organic matter (feces, hair, soil), (3) dissolve readily in water and remain in solution, (4) be nontoxic to animals and humans, (5) penetrate organic matter rapidly, (6) remove dirt and grease, and (7) be economical to use.

The number of available disinfectants is large because there is no ideal universally applicable disinfectant. Table 15-3 summarizes the limitations, usefulness, and strength of some common disinfectants.

When using a disinfectant, always read and follow the manufacturer's directions.

TABLE 15-3
DISINFECTANT GUIDE[1]

Kind of Disinfectant	Usefulness	Strength	Limitations and Comments
Alcohols (ethyl-ethanol, isopropyl, methanol)	Primarily as skin disinfectant and for emergency purposes on instruments.	70% alcohol—the content usually found in rubbing alcohol.	They are too costly for general disinfection. They are ineffective against bacterial spores.
Boric acid[2]	As a wash for eyes, and other sensitive parts of the body.	1 oz in pt water (about 6% solution).	It is a weak antiseptic. It may cause harm to the nervous system if absorbed into the body in large amounts. For this and other reasons, antibiotic solutions and saline solutions are fast replacing it.
Chlorines (sodium hypochlorate, chlomine-T, chlorine dioxide)	Used for equipment and as deodorants. They will kill all kinds of bacteria, fungi, and viruses, providing the concentration is sufficiently high.	Generally used at about 200 ppm for equipment and as a deodorant.	They are corrosive to metals and neutralized by organic materials.
Cresols (many commercial products available)	A generally reliable class of disinfectant. Effective against brucellosis, swine erysipelas, and tuberculosis.	Cresol is usually used as a 2–4% solution (1 cup/2 gal. of water makes a 4% solution).	Cannot be used where odor may be absorbed, and, therefore, not suited for use around milk and meat.
Formaldehyde (gaseous disinfectant)	Formaldehyde will kill anthrax spores, TB organisms, and animal viruses in a 1–2% solution. It is often used to disinfect buildings following a disease outbreak. A 1–2% solution may be used as a footbath to control foot-rot.	As a liquid disinfectant, it is usually used as a 1–2% solution. As a gaseous disinfectant (fumigant), use 1½ lb of potassium permanganate plus 3 pt of formaldehyde. Also, gas may be released by heating paraformaldehyde.	It has a disagreeable odor, destroys living tissue, and can be extremely poisonous. The bacterial effectiveness of the gas is dependent upon having the proper relative humidity (above 75%) and temperature (above 65°F and preferable near 80°F).
Heat (by steam, hot water, burning, or boiling)	In the burning of rubbish or articles of little value, and in disposing of infected body discharges. The steam "Jenny" is effective for disinfection if properly employed, particularly if used in conjunction with a phenolic germicide.	10 minutes exposure to boiling water is usually sufficient.	Exposure to boiling water will destroy all ordinary disease germs but sometimes fails to kill the spores of such diseases as anthrax and tetanus. Moist heat is preferred to dry heat, and steam under pressure is the most effective. Heat may be impractical or too expensive.
Iodine[2] (tincture)	Extensively used as a skin disinfectant, for minor cuts and bruises.	Generally, used as tincture of iodine, either 2% or 7%.	Never cover with a bandage. Clean skin before applying the iodine. It is corrosive to metals.
Iodophors (tamed iodine)	Primarily used for dairy utensils. Effective against all bacteria (both gram-negative and gram-positive), fungi, and moist viruses.	Usually used as disinfectants at concentrations of 50–75 ppm titratable iodine, and as sanitizers at levels of 12.5–25 ppm. At 12.5 ppm titratable iodine, they can be used as an antiseptic in drinking water.	They are inhibited in their activity by organic matter. They are quite expensive.
Lime (quicklime, burnt lime, calcium oxide)	As a deodorant when sprinkled on manure and animal discharges; or as a disinfectant when sprinkled on the floor or used as a newly made "milk of lime" or as a whitewash.	Use as a dust; as "milk of lime"; or as a whitewash, but use fresh.	Not effective against anthrax or tetanus spores. Wear goggles, when adding water to quicklime.
Lye (sodium hydroxide, caustic soda)	On concrete floors; against microorganisms of brucellosis and the viruses of foot-and-mouth disease. In strong solution (5%), effective against anthrax.	Lye is usually used as either a 2% or a 5% solution. To prepare a 2% solution, add 1 can of lye to 5 gal. of water. To prepare a 5% solution, add 1 can of lye to 2 gal of water. A 2% solution will destroy the organisms causing foot-and-mouth disease, but a 5% solution is necessary to destroy the spores of anthrax.	Damages fabrics, aluminum, and painted surfaces. Be careful, for it will burn the hands and face. Not effective against organism of TB or Johne's disease. Lye solutions are most effective when used hot. Diluted vinegar can be used to neutralize lye.

(Continued)

TABLE 15-3 (Continued)

Kind of Disinfectant	Usefulness	Strength	Limitations and Comments
Lysol (the brand name of a product of cresol plus soap)	For disinfecting surgical instruments used in castrating and tatooing. Useful as a skin disinfectant before surgery, and for use on the hands before castrating.	0.5–2.0%.	Has a disagreeable odor. Does not mix well with hard water. Less costly than phenol.
Phenols (carbolic acid): 1. Phenolics—coal tar derivatives 2. Synthetic phenols	They are ideal general-purpose disinfectants. Effective and inexpensive. They are very resistant to the inhibiting effects of organic residue; hence, they are suitable for barn disinfectioin, and foot and wheel dip-baths.	Both phenolics (coal tar) and synthetic phenols vary widely in efficacy from one compound to another. So, note and follow manufacturer's directions. Generally used in a 5% solution.	They are corrosive, and they are toxic to animals and humans. Ineffective on fungi and viruses.
Quaternary ammonium compounds (QAC)	Very water soluble, ultrarapid kill rate, effective deodorizing properties, and moderately priced. Good detergent characteristics and harmless to skin.	Follow manufacturer's directions.	They can corrode metal. Not very potent in combating viruses. Adversely affected by organic matter.
Sal soda	It may be used in place of lye against foot-and-mouth disease.	10½% solution (13½ oz/1 gal. water).	
Sal soda and soda ash (or sodium carbonate)	They may be used in place of lye against foot-and-mouth disease.	4% solution (1 lb/3 gal. water). Most effective in hot solution.	Commonly used as cleansing agents, but have disinfectant properties, especially when used as a hot solution.
Soap	Its power to kill germs is very limited. Greatest usefulness is in cleansing and dissolving coatings from various surfaces, including the skin, prior to application of a good disinfectant.	As commercialy prepared.	Although indispensable to sanitizing surfaces, soaps should not be used as disinfectants. They are not regularly effective; staphylococci and the organisms which cause diarrheal disease are resistant.

[1]For metric conversions, refer to Appendix.

[2]Sometimes loosely classified as a disinfectant but actually an antiseptic and practically useful only on living tissue.

POISONS

Poisonous plants and feeds should be avoided. The swine producer should know the poisonous plants common to the area, and avoid them. Also, the following poisons should be avoided: ergot, mycotoxins, pitch, excessive copper, fluorine, lead, mercury, nitrates, and selenium. Since poisons are generally eaten, a more detailed discussion of potentially poisonous elements is provided in Chapter 9.

FEDERAL AND STATE REGULATIONS RELATIVE TO DISEASE CONTROL

Certain diseases are so devastating that swine producers cannot protect their herds against invasion. Moreover, where human health is involved, the problem is much too important to be entrusted to individual action. In the United States, therefore, certain regulatory activities in animal-disease control are under the supervision of various federal and state organizations. Federally, this responsibility is entrusted to the following agency:

Veterinary Services
Animal and Plant Health Inspection Service
U.S. Department of Agriculture
Federal Center Building
Hyattsville, Maryland 20782

In addition to the federal interstate regulations, each of the states has requirements for the entry of livestock. Generally, these requirements include compliance with interstate regulations. States usually require a certificate of health or a permit, or both, and additional testing requirements depending upon the class of livestock involved.

Detailed information relative to animal disease control can be obtained from federal and state animal health officials, or from accredited veterinarians in all states. Shippers are urged to obtain such information prior to making interstate shipments of livestock.

■ **Quarantine**—Many highly infectious diseases are prevented by quarantine from (1) gaining a foothold in this country, or (2) spreading. *Quarantine involves (1) segregating and confining of one or more animals in the smallest possible area to prevent any direct or indirect contact with animals not so restrained; or (2) regulating movement of animals at points of entry.* When an infectious disease outbreak occurs, drastic quarantine must be imposed to restrict movement out of an area or within areas. The type of quarantine varies from one involving a mere physical examination and movement under proper certification to the complete prohibition against the movement of animals, produce, vehicles, and even human beings.

■ **Indemnities**—Where certain animal diseases are involved, the swine producer can obtain financial assistance in eradication programs through federal and state sources. Information relative to indemnities paid to owners by the federal government may be secured from the USDA agency, listed earlier in this section, by asking for Chapter I, Subchapter B. Title 9, of the Code of Federal Regulations. Similar information pertaining to each state can be secured by writing to the respective state departments of agriculture.

HARRY S. TRUMAN ANIMAL IMPORT CENTER

A Federal Quarantine Center was authorized in Public Law 91-239, signed by the President on May 6, 1970. A 16.1-acre site for the Center was selected at Fleming Key, near Key West, Florida. In 1978, the Center was named the Harry S. Truman Import Center in honor of former President Truman; and in 1979, the Center opened under the administration of the USDA's Animal and Plant Inspection Service.

The Center holds some 400 head of cattle or other animals at one time, for a 5-month quarantine period. This maximum security station enables American livestock producers to import breeding animals from all parts of the world, while at the same time safeguarding our domestic herds and flocks from such diseases as foot-and-mouth disease, rinderpest, piroplasmosis, and others.

QUESTIONS FOR STUDY AND DISCUSSION

1. Why are books and bulletins describing diseases and their control of value to swine producers?

2. Define the following pharmaceutical terms: (a) over-the-counter drugs, (b) prescription drug, and (c) extra-label drugs.

3. What is the normal temperature, pulse rate, and breathing rate of hogs? How would you determine each?

4. Select a specific farm (either your own or one with which you are familiar) and outline (in 1, 2, 3 order) a program of Life Cycle Herd Health.

5. How should feeder pigs be handled in order to get them off to a good start?

6. Detail the provisions for producing specific pathogen-free (SPF) pigs.

7. Define the all-in/all-out system of swine management. Why and how is it used?

8. Define the segregated early weaning (SEW) system. Why and how is it used?

9. Give the (a) symptoms and signs, and (b) the cause, prevention, and treatment of each of the following swine diseases:

    a. Atrophic rhinitis

    b. Enteric colibacillosis (baby pig scours)

    c. Porcine parvovirus

    d. Porcine reproductive and respiratory syndrome (PRRS)

    e. Porcine stress syndrome (PSS)

    f. Transmissible gastroenteritis (TGE)

10. Explain why, even with antibiotics, drugs, and vaccines, a program of sanitation is important to herd health.

11. Some swine diseases have declined in prominence while others have increased. How do you explain this? Give some examples.

12. Assume that a specific parasite (you name it) has become troublesome in your herd. What steps would you take to meet the situation (list in 1, 2, 3 order; be specific)?

13. What is the McLean County System of Sanitation?

14. Why would you expect to see less of certain parasitic infestations with confinement housing on slotted flooring?

15. Recommend a treatment for lice and mites.

16. How are screwworms controlled?

17. Is it likely that, in the United States, a person will contract trichinosis? Justify your answer.

18. Producers have been criticized for their use of chemicals. To eliminate reasons for the criticism, what precautions should producers take when using a pesticide in their herd health program?

19. Explain the difference between natural immunity, active immunity, and passive immunity.

20. Assume that you have, during a period of a year, encountered swine death losses from three different diseases. What kind of disinfectant would you use in each case?

21. Why are certain regulatory activities in animal disease control under the supervision of various federal and state agencies?

22. How can federal and state regulatory officials be of assistance to the individual swine producer?

SELECTED REFERENCES

Title of Publication	Author(s)	Publisher
Animal Diseases, Yearbook of Agriculture, 1956	U.S. Department of Agriculture	Superintendent of Documents, Washington, DC, 1956
Animal Health: A Layman's Guide to Disease Control	J. K. Baker W. J. Greer	Interstate Publishers, Inc., Danville, IL, 1992
Animal Parasitism	C. P. Read	Prentice-Hall, Inc., Englewood Cliffs, NJ, 1972
Animal Science, Ninth Edition	M. E. Ensminger	Interstate Publishers, Inc., Danville, IL, 1991
Diseases of Swine, Seventh Edition	Ed. by A. D. Leman, *et al.*	The Iowa State University Press, Ames, IA, 1992
Diseases Transmitted from Animals to Man	Ed. by W. T. Hubbert, *et al.*	Charles C Thomas, Publisher, Springfield, IL, 1975
Farm Animal Health and Disease Control	J. K. Winkler	Lea & Febiger, Philadelphia, PA, 1982
Keeping Livestock Healthy, Yearbook of Agriculture, 1942	U.S. Department of Agriculture	Superintendent of Documents, Washington, DC, 1942
Merck Veterinary Manual, The, Seventh Edition	Ed. by C. M. Fraser, *et al.*	Merck & Co., Inc., Rahway, NJ, 1991
Pork Industry Handbook		Cooperative Extension Service, University of Illinois, Urbana-Champaign, IL, updated as needed
Stockman's Handbook, The, Seventh Edition	M. E. Ensminger	Interstate Publishers, Inc., Danville, IL, 1992
Swine Diseases, An Outline Of	R. P. Cowart	Iowa State University Press, Ames, IA, 1995

In addition to these selected references, valuable publications on different subjects pertaining to swine diseases, parasites, disinfectants, and poisonous plants can be obtained from the following sources:

1. Division of Publications
   Office of Information
   U.S. Department of Agriculture
   Washington, D.C. 20250

2. Your state agricultural college.

3. Several biological, pharmaceutical, and chemical companies.

Murphy Farms swine buildings and feed storage facilities, with cattle in the foreground grazing pasture irrigated from swine lagoon effluent. Murphy Farms, with headquarters at Rose Hill, North Carolina, the nation's largest pork producer, produces 4 million pigs a year. (Courtesy, Dr. Garth W. Boyd and Ms. Rhonda Campbell, Murphy Farms, Rose Hill, NC)

16

BUILDINGS AND EQUIPMENT FOR SWINE

Fig. 16-1. A complete farrow-to-finish confinement facility. (Courtesy, Kansas State University, Manhattan)

Until about 1960, pastures, dry-lots, and portable houses were an important part of most U.S. swine production systems. At this point and period of time, specialized commercial pork producers evolved. With their advent, came recognition that a good environment for swine is essential for maximum comfort and optimal performance. With their advent also came larger units, confinement production, and specialized swine buildings for specialized systems.

Today, swine producers have a multitude of choices from which they may select the system, along with the buildings and equipment, to match their management, labor, capital, and environment. Then, any of their choices can be engineered and designed to provide the maximum comfort and optimal performance of the animals. A brief relative to each of the most common specialized swine buildings in confinement operations follows.

■ **Farrowing house**—Central farrowing houses are generally environmentally controlled. In comparison with portable houses on pasture, central farrowing houses require more capital and a higher level of management, but the labor requirements are reduced and the comfort and safety of the sow and pigs are enhanced.

■ **Nursery**—At weaning, pigs are usually moved to a nursery, where they remain until they weigh 40 to 60 lb. This facility is kept dry, warm, and free of drafts. Generally, slotted floors are used to help separate the pigs from their waste and to keep the floors dry, thereby improving sanitation and aiding in disease control. Decking nursery pigs by raising them off the floor and allowing the cold air beneath the deck makes for greater comfort.

■ **Growing and finishing facilities**—Pigs are usually moved from the nursery when they weigh from 40 to 60 lb. In a farrow-to-finish operation, they are moved to a growing-finishing facility on the same farm. In an operation that produces feeder pigs, they are marketed when moved from the nursery. Pigs are generally considered to be in the growing phase until they reach 120 lb, and in the finishing phase from 120 lb to market weight of 220 to 250 lb.

The four types of confinement swine buildings used for growing-finishing pigs are: (1) totally enclosed and environmentally controlled, (2) modified open front, (3) double curtain, and (4) open front with an outside concrete apron.

The totally enclosed and environmentally controlled building is the most costly of the four, but it provides the greatest control over temperature, humidity, and insects.

The modified open front is so named because the front (generally the south side) may be opened for

Fig. 16-2. An open front building with concrete apron used for growing-finishing pigs. (Courtesy, E. J. Stevermer, Iowa State University, Ames)

summer ventilation, but completely closed during winter. This type of building is naturally ventilated, alleviating any need for mechanical ventilation.

Double-curtain buildings are the latest design of growing-finishing buildings. They are usually placed perpendicular to prevailing winds. They have automatically controlled curtains on both sidewalls. Double-curtain buildings maintain proper temperatures and provide fresh air by using a combination of mechanical and natural ventilation.

An open-front building with an outside concrete apron is the lowest cost of the four types of units. But the comfort and performance of the growing-finishing pigs that they house may be affected by weather.

■ **Breeding and gestation facilities**—Usually, the breeding herd (gestating sows and gilts, and the herd boars) is the last group to be moved inside. Typically, the order of confinement is as follows: (1) farrowing house, (2) nursery, (3) growing-finishing unit, and (4) breeding/gestating gilts and sows, and herd boars.

In all walks of life, people expect a return on their investments; and swine producers are people. Thus the cost associated with an empty farrowing crate or finishing pen must be recovered by the production from those crates or spaces which are occupied. All buildings must be utilized at or near capacity for maximum return on investment.

Hand-in-hand with confinement operations, waste disposal, and animal welfare have plagued swine producers. To cope with these problems, swine producers and researchers have worked feverishly to evolve with new building and equipment designs to alleviate these problems. In the meantime, outdoor production has taken a new name and a new look; it's called *outdoor intensive swine production.*

ENVIRONMENTAL CONTROL

Most animals are able to control their environment to a large extent. But in terms of coping with heat or cold, pigs are poorly equipped. They possess very little hair for protection against the cold; moreover, piglets—with their sparse hair coat, large surface area to body weight ratio, and limited fat reserves—are very sensitive to cold. High temperatures will reduce performance and lower fertility. In confinement rearing, pigs must rely on their caretakers to modify their environment. Pigs respond quickly to environmental stresses, and often their response is counter-productive.

The primary reason for having swine buildings is to modify the environment. Properly designed barns and other shelters, shades, insulation, ventilation, heating, and air conditioning can be used to approach the environment that producers desire. Naturally, the investment in environmental control facilities must be balanced against the expected increased returns; and there is a point beyond which further expenditures for environmental control will not increase returns sufficiently to justify the added cost. This point of diminishing returns will differ between sections of the country, between animals of different ages, and between operators. For example, higher expenditures for environmental control can be justified for farrowing units than for gestating sow facilities. Also, labor and feed costs will enter into the picture.

Normal metabolic processes of the body result in the production of heat and water vapor. These two products are important considerations when designing swine buildings.

HEAT PRODUCTION OF SWINE

Heat produced by swine varies with age, body weight, diet, activity, ambient temperature, and humidity at high temperatures. However, Table 16-1 may be used as a guide.

As noted, Table 16-1 gives both "total heat production" and "sensible heat production." "Total heat production" includes both sensible heat and latent heat combined. Latent heat refers to the energy involved in a change of state and cannot be measured with a thermometer; evaporation of water and respired moisture from the lungs are examples. Sensible heat is that portion of the total heat, measurable with a thermometer, that can be used for warming air, compensating for building losses, etc.

TABLE 16-1
HEAT PRODUCTION OF SWINE[1]

Heat Source	Unit		Heat Production, Btu/Hr[2]			Heat Production, Kcal/Hr[2]		
			Temperature	Total	Sensible	Temperature	Total	Sensible
	(lb)	*(kg)*	*(°F)*			*(°C)*		
Sow & litter (3 weeks after farrowing)	400	*181.4*	—	2,000	1,000	—	*504.0*	*252.0*
Finishing hog	200	*90.7*	35	860	740	*2*	*216.7*	*186.5*
			70	610	435	*21*	*153.7*	*109.6*

[1]Adapted by the authors from *Agricultural Engineers Yearbook*, St. Joseph, Michigan, ASAE Data Sheet D-249.2, p. 424.

[2]One Btu (British thermal unit) is the amount of heat required to raise the temperature of 1 lb of water 1°F, while 1 kcal (kilocalorie) is the amount of heat required to raise the temperature of 1 kg of water 1°C.

VAPOR PRODUCTION OF SWINE

Pigs give off moisture during normal respiration; and the higher the temperature, the greater the moisture. This moisture should be removed from buildings through the ventilation system. Most building designers govern the amount of winter ventilation by the need for moisture removal. Also, cognizance is taken of the fact that moisture removal in the winter is lower than in the summer; hence, less air is needed. However, lack of heat makes moisture removal more difficult in the wintertime. Fig. 16-3 gives the information necessary for determining the approximate amount of moisture to be removed.

Since ventilation also involves a transfer of heat, it is important to conserve heat in the building to maintain desired temperatures and reduce the need for supplemental heat. In a well-insulated building, mature animals may produce sufficient heat to provide a desirable balance between heat and moisture; but young animals will usually require supplemental heat. The major requirement of summer ventilation is temperature control, which requires moving more air than in the winter.

RECOMMENDED ENVIRONMENTAL CONDITIONS FOR SWINE

The comfort of animals (or humans) is a function of temperature, humidity, and air movement. Likewise, the heat loss from animals is a function of these three items. Additionally, when considering buildings, there are certain general requisites that should always be considered; among them, reasonable construction and maintenance costs, reduced labor, and utility value.

Temperature, humidity, and ventilation recommendations for different classes and ages of swine are given in Table 16-2. This table will be helpful in obtaining a satisfactory environment in confinement buildings, which require careful planning and design. Features of buildings which require emphasis include temperature and humidity control, insulation, vapor barrier, heat, ventilation, and automation.

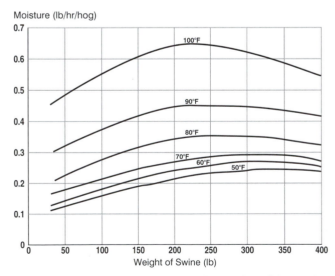

Fig. 16-3. Moisture (vapor) produced by different weights of pigs at various temperatures. To convert pounds to kilograms divide by 2.205, and to convert degrees Fahrenheit to degrees centigrade subtract 32 and multiply by 5/9.

TABLE 16-2
RECOMMENDED ENVIRONMENTAL CONDITIONS FOR SWINE

Hog Unit	Temperature				Acceptable Humidity	Commonly Used Ventilation Rates[1]			
	Comfort Zone		Optimum			Winter[2]		Summer	
	(°F)	(°C)	(°F)	(°C)	(%)	(cfm)	(m³/min)	(cfm)	(m³/min)
Sow and litter	60–70	15–20	65	17	60–85	80	2.2	210	5.9
Newborn pigs (brooder area)	85–95	29–35	90	32	60–85	—	—	—	—
Growing-finishing hogs									
20–40 lb (9–18 kg)	70–75	21–24	75	24	60–85	15	0.4	36	1.0
40-100 lb (18–45 kg)	65–70	18–21	65	18	60–85	20	0.6	48	1.3
100–150 lb (45–68 kg)	45–75	7–16	60	16	60–85	25	0.7	72	2.0
150–210 lb (68–95 kg)	45–75	7–16	60	16	60–85	35	1.0	100	2.8
Gilt, sow, or boar									
200–250 lb (91–113 kg)	45–75	7–16	60	16	60–85	35	1.0	120	3.4
250–300 lb (113–136 kg)	45–75	7–16	60	16	60–85	45	1.1	180	5.0
300–500 lb (136–227 kg)	45–75	7–16	60	16	60–85	45	1.3	250	7.0

[1]Generally two different ventilating systems are provided: one for winter, and an additional one for summer. In practice, in many buildings, added summer ventilation is provided by opening (1) barn doors, and (2) high-up hinged walls.

[2]Provide approximately one-fourth the winter rate continuously for moisture removal.

TEMPERATURE AND HUMIDITY

During cold weather, the necessary added warmth should be provided through properly constructed buildings and artificial means (heated buildings, brooders, etc.); and summer coolness should be enhanced by shades, wallows, and sprinklers.

Rate of gain and feed efficiency are lowered when swine must endure temperatures appreciably below or above the comfort zone. Table 16-3 shows the feed consumption, growth rate, and feed efficiency of growing-finishing pigs housed at temperatures ranging from cold stress to heat stress. Cold pigs eat more and gain less, while heat stressed pigs eat less and gain less. From the standpoint of promoting maximum efficiency, therefore, it is desirable to provide animal housing, shades, wallows, etc., which eliminate extreme changes in environment temperature.

Humidity influences are tied closely to temperature effects. The relative humidity should be within the range of 60 to 85%. The most desirable temperatures and humidities for different classes and ages of hogs are given in Table 16-2.

TABLE 16-3
EFFECT OF TEMPERATURE ON INTAKE, GROWTH RATE, AND
EFFICIENCY OF ENERGY CONVERSION FOR FINISHING PIGS[1]

Temperature		Caloric Intake[2]	Growth Rate		Caloric Value of Gain[3]	Caloric Efficiency[4]
(°F)	(°C)	(kcal/day)	(lb/day)	(kg/day)	(kcal)	(%)
32	0	15,377	1.2	0.54	2,991	19.4
41	5	11,404	1.2	0.53	2,936	25.7
50	10	10,616	1.8	0.80	4,432	41.7
59	15	9,554	1.7	0.79	4,376	45.8
68	20	9,766	1.9	0.85	4,709	48.2
77	25	7,976	1.6	0.72	3,988	50.1
86	30	6,703	1.0	0.45	2,493	37.1
95	35	4,579	0.7	0.31	1,717	37.4

[1]Adapted by the authors from Ames, D. R. "Thermal Environment Affects Livestock Performance," Bioscience, Vol. 30, p. 457.

[2]Digestible energy.

[3]The caloric value for each gram of gain was estimated at 5.54 kcal.

[4]Calculated from the division of the caloric value of the gain by the caloric intake.

INSULATION

The term *insulation* refers to materials which have a high resistance to the flow of heat. Such materials are commonly used in the walls and ceiling of hog houses. Proper insulation makes for a more uniform temperature—cooler houses in the summer and warmer houses in the winter—and makes for a substantial fuel saving in heated houses.

Adequate amounts of insulation should be installed during initial construction because it is easier and cheaper to install at that time. Although it will be more difficult and costly, existing inadequately insulated buildings should be better insulated. The energy savings will be long term.

Fig. 16-4 illustrates the insulative value—the R value—of some construction materials.

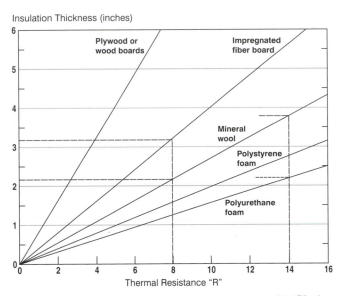

Fig. 16-4. The relationship between insulation thickness and its "R" value. One in. equals 2.54 cm. (Courtesy, USDA)

Polystyrene foam, polyurethane foam, and mineral wool provide the greatest insulation with the least thickness. But the insulating values of the other materials used in wall and ceiling construction must also be considered.

The amount of insulation depends on a number of factors, including climate. In mild and moderate climates, the walls can be about R9 to R12, and the ceilings about R12 to R16. In cold climates, the walls should be around R14 and the ceilings about R23.

VAPOR BARRIER

There is much moisture in hog houses; it comes from waterers, wet bedding, the respiration of the animals, and from the feces and urine. When the amount of water vapor in the house is greater than in the outside air, the vapor will tend to move from inside to outside. The moisture enters the wall and moves outward, condensing when it reaches a cold enough area. Condensed water in the wall greatly reduces the value of the insulation and may damage the wall. Since warm air holds more water vapor than cold air, the movement of vapor is most pronounced during the winter months. The effective way to combat this problem in a hog house is to use a vapor barrier with the insulation. It should be placed on the warm side or inside of the house. Common vapor barriers are 4-mil plastic film and some of the asphalt-impregnated building papers.

HEAT

In many hog-producing areas, buildings need supplemental heat. Table 16-4 shows the amount of heat produced by hogs of different weights and the supplemental heat needs. Even with insulation, about 60% of this body heat is lost by ventilation to remove moisture; without insulation the heat loss is more.

VENTILATION

Ventilation refers to the changing of air. Its purpose is (1) replacement of foul air with fresh air, (2) removal of moisture, (3) removal of odors, and (4) removal of excess heat in hot weather. Hog houses should be well ventilated, but care must be taken to avoid direct drafts and coldness. Good swine house ventilation saves feed and helps make for maximum production.

All swine building ventilation is based on using the difference in static pressure inside and outside the building to bring in fresh air and exhaust foul air.

Five factors are essential for good ventilation: (1) fresh air moving into the hog house (inlets), (2) insulation to keep the house temperatures warm, (3) supplemental heat in the winter in cold areas, (4) vapor barrier, and (5) removal of moist air (outlets). The importance of the latter point becomes evident when it is realized that, through breathing, (1) each sow and litter adds about 1 gal of water per day, and (2) 50 head of 125-lb pigs give off about 1 gal of moisture per hour.

In most hog houses, easily controlled electric fans do the best job of putting air where it is needed.

Fans are rated in cubic feet per minute (cfm) of air they move. By selecting and using two or three fans, each of a different size, the needed variable rate, or flexibility, of ventilation can be achieved.

Provision should be made for one or more of the following, should there be a power or equipment failure in a closed building:

TABLE 16-4
THE HEAT PRODUCTION OF SWINE AND THEIR SUPPLEMENTAL HEAT NEEDS
IN COLD AND MILD CLIMATES[1]

Hog Unit	Building Temperature		Approximate Heat Production[2]	Supplemental Heat Needs			
				Slotted Floor		Bed-scraped Floor	
				Cold	Mild	Cold	Mild
	(°F)	(°C)	- (Btu/hr) -				
Sow & litter							
400 lb (181 kg)	60	16	2,100			2,000[3]	1,400[3]
	80	27		1,500	1,000		
Pigs							
20–40 lb (9–18 kg)	70	21	330	275[3]	125[3]	300[3]	150[3]
40–100 lb (18–45 kg)	45–75	7–24	410	250	100	500	200
100–150 lb (45–68 kg)	45–75	7–24	500	250	100	500	200
150–210 lb (68–95 kg)	45–75	7–24	600	250	100	500	200
Sow or boar, limit-fed							
200–250 lb (91–113 kg)	45–75	7–24	680	250	100	500	200
250–300 lb (113–136 kg)	45–75	7–24	760	250	100	500	200
300–500 lb (136–227 kg)	45–75	7–24	980	250	100	500	200

[1]Mild climates are seldom colder than 0°F (–18°C), while cold climates are often colder than 0°F.

[2]At 60°F (16°C). Depending upon the room temperature, about two-thirds of this total heat production is sensible heat that is available for warming air, building heat losses, and evaporating water.

[3]In addition to this amount, it is necessary to provide brooder heat for small pigs.

1. A battery-powered alarm which sounds in the home of the caretaker if power goes off.

2. A standby generator to supply power.

3. Solenoid or other electrically closed doors which open if power goes off.

4. Manually opened doors and windows.

5. A place to move the hogs outside if the building cannot be opened sufficiently.

The amount of ventilation air required depends upon the (1) inside-of-house temperature desired, (2) outside air temperature, (3) relative humidity, (4) number and size of animals in the building, and (5) amount of insulation. Since size of hog is a factor, and since the comfort zone of baby pigs and finishing hogs differs, it is obvious that at least two ventilation tables are needed; one for the farrowing house, and the other for the finishing house. Also, ventilation needs are different in the winter than in the summer. Recommended ventilation rates are given in Table 16-2.

A common misconception is that heating costs will be reduced when the inside temperature is dropped by lowering the thermostat setting. Losses through the structure will be less, but the ventilation loss will be much greater.

Incoming cold air must be heated in order to increase its moisture-holding capacity (see Fig. 16-5). Cold air has little capacity for picking up moisture. In general, the moisture-holding capacity doubles as the air temperature is raised 20°F.

An example to illustrate the interrelationships between air temperatures, ventilation rate, and heating requirements follows:

A farrowing house maintained at 70°F and 75% relative humidity (RH) while the outside air temperature is at 30°F and 75% RH requires a ventilation rate of approximately 25 cubic feet of air per minute (cfm) to remove the 1 lb of moisture produced per hour by the sow and litter. If the inside temperature is maintained at 50°F and 75% RH while the outside remains at 30°F and 75% RH, the ventilation required to remove the pound of moisture is 70 cfm. To heat the 25 cfm from 30 to 70°F requires 1,160 Btu per hour, but to heat the 70 cfm from 30 to 50°F requires 1,625 Btu per hour. With the minimum ventilation the heat requirement per day is 27,840 Btu or 0.30 gal of LP gas or equivalent. But in the cooler room with the higher ventilation rate the LP gas requirement is 0.42 gal per day, or an increase of 40% in the heating requirement. Hence, the best approach to energy efficiency in enclosed, well-insulated confinement buildings is to maintain the temperature in the 70 to 80°F range and

Pounds of Water per
1,000 lb of Dry Air

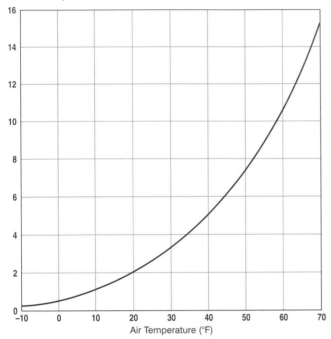

Fig. 16-5. Influence of air temperature on its water-holding capacity. The water-holding capacity of air increases with rising temperature.

Fig. 16-6. A naturally ventilated nursery-finishing unit with nine roof ventilators and adjustable vent doors on all four sides. (Courtesy, *National Hog Farmer*, St. Paul, MN)

provide good air distribution with precise control over the winter ventilation rate for moisture removal. Adjustments may be necessary in the ventilation rate for gas removal depending upon waste management practices employed in the building.

■ **Natural ventilation**—Swine producers continue to have interest in naturally ventilated buildings, as well as in environmentally controlled buildings, primarily because of their significantly lower construction and operating costs. Because no attempt is made to regulate temperature, the costs of heavy insulation, tight fitting doors and windows, and a mechanical ventilation system are averted. Of course, buildings which for management reasons must be maintained at temperatures above winter levels are not suited to natural ventilations; for example, farrowing barns.

Naturally ventilated buildings can be successfully used for swine growing-finishing buildings and gestation units.

Naturally ventilated buildings are mainly a shell to protect animals from rain and snow, and to protect the building contents. Winter inside temperatures will often be within 3 to 10°F of outside temperatures. Thus, such buildings are often referred to as cold confinement livestock buildings.

A naturally ventilated building has a continuous opening at the high point (normally the ridge) of the building for air exhaust and continuous openings or inlets along the long sidewalls of the building for fresh air. The size of these openings is based on rules of thumb or experience. Air entering along the sidewalls (normally under the eaves) of the building is warmed by the heat from the animals in the building and picks up moisture as it rises toward the ridge. The continuous open ridge allows this warm, moist air to escape, thus completing the air exchange process.

During the warm weather, the building should serve mainly to keep rain out and act as a sunshade. Large, continuous openings in the sidewalls allow summer breezes to blow through the building.

Typical naturally ventilated buildings can be divided into three types as follows:

1. **Open front**—These buildings have one long side completely open at least one-half the height of the sidewall. The open side faces away from the direction of prevailing winter winds, normally to the south or southeast.

2. **Enclosed**—These buildings have all sides closed but provide continuous eave openings and large doors or vent panels for summer conditions. Enclosed naturally ventilated buildings offer more protection from wind and precipitation than open-front buildings.

■ **Pressure, exhaust, and tunnel ventilation**—In *pressure ventilation*, fresh air is forced in and stagnant air is allowed to flow out of the building through vents.

In *exhaust ventilation*, stagnant air is forced out of the building by fans and fresh air is drawn into the building through appropriately spaced openings.

In *tunnel ventilation*, air is pushed from one end of the building through curtained openings on the opposite end. Standard designs have curtains on three sides, with a bank of large fans at one end to push air through the building.

AUTOMATION

Automation is a coined word meaning the mechanical handling of materials. Swine producers automate to lessen labor and cut costs. Confinement buildings lend themselves to automating, because of their compactness. Obviously, the feed and water facilities of confinement houses should be automated. The extent and type of automation depends largely on the production system.

PLANS AND SPECIFICATIONS

No attempt will be made in this chapter to present detailed buildings and equipment plans and specifications. Rather, it is desired merely to convey suggestions regarding some of the desirable features of buildings and equipment in use in various parts of the country. For detailed plans and specifications for a particular locality, the swine producer should (1) study successful buildings and equipment on neighboring hog farms; (2) consult the local county agricultural agent, vocational agriculture instructor, or lumber dealer; and/or (3) write to the state college of agriculture.

Nevertheless, the following items should be considered when checking into building plans:

1. Be sure drainage is away from the farm home and other buildings. Pasture or feedlot drainage should not enter waterways leading to streams or lakes.

2. Always leave room for more buildings to be located properly with respect to feed storage and roads.

3. Be sure facilities are located properly for efficient snow control. Locate facilities down-wind from homes to minimize odor problems.

4. Plan for the necessary movement of animals, operators, and equipment.

5. Be certain of the accessibility of water and electricity.

BUILDING SYSTEMS

Swine move through a sequence of production steps; as a result, one building does not work alone but it is part of a system. A complete farrow-to-finish confinement system generally has one or more buildings devoted to each of the following phases (with the number of buildings determined by the size of the buildings and the number of hogs):

1. Farrowing building.
2. Nursery unit.
3. Growing-finishing facility.

4. Breeding/gestation and herd boar facility.

In a feeder pig system, the feeder pigs are marketed at the end of the nursery period, when they weigh 40 to 60 lb. Hence, growing-finishing facilities are not needed.

In a growing-finishing system, feeder pigs are bought; hence, only growing-finishing facilities are needed.

FARROWING BUILDINGS

Fig. 16-7. Interior of modern farrowing house. (Photo by J. C. Allen and Son, West Lafayette, IN)

Farrowing buildings as used in specialized confinement operations are for the care and protection of sows and baby pigs during farrowing and until weaning.

The design and construction of the farrowing building should be such as to provide optimum environmental control, minimum pig losses, maximum labor efficiency, and satisfactory manure handling. To this end, the following features are being incorporated in modern farrowing houses:

1. **Floors.** Slotted floors are commonly used in farrowing houses. Studies have demonstrated that slotted floors save more labor in farrowing than in any other phase of confinement production.

Coated metal, concrete, triangular bars, or woven wire slats are generally used.

A 3-in. slat with $\frac{3}{8}$-in. spacing appears to be most popular for farrowing buildings.

2. **Farrowing crates (or stalls).** A variety of designs and sizes of farrowing crates or stalls are available. Those with a pit under the entire crate are popular. The important thing is that the farrowing crate include a brooder area about 18 to 24 in. wide for the baby pigs.

Alleys should be provided for easy access of sows to and from crates, and of the operator to the brooder area.

Where slats are used, they should be covered for a few days after farrowing. Slatting can be complete or partial.

3. **Temperature.** The brooder temperature for newborn pigs should be maintained at 90 to 95°F with a 2° drop per day to about 70°F.

Winter environment is more critical inside a slotted floor farrowing house than in a house with a solid floor. Air temperatures should be warmer, and drafts are more critical.

Many commercial producers use space heat to maintain up to 80°F at farrowing time.

NURSERIES

Fig. 16-8. Pigs in nursery. (Courtesy, Farmers Hybrid Co., Des Moines, IA)

A swine nursery unit is for weaned pigs, which are generally weaned at 10 days to 4 weeks in the United States.

The following considerations should be given to nursery units:

1. **Temperature.** At weaning, the temperature in the nursery should be maintained at 80 to 90°F for the comfort of the piglets.

2. **Decking.** Decking refers to the housing of weaned pigs in small cages or pens in single, double, or triple tiers. Single decks are called flat or raised decks. Decking was developed for early weaned programs, reduced operating cost, better environmental control, and increased stocking density.

Decking got the pigs off the cold, wet floors. But they were not practical. Producers didn't object to lifting 12 to 15 lb pigs into a deck, but they objected to lifting 40 to 50 lb pigs.

Fig. 16-9. Early-weaned pigs in double-deck nursery pens. (Courtesy, University of Illinois, Champaign-Urbana)

3. **Floors.** Some type of slotted flooring is recommended. This may be either a partially or a totally slotted floor. In pig nurseries, coated metal, triangular bars, or woven wire are desirable because of their cleaning characteristics.

4. **Feeders.** If the nutritious, scientifically formulated, and costly baby pig diet ends up in the manure pit or crammed in inaccessible corners of the feeder instead of in the pig's stomach, growth, feed efficiency, and health will suffer. Researchers at the University of Illinois reported that nursery feeder waste ranges from 2 to 11%.[1] Based on this study, the scientists prepared a checklist of desirable design features of nursery feeders; among which, the following were high on the checklist:

a. The feeding space should allow the pig to adopt its normal eating stance and carry out its normal eating motions.

b. The feed-delivery mechanism should prevent feed buildup in the trough, yet not restrict intake.

c. The delivery mechanism should be easily adjustable.

d. The agitator design should be easily mastered by the weanling pig.

e. The design should allow the pig to swallow meal with its mouth in or above the feed trough.

f. The feed delivered by the agitator should be readily accessible to the pig at or near the floor level.

g. Any feed disturbed by eating movements should automatically return by gravity to the accessible area.

---

[1]Taylor, I. and S. Curtis, *National Hog Farmer*, Vol. 34, No. 6, May 15, 1989, p. 18.

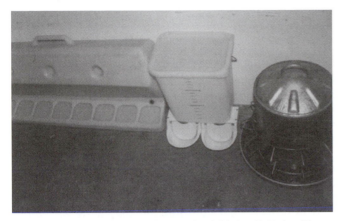

Fig. 16-10. Nursery feeders. (Courtesy, Land O Lakes, Ft. Dodge, IA)

h. There should be no blind corners.

i. Pigs should not be able to climb into, or become trapped by, the feeder.

MODULAR NURSERIES

Modular nurseries are self-contained buildings which are constructed by the manufacturing plant before being delivered to the farm. The producer is generally responsible for preparing the delivery site and installing connections for water, sewer, and electricity. Among the advantages of modulars over conventional facilities built on site, or stick-built, is speedy delivery. Several modular companies will guarantee delivery within six weeks or less from the time a contract is signed. Stick-built may take several months. Additionally, modulars' second strong selling point is that they are portable—they can be moved easily and quickly.

GROWING FACILITIES

Growing is considered to be that span in a pig's life from about 40–60 to 120 lb. The temperature needs of growing pigs are not as critical as those of newborn pigs. Growing pens may be any one of the following:

1. Pens in a separate growing building.

2. Pens in a combination growing-finishing building, with the pigs separated by age and weight.

3. Full sized finishing pens, designed to accommodate both the growing and finishing stages.

Moreover, a variety of building styles and floor plans are available, from totally enclosed or environmentally controlled, to modified open front, to open front-outside apron. Floors may be totally or partially slotted with appropriate manure storage and disposal.

FINISHING FACILITIES

Finishing refers to that time in a pig's life span from about 120 to 250 lb, or market weight. During this time the environmental temperature is less critical than at any other stage. Finishing pigs grow the fastest and with the least feed at about 55°F. Space requirements for finishing pigs vary according to pig size and type of pen floor—bedded, solid, or slotted. The finishing of pigs can be successfully accomplished in an open front building, or in complete confinement. Often, the growing-finishing facility is one unit. When solid floors are used, they should be sloped ½ to 1 in. per foot to aid manure handling and cleanliness. Totally slotted floors with appropriate manure storage and disposal are commonly used in confinement operations.

Fig. 16-11. A modular nursery unit for early-weaned pigs. (Courtesy, Iowa State University, Ames)

Fig. 16-12. Exterior view of an automated curtain opening on the side of a finishing unit. Note the plastic screen over the openings to keep birds out. (Courtesy, Iowa State University, Ames)

GESTATING SOW HOUSING

Total confinement of gestating sows has the following **advantages**:

1. **Less labor.** The labor requirements are reduced in confinement, primarily in handling, cleaning, and feeding.
2. **Higher land use.** Pastures may be used for crop production, thereby (a) saving fencing, and (b) making for higher return from the land.
3. **Better pollution control.** It makes for better control of mud, dust, and manure.
4. **Improved parasite control.** It makes for improved control of internal and external parasites.
5. **Better supervision.** It makes for better supervision of the herd at breeding time.
6. **Improved caretaker comfort.** It makes for improved operator comfort and convenience.
7. **Facilitates individual feeding.** Tying (tethering) individual sows in stalls has been common practice in Europe for many years. Now the practice of keeping gestating sows in total confinement and individually feeding is common in the United States. Among its advantages are:

    a. It allows limited-feeding of each sow.
    b. It discourages bossism and fighting among sows.
    c. It makes it possible to control dosages of feed additives.
    d. It provides facilities (stalls) for artificial breeding and other sow care.

Three types of individual feeding arrangements are being used; namely—

    a. **Loose feeding stalls.** Loose feeding stalls are generally about 20 in. wide and 8 ft long. Usually, the back end is open, although a gate may be used to shut the sows in.
    b. **Tie stalls.** In this arrangement, sows are tied in their individual stalls with a strap. A tie-ring for the tether is centered in the floor under each sow's neck, about 8 in. behind the feed trough.

Tie stalls should be about 24 in. wide, from 3 to 7 ft long, and about 30 in. high. When in the stalls, the sows should be able to see each other. This applies to both loose and tie stalls.
    c. **Individual pens or crates.** This refers to small, individual, enclosed pens, in which one sow is kept during the gestation period.

Not all confinement handled gestating sows are individually fed. Some of them are, and will likely continue to be, group fed—either in a separate feeding area, or in their regular pen.

On some confinement swine establishments, gestating sows are kept in open-front houses; in others, completely enclosed houses are used. A brief discussion of each follows:

1. **Open-front houses for gestating sows.** Some producers provide about 15 sq ft per gestating sow in open-front houses (cold housing), which they divide into pen areas that will hold 10, 20, or 30 sows. The sows require bedding, and manure is handled as solids. Feeders or feed stalls are placed outside the shelter on a concrete floor.
2. **Enclosed houses for gestating sows.** Other producers are building completely enclosed (warm) buildings for gestating sows, which allow year-round climate moderation. Such buildings should be well insulated, have vapor barriers installed in all walls and ceilings, and be equipped with ventilating fans. The inside temperature should be maintained above 50°F, which can usually be achieved by running the fans. During the summer, large ventilator doors in the walls should be opened. Such buildings are divided into pens holding 10 to 30 sows, as in open housing; and the sows are either group fed or individually fed.

In totally confined gestation sow housing, slotted floors are generally used, no bedding is used; and manure is handled as a liquid.

The major **disadvantage** to confinement housing of gestating sows is the high initial investment in buildings and equipment.

HOUSING BOARS

When used in an individual mating system, boars are generally penned separately in individual crates about 28 in. wide and 7 ft long, or in pens 6 ft × 8 ft. Individual housing of boars eliminates fighting, riding, and competition for feed.

Groups of boars used in a pen mating system should be penned together when they are removed from a sow group.

INDOOR VS OUTDOOR INTENSIVE SWINE PRODUCTION

Many changes in U.S. swine production have occurred since the pre-1960 pasture system with portable houses. Beginning in the late 1950s, the industry moved from the low-investment, low-intensity system to the high-investment, high-intensity confinement system.

In the mid-1990s, England led the way with a pasture and portable house system to which they gave the new look and the new name: *outdoor intensive swine production*. Throughout Europe, confinement swine production was, and is, being attacked by pollution and animal welfare activists. Also, the concentration of confinement swine made for disease problems.

But neither European nor United States swine producers are returning to the pasture and portable house system of the old days! Instead the outdoor intensive swine production of the 1990s adapted and applied many of the changes that evolved with specialized confinement systems.

The subject of "Indoor Vs Outdoor Intensive Swine Production" is fully covered in Chapter 14 of this book; hence, the reader is referred thereto.

FLOORS

Floors are a concern when considering buildings for swine. While there are many variations, there are basically two types: solid floor and slotted floor.

The question as to what type of floor to install in a building depends upon how management intends to handle the manure. If manure is to be handled as a solid or semisolid product, the floor will usually be solid. If manure is to be handled as a liquid, the floor will usually be partially or totally slotted.

SOLID FLOORS

Solid floors should slope toward alleys and drains. Annoying puddles accumulate when floors fail to have a proper slope or slope in the wrong direction. A two-slope floor is recommended if waterers are installed in the stall or pen. Alleys should slope ½ in. per ft cross slope to form a crown or ¼ in. per ft to drains. Floors should slope ½ to ¾ in. per ft without bedding and ¼ to ½ in. per ft with bedding.

SLOTTED FLOORS

Four types of slats in swine building floors are illustrated in Figs. 16-13, 16-14, 16-15, and 16-16.

Slotted floors are floors with slots through which the feces and urine pass to a storage area below or nearby. Such floors are not new; they have been used in Europe for over 200 years. Slotted floor buildings should be warm, well ventilated, and flat.

Fig. 16-13. Concrete slats in a finishing facility. (Courtesy, Land O Lakes, Ft. Dodge, IA)

Fig. 16-14. Steel tri-bar slats in a nursery. (Courtesy, Iowa State University, Ames)

Fig. 16-15. Plastic slats in a nursery. (Courtesy, Iowa State University, Ames)

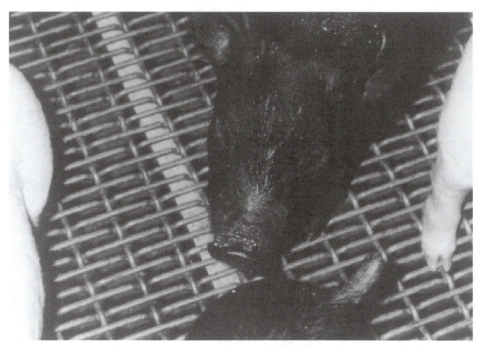

Fig. 16-16. Woven wire floor in a nursery. (Courtesy, Iowa State University, Ames)

The main **advantages** of slotted floors are (1) they facilitate automation and save labor; (2) they lessen or eliminate bedding; (3) they facilitate handling of manure; (4) they necessitate less space per animal; (5) they require less land; (6) they increase sanitation;

(7) they lessen mud, dust, odor, and fly problems; and (8) they lessen pollution.

The chief **disadvantages** of slotted floors are: (1) higher initial cost than conventional solid floors, (2) less flexibility in the use of the building, (3) any spilled feed may be lost through the slots, (4) pigs raised on slotted floors resist being driven over a solid floor, and (5) environmental conditions become more critical.

Studies at the University of Nebraska have shown decreased feed efficiency with an increased percentage of slotted floor area during winter.

■ **Partially slotted floors**— Studies have generally shown no difference in performance when comparing similar buildings with partially slotted floors to those with totally slotted floors, except during the winter months in cold climates.

The initial floor cost increases slightly as the amount of slotted area increases. Despite this fact, many new growing-finishing facilities are totally slotted because many producers want to avoid the messy pens of partially slotted floors. For systems which store manure in under-house pits, partial slots are impractical due to limited manure storage capacity.

■ **Slotted floor materials and design**—Durability, ease of cleaning, pig comfort, and cost should be considered when selecting slotted flooring. Slotted flooring material may be made from the materials listed in Table 16-5. Table 16-5 summarizes pertinent facts pertaining to some of the common types of materials.

Slat width and spacing (slot width) are governed by size of hogs and cleaning efficiency. Narrow slots are usually more effective in farrowing and nursery units. Spacing narrow slats too wide can cause injury to the feet and legs of finishing hogs. On the other

TABLE 16-5
MATERIALS FOR SLOTTED FLOORING

Material	Expected Life	Advantages	Disadvantages	Comments
Cast iron	20 years	High cost.	Cast iron is conductive. It pulls excess heat from the animals.	Avoid sharp edges on cast iron slats.
Concrete slats	20 years	Long life. May be homemade.	Quality control difficult when poured on site.	Should have flat, smooth surface and slightly pencil-rounded edge.
Metal slats (steel or aluminum)	4–8 years	Easily cleaned.	High cost.	Longer life from stainless steel than plain steel. Porcelainized steel slats available. Some steel slats are perforated.
Plastic coated metal	3–5 years	Comfortable. Has diamond-shaped holes.	May become slick.	Generally constructed of welded wire or expanded metal, then coated with a plastic substance. Performs well in farrowing and nursery facilities.
Triangular and T-bar metal slats	5–10 years	Durable. 50% open.	High cost.	Generally fabricated from steel. Performs well in farrowing and nursery facilities.
Wood slats	2–4 years	Low initial cost.	Difficult to maintain spacing. Not too durable.	Make from hardwood, like oak.
Woven and welded wire . . .	3-gauge: 5–10 years 5/16-gauge: 7–15 years	Cleaning ease. Low cost.	Must be supported rather thoroughly with a frame.	Galvanized or coated with plastic. Performs well in farrowing and nursery facilities.

hand, wide slats and narrow openings result in floors that are not completely self-cleaning. In most phases of production, slats placed so that the openings are parallel to the long dimension of the pen or crate appear to favor pig comfort and mobility. Table 16-6 provides recommendations for slot width—slot spacing.

A host of flooring designs and materials is available for constructing slotted floors. Practically all of these materials and designs have their strengths and weaknesses. When choosing a slotted floor, the first consideration should be given to where it will be installed—farrowing, nursery, or the growing-finishing area. Then consideration should be given to the following: pig comfort, durability and longevity, traction, cost, cleanability, and quality.

TABLE 16-6
RECOMMENDED SLOT WIDTH

Production Stage	Slot Width		Comments
	(in.)	*(mm)*	
Farrowing	3/8 for baby pigs	9	In farrowing crate, use larger slot width behind sow and smaller slot width elsewhere. Cover openings behind sow with plywood, sheet metal, or mesh during farrowing.
Nursery (20 to 60 lb or *9 to 27 kg*)	3/8	9	Spacing depends on width of slat; wider slat, wider spacing.
Growing-finishing (60 to 250 lb or *27 to 114 kg*) . .	3/4 to 1	19 to 25	When narrower slats of material other than concrete are used, narrower openings are used.
Gestating sows; boars	3/4 to 1	19 to 25	Slats which are 5 to 8 in. *(13 to 20 cm)* wide need the 1-in. *(2.5-cm)* opening.

SPACE REQUIREMENTS OF BUILDINGS AND EQUIPMENT FOR SWINE

One of the first, and frequently one of the most difficult, problems confronting the producer who wishes to construct a building or item of equipment is that of arriving at the proper size or dimensions. Tables 16-7 and 16-8 contain some conservative average figures which, it is hoped, will prove helpful. In general, less space than indicated may jeopardize the health and well-being of the pigs; whereas, more space may make the buildings and equipment more expensive than necessary.

Note: In the 1990s, hogs were carried to market weights of 240 to 250 lb, 20 to 30 lb heavier than in the 1980s. This prompted concern as to whether adequate space was being provided for these heavier market hogs. In a study of the "Space Requirements for Finishing Pigs," Kansas State University found that, based on average daily gain, average daily feed intake, and feed efficiency, 10 sq ft is adequate for feeding finishing hogs to a market weight of 250 lb.[2]

STORAGE SPACE REQUIREMENTS FOR FEED AND BEDDING

The space requirements for feed storage for the swine enterprise vary so widely that it is difficult to provide a suggested method of calculating space requirements. The amount of feed to be stored depends primarily upon: (1) length of pasture season, (2) method of feeding and management, (3) kind of feed,

---

[2]*Swine Update*, Vol. 11, No. 1, Kansas State University, Manhattan.

TABLE
SPACE REQUIREMENTS OF BUILDINGS

Age and Size of Animal	Swine Buildings					Shades		Pasture or Feeding Floor	
	Inside Sleeping Space or Shelter per Animal[3]	Height of Ceiling[4]	Height of Pen Partition	Hog Door Height	Hog Door Width	Shade per Animal	Shade Height	Good Pasture	Paved Feeding Floor in Addition to Sleeping Space, When Confined, per Animal
	(sq ft)	(ft)	(in.)	(in.)	(in.)	(sq ft)	(ft)	(animals/ acre)	(sq ft)
Sows before farrowing:									
Gilts	15–17	7–8	36	36	24	17	4–6	10–12	15–20
Mature sows	18–20	"	36	"	"	20	"	8–10	15–20
Sows with pigs:									
Gilts	48	"	36	"	"	20	"	6–8	48
Mature sows	64	"	36	"	"	30	"	6–8	64
Herd boars	15–20	"	48	"	"	15–20	"	¼ acre/ boar	15–20
Growing-finishing swine:									
Weaning[7] to 75 lb	5–6[8, 9]	"	30	"	"	4	"	50–100	6–8[10]
75 lb to 125 lb	6–7[8, 9]	"	33	"	"	6	"	50–100	7–9[10]
125 lb to market	8–10[8, 9]	"	36	"	"	6	"	50–100	8–10[10]

[1]With the following information, these requirements may be converted to metric equivalents: 1 sq ft equals *0.093 m²*; 1 ft equals *0.305 m*; 1 in. equals *25 mm*; 1 acre equals *0.405 ha*; 1 lb equals *0.45 kg*.

[2]The drinking water should not fall below 35 to 40°F *(2 to 4°C)* during winter.

[3]When using slotted floors, only half as much floor space per hog is needed as when using conventional solid floors.

[4]Ceiling heights in excess of 7 to 8 ft create cold hog houses in cold climates.

[5]For example, a 6-ft feeder open on both sides has 12 linear ft of feeding space.

(4) climate, and (5) the proportion of feeds produced on the farm in comparison with those purchased. Normally, the storage capacity should be sufficient to handle all feed grain grown on the farm and to hold purchased supplies. Bedding may or may not be stored under cover. Sometimes poled framed sheds or a cheap cover of plastic is used for protection.

Table 16-8 gives the storage space requirements for feed and bedding. This information may be helpful to the individual operator who desires to compute the building space required for a specific operation. This table also provides a convenient means of estimating the amount of feed or bedding in storage.

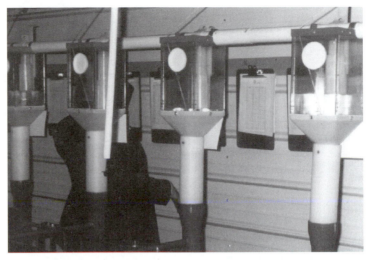

Gestation automated feeding jugs. Feed is delivered via the horizontal auger into jugs for each sow. Then the feed can be automatically dropped to each gestating sow via the vertical pipe. There is one jug per sow. (Courtesy, Iowa State University, Ames)

16-7
AND EQUIPMENT FOR SWINE[1]

Feeding Equipment					Watering Equipment[2]			Comments
Self-feeder Space (Animals/Linear Ft or per Hole)		Percent of Total Self-feeder Space Given to Protein Supplement		Feed Trough Space/ Animal for Hand- Feeding	Water Trough Space Hand- Feeding	Automatic Watering Cups (two openings considered 2 cups)		
Drylot	Pasture	Drylot	Pasture					
(no. animals/ linear ft)[5]	(no. animals/ linear ft)[5]	(% total feeder space)	(% total feeder space)	(linear ft/ animal)	(linear ft/ animal)			
2	3	15	10–15	1½	1½	1 cup/12 gilts		When alfalfa hay is fed in rack, allow 4 sows/linear ft.
3	4	15	10–15	2	2	1 cup/10 sows		
1[6]	1[6]	15	10–15	1½[6]	2	1 cup/4 sows		For the pig creep, provide a minimum of 1 ft of feeder space per 5 pigs; see that the edge of the feeder trough does not exceed 4 in. above the ground floor, and do not allow more than 40 pigs per creep.
1[6]	1[6]	15	10–15	1½[6]	2	1 cup/4 sows		
1	1	15	10–15	2	2	1 cup/2 boars		
4	4–5	25	20–25	¾	¾	1 cup/20 pigs		When salt or mineral is fed free-choice, provide 3 linear ft of mineral box space or 3 self-feeder holes per 100 pigs.
3	3–4	20	15–20	1	1	1 cup/20 pigs		
3	3–4	15	10–15	1¼	1¼	1 cup/20 pigs		

[6]With creep provided for pigs in addition.

[7]For early weaning space requirements, see Table 14-4, under section entitled "weaning pigs" in Chapter 14, Swine Management.

[8]Over the above sleeping space given herein, pigs that are confined from weaning to market should be provided the feeding floor space recommended in the column headed "Paved Feeding Floor in Addition to Sleeping Space, When Confined, per Animal."

[9]The larger area in the summertime.

[10]The larger area when fed from troughs; the smaller area is adequate where self-feeders are used.

TABLE 16-8
STORAGE SPACE REQUIREMENTS FOR FEED AND BEDDING[1]

Kind of Feed or Bedding	Lb/Cu Ft	Cu Ft/Ton	Lb/Bu of Grain	Cu Ft/Bu
Hay—Straw:				
1. Loose				
Alfalfa	4.4–4.0	450–500		
Nonlegume	4.4–3.3	450–600		
Straw	3.0–2.0	670–1,000		
2. Baled				
Alfalfa	10.0–6.0	200–330		
Nonlegume	8.0–6.0	250–330		
Straw	5.0–4.0	400–500		
3. Chopped				
Alfalfa	7.0–5.5	285–360		
Nonlegume	6.7–5.0	300–400		
Straw	8.0–5.7	250–350		
Silage:				
Corn or sorghum in tower silos	40	50		
Corn or sorghum in trench silos	35	57		
Grain:				
Corn, shelled[2]	45	45	56	1.25
Corn, ear	28	72	70	2.50
Corn, shelled, ground	38		48	
Barley	39	51	48	1.25
Barley, ground	28		37	
Oats	26	77	32	1.25
Oats, ground	18	106	23	
Rye	45	44	56	1.25
Rye, ground	38		48	
Sorghum	45	44	56	1.25
Wheat	48	42	60	1.25
Wheat, ground	43	46	50	
Mill Feed:				
Bran	13	154		
Middlings	25	80		
Linseed meal	23	88		
Cottonseed meal	38	53		
Soybean meal	42	43		
Alfalfa meal	15	134		
Miscellaneous:				
Soybeans	48	56		
Salt, fine	50	40		
Pellets, mixed feed	37	58		
Shavings, baled	20	100		

[1]For metric conversion, refer to the Appendix.

[2]These values are for corn with 15% moisture. Corn that is 30% moisture weighs 51 lb/cu ft shelled and 36 lb/cu ft ground or 68 lb/bu shelled and 90 lb/bu ground.

SWINE EQUIPMENT

The successful hog producer must have adequate equipment with which to provide feed, water, shelter, and care for the animals. Suitable equipment saves labor and prevents the loss of many baby pigs—and even the loss of older animals. These are items which should not be overlooked.

Certain features are desirable in swine equipment. Equipment should be convenient and economical; and, as hogs are subject to numerous diseases and parasites, it should be constructed for easy cleaning and disinfection. Equipment should be useful, durable, and yet economical.

There are many types and designs of the various pieces of swine equipment, and some producers will introduce their own ideas to fit the material available to local conditions. It is expected that certain adaptations should be made in designs to meet individual conditions.

BREEDING RACK

A breeding rack is often very necessary in breeding a young gilt to a mature, heavy boar; or in breeding a large, mature sow to a young boar. Satisfactory breeding racks may be either purchased commercially or built in the farm workshop.

CREEP

Research has shown that very little creep feed is consumed before three weeks of age. So, with early weaning, many swine producers no longer creep feed.

If a creep is used, the enclosed area should be of sufficient size to accommodate the number of pigs intended along with the feeders. There should be enough openings in the fence, or panels, through which the little pigs can pass. The openings should be wide enough for the pigs, but narrow enough to keep older hogs out; preferably, they should be adjustable, so that they can be widened as the pigs become older. It is important that the openings be sufficiently high that the pigs do not have to lower their backs when passing through; otherwise, there is a tendency to produce objectionable sway backs.

(See Chapter 9, section headed "Creep Feeding.")

HEAT LAMPS AND BROODERS

Newborn pigs are quite comfortable at temperatures of 85 to 95°F, shiver when standing alone at

70°F, and are quite cold at 60°F. Recent estimates indicate that swine producers can save an average of 1½ pigs per litter by using supplementary heat to keep baby pigs warm.

Brooders of various designs are used. Where sows are farrowed in conventional pens, a triangular-shaped brooder is usually secured in one corner of the pen.

In cold climates, heat may be supplied by infrared-heat lamps, by electric hovers, or by gas-fired ceramic core radiant heaters. Heat lamps should be of hard Pyrex glass (red filter type) that resists breakage when splashed with water. Use heavy porcelain sockets, a metal hood, and suspend the lamp by a chain. **Never suspend the heat lamp by the electrical cord.** A 250-watt lamp should be 24 in. above the floor of the brooder area.

Any type of heating device can cause a fire, so heaters should be installed carefully.

Note: Radiant heating pads and other baby pig heating devices are safer and cheaper than heat lamps.

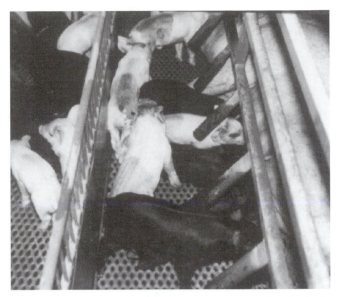

Fig. 16-18. Piglets and sow in farrowing crate. Note excellent flooring for piglets. (Courtesy, Iowa State University, Ames)

FARROWING CRATES
(Stalls)

Adequate facilities for farrowing are important because one-third or more of all death losses before weaning result from over-laying or crushing.

A variety of farrowing crates, stalls, or tethering arrangements of different designs and sizes are available.

The most exciting new systems of gestation housing preserve the advantages of individual living places, while permitting the sow much more freedom of movement than the traditional crates, stalls, or tethering.

The new installations are referred to as turn-around, comfort stalls, or freedom stalls.

It is noteworthy that the United Kingdom's agricultural ministry has banned new sow stalls and tether systems. Of possibly greater significance, the U.K. minister has also decreed that existing stall and tethering systems be phased out by the end of 1998. The rest of the E.C. countries must remove tethers prior to 1999.

(Also see Chapter 14, section headed "Farrowing Management.")

LOADING CHUTES

Loading chutes are desirable on most farms. Chutes make possible loading hogs without injury and with the least amount of effort. Loading chutes may be either fixed or portable. Although slightly more expensive to build, a stairstep loading chute is preferable to the ramp-and-cleat type; hogs find the steps easier to ascend, and will go up more willingly and with less chance of injury. Ramp-type chutes are still the most prevalent, due to their slightly lower initial cost.

The chute should be fairly narrow so animals cannot turn around. Care should be taken so that no cleats, nails, or other projections injure the hog. A portable chute is convenient, because hogs can be loaded from any lot or pasture. The chute should be durably constructed.

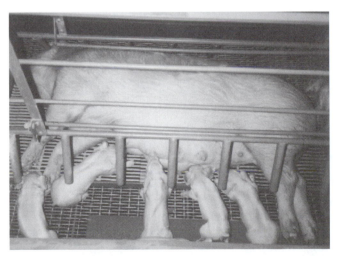

Fig. 16-17. Sow and litter in farrowing crate. (Courtesy, American Diamond Swine Genetics, Prairie City, IA)

Fig. 16-19. Portable stairstep loading chute. Stairsteps are safer and are preferred by hogs. (Courtesy, V. J. Morford, Department of Agricultural Engineering, Iowa State University)

SELF-FEEDER

The design of the self-feeder is extremely important. An adequate self-feeder should be constructed so that:

1. It will be sturdy and durable.
2. The feed will not clog; to this end, the feeder should have means for agitating the feed to keep it from bridging and adjustable throats for controlling the rate of flow.
3. Waste will be reduced to a minimum.

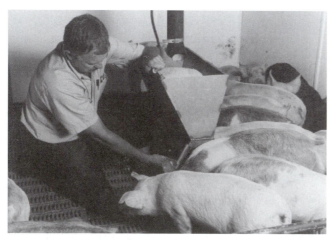

Fig. 16-20. Adjusting a self-feeder. (Courtesy, National Pork Producers Council, Des Moines, IA)

4. The feed will be protected from wind, rain, birds, and rodents.
5. It is large enough to hold several days' feed.

Many satisfactory designs of self-feeders are available, including both commercial and homemade equipment.

When stationed out-of-doors, self-feeders should always be placed on concrete platforms; otherwise, mudholes soon appear around the base in wet weather, and this causes much waste of feed. Also, they should be so designed and located that they can be filled with a self-unloading wagon, an auger, or a blower.

SHADE

Hogs require shade during the hot months of the year. Heat-stressed pigs eat less feed, gain less weight, and the gains are less efficient (see Table 16-3). Studies show that at 85°F it takes 1,200 lb of feed to produce 100 lb of gain; whereas at 60°F, 400 lb of feed will produce 100 lb of gain.

In many cases, houses may double for shade purposes. But it is often desirable to provide additional protection from the sun.

SHIPPING CRATES

Shipping crates are often used for transporting breeding stock. Crates must be durably constructed to prevent animals breaking out during shipment. Crates should be sufficiently light so shipping charges will be held at a minimum. A good crate is an advertisement for the breeder.

SOW WASH

As part of a herd health program, sows should be washed before being placed in farrowing crates or stalls. Often areas for washing sows are located near the entrance of the farrowing house. A variety of designs will work as an area for washing sows, but it is a good idea to include both a hot and a cold water source and a drain with the floor sloping toward it.

TROUGHS

The hog trough, a universally used piece of feedlot equipment, is often carelessly constructed. A good trough should be easy to clean and should be constructed so that hogs cannot lie in it or otherwise contaminate the feed. The little pig trough is similar in construction and is useful for small creep-fed pigs.

VACCINATING AND CASTRATING RACK

A vaccinating and castrating rack makes for convenience in treating young pigs. This is a simple rack, made like a common sawbuck with a V-shaped trough added. A rope or strap is attached at one or both ends of the trough to hold the pigs securely in place. Also, some commercial holders are available which hold piglets upside down by their hind legs.

WALLOWS; SPRINKLERS

Hogs have very few sweat glands; thus they cannot cool themselves by perspiration. In the South, therefore, mature breeding animals and finishing hogs need a wallow during the hot, summer months. Instead of permitting an unsanitary mud wallow, successful swine producers install hog wallows. This equipment, which may be either movable or fixed, will keep the animals cool and clean and make for faster and more economical gains. The wallow should be in close proximity to shade, but in no case should shade be built directly over the wallow. Such an arrangement will cause all the hogs to lie in the water all day. The size of wallow to build will depend upon the number and size of animals. Up to 50 growing-finishing pigs can be accommodated per 100 sq ft of wallow provided shade or shelter is nearby.

Where sprinklers are used for cooling, (1) provide 1 nozzle per 25 to 30 hogs, (2) station the nozzles 4 to 6 ft from the floor or ground and about 8 ft apart, (3) spray about 0.09 gal per hour per pig, (4) control by timer set at 2 minutes on in each hour and set thermostat at 75°F, and (5) provide a fine inline filter. Sprinklers should be limited to concrete lots or sandy soils where mud holes do not develop, and they should be located over the manure collection area of the pens.

WATERER

Swine will consume ¼ to ⅓ gal. of water for every pound of dry feed. The higher the temperature, the greater the water consumption. It is preferable that swine have access to automatic waterers, with cool, clean water available at all times. Otherwise, they should be hand-watered at least twice daily. During winter, the drinking water should not be permitted to fall below 40°F. The estimated water needs for various classes of swine follow:

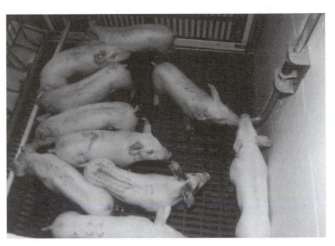

Fig. 16-21. Pigs drinking from a nipple waterer. (Courtesy, National Pork Producers Council, Des Moines, IA)

Class of Swine	Water Consumption
	(gal./head/day)
Gestating sows	2–3
Lactating sows	4–5
Weaned pigs (15–40 lb)	0.5–1
Growing pigs (40–120 lb)	1
Growing-finishing pigs (120–250 lb)	1.5–2

For running water to pastures and other fields, plastic pipe is recommended, because it is easily installed either on top of the ground or in a trench. By locating the waterer in a line fence, two pastures can be supplied from one unit.

MANURE MANAGEMENT

The handling of manure is probably one of the most important problems confronting commercial swine producers today. It must be removed for sanitary reasons, and there is a limit to how much of it can be left to accumulate in pits or other storage areas. Higher labor costs reduce the value of manure as a fertilizer. Also, to the urbanite and city dweller alike, manure odor is taboo. Manure is too costly to burn, too bulky to bury or put in a dump, and, with large operations, there may be too much of it to put in lagoons.

Building types or building systems affect the method of manure management. Slotted floors/pit storage and lagoons are widely used.

Producers should thoroughly evaluate the methods for handling manure and choose a system that will complement their overall production system.

(Also, see Chapter 13, section headed "Manure"; and Chapter 14, section headed "Manure.")

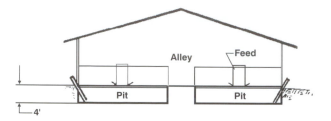

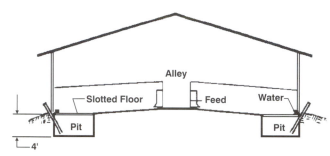

Fig. 16-22. Diagrams of totally slotted pens over pits (top) and partially slotted pens over pits (bottom).

EQUIPMENT FOR HANDLING MANURE

Depending on the production system, there are several phases involved in handling manure, each requiring some different equipment and facilities. The phases of manure handling may be (1) collection and transfer from the building, (2) storage and treatment, and (3) application.

The following are considered as transfer methods:

1. Pits below slotted floors.
2. Flushing—open gutter.

Fig. 16-23. Manure lagoon. (Courtesy, Oklahoma State University, Stillwater)

3. Flushing—below slats
4. Mechanical scrapers.
5. Deep, narrow gutters.

Storage and treatment of manure may be in the following:

1. Pit below slotted floors.
2. Remote storage slurry, above ground, below ground, or in earthen structure.
3. Anaerobic lagoons.
4. Aerated lagoons.
5. Oxidation ditch.
6. Solids separation.

As for application, there are the following methods:

1. Broadcast (top dress) with plow-down or disking.
2. Broadcast without plow-down or disking.
3. Knifing (injection) under the soil surface.
4. Irrigation.

Rapid incorporation of manure will minimize nitrogen loss to the air and will hasten biological activity to release nutrients to plant use. Injecting, chiseling, or knifing liquids beneath the soil surface will minimize odors and nutrient losses to the air and/or runoff. Disking or chiseling land soon after surface application will also reduce odors generated from field-spread wastes and runoff transfer.

Timing manure application can help minimize complaints. If possible, select dry windy days for manure disposal. Pay attention to wind direction with respect to neighboring residences. Direct injection or rapid incorporation can be used advantageously where neighbors live close to a disposal site. Good communication, sincere response to complaints, and other good neighbor techniques can be as important as following good waste management practices.

Since much of the swine wastes is handled as a liquid, the following equipment considerations are presented.

1. **Storage container.** A watertight storage container to which water can be added will be needed. The size of container will depend on the way the swine operation is managed, the length of time between emptyings, and the kind, number, and size of hogs. Large storages have maximum labor advantage.

Approximate daily and yearly manure production figures for swine are given in Table 14-6 in Chapter 14, Swine Management.

a. There are approximately 34 cu ft in a ton of manure.

b. Cleaning swine facilities with high-pressure water may double the volume of wastes.

c. Roof or lot drainage will have an unpredictable effect if allowed to run into the storage.

d. Extra water may have to be added if the manure is to be pumped; anywhere from one-fifth to three-fifths of the storage volume may be needed for extra water. For irrigation, there should be about 95% water and 5% manure; for field spreading the extra water should be held to a minimum.

The needed storage capacity for a given swine unit can be computed as follows: storage capacity = no. of animals × daily manure production (Table 14-6) × desired storage time (days) + extra water.

It is also important that indoor storage containers be well ventilated, that all storage containers (indoor and outdoor) be insulated against freezing, that water be added to the storage container before it is filled with manure (add 3 to 4 in. of water under slotted floors, and 6 to 12 in. of water where manure is scraped into the storage area), that frozen manure never be added to a storage container, that all openings be closed when not in use, and that there be a fly control program (including baits and sprays).

2. **Equipment or facilities to move excrement to storage.** These include a scraper, gutters, slotted floors, drains.

3. **Equipment to stir manure and remove it from storage.** This includes pumps, agitators, augers, etc.

4. **Equipment or facilities to dispose of liquid manure.** These may involve a tank truck or wagon, irrigated fields, available land, a lagoon, etc.

Regardless of the collection, storage, treatment, and application methods, there will be many advantages and disadvantages to consider. Some will work better than others for specific circumstances. Often the end use of swine manure dictates the most appropriate

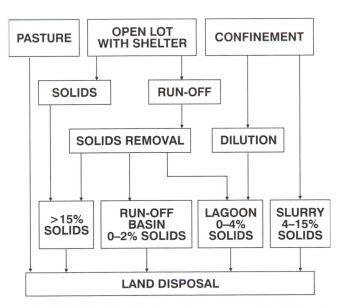

Fig. 16-24. Alternatives for swine waste management, depending on the type of production system. Ultimately the end product is disposed of on the land.

waste disposal system. But before selecting a waste disposal system, all variables should be considered. Furthermore, the least expensive method may not meet regulations or be acceptable to the neighbors.

(Also see Chapter 14, section headed "Manure.")

FENCES FOR SWINE

Good fences (1) maintain farm boundaries, (2) make livestock operations possible, (3) reduce losses to both animals and crops, (4) increase land values, (5) promote better relationships between neighbors, (6) lessen accidents from animals getting on roads, and (7) add to the attractiveness and distinctiveness of the premises.

The discussion which follows will be limited primarily to wire fencing, although it is recognized that such materials as rails, poles, boards, stone, hedge, pipe, and concrete have a place and are used under certain circumstances. Also, where there is a heavy concentration of animals, such as on feeding floors, there is need for a more rigid type of fencing material than wire. Moreover, certain fencing materials have more artistic appeal than others; and this is an especially important consideration on the purebred establishment.

The kind of wire to purchase should be determined primarily by the class of animals to be confined.

The following additional points are pertinent in the selection of wire:

1. **Styles of woven wire.** The standard styles of woven wire fences are designated by numbers as 958, *1155*, 849, *1047*, 741, *939*, *832*, and *726* (the figures in italics are most common). The first one or two digits represent the number of line (horizontal) wires; the last two the height in inches; *i.e.*, 832 has 8 horizontal wires and is 32 in. in height. Each style can be obtained in either (a) 12-in. spacing of stays (or mesh), or (b) 6-in. spacing of stays.

2. **Mesh.** Generally, a close-spaced fence with stay or vertical wires 6 in. apart (6-in. mesh) will give better service than a wide-spaced (12-in. mesh) fence. However, some fence manufacturers believe that a 12-in. spacing with a No. 9 wire is superior to a 6-in. spacing with No. 11 filler wire (about the same amount of material is involved in each case).

3. **Weight of wire.** A fence made of heavier weight wires will usually last longer and prove cheaper than one made of light wires. Heavier or larger size wire is designated by a smaller gauge number. Thus, No. 9 gauge wire is heavier and larger than No. 11 gauge. Woven wire fencing comes in Nos. 9, 11, 12½, and 16 gauges—which refers to the gauge of the wires other than the top and bottom line wires. Heavy barbed

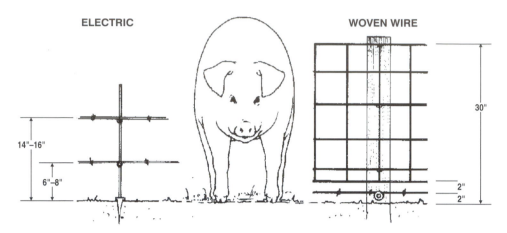

ELECTRIC   WOVEN WIRE

14"–16"

6"–8"

30"

2"
2"

Fig. 16-25. Woven wire fences for hogs may be 30 or 36 in. high with a strand of barbed wire at the bottom to prevent rooting. Electric fences of two strands of barbed wire may be used for temporary fencing. If producers will spill corn along the fence, they can "teach" their pigs about electric fences.

wire is 12½ gauge. But there is a lighter, high tensile barbed wire which comes in 14 and 16 gauge.

Heavier or larger wire than normal should be used in those areas subject to (a) salty air from the ocean, (b) smoke from close proximity industries which may give off chemical fumes into the atmosphere, (c) rapid temperature changes, or (d) overflow or flood. Also, heavier or larger wire than normal should be used in fencing (a) small areas, (b) where a dense concentration of animals is involved, and (c) where animals have already learned to get out.

4. **Styles of barbed wire.** Styles of barbed wire differ in the shape and number of the points of the barb, and the spacing of the barbs on the line wires. The 2-point barbs are commonly spaced 4 in. apart while 4-point barbs are generally spaced 5 in. apart. Since any style is satisfactory, selection is a matter of personal preference.

5. **Standard size rolls or spools.** Woven wire comes in 20- and 40-rod rolls; barbed wire in 80-rod spools.

6. **Wire coating.** The kind and amount of coating on wire definitely affects its lasting qualities. Galvanized coating is most commonly used to protect wire from corrosion. Coatings are specified as Class I, Class II, and Class III. The higher the class number the greater the coating thickness and performance.

Three kinds of material are commonly used for fence posts: wood, metal, and concrete. The selection of the particular kind of posts should be determined by (1) the availability and cost of each, (2) the length of service desired (posts should last as long as the fencing material attached to them, or the maintenance cost may be too high), (3) the kind and amount of livestock to be confined, and (4) the cost of installation.

ELECTRIC FENCES

Where a temporary enclosure is desired or where existing fences need bolstering from roguish or breachy animals, it may be desirable to install an electric fence, which can be done at minimum cost.

The following points are pertinent in the construction of an electric fence:

1. **Safety.** If an electric fence is to be installed and used, (a) necessary safety precautions against accidents to both persons and animals should be taken, and (b) the hog producer should first check into state regulations relative to the installation and use of electric fences. *Remember that an electric fence can be dangerous.* Fence controllers should be purchased from a reliable manufacturer; homemade controllers may be dangerous.

2. **Wire height.** As a rule of thumb, the correct wire height for an electric fence is about three-fourths the height of the animal; with two wires provided for swine—one wire should be 6 to 8 in. from the ground, and the other 14 to 16 in. from the ground.

3. **Posts.** Either plastic or steel posts may be used for electric fencing. Corner posts should be as firmly set and well braced as required for any non-electric fence so as to stand the pull necessary to stretch the wire tight. Line posts (a) need only be heavy enough to support the wire and withstand the elements, and (b) may be spaced 25 to 40 ft apart for hogs.

4. **Wire.** In those states where barbed wire is legal, new 4-point hog wire is preferred. Barbed wire is recommended, because the barbs will penetrate the hair of animals and touch the skin, but smooth wire can be used satisfactorily. Rusty wire should never be used, because rust is an insulator.

5. **Insulators.** Wire should be fastened to the posts by insulators and should not come into direct contact with posts, weeds, or the ground. Inexpensive solid glass, porcelain, or plastic insulators should be used, rather than old rubber or necks of bottles. Plastic or fiberglass posts do not require insulators.

6. **Charger.** The charger should be safe and effective (purchase one made by a reputable manufacturer). There are five types of chargers: (a) *the battery charger*, which uses a 6-volt hot shot battery; (b) *the inductive discharge system*, in which the current is fed to an interrupter device called a circuit breaker or chopper which energizes a current limiting transformer; (c) *the capacitor discharge system*, in which the power line is rectified to direct current and the current is stowed in the capacitor; (d) *the continuous current type*, in which a transformer regulates the flow of current from the power line to the fence; and (e) *the solar (sun) charger.*

7. **Grounding.** One lead from the controller should be grounded to a pipe driven into the moist earth. *An electric fence should never be grounded to a water pipe, because it could carry lightning directly to connecting buildings.* A lightning arrestor should be installed on the ground wire.

Note: Do not string an electric fence across a gate. Use wood or metal gates. If pigs get shocked by an electrified gate, it will be very difficult to get them through the gate even though it is wide open.

QUESTIONS FOR STUDY AND DISCUSSION

1. In the late 1950s, specialized pork producers evolved. With their advent, came confinement production, replacing much of the old system of pastures, dry-lots, and portable houses. Why did this happen?

2. Why are hogs more sensitive to extremes in temperatures—either hot or cold—than other farm animals? What is the most desirable temperature for each class of hogs? What can be done to modify (a) winter and (b) summer temperatures?

3. Why is it necessary to know how much heat the bodies of hogs produce?

4. How is the effect of cold stress and heat stress demonstrated in the performance of growing hogs?

5. The major requirement of winter ventilation is moisture removal, whereas the major requirement of summer ventilation is temperature control. Why the difference?

6. What is natural ventilation? How is natural ventilation accomplished?

7. A complete farrow-to-finish confinement system generally has one or more buildings devoted to each of the following phases: (a) farrowing, (b) nursery, (c) growing-finishing, (d) breeding/gestation and herd boars. Describe each of the types of buildings used for each of these phases.

8. Why is there renewed interest in outdoor swine production, now referred to as *outdoor intensive swine production*?

9. List and discuss the advantages and disadvantages of the various materials for slotted floors.

10. Describe and justify the type of flooring you would use in a nursery.

11. One of the first, and frequently one of the most difficult, problems confronting the swine producer who wishes to construct a building or item of equipment is that of arriving at the proper size or dimensions. In planning to construct new buildings and equipment for swine, what factors and measurements for buildings and equipment should be considered?

12. Why should a swine producer be knowledgeable about the storage space requirements for feed and bedding?

13. Make a critical study of your own swine buildings and equipment, or those on a hog farm with which you are familiar, and determine their (a) desirable and (b) undesirable features.

14. Is a turn-around farrowing crate necessary?

15. What features would you build or buy in a self-feeder?

16. Why are shades and wallows important for swine on pasture?

17. What is meant by "Manure Management"?

18. In modern swine production units, how is manure handled and what equipment is required?

19. Describe the construction of an electric fence for swine.

SELECTED REFERENCES

Title of Publication	Author(s)	Publisher
Confinement Swine Housing	J. E. Turnbull N. A. Bird	Agriculture Canada, Ottawa, Canada, 1979
Farm Builder's Handbook	R. J. Lytle	Structures Publishing Company, Farmington, MI, 1973
Farm Buildings: From Planning to Completion	R. E. Phillips	Doane-Western, Inc., St. Louis, MO, 1981
Livestock Environment	Ed. by E. Collins C. Boon	American Society of Agricultural Engineers, 1993
Pig Housing	D. Sainsbury	Farming Press, Ltd., Ipswich, England, 1972
Pork Facts	Staff	National Pork Producers Council, 1995–96
Pork Industry Handbook		Cooperative Extension Service, Purdue University, West Lafayette, IN
Stockman's Handbook, The, Seventh Edition	M. E. Ensminger	Interstate Publishers, Inc., Danville, IL, 1992
Structures and Environment Handbook		Midwest Plan Service, Iowa State University, Ames, IA, 1980
Swine Housing and Equipment Handbook		Midwest Plan Service, Iowa State University, Ames, IA, 1982

Uniform pork carcasses in a meat packer's cooler. (Courtesy, DeKalb Swine Breeders, Inc., DeKalb, IL)

17

MARKETING AND SLAUGHTERING HOGS[1]

[1]The authors wish to express their appreciation for the authoritative review accorded by, and the helpful suggestions received from, the following persons in the revision of Chapter 17: Jim Wise, Ph.D., USDA, AMS, Livestock and Seed Division, STDZ, Washington, DC; and Ken Johnson, Vice President, and Terence R. Dockerty, Ph.D., Director of Meat Science Programs, National Live Stock and Meat Board, Chicago, IL.

Livestock marketing embraces those operations beginning with loading animals out on the farm and extending until they are sold to go into processing channels.

Marketing—along with breeding, feeding, and management—is an integral part of the modern livestock production process. It is the end of the line; that part which gives point and purpose to all that has gone before. Market receipts constitute the only source of reimbursement to the producer; market day is the producer's payday—hence, it is the most important single day of the operation. The importance of hog marketing is further attested by the following facts:

1. In the past 10 years, an average 87,142,900 hogs were marketed annually in the United States.[2]

2. U.S. farmers receive an average 7.6% of their cash farm marketing income of livestock and products from hogs.[3]

3. Livestock markets establish values of all animals, including those down on the farm. On December 1, 1994, there were 57,938,000 head of hogs in the United States, with an aggregate value of $4.3 billion, or $74.90 per head.[4]

4. The U.S. hog industry loses an estimate $32 million annually from Pale Soft Exudative (PSE) pork, much of it (a) due to excitement and overheating just before stunning at the packing plant, and (b) in normal stress-resistant hogs.[5]

It is important, therefore, that the swine producer know and follow good marketing practices.

THE CHANGING HOG MARKET

In the good old days, farm produce marketing was relatively simple. On Saturdays, the farmer toted to town and sold to the corner produce store a basket of eggs, a gunny sack of old hens or fryers, and a jar of sour cream. Surplus hogs were usually sold to local buyers. The farmer did little figuring as to the best time to sell animals; the chief worry was in growing rather than in selling. The farmer could be successful by knowing how to breed, feed, and manage stock. Today, this is not enough; preconsidered, if not prearranged, markets are essential.

Consumer preference has dictated, and will continue to dictate, changes in market hogs. In recent years, the consumer has been demanding (1) less lard, (2) smaller cuts with less fat and more lean, and

Fig. 17-1. Hogs being driven to market. This was a common scene In the early 1900s. Truck transportation of livestock was first started in 1911. (Photo by J. C. Allen and Son, West Lafayette, IN; courtesy, American Feed Manufacturers Association)

(3) a larger proportion of the valuable cuts. These requirements are met by meat-type hogs (see Fig. 17-2). Further, from the producers' standpoint, there is a saving in feed in marketing meat-type hogs. The feed costs of production of fat are about 2½ times greater than those for lean.

■ **Standard of excellence in hog type**—In order to meet consumer preference, the National Pork Producers Council has provided a description of the ideal market pig (known as Symbol II) for the year 2,000—and beyond. It follows: Marketed at 260 lb, with a carcass weight of 195 lb, grown from a terminal crossbreeding program and maternal line capable of

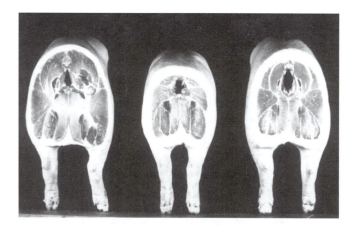

Fig. 17-2. Market hogs have changed over the years! Fat hogs like the center hog are no longer desirable. Most hogs presently going to market are like the hog on the right. But the goal ahead is like the well muscled, meat-type hog on the left. (Courtesy, Dekalb Swine Breeders, Inc., Dekalb, IL)

---

[2]*Agricultural Statistics*, USDA, 1994, p. 235, Table 397.

[3]*Pork Facts 1995/1996*, National Park Producers Council, p. 16.

[4]*Agricultural Statistics*, USDA, 1994, p. 233, Table 393.

[5]Grandin, Temple, *Feedstuffs*, March 25, 1991, p. 20.

weaning 25 pigs per year, free of the halothane or "stress" gene.

(Also see Chapter 8, section on "Growth," Table 8-2.)

MARKET CHANNELS FOR HOGS

Hog producers are confronted with the perplexing problem of determining where and how to market their animals. Usually there is a choice of market outlets, and the one selected often varies between classes and grades of hogs and among sections of the country. Thus, the method of marketing usually differs between slaughter hogs and feeder pigs, and both of these differ from the marketing of purebreds.

Most hogs and pigs are sold through the following channels:

1. Terminal or central market.
2. Auction market.
3. Direct to buyer (including country dealers).
4. Carcass grade and yield basis.
5. Slaughter hog pooling.
6. Packer marketing contract.
7. Feeder pig marketing.
8. Early weaner pig marketing (SEW pigs).

There are other methods of marketing, but the methods listed are the primary ones.

Note: The listing and discussion of market channels for hogs is in the historical order and importance in which they evolved, rather than in the current annual volume of hogs marketed through them.

TERMINAL, OR CENTRAL, MARKET

Terminal markets (also referred to as terminals, central markets, public stockyards, and public markets) are livestock trading centers which generally have several commission firms and an independent stockyards company. Up through World War I, the majority of slaughter livestock in the United States was sold through terminal public markets by farmers or local buyers shipping to them. Since then, the importance of these markets has declined in relation to other outlets.

Formerly, terminal markets were synonymous with private treaty selling. Today, however, many terminal markets operate their own sale ring and all, or almost all, of their animals are sold by auction.

■ **Leading terminal hog markets**—As would be expected, the leading hog markets of the United States are located in or near the Corn Belt—the area of densest hog population. With the advent of the motor

truck and improved highways, the pork packing plants were moved nearer the areas of hog raising. Thus, during the last half of the 1990s, local, or interior packers, increased in numbers. In order to meet this added competition, the large packers at more distant points resorted to direct buying and the purchase of interior plants.

The rank of the major hog markets shifts considerably according to feed supplies and general economic conditions. One drastic example is the Union Stockyards in Chicago which was the leading terminal market for years; in 1970, the Union Stockyards stopped receiving hogs, and, eventually, it closed. The leading terminal hog markets, by rank, and their receipts are listed in Table 17-1.

TABLE 17-1
RECEIPTS AT LEADING TERMINAL HOG MARKETS[1]

Market	Year	
	Average 1989–1993	1993
	- - - - - (thousands) - - - - -	
National Stock Yards, East St. Louis, IL . . .	582	475
South St. Paul, MN	515	443
South St. Joseph, MO	406	411
Omaha, NB	419	372
Sioux City, IA	457	40
U.S. Total[2]	2,224	2,936

[1]*Agricultural Statistics 1994*, USDA, p. 238, Table 400.

[2]The number of stockyards reporting varies from 41 to 68.

AUCTION MARKET

Auction markets (also referred to as sales barns, livestock auction agencies, community sales, and community auctions) are trading centers where animals are sold by public bidding to the buyer who offers the highest price per hundredweight or per head. Auctions may be owned by individuals, partnerships, corporations, or cooperative associations.

This method of selling livestock in this country is very old, apparently being copied from Great Britain where auction sales date back many centuries.

The auction method of selling was used in many of the colonies as a means of disposing of property, imported goods, secondhand household furnishings, farm utensils, and animals. According to available records, the first U.S. public livestock auction sale was held in Ohio in 1836, by the Ohio Company, whose business was importing English cattle.

Fig. 17-3. An auction sale of purebred Yorkshires. (Courtesy, American Yorkshire Club, West Lafayette, IN)

Livestock auction markets had their greatest growth since 1930, both in numbers established and the extensiveness of the area over which they operate. Their most rapid growth occurred between 1930 and 1937.

Several factors contributed to the phenomenal growth in auction markets during the thirties, chief of which were the following:

1. The decentralization of markets, and the improvement and extension of hard surfaced roads accompanied by the increased use of trucks as a means of transporting livestock.

2. The development of more uniform class and grade classifications for livestock.

3. Improvements made by the federal government in providing more extensive collection and dissemination of market news.

4. The greater convenience afforded in disposing of small lots of livestock and in purchasing stockers, feeders, and breeding animals.

5. The recognized educational value of these nearby markets, which enabled producers to keep currently informed of local market conditions and livestock prices.

6. The Depression of 1930-1933, which created the need to lower the cost of marketing and transportation.

7. The abnormal feed distribution caused by the droughts of 1934 and 1936 in the western Corn Belt and range states.

8. The desire to sell near home.

Prior to the advent of community livestock auctions, small livestock operators had two main market outlets for their animals—(1) shipping them to the nearest terminal public market, or (2) selling them to buyers who came to their farms or ranches. Generally, the first method was too expensive because of the transportation distance involved and the greater expense of shipping small lots. The second method pretty much put producers at the mercy of buyers because they had no good alternative to taking the price they offered, and often they did not know the value of their animals.

The auction market method of selling is similar to the terminal market in that both markets (1) are an assembly or collection point for livestock being offered for sale, (2) furnish or provide all necessary services associated with the selling activity, (3) are supervised by the federal government in accordance with the provisions of the National Packers and Stockyards Act, and (4) are characterized by buyers purchasing their animals on the basis of visual inspection.

But there are several important differences between terminals and auctions; among them, auction markets (1) are not always terminal with respect to livestock destination; (2) are generally smaller; (3) are usually single-firm operations; (4) sell by bid, rather than by offer and counter-offer; and (5) are completely open to the public with respect to bidding, and all buyers present have an equal opportunity to bid on all livestock offered for sale, whereas the terminal method is by private treaty (the negotiation is private).

Livestock auctions are really of greatest importance to the small operator. Also, auctions are an important outlet for feeder pigs; thus, a large proportion of the hogs sold through auctions go back to the farm.

DIRECT SELLING
(including country dealers)

Direct selling, including selling to country dealers, refers to producers selling hogs directly to packers or local dealers, without the support of commission firms, selling agents, buying agents, or brokers. Direct selling does not involve a recognized market. The selling usually takes place at the farm or some other nonmarket buying station or collection yard. Some country buyers purchase livestock at fixed establishments similar to packer-owned country buying points.

Direct selling is similar to terminal market selling with respect to price determination; both are by private treaty and negotiation. But it permits producers to observe and exercise some control over selling while it takes place, whereas consignment to distant terminal markets usually represents an irreversible commitment to sell. Larger and more specialized hog producers feel competent to sell their livestock direct.

Prior to the advent of public stockyards in 1865, country selling accounted for virtually all sales of livestock. Sales of livestock in the country declined with the growth of terminal markets until the latter method reached its peak of selling at the time of World

War I. Country selling was accelerated by the large nationwide packers following World War I in order to meet the increased buying competition of the small interior packers.

Improved highways and trucking facilitated the growth of direct selling. Farmers were no longer tied to outlets located at important railroad terminals or river crossings. Livestock could move in any direction. Improved communications, such as the radio and telephone, and an expanded market information service, also aided in the development of direct selling of livestock, especially in sales direct to packers.

CARCASS GRADE AND YIELD BASIS

Carcass selling is commonly called grade and yield selling.

All hogs should be evaluated by a practical, accurate, and objective measure of carcass composition. Any of several measures may be used; among them, the following types: (1) backfat probes, (2) lean meters, (3) ultrasonics, or (4) electrical conductivity. *Note well*: Today's state-of-the-art carcass measures will be obsolete in less than 5 years.

In addition to carcass measures, PSE should be determined and used as part of the price structure.

(Also see Chapter 4, section headed "Selection Based on Mechanical Measure.")

This is the most common method of marketing hogs in Denmark, Sweden, and Canada. The bargaining is in terms of the price to be paid per hundred pounds dressed weight for carcasses that meet certain grade specifications. It is the most accurate and unassailable evaluation of a carcass.

In general, hog producers who market superior animals benefit from selling on the basis of carcass grade and weight, whereas the producers of lower quality animals usually feel that this method unjustly discriminates against them. In countries where rail grading has been used extensively, there has been an unmistakable improvement in the breeding and feeding of swine. Denmark, Sweden, and Canada have effectively followed this type of program in producing high-quality bacon, chiefly for export to the London market.

The factors **favorable** to selling on the basis of carcass grade and weight may be summarized as follows:

1. It encourages the breeding and feeding of quality hogs.
2. It provides the most unassailable evaluation of the product.
3. It eliminates wasteful filling on the market.
4. It makes it possible to trace losses from con-

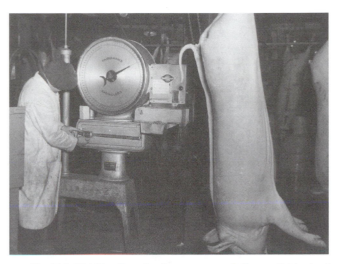

Fig. 17-4. Weighing Canadian hog carcasses. Selling on the basis of carcass grade and weight is the most common method of marketing hogs in Denmark, Sweden, and Canada. (Courtesy, Department of Agriculture, Ottawa, Ontario, Canada)

demnations, bruises, soft pork, and diseases to the producer responsible for them.

5. It is the most effective approach to animal improvement.

Note: It's not live animals, but pounds of lean on the rail that counts.

The factors **unfavorable** to selling on the basis of carcass grade and weight are:

1. The procedure is more time consuming than the conventional basis of buying.
2. From the standpoint of the meat packer, there is less flexibility in the operations.
3. The physical difficulty of handling the vast United States volume of hogs on this basis is great.
4. There is delay in returns to the seller.
5. The seller usually assumes the risk and loss of totally or partially condemned carcasses.

■ **USDA guidelines**—The U.S. Department of Agriculture guidelines, wherein meat packers buy livestock on the basis of carcass grade, carcass weight, or a combination of the two, are:

1. Make known to the seller, before sale, significant details of the purchase contract.
2. Maintain identity of each seller's livestock and carcass.
3. Maintain sufficient records to substantiate settlement for each purchase transaction.
4. Make payment on the basis of actual carcass weight.
5. Use hooks, rollers, gambrels and similar equipment of uniform weight in weighing carcasses of the same species of livestock in each packing plant; and

include only the weight of this equipment in the actual weight of the container.

6. Make payment on the basis of USDA carcass grades or furnish the seller with detailed written specifications for any other grades used in determining final payment.

7. Grade carcasses by the close of the second business day following the day of slaughter.

It is generally agreed that there is need for a system of marketing which (1) favors payment for pounds of lean or percent of lean in the carcass; and (2) the maximum yield in the four primal cuts: the ham, loin, picnic shoulder, and Boston butt, which account for three-quarters of the value of the entire carcass. Selling on the basis of carcass grade fulfills these needs.

Producers need the following information on every hog they sell:

1. Carcass identification.
2. Carcass weight.
3. A quality score.
4. Pounds of lean in each carcass, determined by an objective measure.

SLAUGHTER HOG POOLING

Slaughter hog pooling, which consists of a number of smaller hog producers joining together for the purpose of marketing hogs in truckload lots, is generally prompted because of lack of access to a nearby market.

Typically, a person is hired to manage or coordinate the pooling program.

Slaughter hogs may either be (1) commingled for shipment with all hogs on a single truckload receiving the same price, or (2) compartmentalized on the truck by owners, with each owner receiving a separate check. In either case, the producers are charged a fee by the marketing association for services rendered.

Pooling slaughter hogs from small producers into truckload lots and timing deliveries to fit buyers' needs will increase marketing efficiency and may increase sale prices for hogs sold.

Smaller producers/poolers may earn higher returns by producing the type of hogs that packers prefer, then marketing their hogs on a grade and yield basis.

Those pooling slaughter hogs are charged a commission, or marketing fee, to cover costs. However, the fee may be less than marketing charges at public markets.

Successful slaughter hog pooling must be customer-oriented. Hog producers should understand the needs and wants of packer buyers. Then, they should meet these needs via a pooling program.

PACKER MARKETING CONTRACT

Fig. 17-5. Finishing hogs, nearing market weight. (Courtesy, DeKalb Swine Breeders, Inc., DeKalb, IL)

A hog marketing contract is an agreement between a seller (usually a producer) and a meat packer to sell/buy at a specified future date a specified number of hogs of a specified weight and grade for a certain price.

The most common hog contracts are *packer marketing contracts.*

Typically, the following terms are included in a packer marketing contract:

1. **Quantity.** The quantity to be delivered, with the minimum amount varying anywhere from 5,000 to 30,000 lb (30,000 lb being equivalent to one live hog futures contract).

2. **Location and date of delivery.** The delivery location and date are usually specified with the time of delivery within a specified time interval, and with provision to change time of delivery by mutual agreement.

3. **Acceptable weights and grades.** Additionally, there may be provision for premiums and discounts. For example, the packer may require that all hogs be at least 50% lean. Some contracts now price hogs on a carcass grade and yield basis, which rewards better producers.

4. **Pricing arrangement.** The most popular contract sets a floor and a ceiling for the duration of the contract. For example, the contract might specify a floor of $38 per cwt and a ceiling of $48 cwt. If the cash market stays within the range, the producer gets

the going cash market when the hogs are delivered. But if the market drops to $34, the producer and the packer split the difference between the floor set in the contract and the actual cash market; in this case, the producer would get $36. Should hog prices rise above the $48 ceiling, the same rule holds; the producer and the packer split the difference, so hogs sold on a $52 cash market would return $50 to the producer.

Some packer contracts specify fixed ceiling and floor prices, without provision to split the difference. Thus in the example given above, the producer would never take less than $38 or get more than $48 regardless of how low or high actual hog prices might go.

Both of the above types of contracts often include a formula for recalculating the price if the cost of production changes, especially as feed prices rise or fall. Thus, in the Midwest where corn and soybeans account for about 60% of the cost of production, the contract may call for adjustment in case corn and soybean meal prices go up.

5. **Duration of contracts.** Contracts are usually long-term—5 to 7 years, because bankers and other financiers may require long-term marketing arrangements before financing facilities.

In some respects, a 5- to 7-year contract is a gamble. The producer is betting hog prices will drift lower in the next few years, whereas the packer is betting prices will gradually move higher.

6. **Some financial experts advise producers against putting their entire hog output under contract.** These specialists advise producers to contract enough hogs (a) to be sure a bad market won't ruin them, and (b) to keep their banker satisfied. But they recommend that some hogs (perhaps 40 to 50%) be without a contract in case prices advance. This allows producers some breathing space in case they are hit with unusually large pig losses (for example, an outbreak of TGE) during the contract.

7. **Future of packer contracting.** Packer market contracting will increase. Some knowledgeable market specialists predict that, in the early 21st century, packers will have 75% of their hog kill under contract.

8. **Miscellaneous provisions of packer marketing contracts.** Typically, marketing contracts also make provisions for (a) nonacceptable hogs and carcasses, (b) credit arrangement, (c) buyer inspection of the hogs while on the seller's premises, and (d) breach of contract. **Before signing a contract, producers are admonished to read the stipulations of the contract with care.**

9. **Concerns.** If packers get 50% or more of their kill under contract, what impact will 5 to 7 year contracts have on the cash market?

If a minority of the hogs marketed (*e.g.*, 25%) set the cash market, will they be hogs of lesser quality? Then, would these few hogs of lesser quality be used to calculate prices for the contract hogs?

FEEDER PIG MARKETING

Fig. 17-6. Feeder pigs. (Courtesy, Kansas State University, Manhattan)

Feeder pig marketing is an important part of the swine industry.

Producers of feeder pigs are getting larger. Also, through market price, packers are signaling pork producers that high-quality, consistent, and lean slaughter animals will receive significant premiums. In turn, this places increasing pressure on feeder pig producers to improve the quality, uniformity, and health of feeder pigs.

The most prevalent marketing methods for feeder pigs are the following:

1. **Direct.** This consists of a feeder pig producer marketing to a feeder pig finisher. Price may be either negotiated or determined by long-term contractual arrangement based on a formula price.

2. **Public auctions.** Pigs may be sold as delivered or graded and pooled for sale to buyers. Pigs are sold to the highest bidder, with price determined by demand.

3. **Electronic auctions.** Electronic feeder pig auctions have gained increasing popularity in recent years. Feeder pigs offered for sale are scored on the basis of health programs, feeding programs, and herd management. This information, along with identification of the seller, is made available to prospective buyers via computer modems. Consigned feeder pigs can be reviewed by potential buyers who may participate in an open auction as each pen is offered for sale. No commingling or central collection of the feeder pigs occurs. Pigs move directly from the seller to the buyer. Feeder pigs that are not accurately represented, or that do not meet weight and grade specifications, are either price adjusted at delivery or rejected by the buyer.

Note: Pertinent facts about one successful electronic auction follow. Other electronic auctions are operating in the United States and Canada.

Facts about the successful computerized feeder pig auction, called Bid-Plus, initiated by Central Livestock, a South St. Paul Marketing Agency, follow:

1. **When held.** It is held at 12:30 p.m. every Wednesday.

2. **Producer registries.** In order to participate, producers must (1) receive a herd qualification and rating from the swine specialist at the University of Minnesota, and (b) have a computer and modem compatible with the system.

3. **Entry of pigs.** Producers enter their pigs by lot number, color, weight, health status, and "rating score." Buyers can request further information on genetics, average daily gain, feed efficiency, etc.

4. **Sale charges.** Sale charges range from $1.60 per head up to 99 pigs, to $1.25 per head for 150 or more pigs.

5. **Credit rating of buyers and base price.** Bid-Plus, the feeder pig auction, checks the credit on all potential buyers, and sets the base price on sale day.

6. **Buyers and sellers are linked, and pigs are transferred.** After the sale is over, buyers and sellers are linked and arrangements for the transaction completed. Pigs are usually transferred trailer-to-trailer at elevators or other locations with scales.

Other innovative feeder pig marketing methods include telephone marketing and video marketing.

Because of alleviating exposure to disease when feeder pigs are sold live in a public auction ring or terminal market, it is predicted that electronic auctions of feeder pigs will increase.

■ **Feeder pig price, supply, and demand**—Feeder pig prices are volatile, reacting to numerous economic factors, supply, and/or demand. Thus, it is noteworthy that, during the eight year period 1985 through 1992, feeder pig prices varied from less than $30 per head to more than $60 per head.

Primary determinants of feeder pig prices are shown in Fig. 17-7. Feeder pig demand is influenced by numerous factors. Profit expectations of feeder pig buyers is the major pricing factor. Expected market hog prices at the time the feeder pigs will be slaughtered are an important consideration. Also, the live hog futures price for contracts maturing near the expected slaughter hog marketing date serve as a barometer of future cash slaughter hog market price levels. Thus, increases in deferred live hog futures prices generally signal increased cash feeder pig prices. Any economic factor that affects market hog prices, including pork production, production of competing meats, meat exports and imports, consumer income, and strength of economy exert indirect influences on feeder pig prices.

The second most important factor affecting feeder

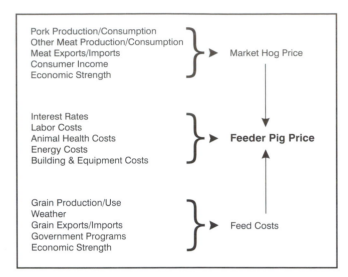

Fig. 17-7. Factors affecting feeder pig prices. (From: *Pork Industry Handbook*, PIH-72, University of Illinois at Urbana-Champaign)

pig demand and prices is feeding costs. Feeding costs are affected both by the cost of the feed itself and the efficiency with which the animal can turn these inputs into salable pork. Increases in feeding costs reduce feeder pig demand driving prices down. As a result, weather affecting feed grain and soybean meal prices can have dramatic influences on feeder pig prices. Characteristics of the pig itself, such as its genetics and health status plus the environment within which the pig will be finished, also affect feeding costs and thereby feeder pig prices. Other costs affecting feeder pig costs include interest rates, labor costs, routine health costs, energy costs, and building and equipment cots. One method to project the expected influence of changes in feed costs and expected slaughter hog prices on feeder pig prices is to develop feeder pig finishing budgets to determine break-even feeder pig purchase prices for different feeding costs and slaughter hog prices. Actual feeder pig prices usually follow break-even prices (perhaps with a profit adjustment).

■ **Feeder pig price information**—Daily feeder pig price information is limited in many regions. Producers often must rely on weekly auction quotations, terminal market sale reports, cooperative market reports, releases from agribusiness firms, electronic wire service reports, and periodic government releases. The percentage of feeder pigs being marketed directly from feeder pig producers to hog finishers in private treaties reduces the availability and, therefore, representativeness of feeder pig price data.

Feeder pig prices usually are quoted on a per head basis or occasionally on per hundredweight (cwt) basis. Interpretation of price reports requires understanding the market conditions, location, terms of the sale, grade standards, number of pigs represented by price quote, feeder pig quality, and preconditioning and vaccination backgrounds of the pigs.

As an alternative to using a competitive bid, some feeder pig producers use a formula to establish the sale price for their pigs. Making a formula pricing system work is very difficult. Because of changing price relationships among slaughter hogs, hog futures, feed prices, interest rates and other production costs, formula prices will rarely exactly match market prices. When the formula price is below the market price, the pig seller often feels short-changed. When the formula price is above the market, the pig buyer is likely to feel disadvantaged. For this reason, formula pricing works best when the feeder pig producer deals with the same buyer(s) over an extended period. The most successful formulas are those based on an established feeder pig market.

■ **Guidelines for selling feeder pigs**—Historically, most feeder pig producers have been able to rely upon organized public markets as an option for selling some or all of their pigs. For a variety of reasons, the number of organized feeder pig auctions has declined greatly in the last decade and is likely to continue to dwindle. This means that direct sales to finishers may be the only option available to many feeder pig producers in the future. Direct sales place importance upon lot size, pig uniformity, genetics, and health status. It is clear that feeder pig buyers place a significant premium upon large lots of uniform pigs from a single producer.

Selling feeder pigs to an individual buyer allows the seller to work with the buyer to refine and customize a health program to a buyer's needs. The seller should work with the buyer and the buyer's veterinarian to insure that the health status of the pigs delivered meets the demands of the buyer.

Some general guidelines for selling feeder pigs follow:

1. Provide healthy pigs of uniform size and quality. Remove the "bottom-sort" or "tail-enders" from the group. All-in, all-out nursery production is critical to overall disease control and helps with uniformity of pigs.

2. Provide pigs with a narrow age spread (2 weeks or less). This helps reduce the transmission of disease organisms, especially respiratory diseases, from older to younger pigs.

3. Dock pigs' tails and castrate male pigs well in advance of delivery, so that they are healed by the time the buyer receives them. Pigs with physical defects, such as hernias or other blemishes, should not be delivered.

4. Establish parasite control practices in the seller's breeding herd to insure the delivery of pigs free of internal and external parasites.

5. Market healthy feeder pigs. The parent herd should have no history of *Actinobacillus pleuropneumonia* (APP) or swine dysentery. Ideally, pigs should be free of pseudorabies (PRV), or status should be known. Some finishers do not discriminate severely against PRV pigs if they are free of other diseases. State laws vary on movement of PRV infected pigs.

6. Vaccinate pigs against erysipelas. If the parent herd has a history, or if the buyer requests, atrophic rhinitis and APP vaccinations may be necessary.

7. Supply the herd history and background information to the buyer and the buyer's veterinarian. Also, provide a health certificate or inspection form executed by the seller's veterinarian.

Some guidelines for selling feeder pigs through a market follow:

Most of the guidelines that apply to selling to an individual buyer apply to selling through a public market, except that usually the communication between seller and buyer is not practical, so a specific program cannot be developed. Most markets have general health requirements concerning weight range and physical appearance of pigs. Uniformity of pigs is still recommended but not critical since they will usually be sorted by weight and grade by market personnel. Most markets now require that pigs be from PRV-monitored herds before they will accept them for sale.

■ **Co-op feeder pig production marketing**—As the pork production industry evolves, pressures for volume production of consistent, high quality pigs are resulting in the reemergence of cooperative feeder pig production. Commonly, several producers pool their resources to develop a feeder pig production site. Each producer may then claim one week's (or some fraction of a week's) production on a rotating basis with excess pigs (if available) contractually sold to third parties. Often a different site is obtained for the sow herd and nursery to take advantage of multiple site health benefits. Pigs may be formula priced to cooperative participants based on cost. Such an arrangement allows specialization, consistency, and the exploitation of economies of size. This trend is expected to increase in the future.

■ **Grades of feeder pigs**—USDA grading standards are available for feeder pig producers. Grade description can provide the uniform terminology for trading nationwide, using new communication technology and accurate formula pricing. Some organizations modify the USDA standards to conform with their own grading systems. All grading standards attempt to relate the feeder pig to the final, finished quality of the animal. USDA grades include U.S. Numbers 1 through 4 and Utility. Additional information about USDA grades can be obtained from the Packers and Stockyards division of USDA or your local Cooperative Extension Office.

(Also see the subsequent section in this chapter headed "Market Classes and Grades of Hogs.")

EARLY WEANER PIG MARKETING (SEW Pigs)

This refers to marketing segregated early weaned (SEW) pigs—pigs weaned under 3 weeks of age, rather than the more conventional 21- to 28-days of age; then, segregating the piglets from their mothers.

Traditionally, feeder pigs have been sold, transported, and mixed at about 30- to 60-lb weight. Removing pigs at less than 3 weeks of age is desirable from a health standpoint; it removes them from the primary source of exposure to infectious organisms (the sow herd) while they still have maternal antibodies (passive immunity) to protect them from many disease organisms. Research is underway to determine the specific logistics of transporting and housing young pigs that originate from several herds. A heated, covered method of transportation and a modern, environmentally-controlled hot nursery in which to be housed are necessary. This technology may allow buyers to acquire pigs from more than one source without the severe disease consequences that sometimes occur when mixing 40- to 60-lb feeder pigs.

In the quest for healthy growing-finishing pigs, more and more producers are turning to the SEW market. In the mid-1990s, these 10-lb pigs were bringing $32 a head in some markets. Under proper feed and management, they reach market weight of 245 lb at 100–105 days of age.

In summary, there are big advantages from SEW pigs: (1) It separates pigs from diseases carried by the sows, and (2) sows return to estrus sooner, in better condition, and farrow more uniform litters.

(Also see Chapter 9, section headed Segregated Early Weaning [SEW]; and Chapter 15, section headed Segregated Early Weaning [SEW].)

CHOICE OF MARKET OUTLETS

Marketing is dynamic. Changes are inevitable in types of market outlets, market structures, and market services. Some outlets have gained in importance; others have declined.

The choice of a market outlet represents the seller's evaluation of the most favorable market among the number of alternatives available. No simple and brief statement of criteria can be given as a guide to the choice of the most favorable market channel. Rather, an evaluation is required of the contributions made by alternative markets in terms of available

services offered, selling costs, the competitive nature of the pricing process, and ultimately the producer's net return. Thus, an accurate appraisal is not simple.

From time to time, producers can be expected to shift from one type of market outlet to another. Because price changes at different market outlets do not take place simultaneously, nor in the same amount, nor even in the same direction, one market may be the most advantageous outlet for a particular class and grade of hogs at one time, but another may be more advantageous at some other time. The situation may differ for different classes and kinds of livestock and may vary from one area to another.

Regardless of the channels through which producers market their hogs, in one way or another, they pay or bear, either in the price they receive from the hogs or otherwise, the entire cost of marketing. Because of this, they should never choose a market because of convenience of habit, or because of personal acquaintance with the market and its operator. Rather, the choice should be determined strictly by the net returns from the sale of hogs; effective selling and net returns are more important than selling costs.

SELLING PUREBRED AND SEEDSTOCK HOGS

Selling purebred and seedstock animals is a highly specialized and scientific business.

Purebred animals are members of a breed which possess a common ancestry and distinctive characteristics and are either registered or eligible for registry.

Seedstock suppliers are competitors of purebred breeders who sell boars and gilts of specialized bloodlines. These lines of breeding, usually called hybrids, originate from crossing two or more breeds, then applying some specialized selection programs.

In general, the vast majority of purebred and seedstock boars saved for breeding purposes go into commercial herds. Only the elite sires are retained with the hope of effecting further improvement in the source herds. On the other hand, the sale of purebred or seedstock females is fairly well restricted to meeting the requirements for replacement purposes in existing herds or for establishing new herds.

Most purebred consignment sales are sponsored by a breed association, local, statewide, or national in character. Such auctions are usually limited to one breed. Purebred auction sales are conducted by highly specialized auctioneers. In addition to being good salespersons, such auctioneers must have a keen knowledge of values and must be familiar with the bloodlines of the breeding stock.

Fig. 17-8. A purebred Hampshire boar. (Courtesy, Lone Willow Farm, Roanoke, IL)

LIVESTOCK MARKET NEWS SERVICES

Accurate market news is essential to the efficient marketing of livestock, both from the standpoint of the buyer and the seller. In the days of trailing, the meager market reports available were largely conveyed by word of mouth. Moreover, the time required to move livestock from the farm or ranch to market was so great that detailed market information would have been of little benefit even if it had been available. With the speed in transportation afforded by trucks, late information on market conditions became important.

The Federal Market News Service was initiated by the U.S. Department of Agriculture beginning in 1916. This service was established for the purpose of providing unbiased and uniformly interpretable market information. It depends on voluntary cooperation in gathering information. There is no legal compulsion for buyers and sellers to divulge purchase and sale information. Re-

ports on direct sales are obtained largely by telephone and teletypewriter, augmented by interviews made at packing plants and farms. Then, for disseminating market reports, the Federal Market News Service relies upon local and privately owned newspapers, radio stations, and TV stations—merely supplying them with the information. Because at least a part of the readers or listeners are interested in this type of information, the local papers and radio stations are usually glad to serve as media for releasing these reports.

Other important sources of market information include farm and trade magazines. Also, many market agencies—such as commission firms, auction markets, and related organizations—prepare and distribute market information. By means of weekly market newsletters or cards, they commonly emphasize the price and market conditions of the particular market they serve.

■ **Terminology of market reports—** Knowledgeable producers must follow market reports in order to determine the best channels through which to market their hogs, as well as to project future trends in supply and demand so that they plan their program accordingly. Table 17-2, briefly describes terms commonly associated with the marketing of commodities.

Fig. 17-9. A seedstock gilt. (Courtesy, Farmers Hybrid Co., Des Moines, IA)

TABLE 17-2
GLOSSARY OF TERMS USED IN FEDERAL-STATE MARKET NEWS REPORTS[1]

Terms	Definitions
Market	1. A geographic location where a commodity is traded. 2. The price, or price level, at which a commodity is traded. 3. To sell.
Market activity:	The pace at which sales are being made.
• Active	Available supplies (offerings) are readily clearing the market.
• Moderate	Available supplies (offerings) are clearing the market at a reasonable rate.
• Slow	Available supplies (offerings) are not readily clearing the market.
• Inactive	Sales are intermittent with few buyers or sellers.
Price trend:	The direction in which prices are moving in relation to trading in the previous reporting period(s).
• Higher	The majority of sales are at prices measurably higher than the previous trading session.
• Firm	Prices are tending higher, but not measurably so.
• Steady	Prices are unchanged from previous trading session.
• Weak	Prices are tending low, but not measurably so.
• Lower	Prices for most sales are measurably lower than the previous trading session.
Supply/Offering:	The quantity of a particular item available for current trading.
• Heavy	When the volume of supplies is above average for the market being reported.
• Moderate	When the volume of supplies is average for the market being reported.
• Light	When the volume of supplies is below average for the market being reported.
Demand:	The desire to possess a commodity coupled with the willingness and ability to pay.
• Very good	Offerings or supplies are readily absorbed.
• Good	Firm confidence on the part of buyers that general market conditions are good. Trading is more active than normal.
• Moderate	Average buyer interest and trading.
• Light	Demand is below average.
• Very light	Few buyers are interested in trading.
Mostly	The majority of sales or volume.
Undertone	Situation or sense of direction in an unsettled market situation.

[1]*Glossary of Terms Used in Federal-State Market News Reports*, Agricultural Marketing Service, USDA.

PREPARING AND SHIPPING HOGS

Improper handling of hogs immediately prior to and during shipment may result in excessive shrinkage; high death, bruise, and crippling losses; disappointing sales; and dissatisfied buyers. Unfortunately, many swine producers who do a superb job of producing hogs, dissipate all the good things that have gone before by doing a poor job of preparing and shipping. Generally speaking, such omissions are due to lack of know-how, rather than any deliberate attempt to take advantage of anyone. Even if the sale is consummated prior to delivery, negligence at shipping time will make for a dissatisfied customer. Buyers soon learn what to expect from various producers and place their bids accordingly.

In addition to the important specific considerations covered in later sections, the following general considerations should be accorded in preparing hogs for shipment and in transporting them to market:

1. **Select the best method of transportation.** Distance of haul is the greatest single factor for consideration in this regard.

All major truckers clean, disinfect, and bed facilities prior to loading, but it is always well that shipper make their own inspection to make sure that these matters have been handled to their satisfaction. Generally, hogs are bedded with about 1 in. of sand, and in the wintertime straw is placed on top of the sand.

To avoid any misunderstanding, it is recommended that all requests for hauling facilities be either requested or confirmed in writing.

2. **Feed and water properly prior to loading.** Never ship hogs on an excess fill. Instead, fast them for 10 hours before shipping. If hogs are in transit longer than 10 hours, feed lightly en route. Although feed is cut off, never withhold water. Access to water is necessary in order to lessen stress in shipment.

Hogs that are too full of feed (especially wet feed) at the time of loading will scour and urinate excessively. As a result, the floors become dirty and slippery and the animals befoul themselves. Such hogs undergo a heavy shrink and present an unattractive appearance when unloaded.

3. **Keep hogs quiet.** Prior to and during shipment, hogs should be handled carefully. Hot, excited animals experience more shrinkage and are more apt to be injured or die.

Although loading may be exasperating at times, take it easy; never lose your temper. Avoid hurrying and striking. Never beat an animal with such objects as pipes, sticks, canes, or forks; instead, use (a) a flat, wide canvas slapper with a handle, or (b) a broom.

4. **Comply with the requirements for health certificates and permits.** When hogs are to be shipped into another state, the shipper should check into and comply with the state regulations relative to health certificates and permits. Usually, the local veterinarian will have this information. Should there be any question about the health regulations, however, the state livestock sanitary board (usually located at the state capital) of the state of destination should be consulted. Knowledge of and compliance with such regulations well in advance of shipment will avoid frustrations and costly delays.

5. **Comply with the federal 28-hour law in rail shipments.**[6] By federal law, passed in 1873, livestock cannot be transported by rail for a longer period than 28 consecutive hours without unloading for the purpose of giving feed, water, and rest for a period of at least 5 consecutive hours before resuming transportation. The period may be extended to 36 hours upon written request from the owner of the animals.

6. **Use partitions in the truck or car when necessary.** When mixed loads (consisting of hogs, cattle, and/or sheep) are placed in the same truck, partition each class off separately. Also, boars, stags, and sows should be properly partitioned.

7. **Avoid shipping during extremes in weather.** Whenever possible, avoid shipping when the weather is either very hot or very cold. During such times, shrinkage and death losses are higher than normal. During warm weather, avoid transporting hogs during the heat of the day; travel at night or in the evening or early morning. In hot weather, wet the sand; and it may even be wise to distribute some ice on the floor of the truck.

---

[6]No such law applies to truck transportation of animals.

PREVENTING BRUISES, CRIPPLING, AND DEATH LOSSES

Losses from bruising, crippling, and death that occur during the marketing process represent a part of the cost of marketing hogs; and, indirectly, the producer foots most of the bill.

The following precautions are suggested as a means of reducing hog market losses from bruises, crippling, and death:

1. Remove projecting nails, splinters, and broken boards from feeding areas and fences.
2. Keep feedlots free from old machinery, trash, and any obstacle that may bruise.
3. Do not feed wet feeds heavily just prior to loading.
4. Use good loading chutes; not too steep.
5. Bed with sand free from stones, to prevent slipping. Cover sand with straw in cold weather, but do not use straw in hot weather. Wet the sand bedding in summer before loading hogs and while enroute. Drench when necessary.
6. Use partitions in trucks that are not fully loaded, to keep animals closer together; and in very long trucks to keep animals from crowding from one location to another.
7. Provide covers for trucks to protect from sun in summer and cold in winter.
8. Always partition mixed loads into separate classes. Partition boars, and stags.
9. Remove protruding nails, bolts, and any sharp objects in truck.
10. Load slowly to prevent crowding against sharp corners and to avoid excitement. Do not overload.
11. Use canvas slappers instead of clubs or canes.
12. Drive trucks carefully; slow down on sharp turns and avoid sudden stops.
13. Back truck slowly and squarely against unloading dock.
14. Unload slowly. Do not drop animals from upper to lower deck; use cleated inclines.
15. Inspect load enroute to prevent trampling of animals that may be down. Get downed animal back on feet immediately.
16. If porcine stress syndrome (PSS) is known to be in a herd, exercise extreme care in all stages of getting hogs to market.

All these precautions are simple to apply, yet all are violated every day of the year.

NUMBER OF HOGS IN A TRUCK

Overcrowding of market animals causes heavy losses. Sometimes a truck is overloaded in an attempt to effect a saving in hauling charges. More frequently, however, it is simply the result of not knowing space requirements. The suggested number of animals should be tempered by such factors as distance of haul, class of livestock, weather, and road conditions.

Truck beds vary in length. The size of the truck and the class and size of animals determine the number of head that can be loaded in a truck. For comfort in shipping, the truck should be loaded heavily enough so that the animals stand close together, but both underloading and overcrowding are to be avoided.

Table 17-3, shows the number of swine for safe trucking.

KIND OF BEDDING TO USE FOR HOGS IN TRANSIT

Among the several factors affecting hog losses, perhaps none is more important than proper bedding and footing in transit.

Footing, such as sand, is required at all times of the year, to prevent the car or truck floor from becoming wet and slick, thus predisposing animals to injury by slipping and falling. Bedding, such as straw, is recom-

mended for warmth in the shipment of swine during extremely cold weather. During warm weather, the sand should be wet down prior to loading, for it has been well said that a hog with a wet belly is a live hog all the way to market. Recommended kinds and amounts of bedding and footing materials are given in Table 17-4.

SHRINKAGE IN MARKETING HOGS

The shrinkage (or drift) refers to the weight loss encountered from the time animals leave the feedlot until they are weighed over the scales at the market. Thus, if a hog weighed 200 lb at the feedlot and had a market weight of 196 lb the shrinkage would be 4 lb or 2.0%. Shrink is usually expressed in terms of percentage. Most of this weight loss is due to excretion, in the form of feces and urine and the moisture in the expired air. On the other hand, there is some tissue shrinkage, which results from metabolic or breakdown changes.

The most important factors affecting shrinkage of hogs are:

1. **Season.** Extremes in temperature, either very

TABLE 17-3
NUMBER OF HOGS FOR SAFE LOADING IN A TRUCK

Floor Length		Weight of Hogs								
		100 lb (45 kg)	150 lb (68 kg)	175 lb (79 kg)	200 lb (91 kg)	225 lb (102 kg)	250 lb (113 kg)	300 lb (136 kg)	350 lb (159 kg)	400 lb (181 kg)
(ft)	(m)									
8	2.4	27	21	19	18	16	14	13	11	9
10	3.1	33	26	24	22	20	18	16	14	12
12	3.7	40	31	28	26	24	22	19	17	14
15	4.6	50	39	36	33	30	27	24	21	17
18	5.5	60	47	43	40	36	33	28	25	21
20	6.1	67	52	48	44	40	35	32	28	24
24	7.3	80	62	57	52	48	44	38	34	28

TABLE 17-4
GUIDE RELATIVE TO BEDDING AND FOOTING MATERIAL WHEN TRANSPORTING SWINE[1, 2, 3]

Class of Livestock	Kind of Bedding for Moderate or Warm Weather; Above 50°F (10°C)	Kind of Bedding for Cool or Cold Weather; Below 50°F (10°C)
Swine	Sand, 0.5 to 2 in. (1.3 to 5.1 cm).	Sand covered with straw.

[1]Straw or other suitable bedding (covered over sand) should be used for protection and cushioning breeding stock that are loaded lightly enough to permit their lying down in the truck.

[2]Sand should be clean and medium-fine, and free from brick, stones, coarse gravel, dirt, and dust.

[3]In hot weather, wet sand down before loading. Never apply water to the backs of hot hogs; it may kill them.

hot or very cold weather, result in higher shrinkage. Shrink is at a minimum between 20 and 60°F. When the temperature is above 80°F, hogs in transit should be sprinkled.

2. **Age and weight.** Young pigs shrink proportionally more than older animals because of less body fat and a greater fill in proportion to liveweight.

3. **Overloading or underloading.** Either overloading or underloading always results in abnormally high shrinkage. Shrinkage from underloading may be prevented by using proper partitions.

4. **Rough ride, abnormal feeding, and mixed loads.** Each of these factors will increase shrinkage.

In a recent study, the University of Missouri found that hogs shrink about 1.7% during a 100-mile shipment from farm to market. Moreover, the majority of the shrink was during the first few miles; and the shrink didn't vary much, no matter how the hogs were raised—in dry lot, in confinement, or in an open-front unit.

AIR TRANSPORTATION

With modern communication and transportation, the world is becoming smaller and smaller. Along with this, there is an increasing need for and desire to move animals efficiently between countries separated by great distances. Ancestors to the present day U.S. hog endured a long sea voyage, which was often a hardship on humans. For many years, ships were considered the only economical method of moving large numbers of animals across the seas, but animals never adapted well to the pitching and tossing of sea travel. In some cases, losses of up to 50% were not uncommon. Then during the late 1960s the concept of air transportation of large numbers of animals was born. Whole planes were adapted for livestock comfort and maximum capacity—literally "flying corrals." With this modern concept, animals can arrive at any destination in the world within hours, thereby minimizing stress and virtually eliminating death losses. Thus, numerous swine have been successfully airlifted worldwide.

The number of animals per load depends on the size of the individual animals and the size of the aircraft, but some of the larger aircraft can carry 900 pigs.

MARKET CLASSES AND GRADES OF HOGS

The generally accepted market classes and grades of hogs are summarized in Table 17-5.

Swine market classes and grades differ from those

TABLE 17-5
THE MARKET CLASSES AND QUALITY GRADES OF HOGS

Hogs or Pigs	Use Selection	Sex Class	Weight Divisions				Commonly Used Grades
			- - - - - - (lb) - - - - - -		- - - - - - (kg) - - - - - -		
Hogs	Slaughter hogs	Barrows and gilts (often called butcher hogs)	120–140 140–160 160–180 180–200 200–220 220–240	240–270 270–300 300–330 330–360 360–400 400 lb up	55–64 64–73 73–82 82–91 91–100 100–109	109–123 123–136 136–150 150–163 163–182 182 kg up	U.S. No. 1, U.S. No. 2, U.S. No. 3, U.S. No. 4, U.S. Utility
		Sows (or packing sows)	270–300 300–330 330–360 360–400	400–450 450–500 500–600 600 lb up	123–136 136–150 150–163 163–182	182–204 204–227 227–272 272 kg up	U.S. No. 1, U.S. No. 2, U.S. No. 3, Medium, Cull
		Stags	All weights				Ungraded
		Boars	All weights				Ungraded
	Feeder hogs	Barrows and gilts	120–140 140–160 160–180		55–64 64–73 73–82		U.S. No. 1, U.S. No. 2, U.S. No. 3, U.S. No. 4, U.S. Utility, Cull
Pigs	Slaughter pigs	Barrows, gilts, and boars	Under 30 30–60		13.6 13.6–27.2		Ungraded
		Barrows and gilts	60–80 80–100 100–120		27.2–36.3 36.3–45.4 45.4–54.5		Ungraded
	Feeder pigs	Barrows and gilts	80–100 100–120		36.3–45.4 45.4–54.5		U.S. No. 1, U.S. No. 2, U.S. No. 3, U.S. No. 4, U.S. Utility, Cull

used in cattle and sheep in that: (1) there are no age divisions by years (*e.g.*, cattle are classified as yearlings and 2-year-olds and over); and (2) rarely are hogs of any kind purchased on the market for use as breeding animals. The class of market hogs indicates the use to which the animals are best adapted, whereas the grade indicates the degree of perfection within the class.

FACTORS DETERMINING MARKET CLASSES OF HOGS

The market classes of hogs are determined by the following factors: (1) hogs or pigs, (2) use selection, (3) sex, and (4) weight.

HOGS OR PIGS

All swine are first divided into two major groups according to age: hogs or pigs. Although actual ages are not observed, the division is made largely by weight in relation to the animal's apparent age. Young animals weighing under 120 lb (under about 3 months of age) are generally known as pigs, whereas those weighing over 120 lb are called hogs.

USE SELECTION OF HOGS AND PIGS

Hogs or pigs are further divided into two subdivisions as slaughter animals or feeders. Slaughter swine are hogs and pigs that are suitable for immediate slaughter. The demand for lightweight slaughter pigs is greatest during the holiday season when they are in demand as roasting pigs for hotels, clubs, restaurants, steamships, and other consumers. Such pigs weigh from 30 to 60 lb, are dressed shipper style (with the head on), and must produce a plump and well-proportioned carcass. Slaughter hogs (the older animals) are in demand throughout the year.

Feeder pigs include those animals that show ability to take on additional weight and finish. Moreover, because of the greater disease hazard with hogs, this class is under very close federal supervision. Before being released for return to the country, feeder swine must be inspected from a health standpoint, and then either sprayed or dipped as a precautionary measure to prevent the spread of disease germs or parasites.

SEX CLASSES

The sex class is used only when it affects the usefulness and selling price of animals. In hogs, this subdivision is of less importance than in cattle. Thus, barrows and gilts are always classed together in the

case of both slaughter and feeder hogs. This is done because the sex condition affects their usefulness so little that a price differentiation is not warranted. In addition, because the carcass is not affected, no sex differentiations are made for slaughter pigs under 60 lb in weight. The terms *barrow, gilt, sow, boar*, and *stag* are used to designate the sex classes of hogs. The definition of each of these terms follows:

■ **Barrow**—*A castrated male swine that was castrated at an early age—before reaching sexual maturity and before developing the physical characteristics peculiar to boars.*

■ **Gilt**—*A female swine that has not produced pigs and which has not reached an evident stage of pregnancy.*

■ **Sow**—*A female swine that shows evidence of having produced pigs or which is in an evident stage of pregnancy.*

■ **Boar**—*An uncastrated male swine of any age.* Mature boars should always be stagged and fed 3 weeks or longer (until the wound heals) before being sent to market. The market value of boars is necessarily low, for a considerable number are condemned as unfit for human consumption, primarily because of odor.

■ **Stag**—*A male swine that was castrated after it had developed the physical characteristics of a mature boar.* Because of relatively thick skins, coarse hair, and heavy bones, stags are subject to dockage. They are usually docked 70 lb, but may be docked from 40 to 80 lb, depending on the market. When marketed direct, stags are usually not docked in weight but are purchased at a price that reflects their true value from a meat standpoint.

WEIGHT DIVISIONS

Occasionally, the terms *light, medium*, and *heavy* are used to indicate approximate weights, but most generally the actual range in weight in pounds is specified both in trading and in market reporting. Moreover, hogs are usually grouped according to relatively narrow weight ranges because variations in weight affect (1) the dressing percentage, (2) the weight and desirability of the cuts of meat, and (3) the amount of lard produced (heavier weights produce more lard). Boars and stags are not usually subdivided according to weights.

THE FEDERAL GRADES OF HOGS

The market grade for swine, as for other kinds of livestock, is a specific indication of the degree of excellence within a given class based upon conformation, finish, and quality. The two chief factors which serve to place a hog in a specific grade are: degree of fatness and amount of muscling. While no official grading of live animals is done by the U.S. Department of Agriculture, market grades do form a basis for uniform reporting of livestock marketings. It is intended that the grade of slaughter hogs on foot be correlated with the carcass grade. However, it takes a great deal of experience and study to correlate live animals with the type of carcass they will produce.

Tentative standards for grades of pork carcasses and fresh pork cuts were first issued by the U.S. Department of Agriculture in 1931. Subsequently, these standards were revised in 1933, 1952, 1955, and 1968. In January, 1985, the standards were further revised, based on backfat thickness of the last rib and muscling.

Grades of barrow and gilt carcasses are based on two characteristics: (1) the quality indications of the lean, and (2) the expected combined yield of the four lean cuts—ham, loin, picnic shoulder, and Boston shoulder. The expected yield of the four lean cuts for each of the four grades are shown in Table 17-6.

TABLE 17-6
EXPECTED YIELDS OF THE FOUR LEAN CUTS BASED ON CHILLED PORK CARCASS WEIGHT, BY GRADE[1]

Grade	Yield
U.S. No. 1	60.4% and over
U.S. No. 2	57.4% to 60.3%
U.S. No. 3	54.4% to 57.3%
U.S. No. 4	Less than 54.4%

[1]These yields will be approximately 1% lower if based on hot carcass weight.

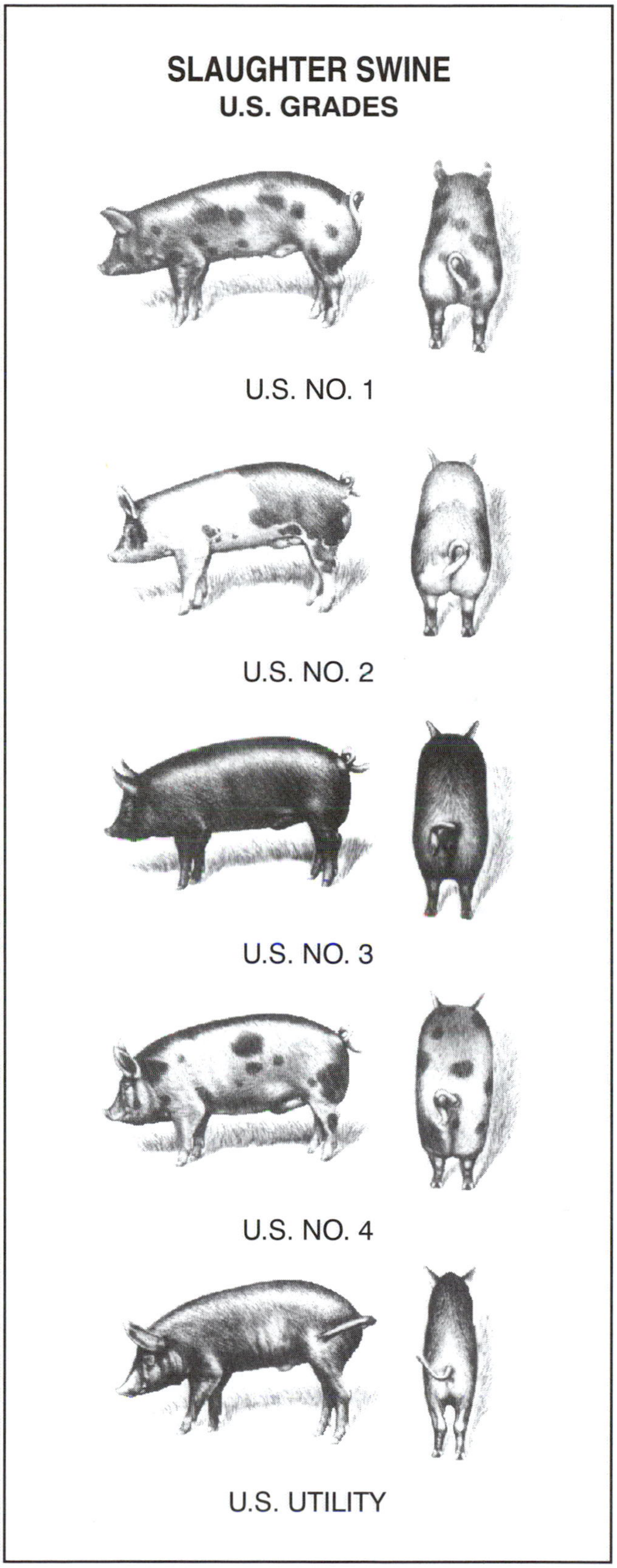

The current federal market grades of slaughter barrows and gilts, which were adopted in 1985, are: U.S. No. 1, U.S. No. 2, U.S. No. 3, U.S. No. 4, and U.S. Utility (see Fig. 17-10). The five grades for slaughter barrows and gilts may be described as follows:

■ **U.S. No. 1**—Slaughter barrows and gilts in this grade will produce carcasses with acceptable lean quality and acceptable belly thickness, and a high percentage of lean cuts (60.4% and over).

■ **U.S. No. 2**—Slaughter barrows and gilts in this grade will produce carcasses with acceptable lean quality and acceptable belly thickness, and a slightly lower percentage of lean cuts (57.4 to 60.3%).

■ **U.S. No. 3**—Slaughter barrows and gilts in this grade will produce carcasses with acceptable lean

Fig. 17-10. The five market grades of slaughter swine. (Courtesy, USDA)

quality and acceptable belly thickness, and a slightly lower percentage of the four lean cuts (54.4 to 57.3%).

■ **U.S. No. 4**—Slaughter barrows and gilts in this grade will produce carcasses with acceptable lean quality and acceptable belly thickness. However, they are fatter and less muscular and will have a lower carcass yield of the four lean cuts than those in the U.S. No. 3 grade (less than 54.4% of four lean cuts).

■ **U.S. Utility**—Barrows and gilts typical of this grade will have a thin covering of fat. The sides are wrinkled and the flanks are shallow and thin. They will produce carcasses with unacceptable lean quality and/or unacceptable belly thickness. Also, all carcasses that are soft and/or oily, or are pale, soft, and exudative (PSE), will be graded U.S. Utility.

The federal grades of slaughter sows, which became effective in 1956, are: U.S. No. 1, U.S. No. 2, U.S. No. 3, medium, and cull. The grades are based on differences in yields of lean cuts and fat cuts and differences in quality of pork. Medium-grade sows have a degree of finish less than the minimum required to produce pork of acceptable palatability. Cull-grade sows have a very low degree of finish.

As a rule, slaughter pigs that weigh under 60 lb are not graded because they have not reached sufficient maturity for variations in their conformation, finish, and quality to affect the market value materially.

For more information regarding the grading of carcasses, refer to Chapter 18, Pork and Byproducts from Hog Slaughter.

The grades of feeder pigs are closely correlated with the standards for slaughter barrows and gilts. These are given in Fig. 17-11. The standards on which these grades are based embrace two general value-determining characteristics of feeder pigs—(1) their logical slaughter potential and (2) their thriftiness. For example, if a feeder pig is graded U.S. No. 1, it has the potential for developing into a U.S. No. 1 slaughter hog that will produce a U.S. No. 1 carcass. Thriftiness indicates the ability of a feeder pig to gain weight rapidly and efficiently.

OTHER HOG MARKET TERMS AND FACTORS

In addition to the rather general terms used in designating the different market classes and grades of hogs, the following terms and factors are frequently of importance.

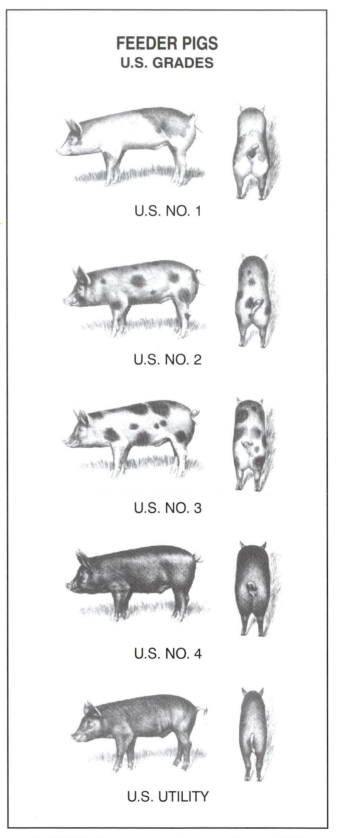

Fig. 17-11. The five market grades of feeder pigs. (Courtesy, USDA)

ROASTERS

Roasters refer to fat, plump, suckling pigs, weighing 30 to 60 lb on foot. These are dressed shipper style (with the head on); and they are not split at the chest or between the hams. When properly roasted and attractively served with the traditional apple in the mouth, roast pig is considered a great delicacy for the holiday season.

SUSPECTS (Governments)

Suspects or governments are suspicious animals that federal inspectors tag at the time of the antemortem inspection to indicate that more careful scrutiny is to be given in the postmortem inspection. If the carcass is deemed unfit for human consumption, it is condemned and sent to the inedible tank.

CRIPPLES

Cripples are hogs that are too crippled to walk. Such animals should never reach a market; they should be dealt with on the farm.

DEAD HOGS

Dead hogs are those that arrive dead at the market. They have practically no salvage value. These carcasses are sent to the tanks for conversion into inedible grease, fertilizer, etc.

SOME HOG MARKETING CONSIDERATIONS

Enlightened and shrewd marketing practices generally characterize the successful hog enterprise. Among the considerations of importance in marketing hogs are those which follow.

SECULAR TRENDS

Secular trends are longtime trends that persist over a period of several cycles. The long-run trend in U.S. hog numbers from 1925 to 1994 have been upward, as shown in Table 17-7. Likewise, the longtime trend in the nation's pork production has been upward due to increased hog numbers, along with improved breeding, feeding and management, which has increased productivity per head and resulted in marketing hogs at younger ages.

TABLE 17-7
ANNUAL COMMERCIAL SLAUGHTER OF HOGS
(Million Head)[1]

Year	Hogs	Year	Hogs	Year	Hogs
1925	65.5	1965	76.5	1981	92.5
1930	67.3	1966	75.4	1982	82.8
1935	46.0	1967	83.4	1983	88.1
1940	77.6	1968	86.4	1984	85.6
1945	71.9	1969	85.0	1985	84.9
1950	79.3	1970	87.1	1986	80.0
1955	81.1	1971	95.6	1987	81.4
1956	85.1	1972	85.9	1988	88.1
1957	78.6	1973	77.9	1989	88.7
1958	76.8	1974	83.1	1990	85.1
1959	87.6	1975	69.9	1991	88.2
1960	84.2	1976	75.0	1992	94.9
1961	82.0	1977	78.4	1993	93.1
1962	83.4	1978	78.4	1994	95.7
1963	87.1	1979	90.2		
1964	86.3	1980	97.2		

[1]From: USDA sources.

CYCLICAL MOVEMENTS

These are movements that follow a pattern that repeats itself. Hog cycles average about 4 years—2 years of expansion, and 2 years of liquidation (see Fig. 17-12). Basically, cycles are the response of producers to prices; and prices reflect supply and demand. In the past, the hog-corn ratio served as a barometer; when it was favorable—above 12—an expansion in hog numbers followed; when it was unfavorable—below 12—a cutback followed. Today, the hog-corn ratio is not as much of a factor as formerly. Also, large swine units are inclined to maintain production near optimum levels for the size unit without regard to cyclical movements.

These cycles are a direct reflection of the rapidity with which the numbers of each class of farm animals can be shifted under practical conditions to meet consumer meat demands. Thus, litter-bearing and early-producing swine can be increased in numbers much more rapidly than either sheep or cattle, whose cycles are 9 to 10 years and 10 to 12 years, respectively.

Normal cycles are disturbed by droughts, wars, general periods of depression or inflation, and federal controls.

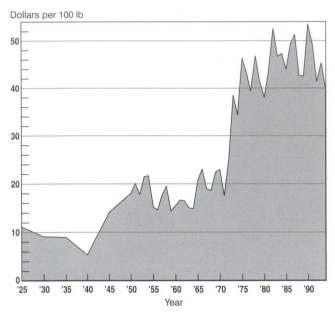

Fig. 17-12. The cyclic trend of the average price received by U.S. producers for hogs. The price cycle for hogs is 3 to 5 years. (*Pork Facts 1995/1996*, National Pork Producers Council, p. 13)

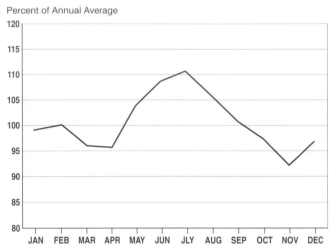

Fig. 17-13. Seasonal price index for barrows and gilts, 7 U.S. markets, 1983–1992.

SEASONAL VARIATIONS

Hog prices vary seasonally (within a year) due to the variation in market receipts. As would be expected, seasons of high market prices are generally associated with light marketings and seasons of low market prices with heavy marketings. It must be realized, however, that the normal seasons of high and low prices may be changed by such factors as (1) federal farm programs and controls, (2) business conditions and general price levels, (3) feed supplies and weather conditions, and (4) wars, etc.

In recent years, seasonal patterns have not been as reliable as they used to be. Year-round farrowing of sows in confinement has made for more uniform marketing throughout the year and lessened seasonality in livestock marketing. Thus, when arriving at livestock forecasts and marketing advice, proper reservation should be exercised in considering seasonal patterns. Anyway, it is not always wise to plan production to hit the highest market, for sometimes that would push up production costs more than enough to offset the gains from higher prices. Nevertheless, a careful study of normal seasonal prices will serve as a useful guide (see Fig. 17-13).

Fig. 17-13 is an average price index for each month. It shows the average relationship of prices in a particular month to the average of all months in the years 1983 to 1992. Primarily, the index reflects the seasonal variation in price since the calculation pro-

cedure eliminated most of the price gyration caused by other factors.

SHORT-TIME CHANGES

Day-to-day variations in swine prices usually are caused by uneven distribution of receipts on a given market because of such factors as weather, interference with transportation, strikes, uncertain or threatened federal policies, and stock market fluctuations. Although such changes are not large, shrewd producers and market specialists are quick to take advantage of fluctuations that are in their financial interest.

DOCKAGE

The value of some market animals is low because dressing losses are high, or because part of the product is of low quality. Some common dockages on hog markets are:

1. **Piggy sows.** Usually docked 40 lb, but it may range from 0 to 50 lb, depending on the market.

2. **Stags (hogs).** Usually docked 70 lb, but it may range from 40 to 80 lb, depending on the market.

PACKER SLAUGHTERING AND DRESSING OF HOGS

Hog slaughtering is unique in that much more pork is cured than is the case with beef or lamb; and pork fat (lard) is not classed as a byproduct, although the surplus fats of beef, veal, mutton, and lamb are in the byproduct category.

Table 17-8 shows the proportion of hogs slaughtered commercially—those slaughtered in federally inspected and in approved state inspected plants. The total figure refers to the number dressed in all establishments and on farms.

TABLE 17-8
PROPORTION OF HOGS SLAUGHTERED COMMERCIALLY
1989 THROUGH 1993[1]

Year	Total Number Slaughtered[2]	Commercial Slaughter	
	(1,000 head)	(1,000 head)	(%)
1989	89,007	88,692	99.6
1990	85,431	85,136	99.7
1991	88,445	88,169	99.7
1992	95,157	94,889	99.7
1993	93,296	93,068	99.8
5-year avg.	90,267	89,991	99.7[3]

[1]*Agricultural Statistics 1994*, USDA, p. 235, Table 397, and p. 240, Table 403.

[2]Includes commercial slaughter in federally inspected plants, in approved state inspected plants, and on farms.

[3]The remaining 0.3% are slaughtered on farms.

FEDERAL MEAT INSPECTION

The federal government requires supervision of establishments which slaughter, pack, render, and prepare meats and meat products for interstate shipment and foreign export. It is the responsibility of the respective states to have and enforce legislation governing the slaughtering, packaging, and handling of meats shipped intrastate. The meat inspection laws do not apply to farm slaughter for home consumption, although all states require inspection if the meat is sold.

The meat inspection service of the U.S. Department of Agriculture was inaugurated and is maintained under the Meat Inspection Act of June 30, 1906. This act was updated and strengthened by the Wholesome Meat Act of December 15, 1967. The latter statute (1) requires that state standards be at least to the levels applied to meat sent across state lines; and (2) assures consumers that all meat sold in the United States is inspected either by the federal government or by an equal state program. The Animal and Plant Health Inspection Service of the U.S. Department of Agriculture is charged with the responsibility of meat inspection.

The purposes of meat inspection are (1) to safeguard the public by eliminating diseased or otherwise unwholesome meat from the food supply, (2) to enforce the sanitary preparation of meat and meat products, (3) to guard against the use of harmful ingredients, and (4) to prevent the use of false or misleading names or statements on labels. The personnel responsible for carrying out the provisions of the act are of two types: Professional or veterinary inspectors who are graduates of accredited veterinary colleges, and nonprofessional food inspectors who are required to pass a Civil Service examination. In brief, the inspections consist of the following two types:

1. **Antemortem (before slaughter)** inspection is made in the pens or as the animals move from the scales after weighing. The inspection is performed to detect evidence of disease or any abnormal condition that would indicate a disease. Suspects are provided with a metal ear tag bearing the notation "U.S. Suspect No. . . . ," and are given special postmortem scrutiny. If in the antemortem examination there is definite and conclusive evidence that the animal is not fit for human consumption, it is "condemned," and no further postmortem examination is necessary.

Fig. 17-14. Antemortem (before slaughter) inspection of hogs being made by a federal veterinarian. Animals that are clearly diseased, emaciated, or otherwise unfit for human food are destroyed. Their carcasses may be used only in making inedible grease, fertilizers, or other nonfood products. Animals that appear slightly abnormal on foot are tagged "U.S. Suspect," and are given special postmortem scrutiny. (Courtesy, Technical Service Staff, Animal and Plant Health Inspection Service, USDA)

2. **Postmortem (after slaughter)** inspection is made at the time of slaughter and includes a careful examination of the carcass and the viscera (internal organs). All healthy carcasses (no evidence of disease) are stamped "U.S. Inspected and Passed," whereas the inedible carcasses are stamped "U.S. Inspected and Condemned." The latter are sent to the rendering tanks, the products of which are not used for human food.

In addition to the antemortem and postmortem

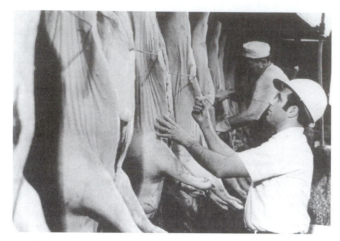

Fig. 17-15. Postmortem (after slaughter) inspection. (Courtesy, USDA)

inspections referred to, the government meat inspectors have the power to refuse the application of the mark of inspection to meat products produced in a plant that is not sanitary. All parts of the plant and its equipment must be maintained in a sanitary condition at all times. In addition, plant employees must wear clean, washable garments, and suitable lavatory facilities must be provided for hand washing.

Meat inspection regulations require the condemnation of all or affected portions of carcasses of animals with various disease conditions, including pneumonia, peritonitis, abscesses and pyemia, uremia, tetanus, rabies, anthrax, tuberculosis, various neoplasms (cancers), arthritis, actinobacillosis, and many others.

Most of the larger meat packers are under federal inspection; hence, they are allowed to ship interstate.

(Also see Chapter 18, section headed "Food Safety," including Irradiation and Hazard Analysis Critical Control Point [HACCP]; and sections headed "Traceback" and "Food Safe Labeling.")

■ **Federally inspected meat plants**—In 1994, there were 968 slaughter plants under federal inspection in the United States, 830 of which slaughtered hogs.

STEPS IN SLAUGHTERING AND DRESSING HOGS

After purchase, hogs are driven from the holding pens to the packing plant where they are given a shower and are held temporarily in a small pen while awaiting slaughter. In large meat packing plants, a series of chutes, and retainer conveyors move the hogs into position for stunning.

The chain method of slaughtering is used in killing and dressing hogs. In this method, the following steps are carried out in rapid succession:

1. **Rendering insensible.** The hogs are rendered insensible[7] by use of a captive bolt stunner, electric current, carbon dioxide, or gun shot. *Note:* The use of a .22 caliber rifle is approved in some locales, but great care must be exercised for human safety when using a rifle. Moreover, the brain from a hog that has been shot with a rifle is not edible because of possible lead or steel contamination by the rifle shell particles.

2. **Shackling and hoisting.** The hogs are shackled just above the hoof on the hind leg and are then hoisted to an overhead rail.

3. **Sticking.** The sticker sticks the hog just under the point of the breast bone, severing the arteries and veins leading to the heart. The animal is allowed to bleed for a few minutes.

4. **Scalding.** The animals are next placed in a scalding vat for about 4 minutes. By means of automatically controlled steam jets, the temperature of the water in the vats is maintained at about 150°F. The scalding process loosens the hair and scurf.

5. **Dehairing.** After scalding, the carcasses are elevated into a dehairing machine which scrapes them mechanically.

6. **Returning to overhead tracks.** As the carcasses are discharged from the dehairing machine, the gam cords of the hind legs are exposed; and gambrel spreaders are inserted in the cords. Then the carcass is again hung from the rail.

7. **Dressing.** A conveyor then moves the carcass slowly along a prescribed course where attendants perform the following tasks:

a. Washing and singeing.

b. Removing the head.

c. Opening the carcass and eviscerating.

d. Splitting or halving the carcass with a cleaver or an electric saw.

e. Removing the leaf fat.

f. Exposing the kidneys for inspection and facing the hams (removing the skin and fat from the inside face or cushion of the ham).

g. Washing the carcass and then sending it to the coolers where the temperature is held at around 34°F. (Rapid chilling is desirable.)

PACKER VS SHIPPER STYLE OF DRESSING

The two common styles of dressing hogs in pack-

---

[7]By federal law (known as the Humane Slaughter Act) passed in 1958 and effective June 30, 1960, unless a packer uses humane slaughter methods, he/she forfeits the right to sell meat to the government. The law lists the following methods as humane: by rendering insensible to pain by a single blow or gunshot or an electrical, chemical, or other means that is rapid and effective, before being shackled, hoisted, thrown, cast, or cut.

Fig. 17-16. Packer (left) vs shipper (right) style of dressing. Note that the head, leaf fat, and kidneys are removed from the packer-style carcass to the left. (Courtesy, Washington State University)

ing plants are: packer style and shipper style. In general, the packer style is followed, and this system is used almost exclusively when carcasses are to be converted into the primal cuts. In packer-style dressing, the backbone is split full length through the center; the head, without the jowl, is removed; and the kidneys and the leaf fat are removed and the hams faced.

The shipper style is ordinarily limited to lightweight slaughter pigs that are sold as entire carcasses to the wholesale trade. In this style of dressing, the carcass is merely opened from the crotch to the tip of the breastbone; the backbone is left intact; the leaf fat is left in; and the entire head is left attached. Roasting pigs are dressed shipper style and prior to cooling are placed in a trough with front legs doubled back from the knee joints and the hind legs extending straight back from the hams.

DRESSING PERCENTAGE OF HOGS

Dressing percentage may be defined as the percentage yield of chilled carcass in relation to the weight of the animal on-foot. For example, a hog that weighed 250 lb on-foot and yielded a carcass weighing 184.2 lb may be said to have a dressing percentage of 73.7

The degree of fatness and the style of dressing are the important factors affecting dressing percentage in hogs. U.S. No. 1 hogs dressed packer style (with head, leaf fat, and kidneys removed) dress about 70% whereas hogs dressed shipper style (head left on, and leaf fat and kidneys in) dress 4 to 8% higher.

Table 17-9 gives the approximate percentages that may be expected from the different grades of barrows and gilts. It is generally recognized that fat, lardy-type hogs give a higher dressing percentage than can be obtained with meat-type or bacon-type animals. Because lard frequently sells at a lower price than is paid for hogs on foot, an excess yield of lard very obviously represents an economic waste of feed in producing the animals and is undesirable from the standpoint of the processor. Accordingly, attaching great importance to the projected dressing percentage of hogs is outmoded. The more progressive buyers are now focusing their attention on the cutout value of the carcass, especially on the maximum yield of the more sought primal cuts of high quality—the production of lean pork.

Hogs have a relatively smaller barrel and chest cavity than cattle and sheep. In addition, they are dressed with their skin and shanks on. Consequently, they dress higher than other classes of slaughter animals.

The average liveweight of hogs, dressed packer style by federally inspected commercial meat packing plants, and their percentage yield in meat are shown in Table 17-10.

TABLE 17-9
APPROXIMATE DRESSING PERCENTAGE OF BARROWS AND GILTS, BY GRADE[1]

Grade	Range	Average
	(%)	*(%)*
U.S. No. 1	68–72	70
U.S. No. 2	69–73	71
U.S. No. 3	70–74	72
U.S. No. 4	71–75	73
Utility	67–71	69

[1]Provided by Livestock Division, Agricultural Marketing Service, USDA.

TABLE 17-10
AVERAGE LIVEWEIGHT, CARCASS YIELD, AND DRESSING PERCENTAGE OF FALL HOGS COMMERCIALLY SLAUGHTERED IN THE UNITED STATES, 1990–1994[1]

Year	Average Liveweight		Average Dressing Weight		Dressing Percentage
	(lb)	*(kg)*	*(lb)*	*(kg)*	*(%)*
1990	249	*113*	181	*82*	72.7
1991	252	*114*	182	*83*	72.2
1992	252	*114*	181	*82*	71.9
1993	254	*115*	183	*83*	72.0
1994	255	*116*	184	*84*	72.0

[1]*Pork Facts 1995/1996*, National Pork Producers Council, p. 21.

QUESTIONS FOR STUDY AND DISCUSSION

1. Define *livestock marketing*.

2. Why is hog marketing important?

3. In recent years, terminal and auction markets have declined in importance, while direct selling, carcass grade and yield selling, and contract selling have increased. Why has this happened?

4. Briefly, describe terminal markets, auction markets, direct selling, carcass grade and yield selling, slaughter hog pooling, packer marketing contracts, feeder pig marketing and early weaner pig marketing.

5. What method of marketing (what market channel) do you consider most advantageous for the hogs sold off your home farm (or a farm with which you are familiar)? Justify your choice.

6. How does selling purebred and seedstock hogs differ from selling slaughter hogs?

7. Some producers and marketing specialists express the following concerns if packers get 50% or more of their kill under contract:

   a. If a minority (25% or fewer) of the hogs marketed set the cash market, will they be hogs of lesser quality?

   b. Would these few hogs of lesser quality be used to calculate prices for the contract hogs?

Do you feel that the above two concerns are justified?

8. Does each market channel give adequate assurance of honesty, of sanitation, and of humane treatment of animals? Justify your answer.

9. Which is the more important to the hog producer: (a) low marketing costs, or (b) effective selling and net returns?

10. List and discuss each of the factors affecting feeder pig prices.

11. What prompted the rise of early weaner pig marketing (SEW pigs)?

12. Why are livestock market news services important?

13. Outline, step by step, how you would prepare and ship hogs to market.

14. Discuss practical ways and means of lessening shrinkage in market hogs.

15. Define on-foot market (a) classes and (b) grades of hogs and tell of their value.

16. List the federal market grades of slaughter barrows and gilts, and brief the specifications of each.

17. What are roasters?

18 Discuss practical ways through which the hog producer can take advantage of cyclical trends and seasonal changes.

19 Describe The federal meat inspection of hogs.

20. Outline the hog slaughtering process.

21. What is the Humane Slaughter Act, and how does it affect hog slaughtering?

22. Describe the packer and shipper styles of dressing.

23. What is the dressing percentage of a hog that weighed 230 lb on-foot and yielded a carcass weighing 150 lb?

SELECTED REFERENCES

Title of Publication	Author(s)	Publisher
Animal Science, Ninth Edition	M. E. Ensminger	Interstate Publishers, Inc., Danville, IL, 1991
Livestock and Meat Marketing	J. H. McCoy	Avi Publishing Co., Westport, CT, 1972
Marketing: Yearbook of Agriculture, 1954	U.S. Department of Agriculture	U.S. Government Printing Office, Washington, DC, 1954
Meat We Eat, The	J. R. Romans, *et al.*	Interstate Publishers, Inc., Danville, IL, 1994
Organization and Competition in the Livestock and Meat Industry, Technical Study No. 1	National Commission on Food Marketing	U.S. Government Printing Office, Washington, DC, June 1966
Pork Facts	Staff	National Pork Producers Council, Des Moines, IA, 1995–1996
Pork Industry Handbook		Cooperative Extension Service, University of Illinois, Urbana-Champaign
Stockman's Handbook, The, Seventh Edition	M. E. Ensminger	Interstate Publishers, Inc., Danville, IL, 1992

Pork on the table. (Courtesy, National Live Stock & Meat Board)

18

PORK AND BYPRODUCTS FROM HOG SLAUGHTER[1]

[1]The authors wish to express their appreciation for the authoritative review accorded by, and the helpful suggestions received from, the following persons in the revision of Chapter 18: Jim Wise, Ph.D., USDA, AMS, Livestock and Seed Division, STDZ, Washington, DC; and Ken Johnson, Vice President, and Torence R. Dockerty, Ph.D., Director of Meat Science Programs, National Live Stock & Meat Board, Chicago, Illinois.

Pork may be defined as the edible flesh of pigs or hogs, while byproducts include all products, both edible and inedible, other than the carcass meat. The edible glands and organs are usually classed as byproducts, but lard is usually grouped along with pork.

Although this chapter is devoted primarily to the final animal product—meat—it must be remembered that the top grades of this food represent the culmination of years of progressive breeding, the best nutrition, vigilant sanitation and disease prevention, superior care and management, and modern marketing, slaughtering, processing, and distribution. Much effort and years of progress have gone into the production of tasty pork chops and hams.

Fig. 18-2 *Pork Quality Assurance* will result in increased consumer confidence and pork consumption. (Courtesy, National Pork Producers Council, Des Moines, IA)

POUNDS OF MEAT MARKETED PER ANIMAL, PER YEAR

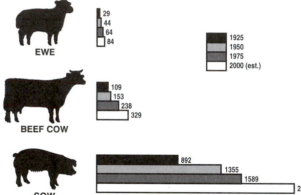

EWE
29
44
64
84

1925
1950
1975
2000 (est.)

BEEF COW
109
153
238
329

SOW
892
1355
1589
2137

Sources: Liveweight marketed per breeding ewe, beef cow, and sow in 1925, 1950, and 1975 from *Food For Animals*, CAST Report No. 82, March 1990, p. 13, Table 2. Pounds of meat marketed per year and dressing percentages for beef cattle and hogs from *Meat and Poultry Facts 1994*, pp. 22–23, American Meat Institute, Washington, DC; for sheep from *Sheep & Goat Science*, fifth ed., p. 216. Year 2000 estimates by the senior author.

Thus, the CAST figures on liveweight marketed per breeding female (pounds) were multiplied by the percentage of edible meat obtained at slaughter from *Meat and Poultry Facts* and *Sheep & Goat Science*, to get the pounds of meat marketed per animal, per year.

Fig. 18-1. Years of progress and much effort have gone into the efficient production of quality meat. The sow has done her part.

PORK OVER THE COUNTER THE ULTIMATE OBJECTIVE

The end product of all breeding, feeding, care and management, marketing, and processing is pork over the counter. It is imperative, therefore, that the progressive hog producer, the student, and the swine scientist have a reasonable working knowledge of pork and of the byproducts from hog slaughter. Such knowledge will be of value in selecting animals and in determining policies relative to their handling.

Of course, the type of animals best adapted to

the production of meat over the counter has changed in a changing world. Thus, in the early history of this country, the very survival of animals was often dependent upon their speed, hardiness, and ability to fight. Moreover, long legs and plenty of bone were important attributes when it came time for animals to be driven to market. The Arkansas razorback was adapted to these conditions.

With the advent of rail transportation and improved care and feeding methods, the ability of animals to travel and fight diminished in importance. It was then possible, through selection and breeding, to produce meat animals better suited to the needs of more critical consumers. With the development of large cities, artisans and craftsmen and their successors in industry required fewer calories than those who were engaged in the more arduous tasks of logging, building railroads, etc. Simultaneously, the American family decreased in size. The demand shifted, therefore, to smaller and less fatty cuts of meats; and, with greater prosperity, high-quality hams, bacons, and chops were in demand. To meet the needs of the consumer, the producer gradually shifted to the breeding and marketing of younger animals with maximum cut-out value of the primal cuts. The need was for meat-type hogs.

Thus, through the years, consumer demand has exerted a powerful influence upon the type of hogs produced. To be sure, it is necessary that such production factors as prolificacy, economy of feed utilization, rapidity of gains, size, etc., receive due consideration along with consumer demands. But once these production factors have received due weight, hog producers—whether they be purebred or commercial operators—must remember that pork over the counter is the ultimate objective.

Now, and in the future, hog producers need to select and feed so as to obtain increased red meat without excess fat. Production testing programs have been reoriented to give greater emphasis to these consumer demands.

QUALITIES IN PORK DESIRED BY THE CONSUMER

The term *pork quality* conveys different messages to different people. To pork processors, it relates primarily to functional properties and color of the muscle. To retailers, it relates to appearance of retail cuts, including fat and bone content, as well as color and juice loss or retention. To consumers, any factor that affects pork eating satisfaction, safety, convenience, and nutritional value falls within the definition of pork quality. Pork producers must recognize all these requirements and use the management practices that maximize pork quality for the entire industry.

In this section, pork qualities which affect consumer preferences are emphasized. In the subsequent section head "Pork Safety and Quality Assurance," pork qualities as affected by the swine producer are emphasized.

As shown in Table 18-1, producers have adapted production to produce more of the valuable retail cuts and less lard, in keeping with consumer demand.

Because consumer preference is such an important item in the production of pork, it is well that the producer, the packer, and the meat retailer be familiar with these qualities, which are summarized as follows:

1. **Quality.** The quality of the lean is based on firmness, texture, marbling, and color.

Fig. 18-3. A pork chop-apple-sauerkraut combination ready for serving. (Courtesy, American Meat Institute, Washington, DC)

2. **Firmness.** Pork muscle should be firm so as to display attractively. Firmness is affected by the kind and amount of fat. For example, pigs that are fed liberally on peanuts produce soft pork.

3. **Texture.** Pork lean that has a fine-grained texture is preferred. Coarse-textured lean is generally indicative of greater animal maturity and less tender meat.

4. **Marbling.** This characteristic contributes to buyer appeal. Feathering (flecks of fat) between the ribs and within the muscles is indicative of marbling.

5. **Color.** Most consumers prefer pork with a white

TABLE 18-1
CHANGES IN THE PRODUCTION OF LARD AND RETAIL MEAT, SINCE 1950[1]

Year	Liveweight		Dressing Percentage	Dressed Weight		Lard Yield		Retail Meat Yield	
	(lb)	(kg)	(%)	(lb)	(kg)	(lb/hog)	(kg/hog)	(lb/hog)	(kg/hog)
1950	240	109	68.9	165.4	75.0	35.4	16.1	137	62.1
1955	237	107	69.5	164.7	74.7	34.9	15.8	136	61.7
1960	236	107	69.5	164.0	74.4	32.2	14.6	139	63.0
1965	238	108	70.1	166.8	75.6	27.9	12.6	147	66.7
1970	240	109	70.3	168.7	76.5	22.8	10.3	155	70.3
1975	240	109	70.6	169.4	76.8	14.8	6.7	166	75.3
1980	242	110	71.7	173.5	78.7	12.8	5.8	172	78.0
1985	245	111	71.4	175.0	79.5	11.0	5.0	136	61.7
1990	249	113	72.7	181.0	82.2	—	—	141	64.0
1994	255	116	72.0	183.6	83.4	—	—	143	64.9

[1]*Pork Facts 1995/1996, Industry Statistics*, National Pork Producers Council, Des Moines, IA, p. 21.

fat on the exterior and a bright reddish lean marbled with flecks of fat.

6. **Maximum muscling; moderate fat.** Maximum thickness of muscling influences materially the acceptability by the consumer. Also, consumers prefer a uniform cover of not to exceed ⅛ in. of firm, white fat on the exterior.

7. **Repeatability.** The consumer wants to be able to secure a standardized product—meat of the same tenderness and other eating qualities as the previous purchase.

8. **Safety.** America has one of the safest food supplies in the world. Yet, the Food and Drug Administration and the Centers for Disease Control and Prevention estimate that about 33 million people, or 14% of the population, become ill each year from microorganisms in food. In most of these cases, the health problem is a mere inconvenience. But it can be life threatening. A total of 9,000 deaths annually are attributed to food-borne diseases. (Also, see the section in this chapter on "Pork Safety and Quality Assurance.")

If these eight qualities are not met by pork, other products will meet them. Recognition of this fact is important, for competition is keen for space on the shelves of a modern retail food outlet.

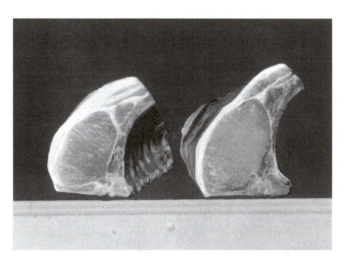

Fig. 18-4. Consumers desire pork with a limited amount of fat and a maximum of lean meat. The pork on the left shows more fat and less lean than the pork on the right. Breeding made the difference!

To this end, we need to conduct more experimental studies on consumer preference and demand; then we need accurately to reflect this information in price differentials at the market place, and, in turn, in production. Factors involved in producing higher quality pork are covered in Chapter 6, Principles of Swine Genetics; Chapter 8, Fundamentals of Swine Nutrition; and Chapter 9, Swine Feeding Standards, Ration Formulation, and Feeding Programs.

FEDERAL GRADES OF PORK CARCASSES

The grade of a pork carcass may be defined as a measure of its degree of excellence based chiefly on quality of lean and the expected yield of trimmed major wholesale cuts. It is intended that the specifications for each grade shall be sufficiently definite to make for uniform grades throughout the country and from season to season, and that on-rail grades shall be correlated with on-foot grades.

Both producers and consumers should know the federal grades of pork and have reasonably clear understanding of the specifications of each grade. From the standpoint of producers—including both purebred and commercial operators—this is important, for, after all, meat over the counter is the ultimate objective. From the standpoint of consumers, this is important, because (1) in these days of self-service prepackaged meats there is less opportunity to secure the counsel and advice of the meat cutter when making purchases, and (2) the average consumer is not the best judge of quality of the various kinds of meats on display in the meat counter.

Because of the relationship between sex in pork and the acceptability of the prepared meat to the consumer, separate standards have been developed for (1) barrow and gilt carcasses and (2) sow carcasses. Only barrow and gilt carcasses will be discussed herein.

The grades of barrow and gilt carcasses are based on two general considerations: (1) the quality-indicating characteristics of the lean and fat, and (2) the expected combined yields of the four primal cuts (ham, loin, picnic shoulder, and Boston butt).

If a carcass qualifies as acceptable in quality of lean and in belly thickness, and is not soft and oily, it is graded U.S. No. 1, 2, 3, or 4, based entirely on projected carcass yields of the four lean cuts. The expected yields of each of the grades in the four lean cuts, based on using the U.S. Department of Agriculture standard cutting and trimming methods, are as given in Table 18-2.

TABLE 18-2
EXPECTED YIELDS OF THE FOUR LEAN CUTS,
BASED ON CHILLED CARCASS WEIGHT, BY GRADE[1]

Grade	Yield
U.S. No. 1	60.4% and over
U.S. No. 2	57.4 to 60.3%
U.S. No. 3	54.4 to 57.3%
U.S. No. 4	less than 54.4%

[1]These yields will be approximately 1% lower if based on hot carcass weight.

Carcasses vary in their yields of the four lean cuts because of variations in their degree of fatness and in their degree of muscling (thickness of muscling in relation to skeletal size).

From the standpoint of quality, two general levels are recognized—"acceptable" and "unacceptable." Acceptability is determined by direct observation of the cut surface and is based on considerations of firmness, marbling, and color, along with the use of such indirect indicators as firmness of fat and lean, and feathering between the ribs. The degree of external fatness is not considered in evaluating the quality of the lean. Suitability of the belly for bacon (in terms of thickness) is also considered in quality evaluation, as is the softness and oiliness of the carcass. Carcasses which have unacceptable quality of lean, and/or bellies that are too thin, and/or carcasses which are soft and oily are graded U.S. Utility.

The grade of a barrow or gilt carcass is determined on the basis of the following math equation: Carcass grade = (4.0 × backfat thickness over the last rib, inches) − (1.0 × muscling score). To apply this equation, muscling should be scored as follows: thin muscling = 1, average muscling = 2, and thick muscling = 3. Carcasses with thin muscling cannot grade U.S. No. 1. The grade may also be determined by calculating a preliminary grade according to the schedule shown in Table 18-3, and adjusting up or down one grade for thick or thin muscling, respectively.

TABLE 18-3
PRELIMINARY CARCASS GRADE, BASED ON BACKFAT THICKNESS OVER THE LAST RIB

Preliminary Grade	Backfat Thickness Range
U.S. No. 1	Less than 1.00 in.
U.S. No. 2	1.00 to 1.24 in.
U.S. No. 3	1.25 to 1.49 in.
U.S. No. 4	1.50 in. and over[1]

[1]Carcasses with last rib backfat thickness of 1.75 in. or over cannot be graded U.S. No. 3, even with thick muscling.

In these standards, the thickness of backfat is the measurement, including the skin, made opposite the last rib.

The second factor considered in barrow and gilt carcass grading is the degree of muscling. The degree of muscling is determined by a subjective evaluation of the thickness of muscling in relation to skeletal size. Since the total thickness of a carcass is affected by both the amount of fat and the amount of muscle in relation to skeletal size, the fatness must also be considered when degree of muscling is evaluated. To best evaluate muscling, primary consideration is given to those parts least affected by fatness, such as the ham. In evaluating the ham for degree of muscling, consideration should be given to both the stifle and back views. The size of lumbar lean area and the relative width through the back of the loin and through the center of the ham are also good indications of muscling.

In barrow and gilt carcass grading, three degrees of muscling—thick (superior), average, and thin (inferior)—are considered.

Thus, the on-foot and carcass federal grades of slaughter barrows and gilts are: U.S. No. 1, U.S. No. 2, U.S. No. 3, U.S. No. 4, and U.S. Utility.

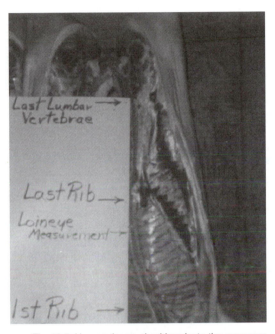

Fig. 18-5. Hog producers should evaluate the carcasses that they produce, especially backfat thickness over the last rib and muscling. (Courtesy, Farmland Industries, Inc., Kansas City, MO)

As a rule, slaughter pigs that weigh under 60 lb are not graded, because they have not reached sufficient maturity for variations in their carcass traits to affect the market value materially.

Unlike meat inspection, government grading is purely voluntary, on a charge basis. Only a very small proportion of the U.S. commercial pork production is federally graded since carcasses are cut and trimmed at the packing plant, and sold as trimmed primals and sub-primals. Graded carcasses are stamped (with an edible vegetable dye) so that the grade will appear on the retail cuts as well as on the carcass and wholesale cuts.

Note: Only negligible quantities of pork carcasses are actually federally graded.

DISPOSITION OF THE PORK CARCASS

Almost all hog carcasses are cut up at the slaughtering plant and are sold in the form of wholesale and retail cuts. In most parts of the country, less than 1% of the pork in large packing plants is sold in carcass form. The whole-carcass trade is largely confined to roasting and slaughter pigs.

The handling of pork differs further from that of beef and lamb in that much of it is cured by various methods, is rendered into lard, or is manufactured into meat products. In general, loins, Boston shoulders, and spareribs are most likely to be sold as fresh cuts. But it must be remembered that practically every pork cut may be cured, and, under certain conditions, is cured. Because pork is well adapted to curing, it has a decided advantage over beef and lamb, which are sold almost entirely in the fresh state. The hog market is stabilized to some extent by this factor.

THE HOG CARCASS AND ITS WHOLESALE CUTS

A minimum of 24 hours chilling at temperatures ranging from 33 to 38°F is necessary to remove the animal heat properly and give the carcasses sufficient firmness to make possible a neat job of cutting. After chilling, the carcasses are brought to the cutting floor where they are reduced to the wholesale cuts. Hot processing, however, is increasing as a means of saving energy.

The method of cutting varies somewhat according to the relative demand for different cuts. Despite some variation, the most common wholesale cuts of pork are: leg (ham), side (bacon), loin, picnic shoulder, Boston shoulder (butt), jowl, and feet, as indicated in Fig. 18-6.

Market hogs weighing from 220 to 240 lb will have about 55 to 60% of liveweight in the five primal cuts: the leg, Boston shoulder, picnic shoulder, loin, and side. Yet, because of the relatively higher value per pound of these cuts, they make up about three-fourths of the value of the entire carcass.

LARD

Lard is the fat rendered (melted out) from fresh, fatty pork tissue. It is considered a primary product of hog slaughter and not a byproduct. The proportion varies with the type, weight, and finish of the hogs.

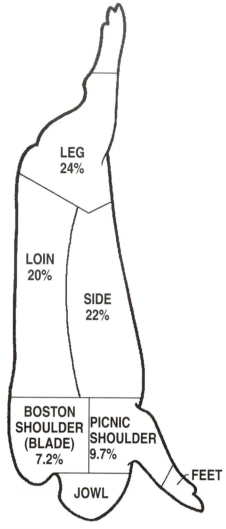

Fig. 18-6. Wholesale cuts of pork and the percentage of the carcass represented by each of five primal cuts.

Lard production per slaughtered hog has decreased sharply in recent years, with the shift away from fat, lardy-type hogs to lean, muscular animals. This is shown by the data in Table 18-1.

KINDS OF LARD

Lard is classified according to the part of the animal from which the fat comes and the method of rendering as follows: kettle-rendered lard, steam-rendered lard, dry-processed-rendered lard, neutral lard, lard substitutes, and lard oil and stearin.

■ **Modern lard**—In an all out attempt to meet consumer demands, meat packers modernized lard. They (1) discolored it—they made it white as snow (natural lard is bluish in color); (2) deodorized it; (3) hydrogenated it—raised the melting point; (4) added an antioxidant—so that it would keep on the shelf; and (5) placed it in a container that would preserve these qualities. But to no avail! The U.S. per capita consumption of lard plummeted from 14.2 lb in 1940 to 1.7 lb in 1993. Today, lard is seldom found in most large retail stores.

PORK RETAIL CUTS AND HOW TO COOK THEM

The method of cutting pork is practically the same in all sections of the United States. Fig. 18-7 illustrates the common retail cuts of pork, and gives the recommended method or methods for cooking each. This informative figure may be used as a guide to wise buying, in dividing the pork carcass into the greatest number of desirable cuts, in becoming familiar with the types of cuts, and in preparing each type of cut by the proper method of cookery.

It is important that meat be cooked at low temperature, usually between 300 and 350°F. At these temperatures, it cooks slowly, and as a result is juicier, shrinks less, and has a better flavor than when cooked at high temperatures.

The methods used in meat cookery depend on the nature of the cut to which it is applied. In general, the types of meat cookery may be summarized as follows:

1. **Dry-heat cooking.** Dry-heat cooking is used in preparing the more tender cuts, those that contain little connective tissue. This method of cooking consists of surrounding the meat by dry air in the oven or under the broiler. The common methods of cooking by dry heat are (a) roasting, (b) broiling, and (c) panbroiling (see Fig. 18-8).

  a. **How to roast:**

    (1) Season with salt and pepper, if desired.

    (2) Place fat side up on rack in open roasting pan.

    (3) Insert meat thermometer so that the end is in the center of the largest muscle.

    (4) Do not add water or cover.

    (5) Roast in preheated oven at 325 to 350°F.

    (6) Roast until the meat thermometer registers medium or well-done, as desired.

For best results, a meat thermometer should be used to test the doneness of roasts (and also for thick steaks and chops). It takes the guesswork out of meat cooking. Allowing a certain number of minutes to the pound is not always accurate, for example, rolled roasts take longer to cook than ones with bones.

The thermometer is inserted into the cut of meat so that the end reaches the center of the largest muscle, and so that it is not in contact with fat or bone. Naturally, frozen roasts need to be partially thawed before the thermometer is inserted, or a metal skewer or ice pick will have to be employed in order to make a hole in frozen meat.

As the oven heat penetrates the meat, the temperature at the center of it gradually rises and registers on the thermometer. Although most meat can be cooked as desired—rare, medium, or well-done, pork should always be cooked well-done—from 160 to 170°F for fresh pork and 160°F for cured pork.

  b. **How to broil:**

    (1) Set the oven regulator for broiling.

    (2) Place the meat on the rack of the broiler pan and cook 2 to 5 in. from heat.

    (3) Broil until the top of meat is brown.

    (4) Season with salt and pepper, if desired.

    (5) Turn the meat and brown the other side.

    (6) Season and serve at once.

  c. **How to panbroil:**

    (1) Place meat in a heavy, uncovered frying pan.

    (2) Do not add fat or water.

    (3) Cook slowly turning at intervals to ensure even cooking.

    (4) If fat accumulates, pour it off.

    (5) Brown meat on both sides.

    (6) Do not overcook. Season, if desired and serve at once.

2. **Moist-heat cooking.** Moist-heat cooking is generally used in preparing the less tender cuts, those containing more connective tissues that require moist heat to soften them and make them tender. In this type of cooking the meat is surrounded by hot liquid or steam. The common methods of moist-heat cooking are: (a) braising, and (b) cooking in water or stewing (see Fig. 8-8).

  a. **How to braise:**

    (1) Brown the meat on all sides in a heavy utensil.

    (2) Season with salt and pepper, if desired.

    (3) Add small amount of liquid, if needed.

    (4) Cover tightly.

    (5) Cook at simmering temperature, without boiling, until tender.

    (6) Make sauce or gravy from the liquid in the pan, if desired.

  b. **How to cook large cuts in water:**

    (1) Brown meat on all sides, if desired.

    (2) Cover the meat with water or stock.

    (3) Season with salt, pepper, herbs,

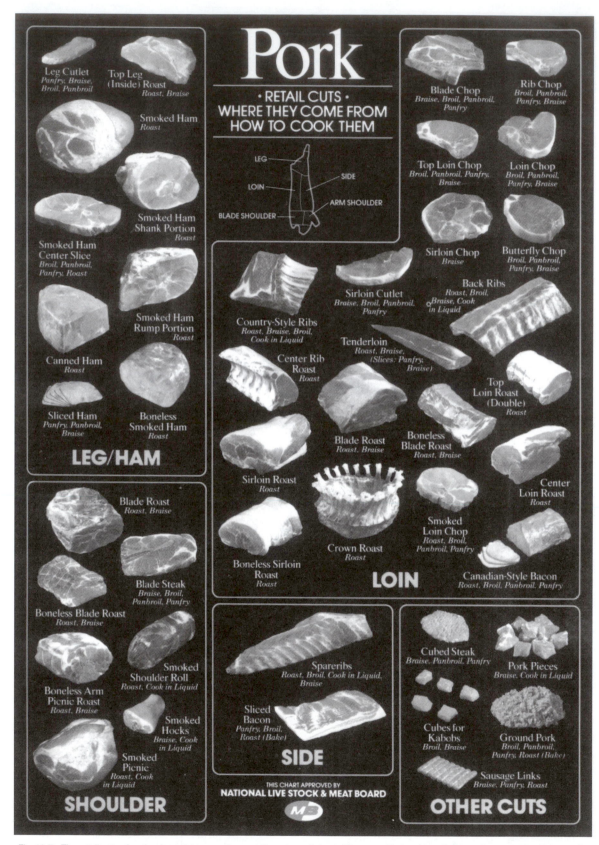

Fig. 18-7. The retail cuts of pork; where they come from and how to cook them. (Courtesy, National Live Stock and Meat Board, Chicago, IL)

DRY-HEAT COOKING

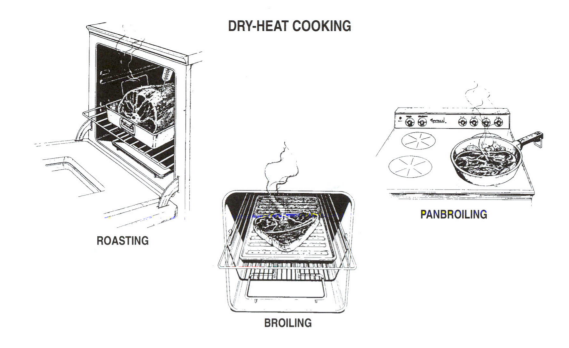

ROASTING

BROILING

PANBROILING

MOIST-HEAT COOKING

BRAISING   COOKING IN LIQUID

Fig. 18-8. Some common methods of meat cookery.

spices, and vegetables, if desired. (Cured or smoked meat does not require salt.)

(4) Cover kettle and simmer (do not boil) until tender.

(5) If the meat is to be served cold, let it cool and then chill in the stock in which it was cooked.

(6) When vegetables are to be cooked with the meat, as in "boiled" dinners, add them whole or in pieces, just long enough before the meat is tender to cook them.

c. **How to cook stews:**

(1) Cut meat in uniform pieces, usually 1- to 2-in. cubes.

(2) If a brown stew is desired, brown meat cubes on all sides.

(3) Add just enough water, vegetable juices, or other liquid to cover the meat.

(4) Season with salt, pepper, herbs and spices, if desired.

(5) Cover kettle and simmer (do not boil) until meat is tender.

(6) Add vegetables to the meat just long enough before serving to be cooked.

(7) When done, remove meat and vegetables to a pan, platter, or casserole and keep hot.

Fig. 18-9. Roast pork loin. Perhaps most people eat meats simply because they like them. For flavor, variety, and appetite appeal, pork is unsurpassed. (Courtesy, National Live Stock and Meat Board, Chicago, IL)

(8) If desired, thicken the cooking liquid with flour for gravy.

(9) Serve the hot gravy (or thickened liquid) over the meat and vegetable or serve separately in a sauce boat.

(10) Meat pies may be made from the stew; a meat pie is merely a stew with a top on it. (The top may be made of pastry, biscuits, or biscuit dough, mashed potatoes, or cereal.)

3. **Frying.** When a small amount of fat is added before cooking, or allowed to accumulate during cooking, the method is called panfrying. This is suitable for preparing comparatively thin pieces of tender meat, or those pieces made tender by pounding, scoring, cubing, or grinding, or for preparing leftover meat. When meat is cooked, immersed in fat, it is called deep-fat frying. This method of cooking is sometimes used for preparing brains, liver, and leftover meat. Usually the meat is coated with eggs and crumbs or a batter, or dredged with flour or cornmeal.

a. **How to panfry:**

(1) Brown meat on both sides in a small amount of fat.

(2) Season with salt and pepper, if desired.

(3) Do not cover the meat.

(4) Cook at moderate temperature, turning occasionally, until done.

(5) Remove from pan and serve at once.

b. **How to deep-fat fry:**

(1) Use a deep kettle and a wire frying basket.

(2) Heat fat to frying temperature.

(3) Place meat in frying basket.

(4) Brown meat and cook it through.

(5) When done, drain fat from meat into kettle and remove meat from basket.

(6) Strain fat through cloth; then cool.

4. **Cooking sausage.** Fresh or uncooked, smoked sausage may be cooked by one of the following methods:

a. **Panfried.** Place links or patties in cold frying pan; add 2 to 4 tsp water; cover tightly and cook slowly 5 to 8 minutes— depending on size and thickness; remove cover and brown slowly; cook until well done.

b. **Cooked in oven.** Arrange sausage in single layer in shallow baking pan; bake in a hot oven (400°F) 20 to 30 minutes, or until well done; turn to brown evenly; pour off drippings as they accumulate.

Cooked, smoked sausage links do not require cooking, but may be heated by one of the following methods:

a. **Simmered.** Drop frankfurters or sausage into boiling water; cover and let water simmer (not boil) until heated through, about 5 to 10 minutes, depending on size.

b. **Panbroiled (or griddle-broiled).** Melt a small amount of fat (1 to 2 tbsp) in a heavy frying pan or on a griddle and brown meat by turning slowly with tongs. Do not pierce with fork.

c. **Broiled.** Brush each frankfurter or sausage link with butter, margarine, or other fat, if desired; broil about 3 in. from the heat; turn to brown evenly, using tongs.

5. **Microwave cooking.** Meat cookery researchers do not recommend cooking fresh pork in the microwave oven. The microwave is, however, very acceptable for cooking cured pork products or reheating cooked fresh pork cuts.

CURING MEAT

In the United States, meat curing is largely confined to pork, primarily because of the keeping qualities and palatability of cured pork products. Considerable beef is corned or dried, and some lamb and veal are cured, but none of these is of such magnitude as cured pork.

Meat is cured with salt, sugar, and certain curing adjuncts (ascorbate, erythrobate, etc.); with sodium nitrite or sodium nitrate (the latter is used only in certain products); and with smoke. Reasons for each of the most common curing additives follow:

■ **Salt**—Sodium chloride is added to cured meats for preservative and palatability reasons. The addition of salt makes it possible to distribute certain perishable meats through what are often lengthy and complicated

Fig. 18-10. Cured meats. (Courtesy, Land O Lakes, Ft. Dodge, IA)

distribution systems. Also, in products such as wieners, bologna, and canned ham, salt solubilizes myosin, the major meat protein, and causes the meat particles to hold together; the meat can then be sliced without falling apart.

■ **Sugar**—Sugar—sucrose, dextrose, or invert sugar—counteracts the harshness of salt, enhances the flavor, and lowers the pH of the cure.

Fig. 18-11. Grilled frankfurters with assorted toppings. (Courtesy, National Live Stock & Meat Board, Chicago, IL)

■ **Sodium nitrite (NaNO$_2$), sodium nitrate (NaNO$_3$), potassium nitrate or saltpeter (KNO$_3$)**—Nitrite and nitrate contribute to (1) the prevention of *Clostridium botulinum* spores in or on meat (*Clostridium botulinum* bacteria produce the botulin toxin that causes botulism, a deadly form of food poisoning); (2) the development of the characteristic flavor and pink color of cured meats; (3) the prevention of "warmed-over flavor" in reheated products; and (4) the prevention of rancidity. Of all the effects of nitrite and nitrate, the antibotulinal effect is by far the most important.

Nitrate is considered essential in cured meats because it performs several important functions, including (1) preventing botulism, (2) retarding liquid oxidation, (3) giving cured meats their characteristic cured flavor, and (4) imparting the characteristic cured pink color.

Nitrite-usage is permitted only in the production of country-cured hams and in some sausages that are cured for several weeks.

■ **Smoking**—Smoking produces the distinctive smoked-meat flavor which consumers demand in certain meats. Many meat packers and processors now use either (1) "liquid smoke," made from natural wood smoke treated so as to remove certain components; or (2) synthetic smoke made by mixing pure chemical compounds found in natural smoke.

■ **Phosphate (PO$_4$)**—Phosphate is added to increase the water-binding capacity of meat; its use increases yields up to 10%. Also, it enhances juiciness of the cooked product. Federal regulations restrict the amount of phosphates to: (1) not to exceed 0.5% in the finished product, and (2) not more than 5% in the pickle solution based on 10% pumping pickle. The addition of phosphate must be declared on the label.

CURING PORK ON THE FARM

Farm meat curing other than freezing is largely confined to pork, primarily because of the keeping qualities and palatability of cured pork products.

The secret of pork curing is to use good sound meat, the correct curing method and formula, clean containers, and to be fortunate enough to secure cool curing weather.

The primary meat curing ingredients are salt, sugar, and saltpeter (potassium nitrate). A combination of 7 lb of salt and 3 lb of white or brown sugar is a basic mixture. In addition to salt or salt and sugar, commercial cures frequently contain spices and flavorings to impart characteristic flavor, appearance, and/or aroma.

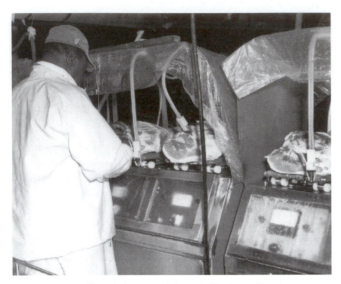

Fig. 18-12. Hams being cured injected with electrical scale computer. (Courtesy, *The National Provisioner*, Chicago, IL)

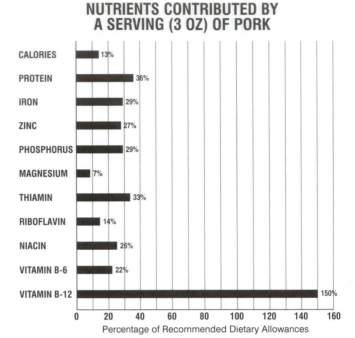

Fig. 18-13. Nutritional value of pork—in percent of RDA contributed by a 3-oz serving. (Based on average values for cooked lean pork and the National Research Council's 1989 *Recommended Dietary Allowances* for a 25- to 50-year-old man)

NUTRITIVE QUALITIES OF PORK

Perhaps most people eat pork simply because they like it. They derive a rich enjoyment and satisfaction therefrom.

But pork is far more than just a very tempting and delicious food. Nutritionally, it contains certain essentials of an adequate diet: high-quality proteins, minerals, and vitamins. This is important, for how we live and how long we live are determined in large part by our diet.

Fig. 18-13 illustrates the contribution of pork toward fulfilling the Recommended Daily Dietary Allowances (RDA) for calories, protein, and certain of the minerals and vitamins.

Effective pork promotion necessitates full knowledge of the nutritive qualities of meats, the pertinent facts of which follow:

1. **Proteins.** The word *protein* is derived from the Greek word *proteios*, meaning *primary*. Protein is needed for growth. Fortunately, meat contains the proper quantity and quality of protein for the building and repair of body tissues. On a fresh basis, pork contains 15 to 20% protein. Also, it contains all of the amino acids, or building blocks, which are necessary for the making of new tissue. Pork is a high-quality protein—a complete protein—because it contains all the essential amino acids.

2. **Calories.** Pork is a good source of energy, the energy value being dependent largely upon the amount of fat it contains.

3. **Minerals.** Minerals are necessary in order to build and maintain the body skeleton and tissues and to regulate body functions. Pork is a rich source of

several minerals, but is especially good as a source of phosphorus and iron. Phosphorus combines with calcium in building the bones and teeth. Phosphorus also enters into the structure of every body cell, helps to maintain the alkalinity of the blood, is involved in the output of nervous energy, and has other important functions.

Iron is necessary for the formation of blood, and its presence protects against nutritional anemia. It is a constituent of the hemoglobin or red pigment of the red blood cells. Thus, it helps to carry the life-giving oxygen to every part of the body. Iron from meat, such as pork, is absorbed most readily by the body, and it also increases the absorption of iron from vegetable sources .

4. **Vitamins.** As early as 1500 B.C., the Egyptians and Chinese hit upon the discovery that eating liver would improve one's vision in dim light. We now know that liver furnishes vitamin A, a very important factor for night vision. In fact, medical authorities recognize that night blindness, glare blindness, and poor vision in dim light are all common signs pointing to the fact that the persons so affected are not getting enough vitamin A in their diets.

Meat is one of the richest sources of the important B group of vitamins, especially thiamin, riboflavin, niacin, and vitamin B-12 (see Table 18-4). Pork is the

TABLE 18-4
VITAMIN CONTENT OF FRESH PORK[1]

Pork Cut[2]	Thiamin	Riboflavin	Niacin	Vitamin B-12
	(mg/100 g)[3]	(mg/100 g)[3]	(mg/100 g)[3]	(mcg/100 g)[3]
Ham	0.64	0.30	5.7	—
Chop	1.10	0.31	6.5	3.0
Roast	0.70	0.30	5.9	—
Spareribs . .	0.43	0.21	3.4	—

[1]Ensminger, A. H., M. E. Ensminger, J. E. Konlande, and J. R. K. Robson, *Foods & Nutrition Encyclopedia*, CRC Press, 1994.

[2]Separable lean, cooked.

[3]One hundred g is approximately equal to 3½ oz.

Fig. 18-14 Low fat, low cholesterol ham. (Courtesy, National Live Stock & Meat Board, Chicago, IL)

leading dietary source of thiamin, containing three times as much as any other food.

These B vitamins are now being used to reinforce certain foods and are indispensable in our daily diet. They are necessary for energy metabolism, synthesis of new tissue (growth), nervous function, and many other functions. A marked deficiency of thiamin causes beriberi. Niacin prevents and cures the disease pellagra. Indeed, one of the reasons for the rapid decline in B vitamin deficiencies in America may well be the increased amount of meat and other B vitamin–containing foods in the daily diet.

5. **Digestibility.** Finally, in considering the nutritive qualities of meats, it should be noted that this food is highly digestible. About 97% of meat proteins and 96% of meat fats are digested. The statement often is heard that "pork is hard to digest." This is not true. Pork, in common with all meats, is well utilized by the body.

We have come to realize, therefore, the important part that pork is playing in the nutrition of the nation.

CONSUMER HEALTH CONCERNS

Much has been written and spoken linking the consumption of meat, including pork, to certain diseases. The primary concerns of consumers about pork pertain to (1) fats and cholesterol, (2) nitrites and nitrosamines, and (3) safety and quality assurance. Each of these concerns is discussed in a separate section that follows.

FATS AND CHOLESTEROL

In 1953, Dr. Ancel Keys of the University of Minnesota first reported a positive correlation between the consumption of animal fat (which is high in cholesterol) and the occurrence of atherosclerosis (heart disease) in humans. Subsequently, other studies correlated high blood levels of cholesterol with increased incidence of atherosclerosis in humans. In 1964, the American Heart Association recommended that the general public reduce cholesterol intake to 300 mg/day. Subsequently, various government and health agencies around the world have followed suit.

Since cholesterol is present in many animal and food products, including pork and lard, it was inevitable that they would be incriminated as causes of heart disease. It is important, therefore, that all members of the pork team—producers, processors, and retailers—along with consumers, know the truth about fats and cholesterol. To this end, the authors present this section on "Fats and Cholesterol," as they perceive them.

Note: From 1983 to 1993, hogs slimmed down so much that, on the average, fresh pork was 31% lower in fat, 14% lower in calories, and 10% lower in cholesterol than 10 years earlier.[2]

■ **About cholesterol**—Cholesterol is found in all body tissues, especially the brain and spinal cord. Further, cholesterol is an essential ingredient for certain biochemical processes, including the production of sex hormones in humans.

Chemically, cholesterol is a fatlike compound—actually an alcohol—which in its pure form appears as pearly flakes. It is composed of 27 carbon atoms which form 3 fused cyclohexane (6-carbon) rings, a cyclopentane (5-carbon) ring and a side chain of 8 carbon atoms.

―――――――

[2]*Pork Facts* 1995–1996, p. 28, National Pork Producers Council.

■ **Blood levels**—Cholesterol is present in both free and esterified forms in the blood. From birth, blood cholesterol increases throughout life as shown in Table 18-5. The moderate risk category in Table 18-5 includes large numbers of people with elevated blood cholesterol due, in part, to their diet. The high risk category in Table 18-5 includes individuals with hereditary forms of high blood cholesterol which require the most aggressive treatment.

TABLE 18-5
AGE AND CHOLESTEROL CONCENTRATION OF MEN
AND WOMEN AT MODERATE AND HIGH RISK[1]

Age	Moderate Risk	High Risk
(years)	*(mg/dl)*	*(mg/dl)*
2 to 19	Greater than 170	Greater than 185
20 to 29	Greater than 200	Greater than 220
30 to 39	Greater than 220	Greater than 240
40 and over	Greater than 240	Greater than 260

[1]National Institute of Health guidelines of moderate and high risk levels of blood cholesterol as measured in milligrams per deciliter (mg/dl).

Since cholesterol is insoluble in a water-based medium such as blood, it is transported in blood as a lipoprotein. Primarily, two types of lipoproteins are involved in cholesterol transport. Low-density lipoproteins—LDL—transport cholesterol from the liver to the cells, while high-density lipoproteins—HDL—transport cholesterol from the tissue cells to the liver. Current research suggests that measuring HDL and LDL may be the most accurate means of assessing one's blood cholesterol level. Regardless of whether total cholesterol, the HDL, and/or LDL are measured in the blood of an individual, one should be aware that just a single determination on a blood sample is far from adequate for evaluating the cholesterol status. Factors such as (1) age, (2) time of day, (3) physical condition, (4) stress, (5) genetic background, and (6) laboratory expertise may all affect the determination.

■ **Metabolism**—Cholesterol in the body arises from two sources: (1) that from the diet—exogenous cholesterol; and (2) that manufactured in the body—endogenous cholesterol. Cholesterol in the blood reflects the overall cholesterol metabolism—that derived from both sources. Pertinent facts about the metabolism of cholesterol in the human body follow:

1. **Digestion and absorption.** The average individual ingests between 500 and 800 mg of cholesterol each day. Dietary fat aids the absorption of cholesterol. Moreover, absorption is dependent upon the availability of bile acids from the liver, and pancreatic cholesterol esterase. Also, the absorption of cholesterol de-

pends upon the amount eaten. Increasing intake decreases the percentage absorbed. At high levels, a person absorbs less than 10%, and the remainder leaves the body via the feces. Initially, dietary cholesterol enters the blood as chylomicrons which are eventually converted to the cholesterol-containing lipoproteins—LDL and HDL. About 2 to 4 hours after eating, the cholesterol in the blood which came from the food is indistinguishable from that synthesized in the body.

2. **Synthesis.** Most all tissues, except possibly the brain, are capable of manufacturing—synthesizing—cholesterol. However, the liver is the major site of synthesis. The entire cholesterol molecule can be synthesized from 2-carbon units called acetate. In the whole metabolic scheme, acetate can be derived from the breakdown of carbohydrates, proteins (amino acids), and of course, fats. Each day the body manufactures 1,000 to 2,000 mg of cholesterol. However, on a day-to-day basis the synthesis and metabolism of cholesterol is controlled by such factors as (1) fasting, (2) caloric intake, (3) cholesterol intake, (4) bile acids, (5) hormones, primarily the thyroid hormones and estrogen, and (6) disorders such as diabetes, gallstones, and hereditary high blood cholesterol—hypercholesterolemia. Control of cholesterol synthesis by cholesterol intake is important, since this means that when intake is high then synthesis is low and vice versa.

3. **Functions.** Cholesterol is vital to the body. Its primary importance concerns tissues, bile acids, and hormones.

4. **Excretion.** Removal from the body occurs primarily via the conversion of cholesterol to bile acids. About 0.8 mg of cholesterol is degraded daily by this method. Also, a minor amount is converted to the above mentioned hormones. Additionally, some cholesterol is never digested, and hence, excreted via the feces, particularly when intake is high.

■ **The risk: heart disease**—High blood cholesterol is one of the three major modifiable risk factors for coronary heart disease (CHD); the other two are high blood pressure and cigarette smoking. Approximately 25% of the adult population 20 years of age and older has high blood cholesterol levels—levels that are high enough to need intensive medical attention. More than half of all adult Americans have a blood cholesterol level that is higher than desirable.

The following seven changes should be made in the diet of persons having high blood cholesterol:

1. **Eat less high-fat food.** The intake of total fat should constitute less than 30% of the calories.
Note: Eating less total fat is an effective way to eat less saturated fat and fewer calories.

2. **Eat less saturated fat.** The intake of saturated fat should constitute less than 10% of the calories.
Note: Saturated fats are found primarily in animal

products. But a few vegetable fats and many commercially processed foods also contain saturated fat. Read labels carefully. Choose foods wisely.

3. **Substitute unsaturated fats for saturated fat.** Unsaturated fats (polyunsaturated and monounsaturated fats) should be substituted for saturated fat to the extent practical.

Note: Unsaturated fats lower blood cholesterol levels when substituted for saturated fats.

4. **Eat less high-cholesterol food.** The intake of cholesterol should be less than 300 mg/day. Dietary cholesterol can raise the blood cholesterol level. Therefore, it is important to eat less food that is high in cholesterol. (See Table 18-7, Cholesterol Content of Some Common Foods.)

Note: There is very little cholesterol in low-fat dairy foods like skim milk and no cholesterol in food from plants, like fruits, vegetables, vegetable oils, grains, cereals, nuts, and seeds.

5. **Substitute complex carbohydrates for saturated fats.** Breads, pasta, rice, cereal, dried peas and beans, fruits, and vegetables are good sources of complex carbohydrates (starch and fiber). They are excellent substitutes for foods that are high in saturated fat and cholesterol.

Note: Foods that are high in complex carbohydrates, if eaten plain, are low in saturated fat and cholesterol as well as being good sources of minerals, vitamins, and fiber.

6. **Maintain a desirable weight.** People who are overweight frequently have higher blood cholesterol levels than people of desirable weight.

Note: To achieve or maintain a desirable weight, caloric intake must not exceed the number of calories the body burns.

7. **Eat foods that are high in soluble fiber.** Among such foods is oat bran.

Note: Basically, there are two types of fibers: water soluble fiber (like oat bran) and non-soluble fiber (like wheat bran). Only soluble fiber is effective in lowering cholesterol. Soluble fiber dissolves in water.

All of the above indicate that the development of atherosclerosis is not just a simple matter of eating too much cholesterol. Moreover, these recommendations encompass accepted measures of good health—cessation of smoking, normal blood pressure, ideal weight, exercise, control of stress, and awareness of family history. Each of the recommendations complement the others and contribute to decreasing the risks of atherosclerosis and heart disease.

■ **Sources**—The following information is presented for those individuals who wish to reduce their cholesterol intake for personal reasons or who require cholesterol restrictions as part of a control measure in cases of hyperlipoproteinemias.

Table 18-6 indicates the foods in the American

TABLE 18-6
AMOUNT OF CHOLESTEROL AVAILABLE PER PERSON PER DAY IN THE UNITED STATES[1]

Food Source	Year	
	1967–1969	1988
	(mg)	*(mg)*
Meat, poultry, fish	183.3	207.2
Eggs	239.6	144.0
Dairy products	73.9	67.0
Fats and oils	28.8	21.0
Animal sources	525.6	440.0
Vegetable sources	0	0

[1] *Agricultural Statistics 1991*, USDA, p. 478.

diet which supply cholesterol and their relative contribution to the total cholesterol available.

Table 18-7 which follows provides a ranking of the cholesterol levels of some common foods. This is provided for informational purposes only—not to encourage or discourage consumption of certain foods by the general public. Rather, each individual should consider his or her own case.

NITRATES/NITRITES AND NITROSAMINES

Nitrates and/or nitrites are considered essential in cured pork because they perform several important functions, the most important of which is their anti-botulism effect.

Both nitrate and nitrite are allowed in meats in specified, limited levels under the Meat Inspection Act, but nitrate usage is permitted only in the production of country-cured hams and in some sausages that are cured for several weeks.

Controversy over the use of nitrates/nitrites, and their production of nitrosamines, stems primarily from nitrosamines producing cancer in test animals.

Generations of consumers have been accustomed to the bright pink color of cured-meat imparted by nitrates and nitrites.

Pork producers, processors, and retailers, along with consumers, need to know the truth about nitrates/nitrites and nitrosamines. So, the authors present the following facts.

■ **About nitrates/nitrites and nitrosamines**—Nitrate refers to the chemical union of one nitrogen (N) and three oxygen (O) atoms, or NO_3, while nitrite refers to the chemical union of one nitrogen (N) and two oxygen (O) atoms, or NO_2. Of prime concern in foods

TABLE 18-7
CHOLESTEROL CONTENT OF SOME COMMON FOODS

Food	Cholesterol
	(mg/100 g)
Egg yolks, chicken	1,602
Kidneys, beef, braised	804
Liver, chicken or turkey, simmered	615
Eggs, fried, poached, or hard cooked	540
Sweetbread (thymus), braised	466
Liver, beef, calf, or pork, fried	438
Kidneys, calf, lamb, or pork, raw	375
Roe, salmon, sturgeon, or turbot, raw	360
Heart, beef, braised	274
Butter	219
Shrimp, canned.	150
Sardines, canned.	140
Whipping cream, 37.6% fat	133
Cream cheese	111
Beef tallow	109
Cheddar cheese	106
Turkey, roasted	105
Veal	101
Beef and lamb, variety of cuts	95
Swiss cheese	92
Chicken, roasted	90
Pork, variety of cuts	89
Frankfurter	62
Ice cream, 12% fat.	60
Tuna, canned	55
Milk, evaporated	31
Milk, whole	14
Milk, 2% fat	8
Yogurt	6
Cottage cheese	4
Milk, 1% fat	4

Fig. 18-15. Polish sausage with peppers and onions—a cured pork product. (Courtesy, National Live Stock & Meat Board, Chicago, IL)

deficiencies of potassium, phosphorus, and calcium or excesses of soil nitrogen, and (6) environmental factors such as drought, high temperature, time of day, and shade. Regardless of the variation of nitrate content in plants, vegetables are the major source of nitrate ingestion as Table 18-8 shows.

Those vegetables which are most apt to contain high levels of nitrates include beets, spinach, radishes, and lettuce. Despite the shift from manure to chemical fertilizers over the years, the overall average concentration of nitrate in plants has remained unchanged. Other natural sources of nitrate are negligible.

Nitrite occurrence in foods is minimal. Interestingly, a naturally occurring source of nitrites appears to be the saliva.

2. **Food additives.** Both nitrates and nitrites are used as food additives, mainly in meat and meat products, according to the guidelines presented in Table 18-9. Their use in meat has been the subject of much publicity, though their use to cure meat is lost in antiquity. The role of nitrates in meats is not clear, though it is believed that they provide a reservoir source of nitrite since microorganisms convert nitrate to nitrite. It is the nitrite which decomposes to nitric oxide, NO, and reacts with heme pigments to form nitrosomyoglobin giving meats their red color. Furthermore, taste panel studies on bacon, ham, hot dogs,

are sodium nitrate (Chile saltpeter), potassium nitrate (saltpeter), sodium nitrite, and potassium nitrite.

■ **Occurence and exposure**—Nitrates and nitrites are common chemicals in our environment whether they come from *natural* or *unnatural* sources. Details follow:

1. **Naturally occurring.** Most green vegetables contain nitrates. The level of nitrates in vegetables depends on (1) species, (2) variety, (3) plant part, (4) stage of plant maturity, (5) soil condition such as

TABLE 18-8
ESTIMATED AVERAGE DAILY INGESTION OF NITRATE AND NITRITE PER PERSON IN THE UNITED STATES[1]

Source	Nitrate (NO_3-)	Nitrite (NO_2-)
	(mg)	(mg)
Vegetables	86.1	0.20
Cured meats	9.4	2.38
Bread	2.0	0.02
Fruits, juices	1.4	0.00
Water	0.7	0.00
Milk and products	0.2	0.00
Total	99.8	2.60
Saliva[2]	30.0	8.62

[1]*Nitrates: An Environmental Assessment*, 1978, National Academy of Sciences, p. 437, Table 9.1.

[2]Not included in the total since the amount of nitrite produced by bacteria in the mouth depends directly upon the amount of nitrate ingested.

and other products have demonstrated that a definite preference is shown for the taste of those products containing nitrites. In addition, nitrites retard rancidity, but more importantly, nitrites inhibit microbial growth, especially *Clostridium botulinum*. Hence, cured meats provide a source of ingested nitrates and nitrites. However, as a source of nitrate, cured meats are minor compared to vegetables. The major dietary source of nitrites is cured meats, but this is small when compared to that produced by the bacteria in the mouth and swallowed with saliva. Currently, there is no other protection from botulism as effective as nitrites. So, for those wishing to eliminate nitrites, possible carcinogens, there is a Catch 22—eliminate the nitrites and increase botulism poisoning, a proven danger.

Note: Since 1990, 90% of the cured meat samples have contained less than 50 ppm nitrate; only 0.1% have contained more than 200 ppm.

3. **Other sources.** Aside from very unusual circumstances, other sources of exposure to nitrates and nitrites are relatively minor. Nitrate concentrations in groundwater used for drinking range from several hun-

TABLE 18-9
FEDERAL NITRATE AND NITRITE ALLOWANCES IN MEAT

Meat Preparation	Level Allowed	
	Sodium or Potassium Nitrate	Sodium or Potassium Nitrite
Finished product	200 ppm or 91 mg/lb (maximum)	200 ppm or 91 mg/lb (maximum)
Dry cure	3.5 oz/100 lb or 991 mg/lb	1.0 oz/100 lb or 283 mg/lb
Chopped meat	2.75 oz/100 lb or 778 mg/lb	0.25 oz/100 lb or 71 mg/lb

dred micrograms per liter to a few milligrams per liter. Nitrates are generally higher in groundwater than in surface water since plants remove the nitrogen from surface water.

■ **Dangers of nitrates and nitrites**—Possibly these chemicals may present a hazard to people through two routes. First, under certain circumstances, nitrates and nitrites can be directly toxic. Second, nitrates and nitrites contribute to the formation of cancer causing nitrosamine.

1. **Toxicity.** Our knowledge of the toxic effects of nitrates and nitrites is derived from its long use in medicine, accidental ingestion, and ingestion by animals. Overall, poisoning by nitrates is uncommon. An accidental ingestion of 8 to 15 g causes severe gastroenteritis, blood in the urine and stool, weakness, collapse, and possibly death. Fortunately, nitrate is rapidly excreted from the adult body in the urine, and the formation of methemoglobin generally is not part of the toxic action of nitrates.

Almost all cases of nitrate-induced methemoglobinemia in the United States have resulted from the ingestion of infant formula made with water from a private well containing an extremely high nitrate level. Overall, it is comforting to note that several hundred million pounds of beets and spinach—nitrate-containing vegetables—are eaten yearly without injury.

2. **The cancer question.** Without doubt, the greatest concern of people is the involvement of nitrates and nitrites in directly causing cancer, or in indirectly producing compounds known as nitrosamines. Since nitrosamines are definitely accepted as carcinogens in test animals, a majority of the furor around nitrates and nitrites stems from this fact.

It cannot be stated that any human cancer has been positively attributed to nitrosamines. However, some nitrosamines have caused cancer in every laboratory animal species tested.

3. **Exposure to nitrosamines.** When nitrosamines are mentioned, foods are the first items which come to mind. However, there are numerous other sources of nitrosamines. Furthermore, people are exposed to such nitrosamines as cosmetics, lotions, and shampoos containing N-nitrosodiethanolamine (NDELA), (the latter is carcinogenic in the rat). Other preformed nitrosamines have been found in tobacco and tobacco smoke. Hence, human exposure can result from breathing or eating preformed nitrosamines, or by applying them to the skin. Foods are not the only source of nitrosamines.

4. **The FDA and the Delaney Clause.** To ban or not to ban the use of nitrates and nitrites in foods is the question. In the summer of 1978, this "fire" received more fuel when a study conducted for the FDA by Dr. Paul Newberne of Massachusetts Institute of Technology (MIT) reported that nitrite alone fed to rats in-

creased the incidence of cancers of the lymphatic system. Immediately, and before the study was properly reviewed, the USDA and the FDA announced they would soon ban nitrites. Tempers flared and pork producers lost money due to the implication of bacon containing a cancer-causing substance. The MIT study has now been reviewed by an independent group, and the research has been shown to be in error. For the time being, the FDA and USDA have backed down from their earlier stand to ban nitrite; they now say that the evidence is insufficient to initiate any action to remove nitrite from foods. However, it is noteworthy that the U.S. Supreme Court has cleared the way for the USDA to approve no-nitrite labels in processed meats, should they wish to do so.

On an individual basis, after carefully considering the issue, it is the old question of benefit versus risk. The risk of botulism in cured meats in the absence of nitrite is both real and dangerous, while the risk of cancer from low levels of nitrosamines and/or nitrites remains uncertain. Thus, the risk of botulism is considered greater than the risk of nitrates, so their use is allowed. Furthermore, no acceptable alternative is as effective as nitrite in preventing botulism. Nevertheless, nitrite should be reduced in all products to the extent protection against botulism is not compromised.

PORK SAFETY AND QUALITY ASSURANCE

In addition to nutritional quality, consumers are concerned about the safety of their food. Food safety and improved meat inspection became public issues in the 1990s. Several pork safety proposals followed; among them, the following:

1. **Irradiation.** The World Health Organization (WHO), American Medical Association (AMA), and the International Atomic Energy Agency have approved the safety of the process. Yet, consumers are squeamish about something they associate with a deadly force.

2. **Hazard Analysis, Critical Control Point (HACCP).** The HACCP plan identifies hazards in food processing, followed by monitoring those crucial points. Problems are fixed as they occur. It is a way of processing safer food.

3. **Rapid microbial test.** This test, which adapts technology already being used in the pharmaceutical and beer industries, takes five minutes and can be used in commercial meat plants. The rapid microbial test provides a means of verifying that meat and poultry plants are operating under appropriate microbiological controls.

4. **Traceback.** This involves a livestock identification system which makes it possible to trace animals back to their origin in order to locate the source of contamination. Some countries already have mandatory identification requirements that allow them to trace animals from birth to slaughter.

5. **Trichinosis prevention.** Trichinosis is caused by a microscopic parasite, *Trichinella spiralis*. In the 1990s, it was estimated that less than 0.1% of the pork in the U.S. was infected with Trichinella.

Note: Trichinella is destroyed by cooking pork to 137°F internal temperature, or by freezing for a continuous period of 20 days at a temperature not higher than 5°F.

(Also see Chapter 15, section headed Trichinosis [*Trichinella spiralis*].)

6. **Food safe labeling.** Effective July 6, 1994, the food labeling requirement became effective. It mandates safe cooking and handling labels for all uncooked meat and poultry products. The labels note that some food products may contain bacteria and can cause illness if mishandled or not cooked properly, and instruct consumers to keep raw meat and poultry refrigerated or frozen. Also, the labels warn that raw meat and poultry should be thawed only in a refrigerator or microwave, kept separate from other foods, cooked thoroughly, and refrigerated immediately or discarded.

7. **The National Pork Producers Council "Pork Quality Assurance Program."** The National Pork Producers Council introduced its Pork Quality Assurance Program in 1989. Its purpose: Enhance consumer confidence in the safety of pork, and increase pork consumption. Figures 18-16 to 18-39, which follow, were selected and adapted by the authors from the Pork Quality Assurance Program of the National Pork Producers Council, Des Moines, Iowa.

Fig. 18-16. Performance of animals is one of their best indicators of health. It follows that production records will help producers spot trends and address potential problems as soon as possible.

Fig. 18-17. It is very important that swine producers prevent the introduction of new diseases into the herd. *Note:* Diseases may enter the farm through several channels.

Fig. 18-19. Isolate new breeding stock, and, if desirable, test and medicate new breeding stock in isolation. Also, vaccinate new breeding stock as needed, based on the vaccination program of the breeding herd that they are joining.

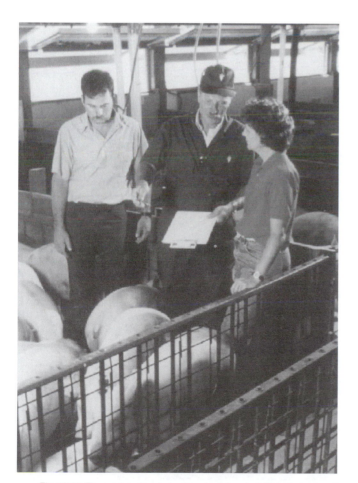

Fig. 18-20. The swine producer should limit the number of visitors, and question them about their last contact with other swine. Also, the producer should not use equipment that has been on other hog farms unless absolutely necessary, and then only after it has been thoroughly cleaned and disinfected.

Fig. 18-18. The producer's veterinarian should contact the veterinarian of the potential source of new stock to discuss the comparative health status of the two herds.

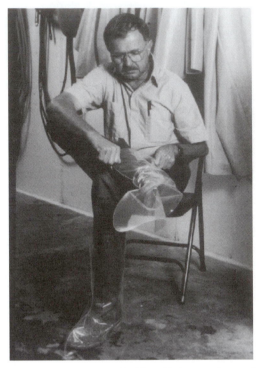

Fig. 18-21. Producers should provide boots and coveralls to all farm visitors. Also, producers should change clothes and shower after visiting other farms, livestock markets, or fairs and shows.

Fig. 18-22. Producers who truck their own hogs to market should not allow pigs to run off the truck, and then return to the truck; should keep separate boots and coveralls in the truck to wear while unloading; and should wash the truck thoroughly before returning to the farm.

Fig. 18-23. Producers should have a good rodent control program consisting of cleaning up feed spills promptly, plugging holes, changing bait regularly, and controlling weeds around buildings.

Fig. 18-24. Waterers and feeders must be easily accessible to hogs and minimize competition. Note pig in nursery drinking from a nipple waterer.

Fig. 18-25. Lagoon for storing swine manure, which may be applied to the land, thereby reducing the need for commercial fertilizers.

Fig. 18-26. Management procedures such as all in, all out pig movement will help prevent disease transmission.

Fig. 18-27. Thoroughly cleaning and disinfecting between groups of pigs results in fewer health problems and improved performance. Some producers have obtained additional improvements by separating age groups by buildings, or by housing them at different sites. Multiple site production is being used by many commercial producers.

Vaccination and Management Schedule

Date Completed _____

Production Stage	Product Name / Procedure	Dosage	Route	When Given / Age Done	Person Responsible	Preslaughter Withd. Ida
Gilts Prebreeding						
Sows Prebreeding						
Boars						
Gilts Prefarrow						
Sows Prefarrow						
Baby Pigs						
Pigs at Weaning						
Grower (50 - 100#)						
Finisher (100# - Market)						

Fig. 18-28. From time to time, producers should review their vaccination and management program with their veterinarians, agricultural extension personnel, and/or agricultural educators. A suggested *Vaccination and Management Schedule* form is presented herewith.

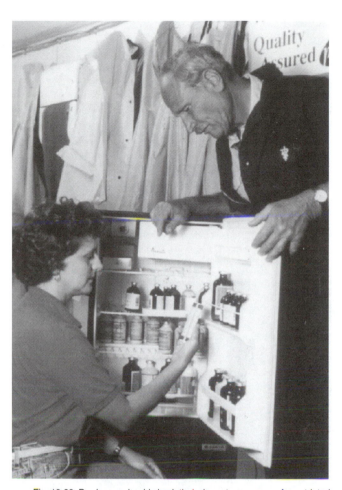

Fig. 18-29. Producers should check their drug storage areas for outdated products. Also, they should identify products requiring refrigeration and check their refrigerator temperature, using a thermometer. As a general rule, penicillin products, oxytocin, and vaccines need refrigeration. Also, all injectable medications should be stored in the refrigerator after they have been opened.

Fig. 18-30. FDA approves animal health products to be labeled for over-the-counter (OTC) use, or prescription (Rx) use. If directions for use cannot be easily written, by federal law the drug is a prescription drug. Prescription products will contain the statement: "Caution: Federal law restricts this drug to be used by, or on the order of, a licensed veterinarian."

Fig. 18-31 The approved uses of swine health products are for only those doses, routes of administration, disease conditions, and species listed on the label. If the product is to be used in any other manner, it needs to take place under a valid veterinarian/client/patient relationship.

Fig. 18-32. There are five main routes of administering injectable medication to pigs: (1) in the muscle (IM), (2) under the skin (SQ), (3) in the abdomen cavity (IP), (4) in the vein (IV), and (5) squirted in the nose (IN). Injections in the muscle should be given in the muscle of the neck, never the ham or loin. Animals should be restrained properly to avoid needle breakage and to deliver the correct dosage.

Fig. 18-33. No person (veterinarian, feed manufacturer, or producer) has "extra-label" drug use privileges for adding drugs to medicated feeds. Drugs in medicated feeds can only be used: (1) if approved for feed use by the FDA; (2) in the manner they were originally FDA approved; (3) as labeled; or (4) as provided for by a form FDA-1900.

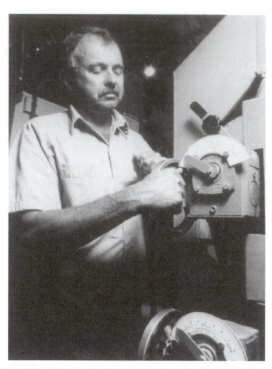

Fig. 18-34. When processing feeds on the farm, it is important that the processor pay close attention to the details of grinding and mixing. Calibrating the mixer and scales helps to ensure accurate addition of animal health products and micronutrients to the feed.

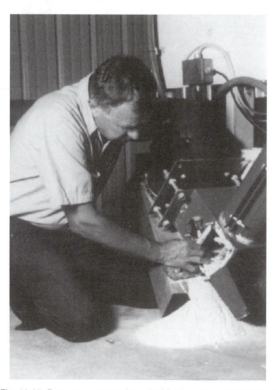

Fig. 18-35. Drug cross-contaminated of feeds can be alleviated or minimized by (1) sequencing and flushing; and (2) cleaning out the equipment, including the bottom of the mixer or auger casing, first making sure that the power to the equipment is off.

Fig. 18-36. Feed bins and feeders should be identified to reduce the possibility of human errors in feed delivery.

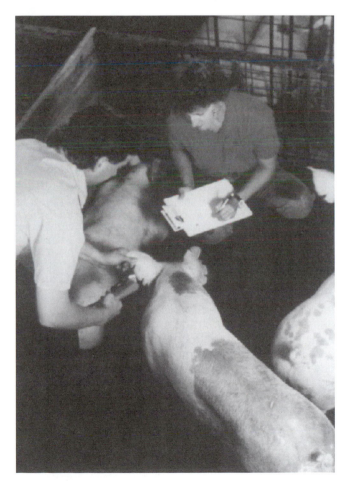

Fig. 18-37. Many producers use paint sticks to mark treated pigs. Ear tags offer individual permanent identification.

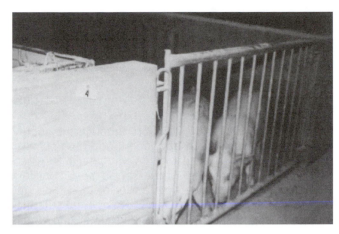

Fig. 18-38. Pen numbers may be helpful in identification and record keeping.

Leading Pork Exporting Countries	Percentage of Production Exporte
1. Denmark	77
2. Netherlands	65
3. Belgium-Luxembourg	50
4. Canada	27
5. Taiwan	26

Fig. 18-39. U.S. pork producers must compete with other pork producing countries. Currently, the U.S. exports only 2.5% of its pork, while Denmark and the Netherlands export 77% and 65%, respectively. To open export markets, U.S. producers need to offer a safe, high quality product.

WHAT DETERMINES PORK PRICES?

During those periods when pork is high in price, especially the choicest cuts, there is a tendency on the part of the consumer to blame any or all of the following: (1) the producer, (2) the packer, (3) the meat retailer, (4) the government; and these four may blame each other. Vent to such feelings is sometimes manifested in political campaign propaganda, consumer boycotts, and sensational news stories.

Who or what is to blame for high meat prices? If good public relations are to be maintained, it is imperative that each member of the meat team—the producer, the packer, and the meat retailer—be fully armed with documented facts and figures with which to answer such questions and to refute such criticisms. Also, the consumer should know the truth of the situation.

Pork prices are determined by the laws of supply and demand; that is, the price of meat is largely dependent upon what the consumers as a group are able and willing to pay for the available supply.

THE AVAILABLE SUPPLY OF PORK

Because pork is a perishable product, the supply of this food is very much dependent upon the number and weight of hogs available for slaughter at a given time. In turn, the number of market animals is largely governed by the relative profitability of the swine enterprise in comparison with other agricultural pursuits. That is to say, swine producers—like other good business people—generally do those things that are most profitable to them. Thus, a short supply of market animals at any given time usually reflects the unfavorable and unprofitable production factors that existed some months earlier and which caused curtailment of breeding and feeding operations.

History, when short pork supplies exist, pork prices rise, and the market price on slaughter hogs usually advances, making hog production more profitable. But, unfortunately, swine breeding and feeding operations cannot be turned on and off like a spigot.

History also shows that if hog prices remain high and feed abundant, producers will step up their breeding and feeding operations as fast as they can within the limitations imposed by nature, only to discover when market time arrives that too many other producers have done likewise. Overproduction, disappointingly low prices, and curtailment in breeding and feeding operations are the result.

Nevertheless, the operations of livestock producers do respond to market prices, bringing about so-called cycles. Thus, the intervals of high production, or cycles, in hogs—which are litter bearing, breed at an early age, have a short gestation period, and go to market at an early age—occur every 3 to 5 years.

THE DEMAND FOR PORK

The demand for pork is primarily determined by buying power and competition from other products. Stated in simple terms, demand is determined by the spending money available and the competitive bidding of millions of homemakers who are the chief home purchasers of meats. On a nationwide basis, a high buying power and great demand for meats exist when most people are employed and wages are high.

Also, it is generally recognized that in boom periods—periods of high personal income—meat purchases are affected in three ways: (1) More total meat is desired; (2) there is a greater demand for the choicest cuts; and (3) because of the increased money

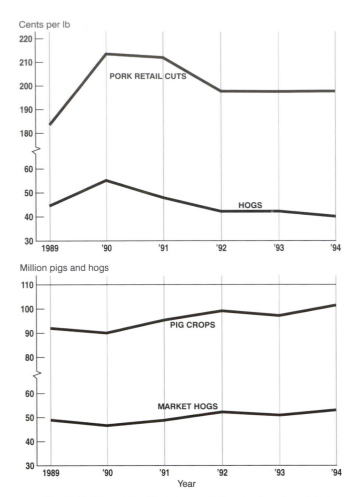

Fig. 18-40. The relationship between retail pork prices, liveweight hog prices, pig crops, and the number of market hogs. Generally, when hog prices are high, producers increase production, and then overproduction lowers the price of hogs, resulting in producers curtailing production. (*Sources: Pork Facts 1995/1996*, National Pork Producers Council, p. 19. *Agricultural Statistics 1995–96*. p. VII-26, Table 405; p. VII-19, Table 394; and p. VII-19, Table 395)

available and shorter working hours, there is a desire for more leisure time, which in turn increases the demand for those meat cuts or products that require a minimum of time in preparation (such as pork chops and hams). In other words, during periods of high buying power, not only do people want more meats, but they compete for the choicer and more easily prepared cuts of meats.

Because of the operation of the old law of supply and demand, when the choicer and more easily prepared cuts of pork are in increased demand, they advance proportionately more in price than the cheaper cuts. This results in a great spread in prices, with some pork cuts very much higher than others. Thus, while pork chops may be selling for 4 or 5 times the cost per pound of the live animal, less demanded cuts may be priced at less than half the cost of the more popular cuts. This is so because a market must be secured for all the cuts.

But the novice may wonder why these choice cuts are so scarce, even though people are able and willing to pay a premium for them. The answer is simple: Hogs are not all pork, and pork is not all chops (see Fig. 18-41). Besides, nature does not make many choice cuts or top grades, regardless of price; a hog is born with only two hams and a limited number of pork chops. It is important, therefore, that those who produce and slaughter animals and those who purchase wholesale and/or retail cuts know the approximate (1) percentage yield of chilled carcass in relation to the weight of the animal on foot, and (2) yield of different retail cuts. For example, the average hog weighing 250 lb on foot will only yield about 139.4 lb of salable pork cuts and 44.8 lb of other products (the balance consists of internal organs, etc.). Thus, only about 55 to 60% of a live hog can be sold as retail cuts of pork. In other words, the price of pork at retail would have to be nearly double

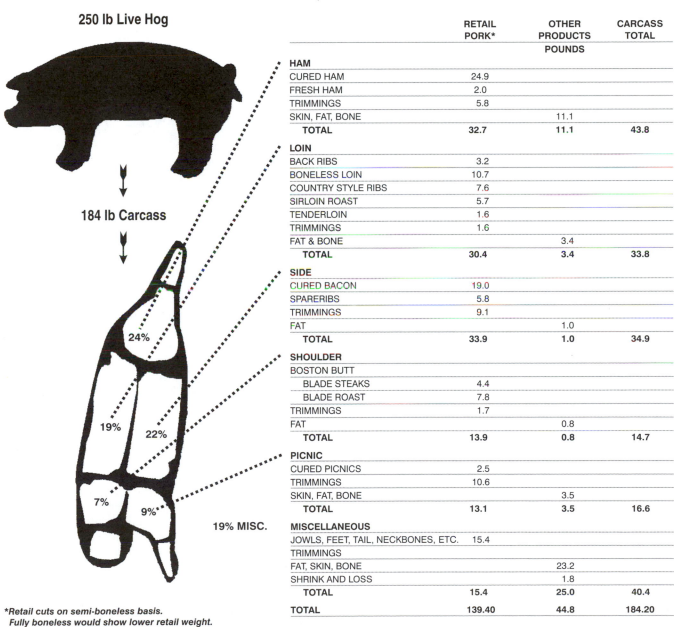

HOGS ARE NOT ALL PORK, AND PORK IS NOT ALL CHOPS!

250 lb Live Hog

184 lb Carcass

19% MISC.

	RETAIL PORK*	OTHER PRODUCTS	CARCASS TOTAL
		POUNDS	
HAM			
CURED HAM	24.9		
FRESH HAM	2.0		
TRIMMINGS	5.8		
SKIN, FAT, BONE		11.1	
TOTAL	**32.7**	**11.1**	**43.8**
LOIN			
BACK RIBS	3.2		
BONELESS LOIN	10.7		
COUNTRY STYLE RIBS	7.6		
SIRLOIN ROAST	5.7		
TENDERLOIN	1.6		
TRIMMINGS	1.6		
FAT & BONE		3.4	
TOTAL	**30.4**	**3.4**	**33.8**
SIDE			
CURED BACON	19.0		
SPARERIBS	5.8		
TRIMMINGS	9.1		
FAT		1.0	
TOTAL	**33.9**	**1.0**	**34.9**
SHOULDER			
BOSTON BUTT			
BLADE STEAKS	4.4		
BLADE ROAST	7.8		
TRIMMINGS	1.7		
FAT		0.8	
TOTAL	**13.9**	**0.8**	**14.7**
PICNIC			
CURED PICNICS	2.5		
TRIMMINGS	10.6		
SKIN, FAT, BONE		3.5	
TOTAL	**13.1**	**3.5**	**16.6**
MISCELLANEOUS			
JOWLS, FEET, TAIL, NECKBONES, ETC.	15.4		
TRIMMINGS			
FAT, SKIN, BONE		23.2	
SHRINK AND LOSS		1.8	
TOTAL	**15.4**	**25.0**	**40.4**
TOTAL	**139.40**	**44.8**	**184.20**

*Retail cuts on semi-boneless basis.
Fully boneless would show lower retail weight.*

Fig. 18-41. Hogs are not all pork, and pork is not all chops—or hams. (Courtesy, American Meat Institute, Washington, DC, from *1994 Meat & Poultry Facts*, p. 23)

the live cost even if there were no processing and marketing charges at all. Secondly, the higher priced cuts make up only a small part of the carcass. Thus, this 139.4 lb will cut out only about 10.7 lb of boneless loin. The other cuts retail at lower prices than do these choice cuts; also, there are bones, fat, and cutting losses which must be considered.

Thus, when the national income is exceedingly high, there is a demand for the choicest but limited cuts of pork from the very top grades. This is certain to make for high prices, for the supply of such cuts is limited, but the demand is great. Under these conditions, if prices did not move up to balance the supply with demand, there would be a marked shortage of the desired cuts at the retail counter.

It must also be remembered that meats must compete with other food products for the consumer's dollar. On the average, U.S. consumers spend about 0.5% of their disposable income, or about 3.5% of their food budget, for pork (1993 figures). In addition to preference, relative prices are an important factor in determining food selection.

WHERE THE CONSUMER'S FOOD DOLLAR GOES

Food is the nation's largest industry. Americans spent $617 billion—almost 10% of the U.S. gross national product—for food in 1993. This vast sum included the bill for about 250,400 (1992 figures) retail food stores; a large majority of the nation's 2,064,930 farms (1993 figures); thousands of wholesalers, brokers, eating establishments, and other food firms; and the transportation, equipment, and container industries. In 1992, food and kindred products employed 12,291,000 people.

In recognition of the importance of food to the nation's economy and the welfare of its people, it is important to know where the consumer's food dollar goes—the proportion of it that goes to the producer, and the proportion that goes to the middleman. Furthermore, it is important to understand the prices and profits of the producer and some of the middlemen—packers and retailers. Often misunderstandings arise and consumers seek to blame someone, frequently the producer, for high prices. For example, some consumers may compare what the packer is paying for hogs on foot to what they are paying for a pound of pork over the counter.

Table 18-10 reveals that of each food dollar in 1992, the farmer's share was only 25¢, the rest—75¢—went for processing and marketing. This means that three-fourths of today's food dollar goes for preparing, processing, packaging, and selling—and not for the food itself.

TABLE 18-10
FARMER'S SHARE OF THE CONSUMER'S DOLLAR[1]

Food	Year			
	1989	1990	1991	1992
	(¢)	(¢)	(¢)	(¢)
Eggs	43	42	41	36
Meat	28	29	26	26
Poultry	42	37	36	38
Dairy products	34	33	29	32
Fats and oils	21	21	19	19
Fruits and vegetables	19	16	16	16
Bakery and cereal products .	10	8	8	7
Average	28	27	25	25

[1] *Farm & Food Facts*, Kiplinger, 1994, p. 34.

The farmer's share of the retail price of eggs and meat is relatively high because processing is simple and inexpensive, and transportation costs are low due to the concentrated nature of the products. On the other hand, the farmer's share of bakery and cereal products is low due to the high processing and container costs. and the bulky, costly transportation.

PACKINGHOUSE BYPRODUCTS FROM HOG SLAUGHTER

The meat or flesh of hogs is the primary object of slaughtering. The numerous other products are obtained incidentally. Thus, all products other than the carcass meat and lard are designated as byproducts, even though many of them are wholesome and highly nutritious articles of the human diet. Yet it must be realized that upon slaughter live hogs yield an average of 40–45% of products other than retail cuts of pork. When meat packers buy hogs, they buy far more than the cuts of meat that will eventually be obtained from the carcass; that is, only about 55–60% of a hog is retail meat.

In the early days of the meat packing industry, the only salvaged animal byproducts were hides, wool, tallow, and tongue. The remainder of the offal was usually carted away and dumped into the river or burned or buried. In some instances, packers even paid for having the offal taken away. In due time, factories for the manufacture of glue, fertilizer, soap, buttons, and numerous other byproducts sprang up in the vicinity of the packing plants. Some factories were company-owned; others were independent industries. Soon much of the former waste material was being converted into materials of value.

Naturally, the relative value of carcass meat and byproducts varies both according to the class of livestock and from year to year. The longtime trend for byproduct values has been downward relative to the value of the live animal, due largely to technological progress in competitive products derived from nonanimal sources.

The complete utilization of byproducts is one of the chief reasons why large packers are able to compete so successfully with local butchers. Were it not for this conversion of waste material into salable form, the price of meat would be higher than under existing conditions.

It is not intended that this book should describe all of the byproducts obtained from hog slaughter. Rather, only a few of the more important ones will be listed and briefly discussed (see Fig. 18-42).

1. **Skins.** Pig skins from domestic sources are virtually nonexistent because, for the most part, they are sold along with the pork cuts. The main source of so-called pigskin leather is the peccary, a pig-like mammal of Central America, which provides leather for wallets, handbags, shoes, clothing, sporting goods, upholstery, saddles, gloves, and razor strops.

Skins from slaughtered hogs are used to make gelatin, which is used for coating pills and making capsules. Also, a porcine collagen product has been developed for stimulating clotting during surgery. Because of its similarity to human skin, specially selected and treated hog skins are used in treating massive burns in humans, injuries that have removed large areas of skin, and in healing persistent skin ulcers.

Note: Some packers now skin, rather than scald, hogs. Scalding makes the skin of pigs unsuited for tanning to leather.

2. **Fats.** Rendered pork fat, known as grease, is used in animals feeds. In addition to energy value, fats reduce dust in feed processing, improve the color and texture of feed, enhance the palatability, increase pelleting efficiency, and reduce machinery wear in the production of animals feeds.

Fatty acids obtained from animal fats through a process referred to as "splitting" are used in the manufacture of a host of products.

Lard oil, made from white grease, is used for making a high grade lubricant which is used on delicate, running machine parts.

During the mid-twentieth century soap making declined significantly, due primarily to the increased use of phosphate-based detergents, powders, and liquids. But soap is biodegradable, whereas phosphate-based detergents are not. So, fat-based cleaning products with detergent-like traits, effective in hard and cold water, have been developed and are in use in various parts of the world. Environmental concerns and fat utilization research have reinstated fat-based materials in the cleaning market.

3. **Variety meats.** The edible byproducts include the liver, brains, kidneys, stomach, ears, pork skins, snout, and weasand (esophagus) meat. All these are sold over the counter as variety meats or fancy meats. They represent about 3.25% of the live weight of a hog; the inedible byproducts represent about 2.5% of the liveweight of a hog.

Some edible hog byproducts which are known as meat and not meat byproducts, but which come under the "variety meats" classification, are head and cheek meat, tongue, heart, tail, and feet.

4. **Hair.** Hog bristles for making brushes were formerly imported from China, but are now produced in the United States in increasing quantities. Proper length bristles are found over the shoulder and back of the hog. The fine hair of most U.S. hogs is not suitable for brush making; it is processed and curled for upholstering purposes.

5. **Heart.** Hog heart valves, specially treated and preserved, are surgically implanted in humans to replace heart valves weakened by disease or injury. Since the first surgery in 1971, thousands of heart

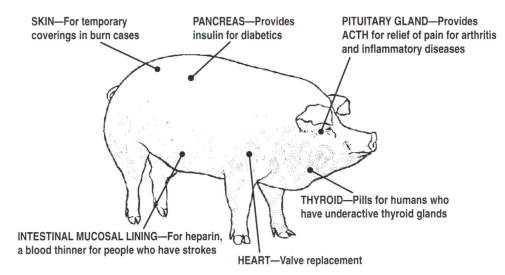

SKIN—For temporary coverings in burn cases

PANCREAS—Provides insulin for diabetics

PITUITARY GLAND—Provides ACTH for relief of pain for arthritis and inflammatory diseases

THYROID—Pills for humans who have underactive thyroid glands

INTESTINAL MUCOSAL LINING—For heparin, a blood thinner for people who have strokes

HEART—Valve replacement

Fig. 18-42. This illustration shows some of the parts of the hog's body which are used in medicine. The list of drugs/pharmaceuticals and industrial/consumer products which follows was adapted by the authors from *Pork Facts* 1995/1996, published by the National Pork Producers Council, Des Moines, IA.

valves have been successfully implanted in human recipients of all ages.

6. **Blood.** Blood albumen from pigs is used in human blood Rh factor typing, and to make amino acids that are part of parenteral solutions for nourishing certain types of surgical patients. Fetal pig plasma is important in the manufacture of vaccines and tissue culture media. *Note:* Fetal blood does not contain any antibodies; so, it is not likely to stimulate immune reactions.

Thrombin, a blood protein, helps create significant blood coagulation.

Plasmin, a hog blood enzyme which has the unique ability to digest fibrin in blood clots, is used to treat patients who have suffered heart attacks.

Hog blood is also used in cancer research, microbiological media, and cell cultures.

7. **Meat scraps and muscle tissue.** After the grease is removed from meat scraps and muscle tissue, they are made into meat meal or tankage.

8. **Bones.** The bones and cartilage are converted into stock feed, fertilizer, glue, crochet needles, dice, knife handles, buttons, toothbrush handles, and numerous other articles.

9. **Intestines and bladders.** Intestines and bladders are used as containers for sausage, lard, cheese, snuff, and putty.

10. **Endocrine glands.** Various endocrine glands of the body, including the thyroid, parathyroid, pituitary, pineal, adrenals, and pancreas, are used in the manufacture of numerous pharmaceutical preparations. Hog pancreas glands are an important source of insulin used to treat diabetics. Hog insulin is especially important because its chemical structure most nearly resembles that of humans.

Proper preparation of glands requires quick chilling and skillful handling. Moreover, a very large number of glands must be collected in order to obtain any appreciable amount of most of these pharmaceutical products.

Until recently, the endocrine glands were the only source of many pharmaceutical preparations. Now, many of these products are made synthetically.

11. **Collagen.** The collagen of the connective tissues—sinews, lips, head, knuckles, feet, and bones—is made into glue and gelatin. The most important uses for glues are in the wood-working industry. Gelatin is used in canning hams and other large cuts, and in baking, ice cream making, capsules for medicine, coating for pills, photography, and culture media for bacteria. Also, a porcine collagen product has been developed for stimulating clotting during surgery.

12. **Contents of the Stomach.** Contents of the stomach are used in making fertilizer.

Thus, in a modern packing plant, there is no waste; literally speaking, "everything but the squeal" is saved.

These byproducts benefit the human race in many ways. Moreover, their utilization makes it possible to slaughter and process pork at a lower cost. But scientists are continually striving to find new and better uses for packinghouse byproducts in an effort to increase their value.

■ **Drugs and pharmaceuticals**—Hogs are a source of nearly 40 drugs and pharmaceuticals, a list of which follows:

Adrenal Glands
Corticosteroids
Cortisone
Epinephrine
Norepinephrine

Blood
Blood fibrin
Fetal pig plasma
Plasmin

Brain
Cholesterol
Hypothalamus

Gall Bladder
Chenodeoxycholic acid

Heart
Heart valves

Intestines
Enterogastrone
Heparin
Secretin

Liver
Desiccated liver

Ovaries
Estrogens
Progesterone
Relaxin

Pancreas Gland
Chymotrypsin
Insulin
Glucagon
Lipase
Pancreatin
Trypsin

Pineal Gland
Melatonin

Pituitary Gland
ACTH—adrenocortico-
   tropic hormone
ADH—antidiuretic
   hormone
Oxytocin
Prolactin
TSH —thyroid stimulating
   hormone

Skin
Dressings
Gelatin
Porcine burn

Spleen
Splenin fluid

Stomach
Intrinsic factor
Mucin
Pepsin

Thyroid Gland
Calcitonin
Thyroglobulin
Thyroxin

■ **Industrial and consumer products**—Hogs make a very significant contribution to industrial and consumer products. Hog byproducts are sources of chemical used in the manufacture of a wide range of products which cannot be duplicated by synthesis.

Blood
Fabric printing & dyeing
Leather treating agents
Plywood adhesive
Protein source in feeds
Sticking Agent

Bones & Skin
Glue
Pigskin garments,
  gloves, & shoes

Bones, Dried
Bone China
Buttons

Bone Meal
Fertilizer
Glass
Mineral source in feed
Porcelain enamel
Water filters

Brains
Cholesterol

Fatty Acids & Glycerine
Antifreeze
Cellophane
Cement fiber
Softeners
Chalk
Cosmetics

Fatty Acids & Glycerine (continued)
Crayons
Floor waxes
Insecticides
Insulation
Linoleum
Lubricants
Matches
Nitroglycerine
Oil polishes
Paper sizing
Phonograph records
Plasticizers
Plastics
Printing rollers
Putty
Rubber
Water-proofing agents
Weed killers

Gall Stones
Ornaments

Hair
Artist brushes
Insulation
Upholstery

Meat Scraps
Commercial feeds
Feeds for pets

PORK PROMOTION

In the 1950s, a voluntary producers organization known as the National Swine Growers Council was formed. At that time, the system was voluntary and only about half of all producers participated. In the beginning, there was a voluntary checkoff of 20¢ on market hogs and 10¢ on feeder pigs. By 1968, 16 state associations were organized. Along the way, the council was renamed the National Pork Producers Council (NPPC). They lobbied hard for congressional approval to activate a market deduction on livestock to fund product promotion. They succeeded in getting amendments to the Packers and Stockyards Act that permitted voluntary checkoff.

In December, 1985, Congress approved a 100% National Legislative Checkoff, with the stated purpose of providing funds for pork promotion and research to enhance the pork producers' opportunity for profit.

On September 7 and 8, 1986, U.S. pork producers voted overwhelming approval (77% of those who voted) for a mandatory checkoff fee on every hog sold in the United States. Following the passage of the referendum, about 95% of all pork producers participated. The program is still voluntary in that those who elect not to participate are given a refund. At the outset, producers were charged 25¢ of every $100 worth of hogs sold in the United States. Imported pork is included at a separate rate determined by U.S. market prices. In 1987, $28,540,336 in gross receipts was obtained from the producer checkoff.

Under the law, if additional funds are needed, not to exceed one-tenth of one percent of the market value of each fiscal year may be levied. But there is a cap of 50¢ for $100 in value under the Act. Currently, the checkoff is 45¢ of every $100 market value.

Effective pork promotion—which should embrace research, education, and sales approaches—necessitates full knowledge of the product.

Research conducted at Land Grant Universities in the United States, and at the Provincial Universities in Canada, provides producers with information to improve production practices and make pork more desirable for consumers.

QUESTIONS FOR STUDY AND DISCUSSION

1. Discuss the significance of the message conveyed by Fig. 18-1.

2. What different messages does the term *pork quality* convey?

3. List the eight qualities which consumers desire in pork.

4. List the federal grades of barrow and gilt carcasses, and the primary factors on which the grades are based. Why are only negligible quantities of pork federally graded?

5. Sketch and label the wholesale cuts of a pork carcass.

6. List and describe the types of meat cookery.

7. In the United States, meat curing is largely confined to pork, with little beef and lamb cured. Why?

8. Discuss the nutritive qualities of pork.

9. Discuss consumer health concerns relative to pork fats and cholesterol.

10. Discuss consumers health concerns relative to nitrates/nitrites and nitrosamines.

11. How are pork prices determined by supply and demand?

12. List and discuss the several pork safety proposals. Why did pork safety and improved meat inspection become public issues in the 1990s?

13. Discuss the National Pork Producers Council "Pork Quality Assurance Program."

14. Fig. 18-41 shows that, on the average, only 139.4 lb of retail pork is obtained from a 250 lb live hog. What accounts for the rest of the original live weight of 250 lb?

15. Why does the farmer get such a small share of the consumer's dollar?

16. List and discuss 10 important packinghouse byproducts from hog slaughter.

17. Trace and discuss the history of pork promotion.

SELECTED REFERENCES

Title of Publication	Author(s)	Publisher
Agricultural Statistics, 1994	Staff	U.S. Department of Agriculture, Washington, DC
Animal Science, Ninth Edition	M. E. Ensminger	Interstate Publishers, Inc., Danville, IL, 1991
Developments in Meat Science—2	Ed. by R. Lawrie	Applied Science Publishers, Inc., Englewood Cliffs, NJ, 1981
Food from Animals	G. C. Smith, Chairman of Task Force	CAST, 1980
Food from Farmer to Consumer	National Commission on Food Marketing	U.S. Government Printing Office, Washington, DC, 1966
Foods & Nutrition Encyclopedia	A. H. Ensminger M. E. Ensminger J. E. Konlande J. R. K. Robson	CRC Press, Boca Raton, FL, 1994
Hides and Skins	National Hide Association	Jacobsen Publishing Co., Chicago, IL, 1970
Lessons on Meat		National Live Stock and Meat Board, Chicago, IL, 1972
Livestock and Meat Marketing	J. H. McCoy	Avi Publishing Co., Westport, CT, 1979
Meat and Poultry Facts	Staff	American Meat Institute, Washington, DC, 1994
Meat Board Meat Book, The	B. Bloch	McGraw-Hill Book Co., New York, NY, 1977
Meat Handbook	A. Levie	Avi Publishing Co., Westport, CT, 1979
Meat, Poultry, and Seafood Technology	R. L. Henderson	Prentice-Hall, Inc., Englewood Cliffs, NJ, 1978
Meat We Eat, The	J. R. Romans, et al.	Interstate Publishers, Inc., Danville, IL, 1994
Pork Facts	Staff	National Pork Producers Council, Des Moines, IA, 1995–96
Practical Meat Cutting and Merchandising, Vol. 2	T. Fabbricante W. J. Sultan	Avi Publishing Co., Westport, CT, 1975
Science of Meat and Meat Products, The	Ed. by J. F. Price B. S. Schweigert	W. H. Freeman and Co., San Francisco, CA, 1971
Statistical Abstracts of the United States	Staff	U.S. Department of Commerce, Washington, DC, 1995
Stockman's Handbook, The, Seventh Edition	M. E. Ensminger	Interstate Publishers, Inc., Danville, IL, 1992

THE NUMBER OF U.S. HOG FARMS DECREASED FROM 2.39 MILLION IN 1954 TO 208,780 IN 1994

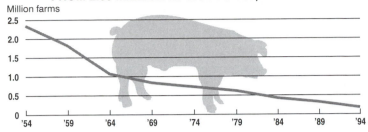

Million farms

... AND THE NUMBER OF HOGS PER FARM INCREASED 10 FOLD

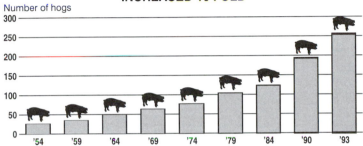

Number of hogs

Pork producers are no longer their father's pig slopper! Today, pork production is big business.

19

BUSINESS ASPECTS OF SWINE PRODUCTION

The great changes that occurred in the U.S. swine industry in the 1980s and 1990s resulted from a shift of our labor-based economy to a knowledge- and capital-based economy. The authors predict that the returns to sweat of brow will continue to shrink, while the returns to knowledge and capital will continue to grow and grow.

But big isn't always better! According to an on-going study at the University of Nebraska, the larger pork producing operations are not always the most profitable. Additionally, the bigger the unit, the more difficult the environmental controls.

COMPUTERS IN THE SWINE BUSINESS

Accurate and up-to-the-minute records and con-trols have taken on increasing importance in all agri-culture, including the swine business, as the invest-ment has risen and profit margins have narrowed. Also, records must be kept current. It no longer suffices merely to know the bank balance at the end of the year.

Big and complex swine operations have outgrown hand record keeping. It is too time consuming, with the result that it does not allow management enough time for planning and decision making. Additionally, it does not permit an all-at-once consideration of the complex interrelationships which affect the economic success of the business. The use of computers is a modern and effective tool in modern swine production, when just a few dollars per head may make the difference between profit and loss.

Today, there are many types of software (pro-grams) available for use in swine production. A *Swine Software Directory* has been prepared by Paul I. Leland and Dr. Jerry Shurson, Department of Animal Science, University of Minnesota, and is available for purchase. The *Table of Contents* and the *Swine Soft-ware Listing* of this publication are herewith presented as a means of showing the magnitude and diversity of available software.

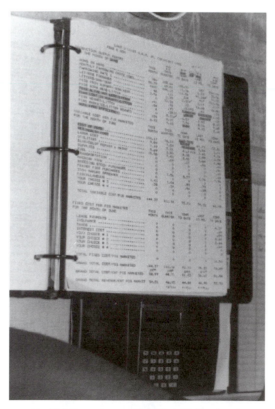

Fig. 19-1. Computers greatly facilitate the keeping of up-to-the-minute records and controls. (Courtesy, Land O Lakes, Ft. Dodge, IA)

Swine Software Directory

prepared by
Paul I. Leland
and
Dr. Jerry Shurson

Department of Animal Science
University of Minnesota

Table of Contents

SWINE SOFTWARE LISTING
PRODUCTION AND MANAGEMENT SYSTEMS

PRODUCT NAME	TAB	COST	SOFTWARE VENDOR
FINPACK	5	$295	University of Minnesota/FINPACK
FIN-PRO	6		Wayne Feeds
Hog Cash Flow Edition	10	$295	Chek-Tech, Inc.
Hog Management*	11	$10,000+	microHELP Services, Inc.
Hog Manager	12	$595	Harvest Computer Systems
HOGS	9		Virginia Tech
Iowa State Pork Production Worksheet Series	13	$150	Iowa State University
KW Supersow	15	$295	KW Software, Inc.
KW SuperFinisher	14	$295	KW Software, Inc.
PigCHAMP®	21	$1200	University of Minnesota/PigCHAMP
PigMON®	23	none	University of Minnesota/PigMON
PigPROPHET	25	$125	Texas A&M University
Smart Breeder™	28	$595	FBS Systems, Inc.
Smart Feeder™	29	$995	FBS Systems, Inc.
SwinePro	31	$200	University of Minnesota/SwinePro
SwineTRAK Recordkeeper	32	$395	Red Wing Business Systems, Inc.
SwineTRAK Plus	32	$795	Red Wing Business Systems, Inc.

A part of the M•A•S Evolution/2 software series by State of the Art, Inc.

PRODUCTION FLOW SCHEDULING

PRODUCT NAME	TAB	COST	SOFTWARE VENDOR
PIGFLOW	22	$15	Purdue University
PigPlan	24	$25	Michigan State University
Swine Breeding & Farrowing Schedule	13	$25	Iowa State University

GROWTH AND SIMULATION MODELS

PRODUCT NAME	TAB	COST	SOFTWARE VENDOR
Fortel	7	$14,900	Heartland Lysine, Inc.
GrowthMaster	8	$90	Prairie Swine Centre, Inc.
NCCISWINE	18	$10	North Central Computer Institute
SLAMSYSTEM	27		US Meat Animal Research Center
TurboSow	35	$1250	ExperCon Inc.

SWINE SOFTWARE LISTING
DIET FORMULATORS

PRODUCT NAME	TAB	COST	SOFTWARE VENDOR
Brill Feed Formulator	2		The Brill Corporation
The Consulting Nutritionist	3	$695	Dalex Computer Systems
Professional NutritionistSwine	26	$300	University of Minnesota/PNSwine
SPARTAN Swine Ration Evaluator	30	$25	Michigan State University
Swine Diet Analysis and Rel. Value	13	$30	Iowa State University
TriLogic Feed Formulator	34		TriLogic Systems

GENETIC/SELECTION PROGRAMS

PRODUCT NAME	TAB	COST	SOFTWARE VENDOR
CrossBreeding Analysis Program for Pigs	4		University of Nebraska
Genetic Improvement in Swine (NCSUGIS)	19	none	North Carolina University
Sow Productivity Index	13	$25	Iowa State University
Ohio Sow Productivity Index (SPI) Service	20	$10	Illinois Cooperative Extension
WISEL 4.0	33		Virginia Tech

WASTE MANAGEMENT PROGRAMS

PRODUCT NAME	TAB	COST	SOFTWARE VENDOR
AMANURE	1	$15	Purdue University
Liquid Manure Storage and Handling	13	$25	Iowa State University
Manure Application Planner (MAP)	16	$120	University of Minnesota/MAP
MSU Nutrient Management	17	$75	Michigan State University

FULL PRODUCT LINE INTEGRATED ACCOUNTING AND AG SYSTEMS

FBS Systems, Inc.   Section V Tab 36
Harvest Computer Systems   Section V Tab 37
Red Wing Business Systems, Inc.   Section V Tab 38

These companies and others provide a full line of integrated accounting packages (general ledger, depreciation, payroll, invoicing), inventory packages, swine production and management packages, and other agricultural products.

COST AND PROFIT/LOSS IN SWINE PRODUCTION

With constant changes in the swine industry, pork producers must continually strive to improve their enterprises. They need to identify their strengths and weaknesses and then determine the opportunities for, and the threats to, their individual swine enterprise.

To analyze the costs and returns from swine production in mid-1995, data from the following three types of swine operations are presented: (1) farrow-to-finish, (2) farrow-to-feeder pigs, and (3) finish-to-market.

Note: In Tables 19-1 and 19-2, in addition to the overall averages for each type of enterprise, averages for the high one-third profit group and low one-third profit group are listed for the farrow-to-finish and farrow-to-feeder pig enterprises.

TABLE 19-1
FARROW-TO-FINISH ENTERPRISES[1]

Item	January 1 to June 30, 1995			July 1, 1994 to June 30, 1995[2]
	Average	High 1/3 Profit	Low 1/3 Profit	
Number of farms . . . . (no.)	37	12	12	20
Profit/cwt pork produced . . . . ($)	4.75	11.32	−2.07	−1.22
Total cost/cwt pork produced . . . . ($)	40.99	36.63	44.20	39.69
Total variable cost/cwt pork produced . . ($)	35.76	33.24	38.46	35.52
Fixed cost/cwt of pork produced . . . . . ($)	5.24	3.39	5.74	4.17
Total feed expense/cwt pork produced . . ($)	24.27	23.12	25.47	24.05
Average cost of diets/cwt . . . . ($)	6.61	6.38	6.79	6.44
Feed/cwt pork produced . . . . (lb)	368	362	377	373
Pigs weaned/female/year . . . . (no.)	17.9	18.3	16.3	17.6
Pigs weaned/crate/year . . . . (no.)	78.4	81.3	79.2	76.6

[1]Kabes, D., M. Brumm, L. Bitney, *Nebraska Swine Report*, "Nebraska Swine Enterprise Records," p. 15, 1996.

[2]Annual data from July 1, 1994 through June 30, 1995.

TABLE 19-2
FARROW-TO-FEEDER ENTERPRISES[1]

Item	January 1 to June 30, 1995			July 1, 1994 to June 30, 1995[2]
	Average	High 1/3 Profit	Low 1/3 Profit	
Number of farms .... (no.)	11	4	4	8
Profit/cwt pork produced .... ($)	2.61	13.38	−7.55	−10.24
Total cost/cwt pork produced .... ($)	63.16	60.39	64.87	66.56
Total variable cost/cwt pork produced . . ($)	53.16	51.32	55.44	54.56
Fixed cost/cwt of pork produced ..... ($)	10.00	9.07	9.43	12.00
Total feed expense/cwt pork produced . ($)	30.80	27.05	34.41	30.64
Average cost of diets/cwt ...... ($)	7.91	7.61	8.20	8.34
Feed/cwt pork produced .... (lb)	389	357	419	368
Pigs weaned/female/year .... (no.)	17.2	18.7	15.3	18.2
Pigs weaned/crate/year .... (no.)	89.6	98.5	81	100.3
Average weight of feeder pig sold . . . . (lb)	50.2	53.9	46.5	49.8

[1]Kabes, D., M. Brumm, L. Bitney, *Nebraska Swine Report*, "Nebraska Swine Enterprise Records," p. 15, 1996.

[2]Annual data from July 1, 1994 through June 30, 1995.

TABLE 19-3
FINISH-TO-MARKET ENTERPRISES FROM 40 TO 230 LB[1]

Hog Prices, Costs and Profit/Loss Margins for Finishing Feeder Pigs

	1993						1994				
Month	Market Hog Price	Feeder Pig Cost	Total Feed Cost	Break-even	Profit/ Loss Margin[2]	Month	Market Hog Price	Feeder Pig Cost	Total Feed Cost	Break-even	Profit/ Loss Margin[2]
	($/cwt)	($)	($)	($/cwt)	($)		($/cwt)	($)	($)	($/cwt)	($)
January	42.40	33.31	38.70	41.14	2.89	January	43.76	41.86	45.29	47.97	−9.68
February	44.31	32.63	38.53	40.76	8.17	February	47.75	40.33	46.49	47.79	−0.09
March	47.01	27.79	38.67	38.57	19.42	March	44.08	35.96	46.83	45.91	−4.20
April	45.59	28.49	39.26	39.15	14.82	April	42.33	32.14	46.26	43.89	−3.58
May	47.19	35.39	39.80	42.60	10.57	May	42.37	34.33	45.62	44.62	−5.18
June	48.48	39.98	39.96	44.80	8.45	June	42.51	45.10	45.29	49.47	−16.01
July	46.21	49.73	40.62	49.58	−7.74	July	42.43	51.14	44.16	51.78	−21.51
August	48.13	48.57	41.38	49.39	−2.86	August	42.22	45.58	42.42	48.44	−14.30
September	49.05	50.57	41.68	50.43	−3.19	September	35.36	43.13	40.76	46.60	−25.85
October	47.04	43.84	41.96	47.43	−0.90	October	31.94	31.68	39.14	40.61	−19.94
November	42.89	50.82	42.95	51.11	−18.91	November	28.01	31.73	37.98	40.13	−27.88
December	40.38	31.54	44.03	42.63	−5.18	December	31.64	29.39	37.73	38.93	−16.77

[1]*Pork Facts*, National Pork Producers Council, Des Moines, Iowa, p. 16, 1995–96. Source: Iowa State University. Data in dollars per head unless otherwise indicated.

[2]Based on selling price required to cover costs of feeding 40 lb feeder pig to 230 lb slaughter hog.

HOG CYCLE—LONGER AND LESS ABRUPT

Concentrating the hog industry in fewer producers is making a difference in the price cycle. Formerly, it averaged about four years (see Chapter 17, section headed "Cyclical Movements").

Fig. 19-2 graph, developed by Dr. Glen Grimes, Agricultural Economist, University of Missouri, based on percentage change in hog slaughter compared to the previous year, indicates that—

1. Increases are not going quite as high as they did in the late 1970s and early 1980s.

2. Decreases are also moderating, compared to prior 1985.

3. The peaks and valleys get broader as time goes on, indicating a longer cycle.

Fig. 19-3. The success of all pork production systems is based on performance tested breeding stock like these performance tested Yorkshire boars. (Courtesy, Lone Willow Farm, Roanoke, IL)

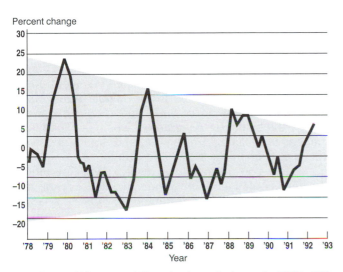

Fig. 19-2. U.S. commercial hog slaughter—*the hog cycle*—1978 to 1993. (Prepared by Dr. Glenn Grimes, University of Missouri, Columbia, MO)

PORK PRODUCTION SYSTEMS

Many different techniques and facilities are being used in hog production. Basically, there are the following three systems of production: (1) farrow-to-finish, (2) farrow-to-feeder pig production, and (3) feeder pig finishing to market. Within each system there are the following variations: high-investment, high-intensity; low-investment, low-intensity; and indoor versus outdoor production.

FARROW-TO-FINISH

This is a complete operation, involving (1) breeding, (2) farrowing-nursery, and (3) growing-finishing-marketing. For success, farrow-to-finish operators must have knowledge of every phase of swine production. Usually, employees are given specialized assignments in one phase only.

(Also see Chapter 14, section headed "Farrow-to-Finish Production.")

FARROW-TO-FEEDER PIG

This system is for the purpose of producing 30 to 60 lb. feeder pigs which are sold to finishers who carry them to slaughter weight. Such operations are usually found on farms, where grain and capital are limited, but where there is adequate labor. Farrowings are scheduled as frequently as possible, within the limitations of disease control and proper breeding and herd management.

(Also see Chapter 14, section headed "Feeder Pig Production.")

FINISH TO MARKET

This system involves the purchase of 30 to 60 lb. feeder pigs which are fed to market weight. Finishing enterprises require rather large sums of operating capital and involve considerable financial risk.

Successful feeder pig finishers tend to be (1) short on labor, but long on feed grain; (2) skilled in buying and selling; and (3) able to withstand periods of financial loss.

(Also see Chapter 14, section headed "Growing-Finishing Production.")

TYPE OF BUSINESS ORGANIZATIONS

The success of today's swine enterprise is very dependent on the type of business organization. No one type of organization is superior under all circumstances; rather, each situation must be considered individually. The size of the operation, the family situation, the enterprises, the objective—all these, and more, are important in determining the best way in which to organize the swine enterprise.

Six types of business organizations are commonly found among swine enterprises: (1) proprietorship (individual), (2) partnership (general partnership), (3) corporations, (4) contracts, (5) co-ops, and (6) networking.

Among the factors which should be considered when deciding which business form best fits a given set of circumstances are the following:

1. Which type of organization is most likely to be looked upon favorably from the standpoint of more credit and capital?
2. How much capital will be required of each individual involved?
3. Are there tax advantages to be gained from the business organization?
4. Is expansion of the business feasible and facilitated?
5. Which type of organization reduces risks and liability most?
6. Which type of organization can be terminated most easily and readily?
7. Which type of ownership provides for the most continuity and ease of transfer?
8. What costs for legal and accounting fees are involved, in setting up the organization and in the preparation of the annual reports required by law?
9. Who will manage the business?

Most hog enterprises are operated as sole proprietorships, not necessarily because this is the best type of organization, but with no effort to form some other type of organization it naturally results. The partnership, the corporation, contracts, co-ops, and networking, which require special planning and effort to bring about, are well suited to the operation of large swine establishments.

PROPRIETORSHIP (Individual)

A sole proprietorship is a business which is owned and operated by one individual.

This is the most common type of business organization in U.S. farming—86% of the nation's farms are individually or family owned.

The **advantages** of a sole proprietorship include ease of formation, direct control over the business, relative freedom from government regulations, allocation of all profits to the owner, and tax deduction of business loans.

In comparison with other forms of organization, the sole proprietorship has three major **limitations**: (1) unlimited personal liability, (2) it may be more difficult to acquire new capital for expansion; and (3) not much can be done to provide for continuity and to keep the present business going as a unit, with the result that it usually goes out of existence with the passing of the owner.

PARTNERSHIP (General Partnership)

A partnership is an association of two or more persons who, as co-owners, operate the business. About 10% of U.S. farms are partnerships.

The basic idea of two or more persons joining together to carry out a business venture can be traced back to the syndicates that were used in major trading centers in western Europe in the Middle Ages. Many of the early efforts to colonize the New World were also partnerships, or "companies" which provided venture capital, ships, provisions, and trade goods to induce settlement of large land grants.

Most swine partnerships involve family members who have pooled land, machinery, working capital, and often their labor and management to operate a larger business than would be possible if each member limited his/her operation to his/her own resources. It is a good way in which to bring a son or daughter, who is usually short on capital, into the business, yet keep the parent in active participation. Although there are financial risks to each member of such a partnership, and potential conflicts in management decisions, the existence of family ties tends to minimize such problems.

In order for a partnership to be successful, the enterprise must be sufficiently large to utilize the abilities and skills of the partners and to compensate them adequately in keeping with their contribution to the business.

A partnership has the following **advantages**:

1. **Ease of formation, low cost of organizing, and little government control.** These are very real advantages.
2. **Combining resources. A partnership often increases returns from the operation due to combining resources.** For example, one partner may contribute labor and management skills, whereas another may provide the capital. Under such an arrangement, it is very important that the partners agree on the value

of each person's contribution to the business, and that this be clearly spelled out in the partnership agreement.

3. **Equitable management.** Unless otherwise agreed upon, all partners have equal rights, regardless of financial interest. Any limitations, such as voting rights proportionate to investment, should be a written part of the agreement.

4. **Tax savings.** A partnership does not pay any tax on its income, but it must file an information return. The tax is paid as part of the individual tax returns on the respective partners, usually at lower tax rates.

5. **Flexibility.** Usually, the partnership does not need outside approval to change its structure or operation—the vote of the partners suffices.

Partnerships may have the following **disadvantages**:

1. **Liability for debts and obligations of the partnership.** In a partnership, each partner is liable for all the debts and obligations of the partnership.

2. **Uncertainty of length of agreement.** A partnership ceases with the death or withdrawal of any partner, unless the agreement provides for continuation by the remaining partners.

3. **Difficulty of determining value of partner's interest.** Since a partner owns a share of every individual item involved in the partnership, it is often very difficult to judge value. This tends to make transfer of a partnership difficult. This disadvantage may be lessened by determining market values regularly.

4. **Limitations on management effectiveness.** This is due to personal differences among partners, and the responsibility of each partner for the acts of the other partners.

The above is what is known as a partnership or general partnership. It is characterized by (1) management of the business being shared by the partners, and (2) each partner being responsible for the activities and liabilities of all the partners, in addition to self activities within the partnership.

LIMITED PARTNERSHIP

A limited partnership is an arrangement in which two or more parties supply the capital, but only one partner is involved in the management. This is a special type of partnership with one or more "general partners" and one or more "limited partners."

The limited partnership avoids many of the problems inherent in a general partnership and has become the chief legal device for attracting outside investor capital into farm ventures. Although this device has been widely used in the oil and gas industry, and for acquiring income producing urban real estate for a number of years, its application to agricultural ven-

tures on a national scale is quite new. As the term implies, the financial liability of each partner is limited to each partner's original investment, and the partnership does not require, and in fact prohibits, direct involvement of the limited partners in management. In many ways, a limited partner is in a similar position to a stockholder in a corporation.

A limited partnership must have at least one general partner who is responsible for managing the business and who is fully liable for all obligations.

The **advantages** of a limited partnership are:

1. It facilitates bringing in outside capital.
2. It need not dissolve with the loss of a partner.
3. Interests may be sold or transferred.
4. The business is taxed as a partnership.
5. Liability is limited.
6. It may be used as a tax shelter.

The **disadvantages** of a limited partnership are:

1. The general partner has unlimited liability.
2. The limited partners have no voice in management.

CORPORATION

A corporation is a device for carrying out a pork enterprise as an entity entirely distinct from the persons who are interested in and control it. Each state authorizes the existence of corporations. As long as the corporation complies with the provisions of the law, it continues to exist—irrespective of changes in its membership.

Until about 1960, few farms and ranches were operated as corporations. In recent years, however, there has been increased interest in the use of corporations for the conducting of farm and ranch business. Even so, only about 3.0% of U.S. farms use the corporate structure.

From an operational standpoint, a corporation possesses many of the privileges and responsibilities of a real person. It can own property; it can hire labor; it can sue and be sued; and it pays taxes.

Separation of ownership and management is a unique feature of corporations. The owners' interest in a corporation is represented by shares of stock. The shareholders elect the board of directors, who, in turn, elect the officers. The officers are responsible for the day-to-day operation of the business. Of course, in a close family corporation, shareholders, directors, and officers can be the same persons.

The major **advantages** of a corporate structure are:

1. It provides continuity despite the death of a stockholder.
2. It facilitates transfer of ownership.

3. It limits the liability of shareholders to the value of their stock.

4. It may make for some savings in income taxes.

The major **disadvantages** of a corporation are:

1. It is restricted to doing only what is specified in its charter.

2. It must register in each state.

3. It must comply with stipulated regulations which involve considerable paperwork and expense.

4. It is subject to the hazard of higher taxes.

5. It is possible to lose control.

FAMILY OWNED (Privately Owned) CORPORATION

Still another type of corporation is family owned (privately owned). It enjoys most of the advantages of its generally larger outside investor counterpart, with few of the disadvantages. The chief **advantages** of the family owned corporation over a partnership arrangement are:

1. **It alleviates unlimited liability.** For this reason, a lawsuit cannot destroy the entire business and all the individual partners with it.

2. **It facilitates estate planning and ownership transfer.** It makes it possible to handle the estate and keep the business in the family and going if one of the partners should die. Each of the heirs can be given shares of stock—which are easy to sell or transfer and can be used as collateral to borrow money—while leaving the management of the enterprise to those heirs interested in operating it, or even to outsiders.

TAX-OPTION CORPORATION
(Subchapter S Corporation)

Instead of paying a corporate tax, a corporation with no more than 35 stockholders may elect to be taxed as a partnership, with the income or losses passed directly to the shareholders, each of whom pays taxes on his/her share of the profits. This special type of corporation is variously referred to as *a tax-option corporation, subchapter S corporation*, pseudo-corporation, or elective corporation.

For income tax purposes, the owners of a tax-option corporation are taxed as if they were a partnership. That is, income earned by the corporation passes through the corporation to the personal income tax returns of the individual shareholders. Thus, the corporation does not pay any income tax. Instead, the shareholders pay tax on their share of corporate income at their individual tax rate; and the shareholders report their share of long-term capital gains and receive their deductions therefor. Although each shareholder's portion of any corporate losses from current

operations is deducted from their personal return, capital losses incurred by the corporation cannot be passed through to the shareholders.

Thus, there are some very real advantages to be gained from a subchapter S or tax-option corporation. However, in order to qualify as a subchapter S corporation, the following requisites must be met:

1. There cannot be more than 35 stockholders.

2. All stockholders must agree to be taxed as a partnership.

3. Nonresident aliens cannot own stock.

4. There can be only one class of stock.

5. Not more than 20% of the gross receipts of the corporation can be from royalties, rents, dividends, interest, or annuities plus gains from sale or exchange of stock and securities; and not more than 80% of the gross receipts can be from sources outside the United States.

ADVANTAGES OF LIMITED PARTNERSHIPS AND CORPORATIONS

In addition to the advantages peculiar to (1) limited partnerships, and (2) corporations, and covered under each, limited partnerships and corporations have the following advantages over individual ownership in the acquisition of capital:

1. They make it possible for several producers to pool their resources and develop an economically-sized operation, which might be too large for any one of them to finance individually.

2. They make it possible for persons outside agriculture to invest through purchase of shares of stock in the business.

3. They can generally borrow money easier because the strength of the loan is not dependent on the financial and management capability of one person.

4. They give assurance that the business will continue, even if one of the owners should die or decide to sell out.

5. They provide built-in management, with continuity; and, generally speaking, they attract very able management.

Thus, those engaged in the livestock business can and do use either of these two business organizations—a limited partnership or a corporation—to develop and maintain an economically sound operation. Actually, no one type of business organization is best suited for all purposes. Rather, each case must be analyzed, with the assistance of qualified specialists, to determine whether there is an advantage to using one of these types of organizations, and, if so, which organization is best suited to the proposed business.

CONTRACTS[1]

A contract is an agreement between two or more persons to do or refrain from doing certain things.

There is increasing interest in hog contracting, due in part to the difficulty for many producers to obtain adequate financing. Contracting also is being used to coordinate pork production from genetics and nutrition to the retail meat counter.

In the mid 1990s, it was estimated that about 20% of U.S. hogs were under production contracts, and a smaller percentage under marketing contracts.

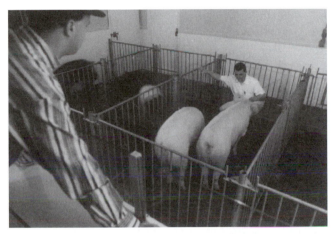

Fig. 19-4. Most contracts specify quality. Many of them pay a bonus for top quality and penalize poor quality. This shows a hog producer selecting breeding stock at a genetic sales center. (Courtesy, DeKalb Swine Breeders, Inc., DeKalb, IL)

Forward pricing (marketing) contracts for market hogs have been available from most major meat packers for a number of years. They are the most commonly used marketing contracts in the industry.

Production contracts for market hog finishing are relatively new but are increasing in the Midwest. However, they have been used for some time in the Southeast where contract hog production is more widely accepted.

The following is an overview of the most common contracts in the pork industry.

MARKETING CONTRACTS FOR SLAUGHTER HOGS

Marketing contracts are of two kinds: (1) marketing

contracts for slaughter hogs, and (2) marketing contracts for feeder pigs. The forward sale contract of market hogs is a contract between a buyer (normally a meat packer or a marketing agent) and a seller (normally a producer), where the producer agrees to sell, at a future date, a specified number of hogs to a buyer for a certain price. The buyer normally will have taken an opposite position in the futures market to offset any price fluctuations between the signing of the contract and the delivery date. To cover margin and commission, the contract price offered by the packer may be lower than futures adjusted for expected basis.

Terms typically found in a forward contract include:

■ The quantity to be delivered, with the minimum amount varying anywhere from 5,000 lb. to 40,000 lb. (40,000 lb. equals one live hog futures contract).

■ The date and location of delivery. The delivery date may normally be changed by mutual agreement. The seller may have the option of selecting the delivery date within a specified time interval.

■ Acceptable weights and grades, including provisions for premiums and discounts.

■ A description of the pricing mechanism, either fixed base price or formula price. Some contracts now price the hogs on a grade and yield basis to reward better producers who would otherwise be less inclined to contract.

■ Provisions for non-deliverable hogs and unacceptable carcasses. The buyer normally deducts from the seller's receipts for unacceptable hogs and carcasses.

■ Provisions outlining the credit requirements of the seller and inspection of the hogs by the buyer. The buyer may request to inspect the hogs while on the seller's premises.

■ A provision dealing with breach of contract. Typically, the seller is liable for all losses incurred by the buyer when the seller is in breach of contract.

The producer retains all production risks, other than the selling price, under a fixed price forward sale contract.

A producer uses a forward sale contract to reduce the risk of price fluctuations and to lock in an acceptable selling price. While the forward sale contract allows the producer to lock in a particular selling price, it may cause the producer to miss out on greater profits if prices rise. Thus, the decision to contract must be based upon each producer's willingness or ability to bear the risk of price uncertainty. Some producers may be forced to contract due to a lack of diversification, indebtedness, or at the request of creditors, while other more financially stable or diversified producers may

[1]This section was adapted by the authors from the excellent section on "Producing and Marketing Hogs under Contract" in the *Pork Industry Handbook*, Cooperative Extension Service, University of Illinois at Urbana-Champaign.

be in a better position to withstand the risk of price movements.

Price risk may be reduced by hedging in the futures market. A marketing contract may be preferred to hedging for the following reasons:

■ Marketing contracts can typically be written for smaller sizes than the 40,000 lb. Chicago Mercantile Exchange contract.

■ A fixed price marketing contract locks in the delivered price. A futures contract hedge locks in the futures price, but the local price differential (basis) may still vary.

■ A marketing contract does not require an initial margin or additional margin calls be paid should the price increase after the contract is signed.

■ A marketing contract is typically made with a local marketing agent or packer rather than dealing with the Chicago Mercantile Exchange. However, unlike a futures contract, the producer is required to deliver the hogs to fulfill a marketing contract.

A floor price contract is a variation of the forward sale contract, however it is not as widely used as the forward sale contract. The seller agrees to deliver a specified number of hogs to a buyer at a future date and the buyer guarantees the seller a minimum price (the floor price) for the hogs. Usually, the seller receives the higher of the floor price or market price at delivery minus a discount. The discount compensates the buyer for the costs (options premiums and other variable costs associated with the contract) of providing the guaranteed minimum price. Both the forward fixed price contract and floor price contract reduce only the risk of hog price fluctuations. The producer must still bear the other risks associated with hog production.

MARKETING CONTRACTS FOR FEEDER PIGS

Typically, feeder pig marketing contracts are between a marketing agency, often a cooperative, and a pig producer, where the marketing agency agrees to market the pigs for the producer in exchange for a fee.

A marketing contract might contain the following provisions:

■ The producer agrees to market all pigs through the marketing agency.

■ The marketing agency prescribes specific management practices to be followed by the producer. These may relate to the weight at which the pigs are to be marketed, health of the animals, and immunization against diseases.

■ Many larger marketing agencies will pool feeder pigs into homogenous groups to increase their marketability and will provide technical assistance to the producer.

Producers are essentially hiring marketing expertise to enhance their market prices and minimize the time and effort of locating buyers for their pigs.

PRODUCTION CONTRACTS

To expand more rapidly their own production, many larger producers use contract production as a way to hold down risk and capital required.

Investors, feed dealers, farmers, and others often are interested in producing hogs, but are unwilling or unable to provide the necessary labor, facilities, and equipment. Therefore, they search out producers who are willing to furnish the labor and equipment in exchange for a fixed wage or share of the profits. The resulting contracts, between owner and producer, vary considerably in form and responsibility of each party involved. These contracting arrangements are attractive to young or financially strapped producers and would-be producers who do not have the capital to invest in a herd, and for producers with underutilized facilities.

FARROW-TO-FINISH CONTRACTS

While base-payment plus bonus contracts are offered in some regions, many farrow-to-finish contracts are on a percentage basis to reflect the relative inputs supplied by each person or form.

Option 1: The producer supplies facilities, labor, veterinary care, utilities, and insurance for an appropriate percentage of gross sales based on input costs. The feed retailer supplies feed, standard feed medications, and receives a predetermined percentage of returns. The capital partner and breeding stock supplier get another percentage. The management firm receives a percentage for supplying computerized records services and management consultation.

Option 2: The current hog inventory is purchased outright by a limited partnership and it will supply sow replacements. The producer supplies facilities, labor, utilities, veterinary costs, repairs, and manure disposal. The feed retailer provides feed and standard feed medications. A management agency supplies production and marketing guidance. Each of the contract participants receives a percentage of the proceeds when hogs are marketed. The remaining percentage is split between the limited partnership and the general partner for managing the partnership.

Option 3: The contractor provides breeding stock,

feed, and a prescribed system of management. The producer provides facilities, labor, utilities, insurance, and disposal of manure. The producer receives fees per head or per pound of hogs marketed plus possibly additional compensation for farrowing and feeding efficiency.

FEEDER PIG PRODUCTION CONTRACTS

Feeder pig production contracts come in several forms.

Option 1: The producer provides everything but the breeding stock and bids what he/she is willing to produce a feeder pig for, based on production criteria such as pigs weaned per litter, etc., with discounts and bonuses based on a target level. Most of the production risk is retained by the producer.

Option 2: A contractor provides breeding stock, feed, management assistance, and supervision, and pays the feeder pig producer a flat fee for each pig. This fee varies according to pig weight and current production costs. In this example most of the risk falls on the person providing breeding stock, feed, and management.

Option 3: The contractor provides breeding stock, feed, facilities, and veterinary costs. The producer provides labor, utilities, maintenance, and manure handling. A fee for each pig produced and a monthly fee for each sow and boar maintained is paid to the manager. This option fits owners who no longer want to be actively involved in production, but have a good manager with limited cash willing to take over the operation.

Option 4: A shared revenue program with revenues divided in proportion to inputs provided. One example would be where the producer supplying facilities, veterinary care, utilities, labor, and insurance would receive a negotiated percentage of gross sales in return for his/her share of production costs for each pig sold. The feed dealer would receive a certain percentage based on his/her share of the total inputs. The remaining percentage would go to the breeding stock supplier and the management firm that supplies computerized records and consultations. Negotiated percentage shares should be based upon inputs provided and risks borne by each participant.

FINISHING CONTRACTS

There are three basic types of hog finishing contracts offered, each with variations on payments and resources provided.

Option 1: A fixed payment contract guarantees the producer a fixed payment per head as well as

bonuses and discounts based on performance. Under a fixed payment contract for finishing hogs, the producer normally provides the building and equipment, labor, utilities, and the necessary insurance. The contractor supplies the pigs, feed, veterinary services and medication, and transportation. The contractor usually provides a prescribed management system and supervises its conduct. The contractor, as the owner of the hogs, does the marketing. The producer often receives an incoming payment based on the weight of the feeder pigs when they come into the producer's facilities. For example, $5 for a 30 lb. pig and $4 for a 40 lb. pig. The remainder of the producer's payment is made when the hogs are sold. The method of calculating base payment varies by contract. Some contracts offer a fixed dollar per head regardless of the weight gained. Other contracts pay a fixed amount per pound of gain based on pay-weights in and out of the facility. Others pay a fixed amount per head per day spent in the facility.

Most contracts contain bonuses for keeping death loss low and improved feed efficiency, as well as penalties for high death losses and unmarketable animals. Producers should have control over factors that impact their bonuses and penalties. For example, the right of refusal on obviously unhealthy pigs, or to negotiate a more lenient bonus schedule for multiple-source pigs. Contract payment methods typically range from a low base payment with high incentive bonuses to a high base with relatively low bonuses.

Option 2: Directed feeding by a cooperative or feed dealer that contracts with a producer to finish-out hogs. The contractor's objective when entering into a directed feeding contract is to increase feed sales and secure a reliable feed outlet.

The contractor provides the feed and some management assistance and typically directs the feeding program. The contracting firm often will purchase the feeder pigs, in which case profits from the sale of the hogs are shared as discussed below, or it will help the producer obtain financing to purchase the pigs. The producer agrees to purchase all feed and related services from the contractor and is responsible for all costs of production. The producer receives all proceeds from the sale of the hogs minus any outstanding balance owed to the contractor.

Option 3: In a profit sharing contract, the producer and contracting firm divide the profit in proportion to the share of the inputs provided by each party.

Typically, the producer provides the facilities, labor, utilities, and insurance for his/her portion of the profit. The contracting firm normally purchases the pigs and is responsible for all feed, the veterinary services, transportation, and marketing expenses. Over the duration of the contract, the contractor's costs are charged to an account. This account balance is then subtracted from the sale proceeds to determine the profit. The contracting firm often will use its own feed and provide management assistance. The producer is normally guaranteed a minimum amount per head as long as death loss is below a set percentage. For instance, depending on contract terms, the producer may receive $5/head if death loss is 3% or less and $3/head if death loss is over 5%. The producer receives this payment regardless of whether a profit is made. The contractor's return depends upon the profit made on the sale of the hogs and the gain received from the markup on feed, pigs, and supplies provided.

Through contracting, producers are able to achieve more stable returns, trading the possibility of large profits for the assurance of a more reliable return. Many producers enter into contracts because they either lack the capital or they do not wish to tie up a large amount of capital in hog production.

BREEDING STOCK LEASING

The popularity of breeding stock leases has declined in recent years and presently they are seldom used. Many contractors were dissatisfied with the care of the breeding herd and sometimes were unable to collect their payments from producers. One lease involves a payment-in-kind for the use of breeding stock. This lease is particularly attractive to producers with limited capital but ample feed, facilities, and labor to produce hogs. The producer pays all production costs and pays the breeding stock owner, for example, one market weight hog per litter.

CHARACTERISTICS OF A GOOD CONTRACT

The relationship of producer and contractor are generally more complex and interdependent for production contracts than for marketing agreements. Hence, production contracts need to be evaluated with special care. When considering contract production, contractors and producers need to evaluate each contract on its own merit. Each party should look for a contract that best fits their operation and management capabilities. Both parties must know their cost of production to make an informed decision. Simply signing a contract will not necessarily improve efficiency or insure a profit. It is doubtful that producers will receive a bonus for feed efficiency better than 2.9 if they have been only achieving 3.5 on their own, for example.

Also, carefully scrutinize the examples used to demonstrate cash flow or producer returns. Unless otherwise stated these are only examples and not guarantees. Producers should consider the impact on cash flow and debt repayment if payments are less than projected. Is there a guarantee of contract length if new facilities or other major capital expenditures are required to obtain the contract? Most contracts guarantee a stated number of turns (groups of hogs) or are in force for a stated length of time. Few, if any, guarantee the number of hogs that will be put through the facility in a set time, say one year. Facilities that sit idle during an unprofitable period in the hog cycle may profit the contractor, but disrupt the producer's debt repayment schedule.

Before considering the details of a contract, one should first consider the reputation and financial stability of the company or individual with whom the contract is to be made. For instance: How long has the company been in business? What has been the company's financial success? How long has the company offered contracts? Do other producers in the area have contracts with the company? Does the company fulfill the terms of its contracts?

Because little can be done after the fact to correct the problem, both parties should be encouraged to gather financial information about the other. This may be best handled on a document separate from the production contract. Problems and risks can arise for both the owner and the feeder due to financial failure of the other. Remember that:

■ Except for the right to remove the hogs, the hog owner is an unsecured creditor of the feeder. The owner has little chance in recovering losses resulting from excessive death loss.

■ The feeder has a statutory lien on the hogs, but this lien is subject to all prior liens of record. This means that the owner's secured creditors can remove the hogs without paying the grower. Once the hogs are removed, the grower has an unsecured claim for his/her contract damages which is probably uncollectible.

■ It is possible for the grower to receive first lien on the hogs if the owner and his/her creditors are willing to give the grower a lien subordination.

■ **Minimal Contract Provisions**

  ■ The contract must be in written form and must be clear and concise.

  ■ The contract should clearly define the rights and responsibilities of both parties involved.

  ■ The contract also should contain the following: number of pigs involved, names of both parties, duration of the contract, method and timing of payment, and definition of who shall supply certain inputs.

  ■ A contract should be thoroughly read and understood before it is signed. Enlisting the advice of a lawyer, farm management specialist, or business consultant is helpful and often essential when evaluating contracts.

- The contract should contain an arbitration clause. Such a clause removes any disputes from the court system. The contract also should define how the arbitrators are chosen.
- Complete records of inventories, deaths, purchases, and sales should be maintained and open to both parties.

- **Other Possible Contract Provisions**
  - The right of the owner to inspect pigs at any time.
  - Designation of responsibility for purchasing and marketing.
  - A procedure for refusing delivery of unhealthy or poor quality pigs.
  - The basis for compensation of feed and non-feed costs.
  - Acceptable weight ranges for incoming feeder pigs and outgoing market hogs.
  - A procedure to use if failure of payment arises.
  - The means and timing of communication by producer to owner when a death loss occurs.
  - Who assumes the risk of death loss.
  - The extent of the producer's responsibility for care of the pigs and record keeping.
  - Designation of who will provide insurance and how much coverage.
  - The brand and quality of feed and supplement that is required if any, and who is responsible for diet formulation.
  - How and when the contract may be terminated by either party.

The key to feeding or producing hogs under contract is finding the type of contract that will allow each individual to profit most from his/her skills, resources, and ability to bear risk associated with hog production. This strength may be record keeping, producing with a low mortality rate, or an ability to maximize herd feed efficiency. Whatever the case, producers should make certain that the contract will reward them appropriately for what they do best.

Once the best contract type has been found, the written contract itself should be carefully read and understood. The responsibilities of both parties should be clearly spelled out and understood as should procedures for dealing with possible disputes. While a well written contract is essential to successful contract production, it is also important that both parties are professional and willing to work out any problems that arise. A contract can never be so complete that every possible problem is anticipated. Individuals interested in contract production should check laws regarding contracting in their state.

COOPERATIVES (Co-ops)

A cooperative is a business formed by a group of people to get certain services for themselves more effectively or more economically than they can get them individually. These people own, finance, and operate the business for their mutual benefit. Often by working together through such a cooperative business, member-owners obtain services not otherwise available to them. Co-ops are engaged in every facet of pork production, including supplies, marketing, processing, and retailing. Like all businesses, their success depends upon management.

NETWORKING

Networking is two or more people working together to achieve individual and group goals that would be more difficult to obtain alone. It's a banding together of small and independent pork producers to compete with the mega-pork producers. Actually, networking is a new name for an old practice. A few years ago, we would have called it cooperating, but some people do not want co-ops raising hogs, so another term evolved.

Networking is not unique to the pork industry. The term was born in the 1960s in the business world, and it flourished through the 1980s. In the business world, the term has many definitions as it does in the pork industry. For example, auto makers may not own a glass factory to manufacture their car windows; instead, they may network with a glass company to make their windows.

Although networking possibilities in the pork industry are unlimited, three categories of networks appear to have merit for pork producers: (1) information, (2) marketing, and (3) production.

- **Information networks**—Information networks are generally informal groups—for example, producers with the same record system—who share information with, and learn from, each other. Some groups charge a membership fee which may be used (1) to bring in consultants and speakers to address the group, and/or (2) to cover all or part of the cost of a study-tour of swine production in other areas.

- **Marketing networks**—Marketing networks require producer commitment relative to the time, number of hogs, and quality of hogs that they will market in the future. Usually, a better price is secured because of both volume and access to additional markets. Marketing networks include input purchasing as well as hog sales.

■ **Production networks**—Production networks may require joint ownership of facilities and/or hogs. Joint sow ownership, gilt multipliers, and congregate nurseries are all examples of production networks. Although joint ownership in centralized production facilities/hogs is typical, it is not a requisite. Networks in which one producer farrows, a second operates a nursery, and third finishes the hogs, all in their own facilities, is possible. Production networks attempt to capture proven technologies which may not be possible for one individual; for example, (1) three sites for segregating groups, and (2) all-in, all-out in cost effective sized buildings.

The key to success of any network is the people involved and their commitment to making it work.

The **benefits** of successful networking are:

1. **Capturing proven technology,** which an individual may not be able to do. For example, by joining together, they may be able to afford multi-site technology; one may farrow, a second may operate the nursery, and a third may finish the hogs. Multi-sites improve hog health, feed efficiency, and rate of gain.

2. **Capturing real economics in volume sales and purchases;** they are able to sell hogs at a higher price, and to buy their supplies at a lower price.

3. **Improving product quality and market access.** Uniform, high quality hogs in sufficient volume will gain market access.

4. **Utilizing production, marketing, and information systems.** A "system" merely implies that the entire pork channel is considered when production and marketing decisions are made, and that actions are taken to improve product quality and efficiency where possible.

The **limitations** of networking are:

1. **Commitment of people.** Successful networking depends upon the commitment of people. *Note:* Networking will seldom make a poor manager better.

2. **Joint responsibility.** Networking depends on the performance of its individual members. Some producers may not wish to assume responsibility for their fellow members.

3. **Formal business procedures.** Tighter control and more formal business procedures are necessary as transactions become more complex. Also, networking requires increased communication between all people involved.

4. **Loss of markets and suppliers.** Typically, networking involves direct negotiation and a likely contract agreement with the supplier and the buyer. This may eliminate local suppliers and buyers.

Summary: The success of a network will depend on the people involved, and their commitment to the network and to the production of high quality pork.

VERTICAL INTEGRATION

Vertical integration in pork refers to a structure in which, through alliances (complete ownership, joint ownership, or contract), an individual or company controls the product through two or more phases; for example, a meat packer owns hogs from birth through slaughter.

The U.S. poultry industry led the way in vertical integration. Currently, the pork industry is following suit. In its 1992 annual report, Smithfield Foods of Virginia, long famous for hams, announced that it was moving to vertical integration faster than any other company in the industry.

In 1995, the level of integration of U.S. animal species was: broilers 100%, turkeys over 90%, hogs 33%, and beef cattle 20%.

(Also see Chapter 2, section headed "Industrialization of U.S. Hog Production," Fig. 2-13.)

HOG FUTURES TRADING[2]

The biggest uncertainty in the hog business is the selling price of market hogs.

Producers know their cost of producing or purchasing feeder pigs; and from their past records, they are usually able to project, with reasonable accuracy, the costs of production to market time. But, unless they contract ahead, they have no assurance of what the hogs will bring when they are ready to market. Moreover, there is little flexibility in market time, for the reason that excess finish is costly, and unwanted by the consumer.

The Chicago Mercantile Exchange first offered live hog futures on February 28, 1966. Since then, trading in hog futures contracts has grown enormously; it increased from 8,063 in 1966 to 1,562,679 in 1994.

The development of hog futures requires that producers and others be knowledgeable of what futures marketing involves and how it applies to the marketing of hogs. This section provides background information.

---

[2]This section on "Hog Futures Trading" was authoritatively reviewed by, and helpful suggestions were received from, Clinton R. Hakes, Director Commodity Marketing, Chicago Mercantile Exchange, Chicago, IL; and R. E. Sheldon, Manager-Agriculture Group, Economic Analysis and Planning Department, Chicago Board of Trade, Chicago, IL.

WHAT IS FUTURES TRADING?

Futures trading is not new. It is a well accepted, century-old procedure used in many commodities; for protecting profits, stabilizing prices, and smoothing out the flow of merchandise. For example, it has long been an integral part of the grain industry; grain elevators, flour millers, feed manufacturers, and others, have used it to protect themselves against losses due to price fluctuations. Also, a number of livestock products—frozen pork bellies, hams, hides, and tallow—were traded on the futures market before the advent of live hog futures. Many of these operators prefer to forego the possibility of making a high speculative profit in favor of earning a normal margin or service charge through efficient operation of their business. They look to futures markets to provide (1) an insurance medium in the marketing field, and (2) the facilities and machinery for underwriting price risks.

A commodity exchange is a place where buyers and sellers meet on an organized market and transact business on paper, without the physical presence of the commodity. The exchange neither buys nor sells; rather, it provides the facilities, establishes rules, serves as a clearing house, holds the margin money deposited by both buyers and sellers, and guarantees delivery on all contracts. Buyers and sellers either trade on their own account or are represented by brokerage firms. Except for dealing in futures and in paper contracts instead of live animals, futures trading on an exchange is very similar to terminal livestock markets. Pork producers need no introduction to the latter.

The unique characteristic of futures markets is that trading is in terms of contracts to deliver or to take delivery, rather than on the immediate transfer of the physical commodity. In practice, however, very few contracts are held until the delivery date. The vast majority of them are canceled by offsetting transactions made before the delivery date.

Many producers contract their feeder pigs for future delivery without the medium of an exchange. They contract to sell and deliver to a buyer a certain number and kind of pigs at an agreed upon price and place. Hence, the risk of loss from a decrease in price after the contract is shifted to the buyer; and, by the same token, the seller foregoes the possibility of a price rise. In reality, such contracting is a form of futures trading. Unlike futures trading on an exchange, however, actual delivery of the pigs is a must. Also, such privately arranged contracts are not always available, the terms may not be acceptable, and the only recourse to default on the contract is a lawsuit. In contrast, futures contracts are readily available and easily disposed.

Fig. 19-5. A view of the trading floor of the Chicago Mercantile Exchange. (Courtesy, Chicago Mercantile Exchange, Chicago, IL)

WHAT CONSTITUTES A FUTURES HOG CONTRACT?

A futures contract is a standardized, legal, binding paper transaction in which the seller promises to make delivery and the buyer promises to take delivery on a specified quantity and type of a commodity at a specified location(s) during a specified future month. The buying and selling are done through a third party (the exchange clearing member) so that the buyer and seller remain anonymous; the validity of the contract is guaranteed by reputable and well financed exchange clearing members; and either buyer or seller can readily liquidate their position by simply offsetting sale or purchase.

On November 10, 1995, the Chicago Mercantile Exchange (CME) began trading futures and options on a new lean hog contract. The lean hog contract upgrades and replaces the live hog contract which began trading on the CME in February 1966.

The new CME lean hog contract makes for several changes in futures trading. For example, it makes it easier for producers to hedge.

The cash settlement price to be used is trademarked by the CME as the "CME Lean Hog Index." The CME Lean Hog Index is a two-day, volume-weighted average of lean value carcass prices from the western Corn Belt, eastern Corn Belt, and mid-South. The prices originate as USDA figures on volume of hogs sold and prices in these major regions.

When the contract expires, the cash settlement price must, by definition, be equal to the futures price. This is expected to reduce, substantially, futures price variability at contract expiration, because it will be explicitly linked to the cash market. This makes it much easier for producers to hedge, because the basis risk should be much lower near contract expiration.

This system also removes the incentive to get out of the futures position before the time of delivery. Liquidation of futures positions prior to final contract expiration can lead to highly variable end-of-contract price movements.

The primary difference between the new lean hog contract and the old live hog contract are as follows:

■ **The Lean Hog Futures Contract is still 40,000 lb, but that amount is on a carcass weight basis**—So, whereas about 167 hogs at 240 lb were necessary to fill a Live Hog Futures Contract, it will now require about 225 (167 divided by 0.74 carcass conversion) hogs to fill the contract on a carcass weight basis. So, it will take more animals to fill the 40,000-lb requirement. The number of hogs is still important, even without delivery, because to form a hedge you will still need to match futures quantities with spot quantities as always. It may be tempting for producers to mismatch the two with no delivery requirement hanging over their heads, but that would constitute increasing risk.

■ **The hogs to be slaughtered must also meet the specifications of 51–52% lean, 0.8–0.99 in. of back-fat at the last rib and weight 170–191 lb on a dressed weight basis**—Of course, all these are new because the old contract was based on live animals' measurements.

■ **The daily price limit is still $1.50/cwt**—But this represents a lower percentage of the total value of the contract than on a live-weight equivalent.

HOW TO GO ABOUT BUYING OR SELLING HOG FUTURES

Here is how to go about buying or selling hog futures:

1. Have good and accurate records of costs.
2. Contact a brokerage house that holds a membership in the commodity exchange.
3. Open up a trading account with the broker, by signing an agreement authorizing the broker to execute trades.
4. Deposit with the broker the necessary perform-

ance money for each contract desired. The broker will then maintain a separate account for the producer. The contract commission fee is due when the contract is fulfilled by either delivery or offsetting purchase or sale of another contract.

SOME FUTURES MARKET TERMS

Futures markets have a jargon and sign language of their own. It is not necessary that producers dealing in futures master many of them, but it will facilitate matters if they at least have a working knowledge of the following:

■ **Basis**—The spread between the cash price and the price of the futures.

■ **Basis movement**—The change in the basis within a particular period of time; in one location, or from location to location at the same time.

■ **Discount to the futures**—When the cash price is under the futures.

■ **Hedgers**—Persons who cannot afford risks, and who try to increase their normal margins through buying and selling of futures contracts.

■ **Limit order**—Placing limitations on orders given the brokerage firm.

■ **Long position**—Buying a contract to sell later.

■ **Pit**—The platform on the trading floor of an exchange where traders and brokers stand while executing futures trades.

■ **Premium**—When the cash price is above the futures.

■ **Short position**—Selling a contract to buy later.

■ **Speculators**—Persons who are willing to accept the risks associated with price changes.

■ **Offset**—Offsetting a position taken earlier.

WORKERS' COMPENSATION[3]

Workers' compensation laws, now in full force in every one of the 50 states, cover on-the-job injuries and protect disabled workers regardless of whether their disabilities are temporary or permanent. Although broad differences exist among the individual states in their workers' compensation laws, principally in their benefit provisions, all statutes follow a definite pattern

---

[3]This section on Workers' Compensation was authoritatively reviewed by Wayman E. Watts, CPA, Fresno, California.

as to employment covered, benefits, insurance, and the like.

Workers' compensation is a program designed to provide employees with assured payment for medical expenses or lost income due to injury on the job. Whenever an employment-related injury results in death, compensation benefits are generally paid to the worker's surviving dependents.

Generally all employment is covered by workers' compensation, although a few states provide exemptions for farm labor, or exempt farm employers of fewer than 10 full-time employees, for example. Farm employers in these states, however, may elect workers' compensation protection. Livestock producers in these states may wish to consider coverage as a financial protection strategy because under workers' compensation, the upper limits for settlement of lawsuits are set by state laws.

This government-required employee benefit is costly for livestock producers. Costs vary among insurance companies due to dividends paid, surcharges, minimum premiums, and competitive pricing. Some companies, as a matter of policy, will not write workers' compensation in agricultural industries. Some states have a quasi-government provider of workers' compensation to assure availability of coverage for small businesses and high risk industries.

For information, contact your area extension farm management or personnel management advisor and an insurance agent experienced in marketing workers' compensation and liability insurance.

MANAGEMENT

Four major ingredients are essential to success in the hog business: (1) good healthy hogs, (2) good feeding, (3) good records, and (4) good management.

Management gives point and purpose to everything else. The skill of the manager materially affects how well hogs are bought and sold, the health of the animals, the results of the diets, the stress of the hogs, the rate of gain and feed efficiency, the performance of labor, the public relations of the establishment, and even the expression of the genetic potential of the hogs. Indeed, a manager can make or break a pork enterprise—a fact often overlooked in the present era with the accent on scientific findings and automation. Management is still the key to success. When in financial trouble, owners should have no illusions on this point.

In manufacturing and commerce, the importance and scarcity of top managers are generally recognized and reflected in the salaries paid to persons in such positions. Unfortunately, agriculture as a whole has lagged; and altogether too many owners still subscribe to the philosophy that the way to make money out of the pork business is to hire a manager cheap, with the result that they usually get what they pay for—a "cheap" manager.

TRAITS OF A GOOD MANAGER

There are established bases for evaluating many articles of trade, including hogs and grain. They are graded according to well-defined standards. Additionally, we chemically analyze feeds and conduct feeding trials. But no such standard or system of evaluation has evolved for managers, despite its acknowledged importance.

The authors have prepared the Hog Manager Check List, given in Table 19-5, which (1) employers may find useful when selecting or evaluating a manager, (2) managers may apply to themselves for self-improvement purposes, and (3) students may use for guidance as they prepare for managerial positions. No attempt has been made to assign a percentage score to each trait, because this will vary among swine establishments. Rather, it is hoped that this check list will serve as a useful guide (1) to the traits of a good manager, and (2) to what the boss wants.

TABLE 19-5
HOG MANAGER CHECK LIST

❏ **CHARACTER—**

Absolute sincerity, honesty, integrity, and loyalty; ethical.

❏ **INDUSTRY—**

Work, work, work; enthusiasm, initiative, and aggressiveness.

❏ **ABILITY—**

Hog know-how and experience, business acumen—including ability to arrive at the financial aspects systematically and to convert this information into sound and timely management decisions; knowledge of how to automate and cut costs; common sense; organized; growth potential.

❏ **PLANS—**

Sets goals, prepares organization chart and job description, plans work, and works plans.

❏ **ANALYZES—**

Identifies the problem, determines pros and cons, then comes to a decision.

❏ **COURAGE—**

To accept responsibility, to innovate, and to keep on keeping on.

❏ **PROMPTNESS AND DEPENDABILITY—**

A self-starter; has "T.N.T.," which means that things are done "today, not tomorrow."

❏ **LEADERSHIP—**

Stimulates subordinates, and delegates responsibility.

❏ **PERSONALITY—**

Cheerful; not a complainer.

ORGANIZATION CHART AND JOB DESCRIPTION

It is important that workers know to whom they are responsible and for what they are responsible; and the bigger and the more complex the operation, the more important this becomes. This should be written down in an organization chart and job descriptions.

AN INCENTIVE BASIS FOR THE HELP

Big farms must rely on hired labor, all or in part. Good help—the kind that everyone wants—is hard to come by; it is scarce, in strong demand, and difficult to keep. And the farm labor situation is going to become more difficult in the years ahead. There is need, therefore, for some system that will (1) give a big assist in getting and holding top-flight help, and (2) cut costs and boost profits. An incentive basis that makes hired help partners in profit is the answer.

Many manufacturers have long had an incentive basis. Executives are frequently accorded stock option privileges, through which they prosper as the business prospers. Laborers may receive bonuses based on piecework or quotas (number of units, pounds produced). Also, most factory workers get overtime pay and have group insurance and a retirement plan. A few industries have a true profit-sharing arrangement based on net profit as such, a specified percentage of which is divided among employees. No two systems are alike. Yet, each is designed to pay more for labor, provided labor improves production and efficiency. In this way, both owners and laborers benefit from better performance.

Family-owned and family-operated farms have a

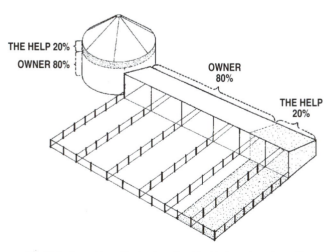

THE HELP 20%
OWNER 80%
OWNER 80%
THE HELP 20%

Fig. 19-6. A good incentive basis makes hired help partners in profit.

built-in incentive basis; there is pride of ownership, and all members of the family are fully cognizant that they prosper as the business prospers.

Many different incentive plans can be, and are, used. There is no best one for all operations.

The incentive basis chosen should be tailored to fit the specific operation; with consideration given to kind and size of operation, extent of owner's supervision, present and projected productivity levels, mechanization, and other factors.

WHAT'S AHEAD FOR U.S. PORK

The U.S. pork team—producers, marketers, processors, and retailers—needs to project what's ahead to the year 2000—and beyond. Here is what the authors' crystal ball shows:

1. **Fewer and bigger hog farms.** This trend will continue. But the growth of mega-producers will be slowed, with a tendency to level off. Their future will be impacted by (a) their ability and willingness to handle wastes in a responsible manner, and (b) the adverse effect of displaced hog producers, displaced agri-businesses, and deteriorating rural communities.

2. **Environmentalists and animal welfarists will become more vocal and more powerful.** These groups will not go away. Environmental control and animal welfare will be major issues in the decades to come.

3. **Small producers will band together.** For survival, small/independent hog producers will band together in co-ops and networks.

4. **Vertical integration of pork will race ahead.** Meat packers will increase their control of pork production and distribution, and share their risk, through ownership (complete or joint) and contracts.

5. **Fat will continue to be an "ugly word."** Markets will pay a premium of $1.00 to $1.50 per cwt for each $1/10$ in. less backfat over the last rib.

6. **Outdoor breeding, gestation, and finishing will remain competitive.** Outdoor swine production will be reinvented as "outdoor intensive swine production," in which some of the efficiencies of complete confinement are incorporated.

7. **Multi-site production will become mainstream.** Having breeding, gestating, and farrowing sows at one location, nursery pigs at a second location, and growing-finishing pigs at a third location, with a mile or more between the three units, will increase for hog health reasons, accompanied by greater rate and efficiency of gains.

8. **Big breeding companies will dominate sales of swine breeding stock.** Mega-hog producers will continue to hook up with a breeding company which can supply them with large numbers of healthy

Fig. 19-7. Big breeding companies, like Farmers Hybrid, will furnish more and more of the breeding stock for mega-swine producers. (Courtesy, Farmers Hybrid Co., Des Moines, IA)

and high quality breeding animals at one time. Smaller producers can survive and compete by forming a co-op or network and producing gilts for a mega-producer, for example.

9. **New standard of excellence.** In 1996, the National Pork Producers Council, Des Moines, Iowa, released a new standard of excellence for their stated purpose "to make U.S. Pork the consumer's meat of choice." This included Symbol II (Fig. 19-8) and the following specifications:

- 260 lb slaughter weight
- 195 lb carcass
- Desirable muscle quality

Fig. 19-8. Symbol II, Standard of Excellence released by the National Pork Producers Council, Des Moines, Iowa, in 1996.

- Minimum loin muscle of 6.5 sq in. barrows (7.1 sq in. gilts), with appropriate
  - Color
  - Water holding capacity
  - Ultimate pH
- Intramuscular fat level greater than or equal to 2.9% barrows (2.5% gilts)
- High health production system
- Produced by an environmentally assured producer on PQA Level III
- Free of the stress gene
- Result of a terminal crossbreeding program
- From a maternal line capable of weaning 25 pigs
- A barrow that is marketed at 156 days of age, with a fat-free lean index of 49.8%
- A gilt that is marketed at 164 days of age, with a fat-free lean index of 52.2%
- Live-weight feed efficiency of 2.40 barrows (2.40 gilts)
- Fat-free lean gain efficiency of 6.4 barrows (5.90 gilts)
- Fat-free lean gain efficiency of 0.78 lb per day
- All achieved on a corn/soy equivalent diet from 60 lb
- Standard Reference backfat of 0.8 barrows (0.6 gilts)

10. **Fewer, but larger, feed companies will sell more complete diets direct to producers, bypassing dealers.** With this trend, more feed companies will be involved in joint financing of hog farms.

11. **More veterinarians will become hog specialists, working with fewer farms but more pigs.** Also, more and more producers will adopt long-term health programs.

12. **Pork will become No. 1 in U.S. per capita meat consumption.** Pork will overtake beef, broilers, and turkeys and become the No. 1 U.S. meat in per capita meat consumption.

13. **The family farm will make a comeback.** Modern environmentally controlled swine buildings use a large variety of high-energy items—for heating, cooling, ventilation, etc. During the first half of the 21st century, fossil fuels will become scarcer and more costly, favoring sustainable agriculture, bio-mass products, labor intensive operations, soil conservation, and the comeback of the family farm.

QUESTIONS FOR STUDY AND DISCUSSION

1. The great changes that occurred in the U.S. swine industry in the 1980s and 1990s resulted from a shift from our labor-based economy to a knowledge- and capital-based economy. Why did this occur?

2. How would you go about selecting the software (program) for a large and modern pork production operation?

3. Cost and profit/loss in swine production are presented in Tables 19-1, 19-2, and 19-3. Based on the data presented in these tables, which type of pork enterprise would you favor if you wish to become a pork producer: (a) farrow-to-finish, (b) farrow-to-feeder pig, or (c) finish to market?

4. Fig. 19-2 shows that hog cycles are becoming longer and less abrupt. What has caused this change?

5. Describe briefly each of the following pork production systems: (a) farrow-to-finish, (b) farrow-to-feeder pig, and (c) finish to market.

6. Describe briefly each of the following types of business organizations: (a) proprietorship (individual), (b) partnership (general partnership), (c) corporations, (d) contracts, (e) co-ops, and (f) networking.

7. What has caused the great increase in contract hog production in recent years?

8. What is vertical integration? Is it good or bad for the pork industry? Justify your answer.

9. If you were becoming a pork producer, would you get involved in hog futures trading? Justify your answer.

10. List the traits of a good manager. How should a graduate fresh out of college train to become the manager of a mega-hog operation?

11. What do you see ahead for U.S. pork?

SELECTED REFERENCES

Title of Publication	Author(s)	Publisher
Beuscher's Law and the Farmer	H. H. Hannah	Springer Publishing Company, Inc., New York, NY, 1975
Complete Guide to Making a Public Stock Offering, A	E. L. Winter	Prentice-Hall, Inc., Englewood Cliffs, NJ, 1962
Contract Farming and Economic Integration	E. P. Roy	The Interstate Printers & Publishers Inc., Danville, IL, 1972
Cooperatives Today and Tomorrow	E. P. Roy	The Interstate Printers & Publishers, Inc., Danville, IL, 1969
Corporation Guide		Prentice-Hall, Inc., Englewood Cliffs, NJ, 1968
Doane's Farm Management Guide		Doane Agricultural Service, Inc., St. Louis, MO, 1965
Economics: Applications to Agriculture and Agribusiness	E. P. Roy F. L. Corty G. D. Sullivan	The Interstate Printers & Publishers, Inc., Danville, IL, 1980
Exploring Agribusiness	E. P. Roy	The Interstate Printers & Publishers, Inc., Danville, IL, 1980
Financial Planning in Agriculture	K. C. Schneeberger D. D. Osburn	The Interstate Printers & Publishers, Inc., Danville, IL, 1977
How to Do a Private Offering—Using Venture Capital	A. A. Sommer, Jr.	Practicing Law Institute, New York, NY, 1970
Lawyer's Desk Book		Institute for Business Planning, Inc., New York, NY, 1979
Pork Industry Handbook	Staff	Cooperative Extension Service, University of Illinois at Urbana-Champaign, current update
Stockman's Handbook, The, Seventh Edition	M. E. Ensminger	Interstate Publishers, Inc., Danville, IL, 1992
Tax-sheltered Investments	W. J. Casey	Institute for Business Planning, Inc., New York, NY, 1973

20

GLOSSARY OF SWINE TERMS

What's that? (Photo by J. C. Allen and Son, West Lafayette, IN)

The mark of distinction of good swine producers is that they "speak the language"—they use the correct terms and know what they mean. Even though swine terms are spoken glibly by people in the business, often they are baffling to the newcomer.

Many terms that are defined or explained elsewhere in this book are not repeated in this chapter. Thus, if a particular term is not listed herein, the reader should look in the Index or in the particular chapter and section where it is discussed.

A

ABATTOIR. A slaughterhouse.

ABSCESS. Localized collection of pus.

ABLACTATION. The act of weaning.

ABORTION. The expulsion or loss of the fetuses before the completion of pregnancy, and before they are able to survive.

ACCLIMATION. This refers to the short-term response of animals to their immediate environment.

ACCLIMATIZATION. Complex of processes of becoming accustomed to a new climate or other environmental conditions.

ACQUIRED IMMUNITY. The immunity that an animal generates during its lifetime, either passively or actively.

ACTIVE IMMUNITY. Immunity created when an animal recovers from natural infection or is vaccinated.

ACUTE. Referring to a disease which has a rapid onset, short course, and pronounced signs.

ADAPTATION. Adjustment of an organism to a new or changing environment.

ADDITIVE. An ingredient or substance added to a basic feed mix, usually in small quantities, for the purpose of fortifying it with certain nutrients, stimulants, and/or medicines.

ADIPOSE TISSUE. Fatty tissue.

AD LIBITUM. Free-choice access to feed.

ADRENAL GLAND. One of the endocrine (ductless) glands of the body, located near the kidney. It secretes hormones needed to utilize nutrients.

AEROBE. In the presence of air. The term usually

applied to microorganisms that require oxygen to live and reproduce.

A-FRAME HOUSE. A portable hog house shaped like an "A," and used by a sow and her litter on pasture.

AFTERBIRTH. The placenta and allied membranes associated with the fetus that are expelled from the uterus at farrowing.

AGALACTIA. Failure to secrete milk following parturition.

AGONISTIC BEHAVIOR. Combat or fighting behavior.

AI. Abbreviation for artificial insemination.

AIR-DRY (approximately 90% dry matter). This refers to feed that is dried by means of natural air movement, usually in the open. It may be either an actual or an assumed dry matter content; the latter is approximately 90%. Most feeds are fed in the air dry state.

ALBUMEN. One of the important proteins of blood, milk, eggs, and other substances.

ALKALI. A soluble salt or a mixture of soluble salts present in some soils of arid or semiarid regions in quantity detrimental to ordinary agriculture.

ALLELOMIMETIC BEHAVIOR. Doing the same thing.

ALL-IN, ALL-OUT (AIAO). In this system, pigs are moved in groups through each of the following stages: (1) farrowing, (2) nursery, and (3) growing-finishing. After each group is moved, the facilities are washed and disinfected. AIAO is being used to control a host of swine diseases.

AMBIENT TEMPERATURE. The prevailing or surrounding temperature.

AMERICAN FEED INDUSTRY ASSOCIATION, INC. (AFIA). The address is 1701 N. Fort Myer Drive, Arlington, Virginia 22209. The AFIA is a nationwide organization of feed manufacturers banded together (1) to improve the quality and promote the use of commercial feeds, (2) to encourage high standards on the part of its members, and (3) to protect the best interests of the feed manufacturer and the producer in legislative programs.

AMINO ACIDS. Nitrogen-containing compounds that constitute the "building blocks" or units from which more complex proteins are formed. They contain both an amino (NH_2) group and a carboxyl (COOH) group.

ANABOLISM. The conversion of simple substances into more complex substances by living cells (constructive metabolism).

ANAEROBE. A microorganism that normally does not require air or free oxygen to live and reproduce.

ANIMAL BEHAVIOR. This refers to the reaction of animals to certain stimuli, or the manner in which they react to their environment.

ANIMAL PROTEIN. Protein derived from meat packing or rendering plants, surplus milk or milk products, and marine sources. It includes proteins from meat, milk, poultry, eggs, fish, and their products.

ANIMAL RIGHTS. Those embracing this concept maintain that humans are animals, too; and that all animals should be accorded the same protection, including the right to live.

ANIMAL WELFARE. The well-being, health, and happiness of animals.

ANOREXIA. A lack or loss of appetite for food.

ANOXIA. Lack of oxygen in the blood or tissues. This condition may result from various types of anemia, reduction in the flow of blood to tissues, or lack of oxygen in the air at high altitudes.

ANTEMORTEM INSPECTION. This is the before slaughter inspection of animals.

ANTHELMINTIC. A drug that kills or expels worms.

ANTIBIOTIC. A compound synthesized by living organisms, such as bacteria or molds, which inhibits the growth of another.

ANTIBODY. A protein substance (modified type of blood-serum globulin) developed or synthesized by lymphoid tissue of the body in response to an antigenic stimulus. Each antigen elicits production of a specific antibody. In disease defense, the animal must have an encounter with the pathogen (antigen) before a specific antibody is developed in its blood.

ANTIGEN. A high molecular-weight substance (usually protein) which, when foreign to the bloodstream of an animal, stimulates formation of a specific antibody and reacts specifically *in vivo* or *in vitro* with its homologous antibody.

ANTIOXIDANT. A compound that prevents oxidative rancidity of polyunsaturated fats. Antioxidants are used to prevent rancidity in feeds and foods.

ANTISEPTIC. A chemical substance that prevents the growth and development of microorganisms.

APPETITE. The immediate desire to eat when food is present. Loss of appetite in an animal is usually caused by illness or stress.

ARCH. The convex curvature of the top line of swine.

ARTHRITIS. Inflammation of a joint.

ARTIFICIAL INSEMINATION. The introduction of semen into the female reproductive tract by a technician, using a pipette.

AS-FED (A-F). This refers to feed as normally fed to animals. It may range from 0 to 100% dry matter.

ASH. The mineral matter of a feed. The residue that remains after complete incineration of the organic matter.

ASSAY. Determination of (1) the purity or potency of a substance, or (2) the amount of any particular constituent of a mixture.

ASSIMILATION. A physiological term referring to the group of processes by which the nutrients in feed are made available to and used by the body; the processes include digestion, absorption, distribution, and metabolism.

ATROPHY. A wasting away of a part of the body, usually muscular, induced by injury or disease.

AUTOSOMES. All chromosomes except the sex chromosomes.

AVERAGE DAILY GAIN (ADG). The average daily liveweight increase of an animal.

AVOIRDUPOIS WEIGHTS AND MEASURES. *Avoirdupois* is a French word, meaning "to weigh." The old English system of weights and measures is referred to as the avoirdupois system, or U.S. Customary Weights and Measures, to differentiate it from the metric system.

B

BACKCROSS. The mating of a crossbred (F_1) animal to one of the parental breeds.

BACON. Cured and smoked meat from the side of the hog.

BACTERIA. Microscopic, single-cell plants, found in most environments, often referred to as microbes; some are beneficial, others are capable of causing disease.

BACTERICIDE. A product that destroys bacteria.

BACTERIN. A suspension of killed bacteria (vaccine) used to increase disease resistance.

BALANCED RATION. One which provides an animal the proper amounts and proportions of all the required nutrients.

BARROW. A male hog whose testicles were removed before reaching breeding age, and before the development of secondary sex characteristics.

BASAL METABOLIC RATE (BMR). The heat produced by an animal during complete rest (but not sleeping) following fasting, when using just enough energy to maintain vital cellular activity, respiration, and circulation, the measured value of which is called the basal metabolic rate (BMR). Basal conditions include thermoneutral environment, resting, postabsorptive state (digestive processes are quiescent), consciousness, quiescence, and sexual repose. It is determined in humans 14 to 18 hours after eating and when at absolute rest. It is measured by means of a calorimeter and is expressed in calories per square meter of body surface.

BATTERY. A series of pens or cages used to house animals in concentrated confinement rearing systems.

BIOAVAILABILITY. The availability of a substance to the animal; for example, the availability of different vitamin sources to pigs.

BIOLOGICAL VALUE OF A PROTEIN. The percentage of the protein of a feed or feed mixture which is usable as a protein by the animal. Thus, the biological value of a protein is a reflection of the kinds and amounts of amino acids available to the animal after digestion. A protein which has a high biological value is said to be of *good quality*.

BIOSYNTHESIS. The production of a new material in living cells or tissues.

BIOTECHNOLOGY. Biotechnology is the use of living organisms or parts of organisms, such as enzymes, to make or modify products.

BLIND TEAT. A small, functionless teat.

BLOOM.

■ Said of an animal that has beauty and freshness. An animal in bloom has a glossy hair coat and an attractive appearance.

■ The condition or time of flowering.

BLUP. This stands for Best Linear Unbiased Prediction. It is a statistical procedure that can be used to analyze swine performance data.

BOAR. A male hog, generally used for breeding purposes.

BOMB CALORIMETER. An instrument used to measure the gross energy content of any material, in which the feed (or other substance) tested is placed and burned in the presence of oxygen.

BRAND NAME. Any word, name, symbol, or device, or any combination of these, often registered as a trademark or name, which identifies a product and distinguishes it from others.

BRED.

■ Refers to an animal that is pregnant.

■ Sometimes used synonymously with the term *mated*.

BREED.

■ Animals that are genetically pure enough to have similar external characteristics of color and conformation, and when mated together produce offspring with the same characteristics.

■ The mating of animals.

BREEDING AND GESTATING FACILITIES. Usually, the breeding herd (gestating sows and gilts, and herd boars) is the last group to be moved inside. So, the breeding and gestating facilities may be inside or outside.

BREED TYPE. The combination of characteristics that makes an animal better suited for a specific purpose.

BRITISH THERMAL UNIT (Btu). The amount of energy required to raise 1 lb of water 1°F; equivalent to 252 calories.

BRIX. A term commonly used to indicate the sugar (sucrose) content of molasses. It is expressed in degrees and was originally used to indicate the percentage by weight of sugar in sucrose solutions, with each degree Brix being equal to 1% sucrose.

BUFFER. A substance in a solution that makes the degree of acidity (hydrogen-ion concentration) resistant to change when an acid or a base is added.

BUSHEL. A unit of capacity equal to 2,150.42 cubic inches (approximately 1.25 cu ft).

BYPRODUCT FEEDS. The innumerable roughages and concentrates obtained as secondary products from plant and animal processing, and from industrial manufacturing.

C

CAKE (presscake). The mass resulting from the pressing of seeds, meat, or fish in order to remove oils, fats, or other liquids.

CALCIFICATION. The process by which organic tissue becomes hardened by a deposit of calcium salts.

CALORIC. Pertaining to heat or energy.

CALORIE (Cal). A small calorie is the amount of heat required to raise the temperature of 1 g of water from 14.5 to 15.5°C. This is equivalent to 4.185 joules.

CALORIMETER. An instrument for measuring the amount of energy.

CANADIAN BACON. The bacon that results from the mild curing and light smoking of pork sirloin muscle.

CANNED HAM. Cured outside the can and cooked inside to reduce shrinkage. A small amount of gelatin is added to set the juices.

CANNIBALISM.
- The habit of one animal pecking at or eating on another animal, such as a fowl pecking at or eating on another fowl or one pig biting the tail of another.
- The eating of young, such as a sow may do after farrowing or a doe rabbit may do if disturbed soon after kindling.

CANOLA. This is an improved rape which was developed by Canadian scientists in the 1970s. It is low in glucosinolates and low in erucic acid (a long-chain fatty acid).

CARCASS. The dressed body of a meat animal, the usual items of offal having been removed.

CARCASS WEIGHT. Weight of the carcass of an animal following slaughter, as it hangs on the rail, expressed either as warm (hot) or chilled (cold) carcass weight.

CARCASS YIELD. The carcass weight as a percentage of the liveweight.

CARRIER.
- A disease-carrying animal.
- A heterozygote for any trait.
- An edible material to which ingredients are added to facilitate their uniform incorporation into feeds.

CARRYING CAPACITY. The number of animal units a property or area will carry on a year-round basis. This includes the land grazed plus the land necessary to produce the winter feed.

CASTRATE.
- To remove the testicles or ovaries.
- An animal that has had its testicles or ovaries removed.

CATABOLISM. The conversion or breaking down of complex substances into more simple compounds by living cells (destructive metabolism).

CATCH. Term used by livestock caretakers to indicate that conception has taken place following breeding.

CENTIGRADE (C). A means of expressing temperature. To convert to Fahrenheit, multiply by $\frac{9}{5}$ and add 32.

CEREAL. A plant in the grass family *(Gramineae)*, the seeds of which are used for human and animal food; *e.g.*, maize (corn), and wheat.

CHELATED MINERAL. A mineral that is bound to a compound such as protein or an amino acid that helps to stabilize it.

CHEMBIOTICS. Compounds similar to antibiotics but they are produced chemically rather than microbiologically.

CHOLESTEROL. A white, fat-soluble substance found in animal fats and oils, bile, blood, brain tissue, nervous tissue, the liver, kidneys, and adrenal glands. It is important in metabolism and is a precursor of certain hormones. Cholesterol is implicated in arteriosclerosis.

CHROMOSOMES. Structures that contain DNA, normally in a double helix.

CHRONIC. Referring to a disease condition that is continuous and long-lasting.

CLINICAL. Referring to direct observation.

CLOSE BREEDING. A form of inbreeding such as brother to sister or sire to daughter.

COEFFICIENT OF DIGESTIBILITY. The percentage value of a food nutrient that is absorbed. For example, if a food contains 10 g of nitrogen and it is found that 9.5 g are absorbed, the digestibility is 95%.

COENZYME. A substance, usually containing a vitamin, which works with an enzyme (protein mainly) to perform a certain function.

COLLAGEN. A white, papery transparent type of connective tissue which is of protein composition. It forms gelatin when heated with water.

COLOSTRUM. The milk secreted by mammalian females for the first few days following parturition, which is high in antibodies and is laxative.

COMBUSTION. The combination of substances with oxygen accompanied by the liberation of heat.

COMMERCIAL FEEDS. Feeds mixed by manufacturers who specialize in the feed business.

COMMODITY EXCHANGE. A place where buyers and sellers meet on an organized market and transact business on paper, without the physical presence of the commodity.

COMPLETE RATION. All feedstuffs (forages and grains) combined in one feed. A complete ration fits well into mechanized feeding and the use of computers to formulate least-cost rations.

CONCENTRATE. A broad classification of feedstuffs which are high in energy and low in crude fiber (under 18%). For convenience, concentrates are often broken down into (1) carbonaceous feeds, and (2) nitrogenous feeds.

CONDITION.
■ The state of health, as evidenced by the coat, and general appearance.
■ The amount of flesh or finish (fat covering).

CONFORMATION. The shape and design of an animal.

CONTAGIOUS. Transmissible by contact.

CONTENTMENT. A stress-free condition exhibited by healthy animals; the cow will stretch on rising, the sheep will stand or lie quietly, the pig will curl its tail, and the horse will look completely unworried when resting.

CONTRACT. An agreement between two or more persons to do or refrain from doing certain things. The most common kinds of hog contracts are:
  1. Marketing contracts.
      a. Marketing contracts for slaughter hogs.
      b. Marketing contracts for feeder pigs.
  2. Production contracts.
      a. Farrow-to-finish contracts.
      b. Feeder pig production contracts.
      c. Finishing contracts.
  3. Breeding stock leasing.

COOKED. Heated to alter chemical or physical characteristics or to sterilize.

COOPERATIVE (co-op). A cooperative is a business formed by a group of people to get certain services for themselves more effectively or more economically than they can get them individually.

CORPORATION. This is a device which may be used for carrying out a pork enterprise as an entity distinct from the persons who are interested in and control it.
■ **Family owned corporation.** This refers to a family owned (privately owned) corporation.

■ **Tax-option corporation (Subchapter S corporation).** This type of corporation cannot have more than 35 members. The income or losses are passed back to the shareholders, each of whom pays his/her share of taxes.

COUNTRY HAM (dry-cured). Ham produced using a dry-cured, slow-smoking, and long drying process. Country hams are heavily salted and may require soaking and simmering before roasting. They may be called Virginia or Georgia (or another state name) ham, depending on the state of origin.

CRACKED. Particle size reduced by combined breaking and crushing action.

CREEP. An enclosure or feeder used for supplemental feeding of nursing young, which excludes their dams.

CROSS BREEDING. The mating of animals of different breeds.

CRUDE FAT. Material extracted from moisture-free feeds by ether. It consists largely of fats and oils with small amounts of waxes, resins, and coloring matter. In calculating the energy value of a feed, the fat is considered to have 2.25 times as much energy as either nitrogen-free extract or protein.

CRYPTORCHID. A male, which is often sterile because one or both testicles are retained in the abdominal cavity.

CULL. An animal taken out of a herd because it is below herd standards.

CUTTING.
■ Removing the testicles.
■ Separating one or more animals from a herd.

CYCLICAL MOVEMENTS. Basically, cycles are the response of producers to prices; and prices reflect supply and demand.

D

DECORTIFICATION.
■ Removal of the bark, hull, husk, or shell from a plant, seed, or root.
■ Removal of portions of the cortical substance of a structure or organ, as in the brain, kidneys, and lung.

DEFECATION. The evacuation of fecal material from the rectum.

DEFICIENCY DISEASE. A disease caused by a lack of one or more basic nutrients, such as a vitamin, a mineral, or an amino acid.

DEHYDRATE. To remove most or all moisture from a substance for the purpose of preservation, primarily through artificial drying.

DERMATITIS. Inflammation of the skin.

DESICCATE. To dry completely.

DIET. Feed ingredient or mixture of ingredients, including water, which is consumed by animals.

DIGESTIBLE NUTRIENT. The part of each feed nutrient that is digested or absorbed by the animal.

DIGESTIBLE PROTEIN. That protein of the ingested food protein which is absorbed.

DIGESTION COEFFICIENT (coefficient of digestibility). The difference between the nutrients consumed and the nutrients excreted expressed as a percentage.

DIRECT SELLING, including sale to country dealers, refers to producers selling hogs directly to packers or local dealers, without the support of commission firms, selling agents, buying agents, or brokers.

DISEASE. Any departure from the state of health.

DISINFECTANT. A chemical capable of destroying disease-causing microorganisms or parasites.

DIURESIS. An increased excretion of urine.

DIURETIC. An agent that increases the flow of urine.

DNA (deoxyribonucleic acid). DNA serves as the genetic information source. The DNA molecule has double helical structure.

DOCK.
- To cut off the tail.
- To reduce the weight and/or price; for example, piggy sows are usually docked 40 lb and stags 70 lb when marketed.

DOMINANT. Describes a gene which, when paired with its allele, covers up the phenotypic expression of that gene.

DRESSING PERCENTAGE. The percentage of the live animal that becomes the carcass at slaughter.

DRIED. Materials from which water or other liquids have been removed.

DRUGS. Substances of mineral, vegetable, or animal origin used in the relief of pain or for the cure of disease.

DRY. Nonlactating female. The dry period is the time between lactations (when a female is not secreting milk).

DRYLOT. A relatively small enclosure without vegetation, either (1) with shelter, or (2) on an open yard, in which animals may be confined.

DRY MATTER BASIS. A method of expressing the level of a nutrient contained in a feed on the basis that the material contains no moisture.

DRY-RENDERED. Residues of animal tissues cooked in open, steam-jacketed vessels until the water has evaporated. Fat is removed by draining and pressing the solid residue.

DUST. A mixture of small particles of different sizes of dry matter.

E

EARLY MATURING. Completing sexual development at an early age.

EARLY WEANER PIG MARKETING (SEW pigs). This refers to marketing segregated early weaned (SEW) pigs—pigs weaned under three weeks of age, rather than the more conventional 21- to 28-days of age; then, segregating the piglets from their mothers.

EARLY WEANING. The practice of weaning young animals earlier than usual; weaning pigs under five weeks of age.

EAR NOTCHING. Making slits or perforations in an animal's ears for identification purposes.

EASY KEEPER. An animal that grows or fattens rapidly on limited feed.

EBV. This stands for *estimated breeding value*. It is the entire worth of the parent as a source of genetic material.

EDEMA. Swelling of a part or all of the body due to the accumulation of excess water.

EFFICIENCY OF FEED CONVERSION. This is expressed as units of feed per unit of meat.

ELECTROLYTE. A chemical compound which in solution dissociates by releasing ions. An ion is an atomic particle that carries a positive (+) or a negative (−) charge.

ELECTRONIC FEEDER PIG AUCTIONS. Feeder pigs offered for sale are scored on the basis of health programs, feeding programs, and herd management. This information, along with the identification of the seller, is made available to prospective buyers via computer modems. Consigned feeder pigs may be reviewed by potential buyers. Buyers participate in an open auction. Pigs move directly from the seller to the buyer, thereby minimizing disease and stress.

ELEMENT. One of the 103 known chemical substances that cannot be divided into simpler substances by chemical means.

EMACIATED. Excessive loss of flesh.

ENDOCRINE. Pertaining to glands and their secretions that pass directly into the blood or lymph instead of into a duct (secreting internally). Hormones are secreted by endocrine glands.

ENDOGENOUS. Originating within the body; *e.g.*, hormones and enzymes.

ENERGY.
- Vigor or power in action.
- Capacity to perform work.

ENERGY FEEDS. Feeds that are high in energy and low in fiber (under 18%), and that generally contain less than 20% protein.

ENTERITIS. Inflammation of the intestines.

ENVIRONMENT. The sum total of all external conditions that affect the life and performance of a pig.

EPD. This stands for *estimated progeny difference*. It is half the genetic worth of each parent and is an estimate of how much of their performance will be passed on to their offspring. EPD is a prediction of the progeny performance of an animal compared to the progeny of an average animal in the population, based on all information currently available. EPDs are available for the following:

    EPD days/230
    EPD backfat
    EPD terminal sire index (TSI)
    EPD number born alive (NBA)
    EPD litter weight (LW)
    EPD sow productivity index (SPI)
    EPD maternal sire index (MSI)

ERGOSTEROL. A plant sterol which, when activated by ultraviolet rays, becomes vitamin D_2. It is also called provitamin D_2 and ergosterin.

ERGOT. A fungus disease of plants.

ESSENTIAL AMINO ACIDS. Those amino acids which cannot be made in the body from other substances or which cannot be made in sufficient quantity to supply the animal's needs.

ESSENTIAL FATTY ACID. A fatty acid that cannot be synthesized in the body or that cannot be made in sufficient quantities for the body's needs.

ESTRUS. The period when the gilt or sow will accept service by the boar.

ETHER EXTRACT (EE). Fatty substances of feeds and foods that are soluble in ether.

EVAPORATED. Reduced to a denser form; concentrated as by evaporation or distillation.

EXCRETA. The products of excretion—primarily feces and urine.

EXOGENOUS. Provided from outside of the organism.

EXPERIMENT. The word *experiment* is derived from the Latin *experimentum*, meaning proof from experience. It is a procedure used to discover or to demonstrate a fact or general truth.

EXTRA-LABEL DRUGS. Use of over-the-counter drugs in therapies or dosages not approved by the labeling constitutes extra-label drug use.

EXTRINSIC FACTOR. A dietary substance which was formerly thought to interact with the intrinsic factor of the gastric secretion to produce the antianemic factor, now known to be vitamin B-12. (Also see INTRINSIC FACTOR.)

F

FARROW. To give birth to piglets.

FARROWING HOUSE. Central farrowing houses are generally environmentally controlled.

FARROW-TO-FINISH. A type of hog farm operation that covers all aspects of breeding, farrowing, and raising pigs to slaughter.

FAT. The term *fat* is frequently used in a general sense to include both fats and oils, or a mixture of the two. Both fats and oils have the same general structure and chemical properties, but they have different physical characteristics. The melting points of most fats are such that they are solid at ordinary room temperatures, while oils have lower melting points and are liquids at these temperatures.

FATTENING. The deposition of energy in the form of fat within the body tissues.

FATTY ACIDS. The key components of fats (lipids). Their degree of saturation (hydrogenation) and the length of their carbon chain determine many of the physical aspects—melting point and stability—of fats (lipids).

FECES. The excreta discharged from the digestive tract through the anus.

FECUNDITY. Ability to produce many offspring.

FEED (feedstuff). Any naturally occurring ingredient, or material, fed to animals for the purpose of sustaining them.

FEED ADDITIVE. An ingredient or a substance added to a feed to improve the rate and/or efficiency of gain of animals, prevent certain diseases, or preserve feeds.

FEED EFFICIENCY. The ratio expressing the number of units of feed required for one unit of production (meat) by an animal. This value is commonly expressed as pounds of feed eaten per pound of gain in body weight.

FEEDER PIG PRODUCTION. This refers to the production and sale of 30 to 60 lb pigs for growing and finishing on other farms.

FEEDER'S MARGIN. The difference between the cost per hundredweight of feeder animals and the selling price per hundredweight of the same animals when finished.

FEED GRAIN. Any of several grains most commonly used for livestock or poultry feed, such as corn, sorghum, oats, and barley.

FEEDLOT. A lot or plot of land on which animals are fed or finished for market.

FEED OUT. To feed a pig until it reaches market weight.

FEEDSTUFF. Any product, of natural or artificial origin, that has nutritional value in the diet when properly prepared.

FERAL. Referring to domesticated animals which have reverted back to their original or untamed state.

FERMENTATION. Chemical changes brought about by enzymes produced by various microorganisms.

FETUS. A young organism in the uterus from the time the organ systems develop until birth.

FIBER CONTENT OF A FEED. The amount of hard-to-digest carbohydrates. Most fiber is made up of cellulose and lignin.

FILL.
■ A term designating the fullness of the digestive tract of an animal.
■ With market animals, the fill refers to the amount of feed and water consumed upon their arrival at the market and prior to selling.

FINISHING PIGS. The phase in the life cycle of market hogs from approximately 120 lb to market weight.

FITTING. The conditioning of an animal for show or sale, which usually involves a combination of special feeding plus exercise and grooming.

FLORA. The plant life present. In nutrition, it generally refers to the bacteria present in the digestive tract.

FLUSHING. The practice of feeding females more generously 1 to 2 weeks before breeding so that they gain in weight from 1.0 to 1.5 lb daily. The beneficial effects attributed to this practice are (1) more eggs (ova) are shed, and this results in more offspring; (2) the females come in heat more promptly; and (3) conception is more certain.

FOLACIN (folic acid/folate). This includes a group of compounds with folic acid activity. Folic acid participates in many enzymatic reactions.

FOLLOWING CATTLE. The former practice of allowing feeder pigs to run behind feedlot cattle so they may glean unused grains and other nutrients from the cattle manure.

FOOD AND DRUG ADMINISTRATION (FDA). The federal agency in the Department of Health and Human Services that is charged with the responsibility of safeguarding American consumers against injury, unsanitary food, and fraud. It protects industry against unscrupulous competition, and it inspects and analyzes samples and conducts independent research on such things as toxicity (using laboratory animals), disappearance curves for pesticides, and long-range effects of drugs.

FOOD SAFETY LABELING. The food labeling requirement became effective June 6, 1994. It mandates safe cooking and handling labels for all uncooked meat and poultry products.

FORTIFY. Nutritionally, to add one or more feeds or feedstuffs.

FREE-CHOICE. Free to eat a feed or feeds at will.

FREEDOM STALLS. This is a system of gestation housing which preserves the advantages of individual living places, but permits more freedom than traditional crates, stalls, or tethering.

FREEZE DRYING. (See LYOPHILIZATION.)

FULL-FEED. The term indicating that animals are being provided as much feed as they will consume safely without going off feed.

FUMIGANT. A liquid or solid substance that forms vapors that destroy pathogens, insects, and rodents.

FUNGI. Plants that contain no chlorophyll, flowers, or leaves, such as molds, mushrooms, toadstools, and yeasts. They may get their nourishment from either dead or living organic matter.

FUTURES CONTRACT. A futures contract is a standardized, legal, binding paper transaction in which the seller promises to make delivery and the buyer promises to take delivery on a specified quantity and type of commodity at a specified location(s) during a specified future month.

FUTURES TRADING. The futures market is a way in which to provide (1) an insurance medium in the marketing field, and (2) the facilities and machinery for underwriting price risks.

G

GASTROINTESTINAL. Pertaining to the stomach and intestines.

GENE. A segment of DNA that carries inherited information.

GENOME. All the inherited information in a cell.

GENOTYPE. An animal's true (genetic) makeup.

GESTATION (pregnancy). Time between breeding and farrowing, about 114 days for swine.

GET. The offspring of a male animal—his progeny.

GILT. Young female, under one year of age, that has not had her first litter.

GIRTH. The circumference of the body of an animal behind the shoulders.

GLUCOSE. A hexose monosaccharide obtained upon the hydrolysis of starch and certain other carbohydrates. Also called dextrose.

GOITROGENIC. Producing or tending to produce goiter.

GOSSYPOL. A toxic yellow pigment found in cottonseed, which is toxic to swine and certain other nonruminants, and which may cause discoloration of egg yolks during cold storage.

GRADE.
■ An animal having parents that cannot be registered by a breed association.
■ A measure of how well an animal or product fulfills the requirements for the class; for example, the federal grades of hogs and their carcasses are a specific indication of the degree of excellence.

GRADING UP. The continued use of purebred sires of the same breed in a grade herd.

GRAIN. Seed from cereal plants.

GRIND. To reduce to small segments by impact, shearing, or attrition (as in a mill).

GROATS. Grain from which the hulls have been removed.

GROUND PORK. At least 70% lean and no seasonings.

GROWING-FINISHING. This refers to the system of buying feeder pigs weighing 30 to 60 lb, then growing and finishing them for market.

GROWING AND FINISHING FACILITIES. Pigs are usually moved from the nursery to the growing facilities where they remain until they weigh 120 lb, then to the finishing facilities from 120 lb to market weight.

GROW OUT. To feed animals so that they attain a certain desired amount of growth with little or no fattening.

GROWTH. May be defined as the increase in size of the muscles, bones, internal organs, and other parts of the body.

GROWTHY. Describes an animal that is large and well developed for its age.

GRUEL. A feed prepared by mixing ground ingredients with hot or cold water.

H

HABITUATION. This is the act or process of making animals familiar with or accustomed to a new environment through use or experience.

HAM. The thigh of a hog prepared for food, or the hind leg of a swine from the hock upwards on the live animal.

HAND-FEEDING. To provide a certain amount of a ration at regular intervals.

HAND-MATING. Controlled breeding with confined boars rather than allowing boars to run loose with groups of unbred sows.

HARD KEEPER. An animal that is unthrifty and grows or fattens slowly regardless of the quantity or quality of feed.

HAZARD ANALYSIS, CRITICAL CONTROL POINT (HACCP). The HACCP plan identifies hazards in food processing, followed by monitoring those crucial points. Problems are remedied as they occur. It is a way of processing safer food.

HEALTH. This is the state of complete well-being, not merely the absence of disease.

HEAT (Estrus). The period when the female will accept service by the boar.

HEAT INCREMENT (HI). The increase in heat production following consumption of feed when the animal is in a thermoneutral environment.

HEAT LABILE. Unstable to heat.

HEDGING. This is an offsetting transaction in which purchases or sales of a commodity are counterbalanced by sales or purchases of an equivalent quantity of futures contracts in the same commodity.

HEMOGLOBIN. The oxygen-carrying, red-pigmented protein of the red blood cells.

HERNIA. The protrusion of some of the intestine through an opening in the body wall—commonly called a rupture.

HETEROSIS (hybrid vigor). Amount the F_1 generation exceeds the P_1 generation for a given trait, or the amount the crossbreds exceed the average of the two purebreeds that are crossed to produce the crossbreds.

HIGH-LYSINE CORN (Opaque-2). Corn that is much higher than normal corn in lysine and tryptophan; hence, it has a better balance of the amino acids for monogastric animals. Also, high-lysine corn is higher in total protein, but lower in leucine than regular corn.

HOG. A large or mature animal of either sex, generally weighing over 120 lb.

HOG/CORN RATIO. This relationship is determined by dividing the price of live hogs per cwt by the price of corn per bushel.

HOG DOWN. Practice of allowing pigs to "harvest" a crop in the field.

HOMOGENIZED. Having particles broken down into evenly distributed globules small enough to remain emulsified for long periods of time.

HOMOLOGOUS CHROMOSOMES. Chromosomes having the same size and shape, and containing the genes affecting the same characteristics.

HORMONE. A body-regulating chemical secreted by an endocrine gland into the bloodstream, then transported to another region within the animal where it elicits a physiological response.

HULL. Outer covering of grain or other seed, especially when dry.

HYDROGENATION. The chemical addition of hydrogen to any unsaturated compound.

HYDROLYSIS. The splitting of a substance into the smaller units by chemically adding water to the material.

HYPERTROPHIED. Having increased in size beyond the normal growth.

HYPERVITAMINOSIS. An abnormal condition resulting from the intake of an excess of one or more vitamins.

HYPOCALCEMIA. Below normal concentration of ionic calcium in blood resulting in convulsions, as in tetany or parturient paresis (milk fever).

HYPOGLYCEMIA. A reduction in concentration of blood glucose below normal.

HYPOMAGNESEMIA. An abnormally low level of magnesium in the blood.

HYPOTHALAMUS. A portion of the brain found in the floor of the third ventricle. It regulates body temperature, appetite, hormone release, and other functions.

I

IDEAL PROTEIN. This refers to a protein that provides a perfect pattern of essential and nonessential amino acids in the diet without any excesses or deficiencies.

IMMUNITY. The ability of an animal to resist or overcome an infection to which most members of its species are susceptible.

IMMUNOGLOBULINS. A family of proteins found in body fluids which have the property of combining with antigens; and, when the antigens are pathogenic, sometimes inactivating them and producing a state of immunity. Also called antibodies.

INBREEDING. The mating of individuals which are more closely related than average individuals in a population. It increases homozygosity.

INDUSTRIALIZATION OF HOG PRODUCTION. The production of hogs in specialized factory-like facilities staffed with specialized labor.

INFLAMMATION. The reaction of tissue to injury, characterized by redness, swelling, pain, and heat.

INGEST. To eat or take in through the mouth.

INGESTA. Food or drink taken into the stomach.

INGESTION. The taking in of food and drink.

INGREDIENT. A constituent feed material.

IN PIG. A pregnant sow or a gilt.

INSULIN. A hormone secreted by the pancreas into the blood, which regulates sugar (glucose) metabolism.

INTRADERMAL. Into, or between, the layers of the skin.

INTRAMUSCULAR. Within the muscle.

INTRAPERITONEAL. Within the peritoneal cavity.

INTRAVENOUS. Within the vein or veins.

INTRINSIC FACTOR. A chemical substance secreted by the stomach which is necessary for the absorption of vitamin B-12. The exact chemical nature of intrinsic factor is not known, but it is thought to be a mucoprotein or mucopolysaccharide. A deficiency of this factor may lead to a deficiency of vitamin B-12, and, ultimately, to pernicious anemia.

IN VITRO. Occurring in an artificial environment, as in a test tube.

IN VIVO. Occurring in the living body.

INVOLUTION. Return of an organ to its normal size and condition after enlargement, as of the uterus after farrowing.

IRRADIATED YEAST. Yeast that has been irradiated. Yeast contains considerable ergosterol, which, when exposed to ultraviolet light, produces vitamin D.

IRRADIATION. Exposure to ultraviolet light.

IU (international unit). A standard unit of potency of a biologic agent (*e.g.*, a vitamin, a hormone, an antibiotic, an antitoxin) as defined by the International Conference for Unification of Formulae. Potency is based on bioassay that produces a particular effect agreed on internationally. Also called a USP unit.

J

JOULE. Proposed international unit (4,184 j = 1 calorie) for expressing mechanical, chemical, or electrical energy, as well as the concept of heat. In the future, energy requirements and feed values will likely be expressed by this unit.

JOWL. Meat from the cheeks of hogs.

K

KERNEL. The whole grain of a cereal. The meat of nuts and drupes (single-stoned fruits).

KILLED VACCINE. A vaccine in which the antigen has been inactivated so that it cannot produce the disease. Toxoids and subunit vaccines also fall in this category.

KJELDAHL. A method of determining the amount of nitrogen in an organic compound. The quantity of nitrogen measured is then multiplied by 6.25 to calculate the protein content of the feed or compound analyzed. The method was developed by a Danish chemist, J. G. C. Kjeldahl, in 1883.

L

LABILE. Unstable. Easily destroyed.

LACTATION. The period in which an animal is producing milk.

LACTOSE (milk sugar). A disaccharide found in milk, having the formula $C_{12}H_{22}O_{11}$. It hydrolyzes to glucose and galactose. Commonly known as milk sugar.

LAGOON. A waste management treatment unit—a digester, for the purpose of biochemical breakdown of organic wastes (manure, straw).

LARD. Fat rendered (melted out) from fresh pork tissue.

LARVA. The immature form of insects and other small animals.

LAXATIVE. A feed or drug that will induce bowel movements and relieve constipation.

LEAN CUTS. Ham, loin, Boston butt, and picnic.

LEAN METER. A large, precisely wound coil of wire through which the carcass passes.

LIMITED-FEEDING. Feeding animals less than they would like to eat. Giving sufficient feed to maintain weight and growth, but not enough for their potential production or finishing.

LIMITED PARTNERSHIP. An arrangement in which two or more parties supply the capital, but only one partner is involved in the management.

LIMITING AMINO ACID. The essential amino acid of a protein which shows the greatest percentage deficit in comparison with the amino acids contained in the same quantity of another protein selected as a standard.

LINEBREEDING. A form of inbreeding which attempts to concentrate the inheritance of some ancestor in the pedigree.

LINECROSS. A cross of two inbred lines.

LIPOLYSIS. The hydrolysis of fats by enzymes, acids, alkalis, or other means to yield glycerol and fatty acids.

LITTER. The pigs farrowed by a sow at one delivery. Such individuals are called *littermates*.

LIVER ABSCESSES. Single or multiple abscesses on the liver, observed at slaughter. Usually the abscess consists of a central mass of necrotic liver surrounded by pus and a wall of connective tissue. At slaughter, those livers affected with abscesses are condemned for human food.

LIVE VACCINE. A vaccine in which the live organism has been altered (attenuated) so that it can no longer cause disease, but can replicate in the animal and stimulate an immune response.

LIVEWEIGHT. Weight of an animal on foot.

LOIN. That portion of the back between the thorax and pelvis.

LOIN EYE AREA. Cross section of the pork chop muscle, usually measured between the 10th and 11th ribs.

LOWER CRITICAL TEMPERATURE. This is the low point of the cold temperature beyond which the animal cannot maintain normal body temperature.

LUMEN. The cavity inside a tubular organ—the lumen of the stomach or intestine.

LYMPH. The slightly yellow, transparent fluid occupying the lymphatic channels of the body.

LYOPHILIZATION. The evaporation of a liquid from a frozen product with the aid of high vacuum. Also called freeze drying.

M

MACROMINERALS. The major minerals—calcium, phosphorus, sodium, chlorine, potassium, magnesium, and sulfur.

MAINTENANCE REQUIREMENT. A ration which is adequate to prevent any loss or gain of tissue in the body when there is no production.

MALNUTRITION. Any disorder of nutrition. Commonly used to indicate a state of inadequate nutrition.

MALTOSE. A disaccharide, also known as malt sugar, having the formula $C_{12}H_{22}O_{11}$. Obtained from the partial hydrolysis of starch. It hydrolyzes to glucose.

MANAGEMENT, SWINE. The art of caring for and handling swine.

MANGY. Infected with a skin disease or parasite so that the skin is dry and scaly.

MANURE. A mixture of animal excrements (consisting of undigested feeds plus certain body wastes) and bedding.

MANURE GASES. The common gases produced by the decomposition of manure are methane, ammonia, hydrogen sulfide, and carbon dioxide. When methane and carbon dioxide displace oxygen, people can be killed.

MARGIN (spread). The difference between the purchase price and the selling price.

MARKER. A part of DNA which provides for the detection of genetic variation from animal to animal.

MARKET CLASS. Animals grouped according to the use to which they will be put, such as slaughter or feeder.

MARKET GRADE. Animals grouped within a market class according to their value.

MARKETING CONTRACT. A marketing contract is an agreement between seller (usually a producer) and a meat packer to sell/buy at a specified future date a specified number of hogs of a specified weight and grade for a specified price.

MASH. A mixture of ingredients in meal form.

MASTICATION. The chewing of feed.

MASTITIS. Inflammation of the mammary gland.

MEAL.
- A feed ingredient having a particle size somewhat larger than flour.
- Mixture of concentrate feeds, usually in which all of the ingredients are ground.

MEATS.
- Animal tissues used as food.
- The edible parts of nuts and fruits.

MECONIUM. Excrement accumulated in the bowels during fetal development.

MEDICATED FEED. Any feed which contains drug ingredients intended or represented for the cure, mitigation, treatment, or prevention of diseases of animals (other than humans).

METABOLISM. Refers to all the changes that take place in the nutrients after they are absorbed from the digestive tract, including (1) the building-up processes in which the absorbed nutrients are used in the formation or repair of body tissues, and (2) the breaking-down processes in which nutrients are oxidized for the production of heat and work.

MICROBE. Same as microorganism.

MICROCHIPS, SWINE. This refers to electronic swine tracking, the goal of which is to assign a number to each farm and each pig. Ultimately, meat retailers will be able to trace meat to its source. Such chips are now being tested.

MICROFLORA. Microbial life characteristic of a region, such as the bacteria and protozoa populating the rumen.

MICROINGREDIENT. Any ration component, such as minerals, vitamins, antibiotics, and drugs, normally measured in milligrams or micrograms per kilogram, or in parts per million.

MICROMINERALS. Minerals required by animals in small amounts—milligrams per pound or smaller units. These are also called *trace elements*, or *trace minerals*.

MICROORGANISM. Any organism of microscopic size, applied especially to bacteria and protozoa.

MILK EJECTION OR "LET-DOWN." The process, controlled by the hormone oxytocin, in which milk is forced from the alveoli, where it is stored, into the larger ducts and cisterns, where it is available to suckling pigs.

MILL BYPRODUCT. A secondary product obtained in addition to the principal product in milling practice.

MINERALS (ash). The inorganic elements of animals and plants, determined by burning off the organic matter and weighing the residue, which is called ash.

MINERAL SUPPLEMENT. A rich source of one or more of the inorganic elements needed to perform certain essential body functions.

MINIATURE SWINE. Genetically small pigs. They are good experimental animals for biomedical studies.

MODULAR NURSERY. These are self-contained portable nurseries which are constructed by the manufacturer before being delivered to the farm.

MOISTURE. A term used to indicate the water contained in feeds—expressed as a percentage.

MOISTURE-FREE (M-F, oven-dry, 100% dry matter). This refers to any substance that has been dried in an oven at 221°F *(105°C)* until all the moisture has been removed.

MOLDS (fungi). Fungi which are distinguished by the formation of mycelium (a network of filaments or threads), or by spore masses.

MORBIDITY. A state of sickness or the rate of sickness.

MORTALITY. Death or death rate.

MULE FOOT. Swine hoof having the shape of a mule's foot—not halved.

MULTIPLE FARROWING. The practice of breeding sows to farrow pigs throughout the year, thereby marketing hogs more frequently.

MUMMIFIED FETUS. A shriveled or dried fetus that has remained in the uterus instead of being expelled or aborted after dying.

MYCOTOXINS. Toxic metabolites produced by molds during growth. Sometimes present in feed materials.

N

NANO. A prefix meaning one billionth (10^{-9}).

NATIONAL PORK PRODUCERS COUNCIL (NPPC). The National Pork Producers Council is the largest commodity organization in the nation with an identified membership. The NPPC is the pork producers' voice and advocate to solve problems efficiently for the industry. Headquartered in Des Moines, Iowa, with an additional office in Washington, DC, the Council's purpose is to enhance the quality, production, distribution, and sale of pork and pork products.

NATIONAL PORK PRODUCERS COUNCIL ASSURANCE PROGRAM. The NPPC first introduced its Pork Assurance Program in 1989. Its purpose is to enhance consumer confidence in the safety of pork, and increase pork consumption.

NATIONAL RESEARCH COUNCIL (NRC). A division of the National Academy of Sciences established in 1916 to promote the effective utilization of scientific and technical resources. Periodically, this private, nonprofit organization of scientists publishes bulletins giving nutrient requirements and allowances of domestic animals, copies of which are available on a charge basis through the National Academy of Sciences, National Research Council, 2101 Constitution Avenue, NW, Washington, DC 20418.

NATIVE DEFENSE SYSTEM. Qualities in a normal, healthy animal that help it fight disease, including the skin and mucous membranes, stomach acid, gut bacteria, enzymes, and types of white blood cells.

NECROPSY. An examination of the internal organs of a dead animal to determine the apparent cause of death—an autopsy or a postmortem.

NECROSIS. Death of tissue.

NEEDLE TEETH. Eight small sharp teeth on the upper and lower corners of a baby pig's mouth.

NEONATE. A newborn.

NEPHRITIS. Inflammation of the nephrons of the kidneys.

NETWORKING. This refers to two or more people working together to achieve individual and group goals that would be more difficult to obtain alone. It's a banding together of small and independent pork producers to compete with the mega-pork producers.

NICK. The result of a certain mating which produces an animal of high order of excellence, sometimes from mediocre parents.

NITRATE/NITRITE. Nitrate refers to the chemical union of one nitrogen (N) and three oxygen (O) atoms, or NO_3, while nitrite refers to the chemical union of one nitrogen (N) and two oxygen (O) atoms, or NO_2. Of prime concern in foods are sodium nitrate, potassium nitrate, sodium nitrite, and potassium nitrite.

NITROGEN. A chemical element essential to life. Animals get it from protein feeds; plants get it from the soil; and some bacteria get it directly from the air.

NITROGEN BALANCE. The nitrogen in the feed intake minus the nitrogen in the feces, minus the nitrogen in the urine.

NITROGEN FIXATION. Conversion of free nitrogen of the atmosphere to organic nitrogen compounds by symbiotic or nonsymbiotic microbial activity.

NITROGEN-FREE EXTRACT (NFE). It consists principally of sugars, starches, pentoses, and nonnitrogenous organic acids. The percentage is determined by subtracting the sum of the percentages of moisture, crude protein, crude fat, crude fiber, and ash from 100.

NITROSAMINES. Nitrates and nitrites contribute to the formation of cancer causing nitrosamines. But foods are not the only source of nitrosamines.

NONPROTEIN NITROGEN (NPN). Nitrogen which comes from other than a protein source but may be used by a ruminant in the building of protein. NPN sources include compounds like urea and anhydrous ammonia, which are used in feed formulations for ruminants only.

NURSERY. The building to which pigs are usually moved at weaning. Normally, they remain in the nursery until they weigh 40 to 60 lb.

NUTRIENT ALLOWANCES. Nutrient recommendations that allow for variations in feed composition; possible losses during storage and processing; day-to-day and period-to-period differences in needs of animals; age and size of animal; stage of gestation and lactation; the kind and degree of activity; the amount of stress; the system of management; the health, condition, and temperament of the animal; and the kind, quality, and amount of feed—all of which exert a powerful influence in determining nutritive needs.

NUTRIENT REQUIREMENTS. This refers to meeting the animal's minimum needs, without margins of safety, for maintenance, growth, fitting, reproduction, lactation, and work. To meet these nutritive requirements, the different classes of animals must receive sufficient feed to furnish the necessary quantity of energy (carbohydrates and fats), proteins, minerals, and vitamins.

NUTRIENTS. The chemical substances found in feed materials that can be used, and are necessary, for the maintenance, production, and health of animals. The chief classes of nutrients are carbohydrates, fats, proteins, minerals, vitamins, and water.

NUTRITION. The science encompassing the sum total of processes that have as a terminal objective the provision of nutrients to the component cells of an animal.

NUTRITIVE RATIO (NR). The ratio of digestible protein to other digestible nutrients in a feedstuff or ration. (The NR of shelled corn is about 1:10.)

O

OFFAL. All organs or tissues removed from the carcass in slaughtering.

OIL. Although fats and oils have the same general structure and chemical properties, they have different physical characteristics. The melting points of oils are such that they are liquid at ordinary room temperatures.

OIL CROPS. Crops grown primarily for oil, including soybeans, cottonseed, peanuts, canola, flaxseed, sunflower seed, safflower, and castor bean.

OILING. The application of oil to the coat of an animal to soften the skin and hair and give the coat a desired gloss.

OPTIMUM TEMPERATURE. This is the temperature at which the animal responds most favorably, as determined by maximum production and feed efficiency.

OPTION. An option is a choice. It is the right, but not the obligation, to buy or sell something, at a specified price on or before a certain expiration date.

OSSIFICATION. The process of bone formation; the calcification of bone with advancing maturity.

OSTEITIS. Inflammation of a bone.

OSTEOMALACIA. A bone disease of adult animals caused by lack of vitamin D, inadequate intake of calcium or phosphorus, or an incorrect dietary ratio of calcium and phosphorus.

OSTEOPOROSIS. Abnormal porosity and fragility of bone as the result of (1) a calcium, phosphorus, and/or vitamin D deficiency, or (2) an incorrect ratio between the two minerals.

OUTCROSS. The introduction of genetic material from some outside and unrelated source, but of the same breed, into a herd which is more or less related.

OUTDOOR INTENSIVE SWINE PRODUCTION. A term in vogue in Britain in the 1990s. It refers to outdoor production which is modernized by incorporating some of the improvements of confinement and environmentally controlled systems.

OVERFEEDING. Excess feeding.

OVERFINISHING. Excess finishing or fatness—a wasteful practice.

OVER-THE-COUNTER DRUGS. These are drugs that may be purchased by the general public for application according to the label.

OXIDATION. The combination with oxygen, or the loss of a hydrogen, or the loss of an electron, all of which render an ion more electropositive. The animal combines carbon from feedstuffs with inhaled oxygen to produce carbon dioxide, energy (as ATP), water, and heat.

OXYTOCIN. The hormone that controls milk letdown.

P

PALATABILITY. The result of the following factors sensed by the animal in locating and consuming feed: appearance, odor, taste, texture, temperature, and, in some cases, auditory properties of the feed (like the sound of pigs eating corn). These factors are affected by the physical and chemical nature of the feed.

PANTOTHENIC ACID. One of the B vitamins. It is a constituent of coenzyme A, which plays an essential role in fat and cholesterol synthesis.

PARASITES. Organisms living in, on, or at the expense of another living organism.

PARTNERSHIP. An association of two or more persons who, as co-owners, operate the business.

PARTS PER BILLION (ppb). It equals micrograms per kilogram or microliter per liter.

PARTS PER MILLION (ppm). It equals milligrams per kilogram or milliliters per liter.

PARTURITION. The act of giving birth—farrowing.

PASSIVE IMMUNITY. Short-lived immunity an animal generates by drinking colostrum, or by receiving hyperimmune serum via injection or orally.

PATHOGENIC. Disease causing.

PEARLED. Dehulled grains which are reduced into smaller and smoother particles by machine brushing, or abrasion.

PEDIGREE. A written statement giving the record of an animal's ancestry.

PERFORMANCE TEST. The evaluation of an animal by its own performance.

PER OS. Oral administration (by the mouth).

PESTICIDE. Any substance that is used to control pests.

pH. A measure of the acidity or alkalinity of a solution. Values range from 0 (most acid) to 14 (most alkaline), with neutrality at pH 7.

PHASE FEEDING. Refers to changes in the animal's diet (1) to adjust for age and stage of production, (2) to adjust for season of the year and for temperature and climatic changes, (3) to account for differences in body weight and nutrient requirements of different strains of animals, or (4) to adjust one or more nutrients as other nutrients are changed for economic or availability reasons.

PHENOTYPE. The characteristics of an animal that can be seen and/or measured.

PHOTOSYNTHESIS. The process whereby green plants utilize the energy of the sun to build up complex organic molecules containing energy.

PHYSIOLOGICAL. Pertaining to the science which deals with the functions of living organisms or their parts.

PHYSIOLOGICAL FUEL VALUES. Units, expressed in calories, used in the United States to measure food energy in human nutrition. Similar to metabolizable energy.

PHYSIOLOGICAL SALINE. A salt solution (0.9% NaCl) having the same osmotic pressure as the blood plasma.

PHYTIN. A form of phosphorus which is poorly utilized.

PICNIC. A shoulder cut often cured and smoked like ham, but contains more internal fat and connective tissue.

PIG. A young swine, generally less than 120 lb and less than 4 months of age.

PIGGY. Refers to a sow that has the appearance of having recently suckled pigs or that is due to farrow soon.

PIGLET. A small pig.

PLANT PROTEINS. This group includes the common oilseed byproducts—soybean meal, cottonseed meal, linseed meal, peanut meal, safflower meal, sunflower seed meal, canola meal.

PNEUMONIA. Inflammation of the lungs.

POLLUTION. Anything that defiles, desecrates, or makes impure or unclean the surrounding environment.

POLYNEURITIS. Neuritis of several peripheral nerves at the same time, caused by metallic and other poisons, infectious disease, or vitamin deficiency. In people, alcoholism is also a major cause of polyneuritis.

POLYUNSATURATED FATTY ACIDS. Fatty acids having more than one double bond. Linoleic acid, which contains two double bonds, is the primary dietary essential fatty acid of humans.

PORCINE. Pertaining to swine.

PORCINE STRESS SYNDROME (PSS). Susceptibility to PSS is caused by a single autosomal recessive gene.

The disease is manifested only in pigs that are homozygous recessive for this gene, which means that they inherited from both the sire and the dam. It causes some pigs subjected to the stress of management or sudden environmental changes to react adversely and even succumb. Pigs with PSS are usually associated with low-quality or pale, soft, and exudative (PSE) pork.

PORCINE REPRODUCTIVE AND RESPIRATORY SYNDROME (PRRS). In 1987, this disease was first reported in North America, at which time it was known as "Mystery Disease." Europeans called it "Blue Ear Disease" because of the bluish coloration of the skin.

PORK. Meat from swine.

PORK CARCASS GRADE. A measure of the degree of excellence based chiefly on quality of lean, amount of backfat, and expected yield of trimmed major wholesale cuts.

PORKER. A young hog (pig).

POSTMORTEM INSPECTION. This is the inspection made at the time of slaughter.

POSTNATAL. Occurring after birth.

POSTPARTUM. Occurring after the birth of the offspring, when referring to the sow.

POT-BELLIED. Designating any individual that has developed an abnormally large abdomen.

PRECURSOR. A compound that can be used by the body to form another compound; for example, carotene is a precursor of vitamin A.

PREHENSION. The seizing (grasping) and conveying of feed to the mouth.

PREMIX. A uniform mixture of one or more microingredients and a carrier, used in the introduction of microingredients into a larger mixture.

PRENATAL. Before birth.

PREPOTENCY. The ability of an individual to transmit its own qualities to its offspring.

PRESCRIPTION DRUGS. These are drugs which are for use by, or on the order of, the veterinarian.

PRESERVATIVES. A number of materials which are available to incorporate into feeds, with claims made that they will improve the preservation of nutrients, nutritive value, and/or palatability of the feed.

PRIMAL CUTS. Ham, loin, Boston butt, picnic, and bacon.

PROBE. A device to measure backfat thickness in pigs.

PROBIOTICS. The term means "in favor of life." They have an opposite effect to antibiotics on the microorganisms of the digestive tract. They increase the population of the desirable microorganisms rather than kill or inhibit undesirable organisms.

PRODUCE. A female's offspring. The produce-of-dam commonly refers to two offspring of one dam.

PROGENY TESTING. An evaluation of an animal on the basis of the performance of its offspring.

PROLAPSE. Abnormal protrusion of a part or organ.

PROPRIETORSHIP (individual). This is a business which is owned and operated by one individual.

PROSTAGLANDINS. A large group of chemically related 20-carbon hydroxy fatty acids with variable physiological effects in the body.

PROTEIN. From the Greek, meaning "of first rank, importance." Complex organic compounds made up chiefly of amino acids present in characteristic proportions for each specific protein. Twenty amino acids generally occur in combinations to form an almost limitless number of proteins. Protein always contains carbon, hydrogen, oxygen, and nitrogen; and, in addition, it usually contains sulfur and frequently phosphorus. Crude protein is determined by finding the nitrogen content and multiplying the result by 6.25. The nitrogen content of proteins averages about 16% ($100 \div 16 = 6.25$). Proteins are essential in all plant and animal life as components of the active protoplasm of each living cell. Feed ingredients that contain more than 20% of their total weight in crude protein are generally classified as protein feeds.

PROTEIN SUPPLEMENTS. Products that contain more than 20% protein or protein equivalent.

PROUD FLESH. Excess flesh growing around a wound.

PROVITAMIN. The material from which an animal may produce vitamins; *e.g.*, carotene (provitamin A) in plants is converted to vitamin A in animals.

PROVITAMIN A. Carotene.

PROXIMATE ANALYSIS. A chemical scheme for evaluating feedstuffs, in which a feedstuff is partitioned into the six fractions: (1) moisture (water) or dry matter (DM); (2) total (crude) protein (CP or TP – N × 6.25); (3) ether extract (EE) or fat; (4) ash (mineral salts); (5) crude fiber (CF)—the incompletely digested carbohydrates; and (6) nitrogen-free extract (NFE)—the more readily digested carbohydrates (calculated rather than measured chemically).

PSE. Pale, soft, and exudative pork. It is related to the porcine stress syndrome (PSS).

PSS. Porcine stress syndrome. Symptoms are extreme muscling, nervousness, easily frightened, tail tremors, and skin blotching.

PUBERTY. The age at which the reproductive organs become functionally operative—sexual maturity.

PUREBRED. An animal of pure breeding, registered or eligible for registration in the herd book of the breed to which it belongs.

PURIFIED DIET. A mixture of the known essential dietary nutrients in a pure form that is fed to experimental (test) animals in nutrition studies.

PURULENT. Consisting of or forming pus.

PUS. A liquid inflammatory product consisting of leukocytes (white blood cells), lymph, bacteria, dead tissue cells. and the fluid derived from their disintegration.

Q

QUALITATIVE TRAITS. Traits in which there is a sharp distinction between phenotypes, usually involving only one or two pairs of genes.

QUALITY. A term used to denote the desirability and/or acceptance of an animal or feed product.

QUALITY OF PROTEIN. A term used to describe the amino acid balance of protein. A protein is said to be of good quality when it contains all the essential amino acids in the proper proportions and amounts needed by a specific animal; and it is said to be poor quality when it is deficient in either content or balance of essential amino acids.

QUANTITATIVE TRAITS. Traits in which there is no sharp distinction between phenotypes, usually involving several genes and the environment. These include such economic traits as gestation length, birth weight, weaning weight, rate and efficiency of gain, and carcass quality.

QUARANTINE.

■ Compulsory segregation of exposed susceptible animals for a period of time equal to the longest usual incubation period of the disease to which they have been exposed.

■ An enforced regulation for the exclusion or isolation of an animal to prevent the spread of an infectious disease.

R

RADIOACTIVE. Giving off atomic energy in the form of alpha, beta, or gamma rays.

RANCID. A term used to describe fats that have undergone partial decomposition.

RANTING. Characteristic behavior of an agitated boar—frothing, chomping, and nervousness.

RATE OF PASSAGE. The time taken by undigested residues from a given meal to reach the feces. (A stained undigestible material is commonly used to estimate rate of passage.)

RATION(S). The amount of feed supplied to an animal for a definite period, usually for a 24-hour period. However, by practical usage, the word *ration* implies the feed fed to an animal without limitation to the time in which it is consumed.

RATIOS. "Weight ratio," "gain ratio," and "conformation score ratio" are used to indicate the performance of an individual in relation to the average of all animals of the same group. Calculated as follows:

$$\frac{\text{Individual record}}{\text{Average of animals in group}} \times 100$$

It is a record or index of individual deviation from the group average expressed in terms of percentage. A ratio of 100 is average for a particular group. Thus, ratios above 100 indicate animals above average, whereas ratios below 100 indicate animals below average.

RED MEAT. Meat that is red when raw, due to the red coloration of myoglobin, the pigment of muscle. Red meats include beef, veal, pork, mutton, and lamb muscle tissue with attendant fat and bone.

REGISTERED. Designating purebred animals whose pedigrees are recorded in the breed registry.

REGURGITATION. The casting up (backward flow) of undigested food from the stomach to the mouth, as by ruminants.

REPLACEMENT. An animal selected to be kept for the breeding herd.

RIDGELING (rig). Any male animal whose testicles fail to descend into the scrotum—a cryptorchid.

RIGOR MORTIS. The stiffness of body muscles that is observed shortly after death.

RING. To place a small ring in the snout of a swine for the purpose of discouraging rooting.

ROASTING PIG (roaster). Fat, plump, suckling pigs weighing 30 to 60 lb dressed with head on and not split at the chest or between the hams.

ROUGHAGE. Feed consisting of bulky and coarse plants or plant parts, containing a high-fiber content and low total digestible nutrients, arbitrarily defined as feed with over 18% crude fiber. Roughage may be classed as either dry or green.

RUNT. A piglet of small size in relation to its litter mates

S

SACCHARIDES. Referring to sugars. The prefixes *mono-*, *di-*, *tri-*, and *poly-* denote the number of sugars contained in the saccharide.

SALIVA. A clear, somewhat viscid solution secreted by glands within the mouth. It may contain the enzymes salivary amylase and salivary maltase.

SALMONELLA. A pathogenic, diarrhea-producing organism, of which there are over 100 known strains, sometimes present in contaminated feeds.

SATIETY. Full satisfaction of desire; may refer to satisfaction of appetite.

SATURATED FAT. A completely hydrogenated fat—each carbon atom is associated with the maximum number of hydrogens; there are no double bonds.

SCOURING. One of the major problems facing livestock producers is scouring (diarrhea) in young animals. It may be due to feeding practices, management practices, environment, or disease.

SCREENED. A feedstuff that has been separated into various sized particles by passing over or through screens.

SECONDARY INFECTION. Infection following an infection already established by other pathogens.

SECTIONED AND FORMED HAM. Ham made from pieces of meat trimmed from the hind leg and after curing formed into a loaf, placed in a casing and cooked and smoked like normal ham.

SECULAR TRENDS. These are longtime trends that persist over a period of several cycles.

SEEDSTOCK SUPPLIERS. These suppliers sell boars and gilts of specialized bloodlines. These lines of breeding, usually called *hybrids*, originate from crossing two or more breeds, then applying some specialized selection programs.

SEGREGATED EARLY WEANING (SEW). This is an infectious disease control procedure the primary objective of which is to improve productivity of the growing/finishing phase by preventing the transfer of disease.

SELECTION. Determining which animals in a population will produce the next generation. Pork producers practice artificial selection while nature practices natural selection.

SELENIUM. An element that functions with glutathione peroxidase, an enzyme which enables the tripeptide glutathione- to perform its role as a biological antioxidant in the body. This explains why deficiencies of selenium and vitamin E result in similar signs—loss of appetite and slow growth.

SELF-FED. Provided with a part or all of the ration on a continuous basis, thereby permitting the animal to eat at will.

SELF-FEEDER. A feed container by means of which animals can eat at will. (See AD LIBITUM.)

SEMEN. The fluid containing the sperm that is ejaculated by the male.

SEPARATE SEX FEEDING. This refers to feeding barrows and gilts separately, due primarily to the higher protein requirement of gilts.

SERUM. The colorless fluid portion of blood remaining after clotting and removal of corpuscles. It differs from plasma in that the fibrinogen has been removed.

SERVICE. Denotes the mating of a female by a male.

SETTLED. Used to indicate that the animal has become pregnant.

SHOAT (shote). A young pig, of either sex, after weaning—synonymous with *pig*.

SHOW BOX (tackbox). A container in which to keep all show equipment and paraphernalia.

SHRINKAGE.

■ A term indicating the amount of loss in body weight when animals are exposed to adverse conditions, such as being transported, severe weather, or shortage of feed.

■ The loss in carcass weight during the aging process.

SIB. A brother or sister.

SIB TESTING. A method of selection in which an animal is selected on the basis of the performance of its brothers or sisters.

SIRE.

■ The male parent.

■ To father or beget.

SLAUGHTER HOG POOLING. This consists of a number of smaller hog producers joining together for the purpose of marketing hogs in truckload lots. Pooling is usually prompted because of lack of access to a nearby market.

SLOTTED FLOORS. Floors with slots through which the feces and urine pass to a storage area below or nearby.

SLOUGHING. A mass of dead tissue separating from a surface.

SOAP. A compound formed along with glycerol from the reaction of fat with alkali.

SOFT PORK. Feed fats are laid down in the body without undergoing much change. Thus, when finishing hogs are liberally fed on high-fat content feeds in which the fat is liquid at ordinary temperatures, soft pork results. This condition prevails when hogs are liberally fed such feeds as soybeans, peanuts, mast, or garbage.

SOLUBLES. Liquids containing dissolved substances obtained from processing animal or plant materials. They may contain some fine suspended solids.

SOLUTION. A uniform liquid mixture of two or more substances molecularly dispersed within one another.

SOW. A female swine that shows evidence of having produced pigs or that is in an evident state of pregnancy.

SOWBELLY. Salt pork; unsmoked fat bacon.

SOY PROTEIN CONCENTRATE. The protein produced by removing the water soluble sugars, ash, and other minor constituents from defatted soy flour. It contains 65 to 70% protein.

SOY PROTEIN ISOLATE. This is the highest protein source. It is produced by removal of the insoluble fibrous material.

SPECIFIC DYNAMIC ACTION (SDA). The increased production of heat by the body as a result of a stimulus to metabolic activity caused by ingesting food.

SPECIFIC GRAVITY. The ratio of the weight of a body to the weight of an equal volume of water.

$$\text{Specific gravity} = \frac{\text{Wt. of body in air}}{\text{Wt. of body in air} - \text{wt. in H}_2\text{0}}$$

SPECIFIC HEAT.

■ The heat-absorbing capacity of a substance in relation to that of water.

■ The heat expressed in calories required to raise the temperature of 1 g of a substance to 1°C.

SPECIFIC PATHOGEN-FREE (SPF). Pigs that are free of disease at birth.

SPECULATING. Risk-taking by anyone who hopes to make a profit in the advances or declines in the price of the futures contract.

SPLIT-SEX FEEDING. This refers to sorting gilts from barrows and feeding each group separate diets.

SPRAY-DRIED BLOOD MEAL. This product contains both the plasma and red-blood cell fractions of blood.

SPRAY-DRIED PORCINE PLASMA. This product is made up of the albumin, globin, and globulin fractions of blood. It contains 68% protein and 6.1% lysine.

STABILIZED. Made more resistant to chemical change by the addition of a particular substance.

STAG. A male that was castrated after the secondary sexual characteristics developed sufficiently to give the appearance of a mature male.

STAGES. STAGES stands for Swine Testing and Genetic Evaluation System. It is a system of the breed registries for recording performance data provided by swine producers.

STANDING HEAT. Period in a sow or gilt's heat (estrus) during which she will stand still when being mounted or when pressure is applied to her back.

STERILE. Incapable of reproducing.

STILLBORN. Born lifeless; dead at birth.

STRAW. The plant residue remaining after separation of the seeds in threshing. It includes chaff.

STRESS. Any physical or emotional factor to which an animal fails to make a satisfactory adaptation. Stress may be caused by excitement, temperament, fatigue, shipping, disease, heat or cold, nervous strain, number of animals together, previous nutrition, breed, age, or management. The greater the stress, the more exacting the nutritive requirements.

SUBCUTANEOUS. Situated or occurring beneath the skin.

SUCKLE. To nurse at the breast or mammary glands.

SUGAR. A sweet, crystallizable substance that consists essentially of sucrose, and that occurs naturally in the most readily available amounts in sugarcane, sugar beet, sugar maple, sorghum, and sugar palm.

SUPPLEMENT. A feed or feed mixture used to improve the nutritional value of basal feeds (*e.g.*, protein supplement—soybean meal). Supplements are usually rich in protein, minerals, vitamins, antibiotics, or a combination of part or all of these; and they are usually combined with basal feeds to produce a complete feed.

SUPPURATION. Formation of pus.

SUSTAINABLE AGRICULTURE. This refers to farming with reduced off-farm purchased inputs of pesticides, herbicides, and fertilizers, along with reduced negative impact on natural resources and improved environmental quality and economic efficiency, while producing and distributing food and fiber.

SWINE. Collective term for all age groups within the species.

SYMMETRY. A balanced development of all parts.

SYNTHESIS. The bringing together of two or more substances to form a new material.

SYNTHETICS. Artificially produced products that may be similar to natural products.

T

TAIL BITING. An abnormal behavior, characterized by one pig biting the tail of another.

TANKAGE. A protein supplement consisting of ground meat byproducts of animals that have been slaughtered.

TATTOO. Permanent identification of animals produced by placing indelible ink under the skin; generally put in the ears of young animals.

TDN. (See TOTAL DIGESTIBLE NUTRIENT.)

TEART. Molybdenosis of farm animals caused by feeding on vegetation grown on soil that contains high levels of molybdenum.

TETANY. A condition in an animal in which there are localized, spasmodic muscular contractions.

TETHER. To tie an animal with a rope or a chain to allow feeding but to prevent straying.

THERMAL. Refers to heat.

THERMOGENESIS. The chemical production of heat in the body.

THERMONEUTRALITY. The state of thermal (heat) balance between an animal and its environment. The thermoneutral zone is referred to as the comfort zone.

THRIFTY. Healthy and vigorous in appearance

THUMPS (thumping). Jerky breathing; a respiratory disturbance in which the pigs breathe with difficulty, rapidly, and spasmodically.

TOCOPHEROL. Any of four different forms of an alcohol also known as vitamin E.

TONIC. A drug, medicine, or feed designed to stimulate the appetite.

TOTAL DIGESTIBLE NUTRIENT (TDN). A term which indicates the energy value of a feedstuff. It is computed by use of the following formula:

$$\% \text{ TDN} = \frac{\text{DCP} + \text{DCF} + \text{DNFE} + (\text{DEE} \times 2.25)}{\text{feed consumed}} \times 100$$

Where DCP = digestible crude protein; DCF = digestible crude fiber; DNFE = digestible nitrogen-free extract; and DEE = digestible ether extract. One lb of TDN = 2,000 kcal of digestible energy.

TOXIC. Of a poisonous nature.

TOXOID. A killed vaccine produced from toxins.

TRACEBACK. This involves an animal identification system which makes it possible to trace animals back to their origin in order to locate the source of contamination.

TRACE ELEMENT. A chemical element used in minute amounts by organisms and held essential to their physiology. The essential trace elements are cobalt, copper, iodine, iron, manganese, selenium, and zinc.

TRACE MINERAL. A mineral nutrient required by animals in micro amounts only (measurable in milligrams per pound or smaller units).

TUBER. A short, thickened, fleshy stem or terminal portion of a stem or rhizome that is usually formed underground, bears minute scale leaves each with a bud capable under suitable conditions of developing into a new plant, and constitutes the resting stage of various plants such as the potato and the Jerusalem artichoke.

TURN-AROUND, COMFORT STALLS. (See FREEDOM STALLS.)

TUSK. An elongated or greatly enlarged tooth.

TWENTY-EIGHT HOUR LAW IN RAIL SHIPMENTS. This law prohibits transporting livestock by rail for a longer period than 28 consecutive hours without unloading, feeding, watering, and resting 5 consecutive hours before resuming transportation. On request of the owner, the period can be extended to 36 hours.

TYPE.
- Physical conformation of an animal.
- All those physical attributes that contribute to the value of an animal for a specific purpose.

U

UDDER. The encased group of mammary glands with each gland provided with a nipple or teat.

ULTRASONICS. This is electronic equipment which employs "pulse echo" technique. Reflected sound waves are used to measure depth of backfat and muscle.

UNDERFEEDING. Usually, this refers to not providing sufficient energy. The degree of lowered production therefrom is related to the extent of underfeeding and the length of time it exists.

UNIDENTIFIED FACTORS. These are referred to as *unidentified* or *unknown* factors because they have not yet been isolated or synthesized in the laboratory. There is evidence that the growth factors exist in dried whey, marine and packinghouse byproducts, distillers' solubles, antibiotic fermentation residues, alfalfa meal, and certain green forages. Most of the unidentified factor sources are added to the diet at a level of 1 to 3%.

UNSATURATED FAT. A fat having one or more double bonds; not completely hydrogenated.

UNSATURATED FATTY ACID. Any one of several fatty acids containing one or more double bonds, such as oleic, linoleic, linolenic, and arachidonic acids.

UNTHRIFTINESS. Lack of vigor, poor growth or development; the quality or state of being unthrifty in animals.

USP (United States Pharmacopoeia). A unit of measurement or potency of biologicals that usually coincides with an international unit. (Also see IU.)

V

VACCINATION (shot). An injection of vaccine, bacterin, antiserum, or antitoxin to produce immunity or tolerance to disease.

VACCINE. A suspension of attenuated or killed microorganisms (bacteria, viruses, or rickettsiae) administered for the prevention, improvement, or treatment of infectious diseases.

VACUUM-PACKED FRESH PORK. Often large boneless cuts of pork, referred to as subprimals, with a shelf life of about 21 days since the oxygen has been excluded. In the package the pork's color is more of a dull purple, but when the package is opened the pink returns.

VARIETY MEATS. Liver, brains, heart, kidney—all excellent sources of many essential nutrients.

VECTORS. Living organisms which carry pathogens.

VERMIFUGE (vermicide). Any chemical substance given to animals to kill internal parasitic worms.

VERTICAL INTEGRATION. In pork, this refers to a structure in which through alliances (complete ownership, joint ownership, or contract), an individual or company controls the product through two or more phases; for example, a meat packer owns hogs from birth through slaughter.

VIETNAMESE POT-BELLIED PIGS. Native Asian pigs, which are normally less than one-fifth the size of traditional U.S. hogs. They have a pot-belly, a swayed back, and a straight tail which they wag when they are happy.

VIRUS. One of a group of minute infectious agents. They lack independent metabolism and can only multiply within living host cells.

VISCERA. Internal organs of the body, particularly in the chest and abdominal cavities.

VITAMIN PRODUCT LABELS. When a product is marketed as a vitamin supplement *per se*, the quantitative guarantees (unit/lb) of vitamins A and D are expressed in USP units; of E in IU; and of other vitamins in milligrams per pound.

VITAMINS. Complex organic compounds that function as parts of enzyme systems essential for the transformation of energy and the regulation of metabolism of the body, and required in minute amounts by one or more animal species for normal growth, production, reproduction, and/or health. All vitamins must be present in the ration for normal functioning, except for B vitamins in the ruminants (cattle and sheep) and vitamin C.

VITAMIN SUPPLEMENTS. Rich synthetic or natural feed sources of one of more of the complex organic compounds, called vitamins, that are required in minute amounts by animals for normal growth, production, reproduction, and/or health.

VOID. To evacuate feces and/or urine.

VOMITING. The forcible expulsion of the contents of the stomach through the mouth.

W

WASTY.
- A carcass with too much fat, requiring excessive trimming.
- Paunchy live animal.

WAXY CORN. The name given to a variety of corn which is sometimes grown for industrial use because of its special type of starch.

WEANER. A pig that has reached the age that it can be weaned or that has been weaned.

WEANING. The stopping of young animals from suckling their mothers.

WHEAT GLUTEN. This is spray-dried protein fraction of wheat remaining after the starch has been extracted for use in human food products.

WHITE MEAT.
- The breast of broilers or turkeys.
- Most cooked pork—"the other white meat."

WILTSHIRE SIDE. The entire half of a dressed pig, minus the head, shank, shoulder bone, and hip bone. All of the side, except the ham and shoulder, is sold as bacon.

WITHDRAWAL PERIOD. The withholding of certain feed additives prior to slaughter in order to insure that drug residues do not occur in carcasses.

WORKERS' COMPENSATION. This is a program designed to provide employees assured payment for medical expenses or lost income due to injury on the job.

Z

ZONING. Ordinances governing the keeping of animals or the type of businesses or residences.

Philippine salt pork container. Historically, pork has been the favored meat in the Philippines. Originally, scavengers, today pigs in the Philippines are produced primarily on home-produced corn and rice byproducts. (Courtesy, Field Museum of Natural History, Chicago, IL)

21

FEED COMPOSITION TABLES

Both nutritionists and swine producers should have access to accurate and up-to-date composition of feedstuffs in order to formulate rations for maximum production and net returns. The ultimate goal of feedstuff analysis, and the reason for feed composition tables, is to be able to predict the productive response of animals when they are fed rations of a given composition. In recognition of this need and its importance, the authors spared no time or expense in compiling the feed composition tables presented in this section. At the outset, a survey of the industry was made in order to determine what kind of feed composition tables would be most useful, in both format and content. Secondly, it was decided to utilize, to the extent available, the monumental work of Lorin Harris,[1] along with the feed composition tables of the National Academy of Sciences. Additional feeds and compositions were provided by the authors with compositions obtained from experimental reports, industries, and other reliable sources.

FEED NAMES

Ideally, a feed name should conjure up the same meaning to all those who use it, and it

---

[1]Professor Emeritus of Nutrition, Department of Animal, Dairy, and Veterinary Sciences, International Feedstuffs Institute, Utah State University, Logan, Utah

should provide helpful information. This was the guiding philosophy of the authors when choosing the names given in the Feed Composition Tables. Genus and species—Latin names—are also included. To facilitate worldwide usage, the International Feed Number of each feed is given. To the extent possible, consideration was also given to source (or parent material), variety or kind, stage of maturity, processing, part eaten, and grade.

MOISTURE CONTENT OF FEEDS

It is necessary to know the moisture content of feeds in ration formulation and buying. Usually, the composition of a feed is expressed according to one or more of the following bases:

1. **As-fed; A-F (wet, fresh).** This refers to feed as normally fed to animals. It may range from 0 to 100% dry matter.

2. **Air-dry (approximately 90% dry matter).** This refers to feed that is dried by means of natural air movement, usually in the open. It may be either an actual or an assumed dry matter content; the latter is approximately 90%. Most feeds are fed in an air-dry state.

3. **Moisture-free; M-F (oven-dry, 100% dry matter).** This refers to a sample of feed that has been dried in an oven at 221°F (105°C) until all the moisture has been removed.

Where available, feed compositions are presented on both **as-fed (A-F)** and **moisture-free (M-F)** bases.

PERTINENT INFORMATION ABOUT DATA

The information which follows is pertinent to the feed composition tables presented in this chapter.

■ **Variations in composition**—Feeds vary in their composition. Thus, actual analysis of a feedstuff should be obtained and used wherever possible, especially where a large lot of feed from one source is involved. Many times, however, either it is impossible to determine actual compositions or there is insufficient time to obtain such analysis. Under such circumstances, tabulated data may be the only information available.

■ **Feed compositions change**—Feed compositions change over a period of time, primarily due to (1) the introduction of new varieties, and (2) modifications in the manufacturing process from which byproducts evolve.

■ **Available nutrients**—The response of animals when fed a feed is termed the available nutrients,

which is a function of its chemical composition and the ability of the animal to derive useful nutrient value from the feed. The latter relates to the digestibility, or availability, of the nutrients in the feed. Thus, soft coal and shelled corn may have the same gross energy value in a bomb calorimeter but markedly different useful energy values when consumed by an animal. Biological tests of feeds are more laborious and costly to determine than chemical analysis, but they are much more accurate in predicting the response of animals to a feed.

■ **Where information is not available**—Where information is not available or reasonable estimates could not be made, no values are shown. Hopefully, such information will become available in the future.

■ **Calculated on a dry matter (DM) basis**—All data were calculated on a 100% dry matter basis (moisture-free), then converted to an as-fed basis by multiplying the decimal equivalent of the DM content times the compositional value shown in the table.

■ **Fiber**—Four values for fiber are given in Table 21-1—crude fiber, cell walls or NDF, acid detergent fiber, and lignin.

Crude fiber, methods for the determination of which were developed more than 100 years ago, is declining as a measure of low digestible material in the more fibrous feeds. The newer method of forage analysis, developed by Van Soest and associates of the U.S. Department of Agriculture, separates feed dry matter into two fractions—one of high digestibility (cell contents) and the other of low digestibility (cell walls)—by boiling a 0.5 to 1.0 g sample of the feed in a neutral detergent solution (3% sodium lauryl sulfate buffered to a pH of 7.0) for 1 hour, then filtering. Also, the amount of lignin in the cell wall is determined.

1. **Crude fiber (CF).** This is the residue that remains after boiling a feed in a weak acid, and then in a weak alkali, in an attempt to imitate the process that occurs in the digestive tract. This procedure is based on the supposition that carbohydrates which are readily dissolved also will be readily digested by animals, and that those not soluble under such conditions are not readily digested. Unfortunately, the treatment dissolves much of the lignin, a nondigestible component. Hence, crude fiber is only an approximation of the indigestible material in feedstuffs. Nevertheless, it is a rough indicator of the energy value of feeds. Also, the crude fiber value is needed for the computation of TDN.

2. **Cell wall (CW) or neutral detergent fiber (NDF).** This is the insoluble fraction resulting from boiling a feed sample in a neutral detergent solution. It contains cellulose, hemicellulose, silica, some protein, and lignin. Cell wall, or NDF, components are of low digestibility and entirely dependent on the micro-

organisms of the digestive tract for any digestion that they undergo; hence, they are essentially undigested by nonruminants. This fraction of a forage affects the volume it will occupy in the digestive tract, a principal factor limiting the amount of feed consumed. Animals fed such forages are often unable to consume enough feed to produce weight gains or milk economically.

The soluble fraction—the cell contents—consists of sugars, starch, fructosans, pectin, protein, nonprotein nitrogen, lipids, water, soluble minerals, and vitamins. This portion is highly digestible (about 98%) by both ruminants and nonruminants.

3. **Acid detergent fiber (ADF).** This involves boiling a 1.0 g sample of air-dry material in a specially prepared acid detergent solution for 1 hour, then filtering. The insolubles, or residue, make up what is known as acid detergent fiber (ADF) and consists primarily of cellulose, lignin, and variable amounts of silica.

Acid detergent fiber differs from neutral detergent fiber in that NDF contains most of the feed hemicellulose and a limited amount of protein, not present in ADF.

ADF is the best predictor of forage digestible dry matter and digestible energy.

4. **Lignin.** This fraction is essentially indigestible by all animals and is the substance that limits the availability of cellulose carbohydrates in the plant cell wall to rumen bacteria.

The acid detergent fiber procedure is used as a preparatory step in determining the lignin of a forage sample. Hemicellulose is solubilized during this procedure, while the lignocellulose fraction of the feed remains insoluble. Cellulose is then separated from lignin by the addition of sulfuric acid. Only lignin and acid-insoluble ash remain upon completion of this step. This residue is then ashed, and the difference of the weights before and after ashing yields the amount of lignin present in the feed.

■ **Nitrogen-free extract**—The nitrogen-free extract was calculated with mean data as: mean nitrogen-free extract (%) = 100 − % ash − % crude fiber − % ether extract − % protein.

■ **Protein values**—Crude protein values are given. Crude protein represents Kjeldahl nitrogen value times 100/16, or 6.25, since protein contains 16% nitrogen on the average.

■ **Energy**—Metabolizable energy, which is considered to be the most accurate evaluation of the energy of feedstuffs for the scientific formulation of poultry feeds, is given.

Metabolizable energy represents that portion of gross energy not lost in feces, urine, and gas (mainly methane). It does not take into account the energy lost as heat, commonly called heat increment.

■ **Minerals**—The level of minerals in forages is largely determined by the mineral content of the soil on which the feeds are grown. Calcium, phosphorus, iodine, and selenium are well-known examples of soil nutrient–plant nutrient relationships.

■ **Carotene**—Where carotene has been converted to vitamin A, the conversion rate of the rat has been used as the standard value, with 1 mg of β-carotene equal to 1,667 IU of vitamin A. Generally speaking, it is unwise to rely on harvested feeds as a source of carotene (vitamin A value), unless the forage being fed is fresh (pasture or green chop) or of a good green color and not over a year old.

TABLE 21-1, COMPOSITION OF FEEDS, DATA EXPRESSED AS-FED AND MOISTURE-FREE

In this table, the commonly used swine feeds are listed alphabetically, and, on both an as-fed and a moisture-free basis, their chemical analysis, TDN, digestible energy, metabolizable energy, mineral and vitamin compositions are given. Note that four pages (two double-page spreads) are devoted to each feed, with presentations as follows:

First page (left-hand): chemical analysis—ash, fiber, fat, nitrogen (N)-free extract, and crude protein.

Second page (right-hand): TDN, digestible energy, metabolizable energy, and macrominerals.

Third page (left-hand): microminerals and fat-soluble vitamins.

Fourth page (right-hand): fat-soluble vitamins (continued) and water-soluble vitamins.

TABLE 21-2, AMINO ACID COMPOSITION OF FEEDS, DATA EXPRESSED AS-FED AND MOISTURE-FREE

This table gives, in alphabetical order, the known amino acid composition of selected feeds.

TABLE 21-3, MINERAL SUPPLEMENTS, COMPOSITION, DATA EXPRESSED AS-FED AND MOISTURE-FREE

This table provides the calcium, phosphorus, sodium, chlorine, magnesium, potassium, sulfur, cobalt, copper, iodine, iron, manganese, selenium, and zinc content of some commonly used mineral sources.

TABLE
COMPOSITION OF FEEDS, DATA

C O M P O S I T I O N   O F   F E E D S

Entry Number	Feed Name Description	International Feed Number[1]	Moisture Basis: A-F (as-fed) or M-F (moisture-free)	Dry Matter (%)	Ash (%)	Crude Fiber (%)	Cell Walls or NDF (%)	Acid Detergent Fiber (%)	Lignin (%)	Ether Extract (Fat) (%)	N-Free Extract (%)	Crude Protein (%)
	ALFALFA (LUCERNE) *Medicago sativa*											
1	-PREBLOOM	2-00-181	A-F	21	2.1	5.3	—	—	—	0.6	8.8	4.5
			M-F	100	9.9	24.9	—	—	—	2.8	41.2	21.2
2	-EARLY BLOOM	2-00-184	A-F	24	2.3	6.7	—	—	—	0.7	9.5	4.6
			M-F	100	9.5	28.0	—	—	—	3.1	40.0	19.4
3	-HAY, SUN-CURED, ALL ANALYSES	1-00-078	A-F	90	8.1	27.2	42.9	32.6	6.3	2.3	36.9	15.9
			M-F	100	9.0	30.1	47.5	36.1	7.0	2.6	40.8	17.6
4	-MEAL, DEHY, 17% PROTEIN	1-00-023	A-F	92	9.7	24.4	—	31.0	—	2.7	38.0	17.5
			M-F	100	10.5	26.4	—	33.7	—	2.9	41.2	18.9
5	-MEAL, DEHY, 20% PROTEIN	1-00-024	A-F	91	10.2	20.4	—	26.9	—	3.2	37.4	20.1
			M-F	100	11.2	22.4	—	29.6	—	3.5	41.1	22.1
	ANIMAL											
6	-BLOOD MEAL, SPRAY, DEHY (BLOOD FLOUR)	5-00-381	A-F	86	7.2	1.4	—	—	—	6.2	−1.3	72.6
			M-F	100	8.4	1.6	—	—	—	7.2	−1.5	84.4
7	-BLOOD, MEAL	5-00-380	A-F	92	5.3	2.4	—	—	—	1.3	3.8	79.1
			M-F	100	5.8	2.7	—	—	—	1.4	4.1	86.0
8	-FAT	4-00-376	A-F	99	—	—	—	—	—	99.0	—	—
			M-F	100	—	—	—	—	—	100.0	—	—
9	-LIVERS, MEAL, DEHY	5-00-389	A-F	93	6.3	1.4	—	—	—	15.8	2.9	66.7
			M-F	100	6.8	1.5	—	—	—	17.0	3.1	71.7
10	-MEAT, MEAL, RENDERED	5-00-385	A-F	93	24.8	2.4	—	—	—	9.0	3.0	53.8
			M-F	100	26.6	2.6	—	—	—	9.7	3.3	57.9
11	-MEAT WITH BLOOD, MEAL, TANKAGE RENDERED	5-00-386	A-F	92	21.8	2.2	—	—	—	9.0	−0.3	59.5
			M-F	100	23.6	2.4	—	—	—	9.7	−0.4	64.7
12	-MEAT WITH BLOOD WITH BONE, MEAL, TANKAGE RENDERED	5-00-387	A-F	92	27.8	2.4	—	—	—	10.5	1.0	50.2
			M-F	100	30.2	2.6	—	—	—	11.4	1.1	54.6
13	-MEAT WITH BONE, MEAL, RENDERED	5-00-388	A-F	93	28.4	2.0	—	—	—	9.9	2.2	50.5
			M-F	100	30.5	2.2	—	—	—	10.6	2.4	54.3
14	-TALLOW	4-08-127	A-F	99	0.1	—	—	—	—	96.9	—	1.6
			M-F	100	0.1	—	—	—	—	97.9	—	1.6
	ANIMAL—POULTRY											
15	-FAT	4-00-409	A-F	19	—	—	—	—	—	98.9	—	—
			M-F	100	—	—	—	—	—	99.9	—	—
	BAKERY WASTE, DEHY											
16	-DRIED BAKERY PRODUCT	4-00-466	A-F	92	4.0	1.5	—	—	—	10.4	65.9	10.2
			M-F	100	4.3	1.6	—	—	—	11.4	71.6	11.1
	BARLEY *Hordeum vulgare*											
17	-GRAIN, ALL ANALYSES	4-00-549	A-F	88	2.3	5.0	—	—	—	1.9	66.6	12.2
			M-F	100	2.6	5.6	—	—	—	2.2	75.7	13.9
18	-GRAIN, GRADE 1, 48 LB/BUSHEL OR *618 G/L*	4-00-535	A-F	88	2.6	5.0	—	—	—	1.8	67.2	11.4
			M-F	100	2.9	5.7	—	—	—	2.1	76.4	13.0
19	-GRAIN, LIGHT, LESS THAN 36 LB/BUSHEL OR *463 G/L*	4-00-566	A-F	88	3.2	10.0	—	—	—	2.0	62.5	10.3
			M-F	100	3.7	11.4	—	—	—	2.3	70.9	11.7
20	-GRAIN, PACIFIC COAST	4-07-939	A-F	90	2.8	6.3	—	—	—	1.7	69.6	9.6
			M-F	100	3.1	7.0	—	—	—	1.9	77.3	10.7
21	-GRAIN SCREENINGS	4-00-542	A-F	89	3.1	8.7	—	—	—	2.3	63.2	11.7
			M-F	100	3.4	9.8	—	—	—	2.6	71.1	13.1
22	-MALT SPROUTS, DEHY	5-00-545	A-F	92	6.4	14.5	—	—	—	1.3	43.7	26.1
			M-F	100	7.0	15.7	—	—	—	1.5	47.5	28.4
	BEAN *Phaseolus vulgaris*											
23	-SEEDS, KIDNEY	5-00-600	A-F	89	3.7	4.2	—	—	—	1.3	57.8	22.0
			M-F	100	4.2	4.7	—	—	—	1.5	64.9	24.7
24	-SEEDS, NAVY	5-00-623	A-F	90	4.7	4.4	—	—	—	1.4	56.7	22.8
			M-F	100	5.2	4.9	—	—	—	1.5	63.0	25.3
25	-SEEDS, PINTO	5-00-624	A-F	90	4.3	4.0	—	—	—	1.3	57.7	22.6
			M-F	100	4.8	4.5	—	—	—	1.4	64.2	25.1

21-1
EXPRESSED AS-FED AND MOISTURE-FREE

Entry Number	TDN	Digestible Energy		Metabolizable Energy		Macrominerals						
						Calcium (Ca)	Phosphorus (P)	Sodium (Na)	Chlorine (Cl)	Magnesium (Mg)	Potassium (K)	Sulfur (S)
	(%)	(kcal)		(kcal)		(%)	(%)	(%)	(%)	(%)	(%)	(%)
		(lb)	(kg)	(lb)	(kg)							
1	12	249	548	228	502	0.48	0.08	0.04	0.08	0.05	0.50	0.13
	58	1,164	2,566	1,067	2,353	2.26	0.35	0.20	0.35	0.25	2.36	0.60
2	—	—	—	—	—	0.55	0.07	—	—	0.01	0.46	—
	—	—	—	—	—	2.33	0.31	—	—	0.03	1.92	—
3	32	645	1,421	523	1,153	1.38	0.20	0.14	0.31	0.30	1.98	0.25
	36	714	1,573	579	1,276	1.53	0.22	0.15	0.34	0.33	2.19	0.28
4	45	891	1,964	540	1,190	1.33	0.24	0.12	0.48	0.29	2.46	0.24
	48	966	2,130	586	1,291	1.44	0.26	0.13	0.52	0.32	2.66	0.26
5	45	892	1,967	761	1,678	1.56	0.28	0.14	0.47	0.33	2.50	0.50
	49	980	2,162	836	1,844	1.70	0.30	0.15	0.51	0.36	2.73	0.55
6	—	—	—	—	—	1.65	0.45	0.31	0.23	0.04	—	0.56
	—	—	—	—	—	1.92	0.52	0.36	0.27	0.04	—	0.65
7	62	1,225	2,701	1,101	2,426	0.27	0.26	0.32	0.32	0.22	0.14	0.34
	68	1,332	2,936	1,196	2,637	0.30	0.29	0.35	0.35	0.24	0.16	0.37
8	198	3,992	8,800	3,583	7,900	—	—	—	—	—	—	—
	200	4,031	8,888	3,620	7,980	—	—	—	—	—	—	—
9	—	—	—	—	—	0.56	1.27	—	—	—	—	—
	—	—	—	—	—	0.61	1.36	—	—	—	—	—
10	63	—	2,046	—	2,379	7.96	4.00	1.31	1.20	0.27	0.57	0.50
	68	—	2,200	—	2,558	8.56	4.30	1.41	1.29	0.29	0.62	0.53
11	68	1,046	2,305	1,020	2,248	5.80	2.99	1.68	1.73	0.34	0.57	0.70
	74	1,137	2,506	1,108	2,443	6.31	3.25	1.82	1.88	0.36	0.62	0.76
12	67	1,344	2,962	1,188	2,618	6.97	4.59	1.71	—	—	0.57	0.26
	73	1,461	3,220	1,291	2,846	7.58	4.99	1.86	—	—	0.62	0.28
13	68	892	1,966	934	2,059	10.16	4.89	0.73	0.74	1.13	1.28	0.26
	73	959	2,114	1,004	2,214	10.92	5.26	0.78	0.80	1.22	1.38	0.28
14	—	—	—	—	—	—	—	—	—	—	—	—
	—	—	—	—	—	—	—	—	—	—	—	—
15	83	3,661	8,071	3,708	8,174	—	—	—	—	—	—	—
	85	3,698	8,153	3,745	8,257	—	—	—	—	—	—	—
16	91	1,871	4,126	1,729	3,813	0.14	0.32	0.98	0.90	0.32	0.44	0.02
	98	2,034	4,484	1,880	4,144	0.16	0.35	1.07	0.98	0.35	0.41	0.02
17	70	1,384	3,052	1,244	2,743	0.04	0.33	0.03	0.18	0.14	0.40	0.15
	80	1,573	3,468	1,414	3,117	0.05	0.37	0.03	0.20	0.15	0.45	0.18
18	73	1,469	3,238	1,372	3,024	0.24	0.36	—	—	—	—	—
	83	1,669	3,680	1,559	3,436	0.27	0.41	—	—	—	—	—
19	62	1,249	2,747	1,165	2,563	—	—	—	—	—	—	—
	71	1,419	3,122	1,324	2,912	—	—	—	—	—	—	—
20	72	1,432	3,156	1,169	2,577	0.05	0.34	0.02	0.15	0.12	0.53	0.15
	80	1,591	3,507	1,299	2,863	0.05	0.38	0.02	0.17	0.13	0.58	0.17
21	70	1,400	3,087	816	1,799	0.23	0.29	—	—	—	1.23	—
	79	1,573	3,468	917	2,021	0.26	0.33	—	—	—	1.38	—
22	35	698	1,538	645	1,422	0.21	0.72	1.35	0.36	0.18	0.21	0.79
	38	758	1,672	701	1,546	0.23	0.79	1.46	0.39	0.20	0.23	0.86
23	—	—	—	—	—	0.11	0.40	0.01	—	—	0.98	—
	—	—	—	—	—	0.12	0.45	0.01	—	—	1.10	—
24	—	—	—	600	1,323	0.13	0.52	0.05	0.04	0.17	1.26	0.23
	—	—	—	667	1,470	0.15	0.58	0.06	0.04	0.19	1.40	0.26
25	—	—	—	—	—	0.13	0.46	—	—	—	—	—
	—	—	—	—	—	0.14	0.51	—	—	—	—	—

(Continued)

TABLE 21-1

Vertical left margin: COMPOSITION OF FEEDS

Entry Number	Feed Name Description	Moisture Basis: A-F (as-fed) or M-F (moisture-free)	Microminerals							Fat-Soluble Vitamins	
			Cobalt (Co)	Copper (Cu)	Iodine (I)	Iron (Fe)	Man-ganese (Mn)	Sele-nium (Se)	Zinc (Zn)	A (1 mg Carotene = 1667 IU Vit. A)	Carotene (Provitamin A)
			(ppm or mg/kg)	(ppm or mg/kg)	(ppm or mg/kg)	(%)	(ppm or mg/kg)	(ppm or mg/kg)	(ppm or mg/kg)	(IU/g)	(ppm or mg/kg)
1	ALFALFA (LUCERNE) *Medicago sativa* -PREBLOOM	A-F	—	—	—	—	5.9	—	—	—	—
		M-F	—	—	—	—	27.7	—	—	—	—
2	-EARLY BLOOM	A-F	—	—	—	—	—	—	—	69.4	41.6
		M-F	—	—	—	—	—	—	—	291.1	174.6
3	-HAY, SUN-CURED, ALL ANALYSES	A-F	0.109	12.6	—	0.018	27.3	—	15.1	68.4	41.1
		M-F	0.120	13.9	—	0.020	30.2	—	16.7	75.8	45.4
4	MEAL, DEHY, 17% PROTEIN	A-F	0.326	9.0	0.149	0.042	30.9	0.336	19.5	201.5	120.9
		M-F	0.353	9.8	0.161	0.046	33.5	0.365	21.1	218.5	131.1
5	-MEAL, DEHY, 20% PROTEIN	A-F	0.258	12.4	0.134	0.038	45.9	0.284	20.1	264.8	158.9
		M-F	0.283	13.6	0.147	0.042	50.2	0.311	22.0	289.7	173.8
6	ANIMAL -BLOOD, MEAL, SPRAY, DEHY (BLOOD FLOUR)	A-F	—	7.6	—	0.257	6.0	—	—	—	—
		M-F	—	8.8	—	0.299	6.9	—	—	—	—
7	-BLOOD, MEAL	A-F	0.089	13.3	—	0.307	5.3	—	4.0	—	—
		M-F	0.096	14.4	—	0.334	5.8	—	4.0	—	—
8	-FAT	A-F	—	—	—	—	—	—	—	—	—
		M-F	—	—	—	—	—	—	—	—	—
9	-LIVERS, MEAL, DEHY	A-F	0.135	89.8	—	0.063	8.9	—	—	—	—
		M-F	0.145	96.5	—	0.068	9.5	—	—	—	—
10	-MEAT, MEAL, RENDERED	A-F	0.129	9.8	—	0.044	9.6	0.420	104.0	—	—
		M-F	0.139	10.5	—	0.047	10.3	0.452	112.0	—	—
11	-MEAT WITH BLOOD, MEAL, TANKAGE RENDERED	A-F	0.154	38.7	—	0.210	19.1	—	—	—	—
		M-F	0.167	42.1	—	0.228	20.8	—	—	—	—
12	-MEAT WITH BLOOD, WITH BONE, MEAL, TANKAGE RENDERED	A-F	0.180	39.7	—	—	19.6	0.258	—	—	—
		M-F	0.196	43.1	—	—	21.3	0.281	—	—	—
13	-MEAT WITH BONE, MEAL, RENDERED	A-F	0.180	1.5	1.313	0.050	13.3	0.262	95.0	—	—
		M-F	0.193	1.6	1.412	0.054	14.3	0.282	102.0	—	—
14	-TALLOW	A-F	—	—	—	—	—	—	—	—	—
		M-F	—	—	—	—	—	—	—	—	—
15	ANIMAL-POULTRY -FAT	A-F	—	—	—	—	—	—	—	—	—
		M-F	—	—	—	—	—	—	—	—	—
16	BAKERY WASTE, DEHY -DRIED BAKERY PRODUCT	A-F	1.009	5.0	—	0.005	56.7	—	15.1	7.7	4.6
		M-F	1.096	5.5	—	0.005	61.6	—	16.4	8.4	5.0
17	BARLEY *Hordeum vulgare* -GRAIN, ALL ANALYSES	A-F	0.099	8.0	0.044	0.008	16.1	0.176	45.2	4.7	2.8
		M-F	0.113	9.1	0.050	0.009	18.3	0.199	51.4	5.4	3.2
18	-GRAIN, GRADE 1, 48 LB/BUSHEL OR *618 G/L*	A-F	—	—	—	—	—	—	—	—	—
		M-F	—	—	—	—	—	—	—	—	—
19	-GRAIN, LIGHT LESS THAN 36 LB/BUSHEL OR *463 G/L*	A-F	—	—	—	—	—	—	—	—	—
		M-F	—	—	—	—	—	—	—	—	—
20	-GRAIN, PACIFIC COAST	A-F	0.088	8.2	—	0.010	16.2	0.102	15.4	—	—
		M-F	0.098	9.1	—	0.012	18.0	0.114	17.1	—	—
21	-GRAIN SCREENINGS	A-F	—	—	—	—	—	—	—	—	—
		M-F	—	—	—	—	—	—	—	—	—
22	-MALT SPROUTS, DEHY	A-F	—	—	—	—	31.7	—	—	—	—
		M-F	—	—	—	—	34.5	—	—	—	—
23	BEAN *Phaseolus vulgaris* -SEEDS, KIDNEY	A-F	—	—	—	0.007	—	—	—	—	—
		M-F	—	—	—	0.008	—	—	—	—	—
24	-SEEDS, NAVY	A-F	—	9.9	—	0.010	21.3	—	—	—	—
		M-F	—	11.0	—	0.011	23.6	—	—	—	—
25	-SEEDS, PINTO	A-F	—	—	—	—	—	—	—	—	—
		M-F	—	—	—	—	—	—	—	—	—

(Continued)

Entry Number	Fat-Soluble Vitamins			Water-Soluble Vitamins								
	D	E (α-tocopherol)	K	B-12	Biotin	Choline	Folic Acid (Folacin)	Niacin (Nicotinic Acid)	Pantothenic Acid	Pyridoxine (B-6)	Riboflavin (B-2)	Thiamin (B-1)
	(IU/kg)	(ppm or mg/kg)	(ppm or mg/kg)	(ppb or mcg/kg)	(ppm or mg/kg)	(ppm or mg/kg)	(ppm or mg/kg)	(ppm or mg/kg)	(ppm or mg/kg)	(ppm or mg/kg)	(ppm or mg/kg)	(ppm or mg/kg)
1	—	—	—	—	—	—	—	—	—	—	—	—
	—	—	—	—	—	—	—	—	—	—	—	—
2	—	—	—	—	—	—	—	—	—	—	—	—
	—	—	—	—	—	—	—	—	—	—	—	—
3	1,447	83.0	15.89	2.0	0.20	—	3.07	38	28.6	5.73	12.0	2.7
	1,602	91.9	17.58	2.2	0.22	—	3.40	42	31.6	6.35	13.3	3.0
4	—	124.3	8.63	—	0.30	1,405	4.50	39	29.4	8.02	13.2	3.4
	—	134.8	9.36	—	0.33	1,523	4.89	43	31.8	8.70	14.4	3.7
5	—	158.9	14.53	10.9	0.33	1,425	3.94	48	35.7	9.11	14.7	5.6
	—	173.8	15.90	12.0	0.36	1,559	4.31	53	39.1	9.97	16.1	6.1
6	—	—	—	—	—	261	—	27	4.9	—	3.9	0.4
	—	—	—	—	—	304	—	31	5.7	—	4.5	0.5
7	—	—	—	44.3	0.08	729	0.10	31	2.9	4.43	2.2	0.4
	—	—	—	48.1	0.09	792	0.10	34	3.1	4.81	2.3	0.4
8	—	—	—	—	—	—	—	—	—	—	—	—
	—	—	—	—	—	—	—	—	—	—	—	—
9	—	—	—	503.7	0.02	11,422	5.59	206	29.3	—	36.4	0.2
	—	—	—	541.6	0.02	12,281	6.01	221	31.5	—	39.1	0.2
10	—	1.0	—	64.2	0.13	2,046	0.37	58	6.4	3.89	5.3	0.2
	—	1.1	—	69.0	0.14	2,200	0.40	63	6.9	4.18	5.7	0.2
11	—	—	—	236.6	—	1,704	1.54	37	2.4	—	2.3	0.4
	—	—	—	257.2	—	1,852	1.67	40	2.6	—	2.5	0.4
12	—	0.8	—	82.4	0.07	2,150	0.57	48	3.7	—	3.6	0.2
	—	0.9	—	89.5	0.08	2,337	0.62	52	4.1	—	3.9	0.2
13	—	0.9	—	109.1	0.10	2,010	0.37	49	4.2	8.73	4.5	0.6
	—	1.0	—	117.3	0.11	2,162	0.40	53	4.5	9.39	4.9	0.7
14	—	—	—	—	—	—	—	—	—	—	—	—
	—	—	—	—	—	—	—	—	—	—	—	—
15	—	7.8	—	—	—	—	—	—	—	—	—	—
	—	7.9	—	—	—	—	—	—	—	—	—	—
16	—	130.6	—	—	0.07	1,008	0.17	25	14.8	17.71	1.3	2.5
	—	141.9	—	—	0.07	1,095	0.19	27	16.1	19.25	1.5	2.7
17	—	15.8	—	—	0.14	903	0.55	85	8.2	6.57	1.6	4.4
	—	18.0	—	—	0.16	1,026	0.62	96	9.3	7.46	1.8	5.0
18	—	—	—	—	—	—	—	—	—	—	—	4.3
	—	—	—	—	—	—	—	—	—	—	—	4.9
19	—	—	—	—	—	—	—	—	—	—	—	—
	—	—	—	—	—	—	—	—	—	—	—	—
20	—	21.1	—	—	0.15	1,003	0.51	48	7.1	2.93	1.6	4.3
	—	23.5	—	—	0.17	1,114	0.56	53	7.9	3.26	1.7	4.7
21	—	—	—	—	—	—	—	—	7.9	—	1.4	—
	—	—	—	—	—	—	—	—	8.8	—	1.6	—
22	—	20.6	—	—	—	1,576	0.20	52	8.6	—	6.7	4.9
	—	22.4	—	—	—	1,713	0.22	56	9.4	—	7.3	5.4
23	—	—	—	—	—	—	—	25	—	—	2.1	5.7
	—	—	—	—	—	—	—	28	—	—	2.4	6.4
24	—	1.0	—	—	0.11	1,675	1.30	25	2.4	0.30	2.0	6.4
	—	1.1	—	—	0.12	1,862	1.44	28	2.7	0.33	2.2	7.1
25	—	—	—	—	—	—	—	22	2.2	—	3.1	8.6
	—	—	—	—	—	—	—	24	2.5	—	3.4	9.6

(Continued)

COMPOSITION OF FEEDS

Entry Number	Feed Name Description	International Feed Number[1]	Moisture Basis: A-F (as-fed) or M-F (moisture-free)	Dry Matter (%)	Ash (%)	Crude Fiber (%)	Cell Walls or NDF (%)	Acid Detergent Fiber (%)	Lignin (%)	Ether Extract (Fat) (%)	N-Free Extract (%)	Crude Protein (%)
26	BEAN, LIMA *Phaseolus limensis* -SEEDS	5-00-613	A-F	90	4.1	4.5	—	—	—	1.4	59.2	20.8
			M-F	100	4.6	5.1	—	—	—	1.5	65.7	23.1
27	BEAN, MUNG *Phaseolus aureus* -SEEDS	5-08-185	A-F	90	3.8	3.9	—	—	—	1.3	57.2	24.0
			M-F	100	4.2	4.3	—	—	—	1.4	63.5	26.6
28	BEET, SUGAR *Beta vulgaris, saccharifera* -MOLASSES, MORE THAN 48% INVERT SUGAR, MORE THAN 79.5° BRIX	4-00-668	A-F	78	8.7	—	—	—	—	0.1	62.2	6.0
			M-F	100	11.2	—	—	—	—	0.2	79.7	7.7
29	-PULP,DEHY	4-00-669	A-F	90	4.8	17.9	—	—	—	0.5	58.1	8.7
			M-F	100	5.3	19.9	—	—	—	0.5	64.5	9.7
30	BLOOD -MEAL	5-00-380	A-F	92	5.3	2.4	—	—	—	1.3	3.8	79.1
			M-F	100	5.8	2.7	—	—	—	1.4	4.1	86.0
31	-MEAL, SPRAY, DEHY (BLOOD FLOUR)	5-00-381	A-F	86	7.2	1.4	—	—	—	6.2	−1.3	72.6
			M-F	100	8.4	1.6	—	—	—	7.2	−1.5	84.4
32	BREWERS' GRAINS -DEHY	5-02-141	A-F	92	3.8	14.4	—	—	—	6.9	42.0	25.0
			M-F	100	4.2	15.6	—	—	—	7.4	45.6	27.2
33	-WET	5-02-142	A-F	22	1.1	3.4	—	—	—	1.4	10.0	5.3
			M-F	100	4.8	15.3	—	—	—	6.5	45.5	24.0
34	BUCKWHEAT, COMMON *Fagopyrum sagittatum* -GRAIN, ALL ANALYSES	4-00-994	A-F	88	2.1	10.3	—	—	—	2.5	62.0	11.1
			M-F	100	2.4	11.7	—	—	—	2.8	70.5	12.6
35	MIDDLINGS	5-00-991	A-F	89	4.9	7.4	—	—	—	7.3	39.5	29.8
			M-F	100	5.5	8.3	—	—	—	8.2	44.4	33.5
36	BUTTERMILK -CONDENSED	5-01-159	A-F	29	3.6	0.1	—	—	—	2.3	12.2	10.7
			M-F	100	12.5	0.3	—	—	—	8.1	42.2	36.9
37	-DEHY	5-01-160	A-F	92	9.0	0.4	—	—	—	4.7	46.4	31.5
			M-F	100	9.8	0.4	—	—	—	5.1	50.4	34.3
38	CASSAVA *Manihot spp* -TUBERS, DEHY, GROUND	4-01-152	A-F	92	2.1	4.5	—	—	—	0.7	82.1	2.6
			M-F	100	2.3	4.9	—	—	—	0.7	89.2	2.9
39	CHEESE -RIND	5-01-163	A-F	36	7.7	0.2	—	—	—	20.0	11.8	46.2
			M-F	100	9.0	0.3	—	—	—	23.3	13.7	53.7
40	CHUFA *Cyperus esculentus* -ROOTS	4-08-374	A-F	27	1.8	2.0	—	—	—	1.8	19.2	2.1
			M-F	100	6.8	7.5	—	—	—	6.8	70.9	7.9
41	CITRUS *Citrus spp* -SYRUP (MOLASSES)	4-01-241	A-F	68	5.0	—	—	—	—	0.2	57.3	4.8
			M-F	100	7.4	—	—	—	—	0.3	84.3	7.1
42	COCONUT *Cocos nucifera* -MEATS, MEAL, MECH EXTD (COPRA MEAL)	5-01-572	A-F	93	6.8	11.5	—	—	—	6.5	47.0	21.2
			M-F	100	7.3	12.4	—	—	—	6.9	50.5	22.8
43	-MEATS, MEAL, SOLV EXTD (COPRA MEAL)	5-01-573	A-F	92	6.6	13.6	—	—	—	4.1	46.3	21.4
			M-F	100	7.2	14.8	—	—	—	4.4	50.3	23.2
44	CORN *Zea mays* -DISTILLERS' GRAINS, DEHY	5-02-842	A-F	94	2.4	11.3	—	—	—	9.2	43.0	28.1
			M-F	100	2.6	12.1	—	—	—	9.8	45.7	29.8
45	-DISTILLERS' SOLUBLES, DEHY	5-02-844	A-F	92	7.3	5.2	—	—	—	8.1	44.5	26.9
			M-F	100	8.0	5.6	—	—	—	8.8	48.4	29.3
46	-GLUTEN, MEAL	5-02-900	A-F	91	3.2	4.5	—	—	—	2.2	38.0	43.1
			M-F	100	3.5	5.0	—	—	—	2.4	41.8	47.3
47	-GLUTEN WITH BRAN (CORN GLUTEN FEED)	5-02-903	A-F	90	6.7	8.7	—	—	—	2.1	48.8	23.6
			M-F	100	7.5	9.7	—	—	—	2.3	54.2	26.2

(Continued)

Entry Number	TDN	Digestible Energy			Metabolizable Energy		Macrominerals						
							Calcium (Ca)	Phosphorus (P)	Sodium (Na)	Chlorine (Cl)	Magnesium (Mg)	Potassium (K)	Sulfur (S)
	(%)	(kcal)			(kcal)		(%)	(%)	(%)	(%)	(%)	(%)	(%)
		(lb)	(kg)	(lb)	(kg)								
26	—	—	—	—	—		0.08	0.38	0.02	0.03	0.18	1.62	0.20
	—	—	—	—	—		0.09	0.42	0.02	0.03	0.20	1.80	0.22
27	—	—	—	—	—		0.13	0.35	0.01	—	—	1.04	—
	—	—	—	—	—		0.14	0.38	0.01	—	—	1.15	—
28	—	—	—	919	2,025		0.12	0.02	1.10	1.50	0.19	4.63	0.47
	—	—	—	1,178	2,597		0.15	0.03	1.41	1.92	0.24	5.94	0.60
29	66	1,316	2,900	1,127	2,485		0.65	0.09	0.21	0.04	0.27	0.17	0.20
	73	1,462	3,223	1,252	2,761		0.72	0.10	0.23	0.04	0.30	0.19	0.22
30	62	1,225	2,701	1,101	2,426		0.27	0.26	0.32	0.32	0.22	0.14	0.34
	68	1,332	2,936	1,196	2,637		0.30	0.29	0.35	0.35	0.24	0.16	0.37
31	—	—	—	—	—		1.65	0.45	0.31	0.23	0.04	—	0.56
	—	—	—	—	—		1.92	0.52	0.36	0.27	0.04	—	0.65
32	56	1.385	3,053	889	1,982		0.30	0.53	0.21	0.12	0.16	0.09	0.30
	61	1,505	3,319	977	2,154		0.32	0.58	0.22	0.13	0.17	0.09	0.33
33	—	—	—	—	—		0.07	0.11	—	—	—	0.02	—
	—	—	—	—	—		0.30	0.51	—	—	—	0.08	—
34	68	1,357	2,991	1,111	2,450		0.10	0.33	0.05	0.04	—	0.45	—
	77	1,542	3,399	1,263	2,784		0.11	0.38	0.06	0.05	—	0.51	—
35	—	—	—	—	—		—	1.02	—	—	—	0.98	—
	—	—	—	—	—		—	1.15	—	—	—	1.10	—
36	—	—	—	—	—		0.44	0.26	0.31	0.12	0.19	0.23	0.03
	—	—	—	—	—		1.51	0.89	1.06	0.41	0.65	0.79	0.10
37	52	1,034	2,280	937	2,066		1.31	0.93	0.80	0.47	0.48	0.89	0.08
	56	1,124	2,478	1,018	2,245		1.42	1.01	0.87	0.51	0.52	0.96	0.09
38	84	1,686	3,717	1,609	3,547		—	0.03	—	—	—	0.24	—
	92	1,833	4,040	1,749	3,856		—	0.03	—	—	—	0.26	—
39	—	—	—	—	—		0.99	0.57	0.82	0.61	0.02	0.28	—
	—	—	—	—	—		1.15	0.66	0.95	0.71	0.03	0.32	—
40	21	422	930	398	878		0.01	0.07	—	—	—	0.14	—
	78	1,563	3,446	1,476	3,253		0.04	0.26	—	—	—	0.53	—
41	—	—	—	—	—		1.09	0.10	0.27	0.07	0.14	0.09	—
	—	—	—	—	—		1.61	0.14	0.40	0.10	0.21	0.13	—
42	—	—	—	—	—		0.20	0.62	0.04	—	0.31	1.54	0.34
	—	—	—	—	—		0.22	0.67	0.04	—	0.33	1.65	0.37
43	—	—	—	1,406	3,099		0.18	0.61	0.04	0.03	0.36	1.21	0.34
	—	—	—	1,528	3,368		0.19	0.66	0.04	0.03	0.39	1.32	0.37
44	65	1,253	2,762	1,030	2,270		0.10	0.41	0.09	0.07	0.08	0.19	0.44
	69	1,333	2,938	1,095	2,415		0.11	0.44	0.10	0.08	0.08	0.20	0.46
45	74	1,460	3,219	1,271	2,802		0.30	1.23	0.26	0.26	0.63	1.58	0.37
	81	1,587	3,499	1,381	3,045		0.33	1.34	0.28	0.28	0.69	1.72	0.40
46	—	—	—	1,447	3,190		0.15	0.47	0.08	0.07	0.05	0.03	—
	—	—	—	1,590	3,505		0.16	0.51	0.09	0.08	0.05	0.03	—
47	74	1,454	3,206	1,149	2,532		0.30	0.73	0.94	0.22	0.35	0.58	0.21
	82	1,616	3,562	1,276	2,813		0.33	0.81	1.04	0.24	0.39	0.64	0.24

(Continued)

Entry Number	Feed Name Description	Moisture Basis: A-F (as-fed) or M-F (moisture-free)	Microminerals							Fat-Soluble Vitamins	
			Cobalt (Co)	Copper (Cu)	Iodine (I)	Iron (Fe)	Man-ganese (Mn)	Sele-nium (Se)	Zinc (Zn)	A (1 mg Carotene = 1667 IU Vit. A)	Carotene (Provitamin A)
			(ppm or mg/kg)	(ppm or mg/kg)	(ppm or mg/kg)	(%)	(ppm or mg/kg)	(ppm or mg/kg)	(ppm or mg/kg)	(IU/g)	(ppm or mg/kg)
26	BEAN, LIMA *Phaseolus limensis* -SEEDS	A-F	—	8.2	—	0.009	16.2	—	—	—	—
		M-F	—	9.1	—	0.010	18.0	—	—	—	—
27	BEAN, MUNG *Phaseolus aureus* -SEEDS	A-F	—	—	—	0.008	—	—	—	—	—
		M-F	—	—	—	0.009	—	—	—	—	—
28	BEET, SUGAR *Beta vulgaris, saccharifera* -MOLASSES, MORE THAN 48% INVERT SUGAR, MORE THAN 79.5° BRIX	A-F	0.374	17.1	—	0.007	4.5	—	—	—	—
		M-F	0.480	22.0	—	0.009	5.7	—	—	—	—
29	-PULP, DEHY	A-F	0.065	12.4	—	0.030	34.4	—	0.7	0.4	0.2
		M-F	0.072	13.8	—	0.033	38.3	—	0.8	0.4	0.2
30	BLOOD -MEAL	A-F	0.089	13.3	—	0.307	5.3	—	4.4	—	—
		M-F	0.096	14.4	—	0.334	5.8	—	4.8	—	—
31	-MEAL, SPRAY, DEHY (BLOOD FLOUR)	A-F	—	7.6	—	0.257	6.0	—	—	—	—
		M-F	—	8.8	—	0.299	6.9	—	—	—	—
32	BREWERS' GRAINS -DEHY	A-F	0.085	21.6	0.065	0.025	37.8	—	27.3	—	—
		M-F	0.093	23.5	0.071	0.027	41.1	—	29.6	—	—
33	-WET	A-F	—	—	—	—	—	—	—	—	—
		M-F	—	—	—	—	—	—	—	—	—
34	BUCKWHEAT, COMMON *Fagopyrum sagittatum* -GRAINS, ALL ANALYSES	A-F	0.049	9.5	—	0.004	33.8	—	8.8	—	—
		M-F	0.055	10.8	—	0.005	38.4	—	10.0	—	—
35	MIDDLINGS	A-F	—	—	—	—	—	—	—	—	—
		M-F	—	—	—	—	—	—	—	—	—
36	BUTTERMILK -CONDENSED	A-F	—	—	—	—	—	—	—	25.3	15.2
		M-F	—	—	—	—	—	—	—	87.2	52.3
37	-DEHY	A-F	—	—	—	0.001	3.4	—	—	—	—
		M-F	—	—	—	0.001	3.7	—	—	—	—
38	CASSAVA *Manihot* spp -TUBERS, DEHY, GROUND	A-F	—	—	—	—	—	—	—	—	—
		M-F	—	—	—	—	—	—	—	—	—
39	CHEESE -RIND	A-F	—	—	—	—	—	—	—	—	—
		M-F	—	—	—	—	—	—	—	—	—
40	CHUFA *Cyperus esculentus* -ROOTS	A-F	—	—	—	—	—	—	—	—	—
		M-F	—	—	—	—	—	—	—	—	—
41	CITRUS *Citrus* spp -SYRUP (MOLASSES)	A-F	0.108	73.4	—	0.034	26.2	—	93.2	—	—
		M-F	0.159	108.0	—	0.050	38.5	—	137.0	—	—
42	COCONUT *Cocos nucifera* -MEATS, MEAL, MECH EXTD (COPRA MEAL)	A-F	0.128	14.1	—	0.132	65.6	—	—	—	—
		M-F	0.138	15.2	—	0.142	70.6	—	—	—	—
43	-MEATS, MEAL, SOLV EXTD (COPRA MEAL)	A-F	0.128	9.5	—	0.069	66.0	—	—	—	—
		M-F	0.139	10.4	—	0.075	71.8	—	—	—	—
44	CORN *Zea mays* -DISTILLERS' GRAINS, DEHY	A-F	0.075	40.7	0.044	0.021	24.6	—	33.1	5.2	3.1
		M-F	0.080	43.3	0.047	0.023	26.2	—	35.3	5.6	3.3
45	-DISTILLERS' SOLUBLES, DEHY	A-F	0.194	82.0	0.107	0.056	73.3	0.332	84.2	1.1	0.7
		M-F	0.211	89.2	0.117	0.061	79.7	0.361	91.5	1.2	0.7
46	-GLUTEN, MEAL	A-F	0.083	28.3	—	0.039	7.3	1.011	—	27.2	16.3
		M-F	0.092	31.1	—	0.043	8.0	1.111	—	29.8	17.9
47	-GLUTEN WITH BRAN (CORN GLUTEN FEED)	A-F	0.205	47.2	0.066	0.043	23.5	0.327	47.6	9.9	5.9
		M-F	0.227	52.4	0.073	0.048	26.1	0.364	52.9	11.0	6.6

(Continued)

Entry Number	Fat-Soluble Vitamins			Water-Soluble Vitamins								
	D	E (α-tocopherol)	K	B-12	Biotin	Choline	Folic Acid (Folacin)	Niacin (Nicotinic Acid)	Panto-thenic Acid	Pyri-doxine (B-6)	Ribo-flavin (B-2)	Thiamin (B-1)
	(IU/kg)	(ppm or mg/kg)	(ppm or mg/kg)	(ppb or mcg/kg)	(ppm or mg/kg)	(ppm or mg/kg)	(ppm or mg/kg)	(ppm or mg/kg)	(ppm or mg/kg)	(ppm or mg/kg)	(ppm or mg/kg)	(ppm or mg/kg)
26	—	—	—	—	—	—	3.33	20	8.4	—	1.7	4.6
	—	—	—	—	—	—	3.69	22	9.4	—	1.9	5.1
27	—	—	—	—	—	—	—	25	—	—	2.1	3.9
	—	—	—	—	—	—	—	27	—	—	2.4	4.3
28	—	—	—	—	—	829	—	41	4.5	—	2.3	—
	—	—	—	—	—	1,063	—	53	5.7	—	2.9	—
29	573	—	—	—	—	810	—	17	1.4	—	0.7	0.4
	637	—	—	—	—	900	—	19	1.5	—	0.8	0.4
30	—	—	—	44.3	0.08	729	0.10	31	2.9	4.43	2.2	0.4
	—	—	—	48.1	0.09	792	0.10	34	3.1	4.81	2.3	0.4
31	—	—	—	—	—	261	—	27	4.9	—	3.9	0.4
	—	—	—	—	—	304	—	31	5.7	—	4.5	0.5
32	—	25.9	—	—	0.96	1,670	7.11	43	8.1	0.66	1.3	0.6
	—	28.1	—	—	1.05	1,815	7.73	47	8.8	0.72	1.4	0.6
33	—	—	—	—	—	—	—	—	—	—	—	—
	—	—	—	—	—	—	—	—	—	—	—	—
34	—	—	—	—	—	441	—	18	11.6	—	5.4	3.7
	—	—	—	—	—	501	—	21	13.2	—	6.2	4.2
35	—	—	—	—	—	—	—	—	—	—	—	—
	—	—	—	—	—	—	—	—	—	—	—	—
36	—	—	—	—	—	—	—	—	—	—	12.4	—
	—	—	—	—	—	—	—	—	—	—	42.8	—
37	—	6.2	—	19.4	0.29	1,665	0.40	9	36.0	2.41	31.0	3.4
	—	6.8	—	21.1	0.32	1,831	0.44	9	39.1	2.62	33.7	3.7
38	—	—	—	—	—	—	—	—	—	—	—	—
	—	—	—	—	—	—	—	—	—	—	—	—
39	—	—	—	—	—	—	—	—	—	—	—	—
	—	—	—	—	—	—	—	—	—	—	—	—
40	—	—	—	—	—	—	—	—	—	—	—	—
	—	—	—	—	—	—	—	—	—	—	—	—
41	—	—	—	—	—	—	—	27	12.6	—	6.2	—
	—	—	—	—	—	—	—	39	18.6	—	9.1	—
42	—	—	—	—	—	1,015	1.39	25	6.2	—	3.2	0.8
	—	—	—	—	—	1,091	1.50	27	6.6	—	3.5	0.8
43	—	—	—	—	—	1,045	0.30	26	6.5	4.40	3.3	0.7
	—	—	—	—	—	1,136	0.33	28	7.0	4.78	3.6	0.7
44	—	—	—	—	0.49	1,261	0.88	38	11.8	4.42	5.3	1.7
	—	—	—	—	0.52	1,341	0.94	41	12.5	4.70	5.6	1.8
45	—	45.5	—	27.9	1.40	4,789	1.32	116	23.4	8.78	21.1	6.7
	—	49.4	—	30.3	1.53	5,206	1.43	127	25.5	9.54	23.0	7.3
46	—	33.9	—	—	0.17	368	0.34	50	9.9	7.98	1.4	0.2
	—	37.2	—	—	0.19	405	0.37	55	10.8	8.77	1.6	0.2
47	—	10.9	—	—	0.32	1,517	0.27	70	13.2	14.93	4.0	2.0
	—	12.1	—	—	0.36	1,685	0.30	78	14.6	16.59	4.4	2.2

(Continued)

TABLE 21-1

COMPOSITION OF FEEDS

Entry Number	Feed Name Description	Inter-national Feed Number[1]	Moisture Basis: A-F (as-fed) or M-F (moisture-free)	Dry Matter	Ash	Crude Fiber	Cell Walls or NDF	Acid Detergent Fiber	Lignin	Ether Extract (Fat)	N-Free Extract	Crude Protein
				(%)	(%)	(%)	(%)	(%)	(%)	(%)	(%)	(%)
	CORN, DENT YELLOW *Zea mays, indentata*											
48	-DISTILLERS' SOLUBLES, DEHY	5-02-844	A-F	92	7.3	5.2	—	—	—	8.1	44.5	26.9
			M-F	100	8.0	5.6	—	—	—	8.8	48.4	29.3
49	-EARS, GROUND (CORN AND COB MEAL)	4-02-849	A-F	86	1.6	8.3	—	—	—	3.2	65.0	7.8
			M-F	100	1.9	9.6	—	—	—	3.7	75.6	9.1
50	-GLUTEN WITH BRAN (CORN GLUTEN FEED)	5-02-903	A-F	90	6.7	8.7	—	—	—	2.1	48.8	23.6
			M-F	100	7.5	9.7	—	—	—	2.3	54.2	26.2
51	-GRAIN, ALL ANALYSES	4-02-935	A-F	88	1.3	2.1	—	—	—	3.9	71.1	9.6
			M-F	100	1.4	2.4	—	—	—	4.5	80.8	10.9
52	-GRAIN, FLAKED	4-02-859	A-F	89	0.9	0.6	—	—	—	2.0	75.6	10.0
			M-F	100	1.0	0.7	—	—	—	2.2	84.9	11.2
53	-GRAIN, GRADE 1, 56 LB/BUSHEL OR 721 G/L	4-02-930	A-F	86	1.3	2.0	—	—	—	4.0	70.2	8.8
			M-F	100	1.5	2.3	—	—	—	4.6	81.4	10.2
54	-GRAIN, GRADE 2, 54 LB/BUSHEL OR 695 G/L	4-02-931	A-F	89	1.3	2.0	—	—	—	3.9	72.7	8.7
			M-F	100	1.5	2.2	—	—	—	4.4	81.7	9.8
55	-GRAIN, GRADE 3, 52 LB/BUSHEL OR 669 G/L	4-02-932	A-F	86	1.2	2.1	—	—	—	3.7	70.3	8.7
			M-F	100	1.4	2.4	—	—	—	4.3	81.8	10.1
56	-GRAIN, GRADE 4, 49 LB/BUSHEL OR 630 G/L	4-02-933	A-F	84	1.2	2.1	—	—	—	3.6	68.5	8.6
			M-F	100	1.4	2.5	—	—	—	4.3	81.6	10.2
57	-GRAIN, GRADE 5, 46 LB/BUSHEL OR 592 G/L	4-02-934	A-F	79	1.1	1.9	—	—	—	3.3	64.7	8.0
			M-F	100	1.4	2.4	—	—	—	4.2	81.9	10.1
58	-GRAIN, GRADE SAMPLE	4-02-929	A-F	82	2.0	1.9	—	—	—	3.0	66.5	8.5
			M-F	100	2.5	2.4	—	—	—	3.7	81.0	10.4
59	-GRAIN, HIGH-MOISTURE	4-20-770	A-F	77	1.2	1.9	—	—	—	3.1	62.5	8.3
			M-F	100	1.5	2.5	—	—	—	4.0	81.2	10.7
60	-HOMINY FEED	4-02-887	A-F	90	2.7	5.0	—	—	—	7.4	64.0	11.0
			M-F	100	3.0	5.6	—	—	—	8.2	71.1	12.2
	CORN GRAIN, OPAQUE-2 *Zea mays*											
61	-HIGH-LYSINE	4-11-445	A-F	87	1.4	2.6	—	—	—	4.0	69.4	9.6
			M-F	100	1.6	3.0	—	—	—	4.6	79.8	11.0
	COTTON *Gossypium* spp											
62	-SEEDS, MEAL, MECH EXTD, 41% PROTEIN	5-01-617	A-F	93	6.2	11.3	—	—	—	4.6	30.0	40.9
			M-F	100	6.6	12.1	—	—	—	5.0	32.3	44.0
63	-SEEDS, MEAL, MECH EXTD, 45% PROTEIN	5-26-100	A-F	94	7.0	10.1	—	—	—	5.5	26.0	45.2
			M-F	100	7.4	10.7	—	—	—	5.9	27.7	48.1
64	-SEEDS, MEAL, PREPRESSED, SOLV EXTD, 41% PROTEIN	5-07-872	A-F	90	6.4	13.6	—	—	—	0.6	28.0	41.4
			M-F	100	7.1	15.1	—	—	—	0.6	31.1	46.1
65	-SEEDS, MEAL, SOLV EXTD, 41% PROTEIN	5-01-621	A-F	91	6.5	12.3	—	—	—	2.2	28.8	41.1
			M-F	100	7.2	13.5	—	—	—	2.5	31.7	45.2
66	-SEEDS, MEAL, PREPRESSED, SOLV EXTD, 48% PROTEIN	5-07-874	A-F	92	7.2	8.5	—	—	—	1.1	26.7	50.0
			M-F	100	7.8	9.2	—	—	—	1.2	28.9	54.0
67	-SEEDS, MEAL, SOLV EXTD, LOW GOSSYPOL	5-01-633	A-F	93	5.8	12.7	—	—	—	1.2	31.6	41.6
			M-F	100	6.3	13.7	—	—	—	1.3	34.0	44.8
68	-SEEDS WITHOUT HULLS, MEAL, PRE-PRESSED, SOLV EXTD, 50% PROTEIN	5-07-874	A-F	93	6.9	8.2	—	—	—	1.1	26.8	50.0
			M-F	100	7.4	8.8	—	—	—	1.2	28.8	53.7
	COTTON, GLANDLESS *Gossypium* spp											
69	-SEEDS, MEAL, SOLV EXTD	5-08-979	A-F	—	—	—	—	—	—	—	—	—
			M-F	100	—	—	—	—	—	—	—	—
	DISTILLERS' PRODUCTS (ALSO SEE SPECIFIC GRAINS)											
70	-GRAINS, DEHY	5-02-144	A-F	93	1.5	12.9	—	—	—	7.4	43.7	27.5
			M-F	100	1.6	13.8	—	—	—	8.0	47.0	29.6
71	-SOLUBLES, DEHY	5-02-147	A-F	92	6.2	3.4	—	—	—	8.9	44.8	28.8
			M-F	100	6.7	3.7	—	—	—	9.6	48.7	31.3
	EMMER *Triticum dicoccum*											
72	-GRAIN	4-01-830	A-F	91	3.5	9.7	—	—	—	2.0	64.1	11.7
			M-F	100	3.9	10.6	—	—	—	2.2	70.4	12.9

(Continued)

Entry Number	TDN	Digestible Energy		Metabolizable Energy		Macrominerals						
						Calcium (Ca)	Phosphorus (P)	Sodium (Na)	Chlorine (Cl)	Magnesium (Mg)	Potassium (K)	Sulfur (S)
	(%)	(kcal)		(kcal)		(%)	(%)	(%)	(%)	(%)	(%)	(%)
		(lb)	(kg)	(lb)	(kg)							
48	74	1,460	3,219	1,271	2,802	0.30	1.23	0.26	0.26	0.63	1.58	0.37
	81	1,587	3,499	1,381	3,045	0.33	1.34	0.28	0.28	0.69	1.72	0.40
49	69	1,410	3,108	1,214	2,676	0.07	0.23	0.04	—	0.12	0.45	0.19
	80	1,639	3,614	1,412	3,112	0.08	0.27	0.05	—	0.13	0.52	0.22
50	74	1,454	3,206	1,149	2,532	0.30	0.73	0.94	0.22	0.35	0.58	0.21
	82	1,616	3,562	1,276	2,813	0.33	0.81	1.04	0.24	0.39	0.64	0.24
51	79	1,579	3,480	1,306	2,880	0.03	0.27	0.01	0.05	0.12	0.31	0.12
	90	1,794	3,955	1,484	3,273	0.04	0.30	0.01	0.05	0.13	0.36	0.14
52	85	1,698	3,744	1,592	3,509	—	—	—	—	—	—	—
	95	1,908	4,206	1,788	3,943	—	—	—	—	—	—	—
53	80	1,595	3,510	1,498	3,296	—	—	—	—	—	—	—
	92	1,848	4,065	1,735	3,818	—	—	—	—	—	—	—
54	82	1,645	3,620	1,544	3,397	0.02	0.30	0.01	0.04	—	0.28	—
	92	1,848	4,065	1,735	3,817	0.02	0.34	0.01	0.04	—	0.31	—
55	73	1,469	3,238	1,380	3,043	0.02	0.25	—	—	—	—	—
	85	1,708	3,766	1,605	3,538	0.02	0.29	—	—	—	—	—
56	72	1,435	3,164	1,348	2,972	0.04	0.29	—	—	0.17	—	—
	85	1,709	3,767	1,605	3,538	0.05	0.34	—	—	0.20	—	—
57	68	1,350	2,977	1,269	2,797	—	—	—	—	—	—	—
	85	1,709	3,768	1,606	3,540	—	—	—	—	—	—	—
58	70	1,393	3,072	1,308	2,884	—	—	—	—	—	—	—
	85	1,699	3,746	1,596	3,518	—	—	—	—	—	—	—
59	66	1,317	2,904	1,236	2,725	—	0.28	—	—	—	0.30	—
	86	1,711	3,722	1,605	3,539	—	0.36	—	—	—	0.39	—
60	76	1,519	3,349	1,421	3,132	0.05	0.53	0.08	0.05	0.23	0.54	0.03
	84	1,688	3,721	1,579	3,481	0.05	0.59	0.09	0.06	0.26	0.59	0.03
61	74	1,485	3,274	1,392	3,070	0.02	0.19	—	—	—	—	—
	85	1,707	3,763	1,600	3,528	0.02	0.22	—	—	—	—	—
62	73	1,465	3,229	1,304	2,875	0.20	1.01	0.06	0.04	0.53	1.28	0.40
	79	1,575	3,472	1,402	3,092	0.22	1.09	0.06	0.04	0.57	1.38	0.43
63	—	—	—	—	—	0.22	1.11	—	—	0.54	—	—
	—	—	—	—	—	0.23	1.18	—	—	0.57	—	—
64	61	1,221	2,692	1,179	2,599	0.15	0.97	0.04	—	0.40	1.22	0.21
	68	1,357	2,992	1,310	2,888	0.17	1.08	0.04	—	0.44	1.36	0.23
65	62	1,245	2,746	1,074	2,368	0.18	1.11	0.05	0.04	0.57	1.45	0.21
	68	1,369	3,017	1,180	2,602	0.19	1.21	0.05	0.05	0.62	1.59	0.23
66	—	—	—	—	—	0.16	1.01	0.05	—	0.46	1.26	—
	—	—	—	—	—	0.17	1.09	0.05	—	0.50	1.36	—
67	—	—	—	—	—	—	—	—	—	—	—	—
	—	—	—	—	—	—	—	—	—	—	—	—
68	—	—	—	—	—	0.19	1.20	0.05	0.05	0.47	1.54	—
	—	—	—	—	—	0.20	1.29	0.06	0.05	0.50	1.65	—
69	—	—	—	—	—	—	—	—	—	—	—	—
	—	—	—	—	—	—	—	—	—	—	—	—
70	—	—	—	—	—	0.14	0.40	0.05	0.05	0.10	0.24	0.46
	—	—	—	—	—	0.15	0.43	0.05	0.05	0.10	0.26	0.49
71	—	—	—	—	—	0.21	1.22	0.15	—	0.46	1.84	—
	—	—	—	—	—	0.22	1.33	0.16	—	0.50	2.00	—
72	70	1,404	3,095	1,311	2,890	0.05	0.36	—	—	—	0.47	—
	77	1,543	3,401	1,441	3,176	0.06	0.40	—	—	—	0.52	—

(Continued)

COMPOSITION OF FEEDS

Entry Number	Feed Name Description	Moisture Basis: A-F (as-fed) or M-F (moisture-free)	Microminerals							Fat-Soluble Vitamins	
			Cobalt (Co)	Copper (Cu)	Iodine (I)	Iron (Fe)	Manganese (Mn)	Selenium (Se)	Zinc (Zn)	A (1 mg Carotene = 1667 IU Vit. A)	Carotene (Provitamin A)
			(ppm or mg/kg)	(ppm or mg/kg)	(ppm or mg/kg)	(%)	(ppm or mg/kg)	(ppm or mg/kg)	(ppm or mg/kg)	(IU/g)	(ppm or mg/kg)
	CORN, DENT YELLOW *Zea mays, indentata*										
48	-DISTILLERS' SOLUBLES, DEHY	A-F	0.194	82.0	0.107	0.056	73.3	0.332	84.2	1.1	0.7
		M-F	0.211	89.2	0.117	0.061	79.7	0.361	91.5	1.2	0.7
49	-EARS, GROUND (CORN AND COB MEAL)	A-F	0.228	6.6	0.022	0.008	24.1	0.074	15.5	1.3	0.8
		M-F	0.265	7.7	0.026	0.010	28.0	0.086	18.0	1.6	0.9
50	-GLUTEN WITH BRAN (CORN GLUTEN FEED)	A-F	0.205	47.2	0.066	0.043	23.5	0.327	47.6	9.9	5.9
		M-F	0.227	52.4	0.073	0.048	26.1	0.364	52.9	11.0	6.6
51	-GRAIN, ALL ANALYSES	A-F	0.033	3.2	—	0.002	4.9	0.069	18.3	3.7	2.2
		M-F	0.037	3.6	—	0.003	5.6	0.079	20.8	4.2	2.5
52	-GRAIN, FLAKED	A-F	—	—	—	—	—	—	—	—	—
		M-F	—	—	—	—	—	—	—	—	—
53	-GRAIN, GRADE 1, 56 LB/BUSHEL OR 721 G/L	A-F	—	4.2	—	0.002	5.5	—	—	—	—
		M-F	—	4.9	—	0.002	6.4	—	—	—	—
54	-GRAIN, GRADE 2, 54 LB/BUSHEL OR 695 G/L	A-F	0.018	—	—	—	5.0	—	10.0	2.9	1.8
		M-F	0.020	—	—	—	5.6	—	11.2	3.3	2.0
55	-GRAIN, GRADE 3, 52 LB/BUSHEL OR 669 G/L	A-F	—	—	—	0.002	5.5	—	—	—	—
		M-F	—	—	—	0.002	6.4	—	—	—	—
56	-GRAIN, GRADE 4, 49 LB/BUSHEL OR 630 G/L	A-F	—	—	—	—	—	—	—	—	—
		M-F	—	—	—	—	—	—	—	—	—
57	-GRAIN, GRADE 5, 46 LB/BUSHEL OR 592 G/L	A-F	—	—	—	—	—	—	—	—	—
		M-F	—	—	—	—	—	—	—	—	—
58	-GRAIN, GRADE SAMPLE	A-F	—	—	—	—	—	—	—	—	—
		M-F	—	—	—	—	—	—	—	—	—
59	-GRAIN, HIGH-MOISTURE	A-F	—	—	—	—	—	—	—	—	—
		M-F	—	—	—	—	—	—	—	—	—
60	-HOMINY FEED	A-F	0.059	13.7	—	0.007	14.5	—	—	15.3	9.2
		M-F	0.066	15.2	—	0.008	16.1	—	—	17.0	10.2
	CORN GRAIN, OPAQUE-2 *Zea mays*										
61	-HIGH-LYSINE	A-F	—	—	—	—	—	—	—	—	—
		M-F	—	—	—	—	—	—	—	—	—
	COTTON *Gossypium* spp										
62	-SEEDS, MEAL, MECH EXTD, 41% PROTEIN	A-F	0.167	18.2	—	0.013	23.0	—	—	0.4	0.2
		M-F	0.179	19.5	—	0.014	24.8	—	—	0.4	0.2
63	-SEEDS, MEAL, MECH EXTD, 45% PROTEIN	A-F	—	11.0	—	0.018	9.2	—	—	—	—
		M-F	—	11.7	—	0.019	9.8	—	—	—	—
64	-SEEDS, MEAL, PREPRESSED, SOLV EXTD, 41% PROTEIN	A-F	—	17.8	—	0.011	20.0	—	62.4	—	—
		M-F	—	19.8	—	0.012	22.2	—	69.3	—	—
65	-SEEDS, MEAL, SOLV EXTD, 41% PROTEIN	A-F	0.151	20.7	—	0.022	20.7	—	—	—	—
		M-F	0.165	22.8	—	0.024	22.8	—	—	—	—
66	-SEEDS, MEAL, PREPRESSED, SOLV EXTD, 48% PROTEIN	A-F	0.093	17.9	—	0.011	22.8	—	73.3	—	—
		M-F	0.100	19.4	—	0.012	24.6	—	79.2	—	—
67	-SEEDS, MEAL, SOLV EXTD, LOW GOSSYPOL	A-F	—	—	—	—	—	—	—	—	—
		M-F	—	—	—	—	—	—	—	—	—
68	-SEEDS WITHOUT HULLS, MEAL, PRE-PRESSED, SOLV EXTD, 50% PROTEIN	A-F	0.042	18.0	—	0.011	23.0	—	73.8	—	—
		M-F	0.045	19.4	—	0.012	24.8	—	79.4	—	—
	COTTON, GLANDLESS *Gossypium* spp										
69	-SEEDS, MEAL, SOLV EXTD	A-F	—	—	—	—	—	—	—	—	—
		M-F	—	—	—	—	—	—	—	—	—
	DISTILLERS' PRODUCTS (ALSO SEE SPECIFIC GRAINS)										
70	-GRAINS, DEHY	A-F	0.092	48.1	—	0.026	35.1	—	—	13.1	7.8
		M-F	0.099	51.7	—	0.028	37.8	—	—	14.0	8.4
71	-SOLUBLES, DEHY	A-F	0.196	71.6	—	0.030	64.1	—	138.0	1.9	1.1
		M-F	0.213	77.9	—	0.033	69.7	—	150.0	2.0	1.2
	EMMER *Triticum dicoccum*										
72	-GRAIN	A-F	—	31.5	—	0.006	78.2	—	—	—	—
		M-F	—	34.6	—	0.006	86.0	—	—	—	—

(Continued)

Entry Number	Fat-Soluble Vitamins			Water-Soluble Vitamins								
	D	E (α-tocopherol)	K	B-12	Biotin	Choline	Folic Acid (Folacin)	Niacin (Nicotinic Acid)	Panto-thenic Acid	Pyri-doxine (B-6)	Ribo-flavin (B-2)	Thiamin (B-1)
	(IU/kg)	(ppm or mg/kg)	(ppm or mg/kg)	(ppb or mcg/kg)	(ppm or mg/kg)	(ppm or mg/kg)	(ppm or mg/kg)	(ppm or mg/kg)	(ppm or mg/kg)	(ppm or mg/kg)	(ppm or mg/kg)	(ppm or mg/kg)
48	—	45.5	—	27.9	1.40	4,789	1.32	116	23.4	8.78	21.1	6.7
	—	49.4	—	30.3	1.53	5,206	1.43	127	25.5	9.54	23.0	7.3
49	—	—	—	—	0.04	357	—	17	4.1	5.54	0.9	2.9
	—	—	—	—	0.05	415	—	19	4.7	6.44	1.0	3.4
50	—	10.9	—	—	0.32	1,517	0.27	70	13.2	14.93	4.0	2.0
	—	12.1	—	—	0.36	1,685	0.30	78	14.6	16.59	4.4	2.2
51	—	22.6	—	—	0.06	536	0.30	30	6.6	5.18	1.3	2.1
	—	25.7	—	—	0.07	609	0.34	34	7.5	5.89	1.5	2.3
52	—	0.8	—	—	—	—	—	—	—	—	—	—
	—	0.9	—	—	—	—	—	—	—	—	—	—
53	—	—	—	—	0.06	—	0.21	—	—	—	—	—
	—	—	—	—	0.06	—	0.24	—	—	—	—	—
54	—	22.0	—	—	0.06	620	0.36	24	4.8	7.00	1.3	3.5
	—	24.7	—	—	0.07	697	0.40	27	5.4	7.87	1.5	4.0
55	—	—	—	—	—	—	—	—	—	—	—	—
	—	—	—	—	—	—	—	—	—	—	—	—
56	—	—	—	—	—	—	—	—	—	—	—	—
	—	—	—	—	—	—	—	—	—	—	—	—
57	—	—	—	—	—	—	—	—	—	—	—	—
	—	—	—	—	—	—	—	—	—	—	—	—
58	—	—	—	—	—	—	—	—	—	—	—	—
	—	—	—	—	—	—	—	—	—	—	—	—
59	—	—	—	—	—	—	—	—	—	—	—	—
	—	—	—	—	—	—	—	—	—	—	—	—
60	—	—	—	—	0.13	993	0.28	47	7.5	10.93	2.1	7.9
	—	—	—	—	0.14	1,104	0.31	53	8.4	12.14	2.3	8.8
61	—	—	—	—	—	500	—	19	4.5	—	1.0	—
	—	—	—	—	—	575	—	22	5.2	—	1.2	—
62	—	32.5	—	—	0.77	2,787	1.80	32	9.9	5.39	4.8	6.6
	—	34.9	—	—	0.83	2,997	1.94	35	10.6	5.79	5.2	7.2
63	—	—	—	—	—	—	—	—	—	—	—	—
	—	—	—	—	—	—	—	—	—	—	—	—
64	—	—	—	—	0.55	2,936	2.67	40	7.0	—	4.0	3.3
	—	—	—	—	0.61	3,263	2.96	45	7.8	—	4.4	3.7
65	—	14.1	—	—	0.60	2,783	0.95	41	14.0	6.37	4.8	6.3
	—	15.5	—	—	0.66	3,058	1.04	45	15.4	6.99	5.3	6.9
66	—	—	—	—	—	—	—	—	—	—	—	—
	—	—	—	—	—	—	—	—	—	—	—	—
67	—	—	—	—	—	—	—	—	—	—	—	20.3
	—	—	—	—	—	—	—	—	—	—	—	21.9
68	—	15.1	—	—	0.10	3,000	1.11	45	12.4	7.04	5.0	—
	—	16.2	—	—	0.11	3,226	1.19	48	13.4	7.57	5.4	—
69	—	—	—	—	—	—	—	—	—	—	—	—
	—	—	—	—	—	—	—	—	—	—	—	—
70	—	—	—	—	—	—	—	47	11.6	—	3.8	2.5
	—	—	—	—	—	—	—	51	12.5	—	4.1	2.6
71	—	—	—	2.9	1.84	4,250	0.73	143	26.0	8.66	13.2	5.8
	—	—	—	3.1	2.00	4,619	0.80	155	28.3	9.42	14.4	6.3
72	—	—	—	—	—	—	—	—	—	—	—	—
	—	—	—	—	—	—	—	—	—	—	—	—

(Continued)

TABLE 21-1

COMPOSITION OF FEEDS

Entry Number	Feed Name Description	International Feed Number[1]	Moisture Basis: A-F (as-fed) or M-F (moisture-free)	Chemical Analysis								
				Dry Matter	Ash	Crude Fiber	Cell Walls or NDF	Acid Detergent Fiber	Lignin	Ether Extract (Fat)	N-Free Extract	Crude Protein
				(%)	(%)	(%)	(%)	(%)	(%)	(%)	(%)	(%)
	FATS AND OILS											
73	-FAT, ANIMAL	4-00-376	A-F	99	—	—	—	—	—	99.0	—	—
			M-F	100	—	—	—	—	—	100.0	—	—
74	-FAT (LARD), SWINE	4-04-790	A-F	99	—	—	—	—	—	99.0	—	—
			M-F	100	—	—	—	—	—	100.0	—	—
75	-TALLOW, ANIMAL	4-08-127	A-F	99	0.1	—	—	—	—	96.9	—	1.6
			M-F	100	0.1	—	—	—	—	97.9	—	1.6
	FEATHERS											
76	-HYDROLYZED MEAL (POULTRY)	5-03-795	A-F	93	3.3	1.2	—	—	—	3.1	0.0	85.6
			M-F	100	3.5	1.3	—	—	—	3.3	0.0	92.0
	FISH											
77	-SOLUBLES, CONDENSED	5-01-969	A-F	50	8.8	0.4	—	—	—	5.1	7.1	28.6
			M-F	100	17.7	0.8	—	—	—	10.1	14.1	57.3
78	-SOLUBLES, DEHY	5-01-971	A-F	92	11.8	1.6	—	—	—	7.4	13.5	57.7
			M-F	100	12.8	1.7	—	—	—	8.1	14.7	62.7
	FISH, ANCHOVY *Engraulis ringen*											
79	-MEAL, MECH EXTD	5-01-985	A-F	92	15.0	1.0	—	—	—	4.1	6.3	65.6
			M-F	100	16.3	1.1	—	—	—	4.5	6.9	71.3
	FISH, MENHADEN *Brevoortia tyrannus*											
80	-MEAL, MECH EXTD	5-02-009	A-F	92	19.2	0.8	—	—	—	9.8	0.8	61.4
			M-F	100	20.9	0.9	—	—	—	10.6	0.9	66.7
	FISH, SARDINE											
81	-MEAL, MECH EXTD	5-02-015	A-F	93	15.8	1.0	—	—	—	5.0	6.1	65.1
			M-F	100	17.0	1.1	—	—	—	5.4	6.5	70.0
	FISH, WHITE *Gadidae* (family), *Lophiidae* (family), *Rafidae* (family)											
82	-MEAL, MECH EXTD	5-02-025	A-F	91	23.9	0.7	—	—	—	4.7	1.1	60.5
			M-F	100	26.3	0.8	—	—	—	5.2	1.2	66.5
	FLAX *Linum usitatissimum*											
83	-SEEDS, MEAL, SOLV EXTD, 33% PROTEIN (LINSEED MEAL)	5-26-089	A-F	90	6.2	8.1	—	—	—	8.4	33.6	33.6
			M-F	100	6.9	9.9	—	—	—	9.4	37.4	37.4
	GARBAGE											
84	-HOTEL AND RESTAURANT, BOILED, WET	4-07-865	A-F	26	1.4	0.7	—	—	—	5.8	13.8	4.3
			M-F	100	5.3	2.7	—	—	—	22.4	53.2	16.3
85	-HOTEL AND RESTAURANT MEAL, BOILED	4-07-879	A-F	54	3.6	1.6	—	—	—	14.6	24.7	9.5
			M-F	100	6.6	2.9	—	—	—	27.1	45.8	17.7
	GRAIN SCREENINGS (SEE SPECIFIC GRAINS)											
86	**HOMINY FEED (CORN)**	4-02-887	A-F	90	2.7	5.0	—	—	—	7.4	64.0	11.0
			M-F	100	3.0	5.6	—	—	—	8.2	71.1	12.2
	MEAT											
87	-MEAL, RENDERED	5-00-385	A-F	93	24.8	2.4	—	—	—	9.0	3.0	53.8
			M-F	100	26.6	2.6	—	—	—	9.7	3.3	57.9
88	-WITH BLOOD, MEAL, TANKAGE RENDERED	5-00-386	A-F	92	21.8	2.2	—	—	—	9.0	-0.3	59.5
			M-F	100	23.6	2.4	—	—	—	9.7	-0.4	64.7
89	-WITH BLOOD WITH BONE, MEAL, TANKAGE RENDERED	5-00-387	A-F	92	27.8	2.4	—	—	—	10.5	1.0	50.2
			M-F	100	30.2	2.6	—	—	—	11.4	1.1	54.6
90	-WITH BONE, MEAL, RENDERED	5-00-388	A-F	93	28.4	2.0	—	—	—	9.9	2.2	50.5
			M-F	100	30.5	2.2	—	—	—	10.6	2.4	54.3
	MILK											
91	-FRESH (COW'S)	5-01-168	A-F	13	0.8	—	—	—	—	3.7	4.9	3.6
			M-F	100	5.9	—	—	—	—	28.8	37.9	27.4
92	-SKIMMED, DEHY (COW'S)	5-01-175	A-F	94	8.0	0.3	—	—	—	0.9	51.3	33.5
			M-F	100	8.5	0.3	—	—	—	0.9	54.6	35.7
93	-SKIMMED, FRESH (COW'S)	5-01-170	A-F	9	0.7	—	—	—	—	0.1	5.2	3.1
			M-F	100	7.2	—	—	—	—	1.0	57.3	34.5

(Continued)

Entry Number	TDN	Digestible Energy		Metabolizable Energy		Calcium (Ca)	Phosphorus (P)	Sodium (Na)	Chlorine (Cl)	Magnesium (Mg)	Potassium (K)	Sulfur (S)
						Macrominerals						
	(%)	(kcal)		(kcal)		(%)	(%)	(%)	(%)	(%)	(%)	(%)
		(lb)	(kg)	(lb)	(kg)							
73	198	3,992	8,800	3,583	7,900	—	—	—	—	—	—	—
	280	4,031	8,888	3,620	7,980	—	—	—	—	—	—	—
74	—	3,485	7,682	3,458	7,623	—	—	—	—	—	—	—
	—	3,520	7,760	3,493	7,700	—	—	—	—	—	—	—
75	—	—	—	—	—	—	—	—	—	—	—	—
	—	—	—	—	—	—	—	—	—	—	—	—
76	62	1,239	2,731	1,018	2,244	0.28	0.66	0.71	—	0.21	0.30	—
	67	1,332	2,936	1,095	2,413	0.30	0.71	0.76	—	0.22	0.33	—
77	43	858	1,892	577	1,273	0.34	0.61	2.34	3.18	0.02	1.54	0.12
	86	1,717	3,784	1,155	2,546	0.68	1.23	4.67	6.37	0.04	3.08	0.24
78	—	—	—	1,298	2,862	1.25	1.96	0.37	—	—	1.90	—
	—	—	—	1,411	3,111	1.36	2.13	0.40	—	—	2.07	—
79	70	1,370	3,021	1,138	2,509	3.76	2.48	0.84	0.30	0.24	0.68	—
	76	1,489	3,283	1,237	2,728	4.08	2.70	0.92	0.32	0.26	0.74	—
80	62	1,486	3,276	1,177	2,595	5.16	2.92	0.37	0.61	0.14	0.69	—
	67	1,615	3,561	1,280	2,821	5.61	3.17	0.40	0.66	0.15	0.74	—
81	67	1,334	2,942	1,146	2,527	4.60	2.68	0.18	0.41	0.10	0.28	—
	72	1,435	3,163	1,233	2,717	4.95	2.88	0.19	0.44	0.11	0.30	—
82	68	1,365	3,009	1,104	2,433	7.33	3.58	0.78	—	0.22	0.79	—
	75	1,500	3,307	1,213	2,674	8.06	3.93	0.85	—	0.24	0.87	—
83	84	1,690	3,719	1,495	3,290	—	—	—	—	—	—	—
	94	1,881	4,139	1,664	3,661	—	—	—	—	—	—	—
84	27	550	1,212	510	1,124	0.11	0.07	—	—	—	—	—
	106	2,115	4,662	1,960	4,322	0.42	0.27	—	—	—	—	—
85	61	1,216	2,680	1,124	2,477	0.32	0.22	—	—	0.17	—	—
	113	2,251	4,963	2,081	4,587	0.60	0.40	—	—	0.32	—	—
86	76	1,519	3,349	1,421	3,132	0.05	0.53	0.08	0.05	0.23	0.54	0.03
	84	1,688	3,721	1,579	3,481	0.05	0.59	0.09	0.06	0.26	0.59	0.03
87	63	928	2,046	1,079	2,379	7.96	4.00	1.31	1.20	0.27	0.57	0.50
	68	998	2,200	1,160	2,558	8.56	4.30	1.41	1.29	0.29	0.62	0.53
88	68	1,046	2,305	1,020	2,248	5.80	2.99	1.68	1.73	0.34	0.57	0.70
	74	1,137	2,506	1,108	2,443	6.31	3.25	1.82	1.88	0.36	0.62	0.76
89	67	1,344	2,962	1,188	2,618	6.97	4.59	1.71	—	—	0.57	0.26
	73	1,461	3,220	1,291	2,846	7.58	4.99	1.86	—	—	0.62	0.28
90	68	892	1,966	934	2,059	10.16	4.89	0.73	0.74	1.13	1.28	0.26
	73	959	2,114	1,004	2,214	10.92	5.26	0.78	0.80	1.22	1.38	0.28
91	16	323	712	292	644	0.12	0.10	0.05	0.20	—	0.14	—
	124	2,484	5,476	2,247	4,954	0.93	0.75	0.39	1.56	—	1.11	—
92	86	1,722	3,796	1,618	3,567	1.29	1.02	0.35	0.90	0.12	1.56	0.32
	92	1,832	4,038	1,721	3,795	1.37	1.09	0.37	0.96	0.13	1.66	0.34
93	9	177	390	157	347	0.12	0.09	—	—	0.01	0.14	0.03
	98	1,964	4,330	1,748	3,855	1.37	1.05	—	—	0.11	1.58	0.32

(Continued)

Entry Number	Feed Name Description	Moisture Basis: A-F (as-fed) or M-F (moisture-free)	Cobalt (Co) (ppm or mg/kg)	Copper (Cu) (ppm or mg/kg)	Iodine (I) (ppm or mg/kg)	Iron (Fe) (%)	Manganese (Mn) (ppm or mg/kg)	Selenium (Se) (ppm or mg/kg)	Zinc (Zn) (ppm or mg/kg)	A (1 mg Carotene = 1667 IU Vit. A) (IU/g)	Carotene (Provitamin A) (ppm or mg/kg)
	FATS AND OILS										
73	-FAT, ANIMAL	A-F	—	—	—	—	—	—	—	—	—
		M-F	—	—	—	—	—	—	—	—	—
74	-FAT (LARD), SWINE	A-F	—	—	—	—	—	—	—	—	—
		M-F	—	—	—	—	—	—	—	—	—
75	-TALLOW, ANIMAL	A-F	—	—	—	—	—	—	—	—	—
		M-F	—	—	—	—	—	—	—	—	—
	FEATHERS										
76	-HYDROLYZED MEAL (POULTRY)	A-F	0.043	6.5	0.043	0.008	8.7	—	54.2	—	—
		M-F	0.047	7.0	0.047	0.008	9.3	—	58.3	—	—
	FISH										
77	-SOLUBLES, CONDENSED	A-F	0.068	42.4	1.099	0.023	11.6	—	33.8	2.2	1.3
		M-F	0.135	84.7	2.198	0.047	23.3	—	67.6	4.3	2.6
78	-SOLUBLES, DEHY	A-F	—	—	—	—	50.0	—	76.0	—	—
		M-F	—	—	—	—	54.3	—	82.6	—	—
79	**FISH, ANCHOVY** *Engraulis ringen* -MEAL, MECH EXTD	A-F	0.173	9.1	0.863	0.021	10.8	1.361	105.6	—	—
		M-F	0.188	9.9	0.938	0.023	11.7	1.480	114.8	—	—
80	**FISH, MENHADEN** *Brevoortia tyrannus* -MEAL, MECH EXTD	A-F	0.153	10.8	1.096	0.046	33.4	2.201	149.0	—	—
		M-F	0.167	11.8	1.191	0.050	36.3	2.392	161.9	—	—
81	**FISH, SARDINE** -MEAL, MECH EXTD	A-F	0.183	20.2	—	0.030	23.1	1.769	—	—	—
		M-F	0.196	21.7	—	0.032	24.9	1.902	—	—	—
82	**FISH, WHITE** *Gadidae* (family), *Lophiidae* (family), *Rafidae* (family) -MEAL, MECH EXTD	A-F	—	5.3	—	0.012	13.0	1.731	79.1	—	—
		M-F	—	5.9	—	0.013	14.3	1.902	87.0	—	—
83	**FLAX** *Linum usitatissimum* -SEEDS, MEAL, SOLV EXTD, 33% PROTEIN (LINSEED MEAL)	A-F	—	—	—	—	—	—	—	—	—
		M-F	—	—	—	—	—	—	—	—	—
84	**GARBAGE** -HOTEL AND RESTAURANT, BOILED, WET	A-F	—	—	—	—	—	—	—	—	—
		M-F	—	—	—	—	—	—	—	—	—
85	-HOTEL AND RESTAURANT, MEAL, BOILED	A-F	—	9.7	—	0.013	4.9	—	—	—	—
		M-F	—	18.0	—	0.024	9.0	—	—	—	—
	GRAIN SCREENINGS (SEE SPECIFIC GRAINS)										
86	**HOMINY FEED (CORN)**	A-F	0.059	13.7	—	0.007	14.5	—	—	15.3	9.2
		M-F	0.066	15.2	—	0.008	16.1	—	—	17.0	10.2
	MEAT										
87	-MEAL, RENDERED	A-F	0.129	9.8	—	0.044	9.6	0.420	104.4	—	—
		M-F	0.139	10.5	—	0.047	10.3	0.452	112.3	—	—
88	-WITH BLOOD, MEAL, TANKAGE RENDERED	A-F	0.154	38.7	—	0.210	19.1	—	—	—	—
		M-F	0.167	42.1	—	0.228	20.8	—	—	—	—
89	-WITH BLOOD WITH BONE, MEAL, TANKAGE RENDERED	A-F	0.180	39.7	—	—	19.6	0.258	—	—	—
		M-F	0.196	43.1	—	—	21.3	0.281	—	—	—
90	-WITH BONE, MEAL, RENDERED	A-F	0.180	1.5	1.313	0.050	13.3	0.262	95.3	—	—
		M-F	0.193	1.6	1.412	0.054	14.3	0.282	102.5	—	—
	MILK										
91	-FRESH (COW'S)	A-F	—	0.0	—	—	—	—	—	1.5	0.9
		M-F	—	0.3	—	—	—	—	—	11.5	6.9
92	-SKIMMED, DEHY (COW'S)	A-F	0.113	11.7	—	0.005	2.2	0.123	41.0	—	—
		M-F	0.120	12.4	—	0.005	2.3	0.131	43.6	—	—
93	-SKIMMED, FRESH (COW'S)	A-F	—	1.0	—	0.001	0.2	—	—	—	—
		M-F	—	11.6	—	0.005	2.3	—	—	—	—

(Continued)

Entry Number	Fat-Soluble Vitamins			Water-Soluble Vitamins								
	D	E (α-tocopherol)	K	B-12	Biotin	Choline	Folic Acid (Folacin)	Niacin (Nicotinic Acid)	Pantothenic Acid	Pyridoxine (B-6)	Riboflavin (B-2)	Thiamin (B-1)
	(IU/kg)	(ppm or mg/kg)	(ppm or mg/kg)	(ppb or mcg/kg)	(ppm or mg/kg)	(ppm or mg/kg)	(ppm or mg/kg)	(ppm or mg/kg)	(ppm or mg/kg)	(ppm or mg/kg)	(ppm or mg/kg)	(ppm or mg/kg)
73	—	—	—	—	—	—	—	—	—	—	—	—
	—	—	—	—	—	—	—	—	—	—	—	—
74	—	22.8	—	—	—	—	—	—	—	—	—	—
	—	23.0	—	—	—	—	—	—	—	—	—	—
75	—	—	—	—	—	—	—	—	—	—	—	—
	—	—	—	—	—	—	—	—	—	—	—	—
76	—	—	—	80.8	0.25	896	0.22	22	9.4	4.42	2.0	0.1
	—	—	—	86.8	0.27	964	0.23	24	10.2	4.75	2.2	0.1
77	—	—	—	273.0	0.14	3,378	0.22	153	32.0	12.04	11.3	4.4
	—	—	—	546.0	0.27	6,757	0.44	305	64.0	24.08	22.6	8.8
78	—	6.0	—	324.0	0.26	5,718	0.43	261	50.0	23.81	11.9	7.9
	—	6.5	—	352.2	0.28	6,216	0.47	284	54.3	25.88	13.0	8.6
79	—	4.5	—	195.8	0.20	3,883	0.18	78	11.3	4.64	7.0	0.5
	—	4.9	—	212.8	0.21	4,221	0.19	84	12.3	5.05	7.6	0.5
80	—	12.0	—	117.8	0.18	3,126	0.44	55	8.8	4.68	4.8	0.6
	—	13.1	—	128.0	0.20	3,398	0.47	60	9.6	5.09	5.2	0.6
81	—	—	—	237.6	0.10	3,272	—	75	11.0	—	5.4	0.3
	—	—	—	255.5	0.11	3,518	—	81	11.8	—	5.8	0.3
82	—	8.9	—	89.3	0.08	5,020	0.20	62	9.6	5.90	9.3	3.8
	—	9.8	—	98.1	0.09	5,516	0.22	68	10.6	6.49	10.2	4.2
83	—	—	—	—	—	—	—	—	—	—	—	—
	—	—	—	—	—	—	—	—	—	—	—	—
84	—	—	—	—	—	—	—	—	—	—	—	—
	—	—	—	—	—	—	—	—	—	—	—	—
85	—	—	—	—	—	—	—	—	—	—	—	—
	—	—	—	—	—	—	—	—	—	—	—	—
86	—	—	—	—	0.13	993	0.28	47	7.5	10.93	2.1	7.9
	—	—	—	—	0.14	1,104	0.31	53	8.4	12.14	2.3	8.8
87	—	1.0	—	64.2	0.13	2,046	0.37	58	6.4	3.89	5.3	0.2
	—	1.1	—	69.0	0.14	2,200	0.40	63	6.9	4.18	5.7	0.2
88	—	—	—	236.6	—	1,704	1.54	37	2.4	—	2.3	0.4
	—	—	—	257.2	—	1,852	1.67	40	2.6	—	2.5	0.4
89	—	0.8	—	82.4	0.07	2,150	0.57	46	3.7	—	3.6	0.2
	—	0.9	—	89.5	0.08	2,337	0.62	52	4.1	—	3.9	0.2
90	—	0.9	—	109.1	0.10	2,010	0.37	49	4.2	8.73	4.5	0.6
	—	1.0	—	117.3	0.11	2,162	0.40	53	4.5	9.39	4.9	0.7
91	—	—	—	—	—	—	—	1	2.9	—	1.8	0.4
	—	—	—	—	—	—	—	10	22.4	—	13.5	2.7
92	419	9.4	—	33.5	0.33	1,391	0.50	11	36.9	4.25	19.1	3.8
	446	10.0	—	35.6	0.35	1,480	0.53	12	39.3	4.52	20.4	4.0
93	—	—	—	—	—	—	—	1	3.3	—	1.9	0.4
	—	—	—	—	—	—	—	12	37.1	—	20.9	4.6

(Continued)

TABLE 21-1

Entry Number	Feed Name Description	International Feed Number[1]	Moisture Basis: A-F (as-fed) or M-F (moisture-free)	Chemical Analysis								
				Dry Matter	Ash	Crude Fiber	Cell Walls or NDF	Acid Detergent Fiber	Lignin	Ether Extract (Fat)	N-Free Extract	Crude Protein
				(%)	(%)	(%)	(%)	(%)	(%)	(%)	(%)	(%)
	MILLET *Setaria* spp											
94	-GRAIN	4-03-098	A-F	90	3.0	6.8	—	—	—	4.0	64.1	12.1
			M-F	100	3.3	7.5	—	—	—	4.5	71.2	13.5
	MOLASSES AND SYRUP											
95	-BEET, SUGAR, MOLASSES, MORE THAN 48% INVERT SUGAR, MORE THAN 79.5° BRIX	4-00-668	A-F	78	8.7	—	—	—	—	0.1	62.2	6.0
			M-F	100	11.2	—	—	—	—	0.2	79.7	7.7
96	-CITRUS, SYRUP (CITRUS MOLASSES)	4-01-241	A-F	68	5.0	—	—	—	—	0.2	57.3	4.8
			M-F	100	7.4	—	—	—	—	0.3	84.3	7.1
97	-SUGARCANE, MOLASSES, DEHY	4-04-695	A-F	90	11.1	4.5	—	—	—	0.9	65.1	8.4
			M-F	100	12.3	5.0	—	—	—	1.0	72.3	9.3
98	-SUGARCANE, MOLASSES (BLACKSTRAP) MORE THAN 46% INVERT SUGAR, MORE THAN 79.5° BRIX	4-04-696	A-F	75	7.7	—	—	—	—	0.1	64.2	3.9
			M-F	100	10.3	—	—	—	—	0.1	85.7	5.2
	OATS *Avena sativa*											
99	-CEREAL BYPRODUCT (FEEDING OAT MEAL, OAT MIDDLINGS)	4-03-303	A-F	91	2.3	4.4	—	—	—	6.5	63.1	14.7
			M-F	100	2.5	4.8	—	—	—	7.2	69.4	16.2
100	-GRAIN, ALL ANALYSES	4-03-309	A-F	89	3.0	10.9	—	—	—	4.9	58.0	12.1
			M-F	100	3.4	12.2	—	—	—	5.6	65.2	13.6
101	-GRAIN, GRADE 1 HEAVY, 36 LB/BUSHEL OR *463 G/L*	4-03-312	A-F	89	2.7	8.9	—	—	—	5.2	59.6	12.6
			M-F	100	3.0	10.0	—	—	—	5.8	67.0	14.2
102	-GRAIN, GRADE 1, 34 LB/BUSHEL OR *438 G/L*	4-03-313	A-F	90	—	11.1	—	—	—	4.5	—	12.1
			M-F	100	—	12.3	—	—	—	5.0	—	13.4
103	-GRAIN, GRADE 2, 32 LB/BUSHEL OR *412 G/L*	4-03-316	A-F	89	3.3	10.9	—	—	—	4.2	59.3	11.4
			M-F	100	3.7	12.2	—	—	—	4.7	66.6	12.8
104	-GRAIN, GRADE SAMPLE	4-03-310	A-F	89	4.1	11.7	—	—	—	4.1	57.6	11.5
			M-F	100	4.6	13.2	—	—	—	4.6	64.7	12.9
105	-GRAIN, LIGHT, LESS THAN 27 LB/BUSHEL OR *347 G/L*	4-03-318	A-F	91	4.2	14.5	—	—	—	4.5	55.9	11.9
			M-F	100	4.6	15.9	—	—	—	4.9	61.5	13.1
106	-GRAIN, PACIFIC COAST	4-07-999	A-F	91	3.8	11.3	—	—	—	4.9	61.8	9.2
			M-F	100	4.2	12.4	—	—	—	5.4	67.9	10.1
107	-GROATS	4-03-331	A-F	89	2.2	2.4	—	—	—	6.3	62.8	15.4
			M-F	100	2.5	2.6	—	—	—	7.0	70.5	17.3
	PEA *Pisum* spp											
108	-SEEDS	5-03-600	A-F	89	3.0	8.2	—	—	—	1.3	54.2	22.4
			M-F	100	3.3	9.2	—	—	—	1.4	60.9	25.2
	PEANUT *Arachis hypogaea*											
109	-KERNELS, MEAL, MECH EXTD, 45% PROTEIN (PEANUT MEAL)	5-03-649	A-F	90	5.3	7.9	—	—	—	7.4	25.3	44.1
			M-F	100	5.9	8.7	—	—	—	8.2	28.1	49.0
110	-KERNELS, MEAL, SOLV EXTD, 47% PROTEIN (PEANUT MEAL)	5-03-650	A-F	92	6.3	9.7	—	—	—	1.3	26.0	48.7
			M-F	100	6.8	10.6	—	—	—	1.5	28.3	52.9
	POTATO *Solanum tuberosum*											
111	-CANNERY RESIDUE, DEHY	4-03-775	A-F	89	3.0	5.8	—	—	—	0.3	72.6	7.2
			M-F	100	3.4	6.6	—	—	—	0.4	81.5	8.1
112	-TUBERS, BOILED	4-03-784	A-F	24	1.3	0.7	—	—	—	0.1	19.7	2.2
			M-F	100	5.3	3.0	—	—	—	0.3	82.3	9.2
113	-TUBERS, DEHY	4-07-850	A-F	91	6.9	2.0	—	—	—	0.5	73.7	7.9
			M-F	100	7.6	2.2	—	—	—	0.5	81.0	8.7
	POULTRY											
114	-BYPRODUCTS, MEAL, RENDERED	5-03-798	A-F	93	14.6	2.3	—	—	—	12.4	5.3	58.4
			M-F	100	15.7	2.5	—	—	—	13.4	5.7	62.8
115	-FEATHERS, HYDROLYZED, MEAL	5-03-795	A-F	93	3.3	1.2	—	—	—	3.1	0.0	85.6
			M-F	100	3.5	1.3	—	—	—	3.3	0.0	92.0
	RAPE *Brassica* spp											
116	-SEEDS, MEAL, SOLV EXTD, 34% PROTEIN	5-26-092	A-F	90	7.0	13.8	—	—	—	1.5	28.5	39.5
			M-F	100	7.8	15.3	—	—	—	1.7	31.6	43.6

(Continued)

Entry Number	TDN (%)	Digestible Energy (kcal) (lb)	(kg)	Metabolizable Energy (kcal) (lb)	(kg)	Macrominerals Calcium (Ca) (%)	Phosphorus (P) (%)	Sodium (Na) (%)	Chlorine (Cl) (%)	Magnesium (Mg) (%)	Potassium (K) (%)	Sulfur (S) (%)
94	73	1,467	3,234	1,225	2,701	0.05	0.28	0.04	0.14	0.16	0.43	0.13
	82	1,630	3,594	1,361	3,001	0.06	0.31	0.04	0.16	0.18	0.48	0.14
95	—	—	—	919	2,025	0.12	0.02	1.10	1.50	0.19	4.63	0.47
	—	—	—	1,178	2,597	0.15	0.03	1.41	1.92	0.24	5.94	0.60
96	—	—	—	—	—	1.09	0.10	0.27	0.07	0.14	0.09	—
	—	—	—	—	—	1.61	0.14	0.40	0.10	0.21	0.13	—
97	67	1,331	2,935	1,074	2,368	0.79	0.26	0.18	—	0.39	3.31	0.41
	74	1,479	3,261	1,194	2,632	0.87	0.29	0.19	—	0.43	3.68	0.46
98	57	1,144	2,523	1,012	2,232	0.78	0.09	0.17	2.78	0.35	2.85	0.35
	76	1,526	3,364	1,350	2,976	1.05	0.11	0.22	3.71	0.47	3.80	0.46
99	79	1,574	3,470	1,459	3,218	0.07	0.44	0.09	0.05	0.16	0.53	0.26
	86	1,729	3,813	1,604	3,536	0.08	0.48	0.10	0.05	0.18	0.59	0.29
100	64	1,285	2,832	1,202	2,648	0.06	0.33	0.16	0.11	0.12	0.39	0.21
	72	1,444	3,182	1,350	2,976	0.07	0.37	0.18	0.12	0.14	0.44	0.23
101	71	1,416	3,121	1,318	2,907	—	—	—	—	—	—	—
	80	1,591	3,507	1,481	3,266	—	—	—	—	—	—	—
102	—	—	—	—	—	0.09	0.32	0.06	0.12	—	0.37	—
	—	—	—	—	—	0.10	0.36	0.07	0.13	—	0.41	—
103	67	1,343	2,960	1,254	2,765	0.06	0.27	—	—	—	—	—
	75	1,509	3,326	1,409	3,107	0.07	0.30	—	—	—	—	—
104	65	1,292	2,849	1,207	2,660	—	—	—	—	—	—	—
	73	1,452	3,201	1,356	2,989	—	—	—	—	—	—	—
105	61	1,227	2,706	1,146	2,526	—	—	—	—	—	—	—
	67	1,349	2,973	1,259	2,776	—	—	—	—	—	—	—
106	69	1,373	3,026	1,112	2,452	0.10	0.31	—	—	—	—	0.21
	75	1,509	3,326	1,222	2,695	0.11	0.34	—	—	—	—	0.23
107	85	1,693	3,731	1,320	2,910	0.08	0.42	0.05	0.09	0.10	0.35	0.20
	95	1,902	4,193	1,483	3,269	0.09	0.47	0.06	0.10	0.11	0.40	0.22
108	76	1,517	3,345	1,379	3,041	0.12	0.42	0.04	0.06	—	0.70	—
	85	1,705	3,759	1,550	3,417	0.13	0.47	0.04	0.07	—	0.79	—
109	84	1,670	3,682	1,438	3,170	0.16	0.56	0.41	0.03	0.32	1.13	0.29
	93	1,856	4,092	1,598	3,523	0.18	0.62	0.45	0.03	0.36	1.25	0.32
110	81	1,622	3,577	1,375	3,031	0.20	0.63	0.42	—	0.04	1.14	—
	88	1,763	3,888	1,495	3,295	0.22	0.69	0.45	—	0.04	1.24	—
111	77	1,542	3,399	1,455	3,207	0.09	0.25	—	—	—	—	—
	87	1,732	3,819	1,635	3,604	0.10	0.28	—	—	—	—	—
112	21	420	925	395	871	—	—	—	—	—	—	—
	87	1,748	3,853	1,646	3,628	—	—	—	—	—	—	—
113	76	1,517	3,344	1,456	3,210	0.06	0.19	0.01	0.36	—	1.99	—
	83	1,667	3,674	1,600	3,527	0.07	0.21	0.01	0.40	—	2.19	—
114	73	1,394	3,073	1,257	2,770	3.88	2.09	0.67	0.48	0.21	0.30	—
	79	1,499	3,305	1,351	2,979	4.17	2.25	0.72	0.52	0.22	0.32	—
115	62	1,239	2,731	1,018	2,244	0.28	0.66	0.71	—	0.21	0.30	—
	67	1,332	2,936	1,095	2,415	0.30	0.71	0.76	—	0.22	0.33	—
116	68	1,357	2,986	1,191	2,621	0.40	0.90	—	—	—	—	—
	75	1,504	3,308	1,320	2,903	0.44	1.00	—	—	—	—	—

(Continued)

TABLE 21-1

Entry Number	Feed Name Description	Moisture Basis: A-F (as-fed) or M-F (moisture-free)	Microminerals							Fat-Soluble Vitamins	
			Cobalt (Co)	Copper (Cu)	Iodine (I)	Iron (Fe)	Man-ganese (Mn)	Sele-nium (Se)	Zinc (Zn)	A (1 mg Carotene = 1667 IU Vit. A)	Carotene (Provitamin A)
			(ppm or mg/kg)	*(ppm or mg/kg)*	*(ppm or mg/kg)*	*(%)*	*(ppm or mg/kg)*	*(ppm or mg/kg)*	*(ppm or mg/kg)*	*(IU/g)*	*(ppm or mg/kg)*
	MILLET *Setaria* spp										
94	-GRAIN	A-F	0.044	21.8	—	0.006	29.9	—	13.9	—	—
		M-F	0.049	24.3	—	0.007	33.3	—	15.4	—	—
	MOLASSES AND SYRUP										
95	-BEET, SUGAR, MOLASSES, MORE THAN 48% INVERT SUGAR, MORE THAN 79.5° BRIX	A-F	0.374	17.1	—	0.007	4.5	—	—	—	—
		M-F	0.480	22.0	—	0.009	5.7	—	—	—	—
96	-CITRUS, SYRUP (CITRUS MOLASSES)	A-F	0.108	73.4	—	0.034	26.2	—	93.2	—	—
		M-F	0.159	108.0	—	0.050	38.5	—	137.0	—	—
97	-SUGARCANE, MOLASSES, DEHY	A-F	1.091	65.5	—	0.021	46.7	—	—	—	—
		M-F	1.213	72.8	—	0.024	51.9	—	—	—	—
98	-SUGARCANE, MOLASSES (BLACKSTRAP), MORE THAN 46% INVERT SUGAR, MORE THAN 79.5° BRIX	A-F	0.908	60.4	1.577	0.019	42.9	—	22.0	—	—
		M-F	1.210	80.5	2.103	0.026	57.1	—	30.0	—	—
	OATS *Avena sativa*										
99	-CEREAL BYPRODUCT (FEEDING OAT MEAL, OAT MIDDLINGS)	A-F	0.044	4.4	—	0.030	43.5	—	140.0	—	—
		M-F	0.049	4.8	—	0.033	47.8	—	153.8	—	—
100	-GRAIN, ALL ANALYSES	A-F	0.057	5.8	0.088	0.008	37.0	0.210	36.6	0.2	0.1
		M-F	0.064	6.5	0.099	0.009	41.6	0.236	41.1	0.2	0.1
101	-GRAIN, GRADE 1 HEAVY, 36 LB/BUSHEL OR *463 G/L*	A-F	—	—	—	—	—	—	—	—	—
		M-F	—	—	—	—	—	—	—	—	—
102	-GRAIN, GRADE 1, 34 LB/BUSHEL OR *438 G/L*	A-F	—	—	—	—	38.2	—	—	—	—
		M-F	—	—	—	—	42.5	—	—	—	—
103	-GRAIN, GRADE 2, 32 LB/BUSHEL OR *412 G/L*	A-F	—	—	—	—	—	—	—	—	—
		M-F	—	—	—	—	—	—	—	—	—
104	-GRAIN, GRADE SAMPLE	A-F	—	—	—	—	—	—	—	—	—
		M-F	—	—	—	—	—	—	—	—	—
105	-GRAIN, LIGHT, LESS THAN 27 LB/BUSHEL OR *347 G/L*	A-F	—	—	—	—	—	—	—	—	—
		M-F	—	—	—	—	—	—	—	—	—
106	-GRAIN, PACIFIC COAST	A-F	—	—	—	—	—	0.076	—	—	—
		M-F	—	—	—	—	—	0.083	—	—	—
107	-GROATS	A-F	—	5.9	—	0.008	46.0	—	—	—	—
		M-F	—	6.6	—	0.009	51.7	—	—	—	—
	PEA *Pisum* spp										
108	-SEEDS	A-F	—	—	—	0.005	—	—	29.3	—	—
		M-F	—	—	—	0.006	—	—	33.0	—	—
	PEANUT *Arachis hypogaea*										
109	-KERNELS, MEAL, MECH EXTD, 45% PROTEIN (PEANUT MEAL)	A-F	—	—	—	—	25.4	—	—	0.4	0.2
		M-F	—	—	—	—	28.2	—	—	0.4	0.2
110	-KERNELS, MEAL, SOLV EXTD, 47% PROTEIN (PEANUT MEAL)	A-F	0.109	15.3	0.065	0.027	26.8	—	32.7	—	—
		M-F	0.119	16.6	0.071	0.029	29.2	—	35.6	—	—
	POTATO *Solanum tuberosum*										
111	-CANNERY RESIDUE, DEHY	A-F	—	—	—	—	—	—	—	—	—
		M-F	—	—	—	—	—	—	—	—	—
112	-TUBERS, BOILED	A-F	—	—	—	—	—	—	—	—	—
		M-F	—	—	—	—	—	—	—	—	—
113	-TUBERS, DEHY	A-F	—	—	—	—	2.3	—	2.0	—	—
		M-F	—	—	—	—	2.5	—	2.2	—	—
	POULTRY										
114	-BYPRODUCTS, MEAL, RENDERED	A-F	0.220	14.0	3.073	0.076	20.0	0.774	568.5	—	—
		M-F	0.236	15.1	3.305	0.082	21.5	0.832	611.3	—	—
115	-FEATHERS, HYDROLYZED, MEAL	A-F	0.043	6.5	0.043	0.008	8.7	—	54.2	—	—
		M-F	0.047	7.0	0.047	0.008	9.3	—	58.3	—	—
	RAPE *Brassica* spp										
116	-SEEDS, MEAL, SOLV EXTD, 34% PROTEIN	A-F	—	—	—	—	—	—	—	—	—
		M-F	—	—	—	—	—	—	—	—	—

(Continued)

Entry Number	Fat-Soluble Vitamins			Water-Soluble Vitamins								
	D	E (α-tocopherol)	K	B-12	Biotin	Choline	Folic Acid (Folacin)	Niacin (Nicotinic Acid)	Panto-thenic Acid	Pyri-doxine (B-6)	Ribo-flavin (B-2)	Thiamin (B-1)
	(IU/kg)	(ppm or mg/kg)	(ppm or mg/kg)	(ppb or mcg/kg)	(ppm or mg/kg)	(ppm or mg/kg)	(ppm or mg/kg)	(ppm or mg/kg)	(ppm or mg/kg)	(ppm or mg/kg)	(ppm or mg/kg)	(ppm or mg/kg)
94	—	—	—	—	—	792	—	53	7.4	—	1.6	6.6
	—	—	—	—	—	880	—	59	8.2	—	1.7	7.3
95	—	—	—	—	—	829	—	41	4.5	—	2.3	—
	—	—	—	—	—	1,063	—	53	5.7	—	2.9	—
96	—	—	—	—	—	—	—	27	12.6	—	6.2	—
	—	—	—	—	—	—	—	39	18.6	—	9.1	—
97	—	5.1	—	—	—	772	—	35	37.5	—	3.3	0.9
	—	5.6	—	—	—	857	—	39	41.6	—	3.7	1.0
98	—	5.0	—	—	0.71	744	0.11	41	39.2	6.50	2.9	0.9
	—	6.7	—	—	0.94	992	0.15	54	52.2	8.67	3.8	1.2
99	—	24.0	—	—	0.22	1,148	0.51	21	18.0	—	1.8	7.0
	—	26.4	—	—	0.24	1,262	0.56	23	19.7	—	1.9	7.7
100	—	12.9	—	—	0.24	1,013	0.34	14	7.1	2.50	1.5	6.4
	—	14.5	—	—	0.27	1,138	0.39	15	8.0	2.81	1.5	7.2
101	—	—	—	—	—	—	—	—	—	—	—	—
	—	—	—	—	—	—	—	—	—	—	—	—
102	—	20.1	—	—	0.11	1,106	0.30	18	13.1	1.31	1.1	—
	—	22.3	—	—	0.12	1,229	0.34	20	14.5	1.45	1.2	—
103	—	—	—	—	—	—	—	—	—	—	—	—
	—	—	—	—	—	—	—	—	—	—	—	—
104	—	—	—	—	—	—	—	—	—	—	—	—
	—	—	—	—	—	—	—	—	—	—	—	—
105	—	—	—	—	—	—	—	—	—	—	—	—
	—	—	—	—	—	—	—	—	—	—	—	—
106	—	20.2	—	—	—	918	—	14	11.7	—	1.2	—
	—	22.2	—	—	—	1,009	—	16	12.8	—	1.3	—
107	—	14.7	—	—	—	1,156	—	16	13.1	3.15	14.9	11.6
	—	16.5	—	—	—	1,298	—	18	14.8	3.54	16.8	13.0
108	—	—	—	—	—	633	0.35	33	9.8	0.98	2.3	101.5
	—	—	—	—	—	712	0.40	37	11.0	1.10	2.6	114.4
109	—	—	—	—	—	1,650	—	166	47.2	—	5.2	7.1
	—	—	—	—	—	1,833	—	184	52.5	—	5.8	7.9
110	—	—	—	—	0.33	1,963	0.65	172	50.2	5.45	10.9	5.7
	—	—	—	—	0.36	2,133	0.71	187	54.5	5.93	11.9	6.2
111	—	—	—	—	—	—	—	—	—	—	—	—
	—	—	—	—	—	—	—	—	—	—	—	—
112	—	—	—	—	—	—	—	—	—	—	—	—
	—	—	—	—	—	—	—	—	—	—	—	—
113	—	—	—	—	0.10	2,620	0.60	33	20.0	14.11	0.7	—
	—	—	—	—	0.11	2,879	0.66	37	22.0	15.50	0.7	—
114	—	2.1	—	299.8	0.19	6,022	0.74	53	11.9	4.39	10.3	0.2
	—	2.3	—	322.3	0.20	6,475	0.79	57	12.8	4.72	11.0	0.2
115	—	—	—	80.8	0.25	896	0.22	22	9.4	4.42	2.0	0.1
	—	—	—	86.8	0.27	964	0.23	24	10.2	4.75	2.2	0.1
116	—	—	—	—	—	—	—	—	—	—	—	—
	—	—	—	—	—	—	—	—	—	—	—	—

(Continued)

COMPOSITION OF FEEDS

TABLE 21-1

Entry Number	Feed Name Description	International Feed Number[1]	Moisture Basis: A-F (as-fed) or M-F (moisture-free)	Chemical Analysis								
				Dry Matter	Ash	Crude Fiber	Cell Walls or NDF	Acid Detergent Fiber	Lignin	Ether Extract (Fat)	N-Free Extract	Crude Protein
				(%)	(%)	(%)	(%)	(%)	(%)	(%)	(%)	(%)
	RAPE, CANADA *Brassica napus*											
117	-SEEDS, MEAL, PREPRESSED, SOLV EXTD, 40% PROTEIN	5-08-135	A-F	92	7.2	9.3	—	—	—	1.1	33.9	40.5
			M-F	100	7.8	10.1	—	—	—	1.2	36.8	44.0
	RICE *Oryza sativa*											
118	-BRAN WITH GERM (RICE BRAN)	4-03-928	A-F	91	12.3	11.6	—	—	—	13.9	40.5	12.8
			M-F	100	13.5	12.7	—	—	—	15.2	44.6	14.0
119	-GRAIN, GROUND (GROUND ROUGH RICE, GROUND PADDY RICE)	4-03-938	A-F	89	4.7	9.1	—	—	—	1.7	65.1	8.4
			M-F	100	5.3	10.3	—	—	—	1.9	73.1	9.4
120	-GROATS (RICE, BROWN)	4-03-936	A-F	88	1.0	0.8	—	—	—	1.7	76.2	8.3
			M-F	100	1.2	0.9	—	—	—	1.9	86.6	9.5
121	-GROATS, POLISHED (RICE, POLISHED)	4-03-942	A-F	89	0.5	0.4	—	—	—	0.4	80.4	7.3
			M-F	100	0.6	0.4	—	—	—	0.5	90.3	8.2
122	-POLISHINGS	4-03-943	A-F	90	7.6	3.2	—	—	—	12.6	54.6	12.1
			M-F	100	8.4	3.6	—	—	—	13.9	60.6	13.4
	RYE *Secale cereale*											
123	-GRAIN, ALL ANALYSES	4-04-047	A-F	87	1.6	2.2	—	—	—	1.5	69.7	12.0
			M-F	100	1.9	2.5	—	—	—	1.7	80.1	13.8
124	DISTILLERS' GRAINS, DEHY	5-04-023	A-F	92	2.3	12.3	—	—	—	7.2	48.6	21.6
			M-F	100	2.5	13.4	—	—	—	7.8	52.8	23.5
125	-DISTILLERS' GRIANS WITH SOLUBLES, DEHY	5-04-024	A-F	91	6.4	8.1	—	—	—	4.1	44.9	27.4
			M-F	100	7.1	9.0	—	—	—	4.5	49.4	30.1
	SAFFLOWER *Carthamus tinctorius*											
126	-SEEDS, MEAL, SOLV EXTD, 20% PROTEIN	5-26-095	A-F	92	4.7	32.3	—	—	—	3.9	29.5	21.4
			M-F	100	5.1	35.2	—	—	—	4.2	32.2	23.3
127	-SEEDS, MEAL, SOLV EXTD, 42% PROTEIN	5-26-094	A-F	90	6.4	8.5	—	—	—	1.7	29.4	44.5
			M-F	100	7.1	9.4	—	—	—	1.9	32.5	49.1
128	-SEEDS WITHOUT HULLS, MEAL, SOLV EXTD, 42% PROTEIN	5-07-959	A-F	90	7.9	15.0	—	—	—	0.9	24.4	41.8
			M-F	100	8.7	16.6	—	—	—	1.0	27.2	46.5
	SESAME *Sesamum indicum*											
129	-SEEDS, MEAL, MECH EXTD	5-04-220	A-F	93	11.2	5.7	—	—	—	7.9	23.8	44.4
			M-F	100	12.1	6.1	—	—	—	8.5	25.5	47.7
	SORGHUM *Sorghum vulgare*											
130	-GRAIN, ALL ANALYSES	4-04-383	A-F	90	1.7	2.4	—	—	—	2.8	71.7	11.4
			M-F	100	1.9	2.7	—	—	—	3.1	79.6	12.6
	SORGHUM, KAFIR *Sorghum vulgare, caffrorum*											
131	-GRAIN	4-04-428	A-F	89	1.5	2.1	—	—	—	2.8	71.6	11.0
			M-F	100	1.7	2.3	—	—	—	3.1	80.5	12.4
	SORGHUM, MILO *Sorghum vulgare, subglabrescens*											
132	-GRAIN	4-04-444	A-F	89	1.7	2.3	—	—	—	2.8	71.6	10.6
			M-F	100	1.9	2.6	—	—	—	3.2	80.4	11.9
	SOYBEAN *Glycine max*											
133	-SEEDS, MEAL, MECH EXTD, 41% PROTEIN	5-04-600	A-F	90	6.0	6.0	—	—	—	4.7	36.1	43.8
			M-F	100	6.7	6.7	—	—	—	5.2	34.6	48.7
134	-SEEDS, MEAL, SOLV EXTD, 44% PROTEIN	5-20-637	A-F	91	6.1	6.3	—	—	—	0.9	37.3	40.3
			M-F	100	6.7	7.0	—	—	—	1.0	41.0	44.3
135	-SEEDS, MEAL, SOLV EXTD, 49% PROTEIN	5-20-638	A-F	91	6.2	4.3	—	—	—	1.3	35.8	43.4
			M-F	100	6.9	4.7	—	—	—	1.5	39.4	47.6
136	-SEEDS, WHOLE	5-04-610	A-F	91	4.9	5.4	—	—	—	17.5	24.0	39.3
			M-F	100	5.4	5.9	—	—	—	19.2	26.3	43.2
137	-SEEDS WITHOUT HULLS, MEAL, SOLV EXTD, 49% PROTEIN	5-04-612	A-F	90	5.8	3.6	—	—	—	1.0	31.5	48.1
			M-F	100	6.4	4.0	—	—	—	1.1	35.0	53.4
	SPELT *Triticum spelta*											
138	-GRAIN	4-04-651	A-F	90	3.5	9.1	—	—	—	1.9	63.5	12.0
			M-F	100	3.9	10.2	—	—	—	2.1	70.5	13.3

(Continued)

Entry Number	TDN	Digestible Energy		Metabolizable Energy		Macrominerals						
						Calcium (Ca)	Phosphorus (P)	Sodium (Na)	Chlorine (Cl)	Magnesium (Mg)	Potassium (K)	Sulfur (S)
	(%)	(kcal)		(kcal)		(%)	(%)	(%)	(%)	(%)	(%)	(%)
		(lb)	(kg)	(lb)	(kg)							
117	—	—	—	—	—	0.66	0.93	—	—	—	—	—
	—	—	—	—	—	0.72	1.01	—	—	—	—	—
118	72	1,433	3,159	1,335	2,943	0.08	1.48	0.07	0.07	0.95	1.73	0.18
	79	1,575	3,472	1,467	3,234	0.09	1.62	0.08	0.08	1.05	1.90	0.20
119	68	1,363	3,005	1,150	2,535	0.06	0.43	0.04	0.08	0.23	0.52	0.05
	77	1,532	3,376	1,292	2,848	0.07	0.48	0.05	0.09	0.25	0.58	0.05
120	77	1,536	3,386	1,445	3,186	0.04	0.22	0.03	0.06	0.06	0.19	0.04
	87	1,745	3,848	1,642	3,620	0.04	0.25	0.03	0.07	0.07	0.21	0.05
121	80	1,601	3,529	1,510	3,329	0.02	0.11	0.02	0.04	0.02	0.10	0.08
	90	1,799	3,965	1,697	3,741	0.02	0.12	0.02	0.04	0.02	0.11	0.09
122	87	1,772	3,907	1,493	3,292	0.05	1.32	0.10	0.11	0.65	2.12	0.17
	97	1,969	4,341	1,659	3,658	0.05	1.47	0.11	0.12	0.72	2.36	0.19
123	76	1,512	3,332	1,280	2,821	0.06	0.32	0.02	0.03	0.12	0.45	0.15
	87	1,737	3,830	1,471	3,243	0.07	0.36	0.03	0.03	0.14	0.52	0.17
124	—	—	—	—	—	0.15	0.48	0.17	0.05	0.17	0.07	0.44
	—	—	—	—	—	0.16	0.52	0.18	0.05	0.18	0.08	0.47
125	—	—	—	—	—	—	—	—	—	—	—	—
	—	—	—	—	—	—	—	—	—	—	—	—
126	—	—	—	—	—	0.34	0.84	—	—	—	—	—
	—	—	—	—	—	0.37	0.92	—	—	—	—	—
127	—	—	—	—	—	0.24	1.66	—	—	—	—	—
	—	—	—	—	—	0.26	1.83	—	—	—	—	—
128	—	—	—	—	—	0.40	1.27	0.04	0.16	1.19	1.19	—
	—	—	—	—	—	0.44	1.41	0.04	0.18	1.33	1.33	—
129	71	1,411	3,110	1,285	2,832	2.02	1.39	0.15	0.07	0.80	1.28	—
	76	1,517	3,344	1,381	3,045	2.17	1.49	0.17	0.07	0.86	1.38	—
130	79	1,538	3,390	1,432	3,156	0.03	0.30	0.03	0.09	0.18	0.35	0.15
	88	1,709	3,767	1,591	3,507	0.03	0.33	0.03	0.10	0.20	0.39	0.16
131	81	1,618	3,567	1,414	3,117	0.03	0.31	0.05	0.10	0.15	0.33	0.16
	91	1,818	4,007	1,589	3,502	0.03	0.35	0.06	0.11	0.17	0.38	0.18
132	78	1,552	3,422	1,273	2,806	0.03	0.28	0.02	0.08	0.20	0.35	0.09
	87	1,744	3,845	1,430	3,153	0.03	0.32	0.02	0.09	0.22	0.39	0.10
133	77	1,535	3,378	1,241	2,731	0.27	0.63	0.24	0.07	0.25	1.71	0.33
	85	1,706	3,753	1,477	3,249	0.30	0.70	0.27	0.08	0.28	1.90	0.37
134	—	—	—	—	—	0.29	0.64	0.29	—	0.27	2.02	0.44
	—	—	—	—	—	0.32	0.71	0.32	—	0.30	2.22	0.48
135	—	—	—	—	—	0.93	0.64	0.38	—	—	2.03	—
	—	—	—	—	—	1.03	0.70	0.41	—	—	2.24	—
136	92	1,836	4,048	1,603	3,533	0.25	0.60	0.12	0.03	0.28	1.61	0.22
	101	2,018	4,449	1,761	3,883	0.28	0.66	0.13	0.03	0.31	1.77	0.24
137	75	1,783	3,931	1,444	3,184	0.25	0.61	0.36	0.05	—	1.77	0.43
	83	1,981	4,368	1,605	3,538	0.27	0.68	0.40	0.05	—	1.97	0.48
138	70	1,394	3,074	1,301	2,869	0.12	0.38	—	—	—	—	—
	77	1,549	3,416	1,446	3,187	0.13	0.42	—	—	—	—	—

(Continued)

COMPOSITION OF FEEDS

Entry Number	Feed Name Description	Moisture Basis: A-F (as-fed) or M-F (moisture-free)	Cobalt (Co)	Copper (Cu)	Iodine (I)	Iron (Fe)	Manganese (Mn)	Selenium (Se)	Zinc (Zn)	A (1 mg Carotene = 1667 IU Vit. A)	Carotene (Provitamin A)
			(ppm or mg/kg)	(ppm or mg/kg)	(ppm or mg/kg)	(%)	(ppm or mg/kg)	(ppm or mg/kg)	(ppm or mg/kg)	(IU/g)	(ppm or mg/kg)
	RAPE, CANADA *Brassica napus*										
117	-SEEDS, MEAL, PREPRESSED, SOLV EXTD, 40% PROTEIN	A-F	—	—	—	—	—	—	—	—	—
		M-F	—	—	—	—	—	—	—	—	—
	RICE *Oryza sativa*										
118	-BRAN WITH GERM (RICE BRAN)	A-F	—	13.1	—	0.019	346.7	—	30.1	—	—
		M-F	—	14.3	—	0.021	381.0	—	33.0	—	—
119	-GRAIN, GROUND (GROUND ROUGH RICE, GROUND PADDY RICE)	A-F	0.044	6.3	0.044	0.010	96.2	—	13.3	—	—
		M-F	0.050	7.1	0.050	0.011	108.1	—	14.9	—	—
120	-GROATS (RICE, BROWN)	A-F	—	3.3	—	0.003	12.7	—	5.6	—	—
		M-F	—	3.7	—	0.003	14.4	—	6.4	—	—
121	-GROATS, POLISHED (RICE, POLISHED)	A-F	—	2.9	—	0.001	11.0	—	2.0	—	—
		M-F	—	3.3	—	0.002	12.4	—	2.2	—	—
122	-POLISHINGS	A-F	—	7.1	0.066	0.012	170.8	—	26.3	—	—
		M-F	—	7.9	0.073	0.013	189.8	—	29.3	—	—
	RYE *Secale cereale*										
123	-GRAIN, ALL ANALYSES	A-F	—	6.7	—	0.006	54.5	0.382	31.4	16.9	10.1
		M-F	—	7.7	—	0.007	62.5	0.439	36.1	19.4	11.6
124	-DISTILLERS' GRAINS, DEHY	A-F	—	—	—	—	18.4	—	—	—	—
		M-F	—	—	—	—	20.0	—	—	—	—
125	-DISTILLERS' GRAINS WITH SOLUBLES, DEHY	A-F	—	—	—	—	—	—	—	—	—
		M-F	—	—	—	—	—	—	—	—	—
	SAFFLOWER *Carthamus tinctorius*										
126	-SEEDS, MEAL, SOLV EXTD, 20% PROTEIN	A-F	—	—	—	—	—	—	—	—	—
		M-F	—	—	—	—	—	—	—	—	—
127	-SEEDS, MEAL, SOLV EXTD, 42% PROTEIN	A-F	—	—	—	—	—	—	—	—	—
		M-F	—	—	—	—	—	—	—	—	—
128	-SEEDS WITHOUT HULLS, MEAL, SOLV EXTD, 42% PROTEIN	A-F	2.000	87.6	—	0.099	39.8	—	184.2	—	—
		M-F	2.222	97.3	—	0.110	44.2	—	204.7	—	—
	SESAME *Sesamum indicum*										
129	-SEEDS, MEAL, MECH EXTD	A-F	—	—	—	—	47.9	—	100.0	0.7	0.4
		M-F	—	—	—	—	51.5	—	107.5	0.8	0.5
	SORGHUM *Sorghum vulgare*										
130	-GRAIN, ALL ANALYSES	A-F	0.264	9.8	0.022	0.005	15.5	0.805	14.5	1.9	1.2
		M-F	0.293	10.8	0.025	0.005	17.3	0.894	16.1	2.1	1.3
	SORGHUM, KAFIR *Sorghum vulgare, caffrorum*										
131	-GRAIN	A-F	0.629	6.9	—	0.006	15.8	0.796	13.5	0.6	0.4
		M-F	0.707	7.8	—	0.007	17.8	0.894	15.2	0.7	0.4
	SORGHUM, MILO *Sorghum vulgare, subglabrescens*										
132	-GRAIN	A-F	0.055	12.9	—	0.004	13.6	—	15.4	0.4	0.2
		M-F	0.062	14.4	—	0.005	15.3	—	17.3	0.4	0.2
	SOYBEAN *Glycine max*										
133	-SEEDS, MEAL, MECH EXTD, 41% PROTEIN	A-F	0.180	18.0	—	0.016	32.3	—	—	0.3	0.2
		M-F	0.200	20.0	—	0.018	35.9	—	—	0.3	0.2
134	-SEEDS, MEAL, SOLV EXTD, 44% PROTEIN	A-F	0.101	37.4	—	0.011	28.0	—	—	—	—
		M-F	0.111	41.2	—	0.012	30.7	—	—	—	—
135	-SEEDS, MEAL, SOLV EXTD, 49% PROTEIN	A-F	—	—	—	—	47.0	—	—	—	—
		M-F	—	—	—	—	51.7	—	—	—	—
136	-SEEDS, WHOLE	A-F	—	15.8	—	0.008	29.9	—	—	1.5	0.9
		M-F	—	17.4	—	0.009	32.8	—	—	1.6	1.0
137	-SEEDS WITHOUT HULLS, MEAL, SOLV EXTD, 49% PROTEIN	A-F	0.065	14.8	0.108	0.011	38.5	0.101	49.7	—	—
		M-F	0.072	16.4	0.120	0.012	42.7	0.112	55.2	—	—
	SPELT *Triticum spelta*										
138	-GRAIN	A-F	—	—	—	—	—	—	—	—	—
		M-F	—	—	—	—	—	—	—	—	—

(Continued)

Entry Number	Fat-Soluble Vitamins			Water-Soluble Vitamins								
	D	E (α-tocopherol)	K	B-12	Biotin	Choline	Folic Acid (Folacin)	Niacin (Nicotinic Acid)	Panto-thenic Acid	Pyri-doxine (B-6)	Ribo-flavin (B-2)	Thiamin (B-1)
	(IU/kg)	(ppm or mg/kg)	(ppm or mg/kg)	(ppb or mcg/kg)	(ppm or mg/kg)	(ppm or mg/kg)	(ppm or mg/kg)	(ppm or mg/kg)	(ppm or mg/kg)	(ppm or mg/kg)	(ppm or mg/kg)	(ppm or mg/kg)
117	—	—	—	—	—	—	—	—	—	—	—	—
	—	—	—	—	—	—	—	—	—	—	—	—
118	—	60.0	—	—	0.42	1,223	—	300	23.1	29.23	2.6	22.6
	—	65.9	—	—	0.46	1,344	—	330	25.4	32.12	2.9	24.8
119	—	9.9	—	—	0.08	925	0.36	35	7.0	4.42	1.0	2.9
	—	11.2	—	—	0.09	1,039	0.41	40	7.8	4.97	1.2	3.2
120	—	8.7	—	—	0.09	—	0.19	43	10.7	6.98	0.6	2.9
	—	9.9	—	—	0.10	—	0.21	49	12.1	7.94	0.7	3.3
121	—	3.6	—	—	—	904	0.15	16	3.6	0.40	0.5	0.7
	—	4.0	—	—	—	1,016	0.17	18	4.0	0.45	0.6	0.7
122	—	71.9	—	—	0.61	1,224	0.44	540	37.9	27.73	1.6	20.0
	—	79.9	—	—	0.68	1,360	0.49	600	42.1	30.81	2.0	22.2
123	—	14.9	—	—	0.32	—	0.61	20	8.8	2.55	1.6	3.0
	—	17.2	—	—	0.37	—	0.70	23	10.2	2.94	1.8	3.4
124	—	—	—	—	—	—	—	17	5.3	—	3.3	1.3
	—	—	—	—	—	—	—	18	5.7	—	3.6	1.4
125	—	—	—	—	—	—	—	63	17.5	—	8.2	3.1
	—	—	—	—	—	—	—	69	19.2	—	9.0	3.4
126	—	—	—	—	—	—	—	—	—	—	—	—
	—	—	—	—	—	—	—	—	—	—	—	—
127	—	—	—	—	—	—	—	—	—	—	—	—
	—	—	—	—	—	—	—	—	—	—	—	—
128	—	0.7	—	—	1.69	3,246	1.59	22	38.8	11.70	2.3	4.6
	—	0.8	—	—	1.88	3,606	1.77	24	43.1	13.00	2.5	5.1
129	—	—	—	—	—	1,547	—	22	6.0	12.50	3.4	2.8
	—	—	—	—	—	1,664	—	23	6.4	13.44	3.7	3.0
130	—	10.9	—	—	0.25	629	0.21	41	11.7	4.68	1.3	4.2
	—	12.1	—	—	0.28	699	0.24	46	13.0	5.19	1.4	4.7
131	—	—	—	—	0.24	436	0.20	38	11.9	6.68	1.3	3.8
	—	—	—	—	0.26	490	0.22	43	13.4	7.50	1.4	4.3
132	—	12.1	—	—	0.87	612	0.21	37	10.9	4.00	1.1	3.9
	—	13.6	—	—	0.98	668	0.23	42	12.2	4.49	1.2	4.4
133	—	6.6	—	—	0.30	2,673	6.60	30	14.9	—	3.5	4.0
	—	7.3	—	—	0.33	2,940	7.73	34	16.6	—	3.9	4.9
134	—	0.7	—	—	—	2,675	—	27	14.5	—	2.9	6.9
	—	0.7	—	—	—	2,939	—	29	15.9	—	3.2	7.6
135	—	0.7	—	—	—	2,731	—	21	13.8	—	2.9	2.5
	—	0.7	—	—	—	3,001	—	24	15.2	—	3.2	2.7
136	—	—	—	—	0.38	2,898	—	22	15.8	—	2.9	11.1
	—	—	—	—	0.42	3,184	—	24	17.4	—	3.2	12.2
137	—	1.7	—	1.9	0.32	2,502	1.69	21	14.9	4.76	2.9	2.9
	—	1.9	—	2.2	0.36	2,780	1.88	24	16.6	5.29	3.3	3.2
138	—	—	—	—	—	—	—	48	—	—	—	—
	—	—	—	—	—	—	—	53	—	—	—	—

(Continued)

TABLE 21-1

Entry Number	Feed Name Description	Inter-national Feed Number[1]	Moisture Basis: A-F (as-fed) or M-F (moisture-free)	Chemical Analysis								
				Dry Matter	Ash	Crude Fiber	Cell Walls or NDF	Acid Detergent Fiber	Lignin	Ether Extract (Fat)	N-Free Extract	Crude Protein
				(%)	(%)	(%)	(%)	(%)	(%)	(%)	(%)	(%)
139	SUGARCANE *Saccharum officinarum* -MOLASSES, MORE THAN 46% INVERT SUGAR, MORE THAN 79.5° BRIX	4-04-696	A-F M-F	75 100	7.7 10.3	— —	— —	— —	— —	0.1 0.1	64.2 85.7	3.9 5.2
140	SUNFLOWER *Helianthus* spp -SEEDS WITHOUT HULLS, MEAL, SOLV EXTD, 44% PROTEIN	5-26-098	A-F M-F	93 100	7.7 8.3	11.0 11.8	— —	— —	— —	2.9 3.1	24.6 26.5	46.8 50.3
141	SWEET POTATO *Ipomoea batata* -CANNERY RESIDUE, DEHY	4-08-535	A-F M-F	90 100	6.0 6.7	9.6 10.6	— —	— —	— —	0.3 0.3	71.6 79.6	2.5 2.8
142	-TUBERS	4-04-788	A-F M-F	31 100	1.1 3.6	1.3 4.2	— —	— —	— —	0.4 1.3	26.5 85.5	1.7 5.4
143	TALLOW, ANIMAL	4-08-127	A-F M-F	99 100	0.1 0.1	— —	— —	— —	— —	96.9 97.9	— —	1.6 1.6
144	TRITICALE *Triticale hexaloide* -GRAIN	4-20-362	A-F M-F	91 100	1.8 2.0	4.0 4.4	— —	— —	— —	1.1 1.2	59.3 65.1	15.0 16.5
145	WHEAT *Triticum aestivum* -BRAN	4-05-190	A-F M-F	89 100	6.2 6.9	10.4 11.6	— —	— —	— —	3.9 4.4	53.4 60.0	15.2 17.1
146	-DISTILLERS' GRAINS, DEHY	5-05-193	A-F M-F	93 100	3.0 3.3	11.7 12.6	— —	— —	— —	6.7 7.2	40.0 43.0	31.5 33.9
147	-ENDOSPERM	4-05-197	A-F M-F	88 100	1.2 1.4	0.3 0.3	— —	— —	— —	1.1 1.3	74.3 84.4	11.1 12.6
148	-GERM, GROUND (WHEAT GERM MEAL)	5-05-218	A-F M-F	88 100	4.1 4.7	3.1 3.5	— —	— —	— —	8.4 9.5	47.9 54.5	24.5 27.8
149	-GRAIN, ALL ANALYSES	4-05-211	A-F M-F	88 100	1.6 1.8	2.5 2.8	— —	— —	— —	1.8 2.0	67.3 76.4	14.9 16.9
150	-GRAIN, HARD RED SPRING	4-05-258	A-F M-F	88 100	1.6 1.8	2.5 2.8	— —	— —	— —	1.8 2.0	66.7 75.8	15.5 17.6
151	-GRAIN, HARD RED WINTER	4-05-268	A-F M-F	88 100	1.7 1.9	2.5 2.8	— —	— —	— —	1.6 1.8	69.6 79.1	12.6 14.4
152	-GRAIN, SOFT RED WINTER	4-05-294	A-F M-F	88 100	1.8 2.1	2.1 2.4	— —	— —	— —	1.6 1.8	70.9 80.6	11.5 13.0
153	-GRAIN, SOFT WHITE WINTER	4-05-337	A-F M-F	88 100	1.7 1.9	2.3 2.6	— —	— —	— —	1.8 2.0	72.4 82.3	9.8 11.2
154	-GRAIN, SOFT WHITE WINTER, PACIFIC COAST	4-08-555	A-F M-F	89 100	1.7 2.0	2.5 2.8	— —	— —	— —	1.8 2.0	72.8 81.9	10.1 11.4
155	-GRAIN SCREENINGS	4-05-216	A-F M-F	89 100	6.0 6.7	7.5 8.4	— —	— —	— —	3.4 3.8	57.9 65.1	14.2 16.0
156	-MIDDLINGS, LESS THAN 9.5% FIBER	4-05-205	A-F M-F	89 100	4.5 5.0	6.9 7.8	— —	— —	— —	4.6 5.2	56.3 63.2	16.7 18.8
157	-MILL RUN, LESS THAN 9.5% FIBER	4-05-206	A-F M-F	90 100	5.4 6.0	8.4 9.3	— —	— —	— —	4.1 4.6	56.5 62.8	15.6 17.3
158	-RED DOG, LESS THAN 4.5% FIBER	4-05-203	A-F M-F	88 100	2.2 2.5	2.4 2.7	— —	— —	— —	3.3 3.8	64.7 73.6	15.3 17.4
159	-SHORTS, LESS THAN 7% FIBER	4-05-201	A-F M-F	88 100	3.9 4.5	6.5 7.4	— —	— —	— —	4.9 5.5	56.3 64.0	16.4 18.6
160	WHEAT, DURUM *Triticum durum* -GRAIN	4-05-224	A-F M-F	87 100	1.6 1.8	2.3 2.6	— —	— —	— —	1.8 2.0	67.5 77.6	13.8 15.9
161	WHEY -CONDENSED	4-01-180	A-F M-F	65 100	6.5 10.0	0.3 0.5	— —	— —	— —	0.6 0.9	48.7 74.9	8.9 13.7
162	-DEHY	4-01-182	A-F M-F	93 100	9.4 10.1	0.2 0.2	— —	— —	— —	0.6 0.7	69.4 74.6	13.4 14.4

(Continued)

Entry Number	TDN	Digestible Energy		Metabolizable Energy		Macrominerals						
						Calcium (Ca)	Phosphorus (P)	Sodium (Na)	Chlorine (Cl)	Magnesium (Mg)	Potassium (K)	Sulfur (S)
	(%)	*(kcal)*		*(kcal)*		*(%)*	*(%)*	*(%)*	*(%)*	*(%)*	*(%)*	*(%)*
		(lb)	*(kg)*	*(lb)*	*(kg)*							
139	57	1,144	2,523	1,012	2,232	0.78	0.09	0.17	2.78	0.35	2.85	0.35
	76	1,526	3,364	1,350	2,976	1.05	0.11	0.22	3.71	0.47	3.80	0.46
140	68	1,369	3,012	1,175	2,586	0.40	1.00	—	0.10	—	1.00	—
	74	1,472	3,239	1,264	2,780	0.43	1.07	—	0.10	—	1.08	—
141	76	1,520	3,351	1,451	3,198	—	—	—	—	—	—	—
	84	1,689	3,723	1,612	3,554	—	—	—	—	—	—	—
142	27	542	1,195	514	1,134	0.03	0.05	0.02	0.02	0.05	0.31	0.04
	87	1,748	3,854	1,659	3,657	0.10	0.15	0.05	0.06	0.16	1.01	0.13
143	—	—	—	—	—	—	—	—	—	—	—	—
	—	—	—	—	—	—	—	—	—	—	—	—
144	—	—	—	—	—	0.05	0.30	—	—	—	—	—
	—	—	—	—	—	0.06	0.33	—	—	—	—	—
145	57	1,075	2,370	1,002	2,208	0.11	1.26	0.03	0.05	0.52	1.46	0.22
	64	1,208	2,663	1,125	2,481	0.12	1.42	0.04	0.06	0.59	1.64	0.25
146	—	—	—	—	—	0.11	0.58	—	—	—	—	—
	—	—	—	—	—	0.12	0.63	—	—	—	—	—
147	76	1,530	3,372	1,429	3,151	—	—	—	—	—	—	—
	87	1,738	3,832	1,624	3,581	—	—	—	—	—	—	—
148	80	1,600	3,527	1,297	2,860	0.05	0.91	0.02	0.08	0.24	0.97	0.24
	91	1,818	4,008	1,474	3,250	0.06	1.04	0.03	0.09	0.28	1.10	0.27
149	80	1,432	3,157	1,478	3,258	0.03	0.38	0.03	0.07	0.15	0.36	0.16
	91	1,627	3,588	1,679	3,702	0.04	0.43	0.03	0.08	0.17	0.41	0.18
150	74	1,411	3,111	1,305	2,877	0.03	0.38	0.03	0.08	0.15	0.36	—
	84	1,604	3,536	1,483	3,269	0.04	0.43	0.03	0.09	0.17	0.41	—
151	75	1,505	3,317	1,401	3,088	0.04	0.38	0.02	0.05	0.10	0.42	0.15
	85	1,710	3,770	1,592	3,509	0.05	0.43	0.02	0.06	0.12	0.48	0.17
152	76	1,510	3,330	1,410	3,109	0.04	0.38	0.01	0.07	0.10	0.41	0.11
	86	1,716	3,784	1,602	3,533	0.05	0.43	0.01	0.08	0.11	0.46	0.12
153	82	1,637	3,609	1,524	3,359	0.06	0.31	0.03	0.08	0.10	0.39	—
	93	1,860	4,101	1,731	3,817	0.07	0.36	0.04	0.09	0.11	0.45	—
154	77	1,534	3,382	1,533	3,379	0.09	0.30	0.05	—	—	0.40	—
	86	1,724	3,800	1,722	3,797	0.10	0.33	0.06	—	—	0.44	—
155	62	1,249	2,754	1,065	2,348	0.15	0.36	—	—	—	—	—
	70	1,404	3,094	1,197	2,638	0.17	0.40	—	—	—	—	—
156	68	1,324	2,918	1,235	2,724	0.11	0.83	0.20	0.03	0.36	0.99	0.16
	76	1,487	3,279	1,388	3,060	0.12	0.93	0.22	0.03	0.40	1.12	0.18
157	68	1,367	3,014	1,300	2,866	0.15	1.03	0.22	—	0.51	1.28	—
	76	1,519	3,349	1,444	3,184	0.17	1.14	0.24	—	0.57	1.42	—
158	72	1,406	3,101	1,270	2,799	0.04	0.49	0.01	0.14	0.14	0.52	0.26
	82	1,598	3,523	1,443	3,181	0.05	0.55	0.02	0.16	0.16	0.59	0.29
159	67	1,343	2,961	1,245	2,746	0.09	0.81	0.02	0.07	0.26	0.94	0.23
	76	1,526	3,364	1,415	3,120	0.10	0.92	0.02	0.08	0.29	1.07	0.26
160	74	1,485	3,273	1,378	3,037	0.08	0.35	—	—	0.14	0.44	—
	85	1,706	3,762	1,583	3,491	0.10	0.41	—	—	0.16	0.51	—
161	52	1,034	2,279	964	2,125	0.39	0.59	—	—	—	—	—
	80	1,591	3,507	1,483	3,270	0.60	0.91	—	—	—	—	—
162	78	1,429	3,151	1,407	3,101	0.78	0.77	0.57	0.07	0.13	0.86	1.04
	84	1,537	3,389	1,512	3,334	0.84	0.83	0.62	0.08	0.14	0.92	1.11

(Continued)

COMPOSITION OF FEEDS

Entry Number	Feed Name Description	Moisture Basis: A-F (as-fed) or M-F (moisture-free)	Microminerals							Fat-Soluble Vitamins	
			Cobalt (Co)	Copper (Cu)	Iodine (I)	Iron (Fe)	Man-ganese (Mn)	Sele-nium (Se)	Zinc (Zn)	A (1 mg Carotene = 1667 IU Vit. A)	Carotene (Provitamin A)
			(ppm or mg/kg)	(ppm or mg/kg)	(ppm or mg/kg)	(%)	(ppm or mg/kg)	(ppm or mg/kg)	(ppm or mg/kg)	(IU/g)	(ppm or mg/kg)
139	**SUGARCANE** *Saccharum officinarum* -MOLASSES, MORE THAN 46% INVERT SUGAR, MORE THAN 79.5° BRIX	A-F M-F	0.908 1.210	60.4 80.5	1.577 2.103	0.019 0.026	42.9 57.1	— —	22.5 30.0	— —	— —
140	**SUNFLOWER** *Helianthus spp* -SEEDS WITHOUT HULLS, MEAL, SOLV EXTD, 44% PROTEIN	A-F M-F	— —	— —	— —	— —	23.0 24.7	— —	— —	— —	— —
141	**SWEET POTATO** *Ipomoea batata* -CANNERY RESIDUE, DEHY	A-F M-F	— —	— —	— —	— —	— —	— —	— —	— —	— —
142	-TUBERS	A-F M-F	— —	1.3 4.2	— —	0.002 0.005	3.4 11.1	— —	— —	222.8 718.8	133.7 431.2
143	**TALLOW**, ANIMAL	A-F M-F	— —	— —	— —	— —	— —	— —	— —	— —	— —
144	**TRITICALE** *Triticale hexaloide* -GRAIN	A-F M-F	— —	— —	— —	— —	— —	— —	— —	— —	— —
145	**WHEAT** *Triticum aestivum* -BRAN	A-F M-F	0.101 0.113	12.7 14.3	0.065 0.073	0.011 0.012	109.9 123.4	0.375 0.422	103.9 116.7	4.4 4.9	2.6 2.9
146	-DISTILLERS' GRAINS, DEHY	A-F M-F	— —	— —	— —	— —	15.0 16.1	— —	— —	1.8 2.0	1.1 1.2
147	-ENDOSPERM	A-F M-F	— —	— —	— —	— —	— —	— —	— —	— —	— —
148	-GERM, GROUND (WHEAT GERM MEAL)	A-F M-F	0.118 0.134	9.3 10.6	— —	0.005 0.006	133.3 151.4	0.339 0.385	119.4 135.7	— —	— —
149	-GRAIN, ALL ANALYSES	A-F M-F	0.118 0.134	5.7 6.5	0.087 0.098	0.006 0.006	36.7 41.7	0.222 0.253	44.7 50.8	16.9 19.3	10.2 11.6
150	-GRAIN, HARD RED SPRING	A-F M-F	0.118 0.134	5.7 6.5	— —	0.006 0.007	36.7 41.7	0.223 0.254	45.5 51.8	16.9 19.3	10.2 11.6
151	-GRAIN, HARD RED WINTER	A-F M-F	0.141 0.160	4.7 5.4	— —	0.003 0.004	29.0 32.9	0.438 0.498	37.7 42.8	— —	— —
152	-GRAIN, SOFT RED WINTER	A-F M-F	0.102 0.116	6.1 6.9	— —	0.003 0.003	31.7 36.0	0.041 0.047	42.0 47.7	— —	— —
153	-GRAIN, SOFT WHITE WINTER	A-F M-F	0.133 0.151	7.4 8.4	— —	0.004 0.005	40.0 45.5	0.051 0.058	22.5 25.6	— —	— —
154	-GRAIN, SOFT WHITE WINTER, PACIFIC COAST	A-F M-F	— —	9.6 10.8	— —	0.010 0.011	50.1 56.3	— —	13.1 14.7	— —	— —
155	-GRAIN SCREENINGS	A-F M-F	— —	— —	— —	— —	14.4 16.2	— —	— —	— —	— —
156	-MIDDLINGS, LESS THAN 9.5% FIBER	A-F M-F	0.093 0.105	18.3 20.6	0.109 0.122	0.008 0.009	112.2 126.1	— —	150.9 169.6	5.1 5.7	3.1 3.4
157	-MILL RUN, LESS THAN 9.5% FIBER	A-F M-F	0.194 0.216	18.7 20.8	— —	0.010 0.011	102.4 113.8	— —	— —	— —	— —
158	-RED DOG, LESS THAN 4.5% FIBER	A-F M-F	0.118 0.134	6.4 7.3	— —	0.004 0.005	55.4 63.0	0.303 0.345	65.0 73.9	— —	— —
159	-SHORTS, LESS THAN 7% FIBER	A-F M-F	0.104 0.118	11.7 13.3	— —	0.007 0.008	116.9 132.8	0.380 0.432	106.3 120.8	— —	— —
160	**WHEAT, DURUM** *Triticum durum* -GRAIN	A-F M-F	— —	6.9 7.9	— —	0.004 0.005	27.9 32.0	0.886 1.019	32.6 37.4	— —	— —
161	**WHEY** -CONDENSED	A-F M-F	— —	— —	— —	— —	— —	— —	— —	— —	— —
162	-DEHY	A-F M-F	0.105 0.113	45.7 49.2	— —	0.015 0.016	5.5 5.9	— —	3.2 3.4	— —	— —

(Continued)

Entry Number	Fat-Soluble Vitamins			Water-Soluble Vitamins								
	D	E (α-tocopherol)	K	B-12	Biotin	Choline	Folic Acid (Folacin)	Niacin (Nicotinic Acid)	Panto-thenic Acid	Pyri-doxine (B-6)	Ribo-flavin (B-2)	Thiamin (B-1)
	(IU/kg)	(ppm or mg/kg)	(ppm or mg/kg)	(ppb or mcg/kg)	(ppm or mg/kg)	(ppm or mg/kg)	(ppm or mg/kg)	(ppm or mg/kg)	(ppm or mg/kg)	(ppm or mg/kg)	(ppm or mg/kg)	(ppm or mg/kg)
139	—	5.0	—	—	0.71	744	0.11	41	39.2	6.50	2.9	0.9
	—	6.7	—	—	0.94	992	0.15	54	52.2	8.67	3.8	1.2
140	—	—	—	—	—	—	—	—	—	—	—	—
	—	—	—	—	—	—	—	—	—	—	—	—
141	—	—	—	—	—	—	—	—	—	—	—	—
	—	—	—	—	—	—	—	—	—	—	—	—
142	—	—	—	—	—	—	—	6	—	—	0.6	1.1
	—	—	—	—	—	—	—	20	—	—	2.0	3.4
143	—	—	—	—	—	—	—	—	—	—	—	—
	—	—	—	—	—	—	—	—	—	—	—	—
144	—	—	—	—	—	468	—	—	—	—	0.4	—
	—	—	—	—	—	514	—	—	—	—	0.5	—
145	—	20.9	—	—	0.57	1,880	1.23	274	31.7	10.18	4.8	6.5
	—	23.4	—	—	0.64	2,113	1.39	308	35.7	11.44	5.4	7.3
146	—	—	—	—	—	—	—	56	8.1	—	3.7	2.0
	—	—	—	—	—	—	—	60	8.7	—	4.0	2.1
147	—	—	—	—	—	—	—	—	—	—	—	—
	—	—	—	—	—	—	—	—	—	—	—	—
148	—	141.1	—	—	0.22	3,056	2.15	72	20.9	11.32	6.1	22.7
	—	160.3	—	—	0.24	3,473	2.45	82	23.7	12.87	6.9	25.8
149	—	13.7	—	0.9	0.10	1,005	0.40	57	9.7	5.00	1.4	4.2
	—	15.6	—	1.0	0.11	1,142	0.45	65	11.6	5.68	1.6	4.8
150	—	12.7	—	—	0.11	1,056	0.41	57	9.6	5.08	1.4	4.2
	—	14.4	—	—	0.13	1,200	0.46	65	10.9	5.77	1.6	4.8
151	—	11.0	—	—	0.11	1,034	0.39	54	9.8	2.99	1.5	4.2
	—	12.5	—	—	0.12	1,175	0.44	61	11.2	3.40	1.7	4.8
152	—	15.6	—	—	—	927	0.41	52	9.6	3.20	1.5	4.5
	—	17.7	—	—	—	1,053	0.46	59	10.9	3.63	1.7	5.1
153	—	13.5	—	—	0.11	950	0.37	53	10.9	4.06	1.2	4.7
	—	15.4	—	—	0.12	1,079	0.43	60	12.4	4.62	1.3	5.3
154	—	13.1	—	—	—	872	—	50	9.8	—	0.9	5.0
	—	14.7	—	—	—	980	—	56	11.0	—	1.0	5.6
155	—	—	—	—	—	—	—	—	—	—	—	6.4
	—	—	—	—	—	—	—	—	—	—	—	7.2
156	—	21.3	—	—	0.11	1,253	0.93	96	18.1	5.66	2.0	14.8
	—	23.9	—	—	0.12	1,408	1.05	107	20.3	6.36	2.2	16.6
157	—	—	—	—	—	989	—	111	13.2	—	1.6	15.2
	—	—	—	—	—	1,099	—	123	14.7	—	1.8	16.9
158	—	32.8	—	—	0.11	1,603	0.76	46	13.4	4.83	2.4	23.0
	—	37.3	—	—	0.12	1,821	0.87	52	15.2	5.49	2.7	26.1
159	—	54.1	—	—	—	2,027	1.57	108	23.3	7.17	4.6	19.7
	—	61.4	—	—	—	2,303	1.78	123	26.4	8.15	5.2	22.4
160	—	—	—	—	—	—	0.38	52	8.8	2.98	1.0	4.6
	—	—	—	—	—	—	0.44	60	10.1	3.42	1.2	5.3
161	—	—	—	—	—	—	—	3	14.3	—	17.2	3.1
	—	—	—	—	—	—	—	5	22.0	—	26.5	4.8
162	—	—	—	17.9	0.35	1,770	0.83	11	45.9	3.34	26.2	4.0
	—	—	—	19.2	0.38	1,904	0.89	11	49.4	3.59	28.2	4.3

(Continued)

A
M
I
N
O

A
C
I
D

C
O
M
P.

TABLE 21-1

Entry Number	Feed Name Description	International Feed Number[1]	Moisture Basis: A-F (as-fed) or M-F (moisture-free)	Chemical Analysis								
				Dry Matter	Ash	Crude Fiber	Cell Walls or NDF	Acid Detergent Fiber	Lignin	Ether Extract (Fat)	N-Free Extract	Crude Protein
				(%)	(%)	(%)	(%)	(%)	(%)	(%)	(%)	(%)
163	-FRESH	4-08-134	A-F	7	0.7	—	—	—	—	0.3	5.1	0.9
			M-F	100	9.4	—	—	—	—	4.3	73.2	13.0
164	-LOW LACTOSE, DEHY (DRIED WHEY PRODUCT)	4-01-186	A-F	93	15.4	0.2	—	—	—	1.0	59.9	16.4
			M-F	100	16.6	0.2	—	—	—	1.1	64.4	17.6
165	YEAST *Saccharamyces cerevisiae* -BREWERS', DEHY	7-05-527	A-F	93	6.7	2.7	—	—	—	1.1	37.8	45.2
			M-F	100	7.2	2.9	—	—	—	1.1	40.4	48.3
166	-IRRADIATED, DEHY	7-05-529	A-F	94	6.2	6.2	—	—	—	1.1	32.4	48.1
			M-F	100	6.6	6.5	—	—	—	1.2	34.5	51.2
167	YEAST, TORULA *Torulopsis utilis* -DEHY	7-05-534	A-F	93	8.7	2.4	—	—	—	2.1	31.0	48.9
			M-F	100	9.3	2.5	—	—	—	2.3	33.3	52.5

TABLE
AMINO ACID COMPOSITION OF FEEDS,

Entry Number	Feed Name Description	International Feed Number	Moisture Basis: A-F (as-fed) or M-F (moisture-free)	Dry Matter	Crude Protein	Arginine	Cystine
				(%)	(%)	(%)	(%)
1	ALFALFA (LUCERNE) *Medicago sativa* -HAY, SUN-CURED	1-00-078	A-F	90	15.9	0.64	0.21
			M-F	100	17.6	0.71	0.23
2	-LEAVES, MEAL, DEHY	1-00-137	A-F	92	20.0	0.96	—
			M-F	100	21.7	1.04	—
3	-LEAVES, SUN-CURED, GROUND	1-00-246	A-F	92	20.5	1.20	0.37
			M-F	100	22.2	1.30	0.40
4	-MEAL, DEHY, 15% PROTEIN	1-00-022	A-F	91	15.4	0.59	0.24
			M-F	100	16.9	0.65	0.26
5	-MEAL, DEHY, 17% PROTEIN	1-00-023	A-F	92	17.5	0.75	0.29
			M-F	100	18.9	0.81	0.32
6	ALFALFA, GRASS *Medicago sativa, grass* -HAY, SUN-CURED	1-08-331	A-F	90	14.0	0.68	0.24
			M-F	100	15.6	0.75	0.27
7	ALFALFA (LUCERNE) *Medicago sativa* -MEAL, DEHY, 20% PROTEIN	1-00-024	A-F	91	20.1	0.94	0.31
			M-F	100	22.0	1.03	0.34
8	-MEAL, DEHY, 22% PROTEIN	1-07-851	A-F	93	22.0	0.99	0.34
			M-F	100	23.7	1.06	0.36
9	BAKERY WASTE -DEHY (DRIED BAKERY PRODUCT)	4-00-466	A-F	92	10.2	0.49	0.17
			M-F	100	11.1	0.53	0.19
10	BARLEY *Hordeum vulgare* -GRAIN	4-00-549	A-F	88	12.2	0.53	0.23
			M-F	100	13.9	0.60	0.26
11	-GRAIN, PACIFIC COAST	4-07-939	A-F	90	9.6	0.44	0.20
			M-F	100	10.7	0.50	0.23
12	-MALT SPROUTS, DEHY	5-00-545	A-F	92	26.1	1.11	0.23
			M-F	100	28.4	1.21	0.25

(Continued)

Entry Number	TDN	Digestible Energy		Metabolizable Energy		Macrominerals						
						Calcium (Ca)	Phosphorus (P)	Sodium (Na)	Chlorine (Cl)	Magnesium (Mg)	Potassium (K)	Sulfur (S)
	(%)	(kcal)		(kcal)		(%)	(%)	(%)	(%)	(%)	(%)	(%)
		(lb)	(kg)	(lb)	(kg)							
163	—	—	—	—	—	0.05	0.05	—	—	—	0.19	—
	—	—	—	—	—	0.72	0.65	—	—	—	2.75	—
164	75	1,064	2,345	1,104	2,434	1.75	1.33	1.58	—	—	3.22	—
	81	1,144	2,521	1,187	2,617	1.88	1.43	1.70	—	—	3.46	—
165	70	1,400	3,086	1,224	2,698	0.14	1.42	0.07	—	0.23	1.73	0.38
	75	1,505	3,319	1,316	2,901	0.15	1.52	0.08	—	0.25	1.86	0.41
166	—	—	—	—	—	0.78	1.42	—	—	—	2.14	—
	—	—	—	—	—	0.83	1.51	—	—	—	2.28	—
167	64	1,277	2,816	1,095	2,415	0.58	1.67	0.01	0.02	0.13	1.88	—
	69	1,373	3,028	1,178	2,597	0.63	1.80	0.01	0.02	0.14	2.03	—

(Continued)

21-2
DATA EXPRESSED AS-FED AND MOISTURE-FREE

Amino Acids

Glycine	Histidine	Isoleucine	Leucine	Lysine	Methionine	Phenyl-alanine	Serine	Threonine	Tryptophan	Tyrosine	Valine
(%)	(%)	(%)	(%)	(%)	(%)	(%)	(%)	(%)	(%)	(%)	(%)
0.60	0.27	0.74	1.15	0.77	0.16	0.69	0.72	0.67	0.22	0.41	0.70
0.67	0.30	0.82	1.27	0.85	0.18	0.77	0.80	0.74	0.24	0.45	0.78
—	0.42	1.00	1.53	1.08	0.30	0.99	—	0.90	0.41	—	1.10
—	0.46	1.08	1.67	1.17	0.33	1.07	—	0.98	0.44	—	1.19
—	0.37	0.92	1.38	1.01	0.37	0.92	—	0.74	0.46	—	1.01
—	0.40	1.00	1.50	1.10	0.40	1.00	—	0.80	0.50	—	1.10
0.71	0.26	0.64	1.03	0.60	0.22	0.62	0.61	0.55	0.39	0.41	0.72
0.78	0.28	0.70	1.13	0.66	0.24	0.68	0.67	0.60	0.43	0.44	0.79
0.90	0.33	0.81	1.28	0.88	0.21	0.79	0.74	0.70	0.36	0.55	0.85
0.98	0.36	0.87	1.39	0.96	0.23	0.86	0.80	0.76	0.39	0.60	0.92
0.50	0.24	0.60	1.26	0.69	0.20	0.69	—	0.61	0.36	0.57	0.77
0.55	0.27	0.67	1.40	0.76	0.22	0.77	—	0.68	0.39	0.63	0.85
0.96	0.36	0.87	1.38	0.90	0.31	0.92	0.89	0.80	0.43	0.62	1.00
1.05	0.40	0.95	1.51	0.99	0.34	1.01	0.97	0.87	0.47	0.68	1.09
1.09	0.44	1.07	1.60	1.00	0.34	1.13	1.02	0.98	0.48	0.65	1.29
1.17	0.48	1.16	1.73	1.08	0.37	1.22	1.10	1.05	0.52	0.70	1.39
0.70	0.15	0.41	0.70	0.31	0.19	0.41	—	0.41	0.10	0.37	0.42
0.76	0.16	0.44	0.76	0.34	0.21	0.44	—	0.44	0.10	0.41	0.45
0.41	0.25	0.47	0.80	0.42	0.15	0.60	0.44	0.38	0.15	0.32	0.60
0.46	0.29	0.54	0.91	0.48	0.18	0.69	0.50	0.43	0.18	0.36	0.68
0.30	0.20	0.41	0.60	0.25	0.14	0.47	0.32	0.30	0.13	0.31	0.47
0.34	0.23	0.45	0.67	0.28	0.16	0.53	0.36	0.34	0.14	0.34	0.52
—	0.53	1.09	1.63	1.22	0.33	0.91	—	1.01	0.41	—	1.45
—	0.57	1.18	1.77	1.32	0.36	0.98	—	1.09	0.44	—	1.58

(Continued)

AMINO ACID COMP.

TABLE 21-1

Entry Number	Feed Name Description	Moisture Basis: A-F (as-fed) or M-F (moisture-free)	Microminerals							Fat-Soluble Vitamins	
			Cobalt (Co)	Copper (Cu)	Iodine (I)	Iron (Fe)	Manganese (Mn)	Selenium (Se)	Zinc (Zn)	A (1 mg Carotene = 1667 IU Vit. A)	Carotene (Provitamin A)
			(ppm or mg/kg)	(ppm or mg/kg)	(ppm or mg/kg)	(%)	(ppm or mg/kg)	(ppm or mg/kg)	(ppm or mg/kg)	(IU/g)	(ppm or mg/kg)
163	-FRESH	A-F	—	—	—	0.002	0.2	—	—	—	—
		M-F	—	—	—	0.029	3.2	—	—	—	—
164	-LOW LACTOSE, DEHY (DRIED WHEY PRODUCT	A-F	—	—	—	—	—	0.051	—	—	—
		M-F	—	—	—	—	—	0.055	—	—	—
165	YEAST *Saccharomyces cerevisiae* -BREWERS', DEHY	A-F	0.184	33.1	—	0.011	5.7	1.250	39.1	—	—
		M-F	0.198	35.6	—	0.012	6.2	1.344	42.0	—	—
166	-IRRADIATED, DEHY	A-F	—	—	—	—	—	—	—	—	—
		M-F	—	—	—	—	—	—	—	—	—
167	YEAST, TORULA *Torulopsis utilis* -DEHY	A-F	—	13.4	—	0.009	12.8	1.227	99.1	—	—
		M-F	—	14.4	—	0.010	13.8	1.320	106.6	—	—

[1]The first digit is the feed class, coded as follows: (1) dry forages and roughages; (2) pasture, range plants, and forages fed green; (3) silages; (4) energy feeds; and (5) protein supplements.

TABLE 21-2

Entry Number	Feed Name Description	International Feed Number	Moisture Basis: A-F (as-fed) or M-F (moisture-free)	Dry Matter	Crude Protein	Arginine	Cystine
				(%)	(%)	(%)	(%)
13	BEAN, NAVY *Phaseolus vulgaris* -SEEDS	5-00-623	A-F	90	22.7	1.19	0.23
			M-F	100	25.3	1.33	0.26
14	BEAN, PINTO *Phaseolus vulgaris* -SEEDS	5-00-624	A-F	90	22.7	1.55	—
			M-F	100	25.1	1.72	—
15	BEET, SUGAR *Beta vulgaris saccharifera* -PULP, DEHY	4-00-669	A-F	90	8.8	0.30	0.01
			M-F	100	9.7	0.33	0.01
16	BLOOD -MEAL	5-00-380	A-F	92	78.9	3.19	1.31
			M-F	100	86.0	3.47	1.43
17	-MEAL, SPRAY, DEHY (BLOOD FLOUR)	5-00-381	A-F	86	72.2	3.15	—
			M-F	100	84.4	3.69	—
18	BREWERS' GRAINS -DEHY	5-02-141	A-F	92	25.0	1.22	0.38
			M-F	100	27.2	1.33	0.41
19	BROOMCORN (MILLET, PROSO) *Panicum miliaceum* -GRAIN	4-03-120	A-F	90	11.6	0.35	—
			M-F	100	12.9	0.39	—
20	BUCKWHEAT *Fagopyrum* spp -GRAIN	4-00-994	A-F	88	11.0	0.98	0.20
			M-F	100	12.6	1.11	0.23
21	BUTTERMILK -DEHY	5-01-160	A-F	92	31.5	1.08	0.39
			M-F	100	34.3	1.18	0.42
22	CASEIN -DEHY	5-01-162	A-F	90	82.0	3.36	0.28
			M-F	100	90.7	3.71	0.31

AMINO ACID COMP.

(Continued)

Entry Number	Fat-Soluble Vitamins			Water-Soluble Vitamins								
	D	E (α-tocopherol)	K	B-12	Biotin	Choline	Folic Acid (Folacin)	Niacin (Nicotinic Acid)	Pantothenic Acid	Pyridoxine (B-6)	Riboflavin (B-2)	Thiamin (B-1)
	(IU/kg)	(ppm or mg/kg)	(ppm or mg/kg)	(ppb or mcg/kg)	(ppm or mg/kg)	(ppm or mg/kg)	(ppm or mg/kg)	(ppm or mg/kg)	(ppm or mg/kg)	(ppm or mg/kg)	(ppm or mg/kg)	(ppm or mg/kg)
163	—	—	—	—	—	—	—	1	5.4	—	0.8	0.3
	—	—	—	—	—	—	—	14	76.7	—	11.7	4.3
164	—	—	—	4,174.1	0.48	4,287	0.98	18	75.2	5.34	45.8	5.2
	—	—	—	4,488.3	0.52	4,609	1.06	19	80.9	5.74	49.2	5.6
165	—	2.2	—	—	0.96	4,075	9.42	458	109.0	43.24	36.4	93.0
	—	2.4	—	—	1.04	4,381	10.13	492	117.2	46.50	39.2	100.0
166	—	—	—	—	—	—	—	—	—	—	18.5	—
	—	—	—	—	—	—	—	—	—	—	19.7	—
167	—	—	—	—	1.39	2,887	22.40	499	83.1	29.45	49.2	6.2
	—	—	—	—	1.50	3,104	24.09	536	89.3	31.66	52.9	6.7

(Continued)

Amino Acids											
Glycine	Histidine	Isoleucine	Leucine	Lysine	Methionine	Phenyl-alanine	Serine	Threonine	Tryptophan	Tyrosine	Valine
(%)	(%)	(%)	(%)	(%)	(%)	(%)	(%)	(%)	(%)	(%)	(%)
0.80	—	—	—	1.29	0.25	—	—	—	0.24	—	—
0.89	—	—	—	1.44	0.28	—	—	—	0.27	—	—
—	0.64	1.14	1.11	1.60	0.26	1.20	—	1.09	0.32	—	1.23
—	0.71	1.26	1.23	1.77	0.29	1.33	—	1.21	0.35	—	1.36
—	0.20	0.30	0.60	0.59	0.01	0.30	—	0.40	0.10	0.40	0.40
—	0.22	0.33	0.66	0.66	0.01	0.33	—	0.44	0.11	0.44	0.44
3.21	3.96	0.90	10.12	5.99	0.91	5.47	7.20	3.47	1.02	1.73	6.41
3.49	4.31	0.98	11.02	6.53	0.99	5.96	7.84	3.78	1.11	1.89	6.99
—	4.55	1.02	9.83	7.70	1.02	5.29	—	3.43	0.93	1.86	6.87
—	5.31	1.19	11.50	9.00	1.19	6.18	—	4.01	1.08	2.17	8.03
1.06	0.49	1.47	1.89	0.87	0.45	1.38	—	0.88	0.37	1.20	1.59
1.15	0.54	1.60	2.05	0.95	0.49	1.50	—	0.95	0.40	1.30	1.73
—	0.20	1.17	1.15	0.26	0.28	0.56	—	0.40	0.17	—	0.58
—	0.23	1.31	1.28	0.29	0.32	0.63	—	0.44	0.19	—	0.64
—	0.26	0.37	0.56	0.62	0.19	0.44	—	0.45	0.18	—	0.54
—	0.30	0.42	0.64	0.70	0.22	0.50	—	0.52	0.21	—	0.61
0.62	0.85	2.37	3.20	2.28	0.71	1.47	1.55	1.52	0.49	1.01	2.56
0.67	0.92	2.58	3.48	2.47	0.77	1.60	1.68	1.65	0.53	1.09	2.78
1.50	2.56	5.56	8.67	7.02	2.76	4.66	5.20	3.86	0.98	4.56	6.62
1.66	2.83	6.15	9.59	7.76	3.05	5.16	5.75	4.27	1.08	5.04	7.32

(Continued)

AMINO ACID COMP.

TABLE 21-2

Entry Number	Feed Name Description	International Feed Number	Moisture Basis: A-F (as-fed) or M-F (moisture-free)	Dry Matter	Crude Protein	Arginine	Cystine
				(%)	(%)	(%)	(%)
23	**CATTLE** -MILK, FRESH	5-01-168	A-F	13	3.4	0.14	—
			M-F	100	27.4	1.09	—
24	**CHICKPEA, GRAM** *Cicer arietinum* -SEEDS	5-01-218	A-F	90	19.5	4.19	—
			M-F	100	21.6	4.63	—
25	**CITRUS** *Citrus* spp -PULP WITHOUT FINES, DEHY (DRIED CITRUS PULP)	4-01-237	A-F	90	6.3	0.23	0.11
			M-F	100	6.9	0.25	0.12
26	**CLOVER, LADINO** *Trifolium repens* -HAY, SUN-CURED	1-01-378	A-F	90	19.1	0.99	0.36
			M-F	100	21.3	1.10	0.40
27	**COCONUT** *Cocos nucifera* -MEATS, MEAL MECH EXTD (COPRA MEAL)	5-01-572	A-F	93	21.2	2.30	0.20
			M-F	100	22.8	2.48	0.22
28	**CORN** *Zea mays* -DISTILLERS' GRAINS, DEHY	5-02-842	A-F	94	28.0	0.97	0.25
			M-F	100	29.8	1.03	0.26
29	-DISTILLERS' GRAINS WITH SOLUBLES, DEHY	5-02-843	A-F	92	27.1	0.93	0.27
			M-F	100	29.5	1.02	0.29
30	-DISTILLERS' SOLUBLES, DEHY	5-02-844	A-F	92	27.0	0.98	0.47
			M-F	100	29.3	1.06	0.50
31	-GLUTEN, MEAL	5-02-900	A-F	91	43.3	1.42	0.65
			M-F	100	47.3	1.56	0.72
32	-GLUTEN WITH BRAN (CORN GLUTEN FEED)	5-02-903	A-F	90	23.6	0.82	0.41
			M-F	100	26.2	0.91	0.46
33	**CORN, DENT YELLOW** *Zea mays, indentata* -EARS, GROUND (CORN AND COB MEAL)	4-02-849	A-F	86	7.8	0.36	0.13
			M-F	100	9.1	0.43	0.15
34	-GERM, MEAL, WET MILLED, SOLV EXTD	5-02-898	A-F	90	20.0	1.30	—
			M-F	100	22.2	1.44	—
35	-GRAIN	4-02-935	A-F	88	9.5	0.41	0.24
			M-F	100	10.9	0.46	0.27
36	-GRAIN, GRADE 2, 54 LB/BUSHEL OR *695 G/L*	4-02-931	A-F	89	8.7	0.50	0.13
			M-F	100	9.8	0.56	0.15
37	-GRAIN, FLAKED	4-02-859	A-F	89	9.9	0.44	0.25
			M-F	100	11.2	0.49	0.28
38	-HOMINY FEED	4-02-887	A-F	90	10.9	0.45	0.18
			M-F	100	12.2	0.50	0.20
39	**CORN GRAIN, DENT WHITE** *Zea mays, indentata*	4-02-928	A-F	91	10.9	0.27	0.09
			M-F	100	12.1	0.30	0.10
40	**CORN GRAIN, OPAQUE-2 (HIGH-LYSINE)** *Zea mays*	4-11-445	A-F	87	9.6	0.60	0.21
			M-F	100	11.0	0.69	0.24
41	**COTTON** *Gossypium* spp -SEEDS, MEAL, MECH EXTD, 36% PROTEIN	5-01-625	A-F	92	38.9	3.56	0.79
			M-F	100	42.3	3.86	0.86
42	-SEEDS, MEAL, MECH EXTD, 41% PROTEIN	5-01-617	A-F	93	40.8	4.05	0.70
			M-F	100	44.0	4.36	0.76
43	-SEEDS, MEAL, PREPRESSED, SOLV EXTD, 41% PROTEIN	5-07-872	A-F	90	41.4	4.59	0.64
			M-F	100	46.1	5.11	0.71
44	-SEEDS, MEAL, SOLV EXTD, 41% PROTEIN	5-01-621	A-F	91	41.3	4.26	0.82
			M-F	100	45.2	4.66	0.90
45	-SEEDS WITHOUT HULLS, MEAL, PREPRESSED, SOLV EXTD, 50% PROTEIN	5-07-874	A-F	93	50.0	4.84	1.07
			M-F	100	53.7	5.20	1.15

AMINO ACID COMPOSITIONS

(Continued)

	Amino Acids											
Glycine (%)	Histidine (%)	Isoleucine (%)	Leucine (%)	Lysine (%)	Methionine (%)	Phenyl-alanine (%)	Serine (%)	Threonine (%)	Tryptophan (%)	Tyrosine (%)	Valine (%)
— —	0.10 0.78	0.32 2.58	0.26 2.03	0.26 2.03	0.07 0.55	0.17 1.33	— —	0.17 1.33	0.05 0.39	— —	0.26 2.03
1.84 2.03	1.17 1.29	1.97 2.18	3.87 4.28	3.07 3.39	0.53 0.59	2.61 2.89	2.58 2.85	1.70 1.88	— —	1.26 1.39	2.05 2.27
— —	— —	— —	— —	0.20 0.22	0.09 0.10	— —	— —	— —	0.06 0.07	— —	— —
0.90 1.00	0.45 0.50	1.08 1.20	1.89 2.10	1.08 1.20	0.27 0.30	1.08 1.20	0.90 1.00	1.17 1.30	0.45 0.50	0.63 0.70	1.17 1.30
1.10 1.19	— —	— —	— —	0.54 0.58	0.33 0.36	— —	— —	— —	0.20 0.22	— —	— —
0.49 0.52	0.58 0.62	1.00 1.06	3.20 3.41	0.82 0.87	0.37 0.40	0.94 1.00	— —	0.40 0.43	0.19 0.21	0.85 0.90	1.19 1.26
0.50 0.55	0.65 0.71	1.39 1.51	2.21 2.40	0.73 0.79	0.50 0.55	1.51 1.64	1.60 1.74	0.94 1.02	0.15 0.17	0.70 0.76	1.50 1.63
1.10 1.19	0.67 0.72	1.31 1.42	2.23 2.42	0.88 0.96	0.58 0.63	1.47 1.59	1.19 1.29	1.00 1.08	0.20 0.22	0.86 0.93	1.53 1.66
1.51 1.65	0.97 1.07	2.24 2.45	7.43 8.13	0.83 0.91	1.07 1.17	2.82 3.09	1.70 1.86	1.43 1.57	0.21 0.23	1.01 1.10	2.24 2.46
0.86 0.96	0.60 0.67	0.94 1.04	2.27 2.52	0.60 0.67	0.37 0.41	0.82 0.91	0.80 0.89	0.78 0.87	0.15 0.17	0.76 0.85	1.14 1.27
0.31 0.36	0.16 0.19	0.34 0.40	0.85 0.99	0.17 0.20	0.14 0.17	0.39 0.45	— —	0.32 0.37	0.07 0.08	0.32 0.37	0.31 0.36
1.10 1.22	— —	— —	— —	— —	— —	— —	— —	— —	0.20 0.22	— —	— —
0.38 0.43	0.26 0.30	0.36 0.41	1.26 1.44	0.26 0.30	0.15 0.17	0.49 0.56	0.52 0.59	0.35 0.40	0.09 0.10	0.38 0.44	0.44 0.50
0.50 0.56	0.20 0.22	0.40 0.45	1.10 1.24	0.20 0.22	0.13 0.15	0.50 0.56	— —	0.40 0.45	0.09 0.11	0.44 0.50	0.38 0.42
0.36 0.40	0.28 0.31	0.34 0.38	1.24 1.40	0.25 0.28	0.15 0.17	0.44 0.50	0.48 0.54	0.35 0.39	— —	0.39 0.44	0.47 0.53
0.49 0.55	0.20 0.22	0.40 0.44	0.84 0.94	0.40 0.44	0.14 0.16	0.35 0.39	— —	0.40 0.44	0.10 0.11	0.50 0.55	0.50 0.55
— —	0.18 0.20	0.45 0.50	0.91 1.00	0.27 0.30	0.09 0.10	0.36 0.40	— —	0.36 0.40	0.09 0.10	0.45 0.50	0.36 0.40
0.43 0.50	0.33 0.38	0.32 0.37	0.96 1.11	0.39 0.45	0.16 0.19	0.37 0.43	0.43 0.50	0.32 0.38	0.10 0.12	0.38 0.44	0.44 0.51
1.83 1.99	0.91 0.99	1.32 1.43	— —	1.22 1.32	0.55 0.60	1.88 2.04	— —	1.12 1.21	0.46 0.50	— —	2.84 3.09
2.06 2.22	1.04 1.12	1.50 1.62	2.30 2.47	1.56 1.68	0.56 0.61	2.04 2.20	— —	1.29 1.39	0.49 0.53	0.77 0.83	1.92 2.07
1.70 1.89	1.10 1.22	1.33 1.48	— —	1.71 1.90	0.52 0.58	2.22 2.47	— —	1.32 1.47	0.47 0.52	— —	1.88 2.09
2.12 2.31	1.11 1.22	1.56 1.70	2.46 2.69	1.69 1.85	0.60 0.66	2.15 2.35	1.71 1.87	1.39 1.52	0.55 0.60	0.83 0.91	2.56 2.80
3.22 3.46	1.21 1.30	1.91 2.05	2.82 3.03	1.84 1.98	0.73 0.79	2.41 2.59	— —	1.61 1.73	0.58 0.62	0.80 0.86	2.31 2.49

(Continued)

AMINO ACID COMPOSITIONS

TABLE 21-2

Entry Number	Feed Name Description	International Feed Number	Moisture Basis: A-F (as-fed) or M-F (moisture-free)	Dry Matter	Crude Protein	Arginine	Cystine
				(%)	(%)	(%)	(%)
46	**DISTILLERS' GRAINS** -DEHY	5-02-144	A-F M-F	93 100	27.4 29.6	1.06 1.14	— —
47	**DISTILLERS' SOLUBLES** -DEHY	5-02-147	A-F M-F	92 100	28.8 31.3	3.55 3.86	0.40 0.44
48	**EMMER** *Triticum dicoccum* -GRAIN	4-01-830	A-F M-F	91 100	11.7 12.9	0.46 0.50	— —
49	**FISH** -MEAL, MECH EXTD	5-01-977	A-F M-F	92 100	64.5 70.4	3.88 4.23	0.70 0.76
50	-SOLUBLES, CONDENSED	5-01-969	A-F M-F	50 100	28.9 57.3	1.62 3.21	0.74 1.47
51	-SOLUBLES, DEHY	5-01-971	A-F M-F	92 100	57.8 62.7	3.15 3.42	0.59 0.64
52	**FISH, ANCHOVY** *Engraulis ringen* -MEAL, MECH EXTD	5-01-985	A-F M-F	92 100	65.6 71.3	3.78 4.10	0.60 0.65
53	**FISH, MENHADEN** *Brevoortia tyrannus* -MEAL, MECH EXTD	5-02-009	A-F M-F	92 100	61.1 66.7	3.73 4.08	0.56 0.61
54	**FISH, SARDINE** *Clupea* spp, *Sardinops* spp -MEAL, MECH EXTD	5-02-015	A-F M-F	93 100	65.3 70.0	2.70 2.90	0.80 0.86
55	-SOLUBLES, CONDENSED	5-02-014	A-F M-F	50 100	29.5 59.4	1.50 3.02	0.20 0.40
56	**FISH, WHITE** *Gadidae* (family), *Lophiidae* (family), *Rajidae* (family) -MEAL, MECH EXTD	5-02-025	A-F M-F	91 100	60.6 66.5	3.99 4.38	0.82 0.90
57	**FLAX** *Linum usitatissimum* -SEEDS, MEAL, MECH EXTD, 33% PROTEIN (LINSEED MEAL)	5-02-045	A-F M-F	91 100	34.3 37.8	2.72 3.00	0.58 0.64
58	-SEEDS, MEAL, SOLV EXTD, 33% PROTEIN (LINSEED MEAL)	5-02-048	A-F M-F	90 100	34.9 38.9	3.07 3.41	0.63 0.71
59	**LIVER** -MEAL, DEHY	5-00-389	A-F M-F	93 100	66.5 71.7	4.11 4.43	0.90 0.97
	MAIZE (SEE **CORN**)						
60	**MEAT** -MEAL, RENDERED	5-00-385	A-F M-F	93 100	53.8 57.9	3.73 4.02	0.65 0.70
61	-WITH BLOOD, MEAL, TANKAGE RENDERED	5-00-386	A-F M-F	92 100	59.5 64.7	3.59 3.91	0.46 0.50
62	-WITH BLOOD, WITH BONE, MEAL, TANKAGE RENDERED	5-00-387	A-F M-F	92 100	50.4 54.6	3.09 3.35	0.30 0.33
63	-WITH BONE, MEAL, RENDERED	5-00-388	A-F M-F	93 100	50.5 54.3	3.53 3.80	0.49 0.53
64	**MILK** -DEHY (COW'S)	5-01-167	A-F M-F	96 100	25.1 26.3	0.92 0.96	— —
65	-FRESH (COW'S)	5-01-168	A-F M-F	13 100	3.4 27.4	0.14 1.09	— —
66	-SKIMMED, DEHY (COW'S)	5-01-175	A-F M-F	94 100	33.4 35.7	1.15 1.23	0.44 0.47

AMINO ACID COMPOSITIONS

(Continued)

						Amino Acids					
Glycine	Histidine	Isoleucine	Leucine	Lysine	Methionine	Phenyl-alanine	Serine	Threonine	Tryptophan	Tyrosine	Valine
(%)	(%)	(%)	(%)	(%)	(%)	(%)	(%)	(%)	(%)	(%)	(%)
—	0.59	1.38	2.95	0.82	0.49	1.13	—	0.90	0.22	0.91	1.37
—	0.63	1.50	3.19	0.89	0.52	1.22	—	0.98	0.23	0.98	1.48
—	0.75	1.10	2.05	0.93	0.50	1.41	0.60	1.01	0.24	0.91	1.61
—	0.82	1.19	2.23	1.01	0.54	1.53	0.66	1.10	0.26	0.99	1.75
—	0.20	0.42	0.67	0.29	0.16	0.46	—	0.38	0.12	—	0.47
—	0.22	0.46	0.74	0.32	0.18	0.50	—	0.42	0.13	—	0.52
4.23	1.54	3.68	4.98	5.87	1.79	2.69	—	2.69	0.76	1.89	3.43
4.61	1.68	4.02	5.43	6.40	1.95	2.93	—	2.93	0.83	2.06	3.74
3.65	1.64	1.00	1.87	1.77	0.83	0.98	0.83	0.85	0.38	0.40	1.18
7.25	3.26	1.99	3.71	3.50	1.64	1.94	1.66	1.69	0.75	0.78	2.34
5.87	1.81	2.09	2.87	3.67	1.17	1.49	2.02	1.40	0.59	0.87	2.16
6.36	1.96	2.27	3.11	3.98	1.26	1.62	2.19	1.52	0.64	0.94	2.35
3.68	1.59	3.12	4.98	5.02	1.99	2.80	2.40	2.76	0.75	2.24	3.51
4.00	1.73	3.39	5.40	5.46	2.16	3.04	2.61	3.00	0.81	2.44	3.81
4.18	1.45	2.88	4.48	4.72	1.75	2.47	2.22	2.50	0.65	1.94	3.22
4.57	1.58	3.15	4.90	5.16	1.91	2.69	2.43	2.73	0.71	2.11	3.52
4.51	1.80	3.34	—	5.91	2.01	2.00	—	2.60	0.50	—	4.10
4.84	1.93	3.59	—	6.34	2.16	2.15	—	2.79	0.54	—	4.40
—	2.00	0.90	1.60	1.60	0.90	0.80	—	0.80	0.10	—	1.00
—	4.02	1.81	3.22	3.22	1.81	1.61	—	1.61	0.20	—	2.01
3.42	1.36	2.97	4.45	4.55	1.68	2.36	—	2.60	0.69	1.98	3.09
3.75	1.49	3.25	4.88	4.99	1.85	2.59	—	2.85	0.75	2.17	3.39
1.52	0.64	1.76	1.88	1.19	0.54	1.41	—	1.12	0.50	0.89	1.57
1.68	0.71	1.94	2.07	1.31	0.60	1.55	—	1.24	0.55	0.99	1.73
1.71	0.69	1.64	2.03	1.20	0.59	1.21	1.96	1.23	0.52	1.09	1.80
1.91	0.77	1.83	2.26	1.34	0.65	1.34	2.18	1.37	0.58	1.21	2.00
5.61	1.50	3.36	5.41	4.81	1.30	2.91	2.50	2.61	0.60	1.70	4.21
6.05	1.62	3.62	5.84	5.19	1.41	3.14	2.70	2.81	0.65	1.84	4.54
5.48	1.06	1.89	3.44	3.43	0.76	1.93	2.14	1.79	0.36	0.87	2.64
5.90	1.14	2.03	3.70	3.69	0.82	2.07	2.30	1.93	0.38	0.94	2.84
6.74	1.90	1.90	5.09	3.73	0.73	2.43	—	2.39	0.72	—	3.75
7.33	2.06	2.06	5.53	4.05	0.80	2.65	—	2.60	0.78	—	4.08
6.04	1.75	1.86	5.23	3.30	0.69	2.27	—	2.17	0.62	—	3.40
6.54	1.90	2.01	5.67	3.57	0.74	2.46	—	2.34	0.67	—	3.68
6.63	0.90	1.66	3.05	2.98	0.66	1.73	—	1.69	0.30	0.77	2.38
7.13	0.97	1.78	3.28	3.20	0.71	1.87	—	1.82	0.32	0.83	2.56
—	0.72	1.33	2.56	2.25	0.61	1.33	—	1.02	0.41	1.33	1.74
—	0.75	1.39	2.67	2.35	0.64	1.39	—	1.07	0.43	1.39	1.81
—	0.10	0.32	0.26	0.26	0.07	0.17	—	0.17	0.05	—	0.26
—	0.78	2.58	2.03	2.03	0.55	1.33	—	1.33	0.39	—	2.03
0.35	0.86	2.19	3.31	2.52	0.90	1.58	1.69	1.59	0.44	1.13	2.31
0.38	0.92	2.34	3.54	2.69	0.96	1.69	1.81	1.69	0.47	1.21	2.46

(Continued)

TABLE 21-2

Entry Number	Feed Name Description	International Feed Number	Moisture Basis: A-F (as-fed) or M-F (moisture-free)	Dry Matter	Crude Protein	Arginine	Cystine
				(%)	(%)	(%)	(%)
	MILLET *Setaria* spp						
67	-GRAIN	4-03-098	A-F	90	12.1	0.35	—
			M-F	100	13.5	0.39	—
	OATS *Avena sativa*						
68	-CEREAL BYPRODUCT, LESS THAN 4% FIBER (FEEDING OAT MEAL, OAT MIDDLINGS)	4-03-303	A-F	91	14.6	0.88	0.25
			M-F	100	16.2	0.97	0.27
69	-GRAIN	4-03-309	A-F	89	12.1	0.65	0.18
			M-F	100	13.6	0.73	0.21
70	-GRAIN, GRADE 1, 34 LB/BUSHEL OR *438 G/L*	4-03-313	A-F	90	12.1	0.80	0.22
			M-F	100	13.4	0.89	0.25
71	-GRAIN, PACIFIC COAST	4-07-999	A-F	91	9.2	0.60	0.17
			M-F	100	10.1	0.66	0.18
72	-GROATS	4-03-331	A-F	89	15.4	0.79	0.20
			M-F	100	17.3	0.89	0.22
	PEA *Pisum* spp						
73	-SEEDS	5-03-600	A-F	89	22.4	1.37	0.17
			M-F	100	25.2	1.54	0.19
	PEA, FIELD *Pisum sativum, arvense*						
74	-SEEDS	5-08-481	A-F	91	23.3	2.11	0.30
			M-F	100	25.5	2.32	0.33
	PEA, GARDEN *Pisum sativum*						
75	-SEEDS	5-08-482	A-F	89	23.8	1.43	—
			M-F	100	26.7	1.60	—
	PEANUT *Arachis hypogaea*						
76	-KERNELS, HULLS ADDED, MEAL, SOLV EXTD	5-03-656	A-F	93	47.4	5.49	0.71
			M-F	100	51.3	5.93	0.77
77	-KERNELS, MEAL, MECH EXTD (PEANUT MEAL)	5-03-649	A-F	90	44.0	4.05	—
			M-F	100	49.0	4.51	—
78	-KERNELS, MEAL, SOLV EXTD (PEANUT MEAL)	5-03-650	A-F	92	48.9	4.62	0.69
			M-F	100	52.9	5.00	0.74
	POULTRY						
79	-BYPRODUCTS, MEAL, RENDERED	5-03-798	A-F	93	58.5	3.85	0.91
			M-F	100	62.8	4.13	0.98
80	-FEATHERS, HYDROLYZED, MEAL	5-03-795	A-F	93	85.7	5.44	3.84
			M-F	100	92.0	5.84	4.12
	RAPE, CANADA *Brassica napus*						
81	SEEDS, MEAL, PREPRESSED SOLV EXTD, 40% PROTEIN	5-08-135	A-F	92	40.5	2.23	—
			M-F	100	44.0	2.42	—
	RICE *Oryza sativa*						
82	-BRAN WITH GERM (RICE BRAN)	4-03-928	A-F	91	12.7	0.85	0.18
			M-F	100	14.0	0.94	0.19
83	-BRAN WITH GERM, MEAL, SOLV EXTD (SOLVENT EXTRACTED RICE BRAN)	4-03-930	A-F	91	13.5	0.85	0.07
			M-F	100	14.9	0.94	0.08
84	-GRAIN, GROUND (GROUND ROUGH RICE, GROUND PADDY RICE)	4-03-938	A-F	89	8.4	0.88	0.14
			M-F	100	9.4	1.00	0.16
85	-GROATS, POLISHED (RICE, POLISHED)	4-03-942	A-F	89	7.2	0.44	0.09
			M-F	100	8.2	0.50	0.11
86	-POLISHINGS	4-03-943	A-F	90	12.1	0.63	0.14
			M-F	100	13.4	0.70	0.16
	RYE *Secale cereale*						
87	-DISTILLERS' GRAINS WITH SOLUBLES, DEHY	5-04-024	A-F	91	27.2	1.00	—
			M-F	100	30.1	1.11	—
88	-GRAIN	4-04-047	A-F	87	12.1	0.54	0.19
			M-F	100	13.8	0.61	0.22
	SAFFLOWER *Carthamus tinctorius*						
89	-SEEDS, WHOLE	4-07-958	A-F	93	18.2	1.60	0.35
			M-F	100	19.5	1.72	0.38

(Continued)

						Amino Acids					
Glycine	Histidine	Isoleucine	Leucine	Lysine	Methionine	Phenyl-alanine	Serine	Threonine	Tryptophan	Tyrosine	Valine
(%)	(%)	(%)	(%)	(%)	(%)	(%)	(%)	(%)	(%)	(%)	(%)
—	0.23	0.49	1.23	0.25	0.30	0.59	—	0.44	0.17	—	0.62
—	0.26	0.54	1.37	0.28	0.33	0.66	—	0.49	0.19	—	0.69
0.65	0.30	0.54	1.08	0.48	0.21	0.70	—	0.49	0.20	0.75	0.75
0.71	0.33	0.60	1.19	0.53	0.23	0.78	—	0.54	0.22	0.82	0.83
0.43	0.19	0.42	0.77	0.39	0.14	0.51	0.45	0.35	0.15	0.36	0.54
0.49	0.21	0.48	0.87	0.43	0.16	0.57	0.50	0.40	0.17	0.40	0.61
0.50	0.20	0.53	0.90	0.50	0.18	0.60	—	0.40	0.16	0.53	0.70
0.56	0.22	0.59	1.01	0.56	0.20	0.67	—	0.45	0.18	0.59	0.78
0.40	0.15	0.37	—	0.33	0.13	0.42	—	0.28	0.12	—	0.48
0.44	0.17	0.41	—	0.36	0.14	0.47	—	0.31	0.13	—	0.53
0.60	0.27	0.55	0.99	0.30	0.20	0.62	—	0.45	0.18	0.60	0.67
0.67	0.31	0.61	1.11	0.33	0.22	0.70	—	0.50	0.20	0.67	0.75
1.07	0.70	1.07	1.76	1.56	0.30	1.27	—	0.92	0.23	—	1.27
1.21	0.79	1.21	1.98	1.76	0.34	1.43	—	1.03	0.26	—	1.43
—	—	—	—	1.21	0.20	—	—	—	0.20	—	—
—	—	—	—	1.32	0.22	—	—	—	0.22	—	—
—	0.63	1.03	1.61	1.47	0.34	1.20	—	0.80	0.23	—	1.18
—	0.71	1.15	1.80	1.65	0.38	1.35	—	0.90	0.26	—	1.32
2.44	1.22	2.03	3.80	1.83	0.44	2.74	2.98	1.52	0.50	1.85	2.84
2.63	1.32	2.19	4.11	1.98	0.47	2.96	3.22	1.65	0.54	2.00	3.07
—	0.85	1.70	2.62	1.33	0.50	1.95	—	1.17	0.47	—	1.89
—	0.94	1.89	2.91	1.48	0.56	2.17	—	1.30	0.53	—	2.11
2.32	0.93	1.69	2.61	1.62	0.42	1.99	—	1.09	0.47	1.52	1.79
2.51	1.01	1.83	2.83	1.75	0.45	2.15	—	1.18	0.51	1.65	1.94
3.85	0.88	2.55	4.14	2.77	1.07	2.31	—	2.11	0.46	0.90	2.99
4.13	0.94	2.73	4.44	2.97	1.15	2.48	—	2.27	0.50	0.97	3.21
6.02	0.51	3.56	6.46	1.63	0.48	3.41	10.04	3.63	0.52	2.29	6.03
6.46	0.54	3.82	6.93	1.75	0.52	3.66	10.78	3.90	0.56	2.45	6.47
1.94	1.09	1.46	2.71	2.15	0.77	1.54	1.70	1.70	0.49	0.85	1.94
2.11	1.19	1.58	2.95	2.33	0.84	1.67	1.85	1.85	0.53	0.92	2.11
0.80	0.32	0.51	0.90	0.57	0.24	0.58	—	0.47	0.10	0.68	0.76
0.88	0.35	0.56	0.99	0.62	0.27	0.64	—	0.52	0.11	0.75	0.83
0.95	0.29	0.45	0.81	0.54	0.21	0.46	—	0.45	0.21	0.72	0.65
1.04	0.31	0.49	0.89	0.59	0.23	0.52	—	0.50	0.23	0.79	0.71
0.84	0.19	0.39	0.72	0.34	0.17	0.47	1.37	0.31	0.11	0.70	0.56
0.94	0.22	0.44	0.82	0.38	0.19	0.53	1.54	0.35	0.12	0.79	0.63
0.74	0.18	0.45	0.71	0.28	0.25	0.53	—	0.36	0.09	0.62	0.53
0.83	0.20	0.50	0.80	0.32	0.28	0.60	—	0.40	0.11	0.70	0.60
0.71	0.19	0.36	0.65	0.53	0.21	0.39	—	0.35	0.10	0.42	0.73
0.78	0.21	0.40	0.72	0.59	0.23	0.44	—	0.39	0.12	0.46	0.81
—	0.70	1.50	2.10	1.00	0.40	1.30	1.20	1.10	0.30	0.50	1.60
—	0.77	1.66	2.32	1.11	0.44	1.44	1.33	1.22	0.33	0.55	1.77
0.51	0.26	0.47	0.71	0.42	0.16	0.56	0.52	0.37	0.13	0.27	0.56
0.58	0.30	0.53	0.82	0.48	0.18	0.65	0.60	0.42	0.14	0.31	0.64
1.00	0.48	0.80	1.20	0.60	0.33	1.00	—	0.64	0.28	—	1.00
1.07	0.52	0.86	1.29	0.64	0.35	1.07	—	0.69	0.30	—	1.07

(Continued)

TABLE 21-2

AMINO ACID COMPOSITIONS

Entry Number	Feed Name Description	International Feed Number	Moisture Basis: A-F (as-fed) or M-F (moisture-free)	Dry Matter	Crude Protein	Arginine	Cystine
				(%)	(%)	(%)	(%)
90	-SEEDS WITHOUT HULLS, MEAL, MECH EXTD	5-08-499	A-F M-F	90 100	42.1 46.8	4.48 4.98	0.67 0.74
91	-SEEDS WITHOUT HULLS, MEAL, SOLV EXTD	5-07-959	A-F M-F	90 100	42.0 46.5	3.70 4.09	0.70 0.77
	SCREENINGS						
92	-REFUSE (CEREAL)	4-02-151	A-F M-F	91 100	12.6 13.8	0.68 0.75	— —
93	-UNCLEANED (CEREAL)	4-02-153	A-F M-F	92 100	13.8 15.0	0.67 0.73	— —
	SESAME *Sesamum indicum*						
94	-SEEDS, MEAL, MECH EXTD	5-04-220	A-F M-F	93 100	44.3 47.7	4.59 4.96	0.60 0.65
	SORGHUM *Sorghum vulgare*						
95	-GRAIN	4-04-383	A-F M-F	90 100	11.3 12.6	0.40 0.44	0.21 0.23
96	-GRAIN, FETERITA	4-04-369	A-F M-F	90 100	12.4 13.8	0.46 0.51	— —
97	-GRAIN, HEGARI	4-04-398	A-F M-F	89 100	10.0 11.1	0.29 0.33	— —
98	-GRAIN, LESS THAN 9% PROTEIN	4-08-138	A-F M-F	89 100	8.9 10.1	0.28 0.32	0.14 0.16
99	-GLUTEN MEAL	5-04-388	A-F M-F	90 100	44.1 49.0	1.27 1.41	0.80 0.89
	SORGHUM, KAFIR *Sorghum vulgare, caffrorum*						
100	-GRAIN	4-04-428	A-F M-F	89 100	11.0 12.4	0.37 0.42	0.16 0.18
	SORGHUM, MILO *Sorghum vulgare, subglabrescens*						
101	-GRAIN	4-04-444	A-F M-F	89 100	10.6 11.9	0.38 0.42	0.17 0.20
102	-GLUTEN WITH BRAN (GLUTEN FEED)	4-08-089	A-F M-F	89 100	23.1 26.0	0.90 1.01	0.20 0.23
	SORGHUM, SHALLU *Sorghum vulgare, roxburghi*						
103	-GRAIN	4-04-456	A-F M-F	90 100	11.5 12.7	0.31 0.34	— —
	SOYBEAN *Glycine max*						
104	-FLOUR BYPRODUCT (SOYBEAN MILL FEED)	5-04-594	A-F M-F	89 100	12.8 14.3	0.81 0.91	0.16 0.18
105	-FLOUR, SOLV EXTD	5-04-593	A-F M-F	93 100	46.8 50.5	3.07 3.32	0.70 0.75
106	-SEEDS, MEAL, SOLV EXTD, 44% PROTEIN	5-20-637	A-F M-F	91 100	40.5 44.3	2.93 3.21	0.80 0.87
107	-SEEDS, MEAL, SOLV EXTD, 49% PROTEIN	5-20-638	A-F M-F	91 100	43.4 47.6	3.14 3.44	0.77 0.84
108	-SEEDS, WHOLE	5-04-610	A-F M-F	91 100	39.4 43.2	7.01 7.68	0.37 0.40
109	-SEEDS WITHOUT HULLS, MEAL, SOLV EXTD	5-04-612	A-F M-F	90 100	48.3 53.4	3.37 3.73	0.76 0.84
	SPELT *Triticum spelta*						
110	-GRAIN	4-04-651	A-F M-F	90 100	12.0 13.3	0.45 0.50	— —
	TRITICALE *Triticale hexaloide*						
111	-GRAIN	4-20-362	A-F M-F	91 100	24.7 27.2	2.04 2.25	0.51 0.56
	WHEAT *Triticum aestivum*						
112	-BRAN	4-05-190	A-F M-F	89 100	15.2 17.1	1.03 1.16	0.35 0.39
113	—DISTILLERS' GRAINS, DEHY	5-05-193	A-F M-F	93 100	31.6 33.9	1.10 1.18	— —

(Continued)

| | | | | | | Amino Acids | | | | | |

Glycine	Histidine	Isoleucine	Leucine	Lysine	Methionine	Phenyl-alanine	Serine	Threonine	Tryptophan	Tyrosine	Valine
(%)	(%)	(%)	(%)	(%)	(%)	(%)	(%)	(%)	(%)	(%)	(%)
2.44	—	—	—	1.29	0.68	—	—	0.79	0.60	—	—
2.71	—	—	—	1.44	0.76	—	—	0.88	0.67	—	—
2.40	1.00	1.70	2.61	1.30	0.69	1.85	—	1.35	0.60	1.05	2.30
2.65	1.11	1.88	2.89	1.44	0.76	2.05	—	1.49	0.66	1.17	2.54
0.59	0.30	0.53	0.98	0.49	0.15	0.64	0.57	0.47	—	0.32	0.63
0.65	0.33	0.58	1.08	0.53	0.16	0.71	0.63	0.51	—	0.35	0.70
0.61	0.30	0.45	0.90	0.42	0.19	0.58	0.67	0.44	—	0.58	0.58
0.66	0.33	0.49	0.98	0.46	0.21	0.63	0.73	0.48	—	0.63	0.63
3.88	1.16	1.97	3.10	1.25	1.37	2.13	2.95	1.60	0.72	1.85	2.36
4.19	1.25	2.12	3.34	1.35	1.47	2.30	3.18	1.72	0.77	1.99	2.54
0.34	0.24	0.45	1.46	0.25	0.13	0.56	0.50	0.37	0.15	0.41	0.53
0.38	0.26	0.50	1.62	0.28	0.14	0.63	0.56	0.41	0.16	0.46	0.59
—	0.26	0.58	1.78	0.20	0.18	0.67	—	0.46	0.17	—	0.67
—	0.29	0.65	1.99	0.23	0.21	0.75	—	0.51	0.19	—	0.75
—	0.18	0.47	1.40	0.17	0.11	0.54	—	0.36	0.11	—	0.55
—	0.21	0.52	1.56	0.20	0.13	0.61	—	0.41	0.12	—	0.61
0.27	0.19	0.46	1.40	0.19	0.12	0.47	—	0.36	0.12	0.60	0.53
0.31	0.21	0.52	1.58	0.21	0.14	0.53	—	0.41	0.14	0.68	0.60
0.80	0.99	2.42	8.00	0.73	0.73	2.73	—	1.47	0.44	—	2.50
0.89	1.10	2.68	8.88	0.81	0.81	3.03	—	1.63	0.49	—	2.77
0.30	0.27	0.55	1.62	0.26	0.19	0.63	—	0.45	0.16	—	0.61
0.33	0.30	0.62	1.81	0.29	0.21	0.71	—	0.51	0.18	—	0.69
0.35	0.22	0.53	1.36	0.18	0.25	0.40	0.53	0.25	0.21	0.48	0.53
0.39	0.24	0.60	1.53	0.21	0.29	0.45	0.60	0.28	0.23	0.55	0.59
0.68	0.60	1.00	2.50	0.70	0.40	1.00	—	0.80	0.20	0.90	1.30
0.76	0.67	1.12	2.81	0.79	0.45	1.12	—	0.90	0.23	1.01	1.46
—	0.19	0.38	0.97	0.19	0.17	0.40	—	0.30	0.10	—	0.46
—	0.21	0.42	1.08	0.21	0.18	0.44	—	0.33	0.11	—	0.51
0.50	0.18	0.41	0.58	0.69	0.14	0.38	—	0.30	0.13	0.23	0.38
0.56	0.20	0.45	0.65	0.77	0.16	0.42	—	0.34	0.15	0.26	0.42
1.60	0.86	1.70	3.05	3.53	0.65	1.89	2.03	1.56	1.00	1.37	0.93
1.73	0.93	1.83	3.29	3.81	0.70	2.04	2.19	1.68	1.08	1.48	1.01
1.66	1.00	1.74	3.07	2.27	0.46	1.94	2.01	1.53	0.66	1.46	1.78
1.82	1.10	1.91	3.36	2.49	0.50	2.12	2.20	1.67	0.72	1.59	1.95
1.97	1.12	1.99	3.48	2.83	0.53	2.23	2.35	1.80	0.71	1.60	1.96
2.16	1.22	2.19	3.82	3.11	0.58	2.45	2.58	1.98	0.78	1.75	2.15
3.31	1.93	1.81	3.67	1.47	0.22	2.90	2.80	1.48	—	2.17	2.29
3.62	2.11	1.98	4.02	1.61	0.24	3.17	3.07	1.62	—	2.38	2.51
2.17	1.15	2.20	3.53	2.96	0.60	2.32	2.33	1.84	0.70	1.62	2.17
2.40	1.27	2.43	3.91	3.28	0.67	2.56	2.58	2.04	0.77	1.79	2.40
—	0.18	0.36	0.63	0.27	0.18	0.45	—	0.36	0.09	—	0.45
—	0.20	0.40	0.70	0.30	0.20	0.50	—	0.40	0.10	—	0.50
1.30	0.83	1.17	2.13	1.69	0.28	1.41	1.56	1.04	0.20	0.88	1.31
1.43	0.92	1.29	2.34	1.86	0.31	1.56	1.72	1.15	0.22	0.97	1.45
0.91	0.40	0.49	0.93	0.60	0.21	0.57	0.70	0.47	0.28	0.44	0.70
1.02	0.46	0.55	1.05	0.68	0.23	0.64	0.79	0.53	0.32	0.49	0.78
—	0.80	2.01	1.71	0.70	—	1.71	—	0.90	—	0.50	1.71
—	0.86	2.15	1.83	0.75	—	1.83	—	0.97	—	0.54	1.83

(Continued)

AMINO ACID COMPOSITIONS

TABLE 21-2

Entry Number	Feed Name Description	International Feed Number	Moisture Basis: A-F (as-fed) or M-F (moisture-free)	Dry Matter	Crude Protein	Arginine	Cystine
				(%)	(%)	(%)	(%)
114	-ENDOSPERM	4-05-197	A-F	88	11.1	0.60	0.30
			M-F	100	12.6	0.68	0.34
115	-FLOUR, LESS THAN 2% FIBER	4-05-199	A-F	87	11.6	0.42	0.31
			M-F	100	13.3	0.48	0.36
116	-GERM, GROUND (WHEAT GERM MEAL)	5-05-218	A-F	88	24.5	1.88	0.47
			M-F	100	27.8	2.14	0.53
117	-GRAIN	4-05-211	A-F	88	14.9	0.58	0.31
			M-F	100	16.9	0.66	0.35
118	-GRAIN, HARD RED SPRING	4-05-258	A-F	88	15.4	0.59	0.25
			M-F	100	17.6	0.67	0.28
119	-GRAIN, HARD RED WINTER	4-05-268	A-F	88	12.6	0.61	0.31
			M-F	100	14.4	0.69	0.35
120	-GRAIN, SOFT RED WINTER	4-05-294	A-F	88	11.5	0.55	0.30
			M-F	100	13.0	0.62	0.34
121	-GRAIN, SOFT WHITE WINTER	4-05-337	A-F	88	9.9	0.46	0.27
			M-F	100	11.2	0.52	0.30
122	-GRAIN, SOFT WHITE WINTER, PACIFIC COAST	4-08-555	A-F	89	10.1	0.45	0.24
			M-F	100	11.4	0.50	0.27
123	-GLUTEN	5-05-221	A-F	89	49.6	2.92	1.71
			M-F	100	56.0	3.30	1.93
124	-MIDDLINGS, LESS THAN 9.5% FIBER	4-05-205	A-F	89	16.7	0.87	0.22
			M-F	100	18.8	0.98	0.25
125	-MILL RUN, LESS THAN 9.5% FIBER	4-05-206	A-F	90	15.6	0.90	0.20
			M-F	100	17.3	1.00	0.22
126	-RED DOG, LESS THAN 4.5% FIBER	4-05-203	A-F	88	15.3	0.96	0.38
			M-F	100	17.4	1.10	0.43
127	-SHORTS, LESS THAN 7% FIBER	4-05-201	A-F	88	16.4	1.21	0.38
			M-F	100	18.6	1.38	0.43
	WHEAT, DURUM *Triticum durum*						
128	-GRAIN	4-05-224	A-F	87	13.8	0.66	—
			M-F	100	15.9	0.76	—
	WHEY						
129	-DEHY	4-01-182	A-F	93	13.4	0.33	0.29
			M-F	100	14.4	0.35	0.31
130	-WHEY, LOW LACTOSE, DEHY (DRIED WHEY PRODUCT)	4-01-186	A-F	93	16.4	0.68	0.49
			M-F	100	17.6	0.73	0.53
	YEAST *Saccharomyces cerevisiae*						
131	-BREWERS', DEHY	7-05-527	A-F	93	502.8	2.22	0.51
			M-F	100	537.8	2.37	0.55
132	-IRRADIATED, DEHY	7-05-529	A-F	94	48.1	2.46	—
			M-F	100	51.2	2.62	—
133	-PRIMARY, DEHY	7-05-533	A-F	93	48.0	2.60	0.50
			M-F	100	51.8	2.81	0.54
	YEAST, TORULA *Torulopsis utilis*						
134	-DEHY	7-05-534	A-F	93	48.8	2.60	0.60
			M-F	100	52.5	2.80	0.65

AMINO ACID COMPOSITIONS

(Continued)

	Amino Acids										
Glycine	Histidine	Isoleucine	Leucine	Lysine	Methionine	Phenyl-alanine	Serine	Threonine	Tryptophan	Tyrosine	Valine
(%)	(%)	(%)	(%)	(%)	(%)	(%)	(%)	(%)	(%)	(%)	(%)
—	0.30	1.10	1.70	0.40	0.20	0.60	—	0.40	0.30	—	0.60
—	0.34	1.25	1.93	0.46	0.23	0.68	—	0.46	0.34	—	0.68
0.44	0.25	0.46	0.88	0.24	0.18	0.60	0.59	0.32	0.10	0.33	0.49
0.51	0.29	0.52	1.01	0.27	0.21	0.68	0.68	0.37	0.11	0.38	0.56
1.47	0.65	0.88	1.56	1.54	0.44	0.94	1.13	0.97	0.30	0.73	1.17
1.66	0.74	1.00	1.77	1.74	0.50	1.07	1.28	1.10	0.34	0.83	1.33
0.57	0.28	0.47	0.87	0.37	0.18	0.61	0.60	0.38	0.16	0.41	0.56
0.64	0.31	0.53	0.98	0.42	0.21	0.69	0.67	0.43	0.18	0.46	0.64
0.67	0.23	0.56	0.87	0.35	0.19	0.66	0.58	0.36	0.14	0.50	0.59
0.77	0.26	0.64	1.00	0.40	0.21	0.75	0.66	0.41	0.16	0.58	0.67
0.57	0.28	0.51	0.88	0.36	0.21	0.63	0.59	0.38	0.18	0.43	0.59
0.64	0.31	0.58	1.00	0.41	0.24	0.71	0.67	0.43	0.20	0.49	0.67
0.54	0.24	0.45	0.90	0.50	0.22	0.64	0.65	0.39	0.26	0.38	0.57
0.62	0.27	0.51	1.02	0.57	0.24	0.72	0.73	0.44	0.30	0.43	0.65
0.49	0.22	0.40	0.65	0.31	0.16	0.45	0.45	0.31	0.12	0.36	0.45
0.55	0.24	0.46	0.73	0.35	0.18	0.51	0.51	0.35	0.14	0.41	0.50
0.50	0.20	0.40	0.59	0.30	0.14	0.42	0.38	0.28	0.12	0.36	0.41
0.56	0.22	0.44	0.67	0.33	0.16	0.47	0.42	0.31	0.13	0.40	0.46
2.72	1.61	3.32	5.43	1.51	1.21	4.12	4.02	2.11	0.70	2.31	3.82
3.07	1.82	3.75	6.14	1.70	1.36	4.66	4.55	2.39	0.80	2.61	4.32
0.37	0.36	0.70	1.09	0.67	0.20	0.64	0.80	0.53	0.22	0.39	0.77
0.42	0.40	0.79	1.22	0.76	0.23	0.72	0.89	0.59	0.24	0.43	0.86
0.40	0.40	0.70	1.20	0.50	0.40	—	—	0.50	0.20	0.50	0.80
0.44	0.44	0.78	1.33	0.56	0.44	—	—	0.56	0.22	0.56	0.89
0.73	0.38	0.54	1.05	0.58	0.24	0.65	0.74	0.50	0.19	0.46	0.72
0.84	0.43	0.62	1.19	0.66	0.27	0.74	0.85	0.57	0.22	0.52	0.82
0.97	0.45	0.57	1.09	0.81	0.28	0.67	0.77	0.60	0.23	0.49	0.83
1.10	0.51	0.65	1.24	0.92	0.32	0.76	0.88	0.68	0.27	0.56	0.94
0.58	0.34	0.58	1.04	0.38	0.18	0.80	0.71	0.42	—	0.39	0.69
0.67	0.39	0.67	1.20	0.43	0.21	0.92	0.81	0.48	—	0.44	0.79
0.47	0.17	0.81	1.18	0.92	0.18	0.35	0.41	0.87	0.17	0.25	0.68
0.50	0.19	0.87	1.27	0.98	0.20	0.37	0.44	0.93	0.18	0.27	0.73
0.81	0.26	0.84	1.17	1.43	0.42	0.51	—	0.82	0.28	0.46	0.76
0.87	0.27	0.90	1.26	1.53	0.45	0.55	—	0.88	0.30	0.50	0.82
1.72	1.11	2.17	3.24	3.11	0.73	1.83	—	2.10	0.51	1.52	2.34
1.84	1.19	2.32	3.46	3.33	0.78	1.96	—	2.25	0.55	1.62	2.50
—	1.00	2.94	3.56	3.70	1.00	2.77	—	2.41	0.73	—	3.06
—	1.06	3.13	3.79	3.94	1.06	2.95	—	2.56	0.78	—	3.26
—	5.60	3.60	3.70	3.80	1.00	2.50	—	2.50	0.40	—	3.20
—	6.05	3.89	4.00	4.10	1.08	2.70	—	2.70	0.43	—	3.46
2.70	1.40	2.90	3.50	3.80	0.80	3.00	—	2.60	0.50	2.10	2.90
2.90	1.51	3.12	3.77	4.09	0.86	3.23	—	2.80	0.54	2.26	3.12

TABLE
MINERAL SUPPLEMENTS, COMPOSITION,

Entry Number	Feed Name Description	International Feed Number	Moisture Basis: A-F (as-fed) or M-F (moisture-free)	Chemical Analysis						Digestible Protein		
				Dry Matter	Ash	Crude Fiber	Ether Extract (Fat)	N-Free Extract	Crude Protein (6.25 × N)	Ruminant	Non-ruminant	Horse
				(%)	(%)	(%)	(%)	(%)	(%)	(%)	(%)	(%)
1	AMMONIUM CHLORIDE	6-08-814	A-F	—	—	—	—	—	—	—	—	—
			M-F	100	—	—	—	—	160.0	—	—	—
2	AMMONIUM PHOSPHATE, MONOBASIC	6-09-338	A-F	97	34.5	—	—	—	68.8	—	—	—
			M-F	100	35.6	—	—	—	70.9	—	—	—
3	AMMONIUM PHOSPHATE, DIBASIC	6-00-370	A-F	97	34.5	—	—	—	112.4	—	—	—
			M-F	100	35.6	—	—	—	115.9	—	—	—
4	AMMONIUM-POLYPHOSPHATE SOLUTION FROM DEFLOURINATED PHOSPHORIC ACID	6-08-042	A-F	60	—	—	—	—	62.5	—	—	—
			M-F	100	—	—	—	—	104.2	—	—	—
5	AMMONIUM-POLYPHOSPHATE SOLUTION FROM FURNACE PHOSPHORIC ACID	6-26-401	A-F	—	—	—	—	—	68.7	—	—	—
			M-F	100	—	—	—	—	—	—	—	—
6	BONE, BLACK, SPENT	6-00-404	A-F	90	—	—	—	—	8.5	—	—	—
			M-F	100	—	—	—	—	9.4	—	—	—
7	BONE, CHARCOAL	6-00-402	A-F	90	—	—	—	—	8.5	—	—	—
			M-F	100	—	—	—	—	9.4	—	—	—
8	BONE MEAL	6-00-397	A-F	95	—	—	9.6	—	17.8	—	—	—
			M-F	100	—	—	10.2	—	18.8	—	—	—
9	BONE MEAL, STEAMED	6-00-400	A-F	97	77.0	1.4	11.3	0.0	12.8	8.7	—	—
			M-F	100	79.3	1.4	11.6	0.0	13.2	9.0	—	—
10	CALCIUM CARBONATE	6-01-069	A-F	99	—	—	—	—	—	—	—	—
			M-F	100	—	—	—	—	—	—	—	—
11	CALCIUM PHOSPHORATE, MONOBASIC, FROM DEFLUORINATED PHOSPHORIC ACID	6-01-082	A-F	97	—	—	—	—	—	—	—	—
			M-F	100	—	—	—	—	—	—	—	—
12	CALCIUM PHOSPHATE, MONOBASIC, FROM FURNACE PHOSPHORIC ACID	6-26-334	A-F	—	—	—	—	—	—	—	—	—
			M-F	100	—	—	—	—	—	—	—	—
13	CALCIUM PHOSPHATE, DIBASIC, FROM DEFLUORINATED PHOSPHORIC ACID	6-01-080	A-F	97	91.0	—	—	—	—	—	—	—
			M-F	100	93.8	—	—	—	—	—	—	—
14	CALCIUM PHOSPHATE, DIBASIC, FROM FURNACE PHOSPHORIC ACID	6-26-335	A-F	97	—	—	—	—	—	—	—	—
			M-F	100	93.8	—	—	—	—	—	—	—
15	CALCIUM PHOSPHATE, TRIBASIC, FROM FURNACE PHOSPHORIC ACID	6-01-084	A-F	—	—	—	—	—	—	—	—	—
			M-F	100	—	—	—	—	—	—	—	—
16	CALCIUM SULFATE ANHYDROUS (GYPSUM)	6-01-087	A-F	85	—	—	—	—	—	—	—	—
			M-F	100	—	—	—	—	—	—	—	—
17	COBALT CARBONATE	6-01-566	A-F	81	—	—	—	—	—	—	—	—
			M-F	100	—	—	—	—	—	—	—	—
18	COBALT OXIDE	6-01-560	A-F	—	—	—	—	—	—	—	—	—
			M-F	100	—	—	—	—	—	—	—	—
19	COBALT SULFATE	6-01-564	A-F	55	—	—	—	—	—	—	—	—
			M-F	100	—	—	—	—	—	—	—	—
20	COLLOIDAL CLAY (SOFT ROCK PHOSPHATE)	6-03-947	A-F	—	—	—	—	—	—	—	—	—
			M-F	100	—	—	—	—	—	—	—	—
21	COPPER (CUPRIC) CHLORIDE	6-01-705	A-F	79	—	—	—	—	—	—	—	—
			M-F	100	—	—	—	—	—	—	—	—
22	COPPER (CUPRIC) OXIDE	6-01-711	A-F	—	—	—	—	—	—	—	—	—
			M-F	100	—	—	—	—	—	—	—	—
23	COPPER (CUPRIC) SULFATE	6-01-720	A-F	64	—	—	—	—	—	—	—	—
			M-F	100	—	—	—	—	—	—	—	—
24	CURACAO PHOSPHATE	6-05-586	A-F	—	—	—	—	—	—	—	—	—
			M-F	100	100.0	—	—	—	—	—	—	—

21-3

DATA EXPRESSED AS-FED AND MOISTURE-FREE

Entry Number	Macrominerals							Microminerals						
	Calcium (Ca)	Phosphorus (P)	Sodium (Na)	Chlorine (Cl)	Magnesium (Mg)	Potassium (K)	Sulfur (S)	Cobalt (Co)	Copper (Cu)	Iodine (I)	Iron (Fe)	Manganese (Mn)	Selenium (Se)	Zinc (Zn)
	(%)	*(%)*	*(%)*	*(%)*	*(%)*	*(%)*	*(%)*	*(ppm or mg/kg)*	*(ppm or mg/kg)*	*(ppm or mg/kg)*	*(%)*	*(ppm or mg/kg)*	*(ppm or mg/kg)*	*(ppm or mg/kg)*
1	—	—	—	—	—	—	—	—	—	—	—	—	—	—
	—	—	—	66.28	—	—	—	—	—	—	—	—	—	—
2	0.50	24.00	0.06	—	0.45	—	0.70	—	80	—	1.200	400	—	300
	0.52	24.74	0.06	—	0.46	—	0.70	—	82	—	1.237	412	—	309
3	0.57	20.00	0.04	—	0.45	0.01	2.50	—	91	—	1.200	400	—	342
	0.59	20.60	0.04	—	0.46	0.01	2.60	—	94	—	1.237	412	—	353
4	0.10	14.50	—	—	—	—	—	—	—	—	—	—	—	—
	0.17	24.20	—	—	—	—	—	—	—	—	—	—	—	—
5	—	16.00	—	—	—	—	—	—	—	—	—	—	—	—
	—	—	—	—	—	—	—	—	—	—	—	—	—	—
6	27.10	12.73	—	—	0.53	0.14	—	—	—	—	—	—	—	—
	30.11	14.14	—	—	0.59	0.16	—	—	—	—	—	—	—	—
7	27.10	12.73	—	—	0.53	0.14	—	—	—	—	—	—	—	—
	30.11	14.14	—	—	0.59	0.16	—	—	—	—	—	—	—	—
8	25.95	12.42	—	—	—	—	—	—	—	—	—	—	—	—
	27.32	13.07	—	—	—	—	—	—	—	—	—	—	—	—
9	29.82	12.49	5.53	—	0.32	0.18	2.44	—	11	33,196	0.085	22	—	126
	30.71	12.86	5.69	—	0.33	0.19	2.51	—	11	34,188	0.088	23	—	130
10	37.62	0.04	0.02	0.04	0.50	0.06	0.09	—	24	—	0.034	277	—	—
	38.00	0.04	0.02	0.04	0.50	0.06	0.09	—	24	—	0.034	280	—	—
11	15.91	20.95	—	—	—	—	—	—	—	—	0.002	—	—	—
	16.40	21.60	—	—	—	—	—	—	—	—	0.002	—	—	—
12	—	—	—	—	—	—	—	—	—	—	—	—	—	—
	22.00	23.00	—	0.00	—	—	—	—	80	—	0.002	—	—	220
13	21.30	18.70	—	—	0.60	0.07	—	—	—	—	—	304	—	—
	22.00	19.30	—	—	0.62	0.07	—	—	—	—	—	313	—	—
14	26.30	18.70	—	—	0.62	0.07	—	—	—	—	—	304	—	—
	27.10	19.30	—	—	0.62	0.07	—	—	—	—	—	313	—	—
15	38.00	19.50	—	—	—	—	—	—	—	—	—	—	—	—
	—	—	—	—	—	—	—	—	—	—	—	—	—	—
16	22.02	0.01	—	—	2.21	—	20.01	—	—	—	0.171	—	—	—
	25.90	0.01	—	—	2.61	—	23.54	—	—	—	0.201	—	—	—
17	—	—	—	—	—	—	—	460,000	—	—	—	—	—	—
	—	—	—	—	—	—	—	570,000	—	—	—	—	—	—
18	—	—	—	—	—	—	—	720,000	—	—	—	—	—	—
	—	—	—	—	—	—	0.20	—	—	—	0.050	—	—	—
19	—	—	—	0.00	0.04	—	11.40	210,000	—	—	0.001	20	—	—
	—	—	—	0.00	0.07	—	20.65	380,000	—	—	0.002	36	—	—
20	17.00	9.00	—	—	—	—	—	—	—	—	—	—	—	—
	—	—	—	—	—	—	—	—	—	—	—	—	—	—
21	—	—	—	41.64	—	—	0.04	—	373,000	—	0.005	—	—	—
	—	—	—	52.71	—	—	0.04	—	472,000	—	0.006	—	—	—
22	—	—	—	—	—	—	—	—	—	—	—	—	—	—
	—	—	—	0.01	—	—	0.13	—	799,000	—	—	—	—	—
23	—	—	—	0.00	—	—	12.84	—	254,000	—	0.003	—	—	—
	—	—	—	0.00	—	—	20.06	—	398,000	—	0.005	—	—	—
24	—	—	—	—	—	—	—	—	—	—	—	—	—	—
	34.00	15.00	—	—	—	—	—	—	—	—	—	—	—	—

(Continued)

M I N E R A L   S U P P L E M E N T S

TABLE 21-3

MINERAL SUPPLEMENTS

Entry Number	Feed Name Description	International Feed Number	Moisture Basis: A-F (as-fed) or M-F (moisture-free)	Chemical Analysis						Digestible Protein		
				Dry Matter	Ash	Crude Fiber	Ether Extract (Fat)	N-Free Extract	Crude Protein (6.25 × N)	Ruminant	Non-ruminant	Horse
				(%)	(%)	(%)	(%)	(%)	(%)	(%)	(%)	(%)
25	DEFLUORINATED PHOSPHATE FROM PHOSPHORIC ACID	6-26-336	A-F	—	—	—	—	—	—	—	—	—
			M-F	100	—	—	—	—	—	—	—	—
26	DIAMMONIUM PHOSPHATE	6-00-370	A-F	97	34.5	—	—	—	112.4	—	—	—
			M-F	100	35.6	—	—	—	115.9	—	—	—
27	DICALCIUM PHOSPHATE (CALCIUM PHOSPHATE, DIBASIC)	6-00-080	A-F	96	90.0	—	—	—	—	—	—	—
			M-F	100	93.8	—	—	—	—	—	—	—
28	DISODIUM PHOSPHATE	6-04-286	A-F	—	—	—	—	—	—	—	—	—
			M-F	100	—	—	—	—	—	—	—	—
29	DOLOMITE LIMESTONE (LIMESTONE, MAGNESIUM)	6-02-633	A-F	99	—	—	—	—	—	—	—	—
			M-F	100	—	—	—	—	—	—	—	—
30	IRON (FERRIC) OXIDE	6-02-431	A-F	—	—	—	—	—	—	—	—	—
			M-F	100	—	—	—	—	—	—	—	—
31	IRON (FERROUS) CHLORIDE	6-01-865	A-F	64	—	—	—	—	—	—	—	—
			M-F	100	—	—	—	—	—	—	—	—
32	IRON (FERROUS) SULFATE	6-01-869	A-F	55	—	—	—	—	—	—	—	—
			M-F	100	—	—	—	—	—	—	—	—
33	LIMESTONE, GROUND	6-02-632	A-F	99	95.9	—	—	—	—	—	—	—
			M-F	100	96.9	—	—	—	—	—	—	—
34	LIMESTONE, MAGNESIUM (DOLOMITE)	6-02-633	A-F	99	—	—	—	—	—	—	—	—
			M-F	100	—	—	—	—	—	—	—	—
35	MAGNESIUM CARBONATE	6-02-754	A-F	81	—	—	—	—	—	—	—	—
			M-F	100	—	—	—	—	—	—	—	—
36	MAGNESIUM OXIDE	6-02-756	A-F	—	—	—	—	—	—	—	—	—
			M-F	100	—	—	—	—	—	—	—	—
37	MAGNESIUM SULFATE (EPSOM SALTS)	6-02-758	A-F	49	—	—	—	—	—	—	—	—
			M-F	100	—	—	—	—	—	—	—	—
38	MANGANESE CARBONATE	6-03-036	A-F	—	—	—	—	—	—	—	—	—
			M-F	100	—	—	—	—	—	—	—	—
39	MANGANESE CHLORIDE	6-03-038	A-F	64	—	—	—	—	—	—	—	—
			M-F	100	—	—	—	—	—	—	—	—
40	MANGANESE DIOXIDE	6-03-042	A-F	—	—	—	—	—	—	—	—	—
			M-F	100	—	—	—	—	—	—	—	—
41	MONOAMMONIUM PHOSPHATE	6-09-338	A-F	97	34.5	—	—	—	68.8	—	—	—
			M-F	100	35.6	—	—	—	70.9	—	—	—
	MONOCALCIUM PHOSPHATE (SEE CALCIUM PHOSPHATE, MONOBASIC)											
42	MONOSODIUM PHOSPHATE, ANHYDROUS	6-04-288	A-F	87	—	—	—	—	—	—	—	—
			M-F	100	—	—	—	—	—	—	—	—
43	ORGANIC IODIDE (ETHYLENEDIAMINE DIHYDROIODIDE)	6-01-842	A-F	—	—	—	—	—	—	—	—	—
			M-F	100	—	—	—	—	—	—	—	—
44	OYSTERSHELL, GROUND (FLOUR)	6-03-481	A-F	99	89.7	—	—	—	1.0	—	—	—
			M-F	100	90.6	—	—	—	1.0	—	—	—
45	PHOSPHATE, ROCK, GROUND (RAW)	6-03-945	A-F	—	—	—	—	—	—	—	—	—
			M-F	100	—	—	—	—	—	—	—	—
46	PHOSPHATE ROCK, DEFLOURINATED	6-01-780	A-F	—	—	—	—	—	—	—	—	—
			M-F	100	100.0	—	—	—	—	—	—	—
47	PHOSPHATE ROCK, LOW FLUORINE	6-03-946	A-F	—	—	—	—	—	—	—	—	—
			M-F	100	—	—	—	—	—	—	—	—

(Continued)

Entry Number	Macrominerals							Microminerals						
	Calcium (Ca)	Phosphorus (P)	Sodium (Na)	Chlorine (Cl)	Magnesium (Mg)	Potassium (K)	Sulfur (S)	Cobalt (Co)	Copper (Cu)	Iodine (I)	Iron (Fe)	Manganese (Mn)	Selenium (Se)	Zinc (Zn)
	(%)	(%)	(%)	(%)	(%)	(%)	(%)	(ppm or mg/kg)	(ppm or mg/kg)	(ppm or mg/kg)	(%)	(ppm or mg/kg)	(ppm or mg/kg)	(ppm or mg/kg)
25	32.00	18.00	—	—	—	—	—	—	—	—	.—	—	—	—
	—	—	—	—	—	—	—	—	—	—	—	—	—	—
26	0.57	20.00	0.04	—	0.45	0.01	2.50	—	91	—	1.200	400	—	342
	0.59	20.60	0.04	—	0.46	0.01	2.60	—	94	—	1.237	412	—	353
27	26.00	18.50	—	—	0.60	0.07	—	—	—	—	—	300	—	—
	27.10	19.30	—	—	0.62	0.07	—	—	—	—	—	313	—	—
28	—	21.50	32.00	0.00	—	—	—	—	—	—	0.001	—	—	—
	—	—	—	—	—	—	—	—	—	—	—	—	—	—
29	22.08	0.04	—	0.12	9.87	0.36	—	—	—	—	0.076	—	—	—
	22.30	0.04	—	0.12	9.99	0.36	—	—	—	—	0.077	—	—	—
30	—	—	—	—	—	—	—	—	—	—	—	—	—	—
	—	—	—	—	—	—	0.13	—	—	—	69.940	—	—	—
31	—	—	—	35.78	—	—	0.03	—	—	—	28.090	—	—	—
	—	—	—	55.91	—	—	0.05	—	—	—	44.628	—	—	—
32	—	—	—	0.00	0.05	—	11.60	—	—	—	20.080	—	—	—
	—	—	—	0.00	0.09	—	21.10	—	—	—	36.709	—	—	—
33	33.66	0.02	0.06	0.03	2.04	0.11	0.04	—	—	—	0.347	269	—	—
	34.00	0.02	0.06	0.03	2.06	0.12	0.04	—	—	—	0.350	270	—	—
34	22.08	0.04	—	0.12	9.89	0.36	—	—	—	—	0.076	—	—	—
	22.30	0.04	—	0.12	9.99	0.36	—	—	—	—	0.077	—	—	—
35	0.02	—	—	0.00	24.96	—	—	—	—	—	0.002	—	—	—
	0.02	—	—	0.00	30.81	—	—	—	—	—	0.002	—	—	—
36	—	—	—	—	—	—	—	—	—	—	—	—	—	—
	—	0.05	—	0.50	0.01	60.31	0.01	—	—	—	—	0.010	—	—
37	0.02	—	0.00	—	9.89	0.00	13.02	—	—	—	0.000	—	—	—
	0.04	—	0.01	—	20.18	0.01	26.58	—	—	—	0.001	—	—	—
38	—	—	—	—	—	—	—	—	—	—	—	—	—	—
	—	—	—	0.02	—	—	0.07	—	—	—	0.002	43	—	—
39	—	—	—	36.04	—	—	0.07	—	—	—	0.001	28	—	—
	—	—	—	56.32	—	—	0.11	—	—	—	0.001	44	—	—
40	—	—	—	—	—	—	—	—	—	—	—	—	—	—
	—	—	—	0.01	—	—	0.02	—	—	—	0.050	632,000	—	—
41	0.50	24.00	0.06	—	0.45	—	0.70	—	80	—	1.200	400	—	300
	0.52	24.74	0.06	—	0.46	—	0.70	—	82	—	1.237	412	—	309
42	—	22.18	16.53	—	—	—	—	—	—	—	0.001	—	—	—
	—	25.50	19.00	—	—	—	—	—	—	—	0.001	—	—	—
43	—	—	—	—	—	—	—	—	—	800,000	—	—	—	—
	—	—	—	—	—	—	—	—	—	—	—	—	—	—
44	37.62	0.07	0.21	0.01	0.30	0.10	—	—	—	—	0.284	133	—	—
	38.00	0.07	0.21	0.01	0.30	0.10	—	—	—	—	0.287	134	—	—
45	35.00	13.00	—	—	—	—	—	—	—	—	—	—	—	—
	—	—	—	—	—	—	—	—	—	—	—	—	—	—
46	—	—	—	—	—	—	—	—	—	—	—	—	—	—
	32.00	16.25	4.00	—	—	0.09	—	—	22	—	0.920	220	—	44
47	36.00	14.00	—	—	—	—	—	—	—	—	—	—	—	—
	—	—	—	—	—	—	—	—	—	—	—	—	—	—

(Continued)

MINERAL SUPPLEMENTS

TABLE 21-3

Entry Number	Feed Name Description	International Feed Number	Moisture Basis: A-F (as-fed) or M-F (moisture-free)	Chemical Analysis						Digestible Protein		
				Dry Matter	Ash	Crude Fiber	Ether Extract (Fat)	N-Free Extract	Crude Protein (6.25 × N)	Ruminant	Non-ruminant	Horse
				(%)	(%)	(%)	(%)	(%)	(%)	(%)	(%)	(%)
48	PHOSPHATE SOFT ROCK (COLLOIDAL CLAY)	6-03-947	A-F	—	—	—	—	—	—	—	—	—
			M-F	100	—	—	—	—	—	—	—	—
49	PHOSPHORIC ACID, DEFLUORINATED, WET PROCESS	6-20-534	A-F	—	—	—	—	—	—	—	—	—
			M-F	100	—	—	—	—	—	—	—	—
50	PHOSPHORIC ACID, FEED GRADE (ORTHO)	6-03-707	A-F	75	—	—	—	—	—	—	—	—
			M-F	100	—	—	—	—	—	—	—	—
51	POTASSIUM CHLORIDE	6-03-755	A-F	—	—	—	—	—	—	—	—	—
			M-F	100	—	—	—	—	—	—	—	—
52	POTASSIUM IODIDE	6-03-759	A-F	92	—	—	—	—	—	—	—	—
			M-F	100	—	—	—	—	—	—	—	—
53	POTASSIUM SULFATE	6-08-098	A-F	—	—	—	—	—	—	—	—	—
			M-F	100	—	—	—	—	—	—	—	—
54	RAW ROCK PHOSPHATE	6-03-945	A-F	—	—	—	—	—	—	—	—	—
			M-F	100	—	—	—	—	—	—	—	—
55	ROCK PHOSPHATE (PHOSPHATE ROCK)		A-F	—	—	—	—	—	—	—	—	—
			M-F	100	—	—	—	—	—	—	—	—
56	SODIUM BICARBONATE	6-04-273	A-F	—	—	—	—	—	—	—	—	—
			M-F	100	—	—	—	—	—	—	—	—
57	SODIUM CHLORIDE	6-04-152	A-F	—	—	—	—	—	—	—	—	—
			M-F	100	—	—	—	—	—	—	—	—
58	SODIUM IODIDE	6-04-279	A-F	—	—	—	—	—	—	—	—	—
			M-F	100	—	—	—	—	—	—	—	—
59	SODIUM PHOSPHATE, MONOBASIC FROM FURNACE PHOSPHORIC ACID, ANHYDROUS	6-04-287	A-F	87	—	—	—	—	—	—	—	—
			M-F	100	—	—	—	—	—	—	—	—
60	SODIUM PHOSPHATE, DIBASIC, FROM FURNACE PHOSPHORIC ACID	6-04-286	A-F	—	—	—	—	—	—	—	—	—
			M-F	100	—	—	—	—	—	—	—	—
61	SODIUM TRIPOLYPHOSPHATE	6-08-076	A-F	96	—	—	—	—	—	—	—	—
			M-F	100	—	—	—	—	—	—	—	—
62	SODIUM SULFATE	6-04-292	A-F	44	—	—	—	—	—	—	—	—
			M-F	100	—	—	—	—	—	—	—	—
63	ZINC CARBONATE	6-05-549	A-F	—	—	—	—	—	—	—	—	—
			M-F	100	—	—	—	—	—	—	—	—
64	ZINC CHLORIDE	6-05-551	A-F	—	—	—	—	—	—	—	—	—
			M-F	100	—	—	—	—	—	—	—	—
65	ZINC OXIDE	6-05-553	A-F	—	—	—	—	—	—	—	—	—
			M-F	100	—	—	—	—	—	—	—	—
66	ZINC SULFATE	6-05-555	A-F	56	—	—	—	—	—	—	—	—
			M-F	100	—	—	—	—	—	—	—	—

(Continued)

Entry Number	Macrominerals							Microminerals						
	Calcium (Ca)	Phosphorus (P)	Sodium (Na)	Chlorine (Cl)	Magnesium (Mg)	Potassium (K)	Sulfur (S)	Cobalt (Co)	Copper (Cu)	Iodine (I)	Iron (Fe)	Manganese (Mn)	Selenium (Se)	Zinc (Zn)
	(%)	(%)	(%)	(%)	(%)	(%)	(%)	(ppm or mg/kg)	(ppm or mg/kg)	(ppm or mg/kg)	(%)	(ppm or mg/kg)	(ppm or mg/kg)	(ppm or mg/kg)
48	17.00 —	9.00 —	— —	— —	— —	— —	— —	— —	— —	— —	— —	— —	— —	— —
49	0.20 —	23.70 —	— —	— —	— —	— —	— —	— —	— —	— —	— —	— —	— —	— —
50	— —	23.70 31.60	0.01 0.03	— —	— —	0.01 0.01	0.05 0.07	— —	— —	— —	0.002 0.003	— —	— —	— —
51	— —	— —	0.01 —	47.30 —	— —	50.50 —	— —	— —	— —	— —	0.000 —	— —	— —	— —
52	— —	— —	0.01 0.01	0.01 0.01	— —	21.67 23.56	— —	— —	— —	699,200 760,000	— —	— —	— —	— —
53	0.15 —	— —	0.09 —	1.52 —	0.60 —	43.10 —	17.70 —	— —	3 —	— —	0.070 —	9 —	— —	4 —
54	35.00 —	13.00 —	— —	— —	— —	— —	— —	— —	— —	— —	— —	— —	— —	— —
55	— —	— —	— —	— —	— —	— —	— —	— —	— —	— —	— —	— —	— —	— —
56	— —	— —	— 27.36	— —	— —	— 0.01	— —	— —	— —	— —	— 0.001	— —	— —	— —
57	— —	— —	39.34 —	60.66 —	— —	— —	— —	— —	— —	— —	— —	— —	— —	— —
58	— —	— —	— 15.33	— 0.01	— —	— —	— —	— —	— —	— 847,000	— 0.001	— —	— —	— —
59	— —	22.18 25.50	16.53 19.00	— —	— —	— —	— —	— —	— —	— —	0.001 0.001	— —	— —	— —
60	— —	— 21.50	— 32.00	— 0.00	— —	— —	— —	— —	— —	— —	— 0.001	— —	— —	— —
61	— —	24.00 25.00	28.80 30.00	— —	— —	— —	— —	— —	— —	— —	0.004 0.004	— —	— —	— —
62	— —	— —	14.27 32.36	0.00 0.00	— —	— —	9.96 22.59	— —	— —	— —	0.000 0.001	— —	— —	— —
63	— —	— —	— —	— 0.00	— —	— —	— 0.14	— —	— —	— —	— 0.002	— —	— —	— 560,000
64	— —	— —	— —	— 52.03	— —	— —	— 0.07	— —	— —	— —	— 0.001	— —	— —	— 480,000
65	— —	— —	— —	— —	— 0.00	— —	— 0.03	— —	— —	— —	— 0.001	— —	— —	— 730,000
66	— —	— —	— —	— —	0.00 0.00	— —	11.15 19.84	— —	— —	— —	0.001 0.001	— —	— —	227,000 404,000

Three pork dishes: Italian, knackwurst, and Polish sausage. (Courtesy, National Livestock & Meat Board)

APPENDIX

Philippine salt pork container. (Courtesy, Field Museum of Natural History, Chicago, IL)

This Appendix is essential to the completeness of *Swine Science*. It provides useful supplemental information relative to (1) animal units, (2) conversions of weights and measures, (3) some uses of weights and measures, (4) swine magazines, (5) breed registry associations, (6) guidelines for the sale of purebred swine, (7) U.S. colleges of agriculture and Canadian provincial universities, and (8) poison information centers.

ANIMAL UNITS

An animal unit is a common animal denominator, based on feed consumption. It is assumed that one mature cow represents an animal unit. Then, the comparative (to a mature cow) feed consumption of other age groups or classes of animals determines the proportion of an animal unit which they represent. For example, it is generally estimated that the ration of 1 mature cow will feed 5 hogs raised to 200 lb. For this reason, the Animal Unit on this class and age of animals is 0.2. The following table gives the animal units for different classes and ages of livestock:

TABLE A-1
ANIMAL UNITS

Type of Livestock	Animal Units
Cattle:	
Cow, with or without unweaned calf at side, or heifer 2 yrs. old or older	1.0
Bull, 2 yrs. old or older	1.3
Young cattle, 1 to 2 yrs. old	0.8
Weaned calves to yearlings	0.6
Horses:	
Horse, mature	1.3
Horse, yearling	1.0
Weanling colt or filly	0.75
Sheep:	
5 mature ewes, with or without unweaned lambs at side	1.0
5 rams, 2 yrs. old or older	1.3
5 yearlings	0.8
5 weaned lambs to yearlings	0.6
Swine:	
Sow	0.4
Boar	0.5
Pigs to 200 lb *(91 kg)*	0.2
Chickens:	
75 layers or breeders	1.0
325 replacement pullets to 6 mo. of age	1.0
650 8-week-old broilers	1.0
Turkeys:	
35 breeders	1.0
40 turkeys raised to maturity	1.0
75 turkeys to 6 mo. of age.	1.0

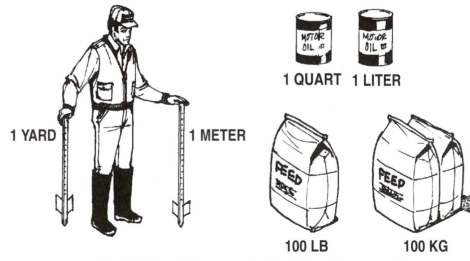

Fig. A-1. Metric vs U.S. customary—length, weight, and volume comparisons.

WEIGHTS AND MEASURES

Weights and measures are standards employed in arriving at weights, quantities, and volumes. Even among primitive people, such standards were necessary; and with the growing complexity of life, they become of greater and greater importance.

Weights and measures form one of the most important parts of modern agriculture. This section contains pertinent information relative to the most common standards used by U.S. pork producers.

METRIC SYSTEM[1]

The United States and a few other countries use standards that belong to the *customary*, or English, system of measurement. This system evolved in England from older measurement standards, beginning about the year 1200. All other countries—including England—now use a system of measurements called the *metric system*, which was created in France in the 1790s. Increasingly, the metric system is being used in the United States. Hence, everyone should have a working knowledge of it.

The basic metric units are the *meter* (length/distance), the *gram* (weight), and the *liter* (capacity). The units are then expanded in multiples of 10 or made smaller by 1/10. The prefixes, which are used in the same way with all basic metric units, follow:

"milli-"	=	1/1,000
"centi-"	=	1/100
"deci-"	=	1/10
"deca-"	=	10
"hecto-"	=	100
"kilo-"	=	1,000

The following tables will facilitate conversion from metric units to U.S. customary, and vice versa:

Table A-2, Weight-Unit Conversion Factors
Table A-3, Weight Equivalents
Table A-4, Weights and Measures
    Length
    Surface or Area
    Volume
    Weight
    Weights and Measures per Unit

---

[1]For additional conversion factors, or for greater accuracy, see *Misc. Pub. 223*, The National Bureau of Standards.

TABLE A-2
WEIGHT-UNIT CONVERSION FACTORS

Units Given	Units Wanted	For Conversion Multiply By
lb	*g*	453.6
lb	*kg*	0.4536
oz	*g*	28.35
kg	lb	2.2046
kg	*mg*	1,000,000
kg	*g*	1,000
g	*mg*	1,000
g	μ*g*	1,000,000
mg	μ*g*	1,000
mg/g	*mg*/lb	453.6
mg/kg	*mg*/lb	0.4536
μ*g/kg*	μ*g*/lb	0.4536
Mcal	*kcal*	1,000
kcal/kg	*kcal*/lb	0.4536
kcal/lb	*kcal/kg*	2.2046
ppm	μ*g/g*	1
ppm	*mg/kg*	1
ppm	*mg*/lb	0.4536
mg/kg	%	0.0001
ppm	%	0.0001
mg/g	%	0.1
g/kg	%	0.1

TABLE A-3
WEIGHT EQUIVALENTS

1 lb	=	*453.6 g*	=	*0.4536 kg*	=	16 oz
1 oz	=	*28.35 g*				
1 kg	=	*1,000 g*	=	2.2046 lb		
1 g	=	*1,000 mg*				
1 mg	=	*1,000* μ*g*	=	*0.001 g*		
1 μ*g*	=	*0.001 mg*	=	*0.000001 g*		
1 μ*g* per *g* or *1 mg* per *kg* is the same as ppm						

TABLE A-4
WEIGHTS AND MEASURES (METRIC AND U.S. CUSTOMARY)

LENGTH

Unit	Is Equal To	
Metric System		(U.S. Customary)
1 millimicron (mμ)	0.000000001 m	0.000000039 in.
1 micron (μ)	0.000001 m	0.000039 in.
1 millimeter (mm)	0.001 m	0.0394 in.
1 centimeter (cm)	0.01 m	0.3937 in.
1 decimeter (dm)	0.1 m	3.937 in.
1 meter (m)	1 m	39.37 in.; 3.281 ft; 1.094 yd
1 hectometer (hm)	100 m	328 ft, 1 in.; 19.8338 rd
1 kilometer (km)	1,000 m	3,280 ft, 10 in.; 0.621 mi
U.S. Customary		(Metric)
1 inch (in.)		25 mm; 2.54 cm
1 hand*	4 in.	
1 foot (ft)	12 in.	30.48 cm; 0.305 m
1 yard (yd)	3 ft	0.914 m
1 fathom** (fath)	6.08 ft	1.829 m
1 rod (rd), pole, or perch	16.5 ft; 5.5 yd	5.029 m
1 furlong (fur.)	220 yd; 40 rd	201.168 m
1 mile (mi)	5,280 ft; 1,760 yd; 320 rd; 8 fur.	1,609.35 m; 1.609 km
1 knot or nautical mile	6,080 ft; 1.15 land mi	
1 league (land)	3 mi (land)	
1 league (nautical)	3 mi (nautical)	

*Used in measuring height of horses.

**Used in measuring depth at sea.

CONVERSIONS

To Change	To	Multiply By
inches	centimeters	2.54
feet	meters	0.305
meters	inches	39.73
miles	kilometers	1.609
kilometers	miles	0.621

(To make opposite conversion, divide by the number given instead of multiplying)

(Continued)

TABLE A-4 (Continued)

SURFACE OR AREA

Unit	Is Equal To	
Metric System		**(U.S. Customary)**
1 square millimeter (mm²)	0.000001 m²	0.00155 in.²
1 square centimeter (cm²)	0.0001 m²	0.155 in.²
1 square decimeter (dm²)	0.01 m²	15.50 in.²
1 square meter (m²)	1 centare (ca)	1,550 in.²; 10.76 ft²; 1.196 yd²
1 are (a)	100 m²	119.6 yd²
1 hectare (ha)	10,000 m²	2.47 acres
1 square kilometer (km²)	1,000,000 m²	247.1 acres; 0.386 mi²
U.S. Customary		**(Metric)**
1 square inch (in.²)	1 in. × 1 in.	6.452 cm²
1 square foot (ft²)	144 in.²	0.093 m²
1 square yard (yd²)	1,296 in.²; 9 ft²	0.836 m²
1 square rod (rd²)	272.25 ft²; 30.25 yd²	25.29 m²
1 rood	40 rd²	10.117 a
1 acre	43,560 ft²; 4,840 yd²; 160 rd²; 4 roods	4,046.87 m²; 0.405 ha
1 square mile (mi²)	640 acres	2.59 km²; 259 ha
1 township	36 sections; 6 mi²	

CONVERSIONS

To Change	To	Multiply By
square inches	square centimeters	6.452
square centimeters	square inches	0.155
square yards	square meters	0.836
square meters	square yards	1.196

(To make opposite conversion, divide by the number given instead of multiplying.)

VOLUME

Unit	Is Equal To		
Metric System Liquid and Dry		**(U.S. Customary)**	
		(Liquid)	**(Dry)**
1 milliliter (ml)	0.001 liter	0.271 dram (fl)	0.061 in.³
1 centiliter (cl)	0.01 liter	0.338 oz (fl)	0.610 in.³
1 deciliter (dl)	0.1 liter	3.38 oz (fl)	
1 liter (l)	1,000 cc	1.057 qt; 0.2642 gal (fl)	0.908 qt
1 hectoliter (hl)	100 liter	26.418 gal	2.838 bu
1 kiloliter (kl)	1,000 liter	264.18 gal	1,308 yd³

(Continued)

TABLE A-4 (Continued)

VOLUME (Continued)

Unit	Is Equal To			
U.S. Customary *Liquid*		(Ounces)	(Cubic Inches)	(Metric)
1 teaspoon (t)	60 drops	0.1666		5 ml
1 dessert spoon	2 t			
1 tablespoon (T)	3 t	0.5		15 ml
1 fl oz		1	1.805	29.57 ml
1 gill (gi)	0.5 c	4	7.22	118.29 ml
1 cup (c)	16 T	8	14.44	236.58 ml; 0.24 l
1 pint (pt)	2 c	16	28.88	0.47 l
1 quart (qt)	2 pt	32	57.75	0.95 l
1 gallon (gal)	4 qt	8.34 lb	231	3.79 l
1 barrel (bbl)	31.5 gal			
1 hogshead (hhd)	2 bbl			
Dry		(Ounces)	(Cubic Inches)	(Metric)
1 pint (pt)	0.5 qt		33.6	0.55 l
1 quart (qt)	2 pt		67.20	1.10 l
1 peck (pk)	8 qt		537.61	8.81 l
1 bushel (bu)	4 pk		2,150.42	35.24 l
Solid **Metric System**		(Metric)	(U.S. Customary)	
1 cubic millimeter (mm³)		0.001 cc		
1 cubic centimeter (cc)		1,000 mm³	0.061 in.³	
1 cubic decimeter (dm³)		1,000 cc	61.023 in.³	
1 cubic meter (m³)		1,000 dm³	35.315 ft³; 1.308 yd³	
U.S. Customary				(Metric)
1 cubic inch (in.³)				16.387 cc
1 board foot (fbm)		144 in.³		2,359.8 cc
1 cubic foot (ft³)		1,728 in.³		0.028 m³
1 cubic yard (yd³)		27 ft³		0.765 m³
1 cord		128 ft³		3.625 m³

CONVERSIONS

To Change	To	Multiply By
ounces (fluid)	cubic centimeters	29.57
cubic centimeters	ounces (fluid)	0.034
quarts	liters	0.946
liters	quarts	1.057
cubic inches	cubic centimeters	16.387
cubic centimeters	cubic inches	0.061
cubic yards	cubic meters	0.765
cubic meters	cubic yards	1.308

(To make opposite conversion, divide by the number given instead of multiplying.)

TABLE A-4 (Continued)

WEIGHT

Unit	Is Equal To	
Metric System		**(U.S. Customary)**
1 microgram (mcg)	0.001 mg	
1 milligram (mg)	0.001 g	0.015432356 grain
1 centigram (cg)	0.01 g	0.15432356 grain
1 decigram (dg)	0.1 g	1.5432 grains
1 gram (g)	1,000 mg	0.03527396 oz
1 decagram (dkg)	10 g	5.643833 dr
1 hectogram (hg)	100 g	3.527396 oz
1 kilogram (kg)	1,000 g	35.274 oz; 2.2046223 lb
1 ton	1,000 kg	2,204.6 lb; 1.102 tons (short); 0.984 ton (long)
U.S. Customary		**(Metric)**
1 grain	0.037 dr	64.798918 mg; 0.064798918 g
1 dram (dr)	0.063 oz	1.771845 g
1 ounce (oz)	16 dr	28.349527 g
1 pound (lb)	16 oz	453.5924 g; 0.4536 kg
1 hundredweight (cwt)	100 lb	
1 ton (short)	2,000 lb	907.18486 kg; 0.907 (metric) ton
1 ton (long)	2,200 lb	1,016.05 kg; 1.016 (metric) ton
1 part per million (ppm)	1 microgram/gram; 1 mg/liter; 1 mg/kg	0.4535924 mg/lb; 0.907 g/ton
	0.0001%; 0.00013 oz/gal	
1 percent (%) (1 part in 100 parts)	10,000 ppm; 10 g/liter	
	1.28 oz/gal; 8.34 lb/100 gal	

CONVERSIONS

To Change	To	Multiply By
grains	milligrams	64.799
ounces (dry)	grams	28.35
pounds (dry)	kilograms	0.4535924
kilograms	pounds	2.2046223
milligrams/pound	parts/million	2.2046223
parts/million	grams/ton	0.90718486
grams/ton	parts/million	1.1
milligrams/pound	grams/ton	2
grams/ton	milligrams/pound	0.5
grams/pound	grams/ton	2,000
grams/ton	grams/pound	0.0005
grams/ton	pounds/ton	0.0022
pounds/ton	grams/ton	453.5924
grams/ton	percent	0.00011
percent	grams/ton	9,072
parts/million	percent	move decimal four places to left

(To make opposite conversion, divide by the number given instead of multiplying.)

WEIGHTS AND MEASURES PER UNIT

Unit	Is Equal To
Volume per Unit Area	
1 liter/hectare	0.107 gal/acre
1 gallon/acre	9.354 liter/ha
Weight per Unit Area	
1 kilogram/cm^2	14.22 lb/in.^{2}
1 kilogram/hectare	0.892 lb/acre
1 pound/square inch	0.0703 kg/cm^2
1 pound/acre	1.121 kg/ha
Area per Unit Weight	
1 square centimeter/kilogram	0.0703 in.^{2}/lb
1 square inch/pound	14.22 cm^2/kg

TEMPERATURE

One centigrade (C) degree is 1/100 the difference between the temperature of melting ice and that of water boiling at standard atmospheric pressure. One centigrade degree equals 1.8°F.

One Fahrenheit (F) degree is 1/180 of the difference between the temperature of melting ice and that of water boiling at standard atmospheric pressure. One Fahrenheit degree equals 0.556°C.

To Change	To	Do This
Degrees centigrade . . .	Degrees Fahrenheit . . .	Multiply by ⁹⁄₅ and add 32
Degrees Fahrenheit . .	Degrees centigrade . . .	Subtract 32, then multiply by ⁵⁄₉

WEIGHTS AND MEASURES OF COMMON FEEDS

In calculating rations and mixing concentrates, it is usually necessary to use weights rather than measures. However, in practical feeding operations it is often more convenient for the producer to measure the concentrates. Table A-5 will serve as a guide in feeding by measure.

TABLE A-5
WEIGHTS AND MEASURES OF COMMON FEEDS

Feed	Approximate Weight[1]	
	Lb per Quart	Lb per Bushel
Alfalfa meal	0.6	19
Barley	1.5	48
Beet pulp (dried)	0.6	19
Brewers' grain (dried)	0.6	19
Buckwheat	1.6	50
Buckwheat bran	1.0	29
Corn, husked ear	—	70
Corn, cracked	1.6	50
Corn, shelled	1.8	56
Corn meal	1.6	50
Corn-and-cob meal	1.4	45
Cottonseed meal	1.5	48
Cowpeas	1.9	60
Distillers' grain (dried)	0.6	19
Fish meal	1.0	35
Gluten feed	1.3	42
Linseed meal (old process)	1.1	35
Linseed meal (new process)	0.9	29
Meat scrap	1.3	42
Milo (grain sorghum)	1.7	56
Molasses feed	0.8	26
Oats	1.0	32
Oats, ground	0.7	22
Oat middlings	1.5	48
Peanut meal	1.0	32
Rice bran	0.8	26
Rye	1.7	56
Sorghum (grain)	1.7	56
Soybeans	1.7	60
Tankage	1.6	51
Velvet beans, shelled	1.8	60
Wheat	1.9	60
Wheat bran	0.5	16
Wheat middlings, standard	0.8	26
Wheat screenings	1.0	32

[1]To convert to metric, refer to Table A-4.

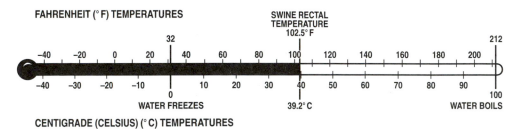

Fig. A-2. Fahrenheit-centigrade (Celsius) scale for direct conversion and reading.

ESTIMATING WEIGHT OF HOGS

Hog weights can be estimated by taking the body measurements, and applying the formula which follows:

Step 1—Measure the circumference (heart girth) of the animal (C in Fig. A-3).

Step 2—Measure the length of body (A-B in Fig. A-3). With the animal standing or restrained in the position shown in Fig. A-3, measure the distance from the poll (between the ears), over the backbone, to the base of the tail.

Step 3—Apply the following formula:

Heart girth × heart girth × length ÷ 400 = weight in pounds

Note: For hogs weighing less than 150 lb, add 7 lb to the weight figure obtained from the formula. For animals weighing 151 to 400 lb, no adjustment is necessary.

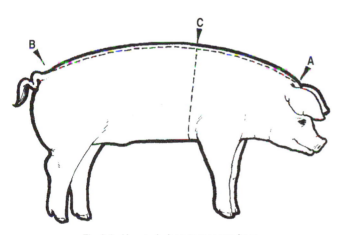

Fig. A-3. How and where to measure hogs.

ESTIMATING WEIGHT OF GRAIN IN A BIN

Sometimes producers need to estimate the weight of grain in storage. Such estimates are difficult to make because of differences in moisture content, depth of material stored, and other factors. However, the following procedure will enable one to figure feed quantities fairly closely.

1. **Corn (shelled) or small grain in rectangular cribs or bins**—Multiply the width by the length by the average depth (all in feet) and multiply by 0.8 to get the number of bushels (multiplying by 0.8 is the same as dividing by 1.25, the number of cubic feet in a bushel).

2. **Ear corn in rectangular cribs or bins**—Multiply the width by the length by the average depth (all in feet) and multiply by 0.4 to get the number of bushels (multiplying by 0.4 is the same as dividing by 2.5, the number of cubic feet in a bushel of ear corn).

3. **Round bins or cribs**—To find the cubic feet in a cylindrical bin, multiply the squared radius by 3.1416 by the depth.

Thus, the volume of a round bin 20 ft in diameter and 10 ft deep is determined as follows:

a. The radius is half the diameter, or 10 feet

b. $10 \times 10 = 100$

c. $100 \times 3.1416 = 314.16$

d. $314.16 \times 10 = 3,141.6$ cubic feet

e. Where shelled corn or small grain is involved, one should multiply $3,141.6 \times 0.8$, which equals 2,513.28 bushels of grain that it would hold if full.

f. Where ear corn is involved, one should multiply $3,141.6 \times 0.4$ which equals 1,256.64 bushels of ear corn that it would hold if full.

SWINE MAGAZINES

The livestock magazines publish news items and informative articles of special interest to swine caretakers.

Also, many of them employ field representatives whose chief duty it is to assist in the buying and selling of animals.

In the compilation of the list herewith presented (see Table A-6), no attempt was made to list the general livestock magazines of which there are numerous outstanding ones. Only those magazines which are devoted exclusively to swine are included.

TABLE A-6
SWINE MAGAZINES

Breed	Publication	Address
General[1]	*Hog Farm Management*	P.O. Box 67 Minneapolis, MN 55440
	National Hog Farmer	Webb Division, Intertec Publishing Corp. 9800 Metcalf Overland Park, KS 66212-2215
	Nebraska Pork Talk	P.O. Box 487 Madison, NE 68748
	Pig Farming	Fenton House, Wharfedale Road Ipswich, Suffolk, England IPI 4LG
	Pigs	Misset International P.O. Box 4, 7000BA Doetinchem, The Netherlands
	Pork	10901 W. 84 Terr. Lenexa, KS 66214
	Southern Hog Producer	P.O. Box 110017 Nashville, TN 37211
	Swine Practitioners	Livestock Division, Vance Publishing Corp. 10901 W. 84th Terr. Lenexa, KS 66214
Joint Publications	*Berkshire News, The* *Chester White Journal* *Poland China Advantage* *Spotted News*	American Berkshire Assn. P.O. Box 2436 West Lafayette, IN 47906 Chester White Swine Record Assn. Poland China Record Assn. National Spotted Swine Record 6320 N. Sheridan Road Peoria, IL 61612-9758
	Seedstock Edge	P.O. Box 2339 1769 U.S. 52 W. West Lafayette, IN 47906-0339
Berkshire	*Berkshire News, The*	American Berkshire Assn. P.O. Box 2436 West Lafayette, IN 47906
Chester White	*Chester White Journal*	6320 N. Sheridan Road Peoria, IL 61612-9758
Duroc	*Duroc News*	1803 W. Detweiller Drive Peoria, IL 61615
Hampshire	*Hampshire Herdsman*	1111 Main Street Peoria, IL 61606
Landrace	*American Landrace, The*	P.O. Box 647 Lebanon, IN 46052
Poland China	*Poland China World, The*	P.O. Box B Knoxville, IL 61448
Spotted	*Spotted News*	6320 N. Sheridan Road Peoria, IL 61612-9758
Tamworth	*Tamworth News, The*	200 Centenary Road Winchester, OH 45697
Yorkshire	*Yorkshire Journal, The*	P.O. Box 2417 West Lafayette, IN 47906

[1]Covers all breeds.

BREED REGISTRY ASSOCIATIONS

A breed registry association consists of a group of breeders banded together for the purposes of: (1) recording the lineage of their animals, (2) protecting the purity of the breed, (3) encouraging further improvement of the breed, and (4) promoting the interest of the breed. A list of the swine breed registry associations is given in Table A-7.

TABLE A-7
BREED REGISTRY ASSOCIATIONS

Breed	Associaton and Address
Berkshire	American Berkshire Assn. P.O. Box 2436 West Lafayette, IN 47906
Chester White	Chester White Swine Record Assn. P.O. Box 9758 6320 N. Sheridan Road Peoria, IL 61612-9758
Duroc	United Duroc Swine Registry P.O. Box 2379 West Lafayette, IN 47906
Hampshire	Hampshire Swine Registry P.O. Box 2807 1769 U.S. 52 W. West Lafayette, IN 47906
Hereford Hog	National Hereford Hog Record Assn. Route 1, Box 37 Flandreau, SD 57028
Landrace	American Landrace Assn., Inc. P.O. Box 2340 West Lafayette, IN 47906
Poland China	Poland China Record Assn. P.O. Box 9758 6320 N. Sheridan Road Peoria, IL 61612-9758
Spotted	National Spotted Swine Record, Inc. P.O. Box 9758 6320 N. Sheridan Road Peoria, IL 61612-9758
Tamworth	Tamworth Swine Assn. 200 Centenary Road Winchester, OH 45697
Yorkshire	American Yorkshire Club, Inc. P.O. Box 2417 1769 U.S. 52 W. West Lafayette, IN 47906

U.S. STATE COLLEGES OF AGRICULTURE AND CANADIAN PROVINCIAL UNIVERSITIES

U.S. pork producers can obtain a list of available bulletins and circulars, and other information regarding swine by writing to (1) their state agricultural college (land-grant institution), and (2) the U.S. Superintendent of Documents, Washington, DC; or by going to the local county extension office (farm advisor) of the county in which they reside. Canadian producers may write to the Department of Agriculture of their province or to their provincial university. A list of U.S. land-grant institutions and Canadian provincial universities follows in Table A-8.

POISON INFORMATION CENTERS

With the large number of chemical sprays, dusts, and gases now on the market for use in agriculture, accidents may arise because operators are careless in their use. Also, there is always the hazard that a child may eat or drink something that may be harmful. Centers have been established in various parts of the country where doctors can obtain prompt and up-to-date information on treatment of such cases, if desired.

Local physicians have information relative to the Poison Information Centers in their area, along with some of the names of their directors, their telephone numbers, and their street addresses. When calling any of these centers, one should ask for the "Poison Information Center." If this information cannot be obtained locally, the U.S. Public Health Service at Atlanta, Georgia, or Wenatchee, Washington, should be contacted.

Also, the *National Poison Control Center* is located at the University of Illinois, Urbana-Champaign. It is open 24 hours a day, every day of the week. The *hot line* number is: 1-800-548-2423. The toxicology group is staffed to answer questions about known or suspected cases of poisoning or chemical contaminations involving any species of animal. It is not intended to replace local veterinarians or state toxicology laboratories, but to complement them. Where consultation over the telephone is adequate, there is no charge to the veterinarian or producer. Where telephone consultation is inadequate or the problem is of major proportion, a team of veterinary specialists can arrive at the scene of a toxic or contamination problem within a short time. The cost of a personal visitation varies according to the distance traveled, personnel time, and laboratory services required.

TABLE A-8
U.S. LAND-GRANT INSTITUTIONS AND CANADIAN PROVINCIAL UNIVERSITIES

State	Address
Alabama	School of Agriculture, Auburn University, Auburn, AL 36830
Alaska	Department of Agriculture, University of Alaska, Fairbanks, AK 99701
Arizona	College of Agriculture, The University of Arizona, Tucson, AZ 85721
Arkansas	Division of Agricutlure, University of Arkansas, Fayetteville, AR 72701
California	College of Agriculture and Environmental Sciences, University of California, Davis, CA 95616
Colorado	College of Agricultural Sciences, Colorado State University, Fort Collins, CO 80521
Connecticut	College of Agriculture and Natural Resources, University of Connecticut, Storrs, CT 06268
Delaware	College of Agricultural Sciences, University of Delaware, Newark, DE 19711
Florida	College of Agriculture, University of Florida, Gainesville, FL 32611
Georgia	College of Agriculture, University of Georgia, Athens, GA 30602
Hawaii	College of Tropical Agriculture, University of Hawaii, Honolulu, HI 96822
Idaho	College of Agriculture, University of Idaho, Moscow, ID 83843
Illinois	College of Agriculture, University of Illinois, Urbana–Champaign, IL 61801
Indiana	School of Agriculture, Purdue University, West Lafayette, IN 47907
Iowa	College of Agriculture, Iowa State University, Ames, IA 50010
Kansas	College of Agriculture, Kansas State University, Manhattan, KS 66506
Kentucky	College of Agriculture, University of Kentucky, Lexington, KY 40506
Louisiana	College of Agriculture, Louisiana State University and A&M College, University Station, Baton Rouge, LA 70803
Maine	College of Life Sciences and Agriculture, University of Maine, Orono, ME 04473
Maryland	College of Agriculture, University of Maryland, College Park, MD 20742
Massachusetts	College of Food and Natural Resources, University of Massachusetts, Amherst, MA 01002
Michigan	College of Agriculture and Natural Resources, Michigan State University, East Lansing, MI 48823
Minnesota	College of Agriculture, University of Minnesota, St. Paul, MN 55101
Mississippi	College of Agriculture, Mississippi State University, Mississippi State, MS 39762
Missouri	College of Agriculture, University of Missouri, Columbia, MO 65201
Montana	College of Agriculture, Montana State University, Bozeman, MT 59715
Nebraska	College of Agriculture, University of Nebraska, Lincoln, NE 68503
Nevada	The Max C. Fleischmann College of Agriculture, University of Nevada, Reno, NV 89507
New Hampshire	College of Life Sciences and Agriculture, University of New Hampshire, Durham, NH 03824
New Jersey	College of Agriculture and Environmental Science, Rutgers University, New Brunswick, NJ 08903
New Mexico	College of Agriculture and Home Economics, New Mexico State University, Las Cruces, NM 88003
New York	New York State College of Agriculture, Cornell University, Ithaca, NY 14850
North Carolina	School of Agriculture, North Carolina State University, Raleigh, NC 27607
North Dakota	College of Agriculture, North Dakota State University, State University Station, Fargo, ND 58102
Ohio	College of Agriculture and Home Economics, The Ohio State University, Columbus, OH 43210
Oklahoma	College of Agriculture and Applied Science, Oklahoma State University, Stillwater, OK 74074

(Continued)

TABLE A-8 (Continued)

State	Address
Oregon	School of Agriculture, Oregon State University, Corvallis, OR 97331
Pennsylvania	College of Agriculture, The Pennsylvania State University, University Park, PA 16802
Puerto Rico	College of Agricultural Sciences, University of Puerto Rico, Mayagüez, PR 00708
Rhode Island	College of Resource Development, University of Rhode Island, Kingston, RI 02881
South Carolina	College of Agricultural Sciences, Clemson University, Clemson, SC 29631
South Dakota	College of Agriculture and Biological Sciences, South Dakota State University, Brookings, SD 57006
Tennessee	College of Agriculture, University of Tennessee, P.O. Box 1071, Knoxville, TN 37901
Texas	College of Agriculture, Texas A&M University, College Station, TX 77843
Utah	College of Agriculture, Utah State University, Logan, UT 84321
Vermont	College of Agriculture, University of Vermont, Burlington, VT 05401
Virginia	College of Agriculture, Viriginia Polytechnic Institute and State University, Blacksburg, VA 24061
Washington	College of Agriculture, Washington State University, Pullman, WA 99163
West Virginia	College of Agriculture and Forestry, West Virginia University, Morgantown, WV 26506
Wisconsin	College of Agricultural and Life Sciences, University of Wisconsin, Madison, WI 53706
Wyoming	College of Agriculture, University of Wyoming, University Station, P.O. Box 3354, Laramie, WY 82070

Canada	Address
Alberta	University of Alberta, Edmonton, Alberta T6H 3K6
British Columbia	University of British Columbia, Vancouver, British Columbia V6T 1W5
Manitoba	University of Manitoba, Winnipeg, Manitoba R3T 2N2
New Brunswick	University of New Brunswick, Federicton, New Brunswick E3B 4N7
Ontario	University of Guelph, Guelph, Ontario N1G 2W1
Québec	Faculty d'Agriculture, L'Université Laval, Québéc City, Québéc G1K 7D4; and Macdonald College of McGill University, Ste. Anne de Bellevue, Québéc H9X 1C0
Saskatchewan	University of Saskatchewan, Saskatoon, Saskatchewan S7N 0W0

To the humble pig

Who, from the remote day of his domestication forward, has artfully mirrored the world around him;

Who has been maligned as unclean, regarded with contempt, fed by the prodigal son, and considered abominable to the Lord;

Who has been people-downgraded as hog wild, as wallowing in it, as a pig in a poke, and as fat as a pig;

Who has been extolled for being the little pig that went to market, for bringing home the bacon, for being the mortgage lifter, and for being in pig heaven;

Whose living legacy to the world is:

    "Root hog or die."

(Courtesy, Cenex/Land O Lakes, Ft. Dodge, IA)

I

INDEX

Producer with micro computer. (Courtesy, Iowa State University, Ames)

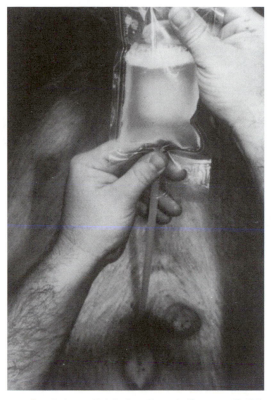

Sow being artificially inseminated. (Courtesy, DeKalb Swine Breeders, Inc., DeKalb, IL)

AI technician. (Courtesy, Lone Willow Farm, Roanoke, IL)

Ultrasound evaluation of backfat and loin area of a live hog. (Courtesy, National Pork Producers Council, Des Moines, IA)

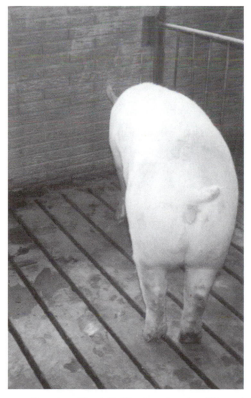

Rear view of bred gilt. (Courtesy, American Diamond Swine, Prairie City, IA)

Imported native Chinese breed—very prolific. (Courtesy, Iowa State University, Ames)

Outside gestation feeding stalls. (Courtesy, Waldo Farms, DeWitt, NE)

XL boars. (Courtesy, Hal Sellers, Director, Research, Farmers Hybrid Co., Des Moines, IA)

Finishing hogs on slatted floors. (Courtesy, Land O Lakes, Ft. Dodge, IA)

Naturally ventilated swine finishing building using "butterfly" doors. (Courtesy, University of Illinois, Urbana)

Environmentally controlled housing. (Courtesy, National Pork Producers Council, Des Moines, IA)

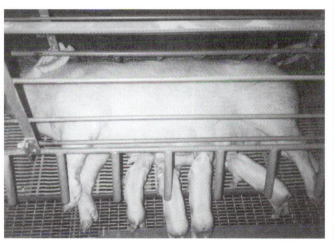

Large White sow in a farrowing crate. (Courtesy, American Diamond Swine, Prairie City, IA)

Finishing hogs in open shed with concrete apron on front. (Courtesy, Iowa State University, Ames)

DELICIOUS END PRODUCT——PORK LOIN. (Courtesy, National Pork Producers Council, Des Moines, IA)